Name	Symbol	Atomic Number	Atomic Mass
Neptunium	Np	93	237.0
Nickel	Ni	28	58.7
Niobium	Nb	41	92.9
Nitrogen	N	7	14.01
Nobelium	No	102	(259)
Osmium	Os	76	190.2
Oxygen	O	8	16.00
Palladium	Pd	46	106.4
Phosphorus	P	15	31.0
Platinum	Pt	78	195.1
Plutonium	Pu	94	(244)
Polonium	Po	84	(209)
Potassium	K	19	39.1
Praseodymium	Pr	59	140.9
Promethium	Pm	61	(145)
Protactinium	Pa	91	231.0
Radium	Ra	88	226.0
Radon	Rn	86	(222)
Rhenium	Re	75	186.2
Rhodium	Rh	45	102.9
Rubidium	Rb	37	85.5
Ruthenium	Ru	44	101.1
Samarium	Sm	62	150.4
Scandium	Sc	21	45.0
Selenium	Se	34	79.0
Silicon	Si	14	28.1
Silver	Ag	47	107.9
Sodium	Na	11	23.0

Name	Symbol	Atomic Number	Atomic Mass
Strontium	Sr	38	87.6
Sulfur	S	16	32.1
Tantalum	Ta	73	180.9
Technetium	Tc	43	(98)
Tellurium	Te	52	127.6
Terbium	Tb	65	158.9
Thallium	Tl	81	204.4
Thorium	Th	90	232.0
Thulium	Tm	69	168.9
Tin	Sn	50	118.7
Titanium	Ti	22	47.9
Tungsten	W	74	183.8
Unnilennium	Une	109	—
Unnilhexium	Unh	106	(263)
Unnilpentium	Unp	105	(262)
Unnilquadium	Unq	104	(261)
Unnilseptium	Uns	107	(262)
Uranium	U	92	238.0
Vanadium	V	23	50.9
Xenon	Xe	54	131.3
Ytterbium	Yb	70	173.0
Yttrium	Y	39	88.9
Zinc	Zn	30	65.4
Zirconium	Zr	40	91.2

CHEMISTRY TODAY 1

Third Edition

Chemistry Today 1, Third Edition
Components: Student Text
                    Laboratory Manual
                    Teacher's Guide

CHEMISTRY TODAY 1

Third Edition

R.L. Whitman,
Vice-Principal, Queen Elizabeth High School

E.E. Zinck,
Head of Chemistry Department, Acadia University

R.A. Nalepa,
Head of Science Department, Halifax West High School

Prentice-Hall Canada Inc., Scarborough, Ont.

To our wives: Gwen, Chérie, Lin.

Canadian Cataloguing in Publication Data

Whitman, R. L. (Ronald Laurie), date
    Chemistry today 1

3rd ed.
Previous eds. published under title: Chemistry today.
For use in secondary schools.
Includes index.
ISBN 0-13-129306-0

1. Chemistry.   I. Zinck, E. E., date.
II. Nalepa, R. A. (Robert Allan), date.
III. Title.

QD31.2.W48 1988   540   C88-093583-9

Supplementary Material: Laboratory Manual and Teacher's Guide

© 1976, 1982, 1988 by Prentice-Hall Canada Inc.

PRENTICE-HALL, INC., Englewood Cliffs, New Jersey
PRENTICE-HALL INTERNATIONAL, INC., London
PRENTICE-HALL OF AUSTRALIA, PTY., LTD., Sydney
PRENTICE-HALL OF INDIA, PVT., LTD., New Delhi
PRENTICE-HALL OF JAPAN, INC., Tokyo
PRENTICE-HALL OF SOUTHEAST ASIA (PTE.) LTD., Singapore
EDITORA PRENTICE-HALL DO BRASIL, LTDA., Rio de Janeiro
PRENTICE-HALL HISPANOAMERICANA, S.A., Mexico

ISBN 0-13-129306-0

Project Editor: Karyn Goldberger/Julia Lee
Production Editor: Elynor Kagan/Ian Chunn
Manufacturing: Pamela Russell
Design: Blue line Productions Inc.

1 2 3 4 5 6   93 92 91 90 89 88

Typesetting by Trigraph Inc.

Printed and bound in Canada by The Bryant Press

Contents

Credits   xv
Acknowledgments   xvi
Preface   xvii

1   The Science Called Chemistry   1

1.1   Why Study Chemistry?   2
1.2   The Place of Chemistry in Science   6
1.3   The Methods of Scientists   10

2   Measuring and Problem Solving   22

Part One: Communicating Chemical Research   23

2.1   Communication and Science   23
2.2   SI: The International System of Units   25
2.3   Exponential Numbers and Scientific Notation   29
2.4   Uncertainty and Significant Digits   32

Part Two: Solving Problems   37

2.5   Problem Solving and Dimensional Analysis   37

3   Matter   47

Part One: Matter and Its Changes   48

3.1   Matter and Energy   48
3.2   The Three States of Matter   52
3.3   Changes in States of Matter   54
3.4   Physical and Chemical Properties   55
3.5   Physical and Chemical Changes in Matter   57
3.6   Classification of Matter   58
3.7   Separating Mixtures   61

Part Two: The Laws of Chemical Change   66

3.8   Law of Conservation of Mass   66
3.9   Law of Constant Composition   66
3.10   Law of Multiple Proportions   68

4   Theories of the Atom   78

Part One: Early Developments of the Atomic Model of Matter   80

4.1   Early Models of Matter   80

4.2 Atomic Mass 83
4.3 Electrons 84
4.4 Protons 87

Part Two: Development of the Nuclear Model of the Atom 88

4.5 Rutherford's Model of the Atom 88
4.6 Isotopes 92
4.7 The Bohr Model of the Hydrogen Atom 96
4.8 Energy Levels and Many-Electron Atoms 103

Part Three: The Present Model of the Atom 105

4.9 Wave Mechanics—The Present Model 105
4.10 Wave Mechanics and Many-Electron Atoms 109
4.11 The Quantum Numbers 114

5 The Elements and the Periodic Law **125**

Part One: The Elements 126

5.1 Discovery of the Elements 126
5.2 "It's Elementary, My Dear..." 128
5.3 Metals, Metalloids, and Nonmetals 131
5.4 Symbols of the Elements 133

Part Two: Periodic Law and the Periodic Table 134

5.5 Periodic Law 134
5.6 The Modern Periodic Table 137
5.7 The Periodic Law and Energy Levels 141
5.8 Electron Dot Symbols 145
5.9 The Periodic Table and Electron Configurations 146

Part Three: Trends in the Periodic Table 150

5.10 Atomic size 150
5.11 Ionization Energies 152
5.12 Electron Affinity 155

6 Chemical Bonding **164**

Part One: The Main Types of Chemical Bonds 165

6.1 The Role of Electrons in Bonding 165
6.2 Ionic Bonding 167
6.3 Covalent Bonding 172

Part Two: Basic Concepts of Chemical Bonding 175

6.4 Polar Covalent Bonding and Electronegativity 175
6.5 The Bonding Continuum 179
6.6 The Octet Rule 182
6.7 Multiple Bonds 185

Part Three: Some Additional Concepts of Chemical
              Bonding   187

    6.8   The Shapes of Molecules—VSEPR Theory   187
    6.9   Coordinate Covalent Bonding   192
    6.10   Chemical Bonding and Quantum Mechanics   195

7   Nomenclature and Formula Writing   210

Part One: Oxidation Numbers   211

    7.1   Assigning Oxidation Numbers   211
    7.2   Using Oxidation Numbers to Write Chemical
                   Formulas   216

Part Two: Chemical Nomenclature   220

    7.3   Naming Chemical Compounds Containing Only Two
                   Elements   220
    7.4   Naming Chemical Compounds Composed of More
                   Than Two Elements   224
    7.5   Nomenclature of Acids   227
    7.6   Reading Chemical Equations   230

8   The Liquid Phase   237

Part One: The Liquid Phase   238

    8.1   The Structure of Liquids   238
    8.2   Attractive Forces in the Liquid Phase   239
    8.3   Physical Properties of Liquids   240
    8.4   Freezing   241
    8.5   Evaporation and Boiling   241
    8.6   Distillation   244

Part Two: Water   247

    8.7   Occurrence of Water   247
    8.8   Physical Properties of Water   247
    8.9   Chemical Properties of Water   249
    8.10   Heavy Water   251
    8.11   Hard Water   252

Part Three: Water Pollution   254

    8.12   Causes of Water Pollution   254
    8.13   Purifying Polluted Water   258
    8.14   Water Pollution in Canada   259

9   Chemical Reactions   269

Part One: The Nature of Chemical Reactions   270

    9.1   Introduction to Chemical Reactions   270
    9.2   Chemical Reactions—The Rearrangement of Atoms   271

9.3 Using the Energy of Chemical Reactions   273

9.4 Factors Affecting the Rate of a Chemical Reaction   276

9.5 Recognizing a Chemical Reaction   279

Part Two: Chemical Reactions and Chemical Equations   282

9.6 Types of Reactions   282

9.7 Balancing Chemical Equations   284

9.8 Information Obtained from a Balanced Chemical Equation   288

Part Three: Reactions Involving Electron Transfer   291

9.9 Identifying Electron Transfer Reactions   291

9.10 The Activity Series of Metals   296

10   Solutions   **306**

Part One: Characteristics of Solutions   308

10.1 General Features of a Solution   308

10.2 Types of Solutions   308

10.3 Applications of Solutions   309

Part Two: Solubility   310

10.4 The Solution Process   310

10.5 Factors Affecting Solubility   312

Part Three: Two Classes of Solutes—Electrolytes and Nonelectrolytes   316

10.6 Electrolytes and Nonelectrolytes   316

10.7 Dissolving Ionic Substances in Water   319

10.8 Dissolving Polar Molecules in Water   320

10.9 Like Dissolves Like   321

Part Four: Reactions of Ions in Solution   324

10.10 Precipitation   324

10.11 Writing Ionic and Net Ionic Equations   325

10.12 Applications of Precipitation Reactions   328

10.13 Qualitative Analysis of Aqueous Solutions   328

11   The Mole and Its Use   **338**

Part One: Introducing the Mole   340

11.1 Avogadro's Number and the Mole   340

11.2 Determining Avogadro's Number   342

11.3 The Mole Triangle   343

11.4 Molecular Mass and Formula Mass   345

11.5 Percentage Composition   346

Part Two: The Mole and Chemical Formulas   349

11.6   Expanding the Mole Concept   349
11.7   Calculating Empirical Formulas   352
11.8   Molecular Formula Determination   354

Part Three: The Mole and Chemical Equations   358

11.9   Relating the Mole to Chemical Equations   358
11.10   Mass Calculations Involving Chemical Equations   359
11.11   Limiting Reagents in Mass Calculations   361
11.12   The Yield of a Chemical Reaction   362
11.13   The Mole and Industrial Chemical Reactions   365

Part Four: Calculations Involving Solutions   368

11.14   Concentration Measured in Moles Per Litre   368
11.15   Concentration Measured in Percentage by Mass   372

12   Acids, Bases, and Acid Rain   384

Part One: The Nature of Acids and Bases   385

12.1   Properties of Acids and Bases   385
12.2   Hydronium and Hydroxide Ions   387
12.3   Neutralization and Salts   390
12.4   Acid-Base Titration   394
12.5   Sources and Uses of Acids and Bases   399

Part Two: Acids and Bases—Additional Concepts   400

12.6   The pH Scale   400
12.7   Determining pH with Indicators   402
12.8   The Brønsted-Lowry Theory: A More General Theory of Acids and Bases   402

Part Three: Acids in the Atmosphere   404

12.9   Sulfur Dioxide: London-Type Smog   404
12.10   Acid Rain   405

13   Gases   418

Part One: Gas Laws   419

13.1   Early Ideas About the Nature of Matter   419
13.2   The Discovery of Air Pressure   421
13.3   Robert Boyle   422
13.4   Boyle's Law   422
13.5   Charles' Law   424
13.6   Application of Boyle's and Charles' Laws   426
13.7   Standard Temperature and Pressure   429
13.8   Dalton's Law of Partial Pressures   430

Part Two: The Kinetic Molecular Theory and the Behaviour of Real Gases   433

13.9 A Model for Gases: The Kinetic Molecular Theory 433
13.10 Further Applications of the Gas Laws 436
13.11 Behaviour of Real Gases 437
13.12 Liquefaction of Gases 439
13.13 Cooling of Gases by Expansion 440
13.14 Heating of Gases by Compression 441

Part Three: Chemistry and the Gaseous Environment 443

13.15 Technology and the Environment 443
13.16 Changes in the Air Environment 444
13.17 The Effects of Solid Particles in the Atmosphere 447
13.18 Carbon Dioxide and the Greenhouse Effect 448
13.19 Photochemical or Los Angeles-Type Smog 450
13.20 Effects of Photochemical Smog 453
13.21 What Can Be Done about Air Pollution? 453

14 **Gases and the Mole** **463**

Part One: The Development of Avogadro's Hypothesis 464

14.1 Dalton's Problem 464
14.2 Gay-Lussac's Law of Combining Volumes 465
14.3 Avogadro's Hypothesis 466

Part Two: The Ideal Gas Law 471

14.4 The Derivation of the Ideal Gas Law 471
14.5 Using the Ideal Gas Law 472
14.6 Molar Mass and the Ideal Gas Law 473

Part Three: Calculations Involving Gases in Chemical Reactions 475

14.7 Volume-Volume Calculations 475
14.8 Mass-Volume Calculations 476

15 **Organic Chemistry** **488**

Part One: Saturated Hydrocarbons 489

15.1 Introduction 489
15.2 The Alkanes 491
15.3 IUPAC Nomenclature of Alkanes 495
15.4 Sources of Alkanes 501
15.5 Chemical Properties of the Alkanes 501
15.6 Cycloalkanes 503
15.7 Physical Properties and Uses of Alkanes 504

Part Two: Unsaturated Hydrocarbons 505

15.8 Alkenes 505
15.9 IUPAC Nomenclature of Alkenes 507

15.10 Sources of Alkenes   509
15.11 Reactions of Alkenes   509
15.12 Alkynes   511
15.13 Ethyne—A Typical Alkyne   511
15.14 Chemical Reactions of Ethyne   512
15.15 Alkadienes   513
15.16 Aromatic Hydrocarbons   514

Part Three: Compounds With Other Functional Groups   517

15.17 Functional Groups   517
15.18 Alcohols   518
15.19 Ethers   520
15.20 Aldehydes   521
15.21 Ketones   522
15.22 Carboxylic Acids   523
15.23 Esters   525
15.24 Other Types of Organic Compounds   526

Part Four: Organic Compounds in Industry   530

15.25 Petroleum   530
15.26 Polymers   532

16   The Chemical Industry, Technology and Society   548

Part One: The Relation of Science and Technology to Industry
            and Society   550

16.1 Science, Technology, and Industry   550
16.2 Technology and Society   552

Part Two: The Chemical Industry   555

16.3 Features of the Chemical Industry   555
16.4 Sources of Raw Materials for the Chemical
        Industry   557
16.5 Production of Important Industrial
        Chemicals   564
16.6 The Chemical Industry in Canada   570
16.7 Factors that Determine the Location of an Industrial
        Plant   570
16.8 Household Chemicals   571

Part Three: Canadian Technology   572

16.9 Canadian Contributions to Technology   572
16.10 Citizens of a Technologically-Based Society   579

17   Nuclear Chemistry   585

Part One: The Principles of Nuclear Chemistry   586

17.1 Radioactivity   586
17.2 Properties of a Radioactive Element— Radium   588

17.3 The Lead Block Experiment 589
17.4 Stability of the Nucleus 590
17.5 Balancing Nuclear Equations 592
17.6 Types of Nuclear Reactions 593
17.7 Radioactive Decomposition 593
17.8 Applications of Natural Radioactivity 595

Part Two: Splitting the Atom 596

17.9 Artificial Transmutation 596
17.10 Particle Accelerators 597
17.11 Neutrons Are Better Bullets 599
17.12 Fission 600
17.13 The Atomic Bomb 602

Part Three: Canada's Nuclear Industry 604

17.14 Nuclear Reactors 604
17.15 Canada and Nuclear Chemistry 604
17.16 AECL Power Projects 606
17.17 CANDU Reactors 606
17.18 Production of Heavy Water 611
17.19 AECL Commercial Products—Uses of Radioactive Isotopes 612

Part Four: The Nuclear Age 613

17.20 Breeder Reactors 613
17.21 Fusion 613
17.22 The Hydrogen Bomb 614
17.23 Ionizing Radiation 614
17.24 Problems of the Nuclear Age 615

Appendices

Appendix A International Atomic Masses 622
Appendix B Periodic Table 624
Appendix C The First Twenty Elements 626
Appendix D Vapour Pressure of Water at Different Temperatures 631

Glossary 632
Index 645

SPECIAL FEATURES

Biographies

Gerhard Herzberg   16
Geraldine Kenney-Wallace   28
Antoine Lavoisier   67
John Dalton   81
Ernest Rutherford   89
Niels Bohr   98
Humphrey Davy   129
Dmitri Mendeleev   135
Gilbert Lewis   165
Linus Pauling   177
Ronald J. Gillespie   190
Dorothy Crowfoot Hodgkin   194
Maud Menten   278
John C. Polanyi   280
Amadeo Avogadro   341
Svante Arrhenius   388
Robert Boyle   423
Jöns Jakob Berzelius   467
Joseph Louis Gay-Lussac   470
August Kekulé   515
Raymond U. Lemieux   522
Alfred Nobel   527
Paul de Mayo   528
Jacqueline Barton   551
Kelvin K. Ogilvie   578
Marie Sklowdowska Curie   587
Lise Meitner   601

Careers

Careers in Chemistry—An Overview   12
The Quality Control Chemist   70
The Chemistry Teacher   154
The Chemical Sales Representative   231
The Science Writer   246
The Analytical Chemist   374
The Chemical Engineer   580

Applications

Chemistry in Our World   5
Plastics: Tailor-Made Chemicals   10
A Chance Discovery   17
Learning from the Aztecs   25
Cherries in Chocolate and Chemical Change   58
Purifying Sea Water   64
From Line Spectra to Light Sources   102
"Seeing" Atoms   117
Chemicals in Space   128
How Many Atoms Make a Metal?   132
Element 109 and Beyond   140
Superconductivity   178
"Seeing" Chemical Bonds   181
The Importance of Molecular Shape   191
The Chemistry of Fireworks   226
Managing Hazardous Waste   262
Chemical Explosives   275
Artificial Blood   315
Antifreeze and Arctic Insects   318
Computers in Chemistry   356
Restoring the Challenger Tapes   398
The Lethal Lake   446
Warnings from Volcanic Gases   455
The Depletion of the Ozone Layer   469
A Strange-Shaped Carbon Molecule   503
New Drug with Sobering Effects   520
Voyager: High-Tech Aircraft   552
Canadian Biotechnology   554
Food Irradiation   573
Synthetic Fuels in Canada   577
Canada's TRIUMFant Particle Accelerator   598
Nuclear Disaster   616

Policy Statement

Prentice-Hall Canada Inc., Secondary School Division, and the authors of *Chemistry Today 1 Third Edition* are committed to the publication of instructional materials that are as bias-free as possible. This text was evaluated for bias prior to publication.

The authors and publisher of this book also recognize the importance of appropriate reading levels and have therefore made every effort to ensure the highest degree of readability in the text. The content has been selected, organized, and written at a level suitable to the intended audience. Standard readability tests have been applied at several stages in the text's preparation to ensure an appropriate reading level.

Research indicates, however, that readability is affected by much more than word or sentence length; factors such as presentation, format and design, none of which are considered in the usual readability tests, also greatly influence the ease with which students read a book. These and many additional features have been carefully prepared to ensure maximum student comprehension. The section on Readability (page xx of the Preface) describes those features. Further information is in the Introduction to the Teacher's Guide.

Photo Credits

Every reasonable effort has been made to find copyright holders of the photographs in this text. The publisher would be pleased to have any errors or omissions brought to its attention. We thank the following for permission to use their material:

Cover photograph: Sherman Hines / Masterfile

AECL: Figs. 17-14, 17-15, 17-16, 17-17, 17-18, 17-19, 17-20, 17-21, 17-22; pp. 573, 612. Paul L. Aird: pp. 444, 448. J. Barton: Fig. 16-3. The Bettman Archive: Figs. 3-20, 4-2, 4-8, 4-13, 5-2, 5-5, 6-1, 6-8, 6-16, 11-2, 12-2, 13-5, 14-3, 14-4, 15-17, 15-20, 17-1, 17-11. British Columbia Ministry of Tourism, Recreation and Culture: Fig. 8-11. British Columbia Parks and Outdoor Recreation Division: Fig. 8-1. Canada Starch Company Inc.: Fig. 15-1. Canadian Pacific: Fig. 16-14. Canadian Salt Ltd.: Fig. 15-2. Canapress: Fig. 17-13. Cape Breton Development Corporation: Fig. 16-7. P. de Mayo: Fig. 15-21. Environment Canada, Atmospheric Environment Service: Figs. 12-1, 12-6(a), 12-7, 13-11, 13-16. Farmers Cooperative Dairy: Fig. 15-1. Fisher Scientific: Figs. 12-5, 13-1. J. Franklin: Figs. 1-10, 9-1. R.J. Gillespie: Fig. 6-15. Hands Fireworks Inc.: p. 78. G. Herzberg: Fig. 1-9. Imperial Oil Ltd.: Figs. 3-1, 3-14, 15-12, 16-4, 16-5, 16-6, 16-8. G. Kenney-Wallace: Fig. 2-3. J. Lee: Figs. 8-10, 16-15; pp. 102, 106. Labatt's Ontario Breweries: p. 71. R.U. Lemieux/C.W. Hill Photography Ltd.: Fig. 15-19. Manitoba Hydro: Fig. 3-3. NASA compliments of SPAR Aerospace Ltd.: p. 552. National Research Council: Fig. 10-1. Nova Scotia Power Corporation: Fig. 11-1. Ontario Cancer Institute/Princess Margaret Hospital: Fig. 1-3. Ontario Ministry of the Environment: Figs. 8-15, 8-18, 11-6, 12-6(b), 13-19, 13-23. Jim Pattison Sign Group: Fig. 4-1. Perfection Foods, Fig. 15-1. J. Polanyi/Sven-Gösta Johansson, Pressens Bild AB: Fig. 9-7. Science North: Fig. 1-2. G.D. Searle: Fig. 7-3. O. Sharman: p. 95. SLAC: Fig. 17-8. M. Stabb: p. 452. Syncrude Canada: Fig. 16-13. TRIUMF: Fig. 17-7. University of Toronto Archives/*Torontonensis* Vol. 9, 1907, p. 174: Fig. 9-5. University of Toronto, Faculty of Forestry: Fig. 1-5. University of Waterloo, Central Photographic: Figs. 3-18, 6-7, 6-10, 6-12, 6-14, 7-2, 8-16, 14-1, 16-1, 16-2. University of Waterloo, Department of Earth Sciences: Fig. 5-3. The Upjohn Company, Kalamazoo, Michigan: Figs. 1-1, 2-1, 3-21. Wadsworth Publishing Company: Fig. 8-14. Uncredited photographs were supplied by the authors.

Acknowledgements

Once again, we would like to begin by thanking our publishers, Prentice-Hall Canada Inc., many of whose staff members have helped us since the inception of *Chemistry Today 1*. These staff members include Steve Lane, John Perigoe, Rob Greenaway, Ian Chunn, Julia Lee, and Pamela Russell. We thank Lesley Wood, Karyn Goldberger, and Elynor Kagan for their editing skills. We especially thank Steve for his many years of valuable assistance and advice. We would also like to acknowledge the work of Susan Howlett and Owen Sharman, who contributed the photo research, and Joe Stevens and Rick Eskins of Blue line Productions for their accomplished design and artwork.

It is an impossible task for us to completely acknowledge our debt to all those who assisted us in this revision of *Chemistry Today 1*. Many teachers and students gave us useful comments and suggestions based on their experience with the first two editions. We thank them for their contributions, and we accept responsibility for any flaws that still remain. We also thank Gwen Whitman for her assistance in preparing the manuscript.

Finally, we give special thanks to our wives, Gwen, Chérie, and Lin and to our children, Ian and Eric Whitman and Jayne and Adam Nalepa. Their patient acceptance of our many absences from home, during the two years it took us to complete this book, was essential to the success of this project.

R.L. Whitman
E.E. Zinck
R.A. Nalepa
Halifax, N.S.

Preface

Chemistry Today 1 Third Edition is an introductory high school program consisting of a student text, a laboratory manual, and a Teacher's Guide. The program is designed to provide students with the fundamental principles and concepts necessary to understand the methods of science and the nature of chemistry. *Chemistry Today 1* makes a special attempt to emphasize that science and chemistry are human pursuits carried on by ordinary people with individual strengths and weaknesses.

The program provides complete coverage of all major introductory chemistry topics. The text and the lab manual provide narratives and lab activities to develop, in a simple, straightforward manner, the concepts, skills, and attitudes helpful to students in their everyday lives.

The text begins with an introduction to the science of chemistry and goes on to deal with problem solving. Chapters 3 to 10 then cover the basic theories and laws of chemistry, including the periodic law, bonding, nomenclature, chemical reactions, and solutions. The following four chapters include the mole concept, acids and bases, and the gas laws. Chapters 15 to 17 form a concluding unit of descriptive chemistry which attempts to relate chemistry to the everyday world, to motivate students and to make chemistry more meaningful by showing that applications of chemistry are all around us.

Rationale for Sequence

The authors have a preferred sequence for an introductory chemistry course, and this is reflected in the scope and sequence for the text. Nonetheless, the authors recognize that teachers do not want to use a completely prescriptive textbook and want the freedom to design their own units of study. Therefore, the authors incorporated much more material than would normally be covered in a one-year chemistry course, and they included certain topics to challenge more capable students. This extra material gives the teacher the flexibility to design a course that is suited to his or her own teaching style, preference for course organization, and the individual abilities, goals, and interests of the students.

Much of the material contained in Chapters 1 to 3 may have been covered in earlier courses. Nevertheless, the authors feel that many of the topics should at least be reviewed at the beginning because they are a necessary prelude to an effective chemistry course. The time spent on these topics could, of course, vary according to the background of the students.

Chapters 4 to 6 constitute a second unit in which the fundamental concepts of chemistry are developed in a relatively traditional pattern. Because of their importance and the sequence in which the concepts are developed, the authors strongly recommend that all three chapters be covered in the order in which they appear in the text.

Chapter 4 traces the development of the modern theory of the atom through several model modifications. The wave mechanical model of the atom is developed in some depth to make it possible for students to describe the structure of the atom in terms of modern theory. However, this wave mechanical model is an optional topic, and the fundamental material presented in following chapters does not depend upon it. This chapters offers more material than some teachers will require their students to know in an introductory program. Individual teachers will decide how much of the modern theory is useful for their students.

Chapter 7 is a rather comprehensive description of the rules of chemical nomenclature. The teacher may decide either to cover it as a unit by itself, or to refer students to the pertinent sections when questions of nomenclature occur in later chapters.

Chapter 8 considers the liquid phase and also includes a discussion of water and water pollution. Some teachers may consider omitting part or all of this chapter.

Chapter 9, a discussion of chemical reactions, is essential for a student who will be continuing on to a second chemistry course, for it will contribute to a better understanding of descriptive chemistry. If this chapter is covered, it should be introduced after chemical bonding but before any descriptive chemistry is introduced.

Chapter 10 deals with solutions, and this chapter can be done as soon as students know how to balance chemical equations.

Chapters 11 to 14 describe the quantitative relationships generally considered relevant for students who will take further courses in chemistry. The chapters could easily be omitted by those teachers who desire a less mathematical approach. Other teachers may elect to cover Chapter 11 on the mole, and then omit Chapters 12 to 14. Some teachers may wish to cover Chapter 12 on acids and bases prior to dealing with the mole. This would be possible, except teachers would have to delay a study of Section 12.4 on acid-base titration until the mole was covered.

Chapters 15 to 17, being largely applications-oriented chapters, can be covered at any time after Chapter 9.

Changes for the Third Edition

Before work on this edition was begun, many teachers were asked for their opinion on how the Second Edition could be improved. *Chemistry Today 1 Third Edition* has been prepared with the resulting comments and suggestions in mind, but following the same basic educational goals as before.

The following are the major changes that have been made for the Third Edition:

1. **Content Coverage**

   - The treatment of the elements (formerly a part of Ch. 3) has been moved to Chapter 5, "The Elements and the Periodic Law."

   - Chapter 8 is now called "The Liquid Phase" and includes a study of water and water pollution (which were previously part of Ch. 11).

- Coverage of "Solutions" (formerly a part of Ch. 9) has been expanded and now covers an entire chapter (Ch. 10).

- The treatment of acids, bases, and acid rain (previously in Ch. 9) has been expanded and now is included in a separate chapter (Ch. 12). One of the new topics included in this chapter is acid-base titration calculations.

- Chapter 16 has been expanded and is now called "The Chemical Industry, Technology, and Society." New topics in the chapter include the relation of science and technology to industry and society, the importance of the chemical industry in Canada, and Canadian contributions to technology.

2. Improved Chapter Organization

Each chapter is divided into "Parts." Each part of a chapter consists of major sections, and these are numbered for easy reference. Within a section, the narrative is often further divided into subsections to help students follow the development of a particular topic more easily. Questions occur at the end of each part. In addition, there are questions and problems at the end of each chapter.

3. More Questions and Problems

The Third Edition of *Chemistry Today 1* contains more questions and problems than the two previous editions.

a) *Review Questions*, at the recall level, appear at the end of each part of a chapter.

b) *Example Problems and Practice Problems*: Example Problems contain step-by-step solutions and are integrated into relevant chapters. Practice problems follow the example problems for immediate reinforcement. The answers are generally provided for these problems as a study aid.

c) *End of Chapter Material*: There has been a major reorganization of the end of chapter assignment material in the Third Edition:

- *Test Your Understanding* are multiple choice exercises at the recall and comprehension levels and help students identify the important concepts of the chapter.

- *Review Your Understanding* are questions at the comprehension and application levels. These items are designed to ensure that students have mastered the essential chapter material. Mathematical problems are also included where appropriate to the content of the chapter.

- *Apply Your Understanding* are questions involving higher level thinking skills (application, synthesis, analysis, and evaluation).

● *Investigations* consist of self-designed and self-directed experiments, research and debate topics, and library assignments. These items are designed to reinforce chapter content and to take the student beyond the material in the chapter.

4. More Emphasis on Applications

Real-world applications continue to be integrated throughout the text to motivate students and make chemistry more meaningful. Examples include: common acids and bases, the pH scale, acid rain, photochemical smog, household detergents, water pollution, the diesel engine, air pollution, the chemical industry in Canada, petroleum, polymers, and Canada's nuclear industry. In addition, the Third Edition provides special features called "Applications." Each feature is about a page long and deals with an interesting application related to the chapter.

5. Addition of Career Profiles

Short descriptions of chemistry-related careers are now included in many chapters. This feature is provided to increase students' understanding of the role of chemistry in our world.

6. Inclusion of New Biographies

A number of new biographies have been added to the Third Edition, and these provide information regarding Canadians who have contributed to the field of chemistry.

7. Reinforcement of New Terms and Concepts

The summary at the end of each chapter is now in point form to enable students to better review the main ideas of the chapter. Key objectives at the beginning and key terms at the end of each chapter are retained in this edition.

8. Improved Format

The new edition features a revised format designed to make the material more readable. In addition, more line drawings and photographs appear in the Third Edition. Furthermore, the revision includes many pages in full colour.

Readability

Every effort has been made in the text to make it a highly readable book. The content was selected, organized, and written at a level appropriate for the intended audience. The readability of *Chemistry Today 1 Third Edition* was controlled for this target audience using standard methods of measuring readability levels.

There is a continuing problem regarding the testing of reading level. Some reading-level tests appear to be more valid for science books than others, and all seem to penalize science texts, which have special vocabulary needs. We have seen from experience that two people performing identical reading tests on the same book often come up with different results.

Research has shown that readability is more than just the number of syllables in a word and/or the length of sentences. Design, format, and motivating content are considered just as important in helping students to read. Thus, the following features were incorporated into the content and design of *Chemistry Today 1 Third Edition* to further enhance readability and understanding.

1. The text consists of 17 chapters. Each chapter, in turn, is divided into parts, and each part consists of several short sections.

2. Chapter parts and sections are listed at the beginning of each chapter. This organizes the chapter at a glance and shows the students how the chapter will develop.

3. A few general objectives are given at the start of each chapter to give the students an idea of what they are expected to be able to do upon completion of the chapter. More detailed learning objectives for each chapter are given in the Teacher's Guide.

4. Each chapter opens with a paragraph or two, providing the students with an overview of what will be covered in the chapter.

5. Narratives are divided into short segments using a consistent system of headings and subheadings. Thus students are able to make a distinction between the various levels of headings and their relative importance.

6. Numerous laboratory activities reinforce the text's content.

7. Whenever possible, the text relates the theory to the everyday world, motivating the students and making the content more meaningful by showing that applications of chemistry are everywhere. In addition, some applications of chemistry are developed in major sections or entire chapters.

8. Review questions occur at the end of every part of a chapter. Most of these are at the recall level and are designed to encourage students to read the material carefully.

9. Example problems, containing step-by-step solutions, and practice problems are integrated into relevant chapters.

10. Each chapter concludes with a listing of the main ideas (Summary) and a list of key terms to assist the students during review.

11. Each chapter concludes with a variety of questions, problems, and investigations.

12. The attractive format is designed to appeal to students and encourage use of the book. The text features an uncluttered single-column format. In addition, many pages are in full colour.

13. The printed word is reinforced with numerous line drawings and photographs placed appropriately in the text to create an inviting and unimposing format.

14. Key terms are boldfaced and defined in context when introduced.

15. Short descriptions of chemistry-related careers are included to increase students' understanding of the role of chemistry in our world.

16. Brief biographies of important chemists are located throughout the text to help show students that chemistry is a human pursuit.

17. The book concludes with a number of useful appendices, a glossary of terms, and a comprehensive index.

Laboratory Manual

The authors believe that an extremely important part of a student's study of chemistry is laboratory work. Chemistry is an experimental science and thus, to understand chemistry fully, a student must be involved in actual investigation. To this end, the Laboratory Manual contains about 50 class-tested experiments related to the student text. These labs are an integral part of the program and are designed to provide students with practical experience in the laboratory as well as to teach them skills in observation, inquiry, and analysis of data.

Each laboratory experiment consists of the following:

- the purpose of the experiment;
- a brief introduction;
- a list of required apparatus and materials;
- the procedure in a step-by-step format;
- questions that elicit observations, generalizations and conclusions.

To assist teachers with their use of the Laboratory Manual, the Teacher's Guide contains a chart for each chapter indicating where the experiments could be placed, in order to give teachers an indication of how the authors saw the lab activities being used. Of course, teachers are free to modify the placement of these experiments.

Teacher's Guide

A comprehensive Teacher's Guide will be available for *Chemistry Today 1 Third Edition*. It will be similar in format to the guide that was prepared for the Second Edition. The Introduction to the Teacher's Guide will offer an explanation of the program's basic methodology and a discussion of the flexibility inherent in the text and lab manual, and ways to organize courses based on the program.

In addition, for each chapter of the text, the Teacher's Guide will provide:

- an **overview** describing the major concepts in the chapter and giving the order in which they appear;

- a list of the **main topics**;
- a **rationale for the sequencing** of the chapter where required;
- a **suggested schedule** including time allotment and possible placement of laboratory experiments;
- comprehensive **learning objectives**;
- **teaching notes** for each part of the chapter. Depending on the chapter, teaching notes provide information such as the following: the concepts that should be emphasized, possible approaches to the treatment of the sections, questions to pose to the students, topics for class discussion, background information on some topics, the relationship of the material covered to other topics, the rationale for including certain sections, suggested demonstrations, and the advantages and disadvantages of various models and theories;
- the **answers** to all questions;
- a **guide to the experiments** which provides the following for each laboratory investigation included in the manual: intent of the investigation, suggested placement, apparatus lists, materials list (including directions for preparation of solutions), experiment notes (including safety precautions), and sample results;
- a bank of **multiple choice test items**.

The Science Called Chemistry

1

CHAPTER CONTENTS

1.1 Why Study Chemistry?
1.2 The Place of Chemistry in Science
1.3 The Methods of Scientists

Thousands of Canadians earn a living working with the science called chemistry. They work for industry, for government, and for schools, colleges, and universities. Others use their knowledge of chemistry in areas such as environmental control, toxicology, pharmacy, medicine, industrial hygiene, food science, library science, photography and in many other fields. Some of these people try to increase the amount of chemical knowledge. Others use their knowledge of chemistry to make items for industries. Still others try to pass their knowledge of chemistry on to their students. These people are scientists, engineers, and educators.

Figure 1-1
Chemists work for industry, for government, and in education.

What is science? What is the science called chemistry? Why do so many people enjoy working with chemistry? What are these people like? Are they different from the rest of us? What skills do they need? What methods do they use?

We attempt to answer these questions in this chapter. Of course, the answers that we can give in this introductory chapter only scratch the surface. Better answers come while you actually work with and study chemistry yourself.

Key Objectives

When you have finished studying this chapter, you should be able to
1. name and describe the type of knowledge upon which science is based.
2. describe the general nature of chemistry and its place among the sciences.
3. list the steps in the traditional scientific method.
4. explain why methods other than the traditional scientific method may be appropriate for the investigations of nature.
5. list and explain three requirements of scientific observation.
6. explain the difference between a law, a theory, and a model, and identify examples of each.

1.1 Why Study Chemistry?

In this section we attempt to answer the question, Is a study of chemistry worth the effort? When students are asked why they take a chemistry course, they usually give reasons that are chosen from a fairly short list. A few students take a chemistry course because they already know something about chemistry, and they enjoy it. Other students study chemistry because they generally do well in science courses, and they expect to succeed in chemistry also.

However, the most common reason is, "I need the course." Chemistry is a necessary course for many areas of study such as nursing, medicine, dentistry, pharmacy, engineering, and some other sciences. Chemistry is also necessary for some programs given at technical schools. Chemistry is required for so many areas of study that it is often referred to as the central science. A few students need another course to round out their high school program and they take chemistry because they prefer it to the other courses that are available.

Even if you do not need chemistry, we think there are some good reasons for taking a chemistry course. In the following subsections, we present some of these reasons.

Chemistry Affects Our Lives

One of the reasons that we need a knowledge of chemistry is that we are surrounded by the products of chemistry. We all make use of these products. Synthetic fibres such as nylon and polyesters are used to make clothing, carpeting and many other useful items. Plastics are used all around us. Grooming products such as deodorants, soaps, and hair sprays are products of chemistry. In fact, the list of chemicals found in our homes seems endless.

The home is not the only place where chemicals affect our daily lives. The chemical changes involved in the refining of petroleum and the additives that are put in gasoline have provided fuels that are much more efficient and less destructive to our automobile engines. In the last decade, chemists have synthesized new drugs that cure or arrest serious diseases. A vast array of pesticides and fungicides has dramatically cut the losses of agricultural crops and contributed to a cheaper, more abundant food supply.

Some beneficial chemicals must be used with care. For example, sodium nitrite is added to cured meats such as hot dogs and ham to improve colour and to stop the growth of the bacterium which causes a dangerous type of food poisoning (botulism). However, sodium nitrite reacts with other substances to produce products which could be harmful. Thus, its use as a food additive must be closely watched. Hexachlorophene is a useful germicide (a substance used to destroy germs) because it kills *Staphylococcus aureus* bacteria 125 times more effectively than phenol, a disinfectant that was formerly used widely in hospitals. (Phenol was responsible for a characteristic "hospital" odour.) However, hexachlorophene is readily absorbed through the skin. When that happens or when it is ingested in high concentrations, it is toxic. Thus, its use is restricted.

Occasionally, a basic knowledge of chemistry could have prevented people from making a tragic mistake. Families have died because they used a charcoal barbecue inside their homes. They did not know that burning charcoal gives off the odourless poisonous gas, carbon monoxide.

Another example in which a knowledge of chemistry can prevent tragedy comes from the farm. The fermentation of pig manure generates large quantities of gaseous methane and hydrogen sulfide. Hydrogen sulfide is an extremely poisonous gas with a characteristic rotten-egg odour. Farmers (and even their would-be rescuers) have died when they entered farm buildings in which there were high concentrations of these gases.

A third example of mistakes arising from chemical ignorance is that people have been injured by mixing bleach with acid-containing substances such as toilet bowl cleaners. They thought that the bleach and the cleaner would do twice as good a cleaning job. They did not know that bleach reacts with acids to produce toxic, irritating chlorine gas.

A knowledge of chemistry may therefore help you to avoid making a bad mistake. It will teach you to be cautious when you use any of the chemical products available to you.

Chemistry Affects Our Environment

For centuries, we have altered our environment to improve our quality of life. Will we be able to continue to modify the natural environment to suit ourselves? It appears that many of the things we have been doing to improve our lives are starting to threaten us.

Products of chemistry such as medicines, pesticides, and plastics have allowed us to live better and longer lives. However, they have not done so without cost. Many of the chemical plants that make these important products have been guilty of poisoning the surrounding environment.

Beneficial chemicals have had other undesirable side effects when used on crops. Pesticides destroy plant and animal pests and improve the productivity of farms, but at a cost. Rain washes the pesticides off the crops and into the ground-water. The pesticides then enter the biological food chain. The use of the insecticide DDT (now banned) vastly increased crop production, but it resulted in the near extinction of several species of birds when it entered the food chain, causing their egg shells to be very soft. Although DDT has not been used for many years, it is decomposing very slowly in the environment. Therefore, the problems caused by its heavy use decades ago still remain.

Plastics, too, have both positive and negative effects. Plastics have greatly increased the convenience of our lives, but they are made from fossil fuels such as petroleum. We use such enormous quantities of plastics that we are placing a further strain on our already-decreasing supplies of petroleum. Furthermore, we tend to throw away plastic products when they are no longer useful to us. Like DDT, plastics decompose in the environment only very slowly. The shorelines of every ocean in the world are littered with plastic objects that have been discarded by our society.

Knowledge Allows Informed Choices

Every day we are exposed to the claims of advertisers: "It helps keep you drier," or "It maintains the hair's normal pH balance." We hear conflicting or confusing statements that are supposed to come from scientists. For example, some scientists have said that we will need the electricity from nuclear power plants. Other scientists have disagreed; they insist that nuclear power stations are not safe enough. Some scientists have said that putting fluorides in drinking water to reduce tooth decay is a good thing. Again, other scientists have disagreed. It is difficult for many citizens to have reasoned opinions regarding the complex, science-related issues of our times.

Nevertheless, decisions are being made which affect the use of our environment every day. If we are to make intelligent decisions, we have to know the facts on each side of the questions. If our knowledge is lacking, we are at a disadvantage, and we may accept decisions which are not in our own interests.

Perhaps it would be pleasant to return to a less complex life style. However, as Canadians, we are citizens of one of the most technologically advanced countries in the world. We owe it to ourselves to become acquainted with the realities of modern life. Science is one of the most important of those realities. The first prime minister of India, Jawaharlal Nehru, said: "The future belongs to science and to those who make friends with science."

Many of the science-related issues that we face have a chemical basis. In order to understand the facts in these chemical issues, you will have to learn the basic principles of chemistry. With this knowledge you will be better able to make decisions and choices.

APPLICATION

Chemistry in Our World

If you were to ask the question, "What has chemistry contributed to the world we live in?", you might be surprised at the answer. You would find that chemistry has influenced nearly every aspect of your life.

Your health is affected by chemistry. Chemical compounds can reduce pain and swelling, relieve headaches, control high blood pressure, and fight cancer. New chemical materials are used to make artificial body parts such as the artificial heart. Chemists are working to produce synthetic blood and skin.

As the world population increases, one of the major problems that we will face is ensuring an adequate food supply. Chemistry is providing some solutions. Many chemical compounds are used as fertilizers to help increase plant growth and crop yield. Chemists are attempting to synthesize plant hormones and growth regulators to increase plant yield. Chemistry also plays an important role in preserving food.

Travel and exploration are influenced by chemistry. Chemistry plays a role in everything from the development of rubber tires and fuels to the materials that go into making the body of an automobile. The mass of the average automobile has decreased steadily in recent years due to the use of new light materials, resulting in increased fuel efficiency. By the year 2000, car bodies may be made largely of light-weight plastics, and the engines may be made of synthetic ceramic materials. These lighter engines will run more efficiently than conventional metal ones. Chemistry has also helped to produce cars which do not rust as easily as they once did.

In the area of space exploration, materials for spacecraft have been designed to be strong, light-weight, and heat-resistant. The spacecraft are propelled by special chemical fuels.

Most of our present energy needs are met by chemical technologies in which wood, coal, oil, and natural gas are used to provide heat. Synthetic fuels such as combustible liquids and gases from coal or oil-bearing shale, and alcohol from sugar cane or corn are being developed to supplement fuels obtained by conventional methods. A process has been developed to convert carbon monoxide and hydrogen gas from coal into gasoline. This process is possible because of chemical agents which speed up chemical reactions.

Many other areas of our lives are affected by chemistry. Our clothing has changed as a result of the production of synthetic fibres such as nylon, rayon, viscose, orlon, and polyesters. Other products of chemistry include cosmetics like make-up, nail polish, lipstick, and perfume, and toiletries like soap, toothpaste, and shampoos. Chemistry has affected sports and recreation through the production of stronger, more durable sports equipment made from plastics or graphite-reinforced epoxy resins. The way we decorate our homes has changed as new synthetic materials for carpets and furniture have been developed. Even the way we communicate has been influenced by chemistry. The components of computers, radios, televisions, telephones, recorders, and copying machines all have a chemical basis. Life without the products of chemistry would certainly be different from life as we know it today.

1.2 The Place of Chemistry in Science

In the last section you learned why it is useful to study chemistry. Now, we will answer two questions: What is science? Where does chemistry fit within science?

The word "science" is derived from the Latin word *scientia*, which means knowledge. **Science** is a human learning activity which is directed toward increasing our knowledge about the composition and behaviour of matter, both living and nonliving. Matter is the material which makes up the universe. Scientific knowledge is mainly empirical knowledge. **Empirical knowledge** is knowledge gained by using the senses. In a situation like the one shown in Figure 1-2 we use our sense of sight to know that a chemical reaction is occurring. Taste, touch, smell, and hearing are the other senses which we could use to obtain information about nature. Often we must use machines and measuring instruments which are more sensitive than any of our five senses to collect the information we need.

Figure 1-2
In this situation, we can easily *see* that a chemical reaction is occurring.

The Types of Knowledge

We have seen that scientific knowledge is mainly empirical knowledge, because it is gained from observation. Other types of knowledge are authoritative knowledge, rational knowledge, and intuitive knowledge.

Authoritative knowledge is knowledge that we gain from experts. For example, when we want to learn about legal matters, we consult lawyers. When we want to install new electrical circuits in our homes, we consult electricians. Other sources of authoritative knowledge include textbooks, journals, and encyclopedias.

Rational knowledge is knowledge based on what we consider to be logical truths. For example, geometry is based on a number of axioms (also called postulates) such as the axiom of equality: "Things that are equal to the same thing are equal to each other." These axioms are not actually proven. They are accepted because they are logically acceptable—they seem obvious to everyone. Thus, geometry is largely a product of rational knowledge. Rational knowledge is as important to science as empirical knowledge is. By using the rules of logic, scientists are able to use their empirical knowledge to develop theories and explanations for the observed behaviour of nature and make predictions about its future behaviour.

Intuitive knowledge is knowledge that people already possess without knowing where it came from. Perhaps a politician intuitively feels that an election should be called. Advisors (authoritative knowledge) may not have suggested that the time is right to call an election. The opinion polls (empirical knowledge) may not suggest that the party is ahead in popularity. An analysis of the strengths and weaknesses of the party and the pros and cons of going to the voters (rational knowledge) may not have led to the conclusion that the election should be called. Years of serving the public simply sharpen one's instincts regarding the public mood. The politician feels that the time is right to call the election but cannot tell you the exact source of this feeling (intuitive knowledge).

Occasionally the different types of knowledge lead to conflicting conclusions regarding a subject or event. For example, ancient philosophers believed that it was rational to expect a large rock to fall faster than a small one. However, empirical observation showed that they fell at the same rate. Some philosophers rejected the empirical evidence. They would not accept empirical knowledge when it conflicted with rational knowledge. However, others did accept the observations. When they accepted the empirical evidence over what many felt to be rational knowledge, these people were behaving as scientists. All scientists believe that careful observation yields accurate information. Therefore, observation is the main activity of science.

There are two types of observation. A *quantitative observation* involves measurement, and therefore numbers are used. For example, an observation that a ball has a diameter of 75 mm is a quantitative observation. The same ball, however, may also be blue in colour and cold to the touch. Observations such as these do not involve numbers. They are called *qualitative observations*.

Pure Science and Applied Science

We have seen that observation is the main activity of scientists. But science is divided into two types, depending on the way in which the scientists use the information obtained. **Pure science** is the search for knowledge by people who want to improve our understanding of the way the universe operates. The scientist who measures the effect of adding sulfur to rubber may be attempting to satisfy a curiosity abut the behaviour of that small portion of the universe. So may the scientist who studies the effect of small amounts of impurities on a crystal of germanium. These people use rational thought based on empirical knowledge to learn more about their surroundings. They are therefore involved in pure science.

On the other hand, science has a practical side, called applied science. **Applied science** (also called **technology**) is the application of scientific discoveries to practical problems. For example, pure scientists have found that the addition of sulfur to rubber improves the durability of the rubber. The applied scientist used this knowledge to develop the process of vulcanization, which gives us longer-lasting automobile tires. Similarly, a pure scientist discovered that the addition of small amounts of impurities to pure germanium converted it to a semiconductor. The applied scientist used this knowledge to develop transistors, which revolutionized the electronics industry. Pure scientists may have studied the effects of radioactivity on living cells, but applied scientists used this knowledge to develop cancer therapy machines (Fig. 1-3).

Figure 1-3
A cancer therapy machine. Atomic Energy of Canada Limited (AECL) has been the pacesetter in the development of radiation equipment for cancer therapy.

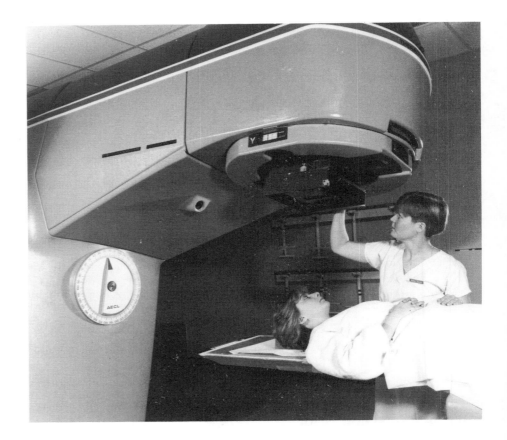

Thus, pure science is an essential basis for applied science (technology). In this book, we will use the word *science* to denote pure science and the word *technology* to denote applied science.

The Disciplines of Science

We have seen that science can be classified as either pure science or technology. Science can also be divided another way—into a number of disciplines. Among these are biology, chemistry, geology, and physics. The division is somewhat artificial. The scientific disciplines are all very much alike and related, because they are all based on empirical knowledge. In fact, sometimes it is difficult to distinguish between physical chemistry and chemical physics, or between biochemistry and molecular biology.

However, the divisions have been made because it is really impossible for one person to be a master of all science. Thus, scientists have been forced to specialize. Many of them specialize out of necessity rather than desire.

Let us examine the place of chemistry in science. It will help if we consider mathematics first. Mathematics is not really a science, since it does not depend on empirical knowledge as the other sciences do. It depends more on rational knowledge. Mathematicians have selected for study only a limited number of concepts, such as number and order. Physics takes the ideas of mathematics and adds further concepts selected from the real world. Matter, energy, and electricity are some of the concepts that physics adds to those of mathematics.

By including more concepts, such as the different kinds of matter and the transformation of one type of matter to another, we arrive at chemistry. Chemistry makes use of the concepts of mathematics and physics. Chemistry and physics are two of the sciences which study nonliving matter. They are called physical sciences. Geology is another physical science. It uses the concepts of physics and chemistry in an attempt to develop an understanding of the processes that occur in the earth's crust and in its interior. If we continue and select the concept of living matter, we leave the physical sciences and enter the area of biology.

Chemistry

Now that we have an idea of the position of chemistry in science, we must ask, What is chemistry? **Chemistry** is the science that deals with the composition, structure, and properties of matter. Consider water as an example. The chemist asks what substances make up water and how they are arranged in a molecule (smallest particle) of water. A chemist also asks about the properties of the water. At what temperature does it freeze? At what temperature does it boil? What happens if a piece of sodium metal is added to the water? Chemistry is concerned with the energy changes involved when water is transformed into other substances. A chemist therefore asks how much heat energy is given off or absorbed when the sodium reacts with the water.

Chemists deal with the smallest particles (molecules) that exhibit the particular properties of substances. Just as architects are people who design buildings and monitor their construction, many chemists can be thought of as molecular architects, since part of their job is to design and make (synthesize) molecules. A good example of the chemist's role as a molecular architect is the fact that aspirin and automobile tires are both derived from crude oil. Crude oil will not cure a headache, nor can you put it on the wheels of your car to provide a smoother ride. The crude oil must first be converted into petrochemicals which are then made into finished products such as aspirin and automobile tires. The chemicals obtained from a single 200 L barrel of crude oil can produce all of the following: 20 polyester shirts, five plastic garbage cans, 19 sweaters, 153 pieces of lingerie, and three inner tubes—with enough left over to fuel the average home for 13 days.

In 1950 there were 1 500 000 known chemical compounds. In 1985 there were approximately ten million known chemical compounds. About two thousand new compounds are discovered daily. We can conclude that there is a rapid growth in chemical knowledge. The rapid growth in chemical knowledge has been aided by new equipment like lasers and computers.

APPLICATION

Plastics: Tailor-Made Chemicals

Many chemists make a career of constructing (synthesizing) new materials with useful properties. The area in which this is most evident is the production of plastics. Common starting materials for the synthesis of plastics are petroleum, salt, and air. Chemists, through their knowledge of the composition, structure, and properties of matter, are able to transform these materials into plastics which are tailor-made to have desired properties. Chemists and chemical engineers have been so successful in the design and manufacture of useful plastics that the annual volume of plastics produced is greater than that of steel.

What properties do plastics have that make them more useful for certain applications than materials such as metal, glass, wood, or paper? Plastics can be designed to be either flexible or rigid. They are waterproof, lightweight and easily shaped. They can be designed to withstand high temperatures, to resist attack by other chemicals, to be nonflammable, or to be good electrical insulators. Plastics can be produced in a number of colours and textures, and they can be transparent, translucent, or opaque. Other properties that chemists can build into plastics include the ability to decompose on exposure to light (photodegradability) or to living organisms (biodegradability). From food containers to bullet-proof vests, the list of uses of plastics in our society seems endless.

1.3 The Methods of Scientists

We have seen that chemistry involves the study of matter. Now we shall examine the way in which chemists, and indeed all scientists, do their work, the rules that they follow, and some of their distinguishing characteristics.

In the past, writers of introductory texts have, in their desire to describe clearly the scientific method, given readers the impression that there is only one acceptable method. The general impression was that all scientists use the same method. This is unfortunate because it makes science look as if it were purely mechanical. Some readers may even feel that using the scientific method is an excuse for not thinking. The traditional form of the scientific method (Fig. 1-4) consists of a number of steps:

1. *State the problem*. For example, how does the volume of a gas depend on the pressure exerted on it?
2. *Make a series of observations*. For example, you might fill a large syringe with a known volume of gas such as air, plug the end of the syringe, and

then exert different pressures on the syringe, measuring the resulting volumes in each case.

3. ***Search for the regularities in the observations***. In your experiment you would use different starting volumes and gases other than air. Is there a pattern that is common to all the observations?

4. ***Formulate a hypothesis***. A **hypothesis** is a tentative explanation of the observed regularities.

5. ***Test the hypothesis***. If it is to be useful, a hypothesis must suggest new experiments that will either verify or refute the hypothesis. Conduct new experiments using other gases under different conditions. Are the results those which the hypothesis predicts? If they are not, the hypothesis must be revised.

6. ***Form a theory***. When enough observations have been made by many scientists working over a long period of time, there may be enough information to provide a general explanation for all of the observations. A theory, then, is a hypothesis that has stood the test of time.

Occasionally, an observation is made that cannot be accounted for by the theory. In such cases, it may be possible to modify the theory to fit the new facts.

This form of the scientific method is a good method, and it is acceptable to many scientists. However, any method which is a combination of curiosity and imagination and which uses careful observation to look for regularities in nature will be acceptable to scientists.

Scientists usually call their investigations research projects. They choose an area to study and conduct a set of experiments. An **experiment** is a planned series of observations. The use of experiments enables scientists to ask specific questions about nature rather then merely making unplanned observations and speculating as to their meaning. Some researchers might plan their projects very carefully. They might contemplate their methods, ideas, and proposed experiments for days. For these people, science is like a chess game— every move is carefully considered. Other scientists might be impatient and wish to get on with the job so that they can test their hunches. Some aspects of their problems may make them so curious that they cannot wait to try and find some answers. Thus, the methods that scientists use really depend on their personalities.

Figure 1-4
Schematic representation of the traditional scientific method

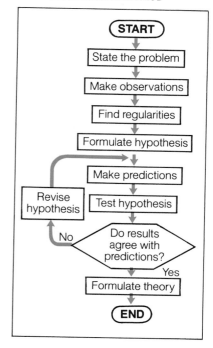

Rules of Experimentation

Scientists must follow certain rules if the results of their research projects are to be acceptable to other scientists. Since observation is the very lifeblood of science, these rules apply particularly to it.

Rule 1. A scientist is expected to be **objective** in making observations. Objectivity requires an unbiased reporting of all observations. We must be careful not to confuse an observation with an interpretation. For example, if two chemicals were mixed and a gas was given off, the recorded observation should be that a gas was evolved. A possible interpretation (explanation of the observation) would be that a chemical reaction had taken place. If five similar experiments give the same answer, and then a sixth gives a different answer, a

CAREER

Careers in Chemistry—An Overview

Figure 1-5 Laboratory technician—one of the many possible careers in chemistry

Chemists study matter and the changes that it undergoes. This is such a large area that chemistry is usually divided into a number of smaller branches. *Organic chemists*, for example, study the compounds of carbon. *Inorganic chemists* study the compounds of the other elements. *Biochemists* study the chemical processes of living systems. *Physical chemists* study the rates and energies of chemical reactions. *Analytical chemists* determine the purity and composition of substances. *Chemical engineers* design and operate the industrial equipment that is used to produce chemicals. The dividing line between these branches is not sharp, and many chemists have extensive knowledge of more than one of these closely related areas.

The chemistry profession provides job opportunities in fields such as pure research, product and process development, technical marketing, market research, instrumentation, environmental protection, forensic chemistry, teaching and information science. In addition, a knowledge of chemistry is valuable in the health sciences, for example, in medicine, dentistry, veterinary medicine, pharmacology, clinical chemistry, and occupational health and safety. It is also useful in areas such as science journalism and publishing, advertising, patent law, and investment counselling.

Industry employs many chemists in areas such as food, metallurgical and petroleum processing, the production of paper, and the manufacture of plastics, pharmaceuticals, and other chemicals. Jobs are available in purchasing, process control, quality control, research, product development, health and safety, environmental control, sales, marketing, public relations, and advertising.

The federal and provincial governments employ chemists to monitor and protect the environment and public health (by studying the effects of acid rain or testing the safety and effectiveness of new drugs, for instance), in forensic laboratories, hospital laboratories, consumer protection agencies, and pollution control agencies. Many chemists work in educational institutions such as universities, community colleges, high schools, and vocational schools, all of which require teachers with a wide knowledge of chemistry.

scientist is not allowed to discard the different or anomalous (irregular) result. Scientists believe that nature does not play tricks. That is, nature does not try to trick people by causing events to happen one way on ninety-nine occasions and another way on the hundredth occasion for no reason at all. There must be a logical explanation for an anomalous event. Time and time again scientists have found that what they thought was an anomaly was really due to a lack of knowledge. Often an anomaly will lead to a new set of observations which might well add to our understanding of nature. Consequently, since anomalous events can turn out to be disguised blessings, it is the scientist's duty to report faithfully all of the observations made (unless, of course, a mistake was made in the experiment).

A number of years ago, a chemist working for a famous company was using a gas called tetrafluoroethylene (TFE). One morning, he discovered that no TFE would come out of its cylinder. He established that no TFE had leaked out. He then discovered that something rattled inside the cylinder. His curiosity led him to cut it open. Inside, he found a white, slippery solid. A new product, Teflon®, was discovered because the chemist did not ignore an anomaly. He felt that there must be TFE in the steel cylinder, and he was determined to find out what had happened to it.

Rule 2. Measurements must be **reproducible**. Scientists believe that all measurements made of the same object or event under the same conditions must agree, regardless of when or by whom the measurements are made. In order for an experiment to be reproducible, all other variables (factors that can change) which *might* also affect the observations must be held constant or "controlled." For example, if we were performing an experiment to determine the effect of pressure on the volume of a gas, we would vary the pressure and observe any changes in the volume of the gas. If we did not keep the temperature or the mass of the gas constant, they might affect the volume of the gas. This would lead to false conclusions about the relation between the pressure and the volume of a gas. The concept of reproducibility follows from the idea that nature does not play tricks.

Rule 3. A scientist must record a measurement so as to convey the level of precision of the measurement. **Precision** is the term used to describe how well a group of measurements made of the same object or event under the same conditions actually do agree with one another (Fig. 1-6). Suppose a thermometer is used to measure the temperature of some hot water. Any number of observers would agree that the temperature is between 85°C and 86°C (Fig. 1-7). However, the observers would probably not all agree on the third digit. Some would record 85.5°C, and others would record 85.4°C or 85.6°C. In terms of the reproducibility of this measurement, the first two digits are *certain*, but the third digit is *uncertain*. Nevertheless, the third digit is useful because it tells us that the temperature is approximately halfway between 85°C and 86°C. The certain digits plus the first uncertain one are always considered to be *significant*.

Suppose the results recorded by seven people are 85.5°C, 85.7°C, 85.6°C, 85.6°C, 85.5°C, 85.4°C, and 85.6°C. The average of the results as displayed by a calculator capable of ten digits is 85.557 142 86°C. However, chemists choose to indicate only the significant digits. The average measurement is precise to

Figure 1-6
Precision is important. Which of these containers measures volume more precisely?

Figure 1-7
What is the temperature reading on this thermometer?

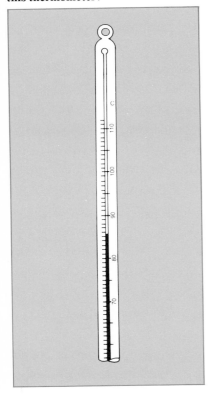

Figure 1-7
What is the temperature reading on this thermometer?

three digits. It is written as 85.6°C and contains three *significant digits*. The number of significant digits written for a measurement informs us of the precision of the measurement. The greater the number of significant digits used for a measurement, the more precise the measurement is.

Rule 4. A scientist must make accurate observations. **Accuracy** represents the closeness of a measurement to the true value. It is important that the scientist measure what is supposed to have been measured, not something else (Fig. 1-8). If a person is trying to determine the mass of a container, it must be clean and dry. If the container holds a few drops of water, the mass of the container plus the mass of the water will be obtained, not the mass of the container alone. In addition, the balance must be in good working order, and it must be properly adjusted so that it gives a zero reading when no object is on the balance pan.

Laws

Once scientists have made their observations, they seek regularities or patterns in these observations. **Laws** are statements of regularities found in observations made on a system. For example, suppose that we observed the relationship between the pressure and the volume of a gas. We noticed that each time the pressure on the gas was increased, the volume of the gas decreased. The observed regularity was that increasing the pressure on a gas decreases its volume. A law can now be stated in words such as: if the pressure on a gas increases, its volume decreases.

A law is simply a concise statement of a relation that is always the same under the same conditions. Often scientists accept a law because it is useful in

Figure 1-8
Accuracy is important. Is the mass of this beaker being determined accurately?

anticipating or predicting what will happen if a certain experiment is performed. Occasionally a law has to be modified to explain some new observation. Scientists are not overly upset that an established law has to be modified. Over the years, they have become used to the necessity of changing or discarding scientific laws as a result of the discovery of new knowledge.

Theories

A law is a rule that nature appears to follow. A law states what happens. It does not state why it happens. Thus, scientists are not satisfied with a law alone. They would like to be able to explain why it is true. They want to know more about the underlying principles associated with the law. A **theory** is an attempt to explain the observed regularity (the law). Experiments can never prove a theory. They can only provide evidence to support a theory. No matter how many experiments are found which yield results to support a theory, there is always a chance additional experiments will demonstrate a flaw in the theory. A theory is not a fact. It is simply a creation of the human mind. Many scientific theories are not perfect. They have known limitations. Nevertheless they are accepted because they lead to useful predictions, even though such predictions may not be accurate in every detail. A theory serves to bring together and unify many ideas, and hence an imperfect theory is generally not discarded until a better one is developed.

For example, the Kinetic Molecular Theory is the modern theory which explains the behaviour of gases. This theory has, as some of its postulates (basic assumptions), the statements

a) Gases are made up of small particles called molecules.
b) These molecules are in rapid random motion.
c) They move in straight lines until their directions are changed by collisions.
d) These molecules are widely separated.

This theory explains many different observations concerning gases, and predicts others that were not known when it was first formulated. It does not, however, explain the fact that gases will condense to liquids if they are cooled to low enough temperatures. Nevertheless, the theory is still widely used with only minor changes.

Models

Often, we must theorize about particles that are too small or too large for us to see. A **model** is a visualization of these particles. The model can be mechanical, as in Figure 1-9. It may be merely a mental picture of a system which is too large or too small for us to see clearly. Or it may be simply a series of equations describing mathematically the behaviour of the system. Models are useful in helping us to understand a theory. For example, the molecules described in the Kinetic Molecular Theory are much to small to be seen. As we cannot see these particles, it may be difficult to visualize them. We may resort to a model. We might visualize a closed room with a few ping-pong balls constantly moving around the room, colliding with the walls of the room and occasionally with one another in a random manner. These never-stopping ping-pong balls

Figure 1-9
This model is used to illustrate the movement of gas molecules. Ball bearings are kept in continuous motion by the vibrating walls of the container.
(a) This represents slower-moving molecules at a lower temperature.

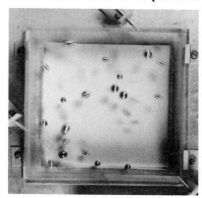

(b) This represents faster-moving molecules at a higher temperature.

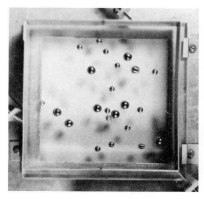

might constitute an acceptable model to help us understand the motion of gas molecules in a container. Of course, the model is not perfect. In reality, the ping-pong balls would fall to the floor and remain there.

Let us be clear about one thing. Theories and models, like laws, must come under constant inspection. There may be events that a certain theory cannot explain. The theory may have to be altered or discarded. Models, too, may need revision.

Moreover, models are only visualizations of the real thing. They are products of the scientists' imaginations. There may be an observation that will not fit a model or theory. We again use the word *anomaly*. The anomalous observation may contradict the accepted theory or model, but it cannot be disregarded. If the experiment was performed correctly, and if the observation was made correctly, this irregular behaviour may serve as the basis for either a new theory or a modified theory. Scientists have their "pet theories," and often they argue strenuously for them; but in the final analysis, they realize that all theories must be put on display for other scientists to examine. Scientists realize that no theory or model will ever be a perfect explanation of the events it was constructed to explain.

Biography

Figure 1-10
Gerhard Herzberg (b. 1904)—
Pioneer in atomic and molecular spectroscopy

GERHARD HERZBERG was born in Hamburg, Germany in 1904. He obtained his doctorate in engineering physics from Darmstadt in 1928. He continued his studies with postdoctoral years at Göttingen and Bristol.

The rise of Nazism in the mid-thirties drove Herzberg from Germany. He went to the University of Saskatchewan. Although the university lacked research equipment and funds at that time, Herzberg had a productive stay there. He slowly obtained equipment and worked with graduate students to produce a number of papers. He also wrote two books on molecular spectroscopy.

In 1945, Herzberg became a professor at the University of Chicago. He built up a laboratory suitable for conducting important experiments. In 1948, Herzberg, a Canadian citizen, accepted an invitation to set up a research laboratory in spectroscopy at the National Research Council (NRC) in Ottawa.

Herzberg was appointed director of the physics division of the NRC in 1949, and he became director of the division of pure physics when it was formed in 1955. In 1969, he became the first person to be appointed Distinguished Research Scientist with the NRC. This appointment enabled him to continue to do research following his retirement as director that year.

Although Herzberg was trained as a physicist, much of his work has been in chemistry. His major area of research has been molecular structure. He has been called the father of atomic and molecular spectroscopy. In 1971, he was awarded the Nobel Prize in Chemistry.

Characteristics of Scientists

Individual scientists do not differ significantly from individuals in other occupations. However, as a group, scientists do show some common characteristics.

Scientists generally have a high level of curiosity (Fig. 1-11). They seem to enjoy looking for answers to unsolved problems. The Nobel Prize-winning American chemist, Linus Pauling, said: "Satisfaction of one's curiosity is one of the greatest sources of happiness in life." Scientists tend to be persistent. They have to be able to cope with the anxiety and discouragement which can result from facing unsuccessful or inconclusive experiments day after day.

Successful scientists, like successful people in other occupations, possess high standards of honesty, a desire to know the truth, open-mindedness, and the ability to concentrate their attention on a problem.

Figure 1-11
Scientists can satisfy their curiosity about how the world works in many ways. This geologist is carrying out research in rock mechanics. She is studying the roughness of rock joints.

APPLICATION

A Chance Discovery

Not all discoveries in science are a result of a strict adherence to the traditional scientific method. Some scientific discoveries are accidental. The following is a good example of the role of chance in research. In 1964, Dr. Barnett Rosenberg, a professor of biophysics at Michigan State University, was studying the effects of electricity on the growth of bacteria. The experimental set-up involved inserting platinum electrodes, conductors of electricity, into a bacteria culture and allowing electricity to pass through it. Within a few hours all of the bacteria had ceased growing.

Chemical analysis of the culture showed that a chemical containing platinum was produced by the passage of the electrical current. Further research showed that this platinum-containing substance was responsible for inhibiting the cell division of the bacteria. Since a characteristic of cancer is uncontrolled cell division, Dr. Rosenberg speculated that this material might be useful as an anticancer drug. His idea was later confirmed, and in 1979 the chemical, known as cisplatin, was approved for use as a cancer chemotherapy agent.

Chapter Summary

- Chemistry affects our lives and has raised our standard of living.
- Care must be taken when using chemicals.
- A knowledge of chemistry can help us with certain science-related issues.
- Science is a human learning activity which is directed toward increasing our knowledge about the composition and behaviour of matter, both living and nonliving.
- Scientific knowledge is mainly empirical knowledge.
- Empirical knowledge is knowledge gained by using the senses.
- Pure science expands our knowledge and understanding of the universe.
- Technology (applied science) is the application of scientific discoveries to practical problems.
- Chemistry is the science which deals with the composition, structure, and properties of matter.
- Not all scientists follow the traditional scientific method.
- Scientists must follow certain rules if the results of their research projects are to be acceptable to other scientists.
- Laws are statements of observed regularities.
- Theories attempt to explain laws.
- Models help us visualize objects which are too small or too big for us to see.
- A hypothesis is a tentative explanation of observed regularities.
- An experiment is a planned series of observations.

Key Terms

science	chemistry	precision
empirical knowledge	hypothesis	accuracy
pure science	experiment	law
applied science	objective	theory
technology	reproducible	model

Test Your Understanding

Each of these statements or questions is followed by four responses. Choose the correct response in each case. (Do not write in this book.)

1. A container of ice cubes is colder than normal body temperature. This is an example of
   a) intuitive knowledge.
   b) rational knowledge.
   c) empirical knowledge.
   d) authoritative knowledge.

2. Scientists are attempting to prepare a powder that is a superconductor of electricity at room temperature. This is an example of
   a) pure science, because it is probably the result of the scientists' curiosity.

b) technology, because it will have useful applications.
c) both of the above.
d) none of the above.

3. The main activity of the scientist is
   a) proposing hypotheses.   b) making observations.
   c) formulating laws.   d) disproving theories.

4. An anomalous result
   a) differs from the expected result and must be discarded.
   b) agrees with the expected result and must be discarded.
   c) differs from the expected result and must not be discarded.
   d) agrees with the expected result and must not be discarded.

5. Which of the following is part of the traditional scientific method?
   a) theory   b) experiment
   c) hypothesis   d) all of these

6. A scientist must make
   a) objective observations.   b) reproducible observations.
   c) accurate observations.   d) all of these.

7. Four students determine the mass of an object. Their results are 16.51 g,
   16.52 g, 16.52 g, and 16.50 g. The actual mass of the object is 17.25 g.
   Which of the following statements best describes their measurements?
   a) Their measurements show poor accuracy but good precision.
   b) Their measurements show good accuracy but poor precision.
   c) Their measurements show good accuracy and good precision.
   d) Their measurements show poor accuracy and poor precision.

8. A scientific model can be
   a) a mechanical device.   b) a mental picture.
   c) a series of mathematical equations.   d) all of these.

9. Which of the following is best considered an example of a law?
   a) Light waves are like water waves.
   b) The total mass of the products of a chemical reaction always equals the
      total mass of the starting materials.
   c) The pressure exerted by a gas is caused by collisions of the gas particles
      with the walls of the container.
   d) All of these are examples of laws.

10. A theory is proved
    a) when enough observations have been made to support the theory.
    b) when scientists are unable to think of any objections to it.
    c) when an anomalous result is obtained.
    d) never.

Review Your Understanding

1. What is empirical knowledge?
2. What are two types of science? What is the purpose of each type? Give
   examples of each.
3. Which of the following are best considered as examples of pure science,
   and which as examples of technology? Why?
   a) The transistor has replaced the vacuum tube in its applications.

    b) It has been found that some of the so-called inert gases will react with fluorine.

    c) Fluoridation of drinking water appears to cause a significant decrease in the incidence of tooth decay in children.

    d) In 1942 Enrico Fermi constructed the first atomic pile in order to prove that nuclear fission could be controlled.

4. What is chemistry? How does chemistry differ from mathematics? How does it differ from biology?

5. Why are chemists referred to as "molecular architects"?

6. What is the traditional form of the scientific method? Are all scientists likely to use the same method? Why or why not?

7. Is there an activity that is common to all scientists?

8. What is an anomaly? Why must an anomaly not be disregarded?

9. Why are scientists forced to assume that nature does not play tricks?

10. List four rules that scientists must obey if their experimental results are to be acceptable to other scientists.

11. If you were walking down a street and noticed smoke coming from behind a building, would it be correct to say that you observed a building on fire? Explain your answer.

12. Distinguish between a law and a theory.

13. Do theories ever change? Explain your answer.

14. Which of the following statements is best considered as a law, and which as a theory?

    a) The pressure exerted by a given volume of a gas is directly proportional to its Kelvin temperature, provided the volume remains constant.

    b) Gases are made up of small particles called molecules.

    c) In all chemical reactions, the total amount of matter remains constant.

    d) A solution of salt in water conducts electricity because the salt has dissociated to form mobile positive and negative particles.

    e) Carbon monoxide is poisonous because it prevents the hemoglobin in the blood from carrying oxygen to the tissues.

15. Why would you use a model when describing a scientific theory?

16. "Even the scientific method, which skeptics suggest hampers creativity, is a mind-freeing method." How is the scientific method a mind-freeing method?

Apply Your Understanding

1. The Nobel Prize-winning chemist Linus Pauling has maintained that massive doses of Vitamin C are effective in preventing colds. Other scientists disagree with this hypothesis. What kind of experiments would you plan in order to prove or disprove Pauling's theory?

2. Suppose you were told by an advertiser that eating a certain product improves your memory. What kind of experiments would you plan in order to prove or disprove this claim?

3. If technology is the application of science to industrial or commercial objectives, can there be a similar technology which applies to either art or music? Support your answer with examples.

4. Explain how a police detective could use a scientific method during an investigation of a crime.

5. Why do you suppose that the public appears to have a more unfavourable view of chemicals and chemistry now than in the past?

6. To what extent is it desirable, or even possible, for governments to manage or direct scientific research completely?

7. Suggest a possible reason why the first prime minister of India said, "The future belongs to science and to those who make friends with science."

8. Gerhard Herzberg, who won the 1971 Nobel Prize in chemistry, has suggested that this country's most pressing problem is the development of Canadian research in pure science. Do you agree? Explain.

Investigations

1. For many years fluorides have been added to municipal water supplies to help in the prevention of tooth decay, yet many communities still refuse to fluoridate their water. Write a research report on the advantages and disadvantages of adding fluorides to drinking water. Do you think that the scientific evidence is greater on one side than on the other?

2. The consequences of scientific discoveries are so far-reaching that science cannot be left entirely to the scientists. The public must have some say in the uses to which scientific discoveries are put. Write a newspaper editorial on this topic.

3. Debate the following topic: More research effort should be put into solving societal problems such as teenage alcoholism than into pure research such as the exploration of space.

4. Write a research report on the effect of chemistry on one of the following:
   a) shelter
   b) national defence
   c) computers
   d) air travel

5. Devise and conduct an experiment to determine the effect that varying quantities of sugar dissolved in water will have on the boiling point of the water. Can you find any regularities in your observations?

2

Measuring and Problem Solving

CHAPTER CONTENTS

Part One: **Communicating Chemical Research**
 2.1 Communication and Science
 2.2 SI: The International System of Units
 2.3 Exponential Numbers and Scientific Notation
 2.4 Uncertainty and Significant Digits

Part Two: **Solving Problems**
 2.5 Problem Solving and Dimensional Analysis

Chemistry requires measurements. There can be no increase in the body of chemical knowledge unless someone carries out well-planned experiments under controlled conditions. Most of these experiments will involve careful measurement of variable quantities such as mass, volume, temperature, energy, and time. In this chapter we will see why it is important to make careful measurements. We will examine the ways in which the results are stated. Then we will examine some of the ways in which these results are manipulated to obtain even more information about a chemical system.

Key Objectives

When you have finished studying this chapter, you should be able to
 1. express the results of a measurement in the correct SI units.
 2. express large and small numbers using scientific notation.
 3. express the result of a measurement to the correct number of significant digits.
 4. identify the significant digits in a given measurement.
 5. express, rounded off to the correct number of significant digits, the results of calculations involving experimental measurements.
 6. solve problems involving measurements using dimensional analysis.

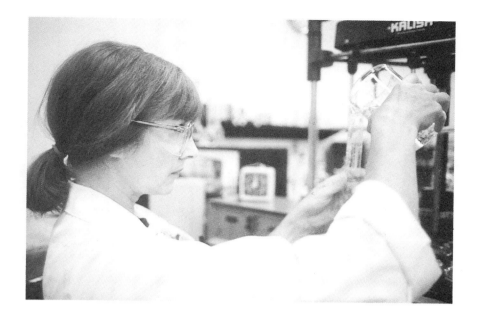

Figure 2-1
Measuring the volume of a liquid in a beaker

Part One: Communicating Chemical Research

2.1 Communication and Science

One of the main reasons for the recent rapid growth of scientific knowledge is that scientists are now able to communicate with one another very quickly. They are able to exchange experimental observations as well as hunches and theories. Obviously, it is quite wasteful for a scientist to spend time and money trying to discover something that has already been discovered by someone else. Also, it is important for a scientist to have the latest knowledge in the field; it may provide the key for making a great discovery. Thus, scientists feel obligated to communicate their findings to other scientists. They do so most often in written descriptions which are published in scientific journals. These journals are published in countries all over the world (Fig. 2-2). They contain descriptions of recent scientific research and are published weekly in some cases, but usually monthly. Because they are published frequently, these journals are reasonably up-to-date. Books are also used to communicate scientific information; however, scientific knowledge is growing so rapidly that books are often out-of-date by the time they are issued. Scientists often communicate their findings in lectures given at annual meetings of the various scientific societies. Occasionally special symposia are organized so that scientists investigating a certain area (for example, cancer research) can meet and discuss their research. Thus, because scientists communicate, scientific knowledge grows.

Figure 2-2
Some international chemistry journals

In addition to learning of new discoveries from journals and lectures, scientists also use knowledge from the past. Ideas from previous generations are used to construct modern theories. Scientists are not required to start from square one. They are able to build on what has already been learned, and there is much to absorb. This is the reason for the long period of training required by a modern scientist.

In order for scientific communication to be effective, it must have **universality**. That is, it must be interpreted the same way in Calgary as it is in Tokyo. A work of art need not have this universality. In fact, it could lose its appeal if everyone interpreted it the same way.

One of the best ways for scientists to communicate is through mathematical equations. Mathematics is the language of science, and mathematics adds to the universality of science. You have probably heard of Einstein's equation, $E = mc^2$. This is understood by scientists throughout the world to mean that energy equals mass times the square of the speed of light.

For such mathematical equations to have universality, scientists must all use the same units of measurement, and they must always report those units. If, for example, you read in a scientific report the statement, "The experiment required ten," you would immediately ask, "Ten what?" Did the experiment require ten cents, ten minutes, or ten grams of salt? With few exceptions, all numbers are quantities of something. It is not sufficient to say only the quantity (number)—a scientist must give the quantity of whatever is being talked about. The "whatever" is called the **units** or *dimensions* of the quantity. Thus, depending on what was being measured in the reported experiment, the units of the measurement were either "cents," "minutes," or "grams of salt." For scientific communication to be meaningful, a number must always have its units written after it.

Some scientific knowledge cannot be communicated by numbers or mathematical equations, and must be communicated by means of analogies. An analogy expresses a likeness. In Chapter 1, we said that gas molecules in motion are like ping-pong balls bouncing in a room. We may say that radio waves are like water waves. Analogies are useful in science, but they do not have the universality that an equation would have. However, there are times when analogies are the only means of communicating scientific ideas.

We have seen that it is important for scientists to communicate their results to others as quickly as possible. They must do so in such a way that no other scientists can possibly misunderstand the results. This is not as easy as it might sound. However, the use of standard units of measurement makes this communication easier.

APPLICATION

Learning from the Aztecs

A search of Spanish literature about plants known to the Aztecs in Mexico has led chemists to the discovery of an intensely sweet substance. In a book written by a Spanish physician around 1750, a sweet-tasting plant was described in such detail that researchers were able to locate it. The sweet-tasting substance, which is more than 1000 times sweeter than table sugar, was found mainly in the leaves and flowers of the plant.

The substance, called hernandulcin, was extracted and purified. Then, using sophisticated equipment, chemists determined its structure. Once the structure was known, chemists were able to prepare hernandulcin in the laboratory.

Hernandulcin shows promise as a sugar substitute. Toxicity tests have so far shown good results. One complaint about hernandulcin is that it tastes less pleasant than table sugar. However, chemists may be able to overcome this problem by slightly altering the structure of the substance. At the least, hernandulcin can be used in studies of the relationship between the sweetness of a substance and its structure.

2.2 SI: The International System of Units

All measurements are made by using or making a comparison with a standard measuring device. The odometer of a car measures distances in terms of a standard of length. A butcher finds the mass of a steak on a scale that measures in terms of a standard of mass. A service station attendant dispenses gasoline with pumps that are calibrated to measure in terms of an accepted standard of volume. In the same ways, chemists use their standards when measuring length, mass, volume, and other properties of a chemical system. Until recently, many standards of measurement were in use. It was necessary for scientists to be familiar with all of them.

Scientists made many efforts to agree on a single set of standards against which all measurements would be made. In 1960 the recognized international authority on units, the General Conference of Weights and Measures, adopted the International System of Units. The abbreviation in all languages is **SI** (for Le Système International d'Unités). Canada has adopted this system.

Base Units

The system is constructed from seven base **units**:

Physical Quantity	Unit	Symbol
Length	metre	m
Mass	kilogram	kg
Time	second	s
Temperature	kelvin	K
Amount of substance	mole	mol
Electric current	ampere	A
Luminous intensity	candela	cd

Of these, length, mass, time, temperature, and amount of substance are the most useful in chemistry.

The SI unit of *length* is the metre. A metre is defined as the distance light travels in a vacuum during 1/299 792 458 of a second.

The SI unit of *mass* is the kilogram, which is defined as the mass of a cylinder of platinum-iridium alloy kept at the International Bureau of Weights and Measures in Paris.

We have defined the metre and the kilogram merely to give you an idea of the care which is taken in defining base units. The other base units are defined just as carefully.

Derived Units

By combining various base units we obtain secondary measuring units, called derived units. Derived units have compound names. Those that have squares or cubes such as "square metre" are derived units. So are those which contain more than one unit in the name, e.g., "metre per second."

The SI unit of *area* is the square metre (m^2). The SI unit of *volume* is the cubic metre (m^3). The litre is now defined as one thousandth of a cubic metre. The use of "litre" as a unit of volume is permitted in SI.

There is also an SI unit of force. A force is any influence on a body which changes the state of rest or motion of that body. If a body moves in a straight line with a constant speed or if it is stationary, no net (unbalanced) force is acting on it. The SI unit of *force* is the newton (N), which is defined as that force which will give a mass of one kilogram a speed of one metre per second when applied for one second. In symbols,

$$1 \text{ N} = \frac{1 \text{ kg} \cdot \text{m/s}}{1 \text{ s}} = 1 \text{ kg} \cdot \text{m} \cdot \text{s}^{-2}$$

Notice that the unit of force is therefore derived from the base units of measurement. A force of about 10 N is required to lift a full one litre carton of milk.

Work is defined as the moving of matter against a force which opposes the motion. Energy is simply the ability to do work. The SI unit for *work* and *energy* is the joule (J), which can be defined as the work done by a force of one newton when acting through a distance of one metre. In symbols,

$$1\,J = 1\,N \cdot m = 1\,N \times 1\,m = 1\,kg \cdot m \cdot s^{-2} \times 1\,m = 1\,kg \cdot m^2 s^{-2},$$

and again the definition is derived from the base units. Lifting a full one litre carton of milk from the kitchen floor to the counter top takes about ten joules of energy.

Pressure results when a force acts perpendicular to a surface. For example, atmospheric pressure is the result of the mass of the atmosphere pressing down on the earth's surface. The pascal (Pa) is the SI unit of *pressure*. It is the force of one newton acting over an area of one square metre. In symbols,

$$1\,Pa = \frac{1\,N}{1\,m^2} = \frac{1\,kg \cdot m \cdot s^{-2}}{1\,m^2} = 1\,kg \cdot m^{-1} \cdot s^{-2}.$$

Again, the definition is derived from the base units. One kilogram of sugar spread evenly over the surface of a card table exerts a pressure of about ten pascals. Atmospheric pressure is about 100 000 Pa.

The commonly used derived units are summarized in Table 2-1.

Table 2-1   Some Derived Units of Measurement

Quantity	Unit	Symbol	Derivation
Area	square metre	m^2	—
Volume	cubic metre	m^3	—
Speed	metre per second	m/s	—
Density	kilogram per cubic metre	kg/m^3	—
Force	newton	N	$kg \cdot m/s^2$
Pressure	pascal	Pa	N/m^2
Energy, Work	joule	J	$N \cdot m$
Power	watt	W	J/s

Table 2-2   Prefixes for Multiples of SI Units and Fractions

Prefix	Symbol	Factor
exa	E	10^{18}
peta	P	10^{15}
tera	T	10^{12}
giga	G	10^{9}
mega	M	10^{6}
kilo	k	10^{3}
hecto	h	10^{2}
deca	da	10^{1}
deci	d	10^{-1}
centi	c	10^{-2}
milli	m	10^{-3}
micro	μ	10^{-6}
nano	n	10^{-9}
pico	p	10^{-12}
femto	f	10^{-15}
atto	a	10^{-18}

Fractions and multiples of SI units are expressed by the appropriate prefixes (Table 2-2).

The prefixes in Table 2-2 tell us how many places we must move the decimal point to get the new unit. For example, moving the decimal point three places to the right (multiplying by 1000) is represented by the prefix *kilo*. Thus a *kilo*metre is a thousand metres, a *kilo*joule is a thousand joules, and a *kilo*pascal is a thousand pascals. A *pico*second is a trillionth (10^{-9}) of a second. Professor Geraldine Kenney-Wallace (Fig. 2-3) of the University of Toronto studies chemical events that take place in less than a picosecond.

SI Style Conventions

In Canada, certain conventions of style must be used when writing measurements in SI units. Among these are the following:

1. Digits are to be grouped in threes about the decimal point without using commas to mark off thousands, millions, and so on. Thus, write 1 234 567.891 23 and not 1,234,567.89123. The space is unnecessary if there are only four digits to the left or right of the decimal point (e.g., 4567.8912), unless the numeral is listed in a column with other numerals of five digits or more.

2. A space is left between the last digit of a numeral and the SI symbol. However, the degree symbol occupies this space when writing Celsius temperatures (25°C). (Although the kelvin is the SI base unit for measurement of temperature, ordinary thermometers measure temperature in degrees Celsius.)

3. There is no period at the end of SI symbols; e.g., write "kg" not "kg.".

4. Units are not pluralized. Thus, a mass of ten kilograms is written as "10 kg", not "10 kgs".

5. Use product dots to avoid ambiguities. For example, ms could mean either "metre-second" or "millisecond." The symbol "m · s" is unambiguous: metre-second.

Biography

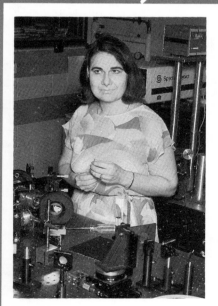

Figure 2-3
Geraldine Kenney-Wallace—
Studying rapid chemical reactions

GERALDINE KENNEY-WALLACE began her study of chemistry in England but obtained her doctorate from the University of British Columbia. Dr. Kenney-Wallace spent a year as a researcher at the University of Notre-Dame and two years as a faculty member at Yale University. She then returned to Canada, joining the Department of Chemistry at the University of Toronto.

Dr. Kenney-Wallace has studied chemical processes that require very little time, in the order of trillionths of a second (picoseconds). She has devised experiments to produce extremely fast laser pulses in order to observe these very short chemical events. Techniques like hers have enabled chemists to probe chemical reactions in time frames that are shorter than the lifetime of even the most reactive chemical substances. These new avenues of study will clarify the factors that determine the rates of chemical reaction. We can expect exciting discoveries from this research in the next decades.

Dr. Kenney-Wallace is an internationally known chemical researcher. In addition to financial support from Canadian sources, funding for her research has come from the American Chemical Society, the Petroleum Research Fund, and the U.S. Office of Naval Research. In 1987, she was appointed to the chair of the Science Council of Canada. In this position, she is the government's chief adviser on science and technology.

6. Words and symbols should not be mixed. One may write "kilograms per cubic metre" or "kg/m^3" but not "kilograms/m^3" or "kg per cubic metre."

7. Symbols should be used in place of full names when units are used in conjunction with numerals. For example, write 180 cm rather than 180 centimetres.

8. Where possible, in the expression of any quantity, a prefix should be chosen so that the numeral lies between 0.1 and 1000. When similar items are compared, however, it is preferable to use the same prefix for all items.

For example, instead of writing 25 000 m, it is preferable to write 25 km. When comparing a series of distances such as 1.5 km, 18 km, 45 Mm, it is preferable to use the same prefix for all items (1.5 km, 18 km, 45 000 km) even though the last item has a numeral that does not lie between 0.1 and 1000.

We have just seen that a measurement consists of a numeral and its SI unit. In Section 2.3, we will see how calculations involving numerals of all sizes (including very large and very small ones) can be performed.

2.3 Exponential Numbers and Scientific Notation

Chemists frequently deal with very large and very small numbers. For example, the average mass of a single carbon atom is 0.000 000 000 000 000 000 000 019 92 g. How many carbon atoms are required to give a total mass of 12.00 g? Imagine how tedious it would be to carry out the division involved in the relation

$$\frac{12.00 \text{ g}}{0.000\ 000\ 000\ 000\ 000\ 000\ 000\ 019\ 92 \text{ g/C atom}}$$

Decimal points have a habit of slipping into the wrong places during such operations, and the answer is a large number indeed.

Exponential Notation

In order to handle both very large and very small numbers, scientists use a technique known as exponential notation. To illustrate this technique, let us consider a simple example. The number 2700 can be written as

$$2700 = (270 \times 10) = (27 \times 100) = (2.7 \times 1000) = (0.27 \times 10\ 000)$$

If we recall that $100 = (10 \times 10) = 10^2$, $1000 = (10 \times 10 \times 10) = 10^3$, and $10\ 000 = (10 \times 10 \times 10 \times 10) = 10^4$, we can write

$$2700 = (270 \times 10^1) = (27 \times 10^2) = (2.7 \times 10^3) = (0.27 \times 10^4)$$

Notice that each of these consists of a number (the coefficient) multiplied by ten raised to an integral power (the exponent). Each is therefore an example of exponential notation.

Scientific Notation

Any measurement which consists of a coefficient multiplied by a power of ten is said to be expressed in exponential notation. Therefore, 2.24×10^1, 224×10^{-1}, and 0.0224×10^3 are all examples of exponential notation. However, if the coefficient is 1 or a number between 1 and 10, the number is said to be expressed in *scientific notation*. In scientific notation, therefore, $2700 = 2.7 \times 10^3$. For a number greater than one, the exponent of the 10 is the same as the number of positions the decimal point must be moved *to the left* to get a coefficient between 1 and 10. E.g., $45\ 000. = 4.5 \times 10^4$.

In scientific notation, we write the number 0.027 as 2.7×10^{-2}. For any number less than one, the exponent of the 10 is a negative integer equal to the number of positions the decimal point must be moved *to the right* to get a coefficient between 1 and 10. For example, $0.000\ 77 = 7.7 \times 10^{-4}$.

EXAMPLE PROBLEM 2-1

Convert the following to scientific notation: **a)** 24 000; **b)** 0.000 000 61.

SOLUTION
a) The decimal point must be moved four places to the left to get a coefficient of 2.4. Therefore $24\ 000 = 2.4 \times 10^4$.
b) The decimal point must be moved seven places to the right to get a coefficient of 6.1. Therefore $0.000\ 000\ 61 = 6.1 \times 10^{-7}$.

PRACTICE PROBLEM 2-1

Convert the following to scientific notation: **a)** 0.000 052 **b)** 186 000

Multiplying Exponential Numbers

The advantage of scientific notation is that it simplifies the manipulation of very large and very small numbers. For example, to *multiply* exponential numbers, we multiply coefficients normally and add the exponents:

$$(3.0 \times 10^3)(2.0 \times 10^{-9}) = (3.0 \times 2.0)(10^3 \times 10^{-9}) = 6.0 \times 10^{3 + (-9)} = 6.0 \times 10^{-6}$$

Dividing Exponential Numbers

To *divide* exponential numbers, we divide the coefficients normally, then subtract the exponent of the denominator from the exponent of the numerator:

$$\frac{4.4 \times 10^6}{8.8 \times 10^{-4}} = \frac{4.4}{8.8} \times \frac{10^6}{10^{-4}} = 0.50 \times 10^{6-(-4)} = 0.50 \times 10^{10}$$

The above answer does not have a coefficient between 1 and 10. We can convert it to scientific notation as follows:

$$0.50 \times 10^{10} = 5.0 \times 10^{10-1} = 5.0 \times 10^9$$

That is, the exponent is decreased by one for each position the decimal point of the coefficient is moved to the right (and is increased by one for each position the decimal point of the coefficient is moved to the left) until the coefficient is a number between 1 and 10. The results of all computations in exponential notation should normally be expressed in scientific notation.

EXAMPLE PROBLEM 2-2

Express in scientific notation the result of the computation

$$\frac{(1.78 \times 10^{-4}) \times 2.41 \times 96.5}{0.724 \times (4.18 \times 10^5)(1.86 \times 10^{-16})}$$

SOLUTION

Treat the coefficients and exponents in groups:

$$\frac{1.78 \times 2.41 \times 96.5}{0.724 \times 4.18 \times 1.86} \times \frac{10^{-4}}{10^5 \times 10^{-16}} = 73.5 \times \frac{10^{-4}}{10^{-11}} = 73.5 \times 10^7$$

Convert to scientific notation: $73.5 \times 10^7 = 7.35 \times 10^8$

PRACTICE PROBLEM 2-2

Express in scientific notation the result of the computation

$$\frac{(3.68 \times 10^{-3}) \times (8.41 \times 10^2) \times 0.248}{1.36 \times (9.24 \times 10^4) \times (5.56 \times 10^{-2}) \times (6.35 \times 10^{-1})}$$

(*Answer*: 1.73×10^{-4})

Addition and Subtraction of Exponential Numbers

Addition and subtraction of exponential numbers is possible only if the exponents are all the same. Thus

$$1 \times 10^{-3} + 2 \times 10^{-3} + 3 \times 10^{-3} = (1 + 2 + 3) \times 10^{-3} = 6 \times 10^{-3}$$

Therefore, if the exponents are different, the numbers must first be converted to a form in which all exponents are the same. It is usually convenient, but not absolutely necessary, to convert all numbers so that they have the same exponent as the largest number in the group. Thus,

$$(1.00 \times 10^{-3}) + (2.00 \times 10^{-4}) + (3.00 \times 10^{-5})$$
$$= (1.00 \times 10^{-3}) + (0.200 \times 10^{-3}) + (0.0300 \times 10^{-3})$$
$$= (1.00 + 0.200 + 0.0300) \times 10^{-3} = 1.23 \times 10^{-3}$$

Similarly,

$$2.0 \times 10^{10} - 6 \times 10^9 = 2.0 \times 10^{10} - 0.6 \times 10^{10}$$
$$= (2.0 - 0.6) \times 10^{10} = 1.4 \times 10^{10}$$

EXAMPLE PROBLEM 2-3

What is the result of these operations?

a) $3.02 \times 10^3 + 1.3 \times 10^2$

b) $5 \times 10^{-6} - 8.2 \times 10^{-5}$

SOLUTION

a) $3.02 \times 10^3 + 1.3 \times 10^2 = 3.02 \times 10^3 + 0.13 \times 10^3 = (3.02 + 0.13) \times 10^3 = 3.15 \times 10^3$

b) $5 \times 10^{-6} - 8.2 \times 10^{-5} = 0.5 \times 10^{-5} - 8.2 \times 10^{-5} = (0.5 - 8.2) \times 10^{-5} = -7.7 \times 10^{-5}$

PRACTICE PROBLEM 2-3

What is the result of these operations?

a) $6.5 \times 10^{-4} - 2.1 \times 10^{-5}$

b) $5.3 \times 10^3 + 8.7 \times 10^4$

(*Answers*: **a)** 6.3×10^{-4}; **b)** 9.2×10^4)

So far, we have seen how scientists must choose the proper units for making a measurement. We have also examined some difficulties the scientist faces when dealing with very large and very small numbers. In the next section, we will see how scientists indicate the degree of confidence or certainty that is implied in their measurements.

2.4 Uncertainty and Significant Digits

A person who is caught for travelling 52 km/h in a 50 km/h zone might well ask the police officer some questions. Was the speed radar recently calibrated? That is, does it actually read 50 km/h when a car is moving at that speed? How reproducible is the radar reading? That is, if the officer took five different measurements of the speed, would they all be 52 km/h? Or would there be a range of values, say from 48 to 52 km/h? Should this person get a ticket? That would depend on how reproducibly the radar can measure the speed, and on how close that measurement is to the actual speed.

Reading Measuring Instruments

Like police officers, scientists do not deliberately make false claims about readings obtained from measuring instruments. However, if they do not pay attention to the degree of reliability of their measurements, they may report results which imply more (or less) certainty than can be justified.

Let us consider again the example of the thermometer (Fig. 2-4). Suppose that the temperature shown is reported by a number of people as 86°C, 85.7°C,

85.6°C, 85.55°C, 85.5°C, 85.4°C, and 85°C. Two people have recorded only two digits in their measurements. This is unsatisfactory because it is reasonably easy to tell that the temperature is about half-way between 85°C and 86°C. These two people have implied less certainty in their temperature readings than seems reasonable. One person has recorded a four digit measurement. Perhaps this person felt the temperature was slightly more than 85.5°C and slightly less than 85.6°C. In any event, a temperature of 85.55°C implies more certainty than seems reasonable. The people that recorded three digits in their measurements reported more reasonable results.

Since the smallest divisions of this thermometer are close together, about the best that can be done is to read it by estimating to fifths (0.2) of the smallest division. The mercury thread in Figure 2-4 is slightly more than half-way between 85°C and 86°C and a temperature of 85.6 ± 0.2°C should be recorded. The ±0.2°C represents the experimental uncertainty in the measurement. The temperature is between 85.4°C and 85.8°C.

What if the mercury thread on a similar thermometer happened to fall exactly (as far as we can tell) on the 86°C mark? Should we record the temperature as "86°C"? The answer is, "No." We can estimate to fifths of a division. If the mercury is right on a division mark we are estimating zero fifths. This is indicated by the zero in the correct reading, 86.0°C. The complete reading would be 86.0 ± 0.2°C.

Whatever instrument we use, the result of an experimental measurement is uncertain. Different laboratory instruments have different uncertainties. For example, a centigram balance typically has an uncertainty of ±0.01 g, and a 50 mL graduated cylinder typically has an uncertainty of ±0.5 mL. However, the precision of a measurement is frequently indicated by the number of digits in the measurement rather than by using the "±" notation.

More About Significant Digits

You must use just enought digits in a measurement to give an idea of the degree of uncertainty in your measurement. As we have seen in Chapter 1, these digits are called the significant digits. The number of **significant digits** in a measurement is usually defined as the number of digits that are known with certainty plus one that is uncertain. Thus, a measurement of 85.6°C contains three significant digits, and a measurement of 85.55°C contains four significant digits. In each of these measurements there is only one uncertain digit (the last one); the others are known with certainty. The larger the number of significant digits, the more precise the measurement.

Leading zeroes are ignored when counting significant digits. For example, how many significant digits are in a measurement of 0.014 g? Since leading zeroes are ignored, there are two significant digits. You will always get the correct number of significant digits if you express the measurement in scientific notation (1.4×10^{-2} g). The number of digits in the coefficient is the same as the number of significant digits in the measurement.

Trailing zeroes may or may not be included among significant digits. Suppose you are told, "The speed of light is 300 000 000 m/s." If the speed had been measured to the nearest metre per second (extremely unlikely), then all

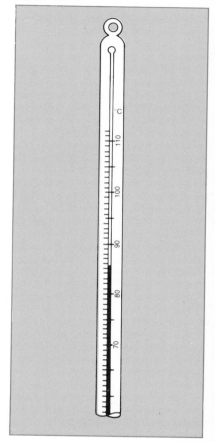

Figure 2-4
What is the uncertainty in the temperature reading of this thermometer?

digits are significant, and the measurement contains nine significant digits. If the measurement is precise only to the nearest million metres per second (i.e., the number is between 299 000 000 and 301 000 000 m/s) then the third digit is uncertain, and there are only three significant digits. This is usually indicated by writing the measurement in scientific notation (3.00×10^8 m/s). Trailing zeros which are not needed to indicate the place of the decimal point are always included when counting significant digits. For example, there are four significant digits in a measurement of 14.50 g. Since the zero is not needed to indicate the position of the decimal point, it must have been written because it is significant.

Two types of numbers in science are considered to be exact and to contain an infinite number of significant digits. These are

a) numbers obtained by counting. If there are 28 students in a classroom, there are not 27 students or 29 students or 28.32 students. There are 28.000... students (the zeroes can be extended to as many decimal places as necessary);

b) numbers obtained by definition. In definitions such as "12 objects = 1 dozen" and "1 m = 100 cm," the numbers 12, 1, and 100 are all considered to be exact and to contain an infinite number of significant digits.

Mathematical Operations Involving Significant Digits

Measurements in chemistry are uncertain. The results of calculations involving these measurements are also uncertain. Consider a line consisting of two segments, one of length 11 ± 1 mm and one of length 17.2 ± 0.1 mm. What is the total length of the line? It is tempting to say that the length is $11 + 17.2 = 28.2$ mm. But is it, really? Both measurements are uncertain. The greatest length the line could have is $12 + 17.3 = 29.3$ mm. The shortest length it could have is $10 + 17.1 = 27.1$ mm. The actual length is somewhere between 27.1 mm and 29.3 mm. Only the first digit (2) is known for certain. The other digits are uncertain. The measurement is better expressed, then, as 28 mm. This result is quickly obtained by using the following procedure. Place a bar over the uncertain digits in each of the numbers to be added or subtracted. An uncertain digit involved in an addition or subtraction results in an answer which is also uncertain. Place bars over all uncertain digits in your answer. The answer should contain all unbarred digits plus the barred digit which is furthest to the left. For example,

$$1\bar{1} \text{ mm}$$
$$17.\bar{2} \text{ mm}$$
$$\overline{28.\bar{2}} \text{ mm}$$

The answer is 28 mm.

EXAMPLE PROBLEM 2-4

To a beaker having a mass of 109.751 g, a student adds 10.23 g of Chemical A, 0.0639 g of Chemical B, and 19.1 g of Chemical C. What is the total mass of beaker plus chemicals?

SOLUTION

Beaker	$109.75\overline{1}$ g
Chemical A	$10.2\overline{3}$ g
Chemical B	$0.063\overline{9}$ g
Chemical C	$19.\overline{1}$ g
Total	$139.\overline{1449}$ g

The total mass is 139.1 g.

PRACTICE PROBLEM 2-4

To a flask having a mass of 115.24 g, a student adds 75.35 g of Chemical A, 8.42 g of Chemical B, and 0.382 g of Chemical C. What is the total mass of flask plus chemicals? (*Answer*: 199.39 g)

The considerations which apply when measurements must be multiplied c divided are also straightforward. Consider the following problem. What is th area of a rectangle of sides 1.14 ± 0.01 cm and 2.3 ± 0.1 cm? The area must fa between 1.13 × 2.2 = 2.486 cm² and 1.15 × 2.4 = 2.760 cm². The answer ca: have only two significant digits. The rule to follow is, "When two or more measurements are multiplied or divided, the answer should contain only as many significant digits as there are in the measurement having the least number of significant digits." Thus the area obtained by multiplying 1.14 cm by 2.3 cm (2.622 cm²) must contain only two significant digits and is written as 2.6 cm².

EXAMPLE PROBLEM 2-5

What is the cross-sectional area of a piece of glass tubing with a diameter of 6.0 mm?

SOLUTION

The area is given by the formula $A = \pi r^2$. The radius is 3.0 mm (two significant digits). π is a constant with an infinite number of significant digits (3.141 592 654...). We shall use at least one more significant digit for π than is in the least precise measurement, giving us 3.14.

$$A = 3.14 \times (3.0 \text{ mm})^2 = 28 \text{ mm}^2$$

The answer must contain only two significant digits.

PRACTICE PROBLEM 2-5

A room is 6.13 m long and 4.1 m wide. Calculate the area of the room. (*Answer*: 25 m²)

Rounding Off

To this point we have simply dropped all the excess uncertain digits in order to get a result with the correct number of significant digits. This process is call ed *truncation* (cutting off). Suppose an area of 2.622 cm² was truncated to 2.6 cm². If the area had been 2.696 cm², would it still have been correct to write the area as 2.6 cm²? The correct value would have been 2.7 cm², because 2.692 is closer to 2.7 than it is to 2.6. Truncation is therefore not always the correct procedure for dropping unnecessary digits. The correct procedure to use is called *rounding off.* The rules for rounding off are:

1. If the first digit to be truncated is less than 5, the preceding digit stays the same. For example, 2.622 cm² becomes 2.6 cm² to two significant digits because the 2 in the hundredths place is less than 5.
2. If the first digit to be truncated is greater than 5, or is 5 followed by numbers other than zeros, the preceding digit is increased by one. Thus, 2.692 cm² becomes 2.7 cm², because 9 is greater than 5.
3. If the digit to be truncated is 5 or is 5 followed only by zeros, the preceding digit stays the same if it is even, and increases by one if it is odd. This is called rounding off to the nearest even number, and is used only when dropping a 5. Thus, to two significant digits, 2.65 cm² is 2.6 cm², 2.75 cm² is 2.8 cm², 2.85 cm² is 2.8 cm², 2.95 cm² is 3.0 cm², and so on.

It might seem strange at first glance to round off fives to the nearest even number. Think for a moment about why it is a reasonable procedure. Suppose you have a group of numbers, all ending in five, to be rounded off and then added together. If you round them all up, the sum will be too high. If you round them all down, the sum will be too low. It is likely that about half of the digits preceding the fives will be odd. The other half will be even. If the numbers with odd digits before the fives are rounded up, and the numbers with even digits before the fives are rounded down, any errors will cancel. The sum will be more nearly correct.

Each time a number is rounded off, a small error is introduced. Thus, if a computation is lengthy and requires the calculation of several intermediate results before a final result is obtained, rounding off after each intermediate step may introduce a significant cumulative error. It is preferable to round off only once—at the very end. The availability of inexpensive hand calculators makes it easy to do this. They will carry eight to ten digits throughout all the steps, allowing intermediate results to be stored temporarily in a memory if necessary. If, for example, we had rounded off all the masses in Example Problem 2-4 before adding, we would have obtained

Beaker	109.8 g
Chemical A	10.2 g
Chemical B	0.1 g
Chemical C	19.1 g
Total	139.2 g

This differs from the answer earlier obtained because of the cumulative errors introduced by rounding off before adding. However, in this case the two answers are identical within experimental uncertainty.

The process of rounding off can often be used to check a computation quickly. All that is necessary is to express each factor to only one significant figure before carrying out the calculations. An approximate answer will result. For example, the cross-sectional area of the glass tubing of Sample Problem 2-5 is approximately given by $A = \pi r^2 \approx 3 \times (3)^2 = 27 \approx 30$ mm^2. The area is somewhere around 30 mm^2 (actually 28 mm^2). If you had obtained 2.8 mm^2 or 280 mm^2 you should have suspected a misplaced decimal point. *Always check your calculations.*

Part One Review Questions

1. Why do scientists feel obligated to communicate their findings to other scientists?
2. Which of the SI base units are most useful in chemistry?
3. What SI units would you use to express the following measurements: **a)** the diameter of a garden hose; **b)** your height; **c)** the distance from your home to school; **d)** the mass of a penny?
4. Make any necessary corrections to the following, using the convention of style for writing measurements with SI units: **a)** 14 kgs; **b)** 42,760 m; **c)** 25 metres/s;   **d)** five grams;   **e)** 44 mL.;   **f)** 75 mg per litre; **g)** 0.00067 kg; **h)** 2.0 L; **i)** 25 °C.
5. Express the following numbers in scientific notation: **a)** 76.43 g; **b)** 4508 mL; **c)** 750 000 m; **d)** 0.000 056 s; **e)** 0.472 kg.
6. Express the results of the following operations in scientific notation:
   a) 2.76×10^{-3} N $\times$ 7.4×10^5 m   **b)** $(5.13 \times 10^{-2}$ N$)/(6.4 \times 10^4$ m$^2)$
   c) 1.25×10^{-2} m^3 + 3.46×10^{-3} m^3 **d)** 2.96×10^{-2} m^3 − 1.72×10^{-3} m^3
7. In what way are precision and significant digits related?
8. How many significant digits are in the following measurements? **a)** 146.20 g; **b)** 14.6 cm; **c)** 0.04 g; **d)** 15 desks; **e)** 54.706 mL.
9. Identify the significant digits in each of the following: **a)** 2.36×10^{-3} km; **b)** 0.0980 g; **c)** 7.82 mL; **d)** 6.217×10^{12} m; **e)** 60 000 L (which is 6.00×10^4 L).
10. Why is the procedure of rounding off necessary?
11. Round off each of the following to two significant digits: **a)** 75.5 mL; **b)** 86.7 J; **c)** 4.32×10^2 kPa; **d)** 24.5 g.

Part Two: Solving Problems

2.5 Problem Solving and Dimensional Analysis

The Problem With Problems

One of the characteristics of a good chemist is the ability to solve problems. Of course, this type of skill is useful to everyone, not just to chemists. We all have to solve problems in our daily lives. Some students give up entirely ("I can do the chemistry, but I can't do the math"). Others adopt a trial and error approach. Yet the solving of problems in chemistry is a skill that can be developed with practice.

Consider a simple example. Students are asked to solve the following problem: A newborn baby has a mass of 3640 g; express its mass in kilograms. Student A knows that the prefix "kilo" has something to do with "a thousand" and multiplies by 1000, to get a result of 3 640 000 kg. Student B has the same knowledge, but divides by 1000, to get a mass of 3.64 kg. Student C follows a more systematic approach.

The reasoning of Student C is something like this:

1. I must convert 3640 g to kilograms. I know that there are a thousand grams in a kilogram. Since a kilogram is much bigger than a gram, I will end up with fewer kilograms than grams.
2. Therefore I will divide the number of grams by 1000.
3. Therefore 3640 g = 3.64 kg.
4. It seems reasonable that a baby should have about the same mass as three or four 1 kg bags of sugar.

Step 4 is an important step, yet it is omitted by most students to their regret. Always check the reasonableness of your answer! Student A, who presumably received a grade of zero for the answer 3 640 000 kg, would have soon realized the silliness of that result after a little thought. A baby with that mass would be huge and equal in mass to about 3000 compact cars.

Dimensional Analysis

A measurement is meaningless unless its SI units are indicated. **Dimensional analysis** is a useful method of calculation in which both the numerals *and* the SI units (dimensions) are carried through the algebraic operations. There are two advantages to this approach. First, the units for the answer will come automatically from the calculations. Second, incorrect units in the answers show that you have made a mistake in setting up the problem. Let us try a problem, using dimensional analysis. How many seconds are in two minutes? "That's easy," you say. "That's not a chemistry problem." But just *how* did you leap to the answer, "120 s," so quickly after reading the question? You probably reasoned, "There are 60 s in 1 min. Therefore, in 2 min there are 2 × 60 s or 120 s."

In this reasoning you used an *equality*, 60 s = 1 min. If you divide both sides of an equality by the same thing, you still have an equality. Thus, division of both sides of the equality

$$60 \text{ s} = 1 \text{ min}$$

by 1 min gives

$$\frac{60 \text{ s}}{1 \text{ min}} = \frac{1 \text{ min}}{1 \text{ min}} = 1$$

The fraction 60 s/1 min is called a *conversion factor*, and it equals 1. *All conversion factors equal 1*. The multiplication of a quantity by 1 does not change its value. That is what you did, unconsciously, when you answered the question. You multiplied 2 min by 1 in the form 60 s/1 min:

$$2 \cancel{\text{min}} \times \frac{60 \text{ s}}{1 \cancel{\text{min}}} = 120 \text{ s}$$

The fraction 60 s/1 min was the conversion factor for converting seconds to minutes. Note that identical units are cancelled in the same way that numbers are cancelled in algebra.

Let us return to the equality, 60 s = 1 min. If we had divided both sides by 60 s, we would have obtained

$$\frac{60 \text{ s}}{60 \text{ s}} = \frac{1 \text{ min}}{60 \text{ s}} = 1$$

Thus, 1 min/60 s is also a conversion factor. There are, then, two conversion factors—60 s/1 min and 1 min/60 s. They are the same except that one is inverted compared with the other. *Any conversion factor can be inverted and it will still be a conversion factor.*

How do we use conversion factors to solve a problem? It is really quite simple. All problems ask a question whose answer will be a number *and its units*. The problems will include information which consists of numbers *and their units*. You will multiply the information given by conversion factors so that all units are cancelled except the one you want in your answer. You may, of course, multiply by as many conversion factors as you wish, since all conversion factors equal 1.

Frequently, a problem contains information given in the form of ratios. Examples include the number of items per unit mass, the distance travelled per unit time, and the mass per unit volume. In each case, the word *per* indicates a ratio, and the number in the denominator is 1 (exactly). Thus, a speed of five kilometres per hour is 5 km/1 h. In this case, 5 km is equivalent (equal in significance) to 1 h. Therefore, speed can be used as a conversion factor for converting distance to time or time to distance. The numerator of a ratio is equivalent to the denominator, and the ratio can be used as a conversion factor.

Consider an example. How long will it take to travel the 546 km between Toronto and Montreal at an average speed of 91 km/h? The question asked is, in effect, "How many hours?" You want to multiply the 546 km by some conversion factor which will have the units of kilometres in the denominator (to cancel the unwanted kilometres) and hours in the numerator (to produce an answer with the required units). How about the conversion factor 91 km/1 h? That won't work because it has kilometres in the numerator. Let's try the conversion factor 1 h/91 km.

$$546 \cancel{\text{km}} \times \frac{1 \text{ h}}{91 \cancel{\text{km}}} = 6.0 \text{ h}$$

We have the required cancellation, and the units of the answer are hours.

If we had used the other conversion factor we would have obtained

$$546 \text{ km} \times \frac{91 \text{ km}}{1 \text{ h}} = 50 \, 000 \text{ km}^2/\text{h}$$

Obviously the km²/h units signal an error. Also, the 50 000 hour answer should give you second thoughts about starting on this trip. Always check your answers for reasonableness.

In this example, we carried both the numerals and their dimensions (units) through the mathematical operations required by the problem. We chose conversion factors so that the appropriate units would cancel to give the correct units for the answer. In other words, we analyzed the dimensions in order to solve the problem. That is why this technique of problem solving is called dimensional analysis.

EXAMPLE PROBLEM 2-6

If light travels 3.00×10^8 m/s, how many minutes are required for light to cover the 1.5×10^8 km between the sun and the earth?

SOLUTION

If we are to use the conversion factor 3.00×10^8 m/1 s or its inverse, we must find a conversion factor relating kilometres and metres. That's easy.

$$1.5 \times 10^8 \ \text{km} \times \frac{1000 \ \text{m}}{1 \ \text{km}} \times \dots$$

Now convert metres to seconds.

$$1.5 \times 10^8 \ \text{km} \times \frac{1000 \ \text{m}}{1 \ \text{km}} \times \frac{1 \ \text{s}}{3.00 \times 10^8 \ \text{m}} \times \dots$$

Finally, convert seconds to minutes.

$$1.5 \times 10^8 \ \text{km} \times \frac{1000 \ \text{m}}{1 \ \text{km}} \times \frac{1 \ \text{s}}{3.00 \times 10^8 \ \text{m}} \times \frac{1 \ \text{min}}{60 \ \text{s}} = 8.3 \ \text{min}$$

∴ 8.3 min are required for light to cover the distance between the sun and the earth.

EXAMPLE PROBLEM 2-7

If a person has a mass of 70 kg, what is this mass in milligrams?

SOLUTION

We can use the conversion factor that relates kilograms to grams.

$$70 \ \text{kg} \times \frac{1000 \ \text{g}}{1 \ \text{kg}} \times \dots$$

The units of kg cancel out. We must now find a conversion factor that relates grams to milligrams.

$$70 \ \text{kg} \times \frac{1000 \ \text{g}}{1 \ \text{kg}} \times \frac{1000 \ \text{mg}}{1 \ \text{g}} = 7.0 \times 10^7 \ \text{mg}$$

∴ The mass of 70 kg in milligrams is 7.0×10^7 mg.

EXAMPLE PROBLEM 2-8

How many seconds are there in 2.0 weeks?

SOLUTION

We can use the conversion factor that relates weeks to days.

$$2.0 \text{ weeks} \times \frac{7 \text{ d}}{1 \text{ week}}$$

We now use a number of conversion factors that relate days to hours, hours to minutes, and finally, minutes to seconds.

$$2.0 \text{ weeks} \times \frac{7 \text{ d}}{1 \text{ week}} \times \frac{24 \text{ h}}{1 \text{ d}} \times \frac{60 \text{ min}}{1 \text{ h}} \times \frac{60 \text{ s}}{1 \text{ min}}$$

$$= 1.2 \times 10^6 \text{ s}$$

∴ In 2.0 weeks there are 1.2×10^6 s.

EXAMPLE PROBLEM 2-9

Suppose you wish to cut a rope into pieces of identical length. Each piece must be 30 cm long. The length of the rope is 540 cm. How many pieces can be obtained?

SOLUTION

We use the conversion factor relating centimetres and length of pieces required by the problem.

$$540 \text{ cm} \times \frac{1 \text{ piece}}{30 \text{ cm}}$$

$$= 18 \text{ pieces}$$

∴ 18 pieces of string 30 cm long can be obtained from a rope of 540 cm.

PRACTICE PROBLEM 2-6

If the price of gasoline is \$0.45/L, how much would it cost to fill a gas tank with a capacity of 73 L? (*Answer*: \$33)

PRACTICE PROBLEM 2-7

How many minutes are there in three weeks? (*Answer*: 30 240 min)

Dimensional analysis is a versatile method of problem solving. Since it is so useful, you should make a point of remembering the steps to be followed:

1. Read the problem carefully to determine what is actually *asked for*, including units, and what is *given*, including units.

2. Decide what conversion factors are needed to make the units cancel and give the desired result. In our example of a trip between Toronto and Montreal, we used 1 h/91 km.
3. Do the arithmetic, and check your answer to see if it is reasonable.
4. Check to make sure that the units cancel properly.
5. Check the number of significant digits.

Part Two Review Questions

Use dimensional analysis to solve each of the following problems:
1. How many kilometres are there in 465 mm?
2. If a car is travelling at 1.00×10^2 km/h, how many minutes will it take to travel 6.5×10^3 m?
3. How many kilograms are there in 400 µg?
4. How many hours are there in 675 000 s?

Chapter Summary

- Scientific knowledge grows, because scientists communicate.
- In order for scientific communication to be effective, it must have universality. That is, it must be interpreted the same way throughout the world.
- Each measurement must have a number followed by a unit.
- All measurements are made by using (or making a comparison with) a standard measuring device.
- Canada has adopted the International System (SI) of measurement.
- In Canada, certain conventions of style must be used when writing measurements in SI units.
- Scientific notation consists of using a coefficient of 1 or a number between 1 and 10 multiplied by a power of ten.
- All measurements involve uncertainty.
- The significant digits in a measurement are all the digits that are known with certainty plus one digit that has been estimated.
- Rounding off is used to provide the proper number of significant digits.
- Dimensional analysis is a useful method for solving numerical problems.
- A conversion factor is a ratio.
- All conversion factors equal 1.

Key Terms

universality	significant digit
SI	dimensional analysis
units	

Test Your Understanding

Each of these statements or questions is followed by four responses. Choose the correct response in each case. (Do not write in this book.)

1. Which of the following is not a derived SI unit?
   a) kelvin   b) pascal   c) joule   d) newton

2. Which of the following measurements does not conform to SI style conventions?
   a) 25.0°C   b) three grams   c) 14.6 grams/mL   d) 5.0 Mm

3. Which of the following is equivalent to 14.6 nm?
   a) 1.46×10^{-10} m   b) 1.46×10^{-12} km   c) 146 pm   d) 1.46×10^{-8} m

4. Convert 0.0047 to scientific notation.
   a) 4.7×10^{-3} b) 4.7×10^3 c) 47×10^4 d) 47×10^{-4}

5. Which of the following equals $(1.40 \times 10^6)/(2.70 \times 10^{-3})$?
   a) 3.78×10^3 b) 5.19×10^8 c) 3.78×10^9 d) 5.19×10^2

6. Which of the following equals $(2.96 \times 10^{-3}) - (1.50 \times 10^{-4})$?
   a) 1.46×10^{-3} b) 1.46×10^{-4} c) 2.81×10^{-4} d) 2.81×10^{-3}

7. The number of significant digits in 0.002 340 m is
   a) four.   b) five.   c) three.   d) six.

8. Which of the following numbers has five significant digits?
   a) $0.092\ 03 \times 10^{-2}$ b) 1.079 73   c) 0.009 203 0   d) 0.7973

9. Which of the following measurements contains the most significant digits?
   a) 0.0014 g   b) 14.75 kg   c) 187°C   d) 6.00×10^2 m

10. When the calculation $(35.43 \times 4.640) + 0.21$ is carried out the number of significant digits in the answer is
    a) two.   b) three.   c) four.   d) five.

11. When the calculation $35.43 \times (4.640 + 0.21)$ is carried out the number of significant digits in the answer is
    a) two.   b) three.   c) four.   d) five.

12. What is the average of the following temperatures, 14.0°C, 9.63°C, 11°C, and 9°C?
    a) 10.9°C   b) 11°C   c) 10°C   d) 10.91°C

13. Which of the following is not correctly rounded off to two significant digits?
    a) 1.45 g becomes 1.5 g   b) 3.58 g becomes 3.6 g
    c) 14.3 g becomes 14 g   d) 1.35 g becomes 1.4 g

14. How many litres equal 5.36 mL?
    a) 5360 L   b) 536 L   c) 0.536 L   d) 0.005 36 L

15. If a liquid has a density of 2.71 g/mL, what volume of the liquid has a mass of 45.0 g?
    a) 0.0602 mL   b) 122 mL   c) 16.6 mL   d) 60.2 mL

16. How many seconds are there in seven days?
    a) 6.05×10^5 s   b) 3.02×10^5 s   c) 2.52×10^4 s   d) 1.01×10^4 s

Review Your Understanding

1. Name three ways in which information is communicated among scientists.
2. The concept of universality illustrates one of the differences between science and art. What is this difference?
3. Why is it necessary to have an international system of units?
4. Why is it necessary to define units of measurement carefully?
5. What is the difference between a base unit and a derived unit? Give an example of each.
6. What are the SI units for the following: a) area; b) volume; c) force; d) pressure; e) work; and f) energy?
7. What SI unit would you use to express the following measurements: a) the diameter of a lead pencil; b) the temperature in your classroom; c) the time required for you to say the word *cheese*; d) your waist measurement; e) the area of your classroom; f) the distance from the earth to the sun?
8. Complete the following table:

   a) 3.15 m =   cm   b) 955 g =   kg
   c) 1630 mL =   L   d) 20.0 Mg =   mg
   e) 178 mm =   cm   f) 15.5 mg =   g
   g) 1620 km =   Mm   h) 144 kg =   mg
   i) 0.0117 mm =   cm   j) 126 mm^3 =   cm^3

9. Make any necessary corrections to the following, using the convention of style for writing measurements with SI units:
   a) 25 gs; b) 10 grams/cm^3; c) 25,000 L; d) fifteen milligrams; e) 65 km.; f) 80 mg per millilitre.
10. Why is it useful to be able to express numbers in scientific notation?
11. Express the following numbers in scientific notation:
    a) 1 003 000 000 000
    b) 0.000 000 000 000 399 8
    c) 52.23
    d) 0.2038
    e) 12 452
12. Convert the following numbers to decimal notation:
    a) 1.776×10^7; b) 2.552×10^{-9}; c) 1.168×10^3; d) 4.44×10^{-1}; e) 1.399×10^0.
13. Express the results of the following operations in scientific notation:
    a) $1.39 \times 10^{-2} + 3.11 \times 10^{-4}$ b) $1.17 \times 10^4 - 3.57 \times 10^2$
    c) $1.34 \times 10^{24} - 2.22 \times 10^2$ d) $2.15 \times 10^5 + 1.56 \times 10^3$
14. Express the results of the following operations in scientific notation:
    a) $(1.81 \times 10^{-3}) \times (1.06 \times 10^{20})$ b) $(5.77 \times 10^{-4})/(1.71 \times 10^{-11})$
    c) $(4.44 \times 10^{-3}) \times (2.252 \times 10^2)$ d) $(7.99 \times 10^{-3})/(1.33 \times 10^6)$
15. Explain why it is important to use the correct number of significant digits in expressing a measurement.
16. Are all experimental measurements uncertain? Explain.
17. How many significant digits are in each of the following measurements?
    a) 133.31 g; b) 0.02 g; c) 24.6 cm^3; d) 109.9457 mL; e) 29 marbles.

18. Identify the significant digits in each of the following:
  a) 6.29 mL; **b)** 0.0990 g; **c)** 42 000 J (which is 4.2×10^4 J);
  d) 1.81×10^{-6} km; **e)** 1.772×10^{10} Pa.
19. **a)** Which of the following three measurements contains the most significant digits: 1057 g, 13 g, or 0.479 g?
  b) Which of the measurements in part **a)** of this question is the least precise?
  c) Find the sum of the three measurements.
20. How many significant digits are there in the answers to the following problems?
  a) 24.4 g + 12.692 g + 14.79 g
  b) 2.229 g − 0.5710 g
  c) 10.6 N × 6.9 m
  d) $(9.93 \times 10^{23}$ s$)(6.9 \times 10^{-2}$ A$)$
  e) 73 mL − 36.9 mL
21. An opened bag of sugar has a mass of 746 ± 3 g.
  a) What is the smallest mass this bag of sugar could have?
  b) What is the largest mass this bag of sugar could have?
22. A person had a mass of 100 ± 1 kg at the start of a diet and 98 ± 1 kg after the first week of the diet.
  a) What is the least amount of mass that could have been lost?
  b) What is the greatest amount of mass that could have been lost?
23. Round off each of the following numbers to two significant digits:
  a) 36.4; **b)** 729; **c)** 0.145; **d)** 8.357; **e)** 0.001 07; **f)** 6.022×10^{23}.
24. Round off each of the numbers in the preceding question to one significant digit.
25. Solve each of the following problems by using dimensional analysis:
  a) What distance is covered in 4.25 h by a car travelling at 95 km/h?
  b) How much does it cost to register a car with a mass of 1800 kg if the registration fee is $2.50/100 kg?
  c) How many grams of alcohol are present in 5.00 L of blood from a person with an alcohol level of 102 mg of alcohol per 100 mL of blood?
26. What was the cost of gasoline for a drive from Banff to Edmonton (428 km) if the car required 10.2 L/100 km and the cost of gasoline was $0.45/L?
27. A recipe using hamburger serves eight people. The recipe calls for 2.0 kg of hamburger. However, you wish to prepare a meal that will serve three people. Use dimensional analysis to determine how many kilograms of hamburger you will need.
28. How many rail cars, each 15.0 m long, are in a freight train which requires 2.00 min to pass a station while the train is travelling at 60.0 km/h?

Apply Your Understanding

1. The league was a unit of distance which varied in length at different periods of time and in different places. In English-speaking countries, it was usually estimated at 4.8 km. However, Jules Verne was probably thinking of a nautical league (5.6 km) when he wrote *Twenty Thousand Leagues Under the Sea*. Compare the radius of the earth (6.4 Mm) with twenty thousand nautical leagues.

2. It has been estimated that a gram of seawater contains 4.0 pg of gold. The oceans of the earth have a total mass of 1.60×10^6 kg. How many grams of gold are present in the oceans?

3. A 9.76 g sample of table sugar is placed in a 25.00 mL flask. The flask is completely filled with benzene, and the sugar and benzene have a total mass of 26.31 g. The sugar does not dissolve in the benzene. If the density of benzene is 0.879 g/mL, what is the density of sugar?

4. The density of ethanol is 0.789 g/mL. The mass of ethanol required to fill a flask is 15.78 g. If this same flask can be filled with 18.34 g of olive oil, what is the density of olive oil?

5. If 4.18 J of energy are required to cause the temperature of one gram of water to increase by one degree Celsius, how many grams of water can be warmed from 15.6°C to 35.9°C by 14.5 kJ?

6. The concentration of pollutants is often expressed in "parts per million" or "ppm." The SI equivalent is "milligrams per kilogram." If one drop of a liquid pollutant has a mass of 50 mg, how many litres of water (density = 1.00 g/mL) are required to dilute the pollutant to a concentration of 1 ppm?

Investigations

1. Debate the following resolution. All systems of measurement other than SI should be made illegal in Canada.

2. Write a review of a scientific report that has appeared in the popular scientific press during the last three months.

3. Write a research report on the historical development of the SI units for length and mass.

4. Devise and conduct an experiment to determine the quantity of dissolved solids in drinking water or lake water.

Matter

3

CHAPTER CONTENTS

Part One: **Matter and Its Changes**
3.1 Matter and Energy
3.2 The Three States of Matter
3.3 Changes in States of Matter
3.4 Physical and Chemical Properties
3.5 Physical and Chemical Changes in Matter
3.6 Classification of Matter
3.7 Separating Mixtures

Part Two: **The Laws of Chemical Change**
3.8 Law of Conservation of Mass
3.9 Law of Constant Composition
3.10 Law of Multiple Proportions

Why does a person weigh less on the moon? Why is a diver able to breathe for such long periods from a small scuba tank? Why are we able to use table salt (sodium chloride) on our food even though it is made from two poisonous substances, sodium and chlorine? How is ultrapure silicon made for the computer industry? You will find the answers to these questions as you study this chapter.

When we look at the world around us, we see that there are always changes taking place. The eruption of volcanoes, the passing of seasons, the breaking of glass, the rusting of metals, are but a few examples of change.

We can understand these changes if we understand the nature of matter. Chemists study the make-up of matter; how matter is put together; its behaviour under certain conditions; and the energy involved when changes in matter occur.

In order to study chemistry, we must become acquainted with its language. The definitions and concepts presented in this chapter serve as basic vocabulary for further study.

Key Objectives

When you have finished studying this chapter, you should be able to

1. distinguish between mass and weight; between work and energy; between kinetic and potential energy.
2. state the characteristics of solids, liquids, and gases.
3. distinguish between physical and chemical properties and identify examples of each.
4. distinguish between physical changes and chemical changes and identify examples of each.
5. distinguish between pure substances and mixtures; elements and compounds; homogeneous and heterogeneous mixtures.
6. describe four methods of separating mixtures.
7. state the Law of Conservation of Mass, the Law of Constant Composition, and the Law of Multiple Proportions.

Part One: Matter and Its Changes

3.1 Matter and Energy

What is the earth composed of? What makes up soil, air, water, plants, and animals? What are sound, electricity, magnetism, heat, and light? We can describe the world around us as being made of matter and energy.

Matter, Mass, and Weight

Matter is defined as anything that has mass and takes up space. Matter possesses **inertia**—the tendency for a moving body to remain in motion and for a stationary body to remain in place. Inertia, then, is resistance to change in state of motion. For example, if you are travelling in a car and the brakes are suddenly applied, you will tend to resist this change in state of motion by continuing to move forward at the same speed, even though the car has stopped.

For the most part, matter is easily recognized: wood, bricks, and water are all examples. Some forms of matter, however, are slightly more difficult to recognize. It is hard to think of air as matter, until you remember that a hovercraft (Fig. 3-1) is supported by a blanket of compressed air or that the high winds of a hurricane can destroy a huge building.

Figure 3-1
A hovercraft, supported on a cushion
of compressed air, moves offshore
and down the Mackenzie River in the
Northwest Territories.

Mass is a measurement of the amount of matter in an object (Fig. 3-2). Since matter has inertia, mass is also defined as the quantity of inertia possessed by an object. The greater the mass of a body, the more inertia it contains. For example, a large object such as a boulder is difficult to move; however, once the boulder is moving, it is difficult to stop. The boulder has a great deal of inertia (mass). The base unit of mass used in chemistry is the kilogram. Since the kilogram is a relatively large unit of mass, the gram (0.001 kg) is more commonly used in practice.

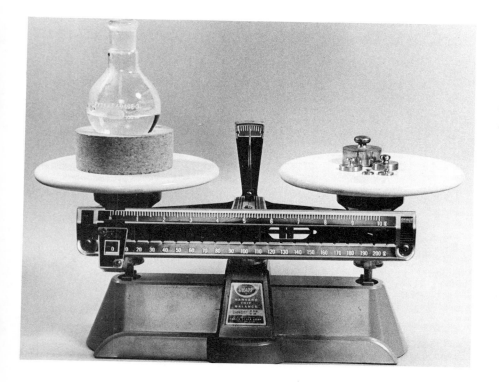

Figure 3-2
The mass of the objects on the left
pan of the balance is equal to the
total mass on the right pan; the
position of the pointer does not alter
with a change in gravitational force.
The force of the earth's attraction
does not influence mass.

Figure 3-3a
This oil-fired thermal generating station located at Dartmouth, Nova Scotia produces 350 MW of power.

Weight is a term which is frequently confused with mass. These two terms are related, but technically they mean different things. **Weight** is a measure of the gravitational force of attraction between the earth and an object. A force is a push or a pull on an object. The gravitational force depends upon the mass of the object and its distance from the centre of the earth. The mass of an object is constant, but the pull of the earth's gravity is not the same everywhere on the earth's surface. Therefore, the weight of a given object does change. If that object is taken far enough from the surface of the earth, the pull of gravity becomes negligible (too small to have any effect), and the object becomes essentially weightless. However, the object still possesses the same amount of mass that it has always possessed.

Energy

A hummingbird hovers, a worm burrows through the earth, oil burns in a furnace, you turn the pages of a book—all these actions involve energy. **Energy** is the ability to do work. **Work** can be defined as the moving of matter against a force which opposes the motion. For example, when you walk up the stairs or lift a book, you are doing work. You are moving matter against a force which opposes the motion—the force of gravity. The energy required to do this work comes from the food you eat.

Energy is subdivided into two main categories: kinetic energy and potential energy. **Kinetic energy** is energy of motion. A moving object such as a stone can pass through a glass window if it has enough kinetic energy to shatter the glass. **Potential energy**, on the other hand, is the energy of position. A book held above a desk has potential energy. If the book is allowed to fall, it acquires kinetic energy. The potential energy grows smaller and the kinetic energy grows larger as the book nears the surface of the desk. When the book hits the desk,

Figure 3-3b
This thermal generating station in Brandon, Manitoba produces 237 MW of power.

its kinetic energy is used to increase the temperature of both the book and the desk and to produce sound energy.

There are many forms of potential energy. The potential energy stored in a tightly wound spring, for example, is a form of *mechanical energy* which is able to drive the hands of a clock. When energy from the sun reaches the earth, a portion of it is absorbed by green plants and converted by photosynthesis into *chemical energy*, another form of potential energy. The chemical energy stored in plants is useful because after long periods of time plants turn into coal and oil. The coal and oil can then be burned to produce heat energy or be used in thermal generating stations to produce *electrical energy*.

The transformation of energy from one form to another is the basis for change in the material world (Fig. 3-4). When the radiant energy of the sun reaches the earth, some of it is converted to heat energy. This heat causes water to evaporate and rise to form clouds. When the water falls to earth again and flows in waterfalls such as the one at Churchill Falls, Labrador, the potential energy of the water is converted to kinetic energy. This energy can be used to drive the blades of a turbine and produce the mechanical energy necessary to keep the generator in motion. The generator converts part of the mechanical energy to electrical energy, which may in turn be used to toast bread (heat energy), to light a room (light energy), or to operate a stereo system (sound energy).

Law of Conservation of Mass and Energy

What is the relationship between matter and energy? At one time it was believed that there was a constant quantity of matter (i.e., mass) in the universe. It was also believed that there was a constant quantity of energy in the universe. These generalizations were known respectively as the Law of Conser-

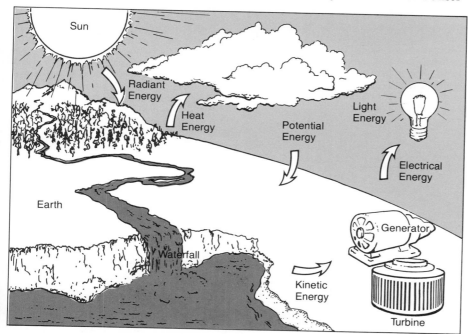

Figure 3-4
Energy transformations

vation of Mass and the Law of Conservation of Energy. Recently, however, the study of nuclear reactions has led to the discovery that in some cases mass can be converted into energy and vice versa. This knowledge gave rise to the **Law of Conservation of Mass and Energy**. This law states that the total quantity of matter and energy in the universe is constant. An increase in the amount of matter must be balanced by a proportionate decrease in the amount of energy, and vice versa.

3.2 The Three States of Matter

Solid

Matter exists in the solid, liquid, and gaseous states. Our own bodies contain examples of all three states of matter. Matter in the **solid** state has a definite shape and volume (Fig. 3-5). A solid object will have a constant size and shape, regardless of the container in which it is placed. Factors such as temperature and pressure have little effect on solids. That is, solids do not expand or contract very much when they are heated or cooled, and they are not easily compressed. These and many other properties of solids can be explained by assuming that the particles which make up the solid are held closely together in fixed positions by strong attractive forces.

Figure 3-5
Solids, liquids and gases
(a) Bulk properties
(b) Relative spacing of particles

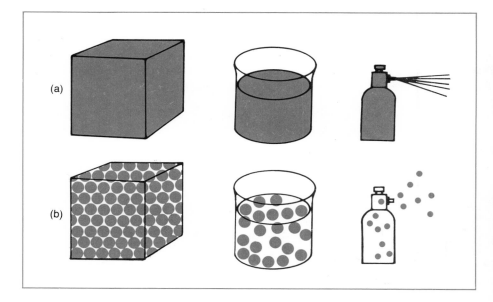

Liquid

Matter in the **liquid** state has a definite volume, but it does not have a definite shape. A liquid takes the shape of the container into which it is poured (Fig. 3-5). Liquids may be compressed slightly, but for most purposes, it is safe to assume that they do have a constant volume. This incompressibility of liquids means that a pressure exerted on a liquid will be transmitted equally in all directions. This property is used in automotive hydraulic brake systems (Fig. 3-6). When the brakes are applied, the pressure is transmitted from the master cylinder by the brake fluid to the slave cylinders which are connected to the brake shoes. The brake shoes press against the revolving brake drums to slow or stop the car. It is easy to explain these properties by assuming that the particles of a liquid are held together by attractive forces which are not quite as strong as those in solids. The particles are not as close to each other as they are in solids, and are able to slide, glide, and slip over one another with relative ease.

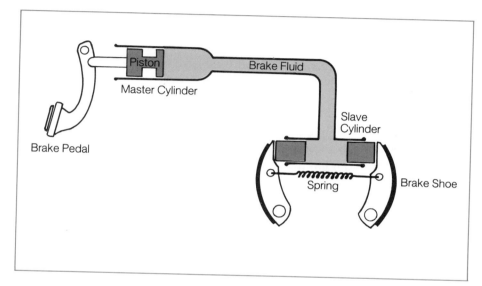

Figure 3-6
Automotive hydraulic brake system

Gas

Matter in the gaseous state has neither definite volume nor definite shape (Fig. 3-5). A **gas** takes both the shape and the volume of the container in which it is placed. Factors such as temperature and pressure have such a large effect on a gas that it is pointless to consider a given volume of gas unless both its temperature and pressure are known. The tank on a scuba diver's back would contain enough air to supply the diver's lungs for only a very short time if it were at normal atmospheric pressure. By forcing more air into the tank at pressures of one hundred to two hundred times that of the atmosphere, the supply of air can be enormously increased with no change in the size of the container. This principle is also used by firefighters, mine rescue personnel, and others who require a self-contained supply of air for work in the presence of toxic gases. These properties are understandable if we assume that the particles of a gas are extremely small and far apart from each other, so that their volume is negligible in comparison with the volume of the container. Furthermore, the attractive forces between the particles are also negligible.

According to modern theories, these gas particles are so small that in a room at normal temperature only about 0.05% of the volume of the room is occupied by the particles. The rest is empty space. This means that the total volume of all the air particles in a room 4 m long, 3 m wide, and 2.5 m high is less than the volume of a standard basketball.

3.3 Changes in States of Matter

As you know, water can exist in three different states. In the form of ice, it exists in the solid state; as water, it is in the liquid state; and when it evaporates or boils away into the atmosphere as water vapour, it is in the gaseous state. Many other substances exhibit similar properties.

How can we explain what is happening when a solid changes to a liquid or when a liquid changes to the gaseous state? Suppose, for example, we remove an ice cube from the freezer and measure its temperature—assume it is –20°C. Now we gradually add heat energy to the cube, and notice that its temperature rises smoothly until it reaches 0°C. At this point the temperature stops rising, even though heat energy is still being added to the ice cube, and we notice that the solid ice is being converted to liquid water. We have reached the melting point of the ice. When all the ice has melted, the water warms until the liquid reaches a temperature of 100°C. The temperature of the water again remains constant as heat is added, but again we notice something happening—bubbles of a gas are forming within the body of the liquid. They rise to the surface and break. The liquid is being converted into a gas, and we have reached the boiling point of the liquid. If the container is closed, the temperature of the gas will again start to rise after the last drop of water has been converted to a gas. These changes are indicated schematically in Figure 3-7.

Figure 3-7
Warming curve for a pure substance (water)

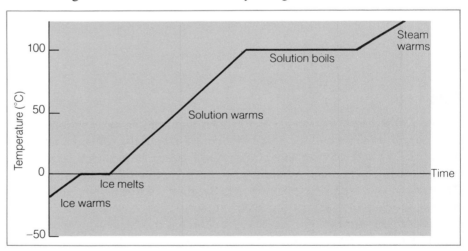

What is happening within the water sample in each of these states as the various changes take place? We have already noted that the particles of a solid are held closely together in a fixed position by strong attractive forces (Fig. 3-5). The various parts of these particles can vibrate back and forth; that is, the particles have vibrational energy (Fig 3-8). The particles are held so rigidly in position in the ice crystal, however, that they can move only minute distances from their original positions. They do not have much sideways motion or translational energy (Latin *trans*, across + *latus*, carried) in any direction.

Translational motion involves the movement of an entire particle from its original position. The situation is analogous to a dance floor in which the couples are crowded together near the band. They can move arms and legs, and perhaps change distances between partners, but the couples do not move far from their starting positions on the floor.

Figure 3-8
Motion of particles in a simple solid crystal

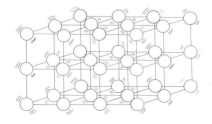

As heat energy is added, the translational energy of the water particles increases, and they begin to move greater distances from their original positions. Eventually some of them are able to overcome the attractive forces of their neighbours. In addition to their increased translational motions, the particles rotate more freely, and their vibrational energy increases. They have increased translational, rotational, and vibrational energies. Figure 3-9 represents translational, vibrational, and rotational motions of a dumbbell-shaped particle. There is so much movement that the crystal starts to fall apart. The ice starts to melt and undergoes a change of state from solid to liquid. At this point the addition of more heat does not cause the temperature to rise. The kinetic energy of the particles remains constant. It is simply used to overcome the attractive forces. That is, it raises the potential energy of the particles. Once all of the ice has melted, the addition of heat to the liquid will increase the kinetic energy of the water particles. The temperature of the water will begin to rise. Similarly, an increase in tempo and loudness of music will cause couples on the dance floor to move more vigorously. The couples will move further apart until eventually some of them will be able to move back and forth through the sea of bodies surrounding them.

Water particles are able to slide past each other, but the attractive forces still hold the particles relatively closely together. As heat energy is added to a liquid, the particles gain ever more translational, vibrational, and rotational energies. Eventually they are able to overcome the restraining attractive forces between particles in the liquid state and escape into the gaseous state as the liquid boils. At this point the temperature will remain constant, because the heat energy is being used to move the particles away from each other in opposition to the attractive forces, thus increasing their potential energy.

Finally, when the last bit of water has boiled away, the temperature of the gas will again begin to rise, because the heat energy is used to increase the total kinetic energy of the particles.

In summary, then, the addition of energy to a substance such as ice will bring about changes by causing the particles to move further apart. The internal energy—that is, the total energy contained in a sample of matter—is increased, largely because the translational kinetic energy of the particles is increased. (The rotational and vibrational energies are increased to a lesser extent.) This increase in internal energy results in either a rise in the temperature or a change in the state (melting or boiling) of the system.

Figure 3-9
Motions of a dumbbell-shaped particle

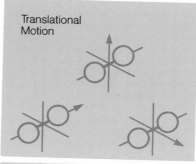
Translational Motion

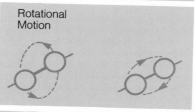
Rotational Motion

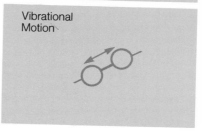
Vibrational Motion

3.4 Physical and Chemical Properties

How do you recognize your father, your house, your clothes? Different kinds of matter are recognized by their properties. A **property** is a quality or characteristic. For example, diamonds are hard; copper wire is a good conductor of electricity; oxygen gas is colourless; iron rusts. The properties of matter can be divided into two classes: physical and chemical.

Physical Properties

The **physical properties** of a substance are those properties which can be determined without changing its composition or make-up. Mercury, for example, is an odourless liquid with a characteristic silvery shine. It does not dissolve in water. It is an excellent conductor of electricity. It freezes at −39°C and boils at 357°C. Properties such as colour, odour, hardness, melting point, boiling point, and electrical conductivity are examples of physical properties. We do not have to change the composition of the mercury in any way to observe these properties. Thus, the density of mercury is also one of its physical properties. **Density** is defined as the mass of an object per unit volume of the object. One cubic metre of mercury has a mass of 13 600 kg. Hence the density of mercury is 13 600 kg/m³. (However, one cubic metre is such a large volume that the density of mercury is usually expressed as 13.6 g/cm³ or 13.6 g/mL.) Since density can be measured using only devices for determining mass and volume, and without changing the mercury into some other substance, it is a physical property.

Chemical Properties

The **chemical properties** of a substance are those properties which can be determined only when the substance undergoes a change in composition. It is difficult to distinguish pyrite (fool's gold) from gold because they are physically so much alike. Yet it is quite easy to distinguish them chemically. Concentrated nitric acid has no effect on gold, but it reacts with pyrite to give a soluble product, iron(III) nitrate, and a residue of insoluble yellow sulfur. Charcoal (carbon) burns in air to give a mixture of two gases, carbon monoxide and carbon dioxide. An iron nail immersed in water gradually becomes coated with a layer of rust. These properties are all chemical properties because they involve changing various kinds of matter into other substances with different compositions, different structures, and different properties. Such changes or transformations are called **chemical reactions**. The formation of iron(III) nitrate and sulfur by the action of nitric acid on pyrite is an example of a chemical reaction. Burning charcoal involves a reaction between carbon and oxygen; the rusting of iron involves a chemical reaction between iron and oxygen.

A substance can often be identified by observing its unique chemical and physical properties. Silver and platinum are both beautiful silvery white metals used extensively in jewelry. Silver melts at 961°C, boils at 2212°C, and has a density of 10.5 g/cm³. Platinum melts at 1769°C, boils at 3827°C, and has a density of 21.4 g/cm³. They can be distinguished from each other and identified by these differences in their physical properties. For a chemist, perhaps the easiest way to distinguish them would be to test a small amount of each metal with nitric acid. Silver dissolves readily in nitric acid with a vigorous chemical reaction, whereas platinum is completely unaffected by the acid.

Quantitative and Qualitative Properties

Both physical and chemical properties can be further subdivided into two categories, quantitative and qualitative properties. A **quantitative property** is one that can be measured. Numbers will be used in expressing a quantitative property. Thus, to say that aluminum has a melting point of 660°C, a boiling point of 2467°C, and a density of 2.7 g/cm³, is really to give a list of some quantitative physical properties of aluminum. Statements to the effect that 1 g of aluminum will react with dilute hydrochloric acid to produce 112 mg of hydrogen gas, or that the burning of 32.1 g of sulfur to form sulfur dioxide releases 297 kJ of heat energy, are examples of quantitative chemical properties of those elements.

Qualitative properties, on the other hand, are those which cannot be expressed numerically, such as appearance and odour. To say that carbon monoxide is a colourless, odourless gas is to state two of its qualitative physical properties. If we state that it is poisonous and burns with a pale blue flame, we are mentioning some of its qualitative chemical properties.

3.5 Physical and Chemical Changes in Matter

Now we shall consider the two types of change that substances undergo: physical and chemical change.

Physical Change

A **physical change** is a change which alters one or more of the properties of the substance with no change in its composition or identity. When liquid water freezes to ice, one of its properties changes: it no longer pours—but it is still water. When water boils and is converted into water vapour, another property changes: it no longer has a definite volume—but again it is still water. Water retains the same chemical composition whether it is in the form of a solid, a liquid, or a gas. Changes of state, such as melting, freezing, or boiling are physical changes because they do not change the composition or chemical identity of the substance. The tungsten filament in a light bulb becomes white hot when electricity is passed through it, but it is still tungsten. This is again a physical change because the tungsten filament is not converted to any other substance during the passage of the electric current. Some other examples of physical change are the breaking of glass, the tearing of a sheet of paper, and the bending of a nail.

Chemical Change

A **chemical change** is a change in a substance which converts it into a different kind (or different kinds) of matter, each with a different composition and new properties. When paper burns, it is converted mainly into two gases, carbon dioxide and water vapour, both of which differ in composition and properties from the original paper. The rusting of iron gives a product, rust, which is different in composition and properties from the original iron. Dramatic proof that new substances with new physical and chemical properties are formed during chemical change is found in the reaction between sodium and chlorine. Taken individually, each is a lethal substance. Yet, when they undergo a

Pyrite (fool's gold) is a compound which resembles the element, gold, in appearance.

Sodium burns vigorously in chlorine to produce the compound, sodium chloride.

chemical change, common table salt is formed. At least one new substance, different in composition and properties from the original substances, must be produced whenever a chemical change occurs. The burning of gasoline in an automobile engine, the preparation of caramel by heating sugar, the bleaching of clothes, the cooking of food, and the digestion of food in our bodies are other examples of chemical changes. When a chemical change occurs, there is usually some evidence such as a change in colour, temperature, or mass of the reacting substances; evolution of a gas; or formation of an odour.

APPLICATION

Cherries in Chocolate and Chemical Change

Did you ever wonder how candy manufacturers get the liquid to surround the cherry in a chocolate-covered cherry? Before being dipped in chocolate, the cherry is coated with a sugary paste containing a substance called invertase. The paste hardens and the cherry is then dipped in chocolate. The candies are then stored for one to two weeks.

During the storage period, the invertase promotes a chemical change in which one form of sugar, table sugar, is changed into two simpler sugars. The result of this chemical reaction is that the hard paste is converted to a syrup which surrounds the cherry.

Chemical reactions like this one are often used in the food processing industry, greatly increasing the variety of foods available to us today.

3.6 Classification of Matter

Pure Substances

A **pure substance** is a homogeneous (uniform) material. It consists of only one particular kind of matter and has the same properties throughout. It always has the same composition. Diamond is a pure substance. It consists only of carbon. Pure substances are classified into two categories: elements and compounds.

Elements are pure substances that cannot be broken down into simpler substances by ordinary chemical methods. Carbon, oxygen, and iron are examples of elements.

Compounds are pure substances, consisting of two or more elements in chemical combination. The preparation of a compound always involves a chemical change. Consider the preparation of the compound, sodium chloride. Sodium (a silvery solid) reacts with chlorine (a greenish-yellow gas) to produce sodium chloride (a white solid). When a compound is prepared, the original components lose their identities, and they can be separated only by chemical means. Using chemical methods, compounds may be decomposed into two or more simpler substances. For example, calcium carbonate decomposes on heating to form calcium oxide and carbon dioxide. Water, to which some acid has been added, decomposes to form hydrogen and oxygen when electricity is passed through it. Water is a *pure substance* by virtue of the fact that it is homogeneous and consists of only one kind of matter. It is a *compound* by virtue of the fact that is contains hydrogen and oxygen in chemical combination.

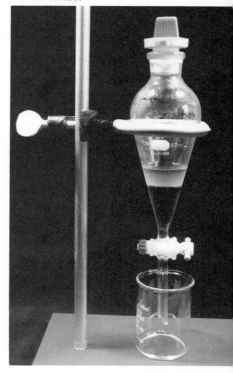

A heterogeneous mixture of two liquid phases. The top layer is mainly water. The bottom layer is a solution of iodine in carbon tetrachloride.

Mixtures

A **mixture** is a combination of two or more pure substances each of which retains its own physical and chemical properties. No chemical change occurs as the mixture is prepared. For instance, we could make a mixture of salt and sugar by merely stirring them together. Each substance would not have changed in composition as a result of forming the mixture. Therefore, a new substance would not be produced. Many mixtures are heterogeneous (nonuniform). A **heterogeneous mixture** consists of more than one phase. The word **phase** is used to mean any portion of a material with a uniform set of properties. For example, a mixture of sand and water is composed of two phases. One phase is the solid sand. The second phase is the liquid water. A heterogeneous mixture is easily identified because the borderline or interface between the different phases is easily seen. A salami, mushroom, and cheese pizza is a heterogeneous mixture with several identifiable components. If you look carefully, you will see that even the salami itself is a heterogeneous mixture. If you had the time and patience, you could manually separate the pizza into its components.

Milk is actually a heterogeneous mixture. If you examine it closely with a microscope, you will see that it consists of small droplets of fat suspended in a water layer. The fat droplets are less dense than water and eventually rise to the top as cream. This process can be speeded up by spinning the milk in a separator. Some other common examples of heterogeneous mixtures include paints, vegetable soup, solid detergents, striped toothpaste, crunchy peanut butter, and most salad dressings.

The important point to note is that heterogeneous mixtures are not uniform throughout. Different portions have different properties. (In our pizza example, salami is different from mushroom.) Furthermore, heterogeneous mixtures can be of variable composition—you can always ask for more salami and less mushroom. A mixture of oil and vinegar can contain any proportion of oil and vinegar. This variability of composition is a feature that distinguishes mixtures from pure substances. Figure 3-10 illustrates a number of heterogeneous mixtures consisting of different numbers of phases.

A **homogeneous** or one-phase mixture is called a **solution**. A homogeneous mixture is uniform throughout. A solution of sugar dissolved in water is an example of a homogeneous mixture. Other examples of homogeneous mixtures include gasoline, air, salt water, and vinegar. What has been said about heterogeneous mixtures can also be said about homogeneous mixtures of solutions but with one caution. A homogeneous mixture of sugar and water

Copper reacts with nitric acid to form a green compound of copper and reddish-brown fumes of nitrogen dioxide.

Figure 3-10
Examples of heterogeneous mixtures
(a) A two-phase system consisting of a solid and a liquid
(b) A four-phase system consisting of four different liquids
(c) A three-phase system consisting of solid ice, liquid water, and water vapour in air

a   b   c

cannot be prepared using *any* quantity of sugar and *any* quantity of water. There is a range of quantities that can be used. One cannot dissolve one kilogram of sugar in a drop of water.

Except for certain solutions (for example, ethyl alcohol and water) the solubility of one substance in another is limited (or has a maximum value). A solution of one metal (or several) in another is called an **alloy**. Brass and bronze are examples of alloys. Solutions are discussed in more detail in Chapter 10. One of the ways in which matter can be classified is summarized in Figure 3-11.

Figure 3-11
Classification of matter

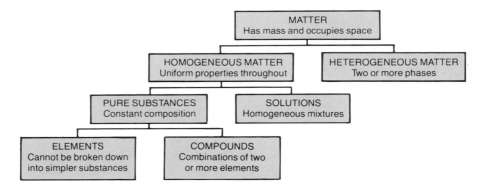

It is fairly easy in principle to tell whether a homogeneous sample of matter is a pure substance or a mixture. If it is a solid we can determine its melting point. We have already seen that if a pure solid substance is heated, it melts at a characteristic temperature (e.g., the melting point of ice is 0°C). As long as any solid remains, the temperature will stay constant (Fig. 3-7). The temperature rises only when the sample is completely melted. Thus, a pure substance has a sharp, characteristic melting point.

An impure solid, on the other hand, does not have a sharp melting point. It melts over a range of temperatures. The impure solid begins to melt at a temperature lower than that of the pure solid, and the temperature rises steadily as more of the solid continues to melt (Fig. 3-12). Thus a mixture usually has a melting point which is lower and covers a wider range of temperatures than a pure substance does. This is a familiar phenomenon to residents of Canada's Atlantic coast. When the temperature is cold enough, both fresh water and salt water will freeze. When the weather warms, however, the impure salt-water ice begins to melt at a much lower temperature than does the ice on fresh-water streams and ponds.

In a similar manner it is possible to test for the purity of a liquid. In this case we look for a constant temperature as the liquid boils. A pure liquid has a characteristic boiling point. For example, at sea level water boils at 100°C. This temperature remains constant as long as any liquid is present. (It is only after the water has completely boiled away that the bottom of a kettle can get hot enough to melt.) If the liquid is impure the temperature usually rises steadily as the liquid boils (Fig. 3-12).

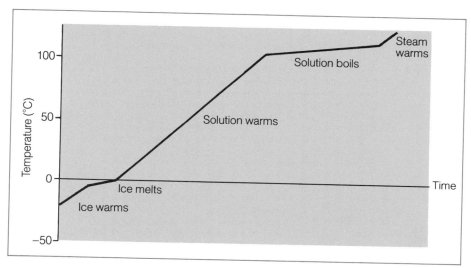

Figure 3-12
Warming curve for an impure substance (salt solution)

3.7 Separating Mixtures

Chemicals found in nature usually consist of mixtures of more than one component—they are not pure substances. How do we separate these mixtures into their components? Different substances have different physical properties. We can make use of this fact to separate the components of mixtures by physical means.

If one component of a mixture is soluble in a liquid such as water and the other is not, the components can be separated by the addition of the liquid. The insoluble component can be separated by allowing it to *settle* to the bottom of the container. The liquid above the solid can then be poured off and evaporated to yield the soluble component.

Insoluble substances can also be separated from soluble components in a liquid mixture by a technique called **filtration**. The mixture is passed through a bed of sand or through a filter paper (Fig. 3-13). Both of these trap solid particles and allow soluble components to pass through. These methods, set-

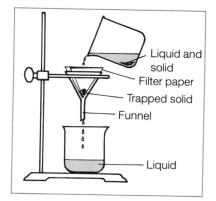

Figure 3-13
Filtration

Figure 3-14
Chemical complex at Redwater, Alberta. Components of crude oil are separated in complexes like this.

An insoluble yellow compound is separated from water by filtration.

tling and filtration, are used in the purification of drinking water (Section 8.13). Filters are also used to remove solid particles from gases as they exit from smokestacks, thus reducing pollution.

Dust particles can be removed from industrial exhaust gases by *electrostatic precipitation*. The dust particles pass through a strong electric field where they become charged. The charged particles are quickly attracted to oppositely-charged plates, where they are neutralized and fall to the bottom of the precipitator. This equipment can remove more than 99% of the dust particles from exhaust gases (Section 13.17).

A substance dissolved in water (or some other liquid) can be separated by a process known as **distillation** (Section 8.6). The solution is heated in a distillation apparatus (Fig. 3-15). If the water vapourizes at a lower temperature than the other substance, the water boils off, leaving the second substance in the distilling flask. The water vapour is liquefied in the condenser and is collected. This process is used by the oil industry (Fig. 3-14) to separate the components of crude oil which boil at different temperatures.

Figure 3-15
Apparatus for distillation

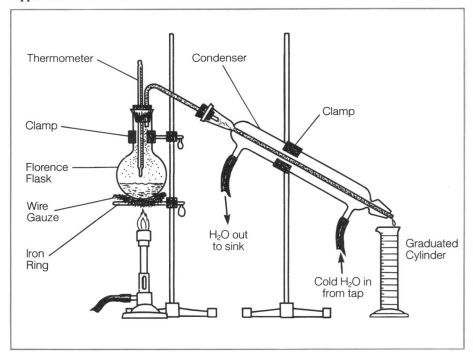

If one of the solids in a mixture is much more soluble in water at high temperatures than at low temperatures, it can be separated from the components by **crystallization** (Section 10.5). The mixture is dissolved in hot water (Fig. 3-16). As the solution cools, the component which is much less soluble in cool water precipitates (forms an insoluble solid) from the solution and can be collected by filtration.

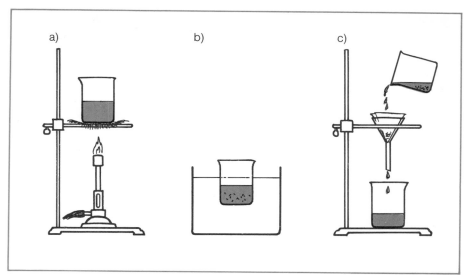

Figure 3-16
Separation by crystallization
(a) Solid mixture is dissolved in hot water.
(b) Crystals of one of the solids in the mixture form as solution cools.
(c) These crystals are then collected by filtration.

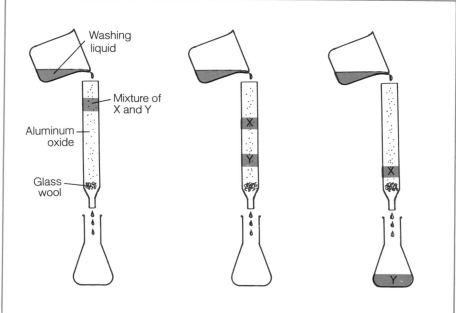

Figure 3-17
Separation of a mixture by column chromatography

Figure 3-18
Working with columns to separate mixtures.

In **chromatography**, separation is achieved by using differences in the degree to which various substances are attracted to the surface of a nonreactive substance. For example, a solution containing several coloured pigments may be washed through a column packed with aluminum oxide. The coloured components, which are strongly attracted to the surface of the aluminum oxide, are washed through more slowly than are the components which are less strongly attracted. The various components are collected one after another as they leave the column (Fig. 3-17).

APPLICATION

Purifying Sea Water

Sea water contains about 3.5% (by mass) of dissolved salts. The removal of these salts from sea water to produce fresh water is called *desalination*. As the world population grows, there will be a shortage of fresh water for drinking. Fresh water is also required for agriculture and industry. Developing efficient methods of desalinating water will therefore become more and more important in the future. There are a number of ways to produce fresh water from sea water.

The method that is used most often to desalinate sea water is *distillation*. This technique accounts for about 90% of the two billion litres of fresh water produced daily by desalination systems. In this process, sea water is boiled until it evaporates. Pure water condenses in the cooling section of the system and is collected. This is an expensive process because it requires energy to heat the sea water. In a technique called multiple-stage distillation, the heat that is obtained from the condensing vapour in one stage is used to heat sea water in the next stage. This method uses energy that would otherwise be wasted. Solar radiation has also been used to reduce the heating costs involved in the evaporation of water. However, there are no large-scale operations using solar radiation.

Research continues into the large-scale production of fresh water by a *freezing* technique. Salt water is partly frozen to form pure ice. The dissolved substances remain in the unfrozen salt water. The ice is then separated and melted to obtain pure water. This method takes less energy than distillation but presents problems with the cleaning and handling of the ice.

Another promising method for purifying water is a technique called *reverse osmosis*. Osmosis is the net movement of water molecules (but not dissolved particles) through a porous membrane from a dilute solution to a more concentrated one. In reverse osmosis (Fig. 3-19) so much pressure is applied to the sea water that the reverse process occurs. Water moves from the more concentrated sea water solution to the other side of the membrane. The low cost of energy required to achieve purification makes this a favourable process. However, some of the major problems to be faced in large scale separations are the tendencies of the membranes to tear under high pressure, to leak dissolved materials, and to become clogged with impurities. The U.S. Navy has developed a portable, hand-operated reverse-osmosis apparatus for use in life rafts. The water produced by this device has a salt content which is low enough for human consumption.

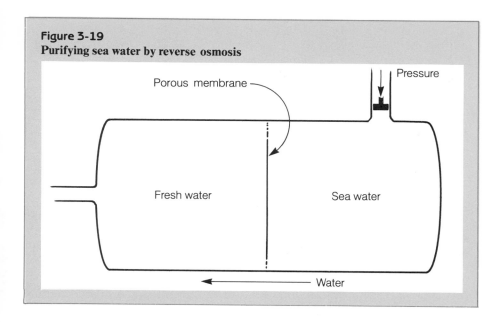

Figure 3-19
Purifying sea water by reverse osmosis

Porous membrane

Pressure

Fresh water

Sea water

Water

Part One Review Questions

1. What are the differences between matter and energy?
2. Why does weight vary with location?
3. State the Law of Conservation of Mass and Energy.
4. How do solids, liquids, and gases differ in the arrangements of their particles?
5. Explain what is meant by translational, rotational, and vibrational motion.
6. Why is melting point considered a physical property?
7. Why is combustion considered a chemical property?
8. Classify each of the following as a physical property or a chemical property. Explain your answers.
    a) density
    b) the boiling point of a compound
    c) the baking of a cake
    d) the filling of a balloon with air
    e) the decomposition of a compound to produce a gas
9. What is meant by a chemical reaction?
10. What is the difference between a physical change and chemical change? Give an example of each.
11. Classify each of the following as a physical or chemical change. Explain your answers.
    a) darkening of silver
    b) fermenting of grapes
    c) formation of snow flakes

d) carving of wood

e) compressing of garbage

12. Give an example of a qualitative property and a quantitative property.

13. Give an example of each of the following: **a)** an element; **b)** a compound; **c)** a homogeneous mixture; and **d)** a heterogeneous mixture.

14. Classify each of the following as an element, compound, heterogeneous mixture, or homogeneous mixture: **a)** ink; **b)** iced tea; **c)** silver; **d)** ice; **e)** grape juice; **f)** garbage.

15. Draw a diagram illustrating the classification of matter.

Part Two: The Laws of Chemical Change

Observations of chemical reactions are summarized in certain generalizations called the laws of chemical change. Three of these are the Law of Conservation of Mass, the Law of Constant Composition, and the Law of Multiple Proportions. They will be used to develop a theory of the atom.

3.8 Law of Conservation of Mass

The **Law of Conservation of Mass** states that during a chemical reaction matter is neither created nor destroyed. Thus, the products of a chemical reaction have the same mass as the starting materials (reactants).

The constancy of mass during chemical reaction can be investigated experimentally by carrying out a reaction in a closed container and obtaining the mass of the system before and after the reaction. A good example of a reaction in a closed system is the ignition of the magnesium wire in a flash bulb. In the bulb, very fine magnesium wire is heated by an electric current supplied by the batteries in the flash unit. The hot magnesium reacts explosively with the oxygen in the bulb to form magnesium oxide, light, and heat. This combustion produces a compound having the same mass as the reactants. If the mass of the used bulb is compared with the mass of the bulb before the flash, it is found that there has been no change in mass. The total mass of the starting materials is the same as the mass of the products. This experiment illustrates the Law of Conservation of Mass.

3.9 Law of Constant Composition

When the French chemist Antoine Lavoisier (Fig. 3-20) showed in 1789 that the total mass remains constant during chemical reactions, he became one of the first people to apply mathematics successfully to chemical processes. Not long afterward, in 1792, the German chemist Jeremias Richter suggested that it should be possible to calculate the composition of the products of a reaction if the composition of the reactants were known. His prose was so difficult to read, however, that chemists paid little attention to this suggestion. Seven years later, in 1799, the French chemist Joseph Proust showed that copper carbonate always had the same chemical composition no matter how it was prepared. In doing so he had discovered the **Law of Constant Compositon**—the percentage composition by mass in particular compounds never varies.

For example, when hydrogen and oxygen are ignited, the product is water. Oxygen and hydrogen can be mixed in any proportion desired: 50/50, 25/75, 75/25, etc. No matter what the proportions of oxygen to hydrogen in the initial mixture, the resulting water will always have the same percentage composition. By mass, the water will contain 11.19% hydrogen and 88.81% oxygen. Any excess hydrogen (or oxygen) will remain unreacted. In the same way, when sodium reacts with chlorine to form sodium chloride, salt, the product will always contain 39.37% sodium and 60.63% chlorine by mass.

Thus, in the formation of a compound, the mass of one element needed to combine with a given mass of another is always the same. This is another way of stating the Law of Constant Composition.

A modern analytical balance

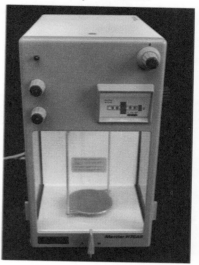

Biography

ANTOINE LAVOISIER was born in Paris, where he studied mathematics and the physical sciences. He was so productive that at the age of 23 he received a gold medal from the Academy of Sciences for his work on the problems of lighting a large town. Two years later he was admitted to the Academy.

Lavoisier became a member of the Ferme Générale, a hated financial company which was allowed to collect taxes on the condition that it pay an annual fee to the state. Lavoisier received a large income which enabled him to equip his laboratory fully.

Lavoisier was always interested in great or far-reaching principles, not in isolated facts. Furthermore, his experiments were usually quantitative—he preferred to weigh and measure. By careful measurements of the masses involved during the reactions of metals and nonmetals with air, he concluded that combustion involves the combination of a substance with oxygen. Lavoisier was a pioneer in the analysis of organic compounds—he burned the compounds and determined the masses of their products to calculate their percentage composition. He also wrote the first modern textbook on chemistry.

In 1793, the French revolution was sweeping the country. All members of the Ferme Générale were arrested and tried. Lavoisier was denounced by a chemist named Fourcroy as a counterrevolutionary, sentenced to death, and executed by guillotine in 1794.

Figure 3-20
Antoine Laurent Lavoisier (1743–1794)--Father of modern chemistry

EXAMPLE PROBLEM 3-1

Lead and sulfur react so that 2.45 g of lead and 0.38 g of sulfur combine to form 2.83 g of lead sulfide. How much sulfur is required to combine with 0.75 g of lead?

SOLUTION

$$0.75 \text{ g of Pb} \times \frac{0.38 \text{ g of S}}{2.45 \text{ g of Pb}} = 0.12 \text{ g of S}$$

EXAMPLE PROBLEM 3-2

Lithium and fluorine always react so that 7.00 g of lithium and 19.2 g of fluorine combine to give 26.2 g of lithium fluoride. If 1.00 g of lithium is reacted with 3.00 g of fluorine to form lithium fluoride, how much fluorine combines with the lithium? How much fluorine remains unreacted?

SOLUTION

$$1.00 \text{ g of Li} \times \frac{19.2 \text{ g of F}}{7.00 \text{ g of Li}} = 2.74 \text{ g of F}$$

∴ 1.00 g of lithium combines with 2.74 g of fluorine.

$$3.00 \text{ g} - 2.74 \text{ g} = 0.26 \text{ g of fluorine}$$

∴ 2.74 g of fluorine react with 1.00 g of lithium, and 0.26 g of fluorine remain unreacted.

PRACTICE PROBLEM 3-1

Sodium and bromine react so that 4.60 g of sodium and 15.98 g of bromine combine to give 20.58 g of sodium bromide. How much sodium is required to combine with 25.00 g of bromine? (*Answer*: 7.20 g)

PRACTICE PROBLEM 3-2

Potassium and iodine react so that 3.00 g of potassium and 9.74 g of iodine combine to give 12.74 g of potassium iodide. If 1.00 g of potassium is reacted with 4.00 g of iodine to form potassium iodide, how much iodine combines with the potassium? How much iodine remains unreacted? (*Answers*: 3.25 g; 0.75 g)

3.10 Law of Multiple Proportions

Some elements can combine with each other in more than one way to produce two or more different compounds. We now know that carbon burns in a plentiful supply of oxygen to produce a gas called carbon dioxide. However, if the supply of oxygen is limited, the carbon burns to form both carbon dioxide and a different gas, carbon monoxide. Analysis of the carbon dioxide reveals that 8.00 g of oxygen are combined with 3.00 g of carbon; in carbon monoxide 8.00 g of oxygen are combined with 6.00 g of carbon according to the following reactions:

$$3 \text{ g of carbon} + 8 \text{ g of oxygen produced carbon dioxide}$$

$$6 \text{ g of carbon} + 8 \text{ g of oxygen produced carbon monoxide}$$

Now let us divide the oxygen-to-carbon mass ratio in carbon dioxide by the oxygen-to-carbon mass ratio in carbon monoxide. That is, we will determine the proportion of oxygen to carbon in each compound and then divide one result by the other.

$$\frac{\text{mass of O in carbon dioxide}}{\text{mass of C in carbon dioxide}} = \frac{8.00 \text{ g of O}}{3.00 \text{ g of C}} = \frac{2.67 \text{ g of O}}{1.00 \text{ g of C}} = 2.67 \text{ g of O/g of C}$$

$$\frac{\text{mass of O in carbon monoxide}}{\text{mass of C in carbon monoxide}} = \frac{8.00 \text{ g of O}}{6.00 \text{ g of C}} = \frac{1.33 \text{ g of O}}{1.00 \text{ g of C}} = 1.33 \text{ g of O/g of C}$$

$$\therefore \frac{2.67 \text{ g of O/g of C}}{1.33 \text{ g of O/g of C}} = \frac{2.01}{1.00} \approx \frac{2}{1}$$

The ratio of the two proportions is a fraction of two whole numbers.

This calculation illustrates the third law of chemical change, which is now known as the **Law of Multiple Proportions**: When two elements, A and B, form more than one compound, the mass of element A divided by the mass of element B in one compound, compared with the mass of element A divided by the mass of element B in the other compound gives a ratio of small whole numbers. In other words, when the proportions (by mass) in which two elements combine to form two different compounds are divided by each other, the result is a fraction composed of small whole numbers.

EXAMPLE PROBLEM 3-3

It is found that 50.0 g of sulfur combines with 73.2 g of oxygen to form a compound (A). It is also found that 100 g of sulfur combines with 97.8 g of oxygen to yield a second compound (B). Show how these data illustrate the Law of Multiple Proportions.

SOLUTION

$$\frac{\text{mass of S in } A}{\text{mass of O in } A} = \frac{50.0 \text{ g of S}}{73.2 \text{ g of O}} = \frac{0.683 \text{ g of S}}{1.00 \text{ g of O}} = 0.683 \text{ g of S/g of O}$$

$$\frac{\text{mass of S in } B}{\text{mass of O in } B} = \frac{100 \text{ g of S}}{97.8 \text{ g of O}} = \frac{1.02 \text{ g of S}}{1.00 \text{ g of O}} = 1.02 \text{ g of S/g of O}$$

$$\therefore \frac{0.683 \text{ g of S/g of O}}{1.02 \text{ g of S/g of O}} = \frac{0.668}{1.00} = \frac{0.668}{1.00} \times \frac{3}{3} = \frac{2.00}{3.00} = \frac{2}{3}$$

Since the result is a ratio of small whole numbers, the Law of Multiple Proportions is obeyed.

CAREER

The Quality Control Chemist

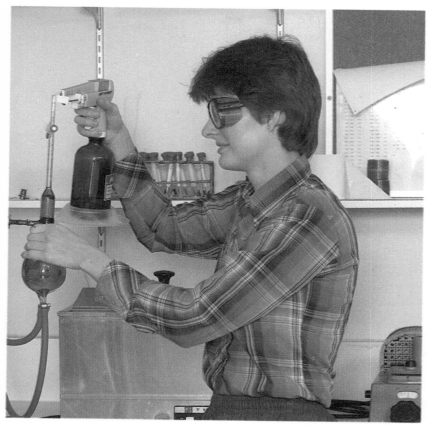

Figure 3-21
A chemist working in a quality control laboratory

Quality control chemists are employed by industry. They control and evaluate all manufacturing materials and operations that affect the quality and reliability of a product. They run continuous tests to check the quality of starting materials and of intermediate substances, to evaluate the effect of changes in production methods, to pinpoint problems in production techniques, and to ensure that the product has the desired characteristics. They improve existing test methods and develop new ones. Because analytical equipment and techniques improve over the years, quality control chemists must evaluate new equipment for future use. They must check products for quality and test returned goods to discover the reasons for their rejection. They are frequently called on to write progress and information reports.

In addition to ensuring the quality and safety of products, some quality control chemists are also responsible for maintaining environmental standards in the vicinity of the plant. They may, therefore, analyze stack gases for atmospheric pollutants, and waste-water effluent for water pollutants.

Quality control chemists often have to modify existing equipment to meet specific needs. Thus, in addition to their basic chemical knowledge, they need some knowledge of electronics and instrumentation. Since a typical quality control laboratory may analyze thousands of samples over a year, quality control chemists must apply statistical methods to ensure accuracy in their sampling and measurement. Therefore, they need some knowledge of mathematical statistics and its applications to sampling. In addition, a knowledge of the use of computers for the processing and management of data is desirable.

The most important requirements for the position of quality control chemist are thoroughness, organization, a concern for precision and detail, and the ability to keep accurate records. A quality control chemist must also have an inquiring mind, which is necessary for conceiving new testing techniques and discovering the sources of problems.

EXAMPLE PROBLEM 3-4

It is found that 12.00 g of carbon combines with 4.00 g of hydrogen to form a compound (A). It is also found that 24.00 g of carbon combines with 6.00 g of hydrogen to produce a second compound (B). Do these results obey the Law of Multiple Proportions?

SOLUTION

$$\frac{\text{mass of C in } A}{\text{mass of H in } A} = \frac{12.00 \text{ g of C}}{4.00 \text{ g of H}} = \frac{3.00 \text{ g of C}}{1.00 \text{ g of H}} = 3.00 \text{ g of C/g of H}$$

$$\frac{\text{mass of C in } B}{\text{mass of H in } B} = \frac{24.00 \text{ g of C}}{6.00 \text{ g of H}} = \frac{4.00 \text{ g of C}}{1.00 \text{ g of H}} = 4.00 \text{ g of C/g of H}$$

$$\therefore \frac{3.00 \text{ g of C/g of H}}{4.00 \text{ g of C/g of H}} = \frac{3}{4}$$

Since the result is a ratio of small whole numbers, the Law of Multiple Proportions is obeyed.

PRACTICE PROBLEM 3-3

One compound of nitrogen and hydrogen is found to contain 83.2 g of nitrogen combined with 2.00 g of hydrogen; a second compound contains 9.28 g of nitrogen combined with 2.00 g of hydrogen. Show that these data are in agreement with the Law of Multiple Proportions. (*Answer*: The proportions of nitrogen to hydrogen in the two compounds are in the ratio of 8.97:1, which equals a ratio of 9:1 within the uncertainty of the measurement.)

PRACTICE PROBLEM 3-4

For one compound of nitrogen and hydrogen it is found that 97.66% of the mass is made up of nitrogen and 2.34% of the mass is made up of hydrogen. A second compound contains 82.3% nitrogen and 17.7% hydrogen. Show that these data are consistent with the Law of Multiple Proportions. (*Answer*: These data are for the same two compounds used in Practice Problem 3-3. The ratio of the N:H proportions in the two compounds is still 9 : 1.)

PRACTICE PROBLEM 3-5

It is found that 118.7 g of tin combines with 71.00 g of chlorine to form a compound (A). It is also found that 118.7 g of tin combines with 142.00 g of chlorine to produce a second compound (B). Show that these data are consistent with the Law of Multiple Proportions. (*Answer*: The proportions of tin to chlorine in the two compounds are in the ratio 2.00 : 1.00, which equals a ratio of 2 : 1 within the uncertainty of the measurement.)

Part Two Review Questions

1. Sodium and chlorine react so that 2.3 g of sodium and 7.1 g of chlorine combine to give 9.4 of sodium chloride. If 1.00 g of sodium is reacted with 4.00 g of chlorine to form sodium chloride, how much chlorine combines with the sodium? How much chlorine remains unreacted?

2. Hydrogen and oxygen react so that 2.00 g of hydrogen and 16.00 g of oxygen combine to produce 18.00 g of water. How many grams of oxygen are required to react completely with 7.00 g of hydrogen? How many grams of water would be formed?

3. It is found that 58.90 g of cobalt combines with 159.80 g of bromine to form a compound (A). It is also found that 58.90 g of cobalt combines with 239.70 g of bromine to produce a second compound (B). Show that these data are consistent with the Law of Multiple Proportions.

4. It is found that 63.5 g of copper combines with 35.5 g of chlorine to form a compound (A). It is also found that 63.5 g of copper combines with 71.0 g of chlorine to produce a second compound (B). Show that these data are consistent with the Law of Multiple Proportions.

Chapter Summary

- The universe consists of matter and energy.
- Matter has mass and occupies space.
- The mass of an object depends upon the amount of matter that is present.
- The weight of an object depends on the force of gravity acting upon it.
- Energy is the ability to do work.
- Kinetic energy is energy of motion.
- Potential energy is energy due to position.
- Energy may be converted from one form to another.
- The Law of Conservation of Mass and Energy states that the total quantity of matter and energy in the universe is constant.
- Mass is conserved in a chemical reaction.
- The three states of matter are solid, liquid, and gas.
- Molecules possess translational, rotational, and vibrational motions.
- The physical properties of a substance are those properties which can be determined without changing its composition or make-up.
- The chemical properties of a substance are those properties which can be determined only when the substance undergoes a change in composition.
- A quantitative property is one that can be measured.
- A qualitative property is one that cannot be expressed numerically.
- A physical change is a change that alters one or more of the properties of a substance without changing its composition or identity.

- A chemical change is a change in a substance that converts it into a different kind (or different kinds) of matter, each with a different composition and new properties.
- A chemical change is also known as a chemical reaction.
- A pure substance consists of only one particular kind of matter and has the same properties throughout.
- Elements are pure substances that cannot be broken down into simpler substances by chemical methods.
- Compounds are pure substances consisting of two or more elements combined chemically.
- Mixtures are combinations of two or more pure substances each of which retains its own physical and chemical properties.
- A phase is any portion of a material with a uniform set of properties.
- A solution is a homogeneous mixture.
- Some methods of separating mixtures include settling, filtration, distillation, crystallization, and chromatography.
- The Law of Constant Composition states that compounds always have the same percentage composition by mass.
- The Law of Multiple Proportions states that when the proportions (by mass) in which two elements combine to form two different compounds are divided by each other, the result is a fraction composed of small whole numbers.

Key Terms

matter	density	heterogeneous
inertia	chemical property	homogeneous
mass	chemical reaction	solution
weight	quantitative property	alloy
energy	qualitative property	filtration
work	physical change	distillation
kinetic energy	chemical change	crystallization
potential energy	pure substance	chromatography
solid	element	Law of Conservation of Mass
liquid	compound	Law of Constant Composition
gas	phase	Law of Multiple Proportions
physical property	mixture	

Test Your Understanding

Each of these statements or questions is followed by four responses. Choose the correct response in each case. (Do not write in this book.)

1. Which of the following is best considered an example of potential energy?
   a) a motorcycle travelling down a highway   b) a ball resting on a table
   c) a person jogging   d) a golfer swinging a club

2. The total mass before a chemical reaction is equal to the total mass after a chemical reaction. Which law does this illustrate?
   a) Conservation of Mass
   b) Constant Composition
   c) Multiple Proportions
   d) Conservation of Energy

3. Which of the following is true?
   a) a liquid has a definite shape
   b) a solid can be compressed easily
   c) the particles of a gas are far apart
   d) a gas has a definite volume

4. Which of the following is a physical property?
   a) carbon reacts with oxygen to produce carbon dioxide
   b) the density of platinum is 2.14 x 10⁴ kg/m³
   c) steel wool rusts
   d) excess stomach acid is neutralized by antacids

5. Which of the following depends on location?
   a) density
   b) weight
   c) mass
   d) inertia

6. Which of the following is a chemical property of water?
   a) it has a boiling point of 100°C
   b) it is soluble in alcohol
   c) it is a colourless liquid
   d) it reacts with potassium to produce hydrogen gas

7. Which of the following is a quantitative property?
   a) oxygen is an odourless gas
   b) a water solution of copper(II) sulfate is blue
   c) 203.9 g of sugar can be dissolved in 100 g of water at 20°C
   d) mercury is a liquid at room temperature

8. Which of the following is a qualitative property?
   a) the density of mercury is 1.36 x 10⁴ kg/m³
   b) silver melts at 961°C
   c) platinum boils at 3827°C
   d) sodium chloride is a white solid

9. Which of the following is a physical change?
   a) the melting of sodium chloride
   b) the burning of rubber
   c) the rusting of an automobile
   d) the rotting of an apple

10. Which of the following is a chemical change?
    a) the stretching of rubber
    b) the burning of a match
    c) the crushing of a grape
    d) the freezing of a liquid

11. Which one of the following is a chemical reaction?
    a) melting
    b) boiling
    c) filtering
    d) burning

12. Which of the following is a pure substance?
    a) salt water
    b) soil
    c) ginger ale
    d) oxygen

13. Which of the following is an element?
    a) air
    b) water
    c) sulfur dioxide
    d) hydrogen

14. Which of the following is a compound?
    a) zinc
    b) aluminum
    c) carbon disulfide
    d) sulfur

15. Which of the following is a heterogeneous mixture?
    a) black coffee
    b) mouth wash
    c) a tossed salad
    d) nail polish remover

16. Which of the following could be separated by filtration?
    a) sawdust and water
    b) sugar and water
    c) milk
    d) soda water

17. Which of the following could be separated by distillation?
    a) mercury   b) sugar and water   c) brass   d) carbon dioxide
18. Which of the following could be separated by column chromatography?
    a) water   b) bronze   c) food colouring   d) salt
19. Thirty-two grams of sulfur react with thirty-two grams of oxygen even if more than thirty-two grams of sulfur are available. Which law does this illustrate?
    a) Conservation of Mass   b) Constant Composition
    c) Multiple Proportions   d) Conservation of Energy
20. It is possible for two elements to join to form two or more different compounds. Which law does this illustrate?
    a) Conservation of Mass   b) Constant Composition
    c) Multiple Proportions   d) Conservation of Energy

Review Your Understanding

1. What is inertia?
2. Discuss the differences that might be found in both the mass and the weight of an astronaut on the earth and on the moon.
3. Why is a moving automobile more difficult to stop than a moving baby carriage if both are travelling at the same speed?
4. Distinguish between a) mass and weight; b) work and energy; c) kinetic and potential energy.
5. What type of energy, kinetic or potential, is illustrated in the following examples: a) a speeding bullet; b) a skier poised at the top of a ski jump; c) water held behind a dam; d) the mainspring in a recently wound watch; e) a bowling ball rolling towards the pins?
6. On which state of matter do temperature and pressure have the greatest effect? Why?
7. What is the difference between a solid and a liquid? What is the difference between a gas and a liquid?
8. Liquid bromine melts at $-7°C$ and boils at $59°C$. Sketch the heating curve that would be obtained if a container of solid bromine at $-20°C$ is placed in a beaker of hot water at $70°C$. Identify on the graph the processes that are occurring at each portion of the curve.
9. Describe the three types of motion that a particle of matter can possess.
10. Trimyristin is a substance which is easily isolated in the pure state from nutmegs. It has a melting point of $56°C$ and a boiling point of $311°C$. Describe the energy changes which occur when a sample of pure trimyristin is heated from $50°C$ to $350°C$.
11. Which of the following are examples of qualitative properties and which are examples of quantitative properties? Give a reason for each answer.
    a) Potassium floats on water, but reacts vigorously with the water.
    b) The melting point of lithium chloride is $610°C$.
    c) Magnesium oxide does not dissolve well in water.
    d) Beryllium oxide has a density of $3.01 g/cm^3$.
    e) Fluorine is a pale yellow gas under ordinary conditions.

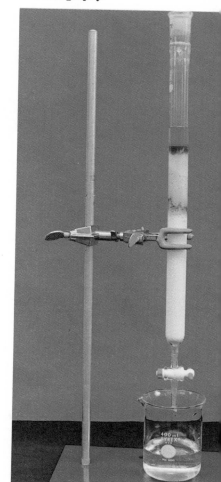

A mixture of three coloured substances is separated into its components by column chromatography.

12. Are the following statements examples of physical properties or chemical properties? Explain your answer in each case.
   a) Gold does not tarnish.
   b) Lead is useful in making weights for scuba divers.
   c) Copper is easily drawn into thin wires.
   d) Laughing gas supports combustion.
   e) Gallium metal melts in your hand.

13. Are the following statements examples of physical changes or chemical changes? State your reasons in each case.
   a) A platinum wire becomes red when held in the flame of a Bunsen burner.
   b) Nitroglycerine explodes on impact or with heating.
   c) Propane burns.
   d) Milk sours when left to stand at room temperature.
   e) Zinc sulfide glows when exposed to gamma radiation.

14. Are the following observations examples of chemical changes or physical changes? Give your reasons in each case.
   a) A firefly glows at night.
   b) Food is digested when it is eaten.
   c) A tree grows in Ottawa.
   d) Ice melts on a hot stove.
   e) A candle gives off light.
   f) A light bulb gives off light.

15. Distinguish between a pure substance and a mixture.

16. How do the properties of a pure substance differ from those of a solution?

17. List two ways in which a solution differs from a compound.

18. Which of the following is heterogeneous? Explain.
   a) a solution of sugar in water
   b) air (a mixture of 80% nitrogen gas and 20% oxygen gas)
   c) a pure sample of silver
   d) a mixture of ice and water

19. Give an example of a homogeneous mixture consisting of **a)** two solids; **b)** two liquids; **c)** two gases; **d)** a solid and a liquid; **e)** a gas and a liquid.

20. Silver chloride is insoluble in water. Sodium chloride is soluble in water. What procedure would you use to separate solid silver chloride from solid sodium chloride?

21. Explain why mixtures do not obey the Law of Constant Composition.

22. Nitrogen and hydrogen react so that 28.00 g of nitrogen and 6.00 g of hydrogen combine to produce 34.00 g of ammonia. How many grams of hydrogen are required to react completely with 17.00 g of nitrogen?

23. If 1.00 g of hydrogen reacts with 8.00 g of oxygen when water forms from these two elements, even if 25.0 g of oxygen are available, which law is illustrated? How many grams of hydrogen are required to react with the 25.0 g of oxygen?

24. Oxygen and hydrogen always react so that 16.0 g of oxygen and 2.00 g of hydrogen combine to give 18.0 g of water. If 10.0 g of oxygen is reacted with 10.0 g of hydrogen to form water, how much hydrogen combines with the oxygen? How much hydrogen remains unreacted?

25. Which of the following obeys the Law of Constant Composition? Explain your reasoning. **a)** a heterogeneous mixture; **b)** a homogeneous mixture; **c)** a solution; **d)** an element; **e)** a compound.

26. Nickel forms several oxides. One contains 71.0% nickel; another contains 78.6% nickel. Are these data consistent with the Law of Multiple Proportions?

27. Iron forms two different compounds with chlorine. There are 44.0 g of iron in 100.0 g of one compound, but only 34.4 g of iron in 100.0 g of the other compound. Show that these data illustrate the Law of Multiple Proportions.

Apply Your Understanding

1. A kind of curve similar to a heating curve and giving the same type of information can be obtained if a solid which has been melted and heated to the boiling point is then cooled. If the temperatures of the sample are taken during cooling, this kind of temperature-versus-time graph is called a cooling curve. Draw a typical cooling curve for water, and explain what processes are occurring at each portion of the curve.

2. An unknown gas was burned in oxygen to produce nitrogen gas and water vapour. Was the unknown gas an element? Explain.

3. When copper wire is heated in the presence of oxygen, a black substance is formed. Is this black substance likely to an element, a mixture, or a compound?

4. Suggest two methods of separating a mixture of sugar and iron filings.

5. You are given two liquids. One is a pure compound, and other the other is a solution. How could you distinguish the two liquids?

6. The labels of two bottles have become so illegible that only the words "m.p. 121°C" can still be read. Each bottle contains a white crystalline solid which does indeed melt at 121°C. How could you demonstrate whether the bottles contain the same substance or two difference substances with the same melting point?

Investigations

1. Write a research report discussing the benefits of reclaiming aluminum from garbage. Include the methods used to separate aluminum from the garbage.

2. Research the use of paper chromatography as a technique for separating mixtures. Design and conduct an experiment which uses paper chromatography to separate the components of washable black ink. Does permanent black ink give different results?

3. Flotation is an important separation technique used in the mining industry. Write a research report on this technique. Indicate some metals that are obtained by using this technique.

4. You are given two separate white solids. One is sugar (sucrose) and the other is salt (sodium chloride). Devise an experiment to distinguish, without tasting, which sample is sugar and which is salt.

Theories of the Atom

CHAPTER CONTENTS

Part One: **Early Developments of the Atomic Model of Matter**
  4.1 Early Models of Matter
  4.2 Atomic Mass
  4.3 Electrons
  4.4 Protons

Part Two: **Development of the Nuclear Model of the Atom**
  4.5 Rutherford's Model of the Atom
  4.6 Isotopes
  4.7 The Bohr Model of the Hydrogen Atom
  4.8 Energy Levels and Many-Electron Atoms

Part Three: **The Present Model of the Atom**
  4.9 Wave Mechanics–The Present Model
  4.10 Wave Mechanics and Many-Electron Atoms
  4.11 The Quantum Numbers

For thousands of years, fireworks have been part of public celebrations.

How do fireworks, X-rays, neon signs, and street lights depend upon the structure of atoms? How are scientists able to identify chemicals in outer space? How are we able to use the properties of electrons to design high-powered microscopes? What property of atoms makes it easier for forensic scientists to identify matter? You will find the answers to these questions as you study this chapter.

The main purpose of this chapter is to trace the development of our view of the atom from the ancient ideas of the Greeks to our current theories of the atom. We will consider the contributions made by a number of important scientists including John Dalton, Ernest Rutherford, and Niels Bohr. We shall learn about those particles in an atom that are important to chemists and see how, as time went on, new discoveries required modifications of existing models of the atom.

Figure 4-1
The structure of atoms allows neon signs to glow.

Key Objectives

When you have finished studying this chapter, you should be able to

1. state the locations of electrons, protons, and neutrons within an atom.
2. state the mass and charge of electrons, protons, and neutrons.
3. describe the behaviour of electrons, protons, and neutrons in an electric field.
4. describe the Rutherford model of the atom.
5. state the number of protons, neutrons, and electrons in an atom with known atomic and mass numbers.
6. calculate the average atomic mass of a mixture of isotopes of known atomic mass.
7. describe the Bohr model of the atom.
8. explain what an atomic line spectrum implies in terms of the structure of an atom.
9. write the energy level populations of the first 20 elements, using only a blank periodic table.
10. state two differences between the Bohr model and the quantum mechanical model of the atom.
11. draw energy level diagrams for the ground states of the first 20 elements.
12. write electron configurations for the ground states of the first 20 elements.

Part One: Early Developments of the Atomic Model of Matter

4.1 Early Models of Matter

The Law of Conservation of Mass, the Law of Constant Composition, and the Law of Multiple Proportions summarize the experimental facts about the behaviour of matter in chemical reactions. Scientists had for centuries tried to develop a theory (or model) of the nature of matter that was consistent with their observations.

The Greek Model

One school of Greek philosophers who lived in the fourth and fifth centuries B.C. believed that matter is composed of tiny indestructible particles which they called atoms (from the Greek word *atomos*, meaning uncut). However, this theory was never verified by experiments and was discarded. More than two thousand years later, in 1803, John Dalton (Fig. 4-2), an English school-teacher, reintroduced the model or theory, and developed it to such an extent that it was able to explain the laws of chemical change.

Dalton's Atomic Theory

To some extent, Dalton's ideas arose from a great controversy between the French chemist J. L. Proust and his fellow countryman C. M. Berthollet. The argument lasted from 1799 to 1806, and it was carried out in the journals with great brilliance and extreme courtesy. For example, Berthollet argued that copper and sulfur would combine in an infinite number of proportions depending on the amounts of each substance used. Proust maintained that the "compounds" of variable composition prepared by Berthollet were simply mixtures of two compounds, each of which had its own constant copper-to-sulfur ratio. Berthollet and his colleagues worked for more than seven years. Yet every time they prepared a "compound" that they thought showed con-tinuously variable composition, Proust and his colleagues were able to demon-strate that the compound was a mixture of two pure substances, each of which had its own composition.

Gradually the tide of opinion shifted in favour of Proust's arguments. One important result, however, was that scientists on both sides of the controversy collected large amounts of data concerning substances, and this information was all available to Dalton. Earlier scientists had occasionally speculated on the atomic nature of matter, but Dalton was able to put all the latest information together and construct a theory which explained the laws of chemical change.

Dalton's theory consisted of several statements:

1. All matter is composed of extremely small particles called atoms.
2. Atoms can be neither subdivided nor changed into one another.
3. Atoms cannot be created or destroyed.
4. All atoms of one element are the same in shape, size, mass, and all other properties.
5. All atoms of one element differ in these properties from atoms of all other elements.
6. Chemical change is the union or separation of atoms.
7. Atoms combine in small whole-number ratios such as $1:1$, $1:2$, $2:3$, etc.

In modern terms, we would say that an **atom** is the smallest particle of an element that has all the chemical properties of that element.

Biography

JOHN DALTON was the son of a poor English weaver. He started keeping meteorological records at the age of 11 and continued until the day before his death. He was the first person to make systematic studies of the weather. He also described the nature of colour blindness, of which he was a victim. At the age of 12 he began his career as a school teacher. He then worked at a boarding school and when he was 19 became its principal. He devoted every spare minute to intellectual pursuits.

In 1793 he became a tutor in mathematics and natural philosophy at New College, Manchester. Six years later he resigned to support himself as a private tutor so he would have more time to spend in scientific pursuits. For financial reasons he was forced to construct his own crude instruments. Furthermore, he was a poor experimenter. Nevertheless, he managed to make discoveries that lie at the very foundations of chemistry.

By 1801 Dalton had discovered several laws involving the behaviour of gases, including his Law of Partial Pressures, all of which stemmed from his meteorological observations. His attempts to explain these observations led to the development of his Atomic Theory, which he published in 1808 in his book *New System of Chemical Philosophy*. He then endeavoured to determine atomic weights and discovered, as a result, the Law of Multiple Proportions. In 1826 he received a Royal Medal from the King.

Dalton's health began to fail in 1837, and in 1844 he died peacefully, having laid the basis for the development of chemistry as a science.

Figure 4-2
John Dalton (1766–1844)—
Proposed atomic theory.

Models of carbon monoxide and carbon dioxide molecules.

Dalton's Atomic Theory Explains the Law of Conservation of Mass

If all chemical reactions are unions or separations of atoms, and if atoms are indestructible, a chemical reaction would not involve the creation or destruction of atoms or mass. Thus, Dalton's theory explains the Law of Conservation of Mass. When oxygen atoms and hydrogen atoms unite to form water, no atoms are created and none are destroyed. Thus, the mass of the atoms before reaction must be the same as the mass of the atoms after reaction.

Dalton's Atomic Theory Explains the Law of Constant Composition

Dalton's atomic theory also explained the Law of Constant Composition. When carbon atoms react with oxygen atoms to form carbon monoxide, each carbon atom reacts with one oxygen atom to form one molecule of carbon monoxide. (A **molecule** is the smallest particle of a compound which shows all the chemical properties of that compound.) If there were 15 oxygen atoms present, only 15 carbon atoms would react to form 15 carbon monoxide molecules, even if 100 carbon atoms were available for reaction with the oxygen. According to the atomic theory, all carbon atoms have the same mass, and all oxygen atoms have the same mass. Therefore, each carbon monoxide molecule will have the same mass—the sum of the atomic masses of carbon and oxygen. The fraction of the molecular mass contributed by carbon will be the mass of a carbon atom (always the same) divided by the mass of the molecule of carbon monoxide (always the same). This fraction will be constant. Thus, when carbon and oxygen unite to form carbon monoxide, the molecule will always have the same fraction of carbon uniting with the same fraction of oxygen. That is, carbon monoxide has a constant composition by mass.

Dalton's Atomic Theory Predicts the Law of Multiple Proportions

It is important to note that Dalton's theory *predicts* the Law of Multiple Proportions, since the Law was not known at the time. The Law of Constant Composition was known, and Dalton's theory explains it. Dalton needed fixed ratios of atoms to obtain the observed constant compositions, and he reasoned that the ratios should be simple ones. If a carbon monoxide molecule contains one oxygen atom per carbon atom, and if a carbon dioxide molecule contains two oxygen atoms per carbon atom, then the ratio of oxygen atoms per carbon atom in the two compounds is exactly 1 : 2.

$$\frac{\text{O atoms per C atom in carbon monoxide}}{\text{O atoms per C atom in carbon dioxide}} = \frac{1}{2}$$

Since there are proportionally half as many oxygen atoms in one compound as in the other, and since each oxygen atom has the same mass according to Dalton's theory, the proportions of mass in each compound that are due to oxygen are also in a 1 : 2 ratio.

Dalton's theory was valuable because it explained the Laws of Constant Composition and Conservation of Mass. It also stimulated chemists to do still more experiments to verify other predicted laws and to try to determine the masses of atoms. Thus began an enormous advance in the knowledge and understanding of chemistry.

4.2 Atomic Mass

One of the most important concepts to arise from Dalton's theory is that all atoms of one element have the same mass and that this mass is different from the masses of the atoms of other elements. Chemists immediately began to try to determine the masses of atoms. It was impossible, of course, for them to determine the masses of individual atoms or even to *prove* that atoms existed at all. But if atoms did exist, they had definite masses, and chemists attempted to assign relative atomic masses that agreed with the known compositions of compounds. There was much confusion, because chemists did not distinguish between atoms and molecules (the concept of a molecule had not yet been invented). Dalton, for example, believed incorrectly that one atom of hydrogen combined with one atom of oxygen to form one atom of water:

$$H + O \longrightarrow HO$$

Since chemical analysis shows that the mass of oxygen in water is eight times the mass of hydrogen, Dalton thought that atoms of oxygen had eight times the mass of hydrogen. Using similar arguments for other compounds, it was possible to construct a table of relative atomic masses of most of the elements then known. There were many errors due to faulty analyses of compounds and wrong guesses as to the number of atoms of each element in a molecule of the compound. Gradually the errors were discovered and corrected, and we now have the modern table of average atomic masses which is reproduced at the front of this book and in Appendix A.

Dalton's atomic theory was a positive contribution to our knowledge of matter. However, some assumptions had to be altered in the light of later discoveries. For example, all atoms of the same element are not identical. Even though their chemical properties are alike, they may differ in mass. These atoms of the same elements with different masses are called **isotopes**. For example, natural carbon consists of two isotopes. One type of carbon atom constitutes 98.90% of natural carbon; another type of carbon atom contains more mass than the first and constitutes 1.10% of natural carbon.

Scientists have chosen the most common isotope of carbon (carbon-12) as a reference standard. An atom of carbon-12 is arbitrarily assigned a mass of 12 atomic mass units (u). The masses of all other atoms are compared with the mass of this type of carbon atom. According to this definition, one **atomic mass unit** (u) is 1/12 the mass of a carbon-12 atom. The mass of an atom expressed in atomic mass units is called its **atomic mass**. The choice of an atomic mass of 12 u for carbon conveniently gives atomic masses for most of the other elements that are very nearly whole numbers.

An atom of ordinary hydrogen, for example, has an atomic mass which is about $\frac{1}{12}$ as great as the mass of carbon-12. Therefore, the mass of an atom of ordinary hydrogen is about 1 u ($12 \times \frac{1}{12} = 1$). A more precise value of the atomic mass of an atom of ordinary hydrogen is 1.007 825 u. The mass of an atom of ordinary oxygen is about $\frac{4}{3}$ as great as the mass of carbon-12. Therefore, the mass of an atom of ordinary oxygen is about 16 u ($12 \times \frac{4}{3} = 16$). A more precise value is 15.994 91 u.

EXAMPLE PROBLEM 4-1

An isotope of neon has a mass which is 3/2 the mass of an atom of carbon-12. What is the atomic mass of this neon isotope?

SOLUTION

$$\text{Mass of carbon-12 atom} = 12.0 \text{ u}$$

$$\frac{3}{2} \times \text{mass of carbon-12 atom} = \frac{3}{2} \times 12.0 \text{ u} = 18.0 \text{ u}$$

$$\therefore \text{ The atomic mass of neon is 18.0 u.}$$

PRACTICE PROBLEM 4-1

An atom of sodium has a mass which is 1.92 times the mass of a carbon-12 atom. What is the atomic mass of sodium? (*Answer:* 23.0 u)

PRACTICE PROBLEM 4-2

An atom of argon has a mass which is 10/3 the mass of an atom of carbon-12. What is the atomic mass of argon?

4.3 Electrons

We have already learned of one flaw in Dalton's atomic theory. His atomic model had to be modified to account for the existence of isotopes: not all atoms of an element are identical. There was still another problem with Dalton's theory. Dalton had suggested that an atom of an element was an extremely small particle which could not be subdivided or changed into an atom of another element. Observations of the behaviour of discharge tubes could be explained, however, only if the atom could be subdivided into even smaller electrically charged particles.

A discharge tube (Fig. 4-3) consists of two metal plates (electrodes) sealed into the ends of a glass tube. These electrodes, called the cathode and the anode, are attached respectively to the negative and the positive sides of a high voltage source. When most of the air is pumped from the tube and the electricity is turned on, current flows between the electrodes, and the air left in the tube begins to glow. If a gas other than air is placed in the discharge tube, a coloured glow characteristic of the gas appears. For example, a discharge tube containing neon gas gives off an orange-red light. Discharge tubes containing neon and other gases are bent to desired shapes and are used to make "neon" signs.

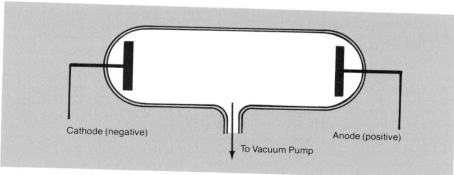

Figure 4-3
A discharge tube

If the air (or any other gas) is almost completely removed from the discharge tube, the glow nearly disappears. However, an invisible ray continues to pass between the electrodes as long as the high voltage is applied to them. For example, a thin piece of metal foil becomes red hot when it is subjected to the invisible rays. Also, some substances, such as zinc sulfide, will emit light when placed between the electrodes of an operating discharge tube. A substance which gives off light when subjected to radiation is called a *phosphor*.

In the case of a discharge tube, the light from the phosphor screen is given off in flashes. This suggests that the electricity is moving through the tube as a stream of individual particles.

Scientists discovered that the charged particles making up the invisible ray originate at the cathode (negative electrode) and travel towards the anode (positive electrode). For example, the path of the ray can be partially blocked by a metal disc containing a slit (Fig. 4-4). A phosphor screen can be placed at

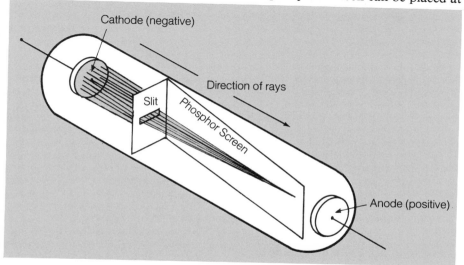

Figure 4-4
Discharge tube designed to show the direction of the invisible rays

a slant beyond the slit. Only the charged particles leaving the cathode which pass through the slit are able to strike the phosphor screen. The result is a straight line of light on the screen. If the invisible rays had come from the anode, the whole screen would have glowed because the rays would not have been blocked by the metal disc. These invisible rays, which originate at the cathode and travel to the anode, are called **cathode rays**.

If electrically charged plates are placed above and below the discharge tube, the cathode ray is deflected towards the positive plate (Fig. 4-5). Since *opposite charges attract and like charges repel*, it seems that cathode rays consist of negative particles. These negative particles, now called electrons, appear to be identical no matter what material the cathode is made of and no matter what gas is present in the tube. These electrons are thought to be part of all matter. In the case of a discharge tube having a platinum cathode, the electrons would come from the platinum atoms. In the case of a discharge tube having a copper cathode, the electrons would come from the copper atoms. "Replacement" electrons are supplied to the cathode from the source of electricity.

Figure 4-5
Deflection of cathode rays

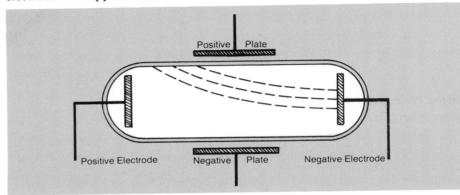

Positive | Plate

Positive Electrode   Negative | Plate   Negative Electrode

A discharge tube. The Maltese cross casts a shadow on the face of the tube.

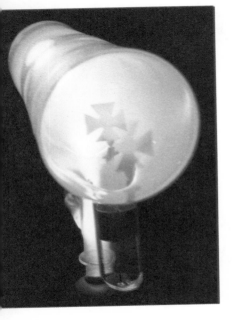

The British physicist J. J. Thomson attempted to discover more about the nature of cathode rays. He studied their deflection by electric and magnetic fields. By 1897 he was able to determine the ratio of the charge, e, to the mass, m, of an electron. He found that this ratio, e/m, had the same value for electrons produced in all discharge tubes. This constant value confirms that all electrons are the same. Since electrons are present in all of the discharge tubes used, it suggests, once again, that electrons are present in all matter.

The e/m ratio told scientists nothing about the actual charge or the actual mass of an electron. If either one of these values could be measured experimentally, the other could then be found. Twelve years later, in 1909, Robert Millikan, an American physicist, determined the charge of an electron. He was therefore able to calculate the mass of an electron. He found that an electron had an atomic mass of 0.000 55 u. This means that electrons contribute very little to the mass of an atom.

The smallest unit of negative charge in an atom is the charge of one electron. For simplicity, the relative charge of an electron is chosen as −1. An **electron** is a particle which has a relative charge of −1 and a mass of 0.000 55 u.

The discovery of the electron changed scientists' views of the atom. The atom could not be an indivisible particle if smaller particles could be removed from it. Scientists began to ask more questions. Did atoms also contain positively charged particles? If electrons contributed little to the mass of an atom, where did the mass of an atom come from? In the next section, we provide answers for these questions.

4.4 Protons

In the late 1800s, several scientists had performed experiments using discharge tubes containing small amounts of several different gases such as hydrogen and oxygen. In 1886, the German physicist Eugen Goldstein made an interesting discovery while using such tubes. His discharge tube had a negative electrode (cathode) in which a hole had been cut. He noticed a bright glow between the electrodes and a faint glow on the other side of the cathode (Fig. 4-6). The glow between the electrodes was due to the collisions of the electrons in the cathode ray with the gas molecules present in the tube. Why did the tube glow on the other side of the cathode? The explanation is that occasionally, as electrons left the cathode and travelled towards the anode, some of them struck neutral gas particles with such force that one or more electrons were knocked off. This loss of electrons converted a gas particle into a positively charged particle called an *ion* which was attracted to the cathode. Most of the ions picked up electrons when colliding with the cathode and were converted back to neutral particles. Occasionally, however, a positive ion passed through the hole in the cathode. Some of these positive ions collided with gas molecules in the tube, producing the glow on the other side of the cathode.

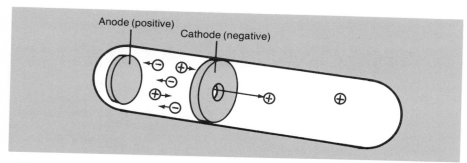

Anode (positive)
Cathode (negative)

Figure 4-6
Goldstein's discharge tube. (A single hole is shown in the cathode for simplicity.)

The charge-to-mass ratio (e/m) was found to be much smaller for positive ions than for electrons. The smallest mass for a positive ion was the mass of the hydrogen ion, 1.0073 u. Furthermore, the charge-to-mass ratio of the positive ions, unlike the charge-to-mass ratio for electrons, varied according to the nature of the gas in the tube. That is, positive ions were not all the same. Eventually, scientists found that the smallest quantity of charge on a positive ion was equal in amount, but opposite in sign, to the negative charge of one electron. This quantity of charge was called the unit positive charge and was assigned a relative value of +1.

The hydrogen ion carries a unit positive charge. It also has the smallest mass of any positive ion. The hydrogen ion is a proton. A **proton** is a particle which has a relative charge of +1 and a mass of 1.0073 u.

The proton, like the electron, is one of the fundamental particles of the atom. Every atom contains one or more protons. Every atom also contains one or more electrons. Where are the protons and electrons located in the atom? The next section contains the answer to this question.

Part One Review Questions

1. Why did scientists accept Dalton's atomic theory more readily than they accepted the atomic theory of the ancient Greeks?
2. Show how Dalton's atomic theory explains the Law of Conservation of Mass and the Law of Constant Composition.
3. What are isotopes?
4. Define the terms *atomic mass unit* and *atomic mass*.
5. A copper atom has a mass which is 5.29 times the mass of a carbon-12 atom. What is the atomic mass of copper?
6. An atom of helium has ⅓ the mass of an atom of carbon-12. What is the atomic mass of helium?
7. What apparatus was used to discover electrons and protons?
8. What are cathode rays?
9. How was it determined that the charge of an electron is negative?
10. What observation led to the discovery of protons?
11. What observation led scientists to believe that protons are positively charged particles?
12. Compare the relative mass and charge of electrons with the relative mass and charge of protons.

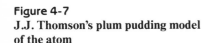

Part Two: Development of the Nuclear Model of the Atom

4.5 Rutherford's Model of the Atom

In 1898, J. J. Thomson proposed a model to explain the arrangement of protons and electrons in the atom. Thomson considered the atom to be a sphere of uniform positive electricity in which negative electrons were embedded like raisins in a plum pudding (Fig. 4-7). It had been shown that protons had more mass than electrons, and therefore most of the atom's mass was thought to be associated with the positive charge.

Meanwhile, the discovery of radioactivity was made by the French physicist Henri Becquerel in 1896. While studying minerals that fluoresce (give off light), he found accidentally that a paper-covered photographic plate became fogged when it was placed near a uranium ore. He wrapped photographic plates in thin sheets of aluminum and copper and placed them near the uranium ore. Again the photographic plates were blackened when they were developed. Becquerel decided that the uranium salts emit invisible rays which, unlike light, are able to pass through black paper and thin sheets of metal. A substance that gives off these invisible rays is said to be *radioactive*.

The New Zealand-born physicist Lord Rutherford (Fig. 4-8) discovered that the invisible rays from radium (a radioactive element) were actually composed

Figure 4-7
J.J. Thomson's plum pudding model of the atom

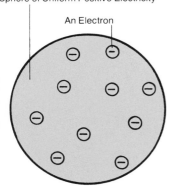

Sphere of Uniform Positive Electricity

An Electron

of three different rays. Some were positively charged, some were negatively charged, and others were uncharged. He named these alpha, beta, and gamma rays, respectively. The experiment which led Rutherford to these conclusions is described in Section 17.3. It is worth noting that Rutherford performed this important experiment while he was head of the physics department at McGill University in Montréal. Later, Rutherford conclusively proved that alpha particles were positive helium ions—helium atoms which had lost both their electrons. These experiments with radioactivity set the stage for an experiment by Lord Rutherford which shed considerable light on the structure of the atom.

Biography

ERNEST RUTHERFORD was born in New Zealand, the fourth in a family of twelve children.

After graduating in mathematics and physics from the University of New Zealand, Rutherford was the first research student to join J.J. Thomson at Cambridge University. One of his research projects at Cambridge involved ingenious methods for measuring the velocities of ions. Rutherford once said: "Ions are jolly little beggars; you can almost see them!"

From 1898 to 1907 Rutherford was professor of physics at McGill University in Montréal, where he worked on radioactivity. In 1903 he published important findings on the nature and properties of alpha particles which led him to believe that they were helium ions.

Rutherford became a professor of physics at the University of Manchester in 1907. It was there that he proved that alpha particles are helium ions. However, Rutherford's greatest contribution to atomic theory came in 1911, when he proposed the nuclear model for the atom.

Rutherford spent the years from 1919 to 1937, when he died, at Cambridge University. In 1919 he became the first person to change one element into another, when he bombarded nitrogen with alpha particles to produce oxygen atoms. Changing given elements into others was the crowning achievement of his life's work and held for Rutherford a deep fascination.

Rutherford was awarded the Nobel Prize in Chemistry in 1908, and he was knighted in 1914. In 1931, he was made Baron Rutherford of Nelson and took his seat in the House of Lords. Sir James Jeans said, "Rutherford was ever the happy warrior—happy in his work, happy in its outcome, and happy in its human contacts." Rutherford had great energy, intense enthusiasm, and an immense capacity for work.

A writer once said to him, "You are a lucky man, Rutherford, always on the crest of the wave." Rutherford laughingly replied, "Well! I made the wave, didn't I?"

Figure 4-8
Ernest Rutherford (1871–1937)—Proposed the nuclear model of the atom.

Figure 4-9
Rutherford's gold foil experiment

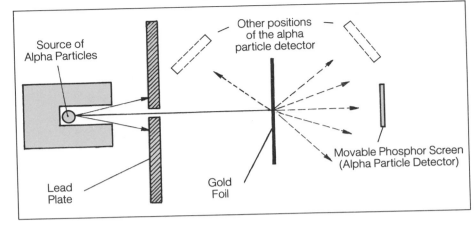

The Gold Foil Experiment

After he had identified alpha particles as positive helium ions, Lord Rutherford next investigated (in 1911) the scattering of these particles by thin sheets of gold metal (Fig. 4-9). Alpha particles from a radioactive source passed through a hole in a lead plate and were allowed to strike a sheet of gold foil. A special detector gave off a flash of light whenever it was struck by an alpha particle.

According to the Thomson model of 1898, atoms are essentially a sphere of uniform positive charge in which the negative particles are embedded. Since alpha particles have high speed, they should penetrate the atoms in the foil. If the positive charge of an atom is uniformly distributed, the positive alpha particle would have little reason to be greatly repelled and therefore deflected from its original path. If negatively charged electrons are spread out in the atom and have little mass, they should not affect the path of an alpha particle. It should pass through the foil without being deflected (Fig. 4-10).

Rutherford found that most of the alpha particles did pass through the gold undeflected, but some were deflected at large angles. In fact, some were *reflected* back towards the lead plate. Rutherford was astonished. He said "it was almost as incredible as if you fired a 15-inch shell at a piece of tissue paper and it came back and hit you."

The Thomson model could not explain this result. According to that model, an alpha particle would never encounter an obstacle large enough to force it to be reflected or to be deflected at a large angle. Rutherford proposed his own model to account for the results of his experiment. He suggested that the mass and the positive charge in the gold atoms must be concentrated in very small regions. Most of the alpha particles could pass through; however, occasionally one came close to a high concentration of positive charge and mass. The large amount of mass in the gold atoms made the positive charge immovable. Thus, alpha particles coming directly at this large concentration of charge and mass were reflected. Alpha particles coming very close to this high concentration of charge were repelled at large angles. Other alpha particles that came not as close were repelled at smaller angles (Fig. 4-11).

Rutherford suggested that an atom had a **nucleus** or centre in which the positive charge and most of the mass were located. This nucleus occupied only a tiny portion of the volume of the entire atom. In other words, most of the

Figure 4-10
Predicted passage of alpha particles through a Thomson model of the atom. (The electrons have been omitted to simplify the diagram.)

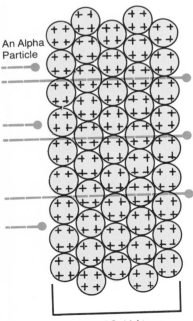

An Alpha Particle

Layer of Gold Atoms

atom was empty space. This explained why most of the alpha particles passed through the atom undeflected. If a large football stadium were to represent the volume of the entire atom, a housefly in the centre of the stadium could represent the size of the nucleus. In the nuclear atom, the electrons, which have only a tiny mass compared with that of the whole atom, occupy all of the volume surrounding the nucleus.

Neutrons

Rutherford had speculated on the existence of a neutral particle, about the same mass as a proton. He thought of it as a hydrogen atom in which the electron had fallen into the nucleus, thus neutralizing the nucleus. He expressed these ideas in a 1920 lecture. However, the third fundamental particle of the atom was not actually discovered until 1932, when the British physicist James Chadwick bombarded atoms of beryllium with high-speed particles. He found that a beam of rays was emitted from the beryllium during the bombardment. Since these rays travelled at only about one-tenth the speed of light, they could not be true radiation. Also, since the rays were unaffected by a magnet, they could not be charged. Chadwick was able to show that they were actually a stream of uncharged particles. These particles were called **neutrons**. The atomic mass of a neutron was found to be 1.0087 u.

Where in an atom are the neutrons found? We can answer this question if we think of a helium atom. The charge on a helium nucleus is twice the charge of the proton. We would therefore expect a helium nucleus to contain two protons. These two protons, however, account for only *half* the mass of a helium nucleus. The other half must be made up of neutral particles. Since neutrons have nearly the same mass as protons but no charge, we conclude that a helium nucleus contains two protons and two neutrons. Neutrons are found in the nucleus of an atom.

Rutherford's gold foil experiment has improved our view of the atom. We believe that the atom has a dense, tiny centre or nucleus. The protons and neutrons are located in this nucleus and are the major contributors to the mass of the atom. The protons contribute the positive charge of the atom. The electrons surround the nucleus, and occupy most of the volume of the atom. They do not contribute very much to the mass of the atom, but they do contribute its negative charge. A summary of the particles found in the atom, that is, of subatomic particles, is shown in Table 4-1.

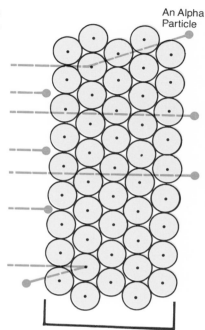

Figure 4-11
Passage of alpha particles through a nuclear atom

An Alpha Particle

Layer of Gold Atoms

Table 4-1   Subatomic Particles

Particle	Location	Relative Mass	Charge
Electron	around nucleus	0.000 55 u	−1
Proton	nucleus	1.007 3 u	+1
Neutron	nucleus	1.008 7 u	0

In the next section we will find out why atoms of the same element are able to have different masses.

4.6 Isotopes

The Nuclear Atom

The nuclear atom is the basis of our present theory of atomic structure. The protons and neutrons are each called **nucleons**. Together they reside in the nucleus, which is a minute but very dense part of the atom. The positive charge of the protons in the nucleus is balanced by the negative charge of the electrons around the nucleus in a neutral atom. The neutrons have no charge. The extranuclear (outside the nucleus) region, which makes up virtually all of the volume of the atom, contains only the electrons. The radii of atoms are around 10^{-10} m, but nuclei have radii in the range of 10^{-15} to 10^{-14} m. The nucleus contributes almost no volume to the atom, but it contains essentially the entire mass of the atom.

Atomic Number

The number of protons in the nucleus is called the **atomic number** (Z). Every atom of each element has its own unique number of protons. That is, all oxygen atoms have 8 protons. If an atom has a number of protons other than 8, it is not an oxygen atom. In a neutral atom, Z is also the number of electrons. Oxygen is element number 8, and a neutral oxygen atom has 8 protons and 8 electrons.

Mass Number

The atomic mass of the proton is 1.0073 u. The atomic mass of the neutron is 1.0087 u. Thus, the approximate masses of the proton and of the neutron are 1 u. The mass of the electron is 0.000 55 u or approximately 0 u compared with the mass of the nucleons. The total number of protons and neutrons in an atom is called the **mass number** (A) and it is approximately equal to the mass of the whole atom. A specific atom may be referred to by using a symbol with both the atomic number and the mass number. For example, the symbol $^{12}_{6}C$ indicates a carbon atom that has $Z = 6$ and $A = 12$. The atom has 6 protons and a total mass of 12 u. The atom contains a total of 12 protons and neutrons, but it contains 6 protons. Therefore, it must also contain 6 neutrons. This atom is the carbon-12 mentioned in Section 4.2.

EXAMPLE PROBLEM 4-2

How many electrons, protons, and neutrons are in an atom for which the atomic number is 92 and the mass number is 235?

SOLUTION

The atomic number is the number of protons in the atom.

$\therefore$ There are 92 protons.

The number of electrons must equal the number of protons.

∴ There are 92 electrons.

The mass number is the sum of the number of protons and the number of neutrons.

∴ The number of neutrons is $235 - 92 = 143$.

PRACTICE PROBLEM 4-3

How many electrons, protons, and neutrons are in an atom for which the atomic number is 103 and the mass number is 257? (*Answer*: 103 electrons, 103 protons, and 154 neutrons)

PRACTICE PROBLEM 4-4

How many electrons, protons, and neutrons are in an atom for which the atomic number is 18 and the mass number is 40?

All atoms of the same element must have the same number of protons, but they need not all have the same number of neutrons. Atoms which have the same number of protons but different numbers of neutrons are called isotopes. Isotopes are atoms that have the same atomic number but different mass numbers. For example, two isotopes of lithium are ^6_3Li and ^7_3Li. The first isotope has three neutrons, and the second one has four neutrons. These isotopes have the same chemical properties, but they have slightly different physical properties.

The atomic mass for lithium found in the table of atomic masses is calculated from the masses of the two isotopes: ^6_3Li makes up 7.4% of all lithium atoms, and ^7_3Li makes up the other 92.6%. Of 1000 lithium atoms, 74 have a mass number of 6 (actual mass 6.015 u) and 926 have a mass number of 7 (actual mass 7.016 u). The average mass is

$$(74 \times 6.015 \text{ u} + 926 \times 7.016 \text{ u}) \div 1000 = 6.94 \text{ u}.$$

In this example, we used 1000 lithium atoms, but we could have chosen 100 atoms or 397 atoms or any other number of atoms. Whatever number of atoms we chose, the result would still have been an average mass of 6.94 u. In this particular case, it is *convenient* to choose 1000 atoms, because that gives us whole numbers of atoms (74 of ^6_3Li and 926 of ^7_3Li). If we had chosen 100 atoms, we would have been finding the average mass of 7.4 atoms of ^6_3Li and 92.6 atoms of ^7_3Li. It is easy to do this calculation mathematically, but we must remember that there is no such thing as a fraction an atom.

The existence of isotopes explains why some elements have average atomic masses that are not close to whole numbers. For example, the atomic mass for chlorine is calculated from the masses of two isotopes: $^{35}_{17}Cl$ makes up 75.5% of all naturally occurring chlorine atoms, and $^{37}_{17}Cl$ makes up the other 24.5%. The average atomic mass of chlorine is 35.5 u. The compositions of the isotopes of a number of elements are given in Table 4-2.

Table 4-2   Table of Isotopes of the First 18 Elements

Atomic Number	Element	Symbol of Isotope	Abundance in Nature (%)	Atomic Mass (u)
1	Hydrogen	^1_1H	99.985	1.007 825
		^2_1H	0.015	2.014 0
2	Helium	^3_2He	0.000 13	3.016 03
		^4_2He	100	4.002 60
3	Lithium	^6_3Li	7.42	6.015 12
		^7_3Li	92.58	7.016 00
4	Beryllium	^9_4Be	100	9.012 18
5	Boron	$^{10}_5B$	19.78	10.012 9
		$^{11}_5B$	80.22	11.009 31
6	Carbon	$^{12}_6C$	98.89	12.000 0
		$^{13}_6C$	1.11	13.003 3
7	Nitrogen	$^{14}_7N$	99.63	14.003 07
		$^{15}_7N$	0.37	15.000 11
8	Oxygen	$^{16}_8O$	99.759	15.994 91
		$^{17}_8O$	0.037	16.999 14
		$^{18}_8O$	0.204	17.999 16
9	Fluorine	$^{19}_9F$	100	18.998 40
10	Neon	$^{20}_{10}Ne$	90.92	19.992 44
		$^{21}_{10}Ne$	0.257	20.993 95
		$^{22}_{10}Ne$	8.82	21.991 38
11	Sodium	$^{23}_{11}Na$	100	22.989 8
12	Magnesium	$^{24}_{12}Mg$	78.70	23.985 04
		$^{25}_{12}Mg$	10.13	24.985 84
		$^{26}_{12}Mg$	11.17	25.982 59
13	Aluminum	$^{27}_{13}Al$	100	26.981 53
14	Silicon	$^{28}_{14}Si$	92.21	27.976 93
		$^{29}_{14}Si$	4.70	28.976 49
		$^{30}_{14}Si$	3.09	29.973 76
15	Phosphorus	$^{31}_{15}P$	100	30.973 76
16	Sulfur	$^{32}_{16}S$	95.0	31.972 07
		$^{33}_{16}S$	0.76	32.971 46
		$^{34}_{16}S$	4.22	33.967 86
		$^{36}_{16}S$	0.014	35.967 09
17	Chlorine	$^{35}_{17}Cl$	75.53	34.968 85
		$^{37}_{17}Cl$	24.47	36.965 90
18	Argon	$^{36}_{18}Ar$	0.337	35.967 55
		$^{38}_{18}Ar$	0.063	37.962 72
		$^{40}_{18}Ar$	99.60	39.962 38

EXAMPLE PROBLEM 4-3

Natural carbon consists of 98.89% carbon-12 and 1.11% carbon-13 (actual mass 13.0033 u). What is the average atomic mass of natural carbon?

SOLUTION

Assume you have 10 000 atoms of carbon. Then there will be 9889 atoms of $^{12}_{6}C$ and 111 atoms of $^{13}_{6}C$.

$$\text{Total atomic mass} = 9889 \times 12.000 + 111 \times 13.0033$$

$$= 120\ 111\ u$$

$$\text{Average atomic mass} = 120\ 111\ u/10\ 000 = 12.0111\ u$$

PRACTICE PROBLEM 4-5

Natural neon contains 90.92% $^{20}_{10}Ne$ (mass 20.0 u), 0.26% $^{21}_{10}Ne$ (mass 21.0 u) and 8.82% $^{22}_{10}Ne$ (mass 22.0 u). Show that this gives an average atomic mass for neon of 20.2 u.

PRACTICE PROBLEM 4-6

Natural argon contains 0.337% $^{36}_{18}Ar$ (mass 35.967 55 u), 0.063% $^{38}_{18}Ar$ (mass 37.962 72 u) and 99.60% $^{40}_{18}Ar$ (mass 39.962 38 u). Show that this gives an average atomic mass for argon of 39.95 u.

There are approximately 2000 known isotopes. Most of these are isotopes of radioactive elements which, as we have seen, emit invisible rays. These elements have unstable nuclei and emit rays in an attempt to become stable. Radioactive isotopes are known as **radioisotopes**. Radioisotopes have many important uses.

Radioactive isotopes have application in medical therapy. Canadians and Canadian reactors have played an important role in this field. Through the effort of Dr. Cipriani at Chalk River, the first radioactive cobalt-60 sources were made for the treatment of malignant tumors. These cobalt-60 sources produce intense beams of gamma rays which destroy tumors. In addition to the cobalt-60 cancer therapy machines (*cobalt bombs*), cesium-137 *Caesatrons* have been made and sold by AECL Commercial Products. Over 700 cancer therapy units have been made by this Canadian company and shipped to 51 different countries.

Many radioactive isotopes have medical uses as "tracers." Since all atoms of the same element have the same chemical properties, they will behave identically in chemical reactions. Thus, a small amount of a radioactive isotope can be mixed with the non-radioactive element. This radioactive "tag" enables doctors to trace the movement of the element through the human body simply by using a Geiger counter, which detects the rays from the radioactive atoms.

A neon laser creates an intense beam of light.

For example, disorders of the thyroid gland (located in the neck) affect a person's metabolism and rate of growth. The disorders are associated with the rate at which iodine appears in the gland. Thus, if a person is given a solution of sodium iodide containing radioactive iodine-131, a Geiger counter placed near the thyroid gland will measure the rate of uptake of iodine-131 by the gland.

Because many pesticides have an adverse effect on the environment, scientists are seeking other methods for controlling destructive insects. They have found that male insects which are exposed to a nonlethal dose of radioactivity become sterile. When these irradiated males are released in the environment they are unable to fertilize the eggs laid by the females, and the insect population is greatly reduced. This method of pest control has been used successfully for several years to control the populations of fruit flies and other pests.

The gamma radiation from cobalt-60 is used to sterilize vegetables, fruit, and grain, as well as medical supplies and wool. Potatoes sterilized by the radiation from cobalt-60 will not sprout, fruit can have a longer shelf life in stores, and medical instruments can be put in packages and sterilized by gamma radiation, ready for use when unpacked. Commercial Products supplied the first commercial units for this type of sterilization to companies in Canada, the United States, India, and New Zealand.

Radioisotopes are used in industry to study the wearing of machine parts. For example, the efficiency of lubricants to diminish wear on moving parts of machines is determined by incorporating radioisotopes into the metal surfaces. After even the slightest amount of metal wear, radioactive atoms will be detected in the lubricants. Radioactive isotopes may be added to a liquid flowing in a pipeline. A Geiger counter could then isolate the position of a leak or obstruction in the pipeline.

In addition, determining the amounts of radioactive isotopes in rocks has enabled scientists to estimate their age, and radioactive carbon dating is used to determine the age of plant and animal specimens. Radioisotopes are also used to study the details of chemical reactions.

We have given only a glimpse into some of the current uses of radioactive isotopes. Future applications in medicine, industry, and research will be limited only be the imaginations of the people who want to use them.

4.7 The Bohr Model of the Hydrogen Atom
Weaknesses of the Rutherford Model

The nuclear model of the atom raised some interesting questions. Why are the negative electrons not pulled into the positive nucleus by the attraction of unlike charges? The electrons must be in some type of motion which prevents them from falling into the nucleus. Rutherford and his colleagues suggested that the electrons move around the nucleus in orbits, much as the planets orbit the sun.

This planetary model of the atom did not solve the problem completely. Physicists had observed that a moving electric charge which changes its direction in space will give off radiant energy. Thus, an electron moving around a nucleus in an orbit would be expected to lose energy. Do electrons lose energy? If so, the loss of energy would cause electrons to slow down and to be more strongly attracted to the nucleus, and they would rapidly spiral into it. (An analogy is found in an orbiting earth satellite which gradually loses energy because of friction with atmospheric molecules and eventually falls back to earth.) The result would be the collapse of the atom. However, atoms do not collapse. Therefore, there must be a flaw in the argument.

A clue to the problem is obtained from a study of the light given off from high energy substances. If a tungsten wire is in a high energy state due to the heating effect caused by an electric current, it releases the extra energy in the form of white light. White light is composed of all colours or frequencies of light. Each frequency has a characteristic energy.

The relationship between energy and frequency was shown by the German physicist, Max Planck. He assumed that light is made up of discrete (separate) little packages of energy. Each package is called a **quantum** (plural *quanta*) or a photon. The relationship between energy and frequency is $E = hv$, where E is the amount of energy possessed by a quantum, h is Planck's constant, and v is the frequency of the quantum. According to this theory, red light consists of a stream of quanta, each of which is a certain amount of energy and has a characteristic frequency. Violet light also consists of a stream of quanta. However, quanta of violet light have more energy and a higher frequency than do quanta of red light.

White light is composed of all frequencies and all energies of visible light. White light can be broken down into its component colours by a prism or a diffraction grating. A spectrum of colours is obtained. It is called a **continuous spectrum** because it consists of all the colours or frequencies of visible light. In order of increasing frequencies (and increasing energies) the colours are red, orange, yellow, green, blue, indigo, and violet.

The Line Spectrum of Hydrogen

When hydrogen gas is excited (forced into a higher energy state) by an electric current passing through it, it will emit energy in the form of visible light. However, the visible light is violet, not white. This violet light can be passed through a prism to show its component colours or frequencies. The visible spectrum of hydrogen atoms consists of four prominent coloured lines (red, bluish-green, indigo, and violet), each of which corresponds to a characteristic energy. This is called a **line spectrum** because it is made up of only certain frequencies (coloured lines) of visible light. When different elements are excited by the passing of an electric current through gaseous samples of the elements, it is found that each element has a characteristic line spectrum. The difference between a continuous spectrum and a line spectrum is shown in Figure 4-12.

Niels Bohr (Fig. 4-13), a Danish physicist working with Rutherford, proposed a model for the hydrogen atom in 1913. The Bohr model explained why the hydrogen atom does not collapse. It also successfully explained the line spectrum of hydrogen. Although it was used by chemists for only about 13 years, several of the postulates of the Bohr model are still considered valid. In the following paragraphs some of Bohr's postulates will be listed and discussed.

Figure 4-12
A comparison between a line spectrum and a continuous spectrum

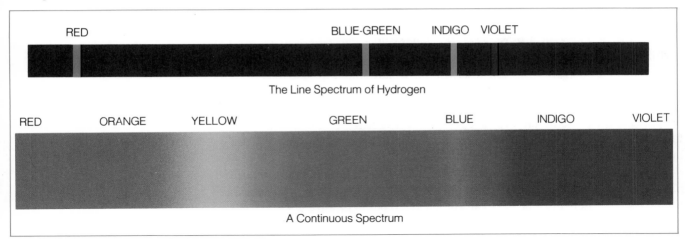

The Line Spectrum of Hydrogen

A Continuous Spectrum

Biography

Figure 4-13
Niels Bohr (1885–1962)—Proposed electron orbits in atoms.

NIELS BOHR was born in Copenhagen, Denmark in 1885, the son of a university professor. He was the second of three children in a wealthy family. He was given much encouragement at home and grew up in an intellectual environment. Bohr showed his skill as an investigator while still a student at the University of Copenhagen, where he won a gold medal from the Academy of Sciences for his precise measurement of the surface tension of water. He received his doctorate in 1911 for his electron theory of metals.

For a time he worked with Rutherford in Manchester. As a result of this contact with Rutherford, Bohr laid the foundations for the quantum theory of the atom in 1913 and 1914. His work on the structure of the atom won him the Nobel Prize in Physics in 1922.

Bohr visited the United States in 1939 and brought with him the news that Hahn and Strassman in Germany had been able to split the uranium atom. This development prompted the Americans to increase their own research efforts in the same direction. Bohr returned to Denmark in 1940. Three years later, he was forced to leave Denmark in order to escape Nazi occupation. He fled to Sweden, then to England, and finally went to the United States. There he acted as an adviser to the physicists working on the development of the atomic bomb. However, he did not work directly on the bomb, and he was opposed to its use. After the war, he returned to Denmark. In 1950, he wrote an open letter to the United Nations, making a plea for world peace. He spent most of the rest of his life working as a director of the Institute for Theoretical Physics in Denmark. He died in Copenhagen in 1962 at the age of 77.

Postulates of the Bohr Theory

First, in every hydrogen atom, there are only certain paths in which an electron is allowed to move. Each of these paths has a definite, fixed energy. Thus, each of these paths corresponds to one of a number of *allowed energy levels* (Fig. 4-14). Bohr suggested that the energy of an electron in a hydrogen atom is **quantized**, which means that the energy is limited to a number of definite quantities that can be calculated by use of the equation

$$E_n = -\frac{R}{n^2}$$

where n is the number of the energy level and R is called the Rydberg constant. The Rydberg constant has a value of 2.18 aJ (2.18×10^{-18} J).

The equation means that the energy of an electron in the first energy level equals $-R/1$. The energy of an electron in the second energy level equals $-R/4$. In the third energy level, the energy of an electron equals $-R/9$. From these energies it can be seen that the larger the value of n, the more energy (less negative value) an electron possesses.

You may be wondering why the energy of an electron in a hydrogen atom is negative. The answer lies in the way scientists have defined the point of zero energy. For example, if you hold a book over a table, the book has potential energy. You could define the table as being the point of zero energy. If you drop the book to the table, the book loses energy and has zero potential energy on the table. However, if the book were to fall from the table to the floor, it would lose more energy. Thus, the book would have a negative potential energy if it were resting on the floor.

Let us return to the case of an electron in a hydrogen atom. The electron in a hydrogen atom has a negative potential energy because scientists have decided to define zero energy as the point at which the electron is separated from the hydrogen nucleus by an infinite distance. As a negative electron approaches a positive nucleus, it feels an increasing force of attraction and gives off energy. Since the electron starts at zero energy outside the energy levels of a hydrogen atom and loses energy as it enters these energy levels, the energy of an electron in a hydrogen atom is negative.

Second, each of these energy levels corresponds to an **orbit**, a circular path in which the electron can move around the nucleus. Thus, Bohr retained the notion of orbits, which was part of the planetary model. (This is the only postulate of the Bohr model listed here which is now considered to be incorrect.) In the model, the first orbit has the smallest radius, the second orbit has a larger radius, and so on (Fig. 4-15). In the first orbit, the energy of an electron equals $-R/1$. In the second, third, and n^{th} orbits, the energy of an electron equals $-R/4$, $-R/9$, and $-R/n^2$, respectively. For these orbits, only certain radii and energies are possible.

If the motion of an electron in an orbit is not limited to a single plane, the three-dimensional path traced by an electron will describe a spherical shell. As with orbits, the shell closest to the nucleus corresponds to the first energy level.

Figure 4-14
Allowed energy levels for a hydrogen atom

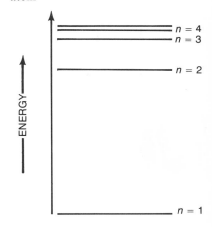

Figure 4-15
The first four Bohr orbits for the hydrogen atom

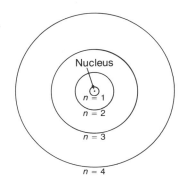

Figure 4-16
Allowed transitions between energy levels

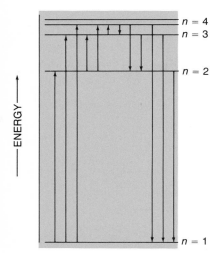

It is called the K shell. Shells farther from the nucleus correspond to higher energy levels: the L shell corresponds to the second energy level, the M shell to the third energy level, the N shell to the fourth energy level, and so on. Thus, an electron in the first energy level is said to be in the first orbit or the K shell. For our purposes, the terms *energy level*, *orbit*, and *shell* are essentially interchangeable.

Third, Bohr assumed that an electron could travel in one of the allowed orbits (energy levels) without a loss of energy.

Fourth, the only way an electron can change its energy value is to "jump" from one allowed energy level (orbit) to another. The jump cannot be gradual—it must occur all at once. To illustrate this point, imagine a person climbing or descending a flight of stairs: the steps can be taken one, two, or even three at a time, but the person can never stop with both feet between the steps. This movement of an electron between allowed energy levels is called a transition.

Bohr's Explanation of Line Spectra

When a hydrogen atom is excited, the electron gains a certain quantity of energy and jumps instantly from a lower allowed energy level to a higher allowed energy level. We say that the electron is promoted from a lower energy level to a higher energy level. The quantity of energy required for this promotion equals the difference in energy between the two energy levels. When the electron drops back to a lower energy level, some or all of the previously gained energy is given off in the form of radiation of a definite frequency (Fig. 4-16). The frequency of the radiation depends on the amount of energy emitted and can be calculated using the equation we saw earlier in this section, $E = h\nu$.

If the electron is already at the lowest energy level, no more energy can be emitted, and thus the atom does not collapse. This is similar to being at the bottom of the flight of stairs. You can descend no further.

In summary, if an electron is promoted from an orbit with a smaller radius (lower energy level) to an orbit with a larger radius (higher energy level), energy is absorbed by the atom. When an electron goes from an orbit with a larger radius to an orbit with a smaller radius, energy is emitted by the atom (Fig. 4-17).

Since there are only certain allowed orbits or energy levels in the atom, the difference in energy between two allowed orbits or energy levels must also be a certain value. Thus, only certain energies can be absorbed or emitted as the electron changes orbits. This fact means that only certain frequencies of radiation can be absorbed or emitted since energy is directly proportional to frequency ($E = h\nu$).

The frequencies of the coloured lines in the visible spectrum of hydrogen agree with the predicted frequencies calculated using the Bohr model of the hydrogen atom. In a hydrogen discharge tube having enormous numbers of hydrogen atoms, all possible jumps from one orbit to another occur simultaneously. That is why we see four coloured lines in the hydrogen spectrum simultaneously.

Figure 4-17
(a) Energy is absorbed, causing the electron to move from the first to the second orbit.
(b) When the electron moves from the second to the first orbit, energy is emitted.

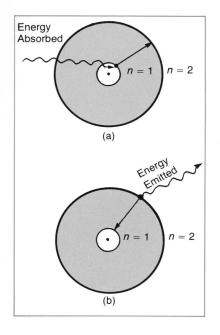

The frequency of the red line agrees with the calculated frequency due to a jump from the third to the second orbit. Also, the frequencies of the bluish-green line, the indigo line, and the violet line agree with the calculated frequencies due to jumps from the fourth to the second orbit, the fifth to the second orbit, and the sixth to the second orbit, respectively (Fig. 4-18).

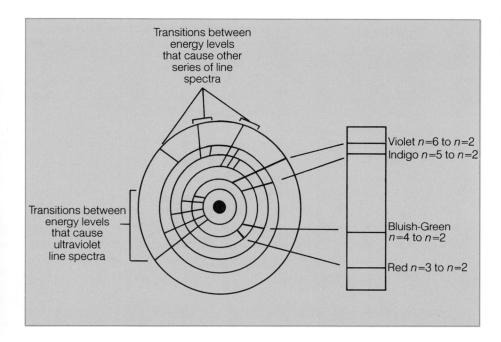

Figure 4-18
The line spectrum of hydrogen and the electron's energy levels

In addition, Bohr was able to predict an undiscovered set of lines corresponding to jumps from the sixth, fifth, fourth, third, and second orbits to the first orbit. He said that these lines would occur in the ultraviolet region of the hydrogen spectrum. These lines were discovered, and their observed frequencies did agree with the predicted frequencies using the Bohr model of the hydrogen atom.

Applications of Electron Transitions

There are a number of applications for electrons changing energy levels in atoms. For example, the light given off when electrons go from higher energy levels to lower levels is used in neon signs, sodium and mercury street lights, red signal flares, fireworks, and coloured fire logs. The spectrum for each element and molecule is unique. It is like a fingerprint which can be used to identify a substance. For example, certain metal ions can be recognized by the colour they impart to a Bunsen burner flame. Forensic scientists are able to identify small traces of chemicals by their spectra. Astronomers are able to identify chemicals in space by their spectra. The gas helium (from Greek, meaning "the sun") was first discovered to be present on the sun as the result of the analysis of its spectrum in 1868. Helium was not found on earth until 1895.

Strengths and Weaknesses of the Bohr Model

Bohr's theory was an important advance, because it introduced the concept that energy is quantized, a concept that is still important in chemistry and physics. Bohr was quite correct in arguing that Newton's laws of motion (classical mechanics), which describe the motion of objects of large mass, do not describe adequately the motion of particles of very small mass. Empirical evidence suggests that these small particles obey a new set of laws of motion. **Quantum mechanics** is the study of the laws of motion which describe the motion of particles of very small mass.

The basic principle of quantum mechanics is that electrons are allowed to have only certain energy values when they are in atoms. No electrons can have energy values that lie between these allowed energy values. When an electron shifts from one allowed energy value (or energy level) to another allowed energy level, the shift must be instantaneous.

However, Bohr was incorrect on a number of points. He wrongly supposed that the electron is a particle whose position and motion can be specified exactly at a given time. He was also wrong in thinking that the electron moves in an orbit at a fixed radius which changes only when the electron "jumps" to another orbit having a different fixed radius. Another difficulty of the Bohr model is that it agrees well with experimental evidence only in the case of a one-electron atom. Thus, the Bohr model had to be improved, although many of its key postulates could be retained.

Sodium vapour street lamps give us a characteristic yellow light.

APPLICATION

From Line Spectra to Light Sources

The movement of excited electrons from higher to lower energy levels produces the characteristic line spectra of the elements. This process has important applications. In the mercury-vapour lamp, bluish-green light is emitted as the excited mercury electrons drop to lower energy levels. In the case of sodium-vapour lamps, mainly yellow light is emitted. There are advantages to using the sodium-vapour lamp. Its yellow light penetrates fog better than normal white light.

A fluorescent lamp is a discharge tube whose inner surface has been coated with a fluorescent material. A fluorescent material is one which emits light when it is exposed to radiant energy, but stops emitting when the energy no longer strikes it. The fluorescent material that is often used is zinc sulfide, the same fluorescent material that Rutherford used in his famous gold foil scattering experiment. The fluorescent lamp is filled with mercury vapour at a low pressure. The electricity passing through the tube causes the electrons in the mercury atoms to become excited and move to higher energy levels. When these electrons return to lower levels, green, blue, and ultraviolet radiation is emitted. When this radiation strikes the inner glass wall, it causes the fluorescent coating to emit a number of different colours which combine to produce white light.

Fluorescent lamps are more efficient to operate than ordinary incandescent light bulbs. In ordinary bulbs, the tungsten filament must be heated to 3000°C to produce the light. Hence much of the energy is wasted as heat. In the fluorescent lamp, the light produced depends only on the transfer of energy from the excited mercury electrons to the fluorescent coating. Thus, little heat is produced, and little energy is wasted.

4.8 Energy Levels and Many-Electron Atoms

We will not use the Bohr model to describe many-electron atoms. However, we will use two principles of quantum mechanics and some of Bohr's terminology to develop a straightforward view of the first twenty elements. The first principle is that electrons exist in energy levels in atoms. Electrons in atoms can have only energies that are calculated by the equation $E_n = - R/n^2$. The number of the energy level, n, is called the principal quantum number. The second principle is that each energy level can hold up to $2n^2$ electrons. Thus the first energy level can hold $2(1)^2 = 2$ electrons; the second energy level can hold $2(2)^2 = 8$ electrons; the third energy level can hold 18 electrons, and so on.

The number of electrons occupying each of the lowest available energy levels in an atom in its ground state is called the **energy level population** of the atom. The energy level populations for the first twenty elements are given in Table 4-3. Normally, as we go from one element to the next, the additional electron goes into the lowest unfilled energy level. In the cases of potassium and calcium, the last electrons go into the fourth energy level even though the third energy level (maximum population = 18) is not filled. We will discuss this exception again in Chapter 5; however, we can say that the third energy level does have eighteen electrons in all elements with thirty or more electrons.

The energy level population of an atom can be written in a short form. For example, the energy level population of element number 11, sodium, can be written 2,8,1. This indicates that two of the sodium electrons occupy the first energy level; eight occupy the second energy level; and the last sodium electron occupies the third energy level. Using Table 4-3, you should be able to verify that the energy level population of phosphorus is 2,8,5. We are going to make use of the simple, but valuable, concept of energy level populations in Chapters 5 and 6.

Table 4-3   Energy Level Populations for the First 20 Elements

Element	Symbol	Number of Protons in the Nucleus	Energy Level Population			
			1st Level (max. = 2)	2nd Level (max. = 8)	3rd Level (max. = 18)	4th Level (max. = 32)
Hydrogen	H	1	1			
Helium	He	2	2			
Lithium	Li	3	2	1		
Beryllium	Be	4	2	2		
Boron	B	5	2	3		
Carbon	C	6	2	4		
Nitrogen	N	7	2	5		
Oxygen	O	8	2	6		
Fluorine	F	9	2	7		
Neon	Ne	10	2	8		
Sodium	Na	11	2	8	1	
Magnesium	Mg	12	2	8	2	
Aluminum	Al	13	2	8	3	
Silicon	Si	14	2	8	4	
Phosphorus	P	15	2	8	5	
Sulfur	S	16	2	8	6	
Chlorine	Cl	17	2	8	7	
Argon	Ar	18	2	8	8	
Potassium	K	19	2	8	8	1
Calcium	Ca	20	2	8	8	2

Part Two Review Questions

1. What results would Rutherford have expected from his gold foil scattering experiment if Thomson's model of the atom had been correct?
2. What did Rutherford's gold foil experiment indicate about the structure of an atom?
3. Which results of the gold foil scattering experiment were unexpected? Why were they unexpected?
4. How does the mass of a neutron compare with the mass of a proton?
5. How does the electrical charge on an electron compare with the charge on a proton?
6. For each of the following indicate the atomic number, mass number, number of electrons, number of protons, and number of neutrons:
   a) $^{170}_{72}Hf$;   b) $^{85}_{36}Kr$;   c) $^{38}_{19}K$.

7. Which of these are acceptable representations for isotopes of strontium: $^{87}_{37}Sr$; $^{87}_{38}Sr$; $^{88}_{38}Sr$; $^{38}_{88}Sr$?
8. What are radioisotopes? What can they be used for?
9. Natural magnesium consists of 78.70% $^{24}_{12}Mg$ (mass 24.0 u), 10.13% $^{25}_{12}Mg$ (mass 25.0 u), and 11.17% $^{26}_{12}Mg$ (mass 26.0 u). What is the average atomic mass of natural magnesium?
10. What is the difference between a continuous spectrum and a line spectrum?
11. What are the postulates of the Bohr model of the atom?
12. In the Bohr model, what is meant by the term *orbit*?
13. State three flaws in the Bohr theory.

Part Three: The Present Model of the Atom

4.9 Wave Mechanics—The Present Model

One of the problems that led to the discarding of the Bohr model was that in some experiments electrons behaved as particles while in other experiments electrons behaved as waves. Bohr supposed that the electron is a particle whose position and motion in an orbit can be specified exactly at any given time. Nevertheless, electrons could be reflected just as light waves can be. In order to describe the motion of electrons in atoms, taking into account both their particle nature and their wave nature, the German mathematician Erwin Schrödinger in 1924 devised a type of mathematics called **wave mechanics**.

The basic idea of wave mechanics is that we cannot measure both the position and the velocity of a body as small as an electron at the same time. The best we can do is to calculate the probability of an electron being in a certain place at a certain time. This may seem rather vague, but much can be learned about an atom by studying electron position probabilities. Many problems of atomic and molecular structure can be solved by use of wave mechanics.

A **wave equation** is a mathematical expression describing the energy and motion of an electron around a nucleus. Such a wave equation exists for the hydrogen atom. It is possible to solve this wave equation and obtain a three-dimensional **wave function**. This wave function is linked to the probability of finding an electron in a certain volume of space around the nucleus. Just as -2 and $+3$ are both solutions for the equation $x^2 - x - 6 = 0$, there are a number of wave functions which are all solutions to the wave equation for the hydrogen atom. They describe regions in space where the electron is most likely to be found. These regions are called **orbitals**. This word was coined to resemble the term *orbit* used in the Bohr model.

Spinning fan blades move so fast that they appear to occupy all the space in front of the fan motor.

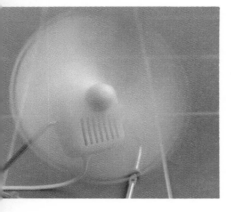

The Orbitals

The **principal quantum number** (n) identifies the energy possessed by an electron in any orbital under study. It is the same as the n used in the equation $E_n = -R/n^2$. Also, as in the Bohr model, n goes from 1 to infinity by units: $n = 1,2,3,\ldots\infty$.

For every value of n, there are n types of orbitals and n^2 actual orbitals. Thus, the first energy level ($n = 1$) has one type of orbital and one (1^2) orbital. This orbital is called the $1s$ orbital. It is spherically shaped. If an electron is in this orbital, it is possible to calculate its energy, but it is not possible to calculate its position or motion. The electron is moving so fast that it seems to occupy all the space in a sphere around the nucleus. It is possible to say only that an electron in the $1s$ orbital of a hydrogen atom spends 95 percent of its time somewhere in a sphere with a radius of about 1×10^{-10} m around the nucleus, and that it is most likely to be at a distance of 5×10^{-11} m from the nucleus.

The second energy level ($n = 2$) has two types of orbitals and four (2^2) orbitals. The first orbital in this level is called the $2s$ orbital. This orbital is much like the $1s$ orbital except that it is larger and corresponds to a higher energy level. In addition, there are three $2p$ orbitals in the second energy level. They all have the same shape, size, and energy, but they point in different directions and are called p_x, p_y, and p_z orbitals. Although it is easy to describe the shape of an orbital, it is impossible to draw a picture of one, since theoretically an orbital extends through all space. The best we can do is draw a shape within which the electron will be found some arbitrary proportion (say 95% or 99%) of the time. In Figure 4-19 the surfaces of the shapes represent boundaries within which the electron can be found 95% of the time. If we wished to use a 99% boundary the shapes would be larger but would have the same appearance.

Figure 4-19
Shapes of s and p orbitals

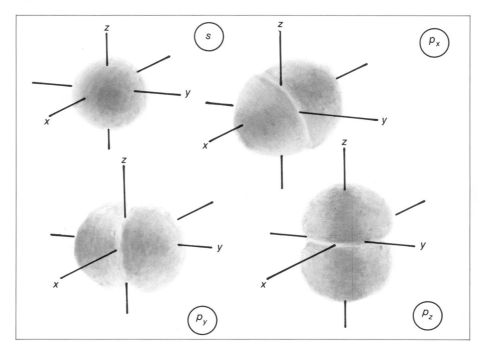

The third energy level ($n = 3$) has three types of orbitals and nine (3^2) orbitals. The first orbital is called the $3s$ orbital. Like all s orbitals, the $3s$ orbital is spherical; however, it is larger than either the $1s$ or the $2s$ orbital. An electron in the $3s$ orbital has more energy than it would have if it were in the $1s$ or $2s$ orbital, and it can move further from the nucleus more often. Next there are three $3p$ orbitals. The $3p$ orbitals are like the $2p$ orbitals except that an electron in a $3p$ orbital is more likely to be further from the nucleus than an electron in a $2p$ orbital. Also it would have more energy than an electron in a $2p$ orbital. Finally there are five $3d$ orbitals. In the third energy level, there are s, p, and d type orbitals, for a total of 9 orbitals (1 s-type + 3 p-type + 5 d-type).

The Hydrogen Energy Level Diagram

The energy level diagram of the orbitals derived for the hydrogen atom is shown in Figure 4-20. Notice that orbitals having the same principal quantum number have the same energy. Orbitals that are at the same level of energy are said to be *degenerate*. For example, the $3s$, the three $3p$, and the five $3d$ orbitals are degenerate in a hydrogen atom. The types of orbitals are also summarized in Table 4-4.

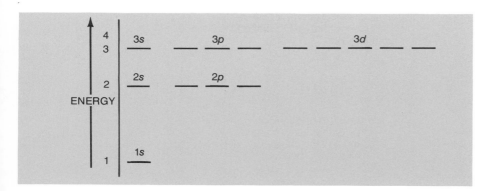

Figure 4-20
The energy level diagram for a hydrogen atom

Table 4-4   Orbitals and Energy Levels in a Hydrogen Atom

Principal Quantum Number (n)	Allowed Orbital Types	Number of Orbital Types	Number of Orbitals
1	s	1	1 ($1s$)
2	s, p	2	4 ($1s, 3p$)
3	s, p, d	3	9 ($1s, 3p, 5d$)
4	s, p, d, f	4	16 ($1s, 3p, 5d, 7f$)
.	.	.	.
.	.	.	.
.	.	.	.
n	$s, p, d, f,\ldots$	n	$n^2(1s, 3p, 5d, 7f,\ldots)$

There are higher energy levels containing more orbitals; however, the trend seems to be clear. A large number of orbitals can be derived mathematically for

Figure 4-21
Summary of models of the atom

Model	Main Postulates	What Model Explains
 GREEK MODEL	An atom is a hard, spherical indivisible particle.	
 DALTON MODEL	Atoms cannot be created or destroyed. Chemical change is the union or separation of atoms.	Laws of chemical change.
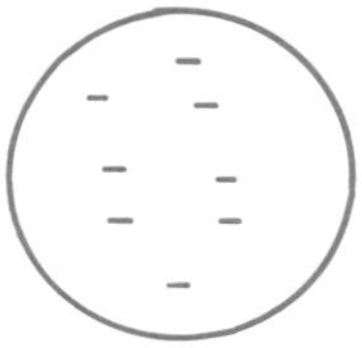 **THOMSON PLUM PUDDING MODEL**	An atom is a sphere of uniformily distributed positive charge with electrons embedded throughout.	Location of protons and electrons.
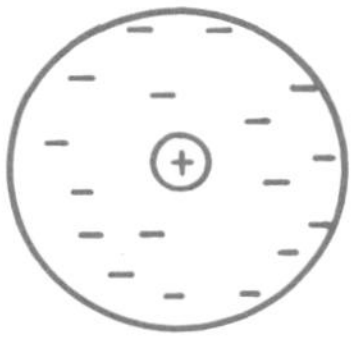 **RUTHERFORD NUCLEAR MODEL**	An atom is mostly empty space with all of the positive charge and most of the mass located in a central part called a nucleus.	Scattering of alpha particles by metal foils.
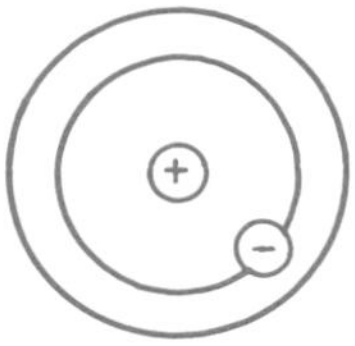 **BOHR MODEL**	The energy of an electron is quantized. Electrons travel around the nucleus in circular paths called orbits.	Atomic line spectrum of hydrogen.
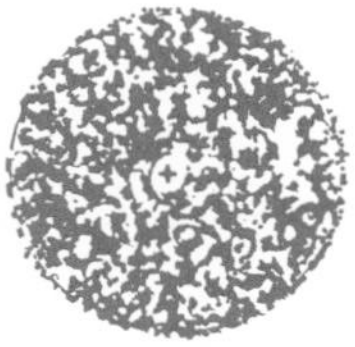 **WAVE MECHANICAL MODEL**	An electron has the properties of both a particle and a wave. Since its location cannot be determined exactly, we can speak only of the region in which it is most likely to be found (an orbital).	Line spectra of many-electron atoms and quantization of the energy of the electron.

the hydrogen atom. This may seem to be a waste of time inasmuch as there is only one electron in the hydrogen atom. This electron can occupy only one orbital at a time while the other orbitals remain empty. It must be remembered that in a large sample of hydrogen atoms most of the atoms would have their one electron in the lowest energy orbital, the $1s$. However, some atoms could be in an excited state, and they would have electrons in the orbitals of the second, third, fourth, or higher energy levels.

Practical Applications of the Wave-Like Properties of Electrons

Although it is not an easy concept to understand, the wave-like properties of electrons do have practical applications. Electron wavelengths can be controlled by applying a high voltage to the electrons. The resulting electron waves can be used in electron microscopes to "see" small objects just as light waves can be used to see small objects with an ordinary microscope. Because electrons have shorter wavelengths than visible light, they can be used to study much smaller objects than an ordinary microscope allows. For example, in 1970, scientists at the University of Chicago used a high-resolution electron microscope to "see" individual atoms of thorium and radium. Electron microscopes are now routine tools in the chemical and biological study of molecular structures.

So far in this chapter, we have examined models of the atom from the early Greek model to the present wave mechanical model. Figure 4-21 summarizes these models.

4.10 Wave Mechanics and Many-Electron Atoms

In a strict sense, the Bohr model applied only to a one-electron atom. For many-electron atoms its predictions were not as reliable as for a hydrogen atom. In the same way, the wave mechanical model applies only to a one-electron atom. For larger atoms the mathematical equations are too complex to solve without making simplifying assumptions. Nevertheless, the theory of orbitals gives us a useful picture of the way in which electrons are distributed in atoms. All we have to do is learn some new rules and modify the energy level diagram slightly.

The **Aufbau principle** states that in going from a hydrogen atom to a larger atom, you add protons to the nucleus and electrons to orbitals having the same shape as those derived for the hydrogen atom. You start at the orbitals having the lowest energy and fill them, in order of increasing energy, until you run out of electrons. Stated briefly, electrons go into the orbital corresponding to the lowest energy level available.

The modified energy level diagram is shown in Fig. 4-22. This diagram shows that the second level is divided into two sublevels. That is, the $2s$ orbital is no longer degenerate with the three $2p$ orbitals. The first sublevel consists of the $2s$ orbital, and the second sublevel consists of the three $2p$ orbitals. In the same way, the third, fourth, and fifth energy levels are divided into sublevels of different energies. Also, the energy levels begin to overlap. For example, the fourth energy level begins before the third level is completed (i.e., the $4s$ sublevel is of lower energy than the $3d$ sublevel). The reasons for having to modify the original energy level diagram, when describing a many-electron system, have been discussed by chemists; however, they are beyond the scope of an introductory course.

Figure 4-22
The energy level diagram for a many-electron atom

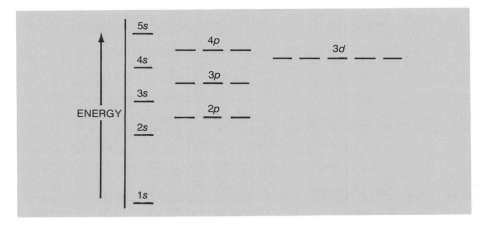

The energies of the orbitals are approximately in the order $1s < 2s < 2p < 3s < 3p < 4s \lesssim 3d < 4p < 5s \lesssim 4d < 5p < 6s \simeq 4f \simeq 5d < 6p$. They fill with electrons in this order. It is possible to remember the order of filling by studying Fig. 4-23.

Figure 4-23
Mnemonic for filling orbitals

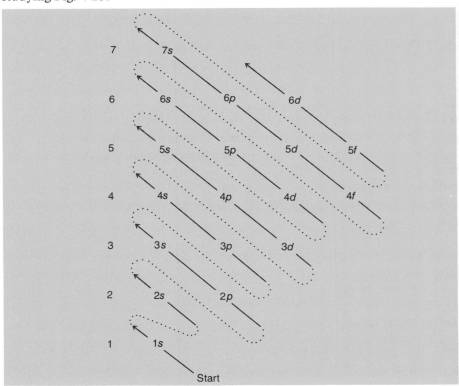

Two rules that must be used in connection with the Aufbau principle are the Pauli exclusion principle and Hund's rule. The **Pauli exclusion principle** states that an orbital can be empty, have one electron, or at most have two electrons. Thus, the electron in a hydrogen atom is placed in the lowest energy level available, a $1s$ orbital. In a helium atom, the second electron can also be placed in the $1s$ orbital, giving a total of two electrons in the orbital. In a lithium atom, however, the third electron cannot be placed in a $1s$ orbital, because an orbital can hold at most two electrons. The third electron must go into the next higher energy level, a $2s$ orbital.

Hund's rule states that electrons in the same sublevel will not pair up (occupy the same orbital) until all the orbitals in the sublevel have been at least half-filled. The filling of orbitals within the $2p$ sublevel is illustrated in Figure 4-24. In this diagram an arrow represents an electron in an orbital. A pair of arrows pointing in opposite directions indicates that there is an electron pair in the orbital.

Number of Electrons in the 2p Sublevel	Distribution		
	$2p_x$	$2p_y$	$2p_z$
1	↑		
2	↑	↑	
3	↑	↑	↑
4	↑↓	↑	↑
5	↑↓	↑↓	↑
6	↑↓	↑↓	↑↓

Figure 4-24
Distribution of electrons in the $2p$ sublevel

An example will show how these rules can be used to describe many-electron atoms. Consider the element sulfur. Sulfur has 16 protons and 16 electrons. It also has 16 neutrons, but the number of neutrons is of no concern to us at this point. Sixteen electrons can be placed into the energy level diagram as shown in Figure 4-25. Each arrow in the diagram represents one electron.

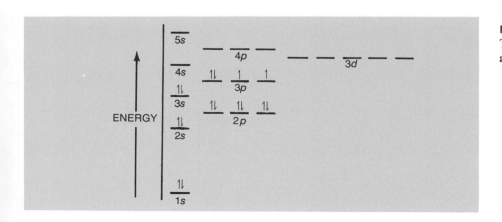

Figure 4-25
The energy level diagram for a sulfur atom

Figure 4-26
The energy level diagram for an iron atom

EXAMPLE PROBLEM 4-4

Draw the energy level diagram for the element iron.

SOLUTION

Iron has 26 electrons. These electrons can be placed into the energy level diagram as shown in Figure 4-26.

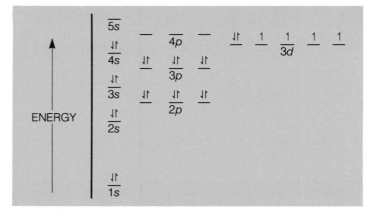

PRACTICE PROBLEM 4-7

Draw an energy level diagram for element 7, nitrogen, and for element 14, silicon. The energy levels for these elements may be checked by consulting Figure 4-27.

PRACTICE PROBLEM 4-8

Draw an energy level diagram for element 10, neon, and for element 20, calcium. The answers are given in Figure 4-27.

PRACTICE PROBLEM 4-9

Draw an energy level diagram for element 28, nickel. The answer is shown in Figure 4-27.

Figure 4-27
Solutions for practice problems 4–7, 4–8, and 4–9

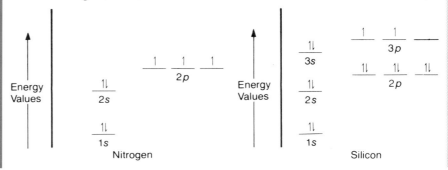

Energy Values

| 2p |
| 2s |
| 1s |

Neon

Energy Values

| 4s |
| 3s | 3p |
| 2s | 2p |
| 1s |

Calcium

Energy Values

4s	4p
3s	3p
2s	2p
1s

3d

Nickel

Table 4-5 Electron Configurations of the First 20 Elements

There is a shorthand method of showing only the distribution of electrons in a given atom. The distribution of electrons among the various orbitals of an atom is called its **electron configuration**. If each electron in an atom is in the orbital of lowest possible energy available to it, the atom is said to be in its *ground state*. The electron configurations given in this text are for the ground states of the atoms, rather than for excited states, in which one or more electrons would be in a higher energy orbital.

The electron configuration of sulfur is written as

$$_{16}S \; 1s^2 \; 2s^2 \; 2p^6 \; 3s^2 \; 3p^4$$

The superscripts indicate the number of electrons occupying each sublevel. Thus, the first and second energy levels are completely filled. However, the third energy level is only partly filled. The $3s$ and one of the $3p$ orbitals are filled. The other two $3p$ orbitals are half-filled. The $3d$ and all orbitals of the fourth, fifth, and higher levels are empty. The electron configurations of the first 20 elements are given in Table 4-5.

Some people use a shortened form of the electron configuration. The shortened form of the electron configuration of sulfur is $_{16}S$ [Ne] $3s^2 \; 3p^4$. The electron configuration of neon is $_{10}Ne \; 1s^2 \; 2s^2 \; 2p^6$, and the [Ne] replaces the $1s^2 \; 2s^2 \; 2p^6$ in the electron configuration of sulfur. When we write the shortened electron configuration of an element, we always start with the symbol of the nearest noble gas that has fewer electrons than the element. The shortened form of the electron configuration of calcium can be written $_{20}Ca$ [Ar] $4s^2$. This indicates that calcium has all of the electrons that argon has plus two electrons in the $4s$ orbital.

Element	Electron Configuration
H	$1s^1$
He	$1s^2$
Li	$1s^2 2s^1$
Be	$1s^2 2s^2$
B	$1s^2 2s^2 2p^1$
C	$1s^2 2s^2 2p^2$
N	$1s^2 2s^2 2p^3$
O	$1s^2 2s^2 2p^4$
F	$1s^2 2s^2 2p^5$
Ne	$1s^2 2s^2 2p^6$
Na	$1s^2 2s^2 2p^6 3s^1$
Mg	$1s^2 2s^2 2p^6 3s^2$
Al	$1s^2 2s^2 2p^6 3s^2 3p^1$
Si	$1s^2 2s^2 2p^6 3s^2 3p^2$
P	$1s^2 2s^2 2p^6 3s^2 3p^3$
S	$1s^2 2s^2 2p^6 3s^2 3p^4$
Cl	$1s^2 2s^2 2p^6 3s^2 3p^5$
Ar	$1s^2 2s^2 2p^6 3s^2 3p^6$
K	$1s^2 2s^2 2p^6 3s^2 3p^6 4s^1$
Ca	$1s^2 2s^2 2p^6 3s^2 3p^6 4s^2$

EXAMPLE PROBLEM 4-5

Predict the electron configuration of element number 24, chromium.

SOLUTION

When electrons are added according to the order given in Figure 4-24, the predicted electron configuration becomes

$$\text{Cr } 1s^2\, 2s^2\, 2p^6\, 3s^2\, 3p^6\, 4s^2\, 3d^4$$

The actual electron configuration of chromium deviates slightly from that predicted. It is found that chromium has the configuration

$$\text{Cr } 1s^2\, 2s^2\, 2p^6\, 3s^2\, 3p^6\, 4s^1\, 3d^5$$

This deviation from the predicted behaviour usually occurs in those elements which are close to filling or half-filling a d or f sublevel. Half-filled and filled sublevels appear to make an atom slightly more stable. When $3d$ sublevels, with energies near those of the $4s$ sublevel are involved, the stability effect is large enough to allow promotion of a $4s$ electron in chromium to the $3d$ sublevel. This process gives chromium a half-filled $3d$ sublevel containing five electrons.

PRACTICE PROBEM 4-10

Predict the electron configuration of element number 35, bromine.
(*Answer*: $1s^2\, 2s^2\, 2p^6\, 3s^2\, 3p^6\, 4s^2\, 3d^{10}\, 4p^5$)

PRACTICE PROBEM 4-11

Predict the electron configuration of element number 60, neodymium.
(*Answer*: $1s^2\, 2s^2\, 2p^6\, 3s^2\, 3p^6\, 4s^2\, 3d^{10}\, 4p^6\, 5s^2\, 4d^{10}\, 5p^6\, 6s^2\, 4f^4$)

4.11 The Quantum Numbers

Bohr's theory was a major step in our understanding of atoms, because it introduced the idea that the energy of an electron is quantized. Bohr simply had to *assume* this fact in order to explain the line spectrum of hydrogen. When the mathematical equations of wave mechanics are solved, however, the idea of quantization arises quite naturally. Solutions exist for the mathematical equations only if the properties described by the equation are quantized. At least three quantum numbers are obtained by solving the wave equation for an atom. We can use these quantum numbers to help us figure out the various orbitals of an atom.

The first quantum number is the principal quantum number, n. We discussed it in Sections 4-7 and 4-9. The principal quantum number identifies the *energy* of an electron in an orbital. Also, n goes from 1 to infinity by whole numbers:

$$n = 1, 2, 3 \ldots \infty$$

The second quantum number, ℓ, is the **secondary quantum number**. It identifies the *shape* of the orbital and the energy sublevel. For any given value of n, ℓ can start at 0 and go to the number which is one unit smaller than n:

$$\ell = 0, 1, 2, \ldots, n - 1$$

If $\ell = 0$, we have an s orbital. If $\ell = 1$, we have a p orbital. Values of ℓ equal to 2 and 3 correspond to d and f orbitals, respectively.

The third quantum number, m_ℓ, is the **magnetic quantum number**. It identifies the *direction of orientation* (pointing) of the orbital in relation to an external magnetic field. For a given value of ℓ, m_ℓ takes every integral value starting at $-\ell$, and going up to $+\ell$. For example, if $\ell = 0$, then $m_\ell = 0$; if $\ell = 1$, then $m_\ell = -1$ or 0 or 1; if $\ell = 2$, then $m_\ell = -2$ or -1 or 0 or 1 or 2. The number of different values of m_ℓ for a given value of ℓ is the same as the number of orbitals of a given type. For example, if $\ell = 1$, we have a p orbital. Since there are three different values for m_ℓ (-1, 0, 1), there must be three orbitals in a set of p orbitals.

We can use the quantum numbers to work out the orbitals for any value of the principal quantum number. As examples, we can consider the orbitals which have n equal to 1, 2, and 3. If $n = 1$, then ℓ must equal 0 and m_ℓ must also equal 0. The values of n and ℓ indicate that this is the $1s$ orbital. The symbol used is $n\ell$, where the numerical value of n is used, but the numerical value of ℓ is not used. Instead, letters are used to replace the numerical value of ℓ. Thus, the designation of a number (n) and a letter (ℓ) gives an approximate location of an electron, just as a postal code consisting of numbers and letters gives an approximate location of a house.

If $n = 2$, ℓ can equal either 0 or 1. If $\ell = 0$, m_ℓ must equal 0, and the orbital is the $2s$ orbital. However, if $\ell = 1$, then there are three acceptable values of m_ℓ (-1, 0, 1), and hence there are three $2p$ orbitals.

If $n = 3$, ℓ can equal 0, 1, or 2. If $\ell = 0$, m_ℓ must equal 0, and the orbital is the $3s$ orbital. If $\ell = 1$, there are three values of m_ℓ (-1, 0, 1), and hence there are three $3p$ orbitals. If $\ell = 2$, the orbitals are d type, and because m_ℓ can have five values (-2, -1, 0, 1, 2), there are five $3d$ orbitals. Table 4-6 summarizes the use of quantum numbers.

Table 4-6   Orbital Names for Different Sets of Quantum Numbers

n	ℓ	m_ℓ	Name of Orbital
1	0	0	$1s$
2	0	0	$2s$
2	1	-1	$2p$
2	1	0	$2p$
2	1	1	$2p$
3	0	0	$3s$
3	1	-1	$3p$
3	1	0	$3p$
3	1	1	$3p$
3	2	-2	$3d$
3	2	-1	$3d$
3	2	0	$3d$
3	2	1	$3d$
3	2	2	$3d$
4	0	0	$4s$
4	1	-1	$4p$
4	1	0	$4p$
4	1	1	$4p$
4	2	-2	$4d$
4	2	-1	$4d$
4	2	0	$4d$
4	2	1	$4d$
4	2	2	$4d$
4	3	-3	$4f$
4	3	-2	$4f$
4	3	-1	$4f$
4	3	0	$4f$
4	3	1	$4f$
4	3	2	$4f$
4	3	3	$4f$

EXAMPLE PROBLEM 4-6

What orbital has quantum numbers $n = 3$, $\ell = 2$, $m_\ell = 1$?

SOLUTION

If $\ell = 2$, the orbital is d type. Since $n = 3$, it must be a $3d$ orbital. We could use the value of m_ℓ to tell us which one of the five $3d$ orbitals is being described. However, it is sufficient for us to say merely that the orbital in question is a $3d$ orbital.

PRACTICE PROBLEM 4-12

Which orbitals have the following quantum numbers: **a)** $n = 5$, $\ell = 3$, $m_\ell = -3$; **b)** $n = 4$, $\ell = 1$, $m_\ell = -1$; **c)** $n = 7$, $\ell = 4$, $m_\ell = 0$? (*Answers*: **a)** $5f$; **b)** $4p$; **c)** $7g$)

According to the Pauli exclusion principle an orbital can contain as many as two electrons. If two electrons are in the same orbital, they must have opposite spins. This introduces the final quantum number m_s. The **spin quantum number**, m_s, can have only two values, $+\frac{1}{2}$ and $-\frac{1}{2}$. That is, an electron behaves as if it were a charged particle that can spin clockwise or counterclockwise. The Pauli exclusion principle means that no two electrons in the same atom can have four identical values for the four quantum numbers. We saw earlier that an arrow represents an electron in an orbital, and a pair of arrows represents an electron pair. We can now add to this by stating that the spin direction of an electron is indicated by the direction of the arrow. Two paired electrons in an orbital are indicated by two arrows, one pointing up and one pointing down—meaning that they have opposite spins. Thus the Pauli exclusion principle has its origin in wave mechanics.

The quantum numbers are not chosen arbitrarily. They are obtained quite naturally in the process of solving a wave equation. The four quantum numbers are summarized in Table 4-7.

Table 4-7   The Four Quantum Numbers

Quantum Number	Name	Values	Meaning
n	principal	1, 2, 3,..., ∞	energy of electron
ℓ	secondary	0, 1, 2,..., $n-1$	shape of orbital
m_ℓ	magnetic	$-\ell$,..., -1, 0, 1,..., $+\ell$	orientation of orbital
m_s	spin	$+1/2$, $-1/2$	spin of electron

EXAMPLE PROBLEM 4-7

Is the set of quantum numbers $n = 2$, $\ell = 0$, $m_\ell = 1$, $m_s = +\frac{1}{2}$ an allowed set for an electron in an atom?

SOLUTION

Since the value of n can be any integer greater than zero, $n = 2$ is allowed. Since ℓ can be any integer from 0 to $n - 1$ (in this case 0 or 1) a value of 0 is allowed. The value of m_ℓ can be any positive or negative integer from $-\ell$ to $+\ell$. In this case, $\ell = 0$, so m_ℓ must also be 0. A value of $m_\ell = 1$ is not allowed. Thus, the set of quantum numbers is not allowed.

PRACTICE PROBLEM 4-13

Which of the following sets of quantum numbers are not allowed?

   a) $n = 2$, $\ell = 1$, $m_\ell = -1$, $m_s = +\frac{1}{2}$
   b) $n = 7$, $\ell = 3$, $m_\ell = 0$, $m_s = -\frac{1}{2}$
   c) $n = 3$, $\ell = 3$, $m_\ell = -2$, $m_s = -\frac{1}{2}$
   d) $n = 4$, $\ell = 2$, $m_\ell = 3$, $m_s = +\frac{1}{2}$
   e) $n = 3$, $\ell = 2$, $m_\ell = -1$, $m_s = 0$

(*Answers*: **c, d, e**)

APPLICATION

"Seeing" Atoms

Why do chemists believe in the existence of atoms? Chemists base their beliefs on the results of studies in areas such as the physical and chemical behaviour of gases, mass relationships in chemical reactions, electrolysis, and radioactivity.

More recently, however, as a result of advances in technology, chemists have obtained more direct evidence of the existence of atoms. For example, in 1970, scientists at the University of Chicago used a high-resolution electron microscope to "see" individual atoms of uranium and thorium. The high-resolution electron micrograph of uranium atoms showed small white dots, indicating individual atoms, and large white dots, indicating clusters of atoms. Further developments by University of Chicago scientists led to motion pictures of individual atoms of metals such as cadmium, palladium, platinum, and silver deposited on a thin film of carbon. An electron beam was fired at the metal. This beam excited outer electrons in the metal atoms and the process was registered on a television screen. The resulting images showed various types of motion for individual metallic atoms and clusters of atoms. Once again, the atoms were depicted as blurred white spots in which the individual atoms were magnified about five million times. The first films of the motion of atoms were in black and white and different kinds of atoms looked almost indistinguishable. A refinement to this method used computers to arbitrarily assign colours to spots of different intensity. Images of different atoms could then be distinguished.

In 1981, a major breakthrough in "seeing" atoms came with the invention of the scanning tunnelling microscope. This microscope produces images which show individual atoms on or near the surface of the sample. The atoms are magnified about ten million times. With these recent advances, chemists have even more reason to believe in the existence of atoms.

Part Three Review Questions

1. What defect in Bohr's theory is remedied by the wave mechanical theory of the atom?
2. What is the basic idea of wave mechanics?
3. Describe the differences between a $2s$ and a $2p$ orbital.
4. State the Aufbau principle, the Pauli exclusion principle, and Hund's rule. Give a specific example of the application of each rule.
5. Draw an energy level diagram for element 17, chlorine.
6. Write the electron configuration for $_{53}$I.
7. Identify the orbital in which electrons with the following quantum numbers are found:
   a) $n = 3$, $\ell = 0$; b) $n = 3$, $\ell = 2$; c) $n = 2$, $\ell = 1$; d) $n = 7$, $\ell = 4$; e) $n = 4$, $\ell = 3$.
8. State what is wrong with each of the following sets of quantum numbers:
   a) $n = 2$; $\ell = 2$; $m_\ell = 0$; $m_s = +\frac{1}{2}$
   b) $n = 3$; $\ell = 1$; $m_\ell = 2$; $m_s = -\frac{1}{2}$
   c) $n = 2$; $\ell = 0$; $m_\ell = 0$; $m_s = 0$
   d) $n = 3$; $\ell = \frac{1}{2}$; $m_\ell = 0$; $m_s = -\frac{1}{2}$
   e) $n = -3$; $\ell = 2$; $m_\ell = 0$; $m_s = +\frac{1}{2}$

Chapter Summary

- The early Greeks pictured matter as being made up of hard, spherical, indivisible particles called atoms.
- John Dalton reintroduced the Greek model of matter and greatly expanded upon it.
- Dalton's atomic theory explains the laws of chemical change.
- An atomic mass unit (u) is $\frac{1}{12}$ the mass of a carbon-12 atom.
- The table of atomic masses of the elements shows the average mass of the atoms of an element relative to the mass of an atom of carbon-12.
- Isotopes are atoms of the same element which have different masses.
- An electron is a particle with a relative charge of -1 and a mass of 0.000 55 u.
- Cathode rays are made up of electrons.
- A proton is a particle with a relative charge of $+1$ and a mass of 1.0073 u.
- Thomson's plum pudding model pictured the atom as a sphere of uniformly distributed positive charge with electrons embedded throughout.
- The Rutherford nuclear model of the atom pictured the atom with a dense central part called the nucleus. The nucleus contains all of the positive charge and most of the mass of the atom. The electrons are at some distance from the nucleus. The atom thus contains mostly empty space.
- Neutrons are particles with no charge and a mass of 1.0087 u.
- Atomic number is the number of protons in the nucleus.
- Mass number is the total number of protons and neutrons in an atom.
- A quantum or photon is a package of energy.
- A continuous spectrum shows light of all colours.
- A line spectrum shows only certain colours of light.

- The Bohr model of the atom pictured an atom made up of a nucleus, with electrons travelling around the nucleus in circular paths called orbits.
- Electrons can jump from one energy level (orbit) to another, giving up or absorbing a quantum of energy in the process.
- The wave mechanical model of the atom treats the electron as both a particle and a wave. It deals with the probability of finding an electron in a certain region of space surrounding a nucleus.
- Orbitals are regions of space in which electrons are most likely to be found.
- An energy level diagram shows the order of the orbitals in terms of energy and the number of electrons present in each orbital.
- The Aufbau principle states that electrons are placed into orbitals in order of increasing energy.
- The Pauli exclusion principle states that an orbital may be empty, half-filled (1 electron) or filled (2 electrons). If an orbital is filled, the electrons must have opposite spins.
- Hund's rule states that a second electron is not added to an orbital until every orbital of the sublevel is half-filled.
- An electron configuration indicates the number of electrons in each energy sublevel.
- The quantum numbers of an electron indicate its approximate location around a nucleus.
- The principal quantum number, n, indicates the size of the orbital and energy of an electron in the orbital ($n = 1, 2, 3, \ldots, \infty$).
- The secondary quantum number, ℓ, indicates the shape of an orbital ($\ell = 0, 1, 2, 3, \ldots, n-1$).
- The magnetic quantum number, m_ℓ, indicates the orientation of an orbital in a magnetic field ($m_\ell = -\ell, \ldots, 0, \ldots, \ell$).
- The spin quantum number, m_s, indicates the direction of electron spin ($m_s = +\frac{1}{2}, -\frac{1}{2}$).

Key Terms

atom	quantized
molecule	orbit
isotope	quantum mechanics
atomic mass unit	energy level population
atomic mass	wave mechanics
cathode ray	wave equation
electron	wave function
proton	orbital
nucleus	principal quantum number
neutron	Aufbau principle
nucleon	Pauli exclusion principle
atomic number	Hund's rule
mass number	electron configuration
radioisotope	secondary quantum number
quantum	magnetic quantum number
continuous spectrum	spin quantum number
line spectrum	

Test Your Understanding

Each of these statements or questions is followed by four responses. Choose the correct response in each case. (Do not write in this book.)

1. Which of the following statements is *not* a part of Dalton's atomic theory?
   a) Atoms of the same element may differ in mass.
   b) All atoms of one element differ from the atoms of every other element.
   c) Chemical change is the union or separation of atoms.
   d) Atoms combine in small whole-number ratios.

2. The modern scale of atomic masses is based on the assignment of
   a) $H = 1.000$ u. b) $C = 12.000$ u.   c) $^{12}C = 12.000$ u.   d) $^{16}O = 16.000$ u.

3. An isotope which has a mass nine times that of an atom of carbon-12 is most likely an atom of
   a) silver.   b) fluorine.   c) boron.   d) element 108.

4. J. J. Thomson's experiments with discharge tubes demonstrated that
   a) alpha particles are the nuclei of helium atoms.
   b) the mass of an atom is concentrated in its nucleus.
   c) cathode rays are streams of negatively charged ions.
   d) the charge-to-mass (e/m) ratio is the same for all cathode ray particles.

5. A proton is a particle with
   a) a relative charge of −1 and a mass of 0.000 55 u.
   b) a relative charge of +1 and a mass of 0.000 55 u.
   c) a relative charge of +1 and a mass of 1.0073 u.
   d) no charge and a mass of 1.0087 u.

6. Rutherford's observation that a gold foil scatters some alpha particles through angles greater than 90° enabled him to conclude that
   a) all atoms are electrically neutral.
   b) the nucleus of the atom contains the positive charge.
   c) an electron has a very small mass.
   d) electrons are a part of all matter.

7. The nucleus of an atom usually consists of
   a) electrons and protons.   b) protons and neutrons.
   c) neutrons and electrons.   d) isotopes.

8. Surrounding the nuclei of atoms are found the
   a) alpha particles.   b) protons.   c) neutrons.   d) electrons.

9. Isotopes of the same element always have the same
   a) number of neutrons.   b) number of protons.
   c) mass number.   d) atomic mass.

10. A nucleus of cobalt-56 contains
    a) 29 protons and 27 neutrons.   b) 27 protons and 29 neutrons.
    c) 29 protons and 27 electrons.   d) 27 protons and 29 electrons.

11. Element X consists of 30.0% of an isotope with mass 24.02 u and 70.0% of an isotope with mass 26.10 u. The average atomic mass of X is
    a) 24.64 u.   b) 25.06 u.   c) 25.48 u.   d) 50.12 u.

12. The spectrum of atomic hydrogen is best described as
    a) a continuous spectrum.
    b) a series of three lines.
    c) a single series of lines with constant line spacings.
    d) several series of lines with varying line spacings.

13. Niels Bohr theorized that
    a) energy is evolved when an electron jumps to a lower energy level.
    b) electrons travel in circular paths called orbitals.
    c) the energy of an electron may have any arbitrary value.
    d) the spectrum produced by hydrogen atoms should be a continuous spectrum.

14. The number of electrons in the second energy level of a nitrogen atom is
    a) 2.   b) 5.   c) 7.   d) 14.

15. The quantum number which determines the shape of an orbital is
    a) n.   b) ℓ c) m_ℓ.   d) m_s.

16. The set of quantum numbers (n, ℓ, m_ℓ, and m_s) which describes an outer-most electron in a strontium atom is
    a) 5, 0, 0, $-\frac{1}{2}$.   b) 5, 0, 1, $+\frac{1}{2}$.   c) 5, 1, 0, $+\frac{1}{2}$.   d) 5, 1, 1, $+\frac{1}{2}$.

17. One allowed set of quantum numbers (n, ℓ, m_ℓ, and m_s) for a 3d electron is
    a) 3, 3, -2, $+\frac{1}{2}$.   b) 2, 1, 1, $-\frac{1}{2}$.   c) 3, 1, 0, $-\frac{1}{2}$.   d) 3, 2, -2, $+\frac{1}{2}$.

Review Your Understanding

1. What is the mass of an atom which has a mass 7/6 times as much as that of a carbon-12 atom?

2. What is the mass of an atom that has 4/13 the mass of a chromium atom? The mass of a chromium atom is 13/3 times the mass of an atom of carbon-12.

3. Where do the rays in a gas discharge tube originate? How do we know?

4. What is an alpha particle? What happens when it comes near a positive charge?

5. Describe the experiment which caused the Thomson model to be replaced by the nuclear model of the atom.

6. Contrast the Thomson model with the Rutherford nuclear model of the atom.

7. What particles are found in the nucleus of the atom? Describe these particles. Describe the third particle found in an atom.

8. How many electrons, protons, and neutrons are in an atom a) for which the atomic number is 38 and the mass number is 90; b) for which the atomic number is 82 and the mass number is 207?

9. How many neutrons, electrons, and protons are found in one atom of each of the following: a) $^{60}_{27}$Co; b) $^{90}_{38}$Sr; c) $^{108}_{47}$Ag; d) $^{207}_{82}$Pb; e) $^{210}_{85}$At; f) $^{238}_{92}$U?

10. What percentage of the total mass of the atom is contributed by the electrons in an atom of a) $^{4}_{2}$He; b) $^{12}_{6}$C; c) $^{35}_{17}$Cl?

11. Natural boron consists of 19.8% $^{10}_{5}$B (mass 10.0 u) and 80.2% $^{11}_{5}$B (mass 11.0 u). What is the average atomic mass of natural boron?

12. Natural potassium consists of 93.1% $^{39}_{19}$K (mass 39.0 u) and 6.9% $^{41}_{19}$K (mass 41.0 u). What is the average atomic mass of natural potassium?

13. Natural thallium consists of 29.50% $^{203}_{81}$Tl (mass 203.0 u) and 70.50% $^{205}_{81}$Tl (mass 205.0 u). What is the average atomic mass of natural thallium?

14. Natural erbium consists of 33.41% $^{166}_{68}Er$ (mass 165.9 u), 22.94% $^{167}_{68}Er$ (mass 166.9 u), 27.07% $^{168}_{68}Er$ (mass 167.9 u), 0.136% $^{162}_{68}Er$ (mass 161.9 u), 1.56% $^{164}_{68}Er$ (mass 163.9 u), and 14.88% $^{170}_{68}Er$ (mass 169.9 u). What is the average atomic mass of natural erbium?

15. What experimental evidence did Bohr have to support his belief that the energy of the electrons in a hydrogen atom is quantized?

16. How is the energy of an electron in a Bohr orbit related to the distance of the electron from the nucleus?

17. What happens to the electron in the atom when a hydrogen atom is excited?

18. Account for the bluish-green line in a hydrogen spectrum in terms of the Bohr model.

19. Give two reasons why the Bohr model was eventually discarded.

20. What would happen to an electron in a hydrogen atom if it absorbed too much energy?

21. How many possible electron transitions are there among the first four energy levels of the hydrogen atom?

22. Write the energy level population of **a)** element number 13; **b)** element number 17.

23. How many types of orbitals are there for $n = 4$? How many actual orbitals are there for $n = 4$?

24. How do the $2s$ and $3s$ orbitals differ? How are they similar?

25. What is one difference between the idea of a Bohr orbit and the idea of an orbital?

26. Contrast the Bohr model of the atom with the wave mechanical model.

27. Use Hund's rule to place **a)** seven electrons in the orbitals of the $3d$ sublevel; **b)** two electrons in the orbitals of the $4p$ sublevel.

28. Draw the energy level diagram for the following elements: **a)** number 5 (boron); **b)** number 9 (fluorine); **c)** number 12 (magnesium); **d)** number 17 (chlorine); **e)** number 33 (arsenic); **f)** number 52 (tellurium); and **g)** number 86 (radon).

29. Write the electron configurations for **a)** $_6C$; **b)** $_{10}Ne$; **c)** $_{15}P$; **d)** $_{19}K$; **e)** $_{35}Br$; **f)** $_{54}Xe$; **g)** $_{84}Po$.

30. Draw the energy level diagram for the following elements: **a)** number 22 (titanium); **b)** number 25 (manganese); **c)** number 32 (germanium); **d)** number 46 (palladium); **e)** number 55 (cesium).

31. Write the electron configurations for **a)** $_{23}V$; **b)** $_{34}Se$; **c)** $_{40}Zr$; **d)** $_{50}Sn$; **e)** $_{56}Ba$.

32. Suggest an atom (other than chromium), with atomic number smaller than 38, which might be expected to deviate from the predicted electron configuration.

33. What are the quantum numbers n, ℓ, and m_ℓ for each of the orbitals of **a)** the $5p$ sublevel; **b)** the $6d$ sublevel?

34. Which orbitals ($1s$, $2s$, $2p$, etc.) have the following quantum numbers: **a)** $n = 2$, $\ell = 1$, $m_\ell = 0$; **b)** $n = 4$, $\ell = 2$, $m_\ell = 2$; **c)** $n = 6$, $\ell = 3$, $m_\ell = 1$?

35. Which of the following are designations for orbitals which are not possible in wave mechanics: **a)** $1d$; **b)** $4f$; **c)** $2s$; **d)** $1p$; **e)** $6d$; **f)** $2f$?

36. Which of the following are sets of quantum numbers for orbitals which are possible in wave mechanics: **a)** $n = 1$, $\ell = 1$, $m_\ell = 1$; **b)** $n = 2$, $\ell = 1$, $m_\ell = 2$; **c)** $n = 2$, $\ell = 0$, $m_\ell = 0$; **d)** $n = 3$, $\ell = 3$, $m_\ell = 1$; **e)** $n = 3$, $\ell = 2$, $m_\ell = -2$; **f)** $n = 4$, $\ell = 3$, $m_\ell = 2$?

37. Why can there be no d orbitals in the second ($n = 2$) energy level?

38. If the magnetic quantum number, m_ℓ, took every integral value starting at 0 and going to $+\ell$, how many orbitals would there be in $n = 5$?

Apply Your Understanding

1. Suppose that in a hypothetical scattering experiment all of the alpha particles went through the gold foil. Propose a model of an atom that would account for this observation.

2. Both protons and neutrons have a mass of 1.67×10^{-27} kg. Suppose that they are spherical in shape with a diameter of 100 fm. What is the density of these nucleons?

3. How many electrons are in each of the following: **a)** $^{23}_{11}Na^+$; **b)** $^{37}_{17}Cl^-$; **c)** $^{24}_{12}Mg^{2+}$; **d)** $^{32}_{16}S^{2-}$?

4. If the diameter of a nucleus is about 100 fm and the diameter of an atom is about 200 pm, what percentage of the total volume of an atom is occupied by the nucleus?

5. The earth has a mass of 5.98×10^{24} kg and a diameter of about 13 Mm. What would be the diameter of the earth if it had the same mass but was composed entirely of nucleons? Assume the density of a nucleon calculated in Question 2.

6. Naturally occurring chlorine is composed of $^{35}_{17}Cl$ with an atomic mass of 34.968 85 u and $^{37}_{17}Cl$ with an atomic mass of 36.965 90 u. The average atomic mass of chlorine is 35.453 u. What are the percentages of each isotope in naturally occurring chlorine?

7. Naturally occurring silver is composed of $^{107}_{47}Ag$ with an atomic mass of 106.9041 u and $^{109}_{47}Ag$ with an atomic mass of 108.9047 u. The average atomic mass of silver is 107.868 u. What are the percentages of each isotope in naturally occurring silver?

8. How is it possible to use atomic emission spectra to determine the elements present in a substance?

9. How could you determine whether the coloured light from an outdoor "neon" sign is a mixture of colours or a single colour?

10. One of the four quantum numbers is not required to describe the orbitals of a hydrogen atom. Which one is it? Why is this quantum number not required to describe an orbital?

11. Suppose the spin quantum number had the values 0, $+\frac{1}{2}$, $-\frac{1}{2}$. How many electrons could one orbital hold? Why?

12. A certain excited state of a carbon atom has all electrons in the first two energy levels, but there are no paired electrons in the second energy level. Use this information to write all the possible sets of quantum numbers for the electrons in the excited atom. What is the electron configuration of the excited atom?

Investigations

1. The effect of magnetic and electric fields on the motion of charged particles has practical application in the mass spectrometer. This instrument is used to determine the masses of molecules and ions. Write a research report on the mass spectrometer.

2. Many different chemicals have been detected in space. List five of these substances. How was each one detected?

3. Large sums of money have already been spent on research into the ultimate building blocks of matter. To determine more information about the smallest particles of matter will require even more money. Debate the usefulness of this type of research.

4. Devise and carry out an experiment to determine whether the light from a street lamp is different from the light from a) a fluorescent lamp; and b) an incandescent lamp.

The Elements and the Periodic Law

5

CHAPTER CONTENTS

Part One: **The Elements**
    5.1   Discovery of the Elements
    5.2   "It's Elementary, My Dear..."
    5.3   Metals, Metalloids, and Nonmetals
    5.4   Symbols of the Elements

Part Two: **Periodic Law and the Periodic Table**
    5.5   Periodic Law
    5.6   The Modern Periodic Table
    5.7   The Periodic Law and Energy Levels
    5.8   Electron Dot Symbols
    5.9   The Periodic Table and Electron Configurations

Part Three: **Trends in the Periodic Table**
    5.10 Atomic Size
    5.11 Ionization Energies
    5.12 Electron Affinity

How many elements are there? What is the only metal that is liquid at room temperature? What is the only nonmetal that is liquid at room temperature? Certain elements have similar physical and chemical properties. How can we arrange a list of the elements so that we can more easily predict the changes in their properties as we proceed from one element to another in the list? Is it possible to predict the properties of elements that have not yet been discovered?

In this chapter we shall learn the answers to these questions. We shall trace the discovery of some elements and note their characteristic properties and uses. Then we shall see how, as time went on, various regularities were recognized in the properties of elements; how this led to the discovery of the periodic law and the development of the modern periodic table; and how the relationship between the properties of elements and their atomic structures finally became clear.

Key Objectives

When you have finished reading this chapter you should be able to

1. state the Periodic Law.
2. locate, in a modern periodic table, the main-group elements, the transition metals, the lanthanides, and the actinides.
3. locate, in a modern periodic table, the metals, nonmetals, and metalloids.
4. locate, in a modern periodic table, the noble gases, the alkali metals, and the halogens.
5. using energy level populations for the first twenty elements, identify pairs of elements which should have similar chemical properties.
6. write electron dot symbols for the first 20 elements.
7. predict trends in atomic size, ionization energy, and electron affinity as you go across a period or down a group in the periodic table.

Part One: The Elements

5.1 Discovery of the Elements

The earliest source of chemical energy was fire. Long before members of the human race had learned to start a fire—thought to be about two million years ago—they had learned how to keep it going. Once they had managed this feat, they were able to practise chemistry.

Then, about eight thousand years ago, people found that shiny materials sometimes melted out of rocks that had been heated in fires. In this way, they discovered soft metallic elements such as copper, silver, and gold. Elements are samples of matter consisting of only one type of atom. For example, a sample of gold consists only of gold atoms. These three elements of copper, silver, and gold were easily hammered into various shapes and could be used to make pieces of jewelry, tools, or weapons. Lead, tin, iron, mercury, carbon, and sulfur had also been discovered by this time. Then, about five thousand years ago, it was found that a little tin added to melted copper changed its properties. The new substance, bronze, was much tougher and made superior tools and weapons. Thus began the Bronze Age. Whole empires were built around this technological breakthrough. There was a thriving tin trade between Phoenicia and the tin mines in Cornwall, England at least as long ago as 1000 B.C., and a prehistoric mine near Salzburg, Germany, is believed to have produced 20 kt (kilotonnes) of copper during this period.

The art of working with metals had interesting effects for chemistry. When the Greek philosophers saw ordinary minerals going into the foundries at Alexandria, Egypt, and large amounts of gold and silver coming out, they naturally assumed that the craftsmen were changing ordinary substances into gold and silver. The craftsmen did not give out their trade secrets. Thus began the search for the elusive "philosopher's stone," a magical substance which could change other substances into gold and silver.

The alchemists (from Arabic *al*, "the" + Greek *chymeia*, "pouring", as of molten metal) continued the search for centuries without success. In the process, however, they made many important discoveries. By the middle of the 18th century they had isolated phosphorus, zinc, arsenic, antimony, and bismuth. The crudeness of their experimental techniques, however, still limited them to discovering only the easily obtainable elements.

The number of known elements increased from 21 to about 50 betweeen 1775 and 1830. During these years new techniques were devised and new theories were proposed. The theories led to predictions about the existence of new elements, and the search for new elements was based on principles rather than chance. Advances in experimental technique were usually responsible for rapid increases in the number of known elements (Fig. 5-1).

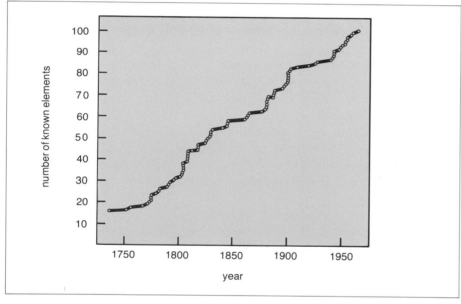

Figure 5-1
Chronology of the discovery of the elements

An example of the importance of new experimental technique lies in the work of Sir Humphrey Davy (Fig. 5-2). At that time the scientific world was interested in the study of "voltaic" electricity. Davy, with a well-equipped laboratory at the Royal Philosophical Institution, enthusiastically joined in these studies. He succeeded in constructing cells (batteries) which were more efficient than previous ones. By 1807 he had succeeded in producing sodium and potassium by passing an electric current through certain of their compounds. The following year in quick succession he isolated boron, magnesium, calcium, strontium, and barium. In less than two years one person had added seven elements to the list, all as the result of an improved experimental technique. Other sharp rises in the curve in Fig. 5-1 are due to similar advances at different periods.

Few new elements were discovered in the first half of this century, and by about 1940 all the 90 naturally-occurring elements had been discovered. Since

Table 5-1	Occurrence of Elements in the Earth's Crust and Atmosphere	
Oxygen	49.20%
Silicon	25.67%
Aluminum	7.50%
Iron	4.71%
Calcium	3.39%
Sodium	2.63%
Potassium	2.40%
Magnesium	1.93%
Hydrogen	0.87%
Titanium	0.58%
Chlorine	0.19%
Phosphorus	0.11%
Manganese	0.09%
Carbon	0.08%
Sulfur	0.06%
Barium	0.04%
Nitrogen	0.03%
Fluorine	0.03%
Strontium	0.02%
All others	0.47%

then work in nuclear science has led to the discovery of at least sixteen new elements, all of which are radioactive.

Only about a quarter of the elements occur in the free or elemental state. The others are all found combined with one another in compounds. The more abundant elements found in the earth's crust and atmosphere are listed in Table 5-1.

APPLICATION

Chemicals in Space

The percentages of the most abundant elements in the earth's crust and atmosphere are listed in Table 5-1. The distribution of elements in the universe is very different from that of elements in the earth. The ten most abundant elements in the universe in order of their decreasing abundance are hydrogen, helium, oxygen, carbon, neon, nitrogen, magnesium, silicon, iron, and sulfur. Hydrogen is the most abundant element in the universe, making up 90-95% of all atoms present. Helium accounts for 5-10% of the composition of the universe. All the other elements combined make up only about 1%.

No elements that are not present on earth have been found in space. Most of the universe is empty space. When matter does accumulate in space, it does so in the form of "clouds." These clouds are thought to be composed of a small percentage of solid matter (about 1% "cosmic dust") with the remainder being gases.

Scientists have identified a large number of molecules in space. Because of the vast difference in conditions such as temperature and density, a number of molecules that don't exist on earth do exist in space. However, common molecules such as water, ammonia, hydrogen sulfide, ethyl alcohol, carbon monoxide, and acetylene, that are found on earth, have also been detected in space. In addition, molecules which are essential to life have been discovered in space. This discovery indicates to some scientists that life may exist beyond earth.

How were elements and molecules identified in space? Elements have been identified by their line spectra which result from movements of electrons between energy levels in atoms. Molecules have been identified by the spectra caused by changes in the movement of valence electrons in molecules or by changes in the energy of rotation of molecules.

5.2 "It's Elementary, My Dear..."

Each of the 109 or more elements is unique. Each has its own fascinating story. You may know a lot about the elements without even realizing it. Just for fun, test your knowledge. See if you can identify the elements for which the following clues are given. The culprits are limited to elements whose atomic numbers are between one and twenty. Feel free to do some library research if you want to improve your score. The answers are given on page 163.

1. I am the metal present in the green pigments in plants which convert light energy to chemical energy.
2. Are you interested in masonry? I am the metal which is an important element in mortar, plaster, and Portland cement.
3. Unlike most metals, I allow X-rays to pass through me quite readily. You are likely to find me in the windows of X-ray tubes.
4. Bang! I am the element formed when a hydrogen bomb explodes.
5. Ring around the rosie! This element can join together in a variety of rings and chains.
6. Need a purgative? You get this metallic element in Epsom salts.
7. I am a colourless, odourless gas. When I burn, I form water.
8. I also am a colourless, odourless gas. I am used in enormous quantities in steel blast furnaces.

Biography

HUMPHREY DAVY was born in Cornwall, England, the eldest of eight children. He did badly at school, mainly because of his love of sport. At the age of 16 he started assisting a local physician with the preparation of remedies, but his taste for startling experiments and explosions soon lost him his job.

At 19, he began his study of chemistry, using materials and apparatus at hand. He made such progress that he was placed in charge of patients at the Pneumatic Institution, which had been established to study the medical effects of the gases discovered in the previous twenty years. He often experimented on himself, inhaling many gases such as nitrous oxide, methane, carbon dioxide, nitrogen, hydrogen, and nitric oxide. Somehow he managed to survive.

Davy prepared nitrous oxide in large quantities. Because it sometimes produces a feeling of exhilaration when inhaled, sniffing nitrous oxide became a fad. Davy's resulting popularity led to his appointment at the age of 22 as an assistant lecturer at the Royal Philosophical Institution of London. Nitrous oxide, also called laughing gas, was used as an anesthetic in dentistry and surgery.

Davy began studies on electrolysis, and soon isolated for the first time the elements potassium, sodium, barium, strontium, calcium, and magnesium. Davy was an inventor, discovering that the copper sheathing of naval vessels could be protected from corrosion by attaching a more reactive metal to the copper. He also invented the safety lamp for miners. At the age of 34 he was knighted, and at 42 he became President of the Royal Society. He was one of the most remarkable men of his time.

Figure 5-2
Humphrey Davy (1778–1829)—
Discovered new elements by
electrolysis.

Emerald crystal

Calcium metal burns with a spectacular flame.

9. Give thanks for me. In the upper atmosphere I prevent ultraviolet rays from the sun from reaching the earth's surface and destroying life on earth.

10. Versatility is my name. There are more than two million compounds containing me.

11. Parlez-vous français? The ore of this metal derives its name from the village of Les Baux in France, where it was discovered in 1821.

12. Be careful! Small amounts of me are needed for plant growth, but large amounts are toxic, so I am found in both fertilizers and weed killers.

13. Guard your jewels! Emerald and aquamarine consist of the oxide of this element.

14. Let there be light! I am the gas that is used to fill electric light bulbs and fluorescent tubes.

15. Don your fireproof suits! I was the gas used to fill the airship *Hindenburg*, which burned at Lakehurst, New Jersey, in less than five minutes.

16. Now put on your mukluks. Many of my compounds are the refrigerants in household refrigerators and the propellants in aerosol sprays.

17. Do I sound like Donald Duck? Mixed with oxygen, I am used as an artificial atmosphere for deep sea divers.

18. Shocking? In one form I am a conductor of electricity; in another form I am an insulator.

19. Use me if you have something to sell. I am the gas that is used in advertising signs—an electrical discharge causes me to glow reddish orange.

20. Stay away from me. I am a pale yellow poisonous gas, the most reactive of all the elements. Even water burns in me with a bright flame! One of my compounds is a poisonous gas which is used to frost electric light bulbs and to etch designs and letters on glass.

21. Brr! I have the lowest melting point of any element.

22. Need a light? I am a wax-like transparent solid that ignites spontaneously in air.

23. I hope you don't mind the heat. Death Valley, California provides 85% of the world's supply of this element.

24. You can't get away from me. I comprise almost as much mass in the earth's crust as do all other elements put together.

25. Watch the birdie! I am used in flash bulbs and flares because I burn with a dazzling white flame.

26. I am another bad actor. My irritating effects and suffocating action were exploited when I was used as a war gas in World War I.

27. Your teacher uses me. I am a metallic element found in chalk, limestone, and marble.

28. I am a pale yellow solid that is obtained in large quantities from Alberta natural gas.

29. Do you drive at night? I am the element responsible for the characteristic yellow colour of many highway lamps.

30. I am a metal that burns with a yellow-red flame. I am an essential constituent of leaves, bones, teeth, and shells.

31. Count your beads. Metal oxides dissolve in a molten bead of one of my

compounds to give characteristic colours which can be used to identify the metals.

32. Do you smoke? A cigarette filter made of this element may not prevent you from getting cancer, but may postpone the day of reckoning.
33. I impart a beautiful crimson colour to a flame, but when I burn strongly the flame is dazzling white.
34. Thinking of a space war? I am the gas that is used in many lasers, where I give a bright orange-red beam of light.
35. I am the most abundant metal in the earth's crust.
36. Let's go spelunking. I am the metallic element present in stalactites and stalagmites.
37. In the liquid form I am a superfluid. I will climb up the inner surface of a cup, over the lip, and down the outer surface, until the cup is empty.
38. More precious stones. I am the second most abundant element on earth. You will find me in agate, amethyst, jasper, and opal.
39. Let's go on a safari. In Africa I am found as pure crystals in what seem to be the necks of old volcanoes.
40. I am a colourless, odourless, generally inert gas, yet many of my compounds are important as explosives.

Amethyst crystals

Quartz crystals contain silicon and oxygen.

5.3 Metals, Metalloids, and Nonmetals

There is much to learn about the elements. It is often helpful to break any topic down into simpler parts. Thus, scientists have subdivided the elements into three groups—metals, nonmetals, and metalloids.

Metals

More than three quarters of the elements known today are **metals** (from Greek *metallon*, "mine"). Not everything that comes out of a mine is a metal, however, so chemists describe metals more carefully. All metals share similar properties (Fig. 5-3). They are good conductors of heat and electricity. They have a characteristic lustre. They are malleable and ductile; that is, they can be hammered into sheets and drawn into wires. They usually have high densities and high boiling points. Many metals have other useful properties such as resistance to stretching and twisting (Fig. 5-4).

Figure 5-3
All metals share similar properties. From left to right are gold, silver, and copper.

Figure 5-4
The suspension cables of this bridge between Halifax and Dartmouth show the resistance to stretching possessed by many metals.

Let us consider a familiar metal—gold. Gold, along with silver and copper, is used for making coins. Gold is valuable, not only because it is the most beautiful element, but because it is both rare and unreactive. Objects made of gold are unaffected by air and most chemical reagents. Gold is used extensively for jewelry, decoration, dental work, and for gold plating. Because gold does not corrode, and because it is a good conductor of electricity, electrical components are often gold plated. Gold is the most malleable metal; one cubic centimetre of gold can be hammered out to 19 m^2 (a square 4.4 m on each edge). Because gold is soft, other metals such as silver and copper are usually added to the gold to give it more strength. The term *karat* is used to indicate the amount of gold present. Pure gold is 24 karat. Thus, 12 karat gold is 50% pure gold.

Silver is more reactive than gold. When exposed to sulfur-containing substances such as eggs, silver quickly tarnishes. Large quantities of silver are used for manufacturing photographic films because some silver compounds are sensitive to light. Copper is one of the most important metals because of its excellent electrical conductivity. Other well-known metals are iron, aluminum, platinum, zinc, and mercury. Mercury is the only metal that is a liquid at room temperature.

Nonmetals

The elements whose properties are not characteristic of metals are called **nonmetals**. They do not usually have a lustre. They are usually poor conductors of heat and electricity. Iodine, phosphorus, and carbon are examples of

APPLICATION

How Many Atoms Make a Metal?

A metal consists of an ordered three-dimensional array of atoms. Each atom contributes one, two or three electrons that are free to move throughout the array. It is this behaviour that gives a metal its characteristic properties. How many atoms are required for the array to exhibit metallic properties?

Researchers in England and the United States have shown that certain magnetic properties characteristic of the bulk metal begin to appear when only ten atoms are present. For example, in a study of osmium metal, they found that when the number of osmium atoms in a cluster was increased from three to ten, the cluster exhibited magnetic properties characteristic of osmium metal.

The researchers plan to study clusters containing up to 40 metal atoms. This could reveal the number of atoms needed for the cluster to provide enough mobile electrons to make it an electrical conductor. With such knowledge it may be possible to design better catalysts, magnetic recording media, and photographic emulsions.

solid nonmetals. Bromine is the only nonmetal that is a liquid at room temperature. Oxygen, nitrogen, chlorine, and fluorine are examples of the gaseous nonmetals. Fluorine is the most reactive of all elements. It is a pale yellow, pungent, corrosive gas which reacts with almost all substances. Metals, glass, ceramics, and even water burn in fluorine with a bright flame. Oxygen is also very reactive and is capable of reacting with most elements. It is a colourless, odourless, tasteless gas which is essential for practically all combustion and respiration. Oxygen forms 21% of air by volume. Nitrogen is a relatively inert gas. It is a colourless, odourless gas and forms 78% of air by volume.

Metalloids

There is no sharp dividing line between metals and nonmetals. Certain elements have some properties of metals and some properties of nonmetals. They are called **metalloids**. Boron, silicon, and arsenic are examples of this type of element.

5.4 Symbols of the Elements

The alchemists of the Middle Ages spent much of their time trying to find methods of transforming iron and lead into gold. They were naturally rather secretive in this hopefully profitable pursuit and developed secret notations to represent the substances with which they experimented. Eventually more or less standard symbols evolved to represent the elements then known. Among these symbols were such diagrams as

☾ or ☽	Silver	♄	Lead	♅ or ○	Gold
♀	Copper	♉	Sulfur	☿	Mercury

The crescent symbolized the silvery colour of the moon; both the horn of plenty and the circle, symbolizing the golden sun, were used to represent gold.

In the early nineteenth century the English schoolteacher John Dalton simplified these symbols by means of drawings related to circular designs. Among his symbols were

⊙	Hydrogen	●	Carbon	◐	Nitrogen
○	Oxygen	⊕	Sulfur	⊙	Mercury

Chemists now use a system of symbols devised by the Swedish chemist Jöns Jakob Berzelius in 1814. The *symbols* usually consist of the first one or two letters in the name of the element. If the second letter is used, it is not capitalized. The symbols are so useful that they are used in all languages, even in Russian and Chinese, which do not use our Roman alphabet. Examples are

H	Hydrogen	B	Boron	C	Carbon
N	Nitrogen	O	Oxygen	F	Fluorine
He	Helium	Be	Beryllium	Ne	Neon

For some elements the symbols consist of the first letter followed by a letter from the middle of the name:

Mg	Magnesium	Mn	Manganese
Cl	Chlorine	Cr	Chromium

The symbols of many elements reveal the time when they were discovered. Many of the elements were known to the Romans or were discovered by the alchemists, at a time when Latin was the language of educated persons. Hence the symbols of many elements are based on their Latin names:

Na	Sodium	(Latin *natrium*)
Cu	Copper	(Latin *cuprum*)
Sn	Tin	(Latin *stannum*)
Au	Gold	(Latin *aurum*)
Pb	Lead	(Latin *plumbum*)
K	Potassium	(Latin *kalium*)
Ag	Silver	(Latin *argentum*)
Sb	Antimony	(Latin *stibium*)
Hg	Mercury	(Latin *hydrargyrum*)
Fe	Iron	(Latin *ferrum*)

One element, tungsten, has a symbol, **W**, derived from its German name, Wolfram.

A list of the elements and their symbols is given on the inside front cover of this book.

Part One Review Questions

1. Why were copper, silver, and gold useful to early society?
2. Why does the discovery of elements appear to go in spurts, with gaps of several years when no elements are discovered, followed by the discovery of several new elements in a short period of time?
3. What are the distinguishing properties of metals, nonmetals, and metalloids? Give three examples of each.
4. The alchemists were the founders of modern chemistry. What motivated them in their work?
5. Why do the symbols of many of the elements not consist of the first one or two letters of their names?

Part Two: Periodic Law and the Periodic Table

5.5 Periodic Law

By the 1800s enough elements had been discovered for some patterns of similarity to be recognized. In 1864 the English chemist John A.R. Newlands arranged the known elements in order of increasing atomic mass. He noticed that similar chemical and physical properties occurred after every eight elements. For example, elements 2, 9, and 16 in his list (lithium, sodium, and potassium) resembled each other chemically. Beryllium, magnesium, and calcium were elements 3, 10, and 17 in his list. They also had similar chemical and physical properties. He found other groupings of closely related elements. Newlands' work was at first ignored or ridiculed. However, he stated more clearly than others had done the periodic reappearance of similar properties among the elements. But there was one flaw in his scheme. A number of elements had not yet been discovered, and Newlands' list left no room for them.

Mendeleev's Periodic Table

This flaw was corrected by two chemists working independently. In about 1869 the Russian chemist, Dmitri Mendeleev (Fig. 5-5), made the exciting discovery that, if he listed the known elements in order of increasing atomic mass, elements with similar *chemical* properties appeared at regular intervals in the list. At the same time Lothar Meyer was working in Germany. He found that when he listed the known elements in order of increasing atomic mass, ele-

Biography

D MITRI MENDELEEV was the fourteenth and youngest child of a Siberian teacher. Dmitri studied at a teacher's training college in St. Petersburg and at the University of St. Petersburg. After he obtained his doctorate in chemistry, he became Professor of General Chemistry in the University of St. Petersburg in 1866.

Mendeleev is best known for his work on the periodic law. He treated the periodic law not only as a system for classifying the elements according to their properties, but as a "law of nature" which could be used to predict new facts. He presented his periodic law in 1869, and two years later gaps in his tables led him to predict the existence of three new elements. He called them eka-boron, eka-aluminum, and eka-silicon, and he predicted their properties. Within 15 years he was proven correct by the discovery of gallium in 1871, scandium in 1879, and germanium in 1886.

Mendeleev was one of the greatest teachers of his day. His lecture room was always crowded with students. His students thought of him as a comrade, and on more than one occasion he supported students in their disputes with the university administration.

In 1890, because of a dispute with the university administration, Mendeleev resigned his professorship at St. Petersburg. A probable cause of this dispute was his outspoken criticism of the classical system of education. However, in 1893 he was appointed Director of the Bureau of Weights and Measures and he retained this position until his death in 1907.

Figure 5-5
Dmitri Mendeleev (1834–1907)—
Proposed the periodic law.

ments with similar *physical* properties appeared at regular intervals in the list. Each had independently discovered the periodic law. Mendeleev, however, is usually given more credit because he presented his ideas first.

Mendeleev's periodic law states that when elements are arranged in order of increasing atomic mass, elements with similar properties occur at regular intervals. Mendeleev prepared a periodic table to illustrate his arguments and made predictions based on it. A version of his table is given in Table 5-2. Mendeleev's table had several gaps. He predicted that new elements would be found to fill these gaps. The discoveries of scandium and gallium filled two of the gaps in the table. These discoveries created interest in Mendeleev's theory, since he had predicted with remarkable accuracy the properties of these elements before they were discovered (Table 5-3). Since Mendeleev's predictions were based upon his periodic table, his work soon gained the recognition that was denied Newlands.

In Mendeleev's table the elements were placed in horizontal rows or **periods**. Elements with similar properties were arranged in vertical columns or **groups**. Mendeleev used Arabic numbers to identify the periods and Roman numerals to identify the groups. The atomic masses of the elements were included with their symbols. Mendeleev's table clearly showed the periodic recurrence of properties noticed by Newlands. Thus, chlorine, bromine, and iodine were all placed together in Group VII. The metals lithium, sodium, and potassium were placed in Group I. The metals beryllium, magnesium, and calcium were in Group II.

Mendeleev's table has stood the test of time with few alterations. The only major modification was required by the discovery of the first noble gas, argon, by Lord Rayleigh and Sir William Ramsay in 1894.

Table 5-2   Mendeleev's Periodic Table

Row	Group I	Group II	Group III	Group IV	Group V	Group VI	Group VII	Group VIII
1	H = 1							
2	Li = 7	Be = 9	B = 11	C = 12	N = 14	O = 16	F = 19	
3	Na = 23	Mg = 24	Al = 27	Si = 28	P = 31	S = 32	Cl = 35.5	
4	K = 39	Ca = 40	Sc = 44	Ti = 48	V = 51	Cr = 52	Mn = 55	Fe = 56 Co = 58.5 Ni = 59
5	Cu = 63	Zn = 65	Ga = 70	Ge = 72	As = 75	Se = 79	Br = 80	
6	Rb = 85	Sr = 87	Y = 89	Zr = 90	Nb = 94	Mo = 96		Ru = 103 Rh = 104 Pd = 106
7	Ag = 108	Cd = 112	In = 113	Sn = 118	Sb = 120	Te = 125	I = 127	
8	Cs = 133	Ba = 137	La = 138	Ce = 140				
9								
10			Yb = 173		Ta = 182	W = 184		Os = 191 Ir = 193 Pt = 196
11	Au = 198	Hg = 200	Tl = 204	Pb = 206	Bi = 208			
12				Th = 232		U = 240		

Table 5-3   **Comparison of Mendeleev's Predictions with Experimental Values for the Properties of Gallium**

Property	Predicted	Observed
atomic mass (u)	69	69.7
density (g/mL)	6.0	5.9
melting point (°C)	low	30
boiling point (°C)	high	2000
formula of oxide	M_2O_3	Ga_2O_3

5.6 The Modern Periodic Table

The discovery of the noble gases meant that a new group had to be added to Mendeleev's table. In 1911, after the discovery of electrons and protons and the development of the nuclear model of the atom, the Dutch physicist A. van den Broek suggested an alteration to Mendeleev's table. He felt that several minor inconsistencies would be removed if the table were arranged according to atomic number instead of atomic mass.

Van den Broek's suggestion was supported by the work of an English physicist, Henry Moseley. In 1913 Moseley studied the wavelengths of the X-rays produced by every element that he could possibly use as a target in his X-ray apparatus (Fig. 5-6). He produced X-rays by focussing high energy cathode rays on the elements and he found that, with one or two exceptions, the wavelengths of the X-rays decreased as the atomic mass of the element increased (Fig. 5-7). The exceptions disappeared when he arranged the elements in order of increasing atomic number. His work confirmed conclusively that the properties of the elements depend on atomic number rather than atomic mass.

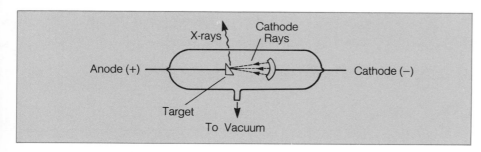

Figure 5-6
X-ray tube. X-rays are produced when high energy cathode rays are focussed on a target.

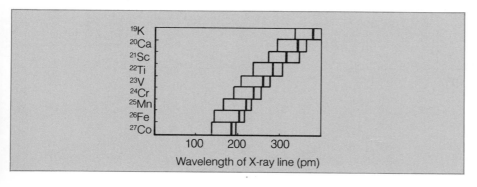

Figure 5-7
X-ray spectra and atomic numbers. Moseley arranged the elements in decreasing order of the most prominent line in their X-ray spectra.

The result of the new arrangement is the modern periodic table which is shown inside the back cover of the book. The revision required a change in the periodic law to its modern form. The modern **periodic law** states that, when the elements are arranged in order of increasing atomic number, elements with similar properties occur at regular intervals. Periodicity is the tendency to recur at regular intervals. Thus, the properties of the elements exhibit periodicity. Another way of stating the periodic law is that the physical and chemical properties of the elements are a periodic function of their atomic numbers.

Features of the Modern Periodic Table

The modern periodic table differs slightly from the arrangements used by Mendeleev and Meyer. In 1920 the Danish physicist Niels Bohr devised a long-form version of the periodic table. There are still horizontal rows or periods, but these periods have different lengths. The first period is short, consisting of only two elements. Periods 2 and 3 contain eight elements each. Periods 4 and 5 are long periods, each containing 18 elements. Period 6 is the longest, with 32 elements, and Period 7 is incomplete. Elements 57 through 70 and 89 through 102 are put into separate groupings below the main table to maintain compactness. Elements having similar properties still appear in vertical columns. As new elements were discovered, all the gaps of Mendeleev's table have become filled. Recently, a committee of the International Union of Pure and Applied Chemists (IUPAC) suggested modifying Bohr's design slightly. In the new format, the groups (vertical columns) are numbered from 1 to 18, thus replacing Roman numerals with Arabic numbers as shown in Table 5-4. These changes are being used more and more frequently in newer chemistry books. Both the new notation and the previous notation are shown in the periodic table inside the back cover of this book; however, we will use only the new notation in this book.

Table 5-4   Arrangement of Elements in the Periodic Table According to Increasing Atomic Number

Group →

Period	1	2	3	4	5	6	7	8	9	10	11	12	13	14	15	16	17	18
1	1																	2
2	3	4											5	6	7	8	9	10
3	11	12				←— Transition Metals —→							13	14	15	16	17	18
4	19	20	21	22	23	24	25	26	27	28	29	30	31	32	33	34	35	36
5	37	38	39	40	41	42	43	44	45	46	47	48	49	50	51	52	53	54
6	55	56	71	72	73	74	75	76	77	78	79	80	81	82	83	84	85	86
7	87	88	103	104	105	106												

6	57	58	59	60	61	62	63	64	65	66	67	68	69	70
7	89	90	91	92	93	94	95	96	97	98	99	100	101	102

Groups 1 and 2, on the left-hand side of the periodic table, and Groups 13 to 17, on the right-hand side, together constitute the **main-group elements**. The middle elements in the periodic table, Groups 3 to 12, are responsible for the extra length of the long periods and are called the **transition metals**. The very long periods, 6 and 7, are compressed in the table by removing 14 of their members and representing them separately below. The elements represented in this portion of period 6 are called the *lanthanides* (rare-earth elements), and the elements represented in this portion of period 7 are called the *actinides*.

The periodic table is useful for organizing chemical knowledge. The elements on the left of the zigzag line in the table are metals. Those on the right side are nonmetals. The most metallic elements are located in the lower left corner of the periodic table. The metallic character of the elements decreases from left to right in a row and increases from top to bottom in a group. The most nonmetallic (least metallic) elements are located in the upper right corner. There is a gradual transition from metallic to nonmetallic properties as one goes from left to right within a period. Since there is no sharp dividing line between metals and nonmetals, those elements near the zigzag line will exhibit properties of both metals and nonmetals. These elements are called metalloids. Boron, aluminum, silicon, germanium, and arsenic fall into this category.

The Noble Gases

When the elements are arranged according to this scheme, elements with atomic numbers 2, 10, 18, 36, 54, and 86 are found in Group 18. These six elements—helium, neon, argon, krypton, xenon, and radon—are the least reactive of all the elements. They are all gases at room temperature and pressure. In chemistry, nobility is equated with chemical inactivity or reluctance to react with other elements. Thus, the elements of Group 18 are called the **noble gas** group. The members of a group in the periodic table are also known as a family. Therefore the elements of Group 18 are also known as the noble gas family. Until 1962, these elements were called the *inert gases*. However, since 1962, compounds involving xenon with fluorine and oxygen and compounds involving krypton with fluorine have been prepared. Therefore, it is no longer correct to consider all the Group 18 elements as inert gases.

The Alkali Metals

The elements with atomic numbers one greater than the noble gases—3, 11, 19, 37, 55, and 87—are lithium, sodium, potassium, rubidium, cesium, and francium. They all exhibit metallic properties and are highly reactive. For example, they all react vigorously with water to release hydrogen. In addition, they all react with chlorine to form colourless compounds which crystallize in cubic shapes and have similar formulas—$LiCl$, $NaCl$, KCl, $RbCl$, $CsCl$, $FrCl$. These chemically similar elements are called the **alkali metals** and are found in Group 1 of the periodic table. The word alkali comes from an Arabic term: the ashes of saltwort. Saltwort was a general term referring to any one of a number of plants from whose ashes the carbonates of sodium and potassium were obtained.

The Halogens

The elements with atomic numbers 9, 17, 35, 53, and 85 are found in Group 17. They are fluorine, chlorine, bromine, iodine, and astatine. These elements all exhibit nonmetallic properties. They react with hydrogen to form compounds which dissolve in water to form acidic solutions. They react with the alkali metals to form compounds with similar formulas and crystal shapes. These chemically similar elements are called the **halogen** family. The word halogen comes from a Greek term: salt-former. The halogens combine readily with metals to form compounds called salts.

Hydrogen

Hydrogen, with atomic number 1, is unique. It is sometimes placed in Group 1 with the alkali metals. Although it is not a metal like the elements of Group 1, hydrogen forms compounds with the halogens in a similar manner to the metals of Group 1. Hydrogen is sometimes placed in Group 17 with the halogens. Hydrogen reacts with metals such as sodium and with nonmetals such as nitrogen in the same way as the elements of Group 17 do. However, hydrogen undergoes other reactions which are not typical of the elements of Group 17. Some versions of the periodic table even place hydrogen in both Groups 1 and 17. Hydrogen does not really fit nicely anywhere in the table.

APPLICATION

Element 109 and Beyond

There are at present 108 known elements. Ninety elements are found in nature and the rest are made in laboratories. The elements above uranium (atomic number 92) are the transuranium elements. These elements are radioactive and are all synthetic. The most recently prepared elements are produced in minute quantities, and because they are highly radioactive, they immediately begin to change into other elements. The term *half-life* is used to describe the length of time it takes for half a given amount of a radioactive element to turn into other elements. The half-lives of some of these newly prepared elements are less than one second. If these new elements are unstable and produced only in minute quantities, why should scientists still be interested in preparing them? The answer is that scientists are curious to learn more about the structure of matter and the forces holding matter together.

The latest element to be prepared, element 109, was synthesized in 1982 by a West German research team. This achievement is even more remarkable since only *one atom* of element 109 was detected. It took a month of analysis before the scientists were certain enough to announce their work. The atom of element 109 was produced by firing a beam of iron-58 nuclei (which contain 26 protons and 32 neutrons) at foil made of bismuth-209 (83 protons and 126 neutrons). The two nuclei fused to produce element 109. This one atom of element 109 lasted for only five thousandths of a second before decaying into another element.

Computer calculations have predicted that certain as yet unmade elements will be stable, with half-lives of up to 1000 years. These elements have atomic numbers around 110, 116, 126, and beyond, to 184. An American research team is attempting to make element 116. They will fire a beam of a rare isotope of calcium (containing 20 protons)—which costs about $300 000 per gram—at a target of curium (containing 96 protons). Only a few grams of this rare isotope of curium exist in the world.

What will be the chemical and physical properties of newly discovered elements such as element 113 and element 114? The periodic table is used to make predictions for these elements. Element 113 should chemically resemble thallium and have an atomic mass of 297 u, a density of 16 g/mL, a melting point of about 430°C, and a boiling point of about 1100°C. Element 114 should chemically resemble lead and have an atomic mass of 298 u, a density of 14 g/mL, a melting point of about 70°C, and boiling point of about 150°C. The properties of these elements are predicted in the same way that Mendeleev predicted the properties of missing elements in his periodic table. We can only hope that, as scientists increase the number of elements in the modern periodic table, their predictions will be as accurate as Mendeleev's were over one hundred years ago.

5.7 The Periodic Law and Energy Levels

We have learned that when the elements are arranged in order of increasing atomic number, elements with similar properties occur at regular intervals. Why does this happen? Is there an explanation for the periodic law? In this section, we shall propose an explanation based on the atomic theory that was introduced in Chapter 4. We will be following the much-used scientific method of *first* observing a regularity (in this case, the periodic law) and *then* proposing an explanation for that regularity.

In Chapter 4, the energy level populations were given for the first 20 elements (Table 4-3). Three of the first 20 elements are members of the alkali metal group. They are lithium, sodium, and potassium. Notice that each of them has one electron in its outer energy level.

Three of the first 20 elements are members of Group 2 (the alkaline earth metals). They are beryllium, magnesium, and calcium. Each of these three elements has two electrons in its outer energy level.

Two of the first 20 elements are halogens. They are fluorine and chlorine. Both of these chemically similar elements have seven electrons in their outer energy level.

It appears that the existence of groups of chemically related elements in the periodic table can be explained by assuming that the properties of atoms depend on the number of electrons present in the outer energy level. Atoms which have the same number of electrons in their outer energy levels are usually members of the same group. Atoms which have different numbers of electrons in their outer energy levels are usually members of different groups.

Three of the first 20 elements (helium, neon, and argon) are noble gases. They are chemically inactive. Two of these (neon and argon) have eight electrons in their outer energy level, but helium has only two electrons in its outer energy level. Except for helium, every member of the noble gas group has eight electrons in its outer energy level. Helium, however, does have a filled first energy level, which accounts for its chemical inactivity. The chemical inactivity of the other noble gases appears to be related to having eight electrons in the outer energy level.

Energy Sublevels

Why do the last electron of potassium and the last two electrons of calcium go into the fourth energy level even though the third energy level is not filled? Why does argon, and each succeeding noble gas, exhibit the same chemical properties as neon, even though each one has only eight electrons in an outer energy level that is capable of holding many more electrons? In this section, we will attempt to answer these questions.

Studies of the spectra of different elements indicate that each shell or energy level actually consists of one or more energy sublevels grouped closely together. The number of energy sublevels equals the number of the energy level. The two electrons of the first energy level have only one possible energy value. However, the eight electrons of the second level can have two possible energy values. That is, the second energy level consists of two energy sublevels. Two electrons can occupy the first sublevel, and the other six electrons occupy the second sublevel. The third energy level consists of three energy sublevels, the

Figure 5-8
Energy levels of a hydrogen atom compared with the energy sublevels of a many-electron atom

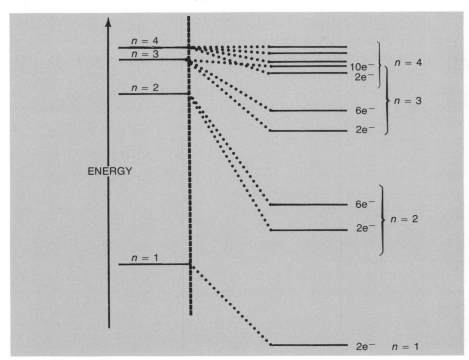

fourth energy level consists of four energy sublevels, and so on.

An energy level diagram is shown in Figure 5-8. To the left of the dashed line are the first four energy levels of a one-electron atom. To the right of the dashed line are the corresponding energy sublevels of a many-electron atom. For any energy level, the first sublevel can hold up to two electrons. For the second and all subsequent energy levels, the second sublevel can hold up to six electrons. For the third and all subsequent energy levels, the third sublevel can hold up to ten electrons. For the fourth and all subsequent energy levels, the fourth sublevel can hold up to 14 electrons.

Thus, the first energy level has a maximum electron population of two, with both electrons found in the first sublevel. The second energy level has a maximum electron population of eight, with two electrons in the first sublevel and six electrons in the second sublevel. The third energy level has a maximum population of 18, with two electrons in the first sublevel, six electrons in the

Table 5-5   Energy Sublevel Populations for the First 20 Elements

Element	Symbol	Number of Protons in Nucleus	1st Level	Energy Sublevel Populations								
				2nd Level		3rd Level			4th Level			
				First Sub-level	Second Sub-level	First Sub-level	Second Sub-level	Third Sub-level	First Sub-level	Second Sub-level	Third Sub-level	Fourth Sub-level
Hydrogen	H	1	1									
Helium	He	2	2									
Lithium	Li	3	2	1								
Beryllium	Be	4	2	2								
Boron	B	5	2	2	1							
Carbon	C	6	2	2	2							
Nitrogen	N	7	2	2	3							
Oxygen	O	8	2	2	4							
Fluorine	F	9	2	2	5							
Neon	Ne	10	2	2	6							
Sodium	Na	11	2	2	6	1						
Magnesium	Mg	12	2	2	6	2						
Aluminum	Al	13	2	2	6	2	1					
Silicon	Si	14	2	2	6	2	2					
Phosphorus	P	15	2	2	6	2	3					
Sulfur	S	16	2	2	6	2	4					
Chlorine	Cl	17	2	2	6	2	5					
Argon	Ar	18	2	2	6	2	6					
Potassium	K	19	2	2	6	2	6		1			
Calcium	Ca	20	2	2	6	2	6		2			

second sublevel, and ten electrons in the third sublevel. The fourth energy level has a maximum population of 32, with successive sublevel populations of 2, 6, 10, and 14 electrons.

In Figure 4-15, we saw that the allowed energy levels for hydrogen (a one-electron atom) are closer together as the numerical value of the energy level increases. For example, the third and fourth energy levels are much closer together than are the first and second energy levels. In fact, when the energy levels are divided into energy sublevels, we observe that the sublevels of the third energy level overlap the sublevels of the fourth energy level. The result is that the highest sublevel of the third energy level is higher than the lowest sublevel of the fourth energy level. This overlapping would be even more complicated if the sublevels of the fifth and sixth energy levels were added to Figure 5-7. The overlapping of the third and fourth energy levels provides an answer for the first of the two questions posed in this section. The last electron of potassium and the last two electrons of calcium go into the fourth energy level before the third energy level is filled, because the lowest (first) sublevel of the fourth energy level is below the highest (third) sublevel of the third energy level.

The existence of sublevels also explains why argon behaves as an unreactive gas with only eight (instead of 18) electrons in its outer energy level. Argon has completely filled first and second energy levels. The first two sublevels of its third energy level are also filled (Table 5-5). Because the next available sublevel is so much higher in energy than the second sublevel of the third energy level, argon behaves as an inert element just as if it had a filled outer energy level. There seems to be a special stability associated with having eight electrons in an outer energy level. Every atom that has eight electrons in its outer energy level is chemically stable and is a member of the noble gas family.

EXAMPLE PROBLEM 5-1

Which one of the first 20 elements should have similar chemical properties to, and be a member of the same family as, phosphorus? Why?

SOLUTION

Phosphorus has five electrons in its outer energy level. The only other element of the first 20 elements to have five electrons in its outer energy level is nitrogen. Thus, nitrogen and phosphorus should be members of a chemically related family.

PRACTICE PROBLEM 5-1

Which one of the first 20 elements should have similar chemical properties to, and be a member of the same family as, oxygen? Why? (*Answer:* sulfur)

PRACTICE PROBLEM 5-2

Which one of the first 20 elements should have similar chemical properties to, and be a member of the same family as, silicon? Why?

5.8 Electron Dot Symbols

In the previous section we discussed how electrons are found in various energy levels. The electrons in the outer energy levels are the ones which usually participate in chemical reactions. The inner electrons are not usually involved. These outer electrons are called **valence electrons**. They allow us to explain the combining powers of atoms. In the case of chlorine, the first and second energy levels contain ten electrons, and those electrons are considered to be inner electrons. The seven electrons in the third energy level are the valence electrons. In the case of potassium, the first, second, and third energy levels contain 18 inner electrons. The one electron in the fourth energy level is the valence electron. These two examples illustrate the point that the *valence electrons of an atom* are those in the *highest energy level*.

In 1916 the American chemist, Gilbert N. Lewis, recognized that the valence electrons of an atom are important in determining its chemical properties. He introduced convenient symbols to represent an atom and its valence electrons. These symbols are now called **electron dot symbols** or Lewis symbols. In an ordinary chemical symbol such as N, Al, or Cl, the letters in the symbol represent the whole atom: nucleus, inner electrons, and valence electrons. In an electron dot symbol, however, the letters represent only the **kernel** of the atom, that is, the *nucleus* and the *inner electrons*. The *outer* or *valence electrons* are written as dots surrounding the kernel. Thus, the electron dot symbol for chlorine is ·C̈l:. The Cl represents the nucleus and the ten inner electrons in the first two energy levels. The seven dots represent the seven valence electrons in the third energy level. The positions of the dots have *no relationship* to the actual positions of the electrons. The main function of the dots is to remind us of the number of valence electrons belonging to the atom. It is common to leave the dots unpaired until there are more than four valence electrons. When there are more than four valence electrons, the dots are paired so as to give the smallest number of pairs (Table 5-6). Note that the following electron dot symbols for oxygen are equally acceptable:

Table 5-6   Electron Dot Symbols of the First 20 Elements

Group							
1	2	13	14	15	16	17	18
H·							·He·
Li·	·Be·	·B·	·C̈·	·N̈:	·Ö:	·F̈:	:Ne:
Na·	·Mg·	·Al·	·Si·	·P̈:	·S̈:	·C̈l:	:Är:
K·	·Ca·						

You will notice that in Groups 1 and 2 the number of electron dots is the same as the number of the group to which the element belongs. For Groups 13 to 18 the number of electron dots is ten less than the number of the group to which the element belongs. The elements of Group 18 therefore have eight electron dots (helium, of course, has only two). The grouping of eight valence electrons is called an **octet**. The chemical inactivity of the noble gases has been attributed to their possessing eight valence electrons.

EXAMPLE PROBLEM 5-2

Write the electron dot symbols for rubidium and strontium.

SOLUTION
Rubidium, in Group 1, has one valence electron. Therefore, Rb· is its electron dot symbol. Strontium, in Group 2, has two valence electrons. Therefore, ·Sr· is its electron dot symbol.

PRACTICE PROBLEM 5-3

Write the electron dot symbols for indium and tin.

PRACTICE PROBLEM 5-4

Write the electron dot symbols for bismuth and polonium.

5.9 The Periodic Table and Electron Configurations

The location of an element in the periodic table is determined by the chemical properties of the element. However, the chemical properties of an element depend on the number of electrons present in the outer energy levels of its atoms. Thus, the location of an element in the periodic table is related to the number of electrons in the outer energy levels of its atoms. We have used these concepts, introduced in Section 5.7, to provide the basis for electron dot symbols. In this section, we go one step further by relating the locations of elements in the periodic table to their electron configurations.

Figure 5-9 shows the order in which 38 electrons are placed in the 19 orbitals of lowest energy. Each orbital is represented by a circle, and each electron is represented by a number in a half-circle. This figure can be used to give the electron configuration of any element below number 39. For example, the energy level diagram for element 16, sulfur, would contain only the first 16 electrons. Thus, the electron configuration of sulfur is $_{16}S\ 1s^2\ 2s^2\ 2p^6\ 3s^2\ 3p^4$. The energy level diagram for element 28, nickel, would contain only the first 28 electrons. The electron configuration of nickel is $_{28}Ni\ 1s^2\ 2s^2\ 2p^6\ 3s^2\ 3p^6\ 4s^2\ 3d^8$.

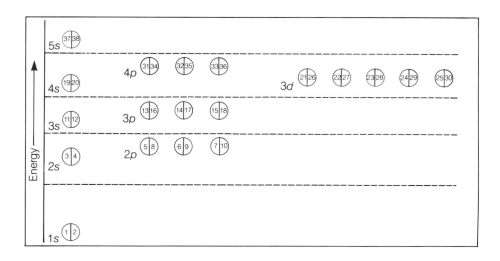

Figure 5-9
Order of placing electrons in the first 19 orbitals

In Figure 5-9, the energy sublevels are grouped into clusters of roughly similar energy by horizontal dotted lines. The $1s$ sublevel stands alone. The $2s$ and $2p$ sublevels are grouped together, as are the $3s$ and $3p$ sublevels. The next grouping, however, consists of $4s$, $3d$, and $4p$ sublevels, and they are filled in the relative order s, then d, then p. Although it is not shown in Figure 5-8, the energy sublevels in the next cluster are also filled in the order s, then d, then p.

We can obtain a table similar to the periodic table by listing the elements according to their electron configurations:

s^1	s^2	d^1	d^2	d^3	d^4	d^5	d^6	d^7	d^8	d^9	d^{10}	p^1	p^2	p^3	p^4	p^5	p^6
1	2																
3	4											5	6	7	8	9	10
11	12											13	14	15	16	17	18
19	20	21	22	23	24	25	26	27	28	29	30	31	32	33	34	35	36
37	38																

The atoms with electron configurations ending in s^1 and s^2 have been placed in two columns at the left side of the listing. Atomic numbers are used in place of the symbols of the atoms. Atoms with electron configurations ending in p^1 to p^6 have been placed in six columns at the right side. Atoms with electron configurations ending in d^1 to d^{10} are placed in ten columns in the middle. These d columns are placed in the middle because the orbitals of the fourth grouping are filled in the order s, then d, then p. The elements are listed in such a way that every atom with an electron configuration ending in s^1 will appear in the first column. Every atom with an electron configuration ending in s^2 will appear in the second column and so on.

The similarity of this listing to the periodic table is not accidental. The periodic table was first constructed to group elements according to their chemical properties. However, it can also be constructed by grouping elements according to their electron configurations. Thus, it is reasonable to conclude that chemical properties are related to the electron configurations of elements. Let us pursue this point.

Consider elements 10, 18, and 36. Their electron configurations are

$_{10}$Ne $1s^2\, 2s^2\, 2p^6$
$_{18}$Ar $1s^2\, 2s^2\, 2p^6\, 3s^2\, 3p^6$
$_{36}$Kr $1s^2\, 2s^2\, 2p^6\, 3s^2\, 3p^6\, 3d^{10}\, 4s^2\, 4p^6$

All of these elements are noble gases. They all have filled outer s and p orbitals. Whenever the outer s and p orbitals are filled with a total of *eight* electrons, the atom is chemically unreactive. (If the d sublevel is filled, as in krypton, the outer s and p electrons can be written at the end of the electron configuration so that the actual number of valence electrons is more obvious.) Helium is also chemically unreactive because two electrons are sufficient to fill the first energy level. Thus, it is included in the noble gas family (Group 18), not in the second column (Group 2).

Consider elements 3, 11, 19, and 37:

$_{3}$Li $1s^2\, 2s^1$
$_{11}$Na $1s^2\, 2s^2\, 2p^6\, 3s^1$
$_{19}$K $1s^2\, 2s^2\, 2p^6\, 3s^2\, 3p^6\, 4s^1$
$_{37}$Rb $1s^2\, 2s^2\, 2p^6\, 3s^2\, 3p^6\, 3d^{10}\, 4s^2\, 4p^6\, 5s^1$

These elements all have electron configurations ending in s^1. That is, their outer (or highest energy) orbitals are half-filled s orbitals. These elements all have similar properties.

Finally, consider elements 9, 17, and 35:

$_{9}$F $1s^2\, 2s^2\, 2p^5$
$_{17}$Cl $1s^2\, 2s^2\, 2p^6\, 3s^2\, 3p^5$
$_{35}$Br $1s^2\, 2s^2\, 2p^6\, 3s^2\, 3p^6\, 3d^{10}\, 4s^2\, 4p^5$

These elements are the halogens, and experiments have shown that they all have similar chemical properties. Their electronic configurations all end in $s^2\, p^5$. That is, they have a total of *seven* electrons in their outer s and p orbitals.

The electron configuration endings for Groups 1 to 18 are summarized in Table 5-7. Note that the electron configurations of the elements in Groups 5 and 6 and Groups 8 to 11 are difficult to predict. The chemical properties of any element depend on the number of electrons in the outermost s and p orbitals. The outer s and p orbitals are called the **valence orbitals**, and they contain the valence electrons. However, partially filled d orbitals are not valence orbitals. The series of elements in which the d orbitals are being filled (Groups 3 - 12) are all metals and have similar chemical properties. Partially filled d orbitals do not contribute as much to the chemical properties of an element as partially filled s and p orbitals.

EXAMPLE PROBLEM 5-3

Consider elements 4, 12, 20, and 38. Show why they should all belong in the same group.

SOLUTION

Their electron configurations are

Be	$1s^2\, 2s^2$
Mg	$1s^2\, 2s^2\, 2p^6\, 3s^2$
Ca	$1s^2\, 2s^2\, 2p^6\, 3s^2\, 3p^6\, 4s^2$
Sr	$1s^2\, 2s^2\, 2p^6\, 3s^2\, 3p^6\, 3d^{10}\, 4s^2\, 4p^6\, 5s^2$

The electron configurations of these atoms all end in s^2, that is, these atoms all have two electrons in their outermost shells.

PRACTICE PROBLEM 5-5

Consider elements 5, 13, and 31. Show why they should belong in the same group.

PRACTICE PROBLEM 5-6

Consider the elements 6, 14, and 32. Show why they belong in the same group.

Table 5-7   Electron Configuration Endings for Elements in a Given Group

Group	Configuration Ending
1	s^1
2	s^2
3	$s^2 d^1$
4	$s^2 d^2$
5	$s^2 d^3$ (usually)
6	$s^1 d^5$ (usually)
7	$s^2 d^5$
8	$s^2 d^6$ (usually)
9	$s^2 d^7$ (usually)
10	varies
11	$s^1 d^{10}$
12	$s^2 d^{10}$
13	$s^2 p^1$
14	$s^2 p^2$
15	$s^2 p^3$
16	$s^2 p^4$
17	$s^2 p^5$
18	$s^2 p^6$

Part Two Review Questions

1. Newlands' Law of Octaves was an early form of the periodic table. What was the main flaw in his scheme?
2. Why did the discoveries of the elements scandium and gallium create interest in Mendeleev's periodic table?
3. State the modern periodic law.
4. What are: **a)** transition metals; **b)** the lanthanides; **c)** the actinides? Give two examples of each.
5. Why are the elements of Group 18 known as the noble gases?
6. Why are the elements of Group 1 known as the alkali metals?
7. Why are the elements of Group 17 known as the halogens?
8. How does the kernel of an atom differ from the nucleus of an atom?
9. What are valence electrons?
10. How is the electron configuration of an atom related to the location of the element in the periodic table?

Part Three: Trends in the Periodic Table

5.10 Atomic Size

If we think of an atom as being a sphere, the size of the atom is a function of its radius. As you move from the left to the right of a period, the number of electrons in the atoms is increasing, and you might expect the radii of atoms to increase. However, this progression does not happen. The radii of atoms decrease as you move from the left hand side of a period to the right hand side. To understand the reasons for this decrease, we must remember that, as you go from the left of a period to the right, each additional electron is going into a sublevel of the same energy level. Electrons having approximately the same energy would be expected to be approximately the same distance from the nucleus. However, as you go from the left to the right side of a period, protons are also being added to the nucleus, increasing the positive nuclear charge. As this happens, electrons are being pulled closer to the nucleus, and the size of the atom decreases.

EXAMPLE PROBLEM 5-4

Of silicon (Si), magnesium (Mg), or sulfur (S), which has the largest atomic radius? Why?

SOLUTION

The atomic radius decreases as one moves from left to right within a period since, as the nuclear charge increases, all the electrons are being pulled closer to the nucleus, and the size of the atom decreases. Thus, the atom with the largest atomic radius should be the farthest left in the periodic table. In this case the element is Mg.

PRACTICE PROBLEM 5-7

Of the elements aluminum (Al), sodium (Na), or chlorine (Cl), which one has the largest atomic radius?

PRACTICE PROBLEM 5-8

Of the elements lead (Pb), barium (Ba) or astatine (At), which one has the smallest atomic radius?

If you go from the top of a group of the periodic table to the bottom of the same group, the atomic radii increase. This happens because, as you move from the top of a group to the bottom, the extra electrons are going into higher energy levels that place the electrons farther from the nucleus. Even though the nuclear charge increases from the top to the bottom of a group, it is less effective at attracting electrons which are much farther away from the nucleus. The atomic radii of the elements are summarized in Figure 5-10.

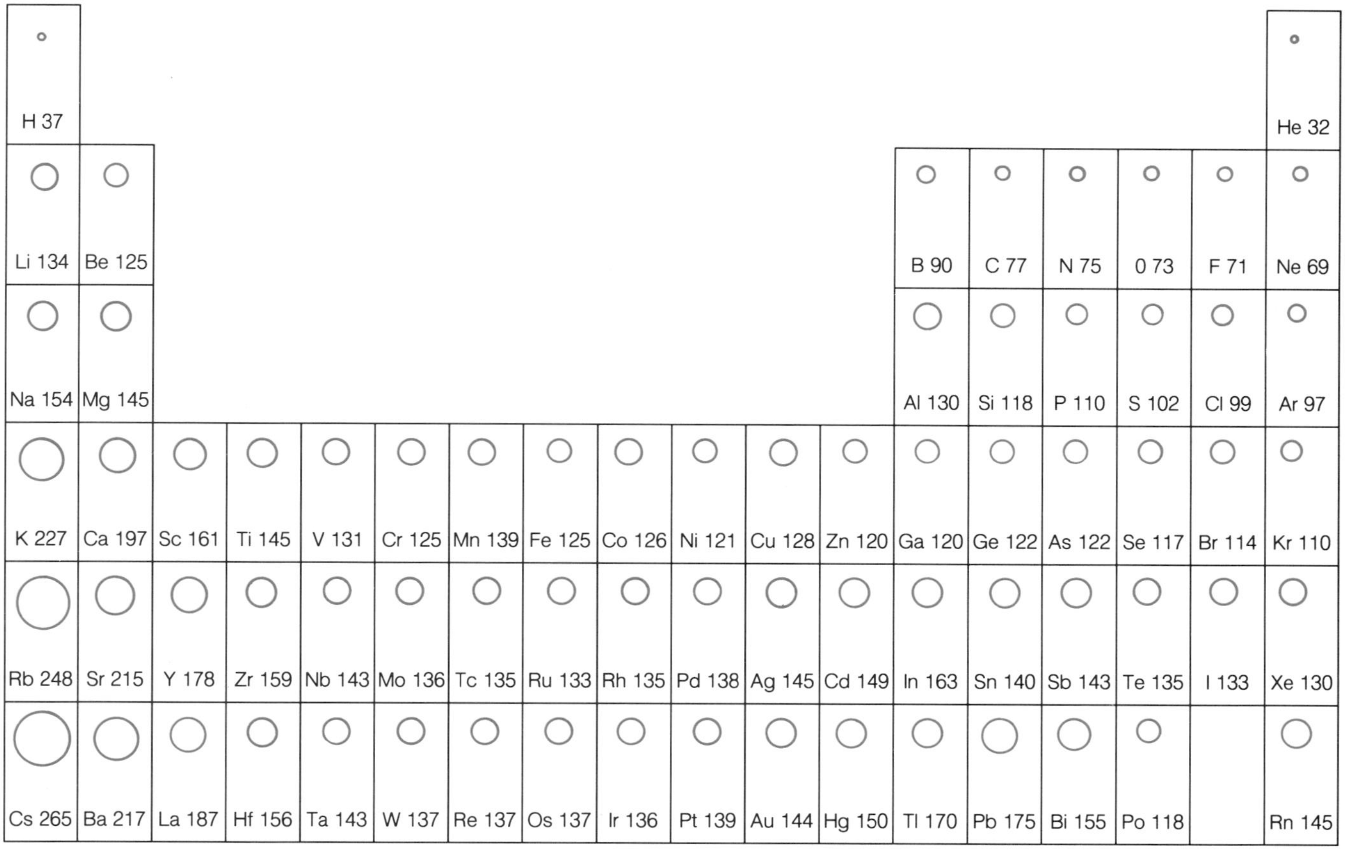

Figure 5-10
Atomic radii of the elements (in picometres)

H 37																	He 32
Li 134	Be 125											B 90	C 77	N 75	O 73	F 71	Ne 69
Na 154	Mg 145											Al 130	Si 118	P 110	S 102	Cl 99	Ar 97
K 227	Ca 197	Sc 161	Ti 145	V 131	Cr 125	Mn 139	Fe 125	Co 126	Ni 121	Cu 128	Zn 120	Ga 120	Ge 122	As 122	Se 117	Br 114	Kr 110
Rb 248	Sr 215	Y 178	Zr 159	Nb 143	Mo 136	Tc 135	Ru 133	Rh 135	Pd 138	Ag 145	Cd 149	In 163	Sn 140	Sb 143	Te 135	I 133	Xe 130
Cs 265	Ba 217	La 187	Hf 156	Ta 143	W 137	Re 137	Os 137	Ir 136	Pt 139	Au 144	Hg 150	Tl 170	Pb 175	Bi 155	Po 118		Rn 145

EXAMPLE PROBLEM 5-5

Which atom in the pair, silicon (Si) and lead (Pb), has the larger atomic radius? Why?

SOLUTION

The atomic radius increases as one goes from the top of a group in the periodic table to the bottom of the same group, because the extra electrons are going into energy levels that place the electrons farther from the nucleus. Both of these elements are in Group 14, and Pb is at the bottom of the group. Therefore Pb is the larger of the two atoms.

PRACTICE PROBLEM 5-9

Which atom in the pair phosphorus (P) and antimony (Sb) has the larger atomic radius?

5.11 Ionization Energies

A skillful player who aims a marble at a group of marbles inside a circle can cause one or more marbles to be knocked outside the circle. In the same way, an electron can be knocked out of an atom such as lithium by hitting it with a high-speed electron aimed at it from outside. The resulting positive particle is an ion (a lithium ion in this case). The process of changing an atom into an ion is called **ionization**. The symbol e^- represents a free electron. The equation for the process as applied to a lithium atom is:

$$\text{Li} \cdot \xrightarrow[\substack{\text{bombard with a high-}\\ \text{speed electron}}]{} \text{Li}^+ + e^-$$

The energy required to remove an electron completely from an atom is called **ionization energy** or the *first ionization energy* of the atom. The additional energy required to remove a second electron from a positive ion is called the *second ionization energy*, and so on. Since atoms are so small, these energies are also small. Ionization energies are therefore expressed in terms of the energy required to remove electrons from a large quantity of atoms, called a mole of atoms (symbol *mol*). The first ionization energy of lithium is 520 kJ/mol. This means that it requires 520 kJ of energy to remove the outermost electron from a mole of lithium atoms. The concept of the mole is discussed fully in Chapter 11.

The first ionization energies of the first 54 elements are arranged in the form of a periodic table in Figure 5-11. The ionization energy increases as one goes from the left side to the right side of a period. Since the positive nuclear charge increases as one moves from the left to the right of a period, it is more difficult to pull electrons from atoms on the right side of the periodic table. As a general rule, the atoms have a large ionization energy if they are on the right-hand side of the periodic table. They have a small ionization energy if they are on the left side of the periodic table.

Figure 5-11
First ionization energies (kJ/mol) of the first 54 elements

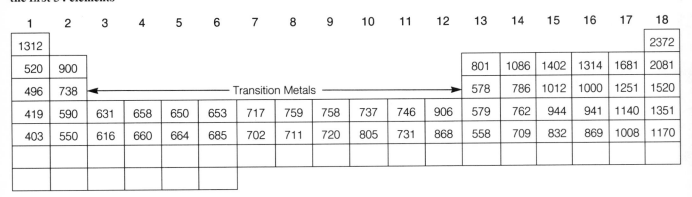

1	2	3	4	5	6	7	8	9	10	11	12	13	14	15	16	17	18
1312																	2372
520	900											801	1086	1402	1314	1681	2081
496	738				Transition Metals							578	786	1012	1000	1251	1520
419	590	631	658	650	653	717	759	758	737	746	906	579	762	944	941	1140	1351
403	550	616	660	664	685	702	711	720	805	731	868	558	709	832	869	1008	1170

The most easily removed electron will be the one furthest from the nucleus. It will also be the one in the highest energy level. Since energy levels depend on the electron populations which are summarized in the periodic table, we should expect to find that ionization energies would vary in a periodic manner. This variation is quite apparent in Figure 5-11. The exceptional stability of the noble gases is readily apparent. Each element in Group 18 has a far higher first ionization energy than any other element in the same row. These gases have filled outer s and p orbitals; that is, they have achieved a completed octet of electrons. An atom with a completed octet is in a relatively stable state, and it is quite difficult to remove an electron from such an octet. Thus, the noble gases have very high ionization energies.

On the other hand, the alkali metals all have very low ionization energies. They all readily lose their single valence electron in order to form a stable positive ion with a completed and exposed electron octet. Thus, Figure 5-10 shows that each alkali metal has a lower first ionization energy than any other element in the same period.

Variations of the ionization energies of the atoms in a group follow a pattern. The ionization energy decreases as you go from the top of a group to the bottom, because the outermost electrons of the small atoms of a group are closer to the nucleus and are more tightly held than the outer electrons of a larger atom in the group. Thus, the first ionization energies of the alkali metals decrease from 520 kJ/mol for lithium (at the top of the periodic table) to 376 kJ/mol for cesium (at the bottom).

EXAMPLE PROBLEM 5-6

Of the elements calcium (Ca), beryllium (Be), or magnesium (Mg), which has the highest ionization energy? Why?

SOLUTION
Since the outermost electrons of the small atoms of a group are closer to the nucleus and are more tightly held than the outer electrons of larger atoms in the group, the ionization energy decreases as you go from the top to the bottom of a group. The atom with the highest ionization energy should be at the top of the group. In this case that element is Be.

PRACTICE PROBLEM 5-10

Of the elements boron (B), aluminum (Al), or gallium (Ga), which has the highest ionization energy?

PRACTICE PROBLEM 5-11

Of the elements argon (Ar), krypton (Kr), or xenon (Xe), which has the highest ionization energy?

CAREER

The Chemistry Teacher

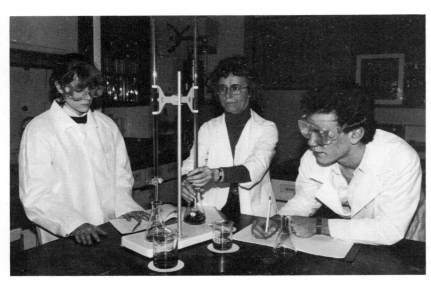

Figure 5-12
A high school science teacher sets up an experiment with her students.

Nearly 20% of all chemists teach chemistry in some type of educational institution, such as a high school or university. People who enjoy learning new topics and sharing what they have learned should consider teaching as a career.

High school chemistry teachers must know something about adolescent psychology, learning theory, student evaluation, and other related education subjects. Thus, they need to take university courses in education as well as in chemistry. Chemistry teachers should also know something of the other sciences, particularly physics and biology. Most have a sufficiently broad mathematical background that they could also teach mathematics if asked to do so. Many chemistry teachers exceed the minimum provincial educational requirements, since continuing an education in chemistry beyond the bachelor's degree is strongly advised for certification requirements and for job security.

A typical high school teacher teaches five or more classes. The teacher also has to plan experiments for the students, make sure that all the equipment and chemicals are set up and ready for use, make up solutions, and generally ensure that the laboratory is efficiently run. There are numerous nonteaching responsibilities. Many teachers serve on committees to discuss changes in the curriculum, recommend purchases,

update safety procedures, and so on. High school teaching demands the ability to work with other people and to communicate knowledge and findings to them. It requires objectivity, accuracy, patience, and the ability to work with factual material.

The short-term outlook for teaching positions at universities is not promising due to the predicted decline in the number of high school graduates. The number of university positions available is less than the number of people with Ph.D.s seeking such positions. Because of the intense competition for appointments, a year or two of postdoctoral research is almost a minimum requirement for applicants. An applicant's chances of appointment are enhanced by having a good academic transcript and degrees from top-rated institutions. The situation will probably change soon, however. Many university professors were hired during the late sixties and early seventies, when Canadian universities were expanding at an enormous rate. About half of the professors now teaching will be retiring in the next ten to fifteen years, and the vacancies created will cause a great increase in the demand for qualified university instructors.

University chemists are normally offered probationary appointments of about three years, which may be renewed once. To receive such appointments, they must be able to develop novel research ideas and make effective presentations of their research. Research is important at universities, and developing a successful research program is the greatest single expectation of a university chemistry professor. Writing grant proposals can be time-consuming, and new faculty members must compete with established researchers for limited research grants from government, industry or private foundations. Another essential part of a chemist's work at a university lies in writing papers for research journals, so that discoveries are shared with others in the same or related fields. Therefore, writing skills are extremely important. University chemists also prepare and give lectures, seminars, tests and examinations, supervise laboratories, and counsel students. In addition, they participate in university government through membership on various committees, faculties, senates, and other governing bodies.

5.12 Electron Affinity

In Section 5.11 we saw that atoms can lose electrons. Now we shall see that some atoms have a tendency to pick up additional electrons. This process results in a lowering of energy of the system; that is, it achieves a more stable state for the added electrons. Energy will be given off when an added electron is moved into the positive field of the nucleus. **Electron affinity** is the energy given off when an electron is added to an atom:

$$:\ddot{X}\cdot\ + e^- \longrightarrow :\ddot{X}:^- + \text{energy}$$

As with ionization energies, electron affinities are usually expressed in units of kilojoules per mole.

In this representation, a neutral atom acquires an electron to become an ion with a negative charge. This is just the reverse of the process in which an electron is removed from a negative ion to form a neutral atom:

$$:\ddot{X}:^- + \text{energy} \longrightarrow :\ddot{X}\cdot\ + e^-$$

Since it is more difficult to remove an electron from a smaller ion (more energy is required because the electrons are closer to the nucleus), the electron affinity increases as the size of the atom decreases. That is, more energy is given off as a smaller atom accepts an additional electron to become an ion.

Because the size of atoms decreases from left to right across a period from Group 1 to 17, the Group 17 elements would be expected to have the highest electron affinities. Although there are many exceptions, the elements at the top of a group generally have a higher electron affinity than do the elements at the bottom. This is also to be expected since the smaller atoms are at the top of a group. The halogens all react to form the halide ions F^-, Cl^-, Br^-, and I^-. The smallest halogen, fluorine, is the most reactive element known.

EXAMPLE PROBLEM 5-7

Which element should have the higher electron affinity, selenium or tellurium? Why?

SOLUTION
The elements at the top of a group generally have a higher electron affinity than do elements at the bottom. Since selenium is closer to the top of Group 16, it has the higher electron affinity.

PRACTICE PROBLEM 5-12

Which element should have the higher electron affinity, arsenic or antimony? Why?

Part Three Review Questions

1. How do atomic radii vary as you go from left to right and from top to bottom among the elements in a periodic table? Why?
2. What is meant by ionization energy?
3. How does ionization energy vary as you go from left to right and from top to bottom among the elements in the periodic table? Why?
4. What is meant by electron affinity?
5. How does electron affinity vary as you go from left to right and from top to bottom among the elements in the periodic table? Why?

Chapter Summary

- The discovery of new elements is often made possible by advances in experimental techniques.
- Metals are substances which are good conductors of heat and electricity, have a characteristic lustre, and are malleable and ductile.
- Nonmetals are elements which are usually brittle, lacking a lustre, poor conductors of heat and electricity, and lacking the characteristics of metals.
- Metalloids are elements which have some of the properties of metals and some of the properties of nonmetals.
- The modern periodic law states that when elements are arranged in order of increasing atomic number, those with similar properties occur at regular intervals.
- The periodic table is an arrangement of the elements in order of increasing atomic number, in such a way that elements with similar properties appear in vertical columns.

- The horizontal rows of a periodic table are called periods.
- The vertical columns of a periodic table are called groups or families.
- The alkali metals are found in Group 1.
- The halogens are found in Group 17.
- The noble gases are found in Group 18.
- The main-group elements are found in Groups 1, 2, and 13 to 18.
- The transition metals are found in Groups 3 to 12.
- The lanthanides are the elements with atomic numbers 57 to 70.
- The actinides are the elements with atomic numbers 89 to 102.
- The metallic character of the elements decreases from left to right in a row and increases from top to bottom in a group.
- The properties of an element depend on the number of electrons present in the outer energy levels of its atoms.
- The valence electrons are the outer electrons of an atom.
- The kernel of an atom consists of the nucleus of the atom and the inner electrons.
- An electron dot symbol is a representation of an atom in which the letters represent the kernel and the dots represent the valence electrons.
- The electron configuration of an atom is related to the location of the element in the periodic table.
- The valence orbitals are the outermost s and p orbitals of an atom.
- The atomic radii of the elements decrease from left to right and increase from top to bottom of the periodic table.
- Ionization of an atom is the addition of one or more electrons to, or the removal of one or more electrons from, the atom.
- The ionization energy of an element is the minimum energy which is required to remove an electron from an atom of the element.
- The ionization energies of the elements increase from left to right and decrease from top to bottom in the periodic table.
- The electron affinity is the energy released when an electron is added to an atom.
- The electron affinities of the elements increase from left to right and decrease from top to bottom of the periodic table.

Key Terms

metal	halogen
nonmetal	valence electron
metalloid	electron dot symbol
period	kernel
group	octet
periodic law	valence orbital
main-group element	ionization
transition metal	ionization energy
noble gas	electron affinity
alkali metal	

Test Your Understanding

Each of these statements or questions is followed by four responses. Choose the correct response in each case. (Do not write in this book.)

1. Which of the following elements was unknown to the alchemists?
   a) copper   b) mercury   c) gallium   d) lead

2. Which of the following properties is not characteristic of metals?
   a) They are malleable.   b) They are good conductors of heat.
   c) They are ductile.   d) They are dull in appearance.

3. Which of the following elements is not a nonmetal?
   a) fluorine   b) oxygen   c) nitrogen   d) antimony

4. Which is the most reactive of all the elements?
   a) fluorine   b) sodium   c) oxygen   d) hydrogen

5. In Mendeleev's periodic table, the horizontal rows are called
   a) groups.   b) periods.   c) families.   d) columns.

6. Which group of the periodic table contains the least reactive elements?
   a) alkali metals   b) halogens   c) actinides   d) noble gases

7. Which of the following is the most metallic element?
   a) oxygen   b) iron   c) potassium   d) neon

8. Which of the following is the most nonmetallic element?
   a) silicon   b) fluorine   c) nitrogen   d) helium

9. Which of the following elements have atoms with four electrons in their outer energy levels?
   a) carbon   b) oxygen   c) neon   d) boron

10. Which of the following elements have atoms that would not have five electrons in their outer energy levels?
    a) phosphorus   b) selenium   c) arsenic   d) antimony

11. The kernel of an atom represents
    a) the nucleus and the valence electrons.
    b) the nucleus and the inner electrons.
    c) the nucleus.
    d) the valence electrons.

12. How many inner electrons does any element of the third period have?
    a) two   b) six   c) eight   d) ten

13. Which of the following should not be written as an electron dot symbol for nitrogen?
    a) $:\!\dot{\underset{.}{N}}\cdot$ b) $\cdot\dot{N}\!:$ c) $\cdot\overset{..}{\underset{.}{N}}\cdot$ d) $\overset{..}{N}\cdot$

14. An element that has an electron configuration of $1s^2\,2s^2\,2p^6\,3s^2\,3p^5$ is one of the
    a) alkali metals.   b) transition metals.
    c) halogens.   d) lanthanides.

15. An element that has an electron configuration of $1s^2\,2s^2\,2p^6\,3s^2\,3p^6\,4s^2\,3d^5$ is one of the
    a) alkali metals.   b) transition metals.
    c) halogens.   d) lanthanides.

16. Which of the following elements has the largest atomic radius?
    a) beryllium   b) carbon   c) nitrogen   d) oxygen

17. Which of the following elements has the smallest atomic radius?
    a) sulfur   **b)** selenium   **c)** oxygen   **d)** tellurium
18. Which of the following elements has the smallest first ionization energy?
    a) strontium   **b)** calcium   **c)** barium   **d)** magnesium
19. Which of the following elements has the largest first ionization energy?
    a) bromine   **b)** potassium   **c)** arsenic   **d)** calcium
20. Which of the following elements has the highest electron affinity?
    a) chlorine   **b)** silicon   **c)** sodium   **d)** phosphorus

Review Your Understanding

1. What are elements that have some metallic and some nonmetallic properties called?
2. How did Mendeleev's presentation of a periodic table allow the discovery of more elements?
3. What would you predict for the atomic mass of the missing element on Group VII, Row 6, of Mendeleev's periodic table?
4. State the modern periodic law. What is meant by periodicity of properties?
5. **a)** Draw a blank outline of the periodic table using Table 5-4 as a guide. Indicate the positions of the representative elements, the transition metals, the lanthanides, and the actinides.
   b) On a similar blank outline indicate the positions of the noble gases, the alkali metals, and the halogens.
6. Which group of the periodic table contains the most metallic element? Which group contains the most nonmetallic element?
7. Find two adjacent elements in the modern periodic table which would appear in a different order when arranged in order of increasing atomic mass.
8. On the basis of their positions in the periodic table, predict which member of each of the following pairs will be more metallic: **a)** silicon or germanium; **b)** arsenic or germanium; **c)** barium or cesium; **d)** beryllium or boron.
9. Arrange the following elements in order of decreasing metallic character: Sc, Fe, Rb, Br, N, Ca, F, Te.
10. In which energy level are the valence electrons of the following elements found: **a)** iodine; **b)** calcium; **c)** gallium; **d)** fluorine; **e)** francium?
11. Using the energy level populations given in Table 4-3, indicate which elements should have similar chemical properties to **a)** carbon; **b)** neon; **c)** sulfur.
12. How many valence electrons are there in **a)** nitrogen; **b)** phosphorus; **c)** arsenic?
13. Write the electron dot symbols for **a)** gallium; **b)** germanium; **c)** arsenic; **d)** selenium.
14. Explain why sodium forms only ions with a +1 charge, and calcium forms only ions with a +2 charge.

15. How many electrons would the following elements tend to gain or lose: **a)** Mg; **b)** Cl; **c)** N; **d)** Al; **e)** S; **f)** Ar? State in each case whether the element gains or loses electrons.

16. To which group in the periodic table do the elements whose electron configurations end in $s^2 p^2$ belong?

17. Using the periodic table, identify each of the following:
    a) an element which has seven electrons in each atom;
    b) an element which has seven electrons in its outer energy level;
    c) an element for which the second energy level is half-filled;
    d) a main-group element which forms ions by losing only s electrons.

18. Without referring to the text or a periodic table, write full electron configurations for elements of the following atomic numbers: **a)** 8; **b)** 14; **c)** 18; **d)** 22.

19. Write the electron configurations for phosphorus and arsenic. Show why they belong to the same group.

20. Consider elements 8, 16, and 34. Show why they should belong in the same group.

21. Without looking at the periodic table, indicate the groups in which elements having the following atomic numbers will be found: **a)** 4; **b)** 7; **c)** 9; **d)** 13; **e)** 16; **f)** 18.

22. Which element of the first 20 elements should belong to the same group of chemically related elements as silicon? Why?

23. The electron configuration of Cr ends in $4s^1 3d^5$. How does this differ from what you might have predicted?

24. Which element in the following sets should have the largest atomic radius and why? **a)** B, Li, or F; **b)** K, Li, or Na.

25. Positive ions are generally smaller than the corresponding neutral atoms, while negative ions are generally larger. Suggest a reason for this fact.

26. The following is a block of elements from the periodic table:

A	B	C	D
E	F	G	H
I	J	K	L

    a) which element has the largest atomic radius?
    b) which element has the smallest atomic radius?

27. Use the data in Figure 5-9 to construct a graph of the atomic radii of the elements plotted against their atomic numbers.

28. Which atom in each of the following pairs has the higher ionization energy? **a)** Cs or Au; **b)** S or P; **c)** Mg or Al; **d)** Rn or At; **e)** Zn or Cu; **f)** Rb or Sr.

29. In each of the following pairs of elements, which has the higher ionization energy? **a)** O or S; **b)** Ge or Se; **c)** Mg or Rb; **d)** Xe or Cs; **e)** Ne or Kr; **f)** P or Si.

30. Which element in the following sets loses an electron most readily, and why? **a)** B, Li, or F; **b)** K, Li, or Na.

31. Arrange the following elements in order of increasing ionization energy: Sr, Cs, S, F, As, N.

32. Elements X, Y, and Z are found in the same group of the periodic table, with X on top and Z on the bottom. Which element will have: **a)** the

largest atomic radius; **b)** the largest ionization energy; **c)** the most metallic character?

33. The ion Na^+ and the atom Ne have the same electron configuration. To remove an electron from gaseous neon atoms requires 2081 kJ/mol. To remove an electron from a gaseous Na^+ ion requires 4562 kJ/mol. Why are these values not the same?

34. The first ionization energy of boron is slightly less than the first ionization energy of beryllium. Use electron configurations of these two elements to provide an explanation for this fact.

35. The electron configurations for six neutral atoms are

    i) $1s^2\ 2s^2\ 2p^6\ 3s^1$ **ii)** $1s^2$

    iii) $1s^2\ 2s^2\ 2p^6\ 3s^2$ **iv)** $1s^2\ 2s^2\ 2p^6$

    v) $1s^2\ 2s^1$ **vi)** $1s^2\ 2s^2\ 2p^3$

    a) Which of these would have the highest first ionization energy?

    b) Which of these would have the lowest first ionization energy?

    c) Which of these would have the lowest second ionization energy?

    d) Which of these would likely have the lowest third ionization energy?

36. The second ionization energy of magnesium is only about twice as great as the first ionization energy. However, the third ionization energy is about ten times as great as the first. Why does it take so much energy to remove the third electron from magnesium?

37. Although the first ionization energy of K is smaller than that of Ca, the second ionization energy of K is much higher than that of Ca. Why is this so?

38. Use the data in Figure 5-10 to construct a graph of the first ionization energies of the elements plotted against their atomic numbers.

39. In what ways would you expect a graph of second ionization energies against atomic number to differ from a similar graph of the first ionization energies? How would you explain these differences?

40. Which element in the following sets has the largest electron affinity, and why? **a)** B, Li, or F; **b)** Br, Cl, or I.

Apply Your Understanding

1. Use the equation $E_n = -R/n^2$, where R = 2.18 aJ (2.18 × 10⁻¹⁸ J), to demonstrate that the energy levels are closer together as the values of the principal quantum number increase.

2. The wavelength, λ, and the frequency, v, of electromagnetic radiation are related by the equation $v\lambda = c$ where c is the speed of light (3.00 x 10⁸ m·s⁻¹). Moseley discovered that the frequency of a characteristic line of the X-ray spectrum of an element is related to the element's atomic number, Z, by the following formula:

$$\sqrt{v} = a(Z - b)$$

In this formula a is 5.00 x 10⁷ s⁻¹ᐟ² and b is 1.00.

    a) If the characteristic line in the X-ray spectrum of an element has a wavelength of 164 pm, what is the atomic number of the element?

b) What is the wavelength of the characteristic line in the X-ray spectrum of chromium?

3. You have landed on a strange planet and have been taken prisoner. You are told that you life will be spared if you are able to construct a periodic table for your captors. Their atomic theory is the same as ours except that their spin quantum numbers are $+\frac{1}{2}$, 0, and $-\frac{1}{2}$ and their magnetic quantum numbers take the values 0, 1, 2,...ℓ.

a) Construct an energy level diagram for the planet's inhabitants. It will differ from ours only because of the changes in the spin quantum numbers and the magnetic quantum numbers. Everything else, including the ordering of the orbitals and the naming of the orbitals, will be the same.

b) Construct a periodic table. Use atomic numbers instead of symbols and go as far as element number 35.

c) What are the atomic numbers of their first three noble gases?

d) What is the electron configuration of their element number 24?

e) Which of the following of their elements should have similar chemical properties: 4, 7, 15, 16, 20, 27, 34, 35?

f) Which of the following of their elements should have the highest electron affinity: 11, 12, 13, 27?

4. Are there any atoms for which the second ionization energy is smaller than the first? Explain.

5. The first, second, and third ionization energies (kJ/mol) for a number of elements are listed:

Element	I_1	I_2	I_3
Aa	531	1087	6270
Bb	2090	3135	4180
Cc	627	1045	5016
Dd	523	8360	11704
Ee	577	1823	2675

a) Which of these elements belong in the same group of the periodic table?

b) Which of these elements would be most like an alkali metal?

c) Which of these elements would be most like a noble gas?

6. As we go across a period from Group 1 to 17, we expect Group 17 elements to have the highest electron affinities. Why do you think the Group 18 elements were omitted from this statement?

7. A significant amount of energy is given off when an electron is added to a chlorine atom. On the other hand, the addition of an electron to argon requires energy. Suggest a reason for the difference.

Investigations

1. A number of elements have been included in lists of controversial chemicals. Do some library research to identify these elements from the following clues:

a) This element has the reputation of being the classical poison used by murderers. Today poisoning involving this element occurs mainly in industry and agriculture among workers who use chemicals which contain this element as an impurity.

b) The name of this element is derived from the ancient name for cala-mine (zinc carbonate). This element often occurs in ores commonly mined for extraction of zinc. Today, we import products such as paint pigments and batteries that contain this element.

c) The increased emphasis on energy conservation has caused people to improve the insulation of their homes. Unfortunately, this improve-ment can result in insufficient ventilation and a potentially hazardous buildup of this gaseous element.

d) This element is still used in gasoline additives. There is speculation about the effect of this element on people.

e) This element was one of the gases used in chemical warfare during the First World War. Today, the addition of this element to water has been a great move forward in public health, but it still causes some concern.

f) Workers in the felt hat industry used compounds of this element. The fact that hatters suffered psychotic disorders because of poisoning by this element may have been the inspiration for the "Mad Hatter." Today, this element is used in the production of electrical goods such as switches and batteries.

2. Write research reports on the potential hazards of at least two of the elements described in the previous question.

3. Trace the history of the development of the periodic table.

4. Write a research report on the commercial uses of chlorine, bromine, and iodine.

Answers for Section 5.2

1. Magnesium (in chlorophyll)
2. Calcium (in quicklime and limestone)
3. Beryllium
4. Helium
5. Carbon (in hydrocarbons)
6. Magnesium (in magnesium sulfate)
7. Hydrogen
8. Oxygen
9. Oxygen (as ozone)
10. Carbon
11. Aluminum (in bauxite)
12. Boron
13. Beryllium
14. Argon
15. Hydrogen
16. Fluorine (in freons)
17. Helium
18. Carbon (as graphite and diamond)
19. Neon
20. Fluorine

21. Helium
22. Phosphorus
23. Boron (in borax)
24. Oxygen
25. Magnesium
26. Chlorine
27. Calcium (in calcium carbonate)
28. Sulfur
29. Sodium
30. Calcium
31. Boron (in borax)
32. Carbon (as charcoal)
33. Lithium
34. Neon
35. Aluminum
36. Calcium (in calcium carbonate)
37. Helium
38. Silicon (in silicates)
39. Carbon (as diamond)
40. Nitrogen

6

Chemical Bonding

CHAPTER CONTENTS

Part One: **The Main Types of Chemical Bonds**
    6.1 The Role of Electrons in Bonding
    6.2 Ionic Bonding
    6.3 Covalent Bonding

Part Two: **Basic Concepts of Chemical Bonding**
    6.4 Polar Covalent Bonding and Electronegativity
    6.5 The Bonding Continuum
    6.6 The Octet Rule
    6.7 Multiple Bonds

Part Three: **Some Additional Concepts of Chemical Bonding**
    6.8 The Shapes of Molecules - VSEPR Theory
    6.9 Coordinate Covalent Bonding
    6.10 Chemical Bonding and Quantum Mechanics

Why do atoms join together? Why does a carbon atom join with four hydrogen atoms in methane, but with only two oxygen atoms in carbon dioxide? Why is the element nitrogen, N_2, so unreactive? Is it possible to predict the formulas of compounds? Why is sodium chloride (table salt) different from sucrose (sugar)? How does the way in which atoms join together affect the properties of their compounds? How do the shapes of compounds affect their properties? The answers to these questions are found in this chapter.

Individual elements possess the ability to unite with each other to form compounds. We know this because, although there are only about 100 elements, there are millions of compounds. Some compounds consist of ions (charged atoms) attracted to each other. Other compounds consist of molecules. A **molecule** is any electrically neutral group of atoms held together tightly enough to be considered a single particle. **Chemical bonds** are the attractions between the atoms or ions of a substance. Using their knowledge of chemical bonding and molecular structure, chemists have manipulated elements and compounds to make products which have enhanced our comfort, convenience, and health.

Key Objectives

When you have finished studying this chapter, you should be able to
  1. explain the role of electrons in bonding.
  2. predict whether a given bond will be ionic or covalent.

3. write equations, using electron dot symbols, for the formation of ionic compounds.
4. predict the formulas of ionic compounds.
5. write Lewis structures to represent the formation of covalent bonds in simple molecules.
6. contrast the properties of ionic compounds and covalent compounds.
7. use VSEPR Theory to predict the shapes of simple molecules.
8. draw orbital diagrams for simple covalent molecules.

Part One: The Main Types of Chemical Bonds

6.1 The Role of Electrons in Bonding

The basis of the present theories of chemical bonding was presented in 1916 by W. Kossel (a German physicist) and G. N. Lewis (Fig. 6-1) in separately published papers.

Biography

GILBERT LEWIS was born in West Newton, Massachusetts, in 1875, the son of a lawyer. He received his early education at home from his parents. He was able to read at the age of three and attended university preparatory school in Nebraska in 1889. He later went to the University of Nebraska and after two years transferred to Harvard University. In 1899, he obtained his doctorate, doing research into electrochemistry. After studying in Germany, he returned to teach at Harvard and then at the Massachusetts Institute of Technology. His early research was in the area of thermodynamics.

In 1912, Lewis became dean and chairman of the College of Chemistry at the University of California at Berkeley. There he recruited many people who would later make great contributions to chemistry and to chemical education. Lewis' work in chemical education set the standard for other chemists to follow and is one of his greatest achievements.

In 1916, he proposed a theory of chemical bonding in which electron pairs were shared between atoms and an octet of valence electrons resulted in stability for an atom.

In 1918, Lewis became a major in the Chemical Warfare Service in France. He was later decorated for his war efforts. Lewis went on to work in the areas of acid-base theory, heavy water, and photochemistry. He died in 1946 in Berkeley, California.

Figure 6-1
Gilbert Newton Lewis (1875–1946)
—Proposed a theory of chemical bonding.

The electrons in the outer (highest) energy level of an atom are the ones which usually participate in chemical bonding. These electrons are called valence electrons. The number of valence electrons possessed by an atom determines the number of other atoms with which that atom can combine.

The number of electrons in the highest energy level of an energy level population equals the number of valence electrons. The energy level populations (and their abbreviations) for the first 20 elements are given in Table 6-1. For example, the sodium atom has two electrons in the first energy level, eight more electrons in the second energy level, and one electron in the third energy level. Sodium's energy level population is 2, 8, 1. Thus, sodium has one valence electron.

Except for helium, the noble gases each have eight valence electrons. Helium has only two valence electrons. However, since its outer level (the first energy level) is capable of holding only two electrons, helium also behaves like a noble gas. The noble gases are stable and virtually unreactive. In fact, it was not until 1962 that chemists were able to unite a noble gas with another element to form

Table 6-1   Energy Level Populations for the First 20 Elements

| Element | Symbol | Atomic Number | Energy Level Populations | | | | Energy Level Population Abbreviations |
			1st Level	2nd Level	3rd Level	4th Level	
Hydrogen	H	1	1				1
Helium	He	2	2				2
Lithium	Li	3	2	1			2,1
Beryllium	Be	4	2	2			2,2
Boron	B	5	2	3			2,3
Carbon	C	6	2	4			2,4
Nitrogen	N	7	2	5			2,5
Oxygen	O	8	2	6			2,6
Fluorine	F	9	2	7			2,7
Neon	Ne	10	2	8			2,8
Sodium	Na	11	2	8	1		2,8,1
Magnesium	Mg	12	2	8	2		2,8,2
Aluminum	Al	13	2	8	3		2,8,3
Silicon	Si	14	2	8	4		2,8,4
Phosphorus	P	15	2	8	5		2,8,5
Sulfur	S	16	2	8	6		2,8,6
Chlorine	Cl	17	2	8	7		2,8,7
Argon	Ar	18	2	8	8		2,8,8
Potassium	K	19	2	8	8	1	2,8,8,1
Calcium	Ca	20	2	8	8	2	2,8,8,2

a compound. In 1962, Neil Bartlett, then of the University of British Columbia, produced the first noble gas compound. This compound consisted of the elements xenon, platinum, and fluorine. A number of noble gas compounds have since been prepared. However, there are no known compounds of helium or neon. Thus, there appears to be a special stability associated with an atom's having eight valence electrons. The term **stable octet** draws attention to this special stability.

Kossel believed that some atoms could acquire a stable octet by the loss or gain of electrons. Lewis suggested that other atoms might achieve a stable octet and at the same time be held together by sharing electrons between them, that is, by chemically bonding to one another. In the following sections, the ideas of Kossel and Lewis will be described in some detail. We will see that atoms can acquire a stable octet and an energy level population similar to that of the noble gases. Thus, atoms gain stability by losing, gaining, or sharing valence electrons.

6.2 Ionic Bonding

The most chemically reactive metals are found in Groups 1, 2, and to a lesser extent 13 of the periodic table. These metals have low ionization energies. That is, a relatively small amount of energy is required to cause them to lose electrons. The most active nonmetals are found in Groups 16 and 17. These nonmetals have relatively large electron affinities. That is, they tend to gain electrons readily. When an active metal reacts with an active nonmetal, electrons are transferred from the metal to the nonmetal. The atoms of the nonmetal, having gained electrons, become negatively charged ions. The atoms of the metal, having lost electrons, become positively charged ions. These oppositely charged ions attract each other. **Ionic bonding** is the type of chemical bonding resulting from the attraction between oppositely charged ions formed when metallic atoms transfer electrons to nonmetallic atoms.

The sodium atom has an energy level population of 2,8,1 and could become **isoelectronic** with (that is, have the same energy level population as) the noble gas neon (2,8) by losing its valence electron to a chlorine atom. The chlorine atom has an energy level population of 2,8,7 and could become isoelectronic with the noble gas argon (2,8,8) by gaining one electron from the sodium atom:

$$\text{Na} \cdot + \cdot \ddot{\underset{..}{\text{Cl}}} : \longrightarrow \underbrace{\text{Na}^+ + : \ddot{\underset{..}{\text{Cl}}} :^-}_{\downarrow}$$
$$\text{NaCl}$$

In this way, a positive and a negative ion are produced when a sodium atom collides with a chlorine atom. These ions attract one another because they have opposite charges, and an ionic bond is formed. The ionic compound is sodium chloride, also known as table salt. Sodium chloride occurs in nature as the mineral halite. It is obtained by mining rock salt or by the evaporation of sea water, normally in the form of white cubic crystals. Sodium chloride melts at a high temperature (804°C). It readily dissolves in water, and aqueous solutions of sodium chloride conduct electricity. Sodium chloride is the source of chlorine gas and sodium metal.

EXAMPLE PROBLEM 6-1

Why does each lithium ion require one fluoride ion in the compound lithium fluoride?

SOLUTION

The energy level populations are Li (2,1) and F (2,7). Each lithium atom tends to lose its valence electron to form a lithium ion which is isoelectronic with the noble gas helium (2). Each fluorine atom tends to gain one electron from the lithium atom to form a fluoride ion which is isoelectronic with the noble gas neon (2,8). The lithium ion attracts the fluoride ion because they have opposite charges:

$$\text{Li}\cdot + \;\cdot\ddot{\text{F}}: \longrightarrow \text{Li}^+ + \;:\ddot{\text{F}}:^-$$
$$\underbrace{\qquad\qquad}_{\downarrow}$$
$$\text{LiF}$$

PRACTICE PROBLEM 6-1

Why does each rubidium ion require one bromide ion in the compound rubidium bromide?

When an ionic bond is formed by electron transfer between a metal and a nonmetal, all of the electrons lost by the metal must be gained by the nonmetal. When ionic bonds are formed, enough electrons must be transferred so that each ion produced is isoelectronic with a noble gas.

A Group 2 metal (such as calcium) has two valence electrons located in its highest energy level. Calcium (2,8,8,2) can become isoelectronic with the noble gas argon (2,8,8) if it loses two valence electrons. Therefore, calcium atoms lose their two valence electrons. If calcium is reacting with chlorine (2,8,7), two chlorine atoms will be required to accept the two electrons from the calcium atom. Each chlorine atom will accept one electron from the calcium atom and will become isoelectronic with the noble gas argon (2,8,8). Two chloride ions will be formed, and both will be attracted to the calcium ion:

$$\text{Ca}: + \;\; 2\;\cdot\ddot{\text{Cl}}: \longrightarrow :\ddot{\text{Cl}}:^- + \text{Ca}^{2+} + :\ddot{\text{Cl}}:^-$$
$$\underbrace{\qquad\qquad\qquad\qquad}_{\downarrow}$$
$$\text{CaCl}_2$$

The subscript 2 indicates that there are two chloride ions bonded to a calcium ion. Calcium chloride is used as a drying agent and is spread on roads to melt ice and snow.

Calcium could also react with an element from Group 16. Oxygen (2,6) is such an element. Oxygen atoms have six valence electrons. Oxygen atoms would have to gain two electrons to become isoelectronic with the noble gas neon (2,8). Thus, one calcium atom could transfer two electrons to a single oxygen atom, and the ionic compound calcium oxide would form:

$$\text{Ca}\!: + \; \cdot\ddot{\text{O}}\!: \longrightarrow \text{Ca}^{2+} + \; :\ddot{\text{O}}\!:^{2-}$$

$$\underbrace{\qquad\qquad}_{\downarrow}$$

$$\text{CaO}$$

Calcium oxide is also known as lime or quicklime. It is a white solid used as a component of cement and fertilizers.

EXAMPLE PROBLEM 6-2

Predict the formula of the compound formed when beryllium reacts with oxygen.

SOLUTION

The energy level populations are Be (2,2) and O (2,6). Each beryllium atom can give up its two valence electrons to form a beryllium ion which is isoelectronic with the noble gas helium (2). Each oxygen atom gains two electrons from the beryllium atom to form an oxide ion which is isoelectronic with the noble gas neon (2,8).

$$\text{Be}\!: + \; \cdot\ddot{\text{O}}\!: \longrightarrow \text{Be}^{2+} + \; :\ddot{\text{O}}\!:^{2-}$$

$$\underbrace{\qquad\qquad}_{\downarrow}$$

$$\text{BeO}$$

PRACTICE PROBLEM 6-2

Predict the formula of the compound formed when magnesium reacts with sulfur.

Let us consider the example of the Group 13 metal aluminum (2,8,3) reacting with oxygen (2,6):

$$2\;\text{Al}\!: + \; 3\;\cdot\ddot{\text{O}}\!: \longrightarrow \; :\ddot{\text{O}}\!:^{2-} + \text{Al}^{3+} + :\ddot{\text{O}}\!:^{2-} + \text{Al}^{3+} + :\ddot{\text{O}}\!:^{2-}$$

$$\underbrace{\qquad\qquad\qquad\qquad\qquad}_{\downarrow}$$

$$\text{Al}_2\text{O}_3$$

Two aluminum atoms have lost a total of 6 electrons to become isoelectronic with the noble gas neon (2,8). Three oxygen atoms have gained a total of 6 electrons to become isoelectronic with the noble gas neon (2,8). The formula Al_2O_3 indicates that there are three oxide ions for every two aluminum ions in aluminum oxide. Figure 6-2 shows schematically how the two to three ratio of aluminum ions to oxide ions can be obtained for aluminum oxide.

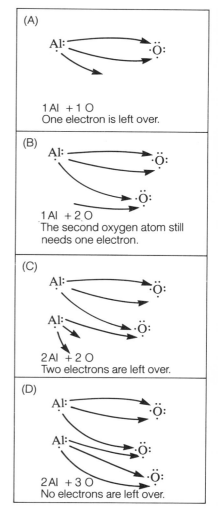

Figure 6-2
Establishing the 2:3 ratio of ions in Al_2O_3

(A)

1 Al + 1 O
One electron is left over.

(B)

1 Al + 2 O
The second oxygen atom still needs one electron.

(C)

2 Al + 2 O
Two electrons are left over.

(D)

2 Al + 3 O
No electrons are left over.

EXAMPLE PROBLEM 6-3

When magnesium burns in air it reacts not only with oxygen to form magnesium oxide, but also with nitrogen to form magnesium nitride. What is the formula for magnesium nitride?

SOLUTION

The energy level populations are Mg (2,8,2) and N (2,5). Each magnesium atom can give up its two valence electrons to form a magnesium ion which is isoelectronic with the noble gas neon (2,8). However, each nitrogen atom must gain three electrons to form a nitride ion which is also isoelectronic with the noble gas neon (2,8).

$$3\,Mg: + \; 2\cdot\ddot{N}: \longrightarrow Mg^{2+} + :\ddot{N}:^{3-} + \; Mg^{2+} + :\ddot{N}:^{3-} + \; Mg^{2+}$$

$$Mg_3N_2$$

PRACTICE PROBLEM 6-3

Predict the formula of the compounds formed **a)** when lithium reacts with nitrogen; **b)** when gallium reacts with sulfur. (*Answers*: **a)** Li_3N; **b)** Ga_2S_3).

The atoms and the ions of an element have different chemical properties. Sodium is a soft, silvery-white metal and chlorine is a poisonous, greenish-yellow gas. However, sodium chloride (sodium ions and chloride ions) is white, solid table salt and, as we have already learned, is not poisonous.

The periodic table can be used to help us predict the formula of an ionic compound. For example, the metals of groups 1, 2, and 13 tend to lose 1, 2, and 3 electrons respectively to become isoelectronic with a noble gas. The elements of groups 16 and 17 tend to gain 2 and 1 electrons respectively to become isoelectronic with a noble gas.

Figure 6-3 can also be used to obtain the formulas of many ionic compounds. Suppose we wish to know the formula of aluminum selenide. Aluminum is a Group 13 metal and selenium is a Group 16 nonmetal. The formula must have the form M_2X_3. The formula is Al_2Se_3. Six electrons have been transferred from the two aluminum atoms to the three selenium atoms.

Figure 6-3
Formulas of ionic compounds. Each rectangle contains the formula of the compound that results when the metal (M) from the indicated group combines with the nonmetal (X) from the indicated group to form an ionic compound. The number in parentheses indicates the total number of electrons transferred.

	Group 1 Metals	Group 2 Metals	Group 13 Metals
Group 17 Nonmetals	MX (1)	MX$_2$ (2)	MX$_3$ (3)
Group 16 Nonmetals	M$_2$X (2)	MX (2)	M$_2$X$_3$ (6)

One point should be made regarding the naming of the ionic compounds which we have considered. The name of the metallic ion is the same as the name of the metallic atom. Hence, during the formation of an ionic compound, a sodium atom becomes a sodium ion. However, the name of the nonmetallic ion differs from the name of the nonmetallic atom. The ending of the name of the nonmetallic atom is dropped and replaced with the suffix *-ide*. During the formation of an ionic compound, a chlorine atom becomes a chloride ion, an oxygen atom becomes an oxide ion, and a sulfur atom becomes a sulfide ion. When an ionic compound is named, the name of the metallic ion is written before the name of the nonmetallic ion. Thus, Na_2S is sodium sulfide.

Properties of Ionic Compounds

Ionic compounds are formed when the metals of Groups 1, 2, and 13 react with nonmetals such as the elements of Groups 16, 17, and nitrogen. Ionic compounds have a number of distinguishing properties, which are determined by their particular crystal structure. In an ionic crystal, positive and negative ions are arranged so that attractive forces between oppositely charged ions are maximized and repulsive forces between ions of the same charge are minimized. The sodium chloride crystal structure is shown in Figure 6-4.

Ionic compounds are solids at room temperature. As solids they are nonconductors of electricity; however, when melted they conduct electricity quite well. The electrical conductivity of molten ionic compounds is due to the presence of ions which are able to move under the influence of an electrical charge. In ionic solids, the ions are tightly held in the crystal structure, so they are not free to move and carry an electric current.

Ionic compounds have high melting points and boiling points, indicating strong bonding. Let us consider the energy change that occurs in the reaction:

$$Na(g) + Cl(g) \longrightarrow NaCl(g)$$

The symbol (g) indicates that a substance is in the gas state.

In order to consider the energy change, we can think of the reaction as if it occurred in three steps. First, energy must be added to a sodium atom in order to remove its electron. This energy is called the ionization energy:

$$Na(g) + Energy \longrightarrow Na^+(g) + e^-$$

Second, the electron is added to the chlorine atom, and energy is given off in the process. This energy is called the electron affinity of chlorine:

$$Cl(g) + e^- \longrightarrow Cl^-(g) + Energy$$

In this case, the electron affinity is smaller than the ionization energy. Third, energy is given off when the sodium ion and the chloride ion are brought together:

$$Na^+(g) + Cl^-(g) \longrightarrow NaCl(g) + Energy$$

The energy given off in this step is larger than either the ionization energy or the electron affinity. Thus, energy is given off in the overall reaction:

$$Na(g) + Cl(g) \longrightarrow NaCl(g) + Energy$$

Figure 6-4
The sodium chloride crystal structure

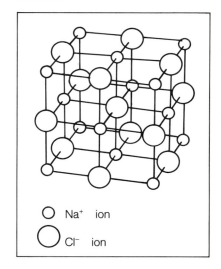

$\bigcirc$ Na^+ ion

$\bigcirc$ Cl^- ion

Figure 6-5
Energy changes which accompany the reaction
$Na(g) + Cl(g) \rightarrow NaCl(g)$

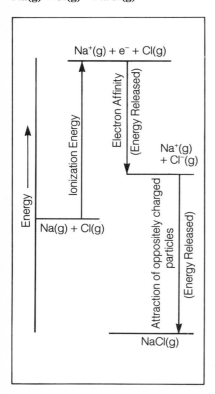

The energy changes which occur in this process are illustrated in Figure 6-5. Since the gaseous ion pair has a lower energy, it is stable when compared with the gaseous atoms.

The energy given off in the formation of solid NaCl is greater than for gaseous NaCl because each ion in the solid form is surrounded by six ions of the opposite charge. The high melting point of solid sodium chloride results from the strong attractions which occur in the crystal structure. Each sodium ion attracts six chloride ions, each of which, in turn, attracts six sodium ions, and so on, throughout the crystal.

Ionic substances are crystalline; that is, they have flat surfaces that make characteristic angles with one another. Ionic substance are hard but brittle. The hardness results from strong attractions which occur in the crystal structure. However, if we apply enough force to move the ions slightly, the attractive forces can become repulsive as sodium ions contact sodium ions and chloride ions contact chloride ions. In this case, the crystal cleaves (breaks apart) along smooth, flat surfaces.

Ionic compounds are often soluble in water. The resulting aqueous solutions conduct electricity, indicating that ions are free to move in solution.

In this section we learned that atoms gain stability by losing or gaining electrons. The ions that result have stable octets and are isoelectronic with noble gases. These oppositely charged ions attract one another and form ionic bonds. In the next section, we shall look at another means by which atoms achieve stability and become isoelectronic with the noble gases.

6.3 Covalent Bonding

In the last section, we discussed ionic bonding. However, the transfer of electrons from one atom to another does not adequately represent all bond types. For example, when two hydrogen atoms collide, there is no reason for one hydrogen atom to give up its electron to the other hydrogen atom. We believe that there is no electron transfer in this case. The electrons are, instead, shared equally between the two hydrogen atoms, since there is no reason for one hydrogen atom to attract the electrons more than the other. A **covalent bond** is formed when a pair of electrons is shared between two atoms. If the electrons are shared equally, the bond is considered to be a pure covalent bond.

Why do the two hydrogen atoms bond together? The formation of a chemical bond indicates that a molecule is more stable; that is, it has less energy than the isolated atoms from which it is formed. This answer leads to another question: Why is a molecule more stable than the isolated atoms?

When any two atoms come close together, the electrons of each atom come under the influence of the nucleus and the electrons of the other atom. We can consider two distant hydrogen atoms (Fig. 6-6a) and two hydrogen atoms which are near to one another (Fig. 6-6b). In the distant hydrogen atoms, the only force is the attraction of the electron of each atom to the proton of the same atom. In the close hydrogen atoms, there are both attractive and repulsive forces. First, there are the same proton-electron attractions which existed for each of the two distant hydrogen atoms. In addition, *between* the two hydrogen atoms there is an electron-electron repulsion; there is a proton-

proton repulsion; and the electron of each atom is attracted to the proton of the other atom (two new attractions).

The fact that the two hydrogen atoms do bond together to form a molecule indicates that the hydrogen molecule is more stable than the distant hydrogen atoms. Obviously, the proton-electron attractions which were originally present in the distant hydrogen atoms cannot account for the stability of the new hydrogen molecule. Furthermore, the electron-electron repulsion and the proton-proton repulsion act against the increase in stability which results from bond formation. Thus, the two hydrogen atoms must bond together because the electron of each atom is attracted to the proton of the other atom. Therefore, the total attractive forces are greater than the total repulsive forces.

Each hydrogen atom has one valence electron ($H\cdot$).When two hydrogen atoms collide, an electron pair is shared equally between the two hydrogen nuclei:

$$H\cdot + \cdot H \rightarrow H\!:\!H$$

The pair of electrons is now simultaneously attracted to the two hydrogen nuclei. This simultaneous attraction of the electron pair to the two nuclei produces a more stable state. The two atoms are held together because they share an electron pair between them. When the hydrogen atoms share a pair of electrons, each has become isoelectronic with the noble gas helium. Because *one* pair of electrons (called the *shared pair*) is shared between the atoms, the bond is a **single covalent bond**.

The representation $H\!:\!H$ is called an electron dot formula or a **Lewis structure**. However, in writing Lewis structures the usual practice is to use single lines to represent each shared pair of electrons and to use dots to represent valence electrons that are not involved in bonding. Hence, the Lewis structure of H_2 is generally written $H-H$.

In the past, we have discussed hydrogen as if were normally found as individual atoms. In fact, a sample of hydrogen gas is actually made up of hydrogen molecules. Each hydrogen molecule consists of two hydrogen atoms bonded together. Thus, hydrogen gas consists of diatomic (two atom) H_2 molecules. The subscript indicates the number of atoms in one molecule of hydrogen. A hydrogen atom with a single unpaired electron is too reactive to remain unbonded for any length of time. If no other suitable type of atom is available for bonding, hydrogen atoms will bond to one another.

Suppose two fluorine atoms (2,7) come together. Will they form a bond? Each fluorine atom has seven valence electrons. Therefore, each fluorine atom requires *one more* electron to become isoelectronic with the noble gas neon (2,8). When the fluorine atoms collide, each tries to gain an eighth valence electron, but neither fluorine atom can pull an electron from the other. Instead, they share a pair of electrons, enabling both atoms to become isoelectronic with the noble gas neon:

$$:\!\ddot{F}\!\cdot\; + \;\cdot\!\ddot{F}\!: \longrightarrow :\!\ddot{F} - \ddot{F}\!:$$

Thus, fluorine, like hydrogen, exists as diatomic (F_2) molecules.

In the same way that two fluorine atoms form a single covalent bond, two chlorine atoms form a single covalent bond, producing diatomic (Cl_2) molecules. The Lewis structure for a chlorine molecule is $:\!\ddot{C}l - \ddot{C}l\!:$. Bromine

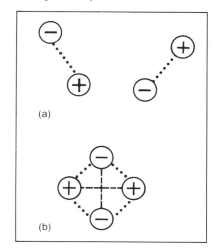

Figure 6-6
(a) Two distant hydrogen atoms
(b) Two near hydrogen atoms. (The dotted lines represent attractions, and the dashed lines represent repulsions.)

and iodine also belong to Group 17, and they form diatomic molecules (Br_2 and I_2).

We have learned so far that five elements exist as diatomic molecules rather than as individual atoms. They are hydrogen (H_2), fluorine (F_2), chlorine (Cl_2), bromine (Br_2), and iodine (I_2). The first three of these are gaseous elements. Bromine is one of the two elements which are liquids at room temperature (20-25°C). (The other liquid element is mercury.) Iodine is a solid element. Two other gaseous elements exist as diatomic molecules rather than as individual atoms, oxygen (O_2) and nitrogen (N_2). In fact, the only gaseous elements which exist as individual atoms are the noble gases. Thus, a total of seven elements exist as diatomic molecules rather than as individual atoms.

PRACTICE PROBLEM 6-4

Draw the Lewis structures for a bromine molecule and for an iodine molecule.

Figure 6-7
Models of water (H_2O) and methane (CH_4)

Covalent bonding also occurs in molecules other than elements that are diatomic. Molecules such as water, carbon dioxide, and methane contain atoms which are also joined by a type of covalent bonding. Water consists of molecules, each containing two hydrogen atoms and one oxygen atom (H_2O). Carbon dioxide gas is found in soda water and in fire extinguishers, is used by plants in photosynthesis, and is produced when food is metabolized in our bodies. It consists of molecules, each containing one carbon atom and two oxygen atoms (CO_2). Methane molecules are the most abundant molecules in natural gas. Methane is also found in coal mines, bogs, and marshes. Each methane molecule consists of one carbon atom and four hydrogen atoms (CH_4). Figure 6-7 shows models of water and methane.

Part One Review Questions

1. What accounts for the increased stability of a molecule formed from isolated atoms?
2. What are valence electrons?
3. What is meant when we say that an atom or ion is isoelectronic with a noble gas?
4. Describe what happens to the valence electrons when ionic and covalent bonds are formed.
5. What are four properties of ionic solids?
6. What is meant by the term *Lewis structure*?
7. What are the seven elements that exist as diatomic molecules?

Part Two: Basic Concepts of Chemical Bonding

6.4 Polar Covalent Bonding and Electronegativity

In the previous sections we have discussed ionic bonding, in which electrons are transferred, and covalent bonding, in which electrons are shared, sometimes equally. In this section, we shall discuss polar covalent bonding, in which electrons are always shared unequally.

Polar Covalent Bonding

When a hydrogen atom and a chlorine atom unite, a hydrogen chloride molecule forms. A pair of electrons is shared between the two atoms and a type of covalent bond is formed:

$$H\cdot + \cdot\ddot{\underset{\cdot\cdot}{Cl}}: \longrightarrow H - \ddot{\underset{\cdot\cdot}{Cl}}:$$

By sharing the pair of electrons, the hydrogen has become isoelectronic with the noble gas helium. The chlorine has become isoelectronic with the noble gas argon.

In the hydrogen chloride molecule, a pair of electrons is shared between two different types of atoms. Will the pair of electrons be shared equally? Will either the chlorine atom or the hydrogen atom have a stronger attraction for the shared pair of electrons?

Chemists have shown that the hydrogen chloride molecule is a *polar molecule*. It has a slightly negative (δ^-, read delta negative) and a slightly positive (δ^+, read delta positive) end. Further work has indicated that the chlorine end of the molecule is slightly negative and the hydrogen end of the molecule is slightly positive:

$$^{\delta+}\,H\,:\ddot{\underset{\cdot\cdot}{Cl}}:\,^{\delta-}\quad\textbf{or}\quad ^{\delta+}\,(H\,Cl)\,^{\delta-}$$

The molecule as a whole is electrically neutral because it has an equal number of protons and electrons. However, the bonding electron pair is unequally shared. The chlorine atom has a greater attraction for the shared pair of electrons than does the hydrogen atom. Therefore, the shared electrons will be closer, on average, to the chlorine atom. The chlorine atom does not attract the shared pair strongly enough to gain complete possession of it. Therefore, this is *not* an ionic bond. It is a covalent bond in which there is unequal sharing of electrons: a **polar covalent bond**.

Another example of a polar molecule is water.

$$2\,H\cdot + \cdot\ddot{\underset{\cdot\cdot}{O}}\cdot \longrightarrow :\ddot{O}\!\!\diagup^{H}_{\diagdown H}$$

Each oxygen-hydrogen bond is a polar covalent bond. The molecule as a whole is polar because the positions taken up by the hydrogen atoms around the oxygen atom result in one side of the water molecule being slightly negative and the opposite side being slightly positive:

$$\delta- \quad \overset{..}{\underset{..}{O}} \overset{\displaystyle H}{\underset{\displaystyle H}{<}} \quad \delta+$$

If water had the following shape:

$$\overset{\delta+}{H} - \overset{\delta-}{\overset{..}{\underset{..}{O}}} - \overset{\delta+}{H}$$

it would be a *nonpolar molecule* because it would not have opposite $\delta+$ and $\delta-$ sides. Thus, the polarity of the bonds and their arrangement in a molecule determine whether the molecule will be polar or nonpolar.

In a polar covalent bond, the shared pair spends more of its time near one of the atoms than it does near the second atom. This makes the first atom slightly negative, and the other atom becomes slightly positive. Generally, polar covalent bonding occurs when atoms of two different elements share a pair of electrons. Is it possible to predict which end of the polar covalent bond will be slightly negative and which end will be slightly positive?

Electronegativity

A quantitative measure of the electron-attracting ability of the atoms in a molecule is **electronegativity**. By measuring various properties such as ionization energies and electron affinities of the atoms making up the molecule, how polar the molecule is, and the energy required to break the bond, it is possible to construct a table of electronegativities. Numerical values assigned for electronegativities of the various elements are included in the periodic table at the back of this book. These values were calculated by the American chemist Linus Pauling (Fig. 6-8). The electronegativity values are such that the bigger the number is, the greater is the tendency of an atom to attract a shared pair of electrons to itself.

Fluorine has been assigned the highest electronegativity (4.0). Fluorine has a high ionization energy, and it has a high electron affinity. That is, it takes a relatively large amount of energy to force a fluorine atom to lose an electron, but a fluorine atom is ready to accept an electron in order to change its energy level population from 2,7 to 2,8 and to become isoelectronic with neon. These two factors both indicate that in any bond between fluorine and another element, it will be more difficult to transfer an electron from fluorine to the other element than it will be to transfer an electron from the other element to fluorine.

Cesium and francium both have the smallest electronegativity (0.7). They both have low ionization energies and it is relatively easy to transfer an electron from cesium or francium to another element.

In general, the elements with the lowest ionization energies have the lowest electronegativities and the elements with the highest ionization energies have the highest electronegativities. Thus, the variation of electronegativity parallels the variation of ionization energies. As we move from left (metals) to right (nonmetals) across a period, the electronegativity increases. As we go from top to bottom in a family or group, the electronegativity decreases.

Another example of a polar covalent bond is the bond between chlorine and bromine in the molecule, BrCl. Each of these atoms is in Group 17 and each has seven valence electrons. By sharing a pair of electrons each becomes isoelectronic with a noble gas:

$$:\ddot{\text{Br}}\cdot + \cdot\ddot{\text{Cl}}: \longrightarrow :\ddot{\text{Br}} — \ddot{\text{Cl}}:$$

The electronegativity of chlorine is 3.0, and it is 2.8 for bromine. A chlorine atom is more electronegative than a bromine atom. Therefore, the chlorine atom will have a greater ability to attract the shared pair of electrons. The shared electrons spend more time near the chlorine nucleus, and the chlorine end of the molecule becomes negative with respect to the bromine end:

$$^{\delta+} :\ddot{\text{Br}} — \ddot{\text{Cl}}: {}^{\delta-}$$

Biography

LINUS PAULING was born in Portland, Oregon in 1901. He entered Oregon State College in 1917 and received his B.Sc. in chemical engineering in 1922. He was appointed a Teaching Fellow in Chemistry in the California Institute of Technology while a student there from 1922 to 1925. In 1925 he received his doctorate (*summa cum laude*) in chemistry, with minors in physics and mathematics.

Pauling's approach to research was marked by intuition and intelligent guesses. In 1922, he began the experimental determination of the structures of certain crystals and also started theoretical work on the nature of the chemical bond. He was one of the first chemists to use quantum mechanics to explain the chemical bond. His work also included the areas of molecular structure determination, hydrogen bonding, metallic bonding, electronegativity, and the structure of proteins. He attempted to explain the chemical nature of sickle cell anemia and general anesthesia. Pauling is well known for his theory that large doses of Vitamin C can prevent or lessen the severity of the common cold.

He has published many books and articles. In 1931, he became the first recipient of the American Chemical Society Award in Pure Chemistry. He was awarded the 1954 Nobel Prize in Chemistry for his research into the nature of the chemical bond and the application of his bonding theory to the structural determination of complex substances. Pauling was awarded the Nobel Peace Prize in 1963 for his efforts on behalf of a nuclear test ban treaty. He became at that time only the second person to win the Nobel Prize twice. (Marie Curie was the first.)

Figure 6-8
Linus Pauling (b. 1901)—Theorized on the nature of the chemical bond.

The greater the difference in electronegativity between the two atoms involved in a polar covalent bond, the more polar the bond will be. Thus, the bond between H and Cl is more polar than the bond between Br and Cl.

PRACTICE PROBLEM 6-5

Draw a Lewis structure for BrF. Is this a polar or a nonpolar molecule? If it is polar, which end of the molecule is the negative end?

PRACTICE PROBLEM 6-6

Which of the bonds in each of the following is more polar: **a)** B-Br or C-Br; **b)** N-F or N-Cl? In each case identify which atom is slightly negative.

Properties of Covalent Compounds

In Section 6.2 we discussed the properties of ionic compounds. What are the properties of covalent compounds? Covalent compounds exist as gases, liquids, and solids at room temperature. Solid covalent compounds are usually soft. Compared with ionic compounds, covalent compounds usually evaporate readily (mothballs and perfume are both examples of covalent compounds) and have low melting and boiling points. Many covalent compounds are not soluble in polar substances such as water, but they are soluble in nonpolar substances such as gasoline or carbon tetrachloride. Covalent compounds do not conduct electricity in either the liquid or the solid state.

APPLICATION

Superconductivity

Superconductivity is the ability of a material to conduct electricity with no resistance. When superconductivity was discovered, about 75 years ago, this phenomenon would occur only at temperatures near −273°C. From 1911 to 1973 the temperature at which a material would be a superconductor was raised to −250°C by the discovery of new superconductive materials. In 1986 a major technological breakthrough pushed the temperature to −183°C. The importance of this advancement is that a temperature of −183°C is easy to achieve. The development spurred much research into superconductivity and a flood of published papers resulted. It has been said that work on superconductivity could be as important as the development of the light bulb, the laser, or the transistor.

Scientists now believe that it is possible to develop materials that are superconductors at room temperatures. Originally, liquid helium (which boils at −269°C) was used to cool a material to its temperature of super-conductivity. The latest breakthrough means that liquid nitrogen (which boils at −196°C) can be used as a coolant. Since liquid nitrogen is a much cheaper and more efficient coolant than liquid helium, the cost of super-conductivity is now very much lower. The development of materials which are superconductors at room temperature eliminates the need for expensive coolants.

The new high-temperature superconductive materials are metal oxide ceramics. One of the most important of these has the formula $YBa_2Cu_3O_x$, where x may vary, but is approximately seven. The potential benefits of these new superconducting ceramics are enormous. Electrical transmission lines could be made of superconductive materials. This would enable them to carry electricity over long distances without pro-ducing waste heat. Superconductors would also allow us to build smaller, faster, more powerful computers.

Certain pieces of medical diagnostic equipment use superconducting magnets. The availability of materials which are superconductive at more economical temperatures means that more hospitals will be able to afford this type of equipment. The new superconductors could be used to build more powerful particle accelerators, which are used to study the ultimate composition of matter. They could also become important in the devel-opment of nuclear fusion reactors. In fact, any electrically driven machine or process could potentially benefit from these superconductive materials.

Finally, an exciting property associated with superconductive materials is that the magnetic field of an ordinary magnet cannot penetrate the magnetic field produced by electricity flowing through a superconductor. A magnet placed above the superconducting material will therefore levi-tate (rise and float in the air). This would enable us to build frictionless trains capable of speeds of up to 500 km/h, levitated above a supercon-ducting track.

6.5 The Bonding Continuum

There is no sharp distinction between ionic and covalent bonding. In a chemi-cal bond between atoms M and X, the polarity of the bond depends on the natures of M and of X. If they each have the same ability to attract electrons, the bond will be covalent and nonpolar. If X can attract electrons better than M, the bond will be polar covalent. If X attracts electrons so strongly that one can say the electron pair spends essentially all of its time in X, the bond is ionic, and there is an M^+ ion and an X^- ion. In fact, an ionic bond is simply an *extreme case* of polar covalent bonding. In these extreme cases, chemists find it easier to think of the bond as being 100% ionic, but they keep in mind that a 100% ionic bond is not always entirely realistic.

Electronegativities help us to predict which bonds are ionic and which bonds are covalent. Two elements of very different electronegativities such as Na (0.9) and Cl (3.0) are expected to form ionic bonds. Two elements of slightly different electronegativities such as C (2.5) and H (2.1) are expected to form only slightly polar covalent bonds. The greater the difference in electronegativity, the more polar the bond becomes. If the difference in electronegativity between two bonding atoms exceeds 1.7, it is better to consider the bond as being ionic.

In any event, the transition between polar covalent bonding and ionic bonding is indefinite. It is better to think of a continuum with pure covalent (equal sharing) bonding at one end and ionic (electron transfer) bonding at the other end. In between is polar covalent bonding (Fig. 6-9).

Figure 6-9
The bonding continuum

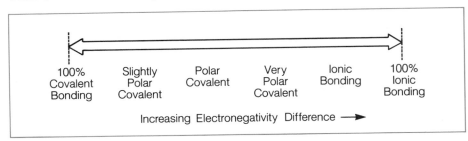

PRACTICE PROBLEM 6-7

Which pair in each of the following groups of atoms will likely form an ionic compound: **a)** H, Na, C, and Cl; **b)** K, Ga, B, and Br; **c)** Ca, As, O, and Ne?

Ionic Solids Do Not Consist of Molecules

One other point should be made in any comparison of ionic, polar covalent, and covalent bonding. It is the use of the word *molecule*. A molecule is any electrically neutral group of atoms that is bonded together tightly enough to be considered a single particle. We are using the word *bond* to describe the linkage between a pair of atoms. In the case of HCl, a hydrogen atom is bonded to a chlorine atom to form a molecule, and hydrogen chloride exists as molecules in the gaseous, liquid, or solid state. In the case of Cl_2, one chlorine atom is bonded to a second chlorine atom to form a molecule, and chlorine exists as molecules in the gaseous, liquid, or solid state.

However, the word *molecule* does not apply very well to ionic substances such as sodium chloride. When sodium reacts with chlorine, a sodium atom gives up its valence electron and becomes a sodium ion. A chlorine atom accepts the electron and becomes a chloride ion. In the gas phase, it is possible to have one positive sodium ion attracted to one negative chloride ion, and the result could perhaps be called a molecule. However, ionic substances exist as solids (not gases) at room temperature. We can say that an ionic solid consists of a stationary array or cluster of positive and negative ions held together by the attractive forces of oppositely charged ions. That is, solid sodium chloride does not consist of NaCl molecules. A crystal of sodium chloride consists of a large number of sodium ions and an equal number of chloride ions, none of which has a specific bonding partner. In the liquid phase, ionic substances consist of a fluid cluster of positive and negative ions in which each ion is free to move slowly throughout the liquid. Thus, liquid sodium chloride does not consist of NaCl molecules, either.

We can speak of molecules of covalently or polar covalently bonded substances. The formulas Cl_2 and HCl refer to a molecule of chlorine which is made up of two chlorine atoms, and to a molecule of hydrogen chloride which is made up of one hydrogen atom and one chlorine atom (Fig. 6-10). With the possible exception of the gas phase, we should not speak of molecules of ionically bonded substances. The formulas for the ionically bonded substances NaCl and $CaCl_2$ merely tell us that for every sodium ion there is one chloride ion in sodium chloride, and that for every calcium ion there are two chloride ions in calcium chloride.

Now that we have discussed ionic bonding and covalent bonding, we can compare the properties of ionic compounds with those of covalent compounds (Table 6-2).

Figure 6-10
A model of hydrogen chloride (HCl)

Table 6-2   A Comparison of Ionic and Covalent Compounds

	Ionic Compounds	Covalent Compounds
Bond type	ionic	covalent
Component particles	ions	molecules
State at room temperature	solid	solid, liquid, or gas
Melting point	high	usually low
Boiling point	high	usually low
Hardness	hard but brittle	soft
Electrical conductivity -of liquid -of solid	conducts does not conduct	does not conduct does not conduct
Solubility in water	usually soluble	usually insoluble

APPLICATION

"Seeing" Chemical Bonds

In 1981 a device for viewing individual atoms on or near the surface of a sample was developed by IBM researchers. This device is called a scanning tunnelling microscope (STM). The microscope has a needlelike probe tip which is moved over, and close to, a surface. Both the distance separating the sharply-pointed metal tip from the surface and a voltage applied to the tip are adjusted so that electrons "tunnel" from the tip to the surface. The tiny electric current that flows between the tip and the surface of a sample is used to trace a contour map of the sample's surface. The motion of the probe across the surface is monitored by a computer.

A study of the surface of a sample of silicon using the STM showed not only the position of the silicon atoms on the surface, but also images of the bonds holding the atoms in place. This technique gave the locations of bonding orbitals and those partially-filled orbitals that were not involved in bonding. For instance, it showed partially filled nonbonding orbitals of the top layer of silicon. It also showed partially-filled nonbonding orbitals that poke up between silicon atoms in the top layer from silicon atoms underneath (in the second layer). In addition, it showed chemical bonds involving orbitals that laterally connect neighbouring atoms in the second layer. The scanning tunnelling microscope will give valuable information about chemical bonding and the surface of matter. This technique will also be used to study chemical reactions that occur on the surface of catalysts.

6.6 The Octet Rule

In our discussion of covalent bonding, we used the concept of atoms forming bonds so as to become isoelectronic with noble gases. In this section we will consider some molecules made up of more than two atoms per molecule. These compounds will involve elements from the second period of the periodic table. We could use the same concept again to determine the formula of each compound. However, instead we will use a more convenient method—the octet rule.

All noble gases (except He) have eight valence electrons. Many atoms react with other atoms in such a way as to end up with eight valence electrons. The **octet rule** states that when atoms combine, bonds form so that each atom finishes with an octet of valence electrons.

There is no clear explanation for an atom's acquiring an octet of valence electrons. In fact, there are some cases where an atom involved in bonding finishes with fewer than an octet of valence electrons. For example, hydrogen stops at two valence electrons, making it isoelectronic to the noble gas helium, which also has two electrons. There are other cases where an atom finishes with more than eight valence electrons. However, for many molecules, there is a special stability associated with atoms having completed octets, and the octet rule is a useful generalization.

Suppose that an oxygen atom ($\cdot \ddot{\text{O}} :$) were to form a bond with one hydrogen atom (H$\cdot$) :

$$\text{H} \cdot + \cdot \ddot{\text{O}} : \longrightarrow \text{H} - \ddot{\text{O}} :$$

By sharing a pair of electrons with the oxygen atom, the hydrogen atom has become isoelectronic with the noble gas helium. However, the oxygen atom has not completed an octet of valence electrons. It still requires one more electron. An H—O molecule is reactive, and the oxygen atom has the ability to form a second bond to another hydrogen atom:

$$H\cdot\ +\ H\!-\!\overset{\cdot\cdot}{\underset{\cdot}{O}}\!: \ \longrightarrow\ H\!-\!\overset{\cdot\cdot}{\underset{\underset{\displaystyle H}{|}}{O}}\!:$$

This compound, formed by the reaction of hydrogen and oxygen, is water, H_2O.

A nitrogen atom ($\cdot\overset{\cdot\cdot}{\underset{\cdot}{N}}\cdot$) has five valence electrons. It requires three more electrons to complete an octet, and we would expect it to bond to three hydrogen atoms:

$$3H\cdot\ +\ \cdot\overset{\cdot\cdot}{\underset{\cdot}{N}}\cdot \longrightarrow\ H\!-\!\overset{\cdot\cdot}{\underset{\underset{\displaystyle H}{|}}{N}}\!-\!H$$

The compound NH_3 is called ammonia. It is found in household cleaners and is used to make fertilizers.

A carbon atom ($\cdot\overset{\cdot}{\underset{\cdot}{C}}\cdot$) has four valence electrons. It requires four more electrons to complete an octet, and we would expect it to bond to four hydrogen atoms:

$$4H\cdot\ +\ \cdot\overset{\cdot}{\underset{\cdot}{C}}\cdot \longrightarrow\ H\!-\!\overset{\overset{\displaystyle H}{|}}{\underset{\underset{\displaystyle H}{|}}{C}}\!-\!H$$

The compound CH_4 is called methane.

We will finish this section by showing a method for writing the Lewis structures for molecules whose formulas are known. First, what is the Lewis structure of the molecule whose formula is N_2F_4? A nitrogen atom ($\cdot\overset{\cdot\cdot}{\underset{\cdot}{N}}\cdot$) has five valence electrons and requires three more. A fluorine ($\cdot\overset{\cdot\cdot}{\underset{\cdot\cdot}{F}}\!:$) atom has seven valence electrons and requires one more. We will attempt to draw the Lewis structure so that the four fluorine atoms and the two nitrogen atoms obey the octet rule. As a rule it is preferable to start with atoms which require the most electrons to complete their octets. In this case, the nitrogen atoms each require three electrons. We start by indicating one shared pair of electrons between the two nitrogen atoms:

$$\cdot\overset{\cdot\cdot}{\underset{\cdot}{N}}\cdot\ +\ \cdot\overset{\cdot\cdot}{\underset{\cdot}{N}}\cdot \longrightarrow \cdot\overset{\cdot\cdot}{\underset{\cdot}{N}}\!-\!\overset{\cdot\cdot}{\underset{\cdot}{N}}\cdot$$

Now, each nitrogen atom requires two electrons, and can form two bonds (a total of four). Since each of the four fluorine atoms requires one electron and can form one bond, the Lewis structure can be completed:

$$\cdot\overset{\cdot\cdot}{\underset{\cdot}{N}}\!-\!\overset{\cdot\cdot}{\underset{\cdot}{N}}\cdot\ +\ 4\cdot\overset{\cdot\cdot}{\underset{\cdot\cdot}{F}}\!: \longrightarrow\ :\overset{\cdot\cdot}{\underset{\cdot\cdot}{F}}\!-\!\overset{\cdot\cdot}{\underset{\underset{\displaystyle :\overset{\cdot\cdot}{\underset{\cdot\cdot}{F}}:}{|}}{N}}\!-\!\overset{\cdot\cdot}{\underset{\underset{\displaystyle :\overset{\cdot\cdot}{\underset{\cdot\cdot}{F}}:}{|}}{N}}\!-\!\overset{\cdot\cdot}{\underset{\cdot\cdot}{F}}\!:$$

In this structure each atom *does* obey the octet rule.

Next, what is the Lewis structure of BF_3? A boron atom ($\cdot\overset{}{\underset{\cdot}{B}}\cdot$) has three valence electrons, and a fluorine atom ($\cdot\overset{\cdot\cdot}{\underset{\cdot\cdot}{F}}\!:$) has seven valence electrons. If we start with the atom that requires the most electrons to complete an octet ($\cdot \ddot{B} \cdot$), we can bond all three fluorine atoms to the one boron atom:

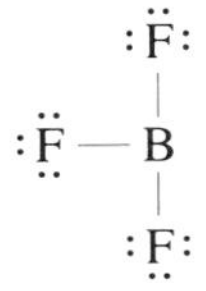

$$
\begin{array}{c}
:\ddot{F}: \\
| \\
:\ddot{F} - B \\
| \\
:\ddot{F}:
\end{array}
$$

In this Lewis structure, each fluorine atom does obey the octet rule; however, the boron atom does not achieve an octet.

What is the Lewis structure of a molecule whose formula is CH_4O? A carbon atom ($\cdot \dot{C} \cdot$) has four valence electrons; an oxygen atom ($:\ddot{O}\cdot$) has six valence electrons; and a hydrogen atom ($H\cdot$) has one valence electron. We start by bonding the atom that requires the most electrons to complete an octet ($\cdot \dot{C} \cdot$) to the atom that requires the next largest number of electrons ($:\ddot{O}\cdot$):

$$
\cdot \dot{C} - \ddot{O}:
$$

The carbon atom has three remaining valence electrons that can bond with three of the four hydrogen atoms. The oxygen atom has one remaining valence electron that can bond with the remaining hydrogen atom:

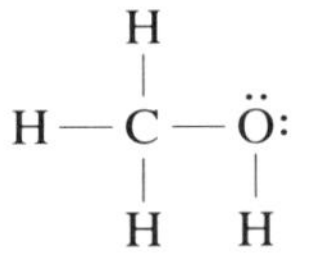

$$
\begin{array}{c}
H \\
| \\
H - C - \ddot{O}: \\
| \quad | \\
H \quad H
\end{array}
$$

What is the Lewis structure of the O_2^{2-} ion? An oxygen atom ($:\ddot{O}\cdot$) has six valence electrons. If two oxygen atoms bond together, they each have only seven valence electrons:

$$
:\ddot{O} - \ddot{O}:
$$

However, we are dealing with an ion which has a charge of 2−. Thus, two extra electrons must be incorporated into the Lewis structure:

$$
(:\ddot{\underset{..}{O}} - \ddot{\underset{..}{O}}:)^{2-}
$$

Each oxygen atom now obeys the octet rule.

PRACTICE PROBLEM 6-8

Draw the Lewis structure of a molecule whose formula is H_2O_2.

PRACTICE PROBLEM 6-9

Draw the Lewis structure of a molecule whose formula is NH_3O.

6.7 Multiple Bonds

Double Covalent Bonds

So far, we have drawn Lewis structures only for molecules having single covalent bonds. In this section, we consider molecules with bonds in which more than one pair of electrons are shared.

A carbon dioxide (CO_2) molecule has one carbon atom bonded to two oxygen atoms. The carbon atom ($\cdot\overset{\cdot}{C}\cdot$) requires four electrons and the oxygen atoms ($\cdot\overset{\cdot\cdot}{O}\colon$) each require two electrons. We start by indicating one shared pair between the carbon atom and an oxygen atom:

$$\cdot\overset{\cdot}{C} - \overset{\cdot\cdot}{O}\colon$$

We continue by indicating one shared pair between the carbon atom and a second oxygen atom:

$$\colon\overset{\cdot\cdot}{O} - \overset{\cdot}{C} - \overset{\cdot\cdot}{O}\colon$$

The carbon atom requires two more electrons, and each oxygen atom requires one more electron. At present, none of the atoms obeys the octet rule. However, if two pairs of electrons are shared between the carbon atom and each of the two oxygen atoms, the three atoms in the molecule will each obey the octet rule:

$$\colon\overset{\cdot\cdot}{O} - \overset{\cdot}{C} - \overset{\cdot\cdot}{O}\colon \longrightarrow \colon O = C = O\colon$$

When two pairs of electrons are shared between two atoms, the result is a **double covalent bond**.

EXAMPLE PROBLEM 6-4

Draw the Lewis structure for formaldehyde, CH_2O, a compound used to preserve biological specimens.

SOLUTION

The carbon atom ($\cdot\overset{\cdot}{C}\cdot$) requires four electrons; the oxygen atom ($\cdot\overset{\cdot\cdot}{O}\colon$) requires two electrons; and the hydrogen atoms ($\cdot H$) each require one electron. We will start by indicating one shared pair between the carbon atom and the atom which requires the next greatest number of electrons, oxygen:

$$\cdot\overset{\cdot}{C} - \overset{\cdot\cdot}{O}\colon$$

Since the carbon atom still requires the most electrons, we will bond each of the two hydrogen atoms to it:

$$H - \overset{\overset{\displaystyle H}{|}}{\underset{\cdot}{C}} - \overset{\cdot\cdot}{O}\colon$$

This results in both the carbon atom and the oxygen atom being one electron short of an octet. This problem can be solved by having the carbon atom and the oxygen atom share two pairs of electrons:

$$H - \overset{\overset{\displaystyle H}{|}}{\underset{\cdot}{C}} - \overset{\cdot\cdot}{O}\colon \longrightarrow H - \overset{\overset{\displaystyle H}{|}}{C} = O\colon$$

Now the carbon atom and the oxygen atom both obey the octet rule.

PRACTICE PROBLEM 6-10

Draw a Lewis structure for carbon disulfide, CS_2.

Triple Covalent Bonds

The nitrogen molecule (N_2) is formed when two nitrogen atoms share electrons. If the two nitrogen atoms were to share a pair of electrons, neither of them would obey the octet rule:

$$:\overset{..}{\underset{..}{N}} - \overset{..}{\underset{.}{N}}:$$

If two pairs of electrons were shared between the two nitrogen atoms, each atom would still have only seven valence electrons:

$$:\underset{.}{N} = \underset{.}{N}:$$

However, if three pairs of electrons were shared between the two nitrogen atoms, each would obey the octet rule:

$$:\overset{.}{\underset{.}{N}}\cdot + \cdot\overset{.}{\underset{.}{N}}: \longrightarrow :N \equiv N:$$

When three pairs of electrons are shared between two atoms, the result is a **triple covalent bond**.

EXAMPLE PROBLEM 6-5

Draw the Lewis structure for hydrogen cyanide, HCN, a poisonous gas which paralyzes the central nervous system and stops respiration.

SOLUTION

The carbon atom ($\cdot\overset{.}{C}\cdot$) requires four electrons; the nitrogen atom ($:\overset{..}{N}\cdot$) requires three electrons; and the hydrogen atom ($\cdot H$) requires one electron. We will select carbon as the central atom, and we will indicate one shared pair between the carbon atom and the nitrogen atom:

$$\cdot\overset{.}{C} - \overset{..}{\underset{.}{N}}:$$

Since the carbon atom still requires the most electrons, we will bond the hydrogen atom to it:

$$H - \overset{.}{\underset{.}{C}} - \overset{..}{\underset{.}{N}}:$$

This means both the carbon atom and the nitrogen atom are two electrons short of an octet. The problem can be solved by having the carbon atom and the nitrogen atom share three pairs of electrons:

$$H - \overset{.}{\underset{.}{C}} - \overset{..}{\underset{.}{N}}: \longrightarrow H - C \equiv N:$$

Now the carbon atom and the nitrogen atom both obey the octet rule.

PRACTICE PROBLEM 6-11

Draw the Lewis structure for the ethyne molecule, C_2H_2.

Part Two Review Questions

1. What is a polar covalent bond? How does it differ from a pure covalent bond?
2. Define the term *electronegativity*.
3. Which element has the largest electronegativity? Which has the smallest electronegativity?
4. How do electronegativity values help us predict whether a bond is pure covalent, polar covalent, or ionic?
5. Why does the word molecule not apply to ionic solids?
6. State the octet rule. Does it always apply?
7. What is a double covalent bond? What is a triple covalent bond?
8. Draw the Lewis structures for H_2S and C_2HBr.

Part Three: Some Additional Concepts of Chemical Bonding

6.8 The Shapes of Molecules—VSEPR Theory

Molecules which are made up of one atom bonded to only one other atom are linear molecules. In Sections 6.3 and 6.4 we discussed some molecules of that type; however, in Sections 6.6 and 6.7 we discussed molecules which have other shapes. Lewis structures do not indicate the true shapes of molecules. In this section, we will use a simple, yet useful, theory to predict the shapes of some of the molecules that we have already discussed.

Shapes of Molecules with Single Bonds

Consider the methane molecule:

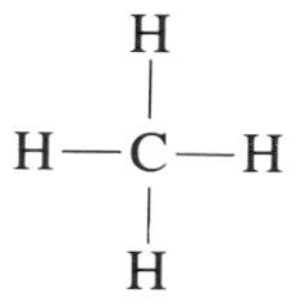

$$
\begin{array}{c}
\text{H} \\
| \\
\text{H}-\text{C}-\text{H} \\
| \\
\text{H}
\end{array}
$$

Figure 6-11
Arrangement of orbitals
(a) Linear arrangement
(b) Trigonal planar arrangement
(c) Tetrahedral arrangement

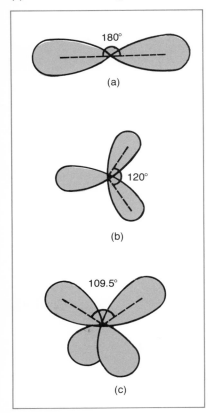

(a)

(b)

(c)

The carbon atom (the central atom) is surrounded by four shared pairs of electrons. These electrons make up an octet in what is called the valence shell of the carbon atom. The **valence shell** is the occupied energy level with the highest principal quantum number. The valence electrons of an atom occupy the valence shell.

The **Valence-Shell Electron-Pair Repulsion (VSEPR) Theory** proposes that the arrangement of atoms around a central atom in a molecule depends upon the repulsions between all of the electron pairs in the valence shell of the central atom. These valence shell electrons are regarded as occupying localized regions of space, called orbitals, which are located around the filled inner shells of the atom so that their average distance from each other is maximized. The basic postulate of VSEPR Theory is that electron pairs in the valence shell repel each other and, therefore, reside in orbitals which are as far apart as possible.

The most probable arrangements of two, three, and four electron pairs are linear, trigonal planar, and tetrahedral, respectively (Fig. 6-11). In each of these arrangements, the electron pairs (residing in orbitals) are as far apart from each other as possible. Stated another way, the angles between the orbitals in each arrangement are maximized. The angle between two linear orbitals is 180°; the angles between three trigonal planar orbitals are each 120°; and the angles between four tetrahedral orbitals are each 109.5°.

In the case of the carbon atom in methane, there are four pairs of valence shell electrons occupying four orbitals. Since each pair of valence electrons involves a carbon electron and a hydrogen electron, they are called shared pairs. According to the basic postulate of VSEPR Theory, these *four* orbitals should be as far apart as possible, giving a tetrahedral arrangement of the four shared pairs of electrons. Thus, VSEPR Theory predicts that the four hydrogen atoms will be arranged tetrahedrally around the carbon atom in the methane molecule (Fig. 6-12).

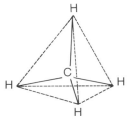

Figure 6-12
Model of methane (CH$_4$)

There is a great deal of evidence to support this view of the structure of the methane molecule. For example, experiments indicate that all bond angles in methane are 109.5°.

Shapes of Molecules with Single Bonds and Lone Electron Pairs

Consider the ammonia molecule:

$$H—\ddot{N}—H$$
$$|$$
$$H$$

The nitrogen atom (the central atom) is surrounded by four pairs of electrons. Three of the pairs are shared pairs, but the fourth is a nonbonding (lone) pair of electrons. How does the presence of a lone pair of electrons in the valence shell of nitrogen affect the shape of the ammonia molecule?

Because they are under the influence of only one nucleus, the two electrons of a lone pair occupy a bigger, more puffy orbital than the electrons of a shared pair which are attracted to two nuclei (Fig. 6-13). Thus, a lone-pair orbital will overlap neighbouring orbitals more extensively, and the electrons in it will repel the electrons in these neighbouring orbitals more strongly than the electrons in a shared-pair orbital would. The repulsions between electron pairs in a valence shell decrease in this order: lone pair—lone pair, then lone pair—shared pair, and finally shared pair—shared pair.

In ammonia, the four pairs of electrons are arranged in a tetrahedral fashion; however, the lone-pair electrons push the three shared electron pairs together slightly:

No atom is bonded to the lone pair in ammonia, and the molecule has a trigonal pyramidal shape:

Furthermore, we would expect that the bond angles would be slightly less than the 109.5° angles of a tetrahedron, because the lone-pair electrons push the other electron pairs together slightly. Experiments indicate that ammonia *is* a trigonal pyramidal molecule (Fig. 6-14) and that the bond angles are 107.3°.

We will next use VSEPR Theory to predict the shape of the water molecule:

$$H - \overset{\cdot\cdot}{O} :$$
$$|$$
$$H$$

The oxygen atom (the central atom) is surrounded by two shared pairs of electrons and two lone pairs of electrons. These four pairs of electrons are arranged in a tetrahedral fashion; however, the two lone pairs are expected to push the two shared pairs together even more than was the case with ammonia:

No atoms are bonded to the lone pairs in water, and the molecule has an angular shape:

Furthermore, we would expect that the bond angle is even less than 107.3°. Experiments indicate that water *is* an angular molecule and that the bond angle is 104.5°.

Figure 6-13
Representation of the approximate shapes of (a) a shared-pair orbital and (b) an unshared-pair orbital

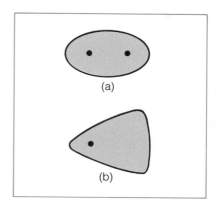

Figure 6-14
Model of ammonia (NH$_3$)

Shapes of Molecules with Multiple Bonds

Let us conclude this section by considering two molecules which have multiple (in these examples, double) bonds—formaldehyde and carbon dioxide. For the purposes of VSEPR Theory, the two shared pairs of a double bond can be considered as occupying one four-electron orbital, and the three shared pairs of a triple bond can be considered as occupying one six-electron orbital. Triple-bond orbitals, because they contain three shared pairs of electrons, are larger than double-bond orbitals, which contain two shared pairs of electrons, and double-bond orbitals are larger than single-bond orbitals. Because they are larger, multiple-bond orbitals repel other orbitals more strongly than single-bond orbitals do. Thus, the repulsions exerted by bonding orbitals decrease in this order: triple-bond orbital, then double-bond orbital, and finally single-bond orbital.

The Lewis structure of formaldehyde is

$$\begin{array}{c} H \\ | \\ H - C = \ddot{O} \colon \end{array}$$

The carbon atom is surrounded by four pairs of electrons occupying three orbitals. The three orbitals are expected to have a trigonal planar arrangement. Since the double-bond orbital is a larger orbital requiring more space, it should push the two single-bond orbitals together slightly. Thus, the H—C—H bond angle should be slightly less than 120°; the two H—C—O bond angles should be

Biography

RONALD J. GILLESPIE was born in London, England in 1924. He studied at University College, London, where he obtained his B. Sc. in 1944 and his Ph. D. in 1947. He now lives in Canada and is a professor of chemistry at McMaster University.

Dr. Gillespie has done research on nonaqueous strong acids, vibrational spectroscopy, and nuclear magnetic resonance spectroscopy, but he is perhaps best known for his important contributions to the Valence-Shell Electron-Pair Repulsion (VSEPR) Theory. He is the author of several chemistry books and has written more than 300 articles in professional journals dealing with inorganic chemistry and with chemistry in education.

Ronald Gillespie has received numerous awards in recognition of his outstanding achievement in inorganic chemical research and in teaching. In 1977 he was the winner of the Chemical Institute of Canada Medal. In 1987, he was awarded the prestigious Izaak Walton Killam Memorial Prize which acknowledges his outstanding contribution to the advancement of knowledge in chemistry.

Figure 6-15
Ronald J. Gillespie (b. 1924)—
Promoted VSEPR Theory.

slightly more than 120°. Nevertheless, the structure of formaldehyde should be basically trigonal planar:

$$
\begin{array}{c}
\ddot{O} \\
\parallel \\
C \\
\diagup \quad \diagdown \\
H \qquad H
\end{array}
$$

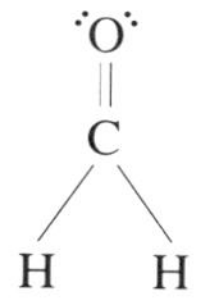

Experiments indicate that formaldehyde *is* basically trigonal planar. The H—C—H bond angle is 118°, and the two H—C—O bond angles are 121°.

Now let us use VSEPR Theory to predict the shape of the carbon dioxide molecule. The Lewis structure of carbon dioxide is

$$
\ddot{O}\!\!=\!\!C\!\!=\!\!\ddot{O}
$$

The carbon atom is surrounded by four pairs of electrons occupying two orbitals. The two orbitals are expected to have a linear arrangement. Thus, carbon dioxide should be a linear molecule, and experiments confirm this prediction.

PRACTICE PROBLEM 6-12

Use VSEPR Theory to predict the shapes of the following molecules:
a) CCl_4; b) NF_3; c) H_2S.

We have shown that the shape of a molecule depends upon the repulsions among all of the electron pairs in the valence shell of the central atom. The most probable arrangements of two, three, and four electron pairs are linear, trigonal planar, and tetrahedral, respectively.

APPLICATION

The Importance of Molecular Shape

In this chapter we used VSEPR theory to predict the shapes of some molecules. Why should we be concerned with the shapes of molecules? As we shall learn from the following examples, the shape of a molecule can influence such fundamental properties as the melting and boiling points of compounds. For example, the boiling point of butane, a rod-shaped molecule, is 0°C. Isobutane has the same number of carbon and hydrogen atoms but a more spherical shape. Its boiling point is −12°C. Let's look at the molecular shapes of butane and isobutane:

$$
CH_3\!-\!CH_2\!-\!CH_2\!-\!CH_3 \qquad\qquad
\begin{array}{c}
CH_3\!-\!CH\!-\!CH_3 \\
\mid \\
CH_3
\end{array}
$$

butane   isobutane

The rod shape of butane molecules allows them to pack together more compactly, so that the attractive forces between them can act more efficiently. Therefore, a higher temperature is required in order to separate the molecules from each other.

Sulfa drugs such as sulfanilamide are important antibiotics. The shapes of their molecules are important factors in determining their activities against bacteria. For example, bacteria need para-aminobenzoic acid (PABA) to synthesize folic acid, which is necessary for the growth of the bacteria. When sulfanilamide is administered to a patient, the bacteria incorporate it into their cells, mistaking it for PABA which has a similar shape. The result is that the bacteria lack an essential growth-producing compound, so they cease to grow.

Molecules can transmit messages within living organisms by fitting into appropriate receptor sites in a very specific way that is determined by their shape. Appropriate chemical processes are stimulated when a molecule with the correct shape occupies a specific receptor site. Current theories of smell and taste are based on this type of interaction. For example, the odour of many substances depends upon how a molecule fits into a receptor site in the nasal cavity. Thus, molecules may have similar odours if they have the same shape, even if they contain different types of atoms and have different chemical properties. Our sense of taste also depends upon how molecules fit receptor sites on the tongue. It is possible to trick these receptors. The molecules of artificial sweeteners fit the sites of taste buds that stimulate a "sweet" response in the brain. In this way, these molecules provide the sensation of a sweet taste but do not contain the unwanted energy of "natural" sugars. Studies of molecular shapes are therefore important in the food and perfume industries.

Plants and animals use chemical communication; that is, they relay messages by releasing chemicals that are picked up by other plants or animals. For example, ants lay a chemical trail to help other ants find a particular food supply. They also warn each other of anticipated danger by emitting certain chemicals. Female insects release chemicals that act as sex attractants for males. In many cases, it is possible to determine the chemical structure of the attractant, and to synthesize it or a molecule with a similar shape. The synthetic attractant can then be used to bait traps for the appropriate insects.

6.9 Coordinate Covalent Bonding

A coordinate covalent bond is another type of covalent bond. In a coordinate covalent bond a pair of electrons is again shared between a pair of atoms; however, both of the electrons of the shared pair were originally part of *one* of the two atoms. A **coordinate covalent bond** is a covalent bond in which both electrons of the shared pair come from only one of the bonded atoms.

The Lewis structure of BF_3 reveals that the boron atom does not have an octet in its valence shell:

One of the valence orbitals of boron is empty, and the chemical behaviour of this molecule reflects the vacancy. It is attracted to molecules that have atoms with unshared pairs of electrons which can be donated to the boron. For example, BF_3 reacts with NH_3 to form BF_3NH_3:

$$
\begin{array}{cc}
\text{H} & :\ddot{\text{F}}: \\
| & | \\
\text{H}-\text{N}: \;+\; \text{B}-\ddot{\text{F}}: \\
| & | \\
\text{H} & :\ddot{\text{F}}:
\end{array}
\;\longrightarrow\;
\begin{array}{cc}
\text{H} & :\ddot{\text{F}}: \\
| & | \\
\text{H}-\text{N}-\text{B}-\ddot{\text{F}}: \\
| & | \\
\text{H} & :\ddot{\text{F}}:
\end{array}
$$

The bond between N and B is a coordinate covalent bond. Both electrons in the N-B bond were originally part of the nitrogen atom. However, once the bond has formed between the nitrogen atom and the boron atom, it is impossible to distinguish it from any bond in which a pair of electrons is shared between two atoms. That is, no experiment can distinguish a coordinate covalent bond from a covalent bond after the bonding is completed. Electrons are the same whether they were originally found in the orbitals of a nitrogen atom or a boron atom.

The BF_4^- ion can be thought of as a BF_3 molecule which has been bonded to a fluoride ion (a fluorine atom which has picked up an eighth valence electron from some outside source):

$$
\begin{array}{c}
:\ddot{\text{F}}-\text{B}-\ddot{\text{F}}: \\
| \\
:\ddot{\text{F}}:
\end{array}
\;+\; :\ddot{\text{F}}:^- \;\longrightarrow\;
\left[
\begin{array}{c}
:\ddot{\text{F}}: \\
| \\
:\ddot{\text{F}}-\text{B}-\ddot{\text{F}}: \\
| \\
:\ddot{\text{F}}:
\end{array}
\right]^-
$$

We know, by following the process through which the Lewis structure of BF_4^- is established, that one of the bonds must be a coordinate covalent bond. However, the four bonds in BF_4^- are all identical after bonding has occurred. In each case a pair of electrons is shared between a fluorine atom and the boron atom. Also, BF_4^- is a negative ion because one of the fluorine atoms bonded to the boron atom was a negative fluoride ion.

PRACTICE PROBLEM 6-13

While the BH_3 molecule does not exist, the BH_4^- ion does exist. Write the Lewis structure for BH_4^- assuming that a hydride ion ($H:^-$)combines with BH_3. What is the shape of the BH_4^- ion?

PRACTICE PROBLEM 6-14

Write the Lewis structure for NH_4^+ assuming that a hydrogen ion (H^+) combines with NH_3. What is the shape of the NH_4^+ ion?

Biography

Figure 6-16
Dorothy Crowfoot Hodgkin
(b. 1910)—Determined structures of
complex molecules.

DOROTHY CROWFOOT was born in Cairo, Egypt, in 1910, and went to school in England. Soon after she entered Oxford University in 1928, she became interested in the exciting structures chemists had proposed for large, complex molecules. She wondered if it would be possible to "see" whether complicated molecules had the structures suggested. By her second year, she was already using X-rays to discover the structures of the molecules of complex substances. She graduated from Oxford in 1932 and received her Ph. D. in 1936 from Cambridge University. In 1937, she married Thomas Hodgkin.

Dorothy Crowfoot Hodgkin spent most of her career teaching and researching at Oxford University. Her research continued to focus on the use of X-rays to determine the structures of complex molecules from natural sources.

In 1942, information on the structure of penicillin was vital for the synthesis of the large quantities of the antibiotic needed during the war. She began the structural analysis of penicillin, and by 1946 had determined its structure. Later, she determined the structure of cephalosporin C, a drug closely related to the penicillin family but able to wipe out many of the infections that resisted penicillin.

From 1948 to 1954 she and her colleagues worked on the structure of vitamin B_{12}, one of the most complicated molecules so far found in nature. Vitamin B_{12} is essential for the building of red blood cells. The structure of vitamin B_{12} was finally determined after almost a dozen years of work on both sides of the Atlantic. Hodgkin and her co-workers found that the vitamin B_{12} molecule was almost spherical in form, with the most chemically reactive groups on its surface. This knowledge shed new light on theories of the blood-building process and demonstrated that it is possible to think about these huge molecules in three dimensions.

Dorothy Hodgkin has made two scientific reputations for herself, one under her maiden name and one under her married name. Many chemists were suprised to learn that the Crowfoot of penicillin fame and the Hodgkin of vitamin B_{12} fame were the same person. For her work on penicillin, vitamin B_{12}, and other molecules, Dr. Hodgkin became, in 1964, the third woman to win the Nobel Prize in Chemistry. (The first two were Marie Curie and her daughter, Irène Joliot-Curie.) Her citation praised "her exceptional skill in which chemical knowledge, intuition, imagination, and perseverance have been conspicuous."

6.10 Chemical Bonding and Quantum Mechanics

Many chemists have made extensive use of quantum mechanics in their descriptions of the chemical bond. There are two general methods of description. First, we can consider the entire molecule as a unit with each electron moving under the influence of all the nuclei and all the other electrons in the molecule. This approach suggests that every electron belongs to the molecule as a whole and can move anywhere throughout the molecule. Thus, we say that the electrons occupy *molecular orbitals*. Second, we can consider that the atoms in molecules behave as if they were isolated atoms, except that one or more electrons in the outer *s* and *p* orbitals of one atom can move to the outer *s* and *p* orbitals of another atom. This description is called the *atomic orbital* or *valence bond* approach. It has its shortcomings, but its chief advantage is that it is easier to visualize than the molecular orbital approach. The valence bond approach will be used in this section as we describe chemical bonding in terms of quantum mechanics.

Covalent Bonding

The main consideration of the valence bond approach is that stable compounds result from a tendency to fill all the orbitals of the valence shells with electron pairs—shared pairs or unshared (lone) pairs. A covalent bond may be described as the result of the overlapping of two atomic orbitals, one on each of two atoms, so that the two orbitals can be occupied by a shared pair of electrons. Let us take a second look at the formation of a bond between two hydrogen atoms. Each hydrogen atom has one electron in a 1s orbital.

As two hydrogen atoms approach one another, it is possible for the two orbitals to overlap:

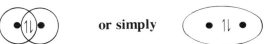

The pair of electrons is now simultaneously attracted to two nuclei and a covalent bond forms. This simultaneous attraction of the electron pair to two nuclei produces a more stable state, and a covalent bond is formed. A simple method of showing the orbitals that are involved in bonding is called the *orbital diagram*. This method represents the outer *s* and *p* (i.e., valence) orbitals by small squares and their electron populations by arrows. A rectangle is drawn around orbitals that are involved in bonding. For the hydrogen molecule the orbital diagram is

Chapter 6

A single line drawn between two atoms (e.g., H—H) indicates that two electrons are shared between the two atoms. This is a single covalent bond.

Suppose two fluorine atoms ($1s^2 2s^2 2p^5$) come together and form a bond. Each fluorine atom has seven electrons in its outer s and p orbitals. Therefore, each fluorine requires one more electron to achieve the electron configuration of the noble gas, neon. When the fluorine atoms collide, each tries to gain an eighth valence electron but neither fluorine atom can pull an electron from the other. They share a pair of electrons:

$$:\ddot{F}\cdot + \cdot\ddot{F}: \longrightarrow :\ddot{F}:\ddot{F}:$$

Each fluorine has a half-filled $2p$ orbital:

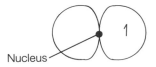

Nucleus

As the two fluorine atoms approach one another, it is possible for the two orbitals to overlap:

For the fluorine molecule, the orbital diagram is

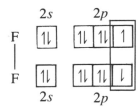

In the same way that two fluorine atoms form a single covalent bond, two chlorine atoms also form a single covalent bond. The electron dot formula for a chlorine molecule is ($:\ddot{C}l:\ddot{C}l:$). The orbital diagram is the same as the one shown for fluorine except that the orbitals involved are the $3s$ and $3p$, not the $2s$ and $2p$.

PRACTICE PROBLEM 6-15

Describe the bonding which occurs in a molecule of iodine. Draw an orbital overlap picture and an orbital diagram. What orbitals are involved in the bonding?

Polar Covalent Bonding

When a hydrogen atom and a chlorine atom react, the H—Cl molecule forms. The bonding orbitals involved are the 1s orbital of the hydrogen atom and a 3p orbital of the chlorine atom:

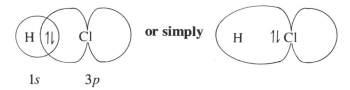

1s   3p

For the hydrogen chloride molecule, the orbital diagram is

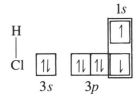

Because the chlorine atom is more electronegative than the hydrogen atom, the shared pair of electrons spends more of its time near the chlorine end of the molecule. Thus, the overlap of the 1s orbital of the hydrogen with the 3p orbital of the chlorine results in a polar covalent bond. The chlorine end of the molecule is slightly negative, and the hydrogen end of the molecule is slightly positive.

Another example of a polar covalent bond is the bond between the chlorine atom and the bromine atom in the molecule BrCl. Each of these atoms is in Group 17 and has seven valence electrons in its outer s and p orbitals. By sharing a pair of electrons each completes its outer s and p orbitals. The orbital diagram is

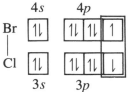

Since a chlorine atom is more electronegative than a bromine atom, the shared pair of electrons spends more time near the chlorine nucleus. The chlorine end of the molecule is slightly negative, and the bromine end of the molecule is slightly positive.

PRACTICE PROBLEM 6-16

Draw an orbital diagram for ICl. Is this a polar molecule? If it is polar, which end of the molecule is slightly negative?

Let us look at an example of a molecule with more than one polar covalent bond. Suppose that oxygen ($1s^2\ 2s^2\ 2p^4$) were to form a bond with hydrogen:

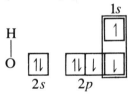

The oxygen still has one unpaired electron in a $2p$ orbital. This is closer to fluorine ($1s^2\ 2s^2\ 2p^5$) than it is to a noble gas electron configuration. The H—O molecule is reactive since the oxygen has the ability to form a second bond to another hydrogen atom:

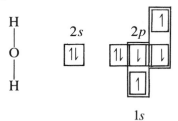

The bonding orbitals of the oxygen are two of the $2p$ orbitals, and these $2p$ orbitals are at right angles to each other:

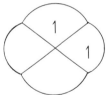

This diagram is a good representation of the two $2p$ orbitals, since it correctly shows that the orbitals partially penetrate one another. However, it is frequently simplified to show the directional nature of the two $2p$ orbitals better:

When the two hydrogen atoms bond to the oxygen atom, their $1s$ orbitals must overlap with the two $2p$ orbitals:

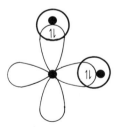

Thus, we might expect the two hydrogen-oxygen bonds to be at 90° to one another. Therefore, this study of the orbitals involved in bonding suggests that water will be a bent molecule:

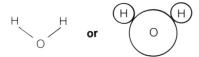

Since the electronegativity of the oxygen atom (3.5) is larger than the electronegativity of the hydrogen atoms (2.1), the bonding will be polar covalent. The oxygen end of the water molecule should be more negative than the hydrogen end:

$$H^{\delta+} \qquad {}^{\delta+}H$$
$$O_{\delta-}$$

Experiments confirm that water is a bent, polar molecule. The $H-O-H$ bond angle is greater than 90°, and this would be expected since the two hydrogen nuclei are likely to repel each other slightly.

Let us now examine the bonding in the ammonia molecule, NH_3. Nitrogen $(1s^2\,2s^2\,2p^3)$ has three unpaired electrons and we would expect a nitrogen atom to bond to three hydrogen atoms:

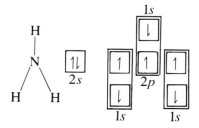

The bonding orbitals of the nitrogen are the three $2p$ orbitals, and these orbitals are all at right angles to each other:

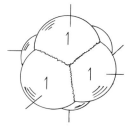

This diagram correctly indicates that the three $2p$ orbitals partially penetrate one another. However, it is frequently simplified to show the directional nature of these orbitals better:

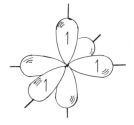

When the three hydrogen atoms bond to the nitrogen their $1s$ orbitals must overlap the three $2p$ orbitals:

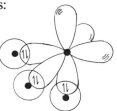

This study of the orbitals involved in bonding suggests that ammonia is a pyramidal molecule (Fig. 6-14). Since the electronegativity of the nitrogen atom (3.0) is larger than the electronegativity of the hydrogen atoms (2.1), the bonding will be polar covalent. The nitrogen end of the molecule should be slightly more negative than the hydrogen end:

Chemists have confirmed these predictions. The study also predicts that the H—N—H bond angles in ammonia should be greater than 90° due to repulsions among the positive hydrogen nuclei. This prediction has also been confirmed.

Now let us consider the bonding in a molecule consisting of carbon and hydrogen. It has been demonstrated that carbon ($1s^2\ 2s^2\ 2p^2$) bonds with four hydrogen atoms to form the molecule CH_4 (methane). In order to have four half-filled orbitals available for bonding, we must postulate that during bonding one of the $2s$ electrons is promoted to a vacant $2p$ orbital:

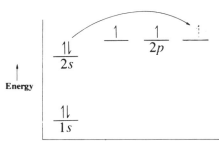

When the promotion is complete, carbon has four half-filled valence orbitals:

A carbon atom is then able to bond to four hydrogen atoms:

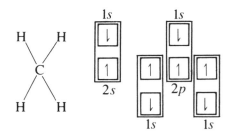

Chemists have shown that the methane molecule consists of a central carbon atom with four hydrogen atoms arranged symmetrically around it. All four C—H bonds are the same, and they point toward the corners of a regular tetrahedron. Methane is a tetrahedral molecule (Fig. 6-12). How do we explain the tetrahedral shape and the four identical bonds in terms of the one $2s$ and three $2p$ orbitals of the carbon atom? Such a model would predict three identical bonds and one which is different.

Orbital Hybridization

In order to describe the shape of the methane molecule, we must add another consideration to the valence bond theory. The added consideration is called *hybridization*.

We learned that orbitals are regions of space where electrons are likely to be. These regions of space are described by mathematical equations. If the $2s$ orbital and the three $2p$ orbitals of a carbon atom are mathematically averaged or hybridized, the result will be four equivalent orbitals which point toward the corners of a tetrahedron:

These four equivalent orbitals are called sp^3 hybrid orbitals, indicating that one s orbital and three p orbitals were used for hybridization. Each C—H bond is formed by the overlap of the $1s$ orbital of a hydrogen atom with one of the four sp^3 hybrid orbitals of the carbon atom. This results in a tetrahedral molecule:

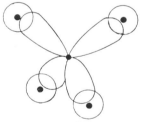

EXAMPLE PROBLEM 6-6

The gas silicon tetrafluoride, SiF_4, is formed by the action of hydrofluoric acid on glass. Describe as completely as possible the bonding in a molecule of silicon tetrafluoride.

SOLUTION

Silicon has the electron configuration

$$Si\ 1s^2 2s^2 2p^6 3s^2 3p^2$$

Each fluorine atom has a half-filled $2p$ orbital:

$$F\ 1s^2 2s^2 2p^5$$

Prior to bonding, a $3s$ electron of silicon is promoted to the $3p$ level, resulting in four half-filled valence orbitals. The hybridization of the $3s$ orbital and the three $3p$ orbitals results in four equivalent sp^3 orbitals which point toward the corners of a tetrahedron and which contain one unpaired electron each. The unpaired $2p$ electrons of fluorine can bond with the four unpaired electrons of silicon to form four silicon-fluorine bonds each pointed towards the corners of a regular tetrahedron.

PRACTICE PROBLEM 6-17

Describe as completely as possible the bonding in a molecule of germanium tetrabromide, $GeBr_4$.

Multiple Bonding

Orbital diagrams can be drawn for molecules with multiple bonds. The orbital diagram for CO_2 involves the promotion of a $2s$ electron of the carbon atom ($1s^2\ 2s^2\ 2p^2$) to the vacant $2p$ orbital, followed by the formation of double bonds between the carbon atom and each of the two oxygen atoms ($1s^2\ 2s^2\ 2p^4$):

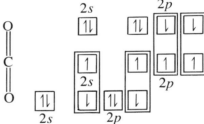

Another example of a molecule with a multiple bond is nitrogen. The orbital diagram for N_2 indicates that the three unpaired $2p$ electrons of each nitrogen atom ($1s^2\ 2s^2\ 2p^3$) are involved in triple bonding.

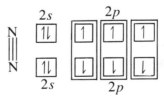

PRACTICE PROBLEM 6-18

Draw orbital diagrams for CH_2O, CS_2, and HCN.

Ionic Bonding

Quantum mechanics does not change the concept of ionic bonding presented in Section 6.2 very much. A common example of ionic bonding is the reaction between sodium atoms ($1s^2\ 2s^2\ 2p^6\ 3s^1$) and chlorine atoms ($1s^2\ 2s^2\ 2p^6\ 3s^2\ 3p^5$). Suppose these two atoms collide. Sodium is a Group 1 element, and it has one outer electron in the $3s$ orbital. If sodium gains some energy from the collision, the $3s$ electron leaves the sodium atom. A sodium ion is produced, and this ion is isoelectronic with the noble gas neon ($1s^2\ 2s^2\ 2p^6$). The electron which the sodium atom loses is picked up by a chlorine atom and enters the half-filled $3p$ orbital of the chlorine atom and a Cl^- ion ($1s^2\ 2s^2\ 2p^6\ 3s^2\ 3p^6$) is produced. The Cl^- ion is called a *chloride ion* and it has the same electron configuration as the noble gas argon. The sodium ion and the chloride ion are held together because they have opposite charges.

Part Three Review Questions

1. What is the valence shell of an atom?
2. State the basic postulate of VSEPR Theory.
3. According to VSEPR Theory, what are the most probable arrangements of two, three, and four electron pairs?
4. What are the bond angles if there are: *a)* four pairs of valence-shell electrons; *b)* three pairs of valence-shell electrons; *c)* two pairs of valence-shell electrons?
5. Explain what is meant by the term *coordinate covalent bond*.
6. What is meant by an orbital diagram?
7. What is meant by the term *hybridization of orbitals*?

Chapter Summary

- A molecule is any electrically neutral group of atoms that is held together tightly enough to be considered a single particle.
- Chemical bonds are the attractions between the atoms or ions of a substance.
- A molecule is more stable than the isolated atoms from which it is formed. The increased stability is caused by electrons being simultaneously attracted to two nuclei.
- Valence electrons are the electrons which occupy the highest energy level of an atom.
- The term *stable octet* draws attention to the special stability associated with an atom's having eight valence electrons.
- Ionic bonding is the type of chemical bond which results from the attraction between oppositely charged ions that forms when metallic atoms transfer electrons to nonmetallic atoms.
- Ionic compounds are formed when the metals of Groups 1, 2, and 3 react with nonmetals such as the elements of Groups 16 and 17 and nitrogen.
- Ionic compounds are nonconductors of electricity as solids, but they are conductors when melted. They have high melting points and are hard but

brittle. They are often soluble in water, producing solutions which conduct electricity.

- A covalent bond results when two atoms share a pair of electrons.
- Lewis structures are formulas which show the shared and unshared valence electrons in molecules.
- The seven elements which exist as diatomic molecules are H_2, O_2, N_2, F_2, Cl_2, Br_2, and I_2.
- A polar covalent bond is a covalent bond in which there is an unequal sharing of electrons.
- Electronegativity is a measure of the ability of an atom to attract a shared pair of electrons to itself.
- If the difference in electronegativity between two bonding atoms exceeds 1.7, it is better to consider the bond as being ionic. If the difference is less than 1.7 but greater than 0, the bond is polar covalent. If the difference is 0, the bond is pure covalent.
- The octet rule states that when atoms combine, bonds form so that each atom finishes with an octet of valence electrons.
- When one pair of electrons is shared between two atoms, the result is a single covalent bond. When two pairs are shared, the result is a double covalent bond. When three pairs are shared, the result is a triple covalent bond.
- The valence-shell electron-pair repulsion (VSEPR) theory states that the arrangement of atoms around a central atom in a molecule depends upon the repulsions between the electron pairs in the valence shell of the central atom. These valence-shell electron pairs reside in orbitals which are as far apart as possible.
- A coordinate covalent bond is a covalent bond in which both electrons of the shared pair come from only one of the bonded atoms.
- In the valence bond theory of bonding, a covalent bond is described as the result of the overlapping of two atomic orbitals, one on each of two atoms, so that the resulting orbital can be occupied by a shared pair of electrons.
- In valence bond theory, the shape of a molecule is accounted for by the directional nature of the bonding orbitals.

Key Terms

molecule	electronegativity
chemical bond	octet rule
stable octet	double covalent bond
ionic bonding	triple covalent bond
isoelectronic	valence shell
covalent bond	Valence-Shell Electron-Pair
single covalent bond	Repulsion Theory
Lewis structure	coordinate covalent bond
polar covalent bond	

Test Your Understanding

Each of these statements or questions is followed by four responses. Choose the correct response in each case. (Do not write in this book.)

1. Chemical bonds are the result of
   a) electron-electron repulsion.
   b) proton-proton attractions.
   c) attraction of an atom's nucleus for its valence electrons.
   d) simultaneous attractions of shared electrons for two nuclei.

2. Ionic bonds are usually formed between two elements with the following combination of properties:
   a) low ionization energy and low electron affinity.
   b) low ionization energy and high electron affinity.
   c) high ionization energy and low electron affinity.
   d) high ionization energy and high electron affinity.

3. An ionic bond is usually formed between
   a) two nonmetals.
   b) two metals.
   c) a metal and a nonmetal.
   d) a metalloid and a nonmetal.

4. What is the correct formula for the ionic compound formed when magnesium reacts with fluorine?
   a) MgF
   b) Mg_2F
   c) MgF_2
   d) Mg_2F_3

5. What is the formula of the ionic compound formed when element X (three valence electrons) reacts with element Y (six valence electrons)?
   a) X_2Y_3
   b) XY
   c) X_2Y
   d) X_3Y_2

6. Which of the following is a property of ionic compounds?
   a) Their solids conduct electricity.
   b) They are usually gases or liquids at room temperature.
   c) They have high melting points.
   d) They are soft and waxy.

7. Which of the following is isoelectronic with a noble gas?
   a) F^{2-}
   b) S^-
   c) Al^{2+}
   d) Ca^{2+}

8. Which of the following has pure covalent bonding?
   a) N_2
   b) HBr
   c) MgF_2
   d) CH_4

9. Which of the following is a property of covalent compounds?
   a) They have low boiling points.
   b) They are hard and brittle.
   c) Their molten state conducts electricity.
   d) They have high melting points.

10. Which of the following has polar covalent bonding?
    a) F_2
    b) H_2Se
    c) KCl
    d) O_2

11. Element X has an electronegativity of 3.0 and element Y has an electronegativity of 1.0. The most probable type of bond between X and Y is
    a) pure covalent.
    b) polar covalent.
    c) ionic.
    d) unpredictable.

12. Which of the following contains an atom that is isoelectronic with argon?
    a) $AlBr_3$
    b) HF
    c) MgI_2
    d) H_2S

13. Which of the following contains an atom or ion that does not obey the octet rule?
    a) BCl_3
    b) RbF
    c) CaI_2
    d) PCl_3

14. Which of the following contains an atom that does not obey the octet rule?

a) MgO   b) IF_5 c) CCl_4 d) KCl

15. The Lewis structure for H_2Te is

a)

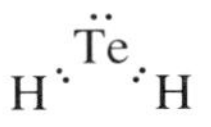

H : Te : H

b)

Te
H   H

c)

Te
H : : H

d)

Te
H   H

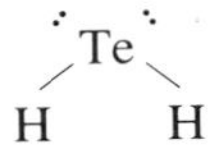

16. Which of the following is a Lewis structure for CBr_4?

a)

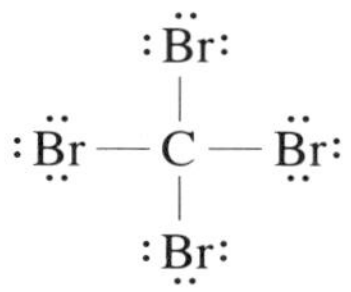

:Br:
|
:Br — C — Br:
|
:Br:

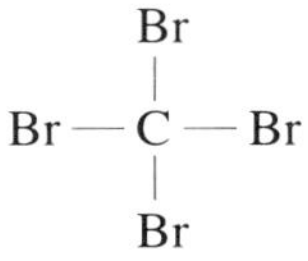

b)

Br
|
Br — C — Br
|
Br

c)

Br
‖
Br = C = Br
‖
Br

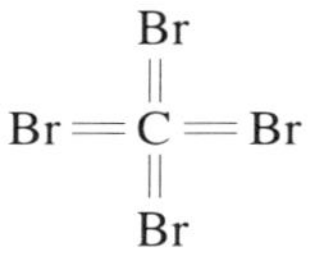

d)

Br
Br :C: Br
Br

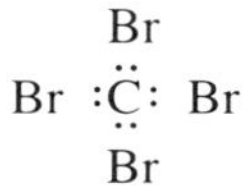

17. Which of the following contains double bonding?

a) NH_2Cl b) CS_2 c) C_2Cl_2 d) H_2Te

18. Which of the following contains triple bonding?

a) NBr_3 b) C_2Br_2 c) CO_2 d) C_2Cl_4

19. What is the predicted shape of PH_3?

a) tetrahedral   b) trigonal planar
c) trigonal pyramidal   d) linear

20. What is the predicted shape of HCN?

a) trigonal pyramidal   b) trigonal planar
c) tetrahedral   d) linear

Review Your Understanding

1. What is meant by stable octet?
2. What is the reason for the stability of the noble gases?
3. What type of elements are involved in ionic bonding?
4. Why will each oxide ion require two potassium ions in the compound potassium oxide?
5. Why will each barium ion require two chloride ions in the compound barium chloride?

6. During the formation of ionic bonds, with which noble gas does each of the following become isoelectronic: **a)** beryllium; **b)** aluminum; **c)** potassium; **d)** sulfur; **e)** fluorine?

7. Assuming that the following elements combine to form ionic compounds, write the formulas of the compounds: **a)** Sr and Cl; **b)** Rb and S; **c)** Al and S; **d)** Ca and N.

8. Describe what occurs when an ionic bond is formed in the case of **a)** potassium bromide; **b)** magnesium iodide; **c)** aluminum sulfide.

9. Why do ionic compounds tend to have high melting points?

10. Why do ionic compounds tend to be hard but brittle?

11. What is the main difference between an ionic bond and a covalent bond?

12. Draw the Lewis structure for IBr. Is the bond between the iodine and the bromine polar? If it is, which end of the molecule is slightly negative?

13. Draw the Lewis structure for IAt. Is the bond between the iodine and the astatine polar? If it is, which end of the molecule is slightly negative?

14. Which, in each of the following pairs of bonds, is more polar: **a)** Si—Cl or P—Cl; **b)** O—F or S—O; **c)**Se—Cl or H—Se? In each case state which atom is slightly negative.

15. What is a molecule? Using examples, distinguish between a molecule and an ion.

16. A cesium atom has one electron in its outer shell and a selenium atom has six valence electrons. What is the formula for cesium selenide? What type of bonding is involved in cesium selenide?

17. What are the differences in physical properties between ionic and covalent substances?

18. Predict the most probable type of bonding (ionic or covalent) found in each of the following:
    a) a substance that melts at relatively low temperatures
    b) a substance that boils at relatively high temperatures
    c) a hard but brittle substance
    d) a substance that is made up of molecules

19. A sulfur atom has six electrons in its outer shell and a hydrogen atom has one valence electron. What is the formula for hydrogen sulfide? Will the molecule be polar or nonpolar? Why?

20. Draw the Lewis structures for **a)** SnH_4; **b)** NF_3; **c)** $TeCl_2$; **d)** CH_5N; **e)** HOBr.

21. Draw the Lewis structure for the OH^- ion.

22. Draw the Lewis structures for **a)** BCl_3; **b)** GeH_4; **c)** SCl_2; **d)** ClO^-; **e)** H_2Se_2.

23. Which of the following would you predict to be extremely reactive: OBr, Br_2O, CH_3, CH_4?

24. Draw the Lewis structures for **a)**N_2H_2; **b)** NOCl.

25. Draw the Lewis structures for **a)** C_2HCl; **b)** C_2Cl_4.

26. Draw Lewis structures for two compounds with the molecular formula $C_2H_2Br_2$.

27. Draw Lewis structures for **a)**PH_3; **b)** SiH_4; **c)**H_2Se; **d)**CCl_4. Use VSEPR Theory to predict the shapes of these molecules.

28. Draw Lewis structures for **a)** CH_3Cl; **b)** Cl_2O; **c)** H_2S_2; **d)** P_2H_4; **e)** PCl_3; **f)** C_2Cl_2. Use VSEPR Theory to predict the shapes of these molecules.

29. Draw the Lewis structure for H_3O^+.
30. Draw the Lewis structure for the ClO^- ion and the ClO_2^- ion. Which of these ions has a coordinate covalent bond?
31. Describe the bonding which occurs in the chlorine molecule. Which orbitals are involved in the bonding?
32. Draw the orbital diagram for a bromine molecule.
33. Draw the orbital diagram for a hydrogen sulfide (H_2S) molecule.
34. Draw the orbital diagram for the HClO molecule.
35. Draw the orbital diagrams for **a)** CH_3Cl; and **b)** Cl_2O.
36. Double covalent bonding occurs in the C_2H_4 molecule. Draw the orbital diagram for C_2H_4.
37. Triple covalent bonding occurs in the C_2H_2 molecule. Draw the orbital diagram for C_2H_2.
38. How many orbitals are averaged to form a set of sp^2 hybrid orbitals?
39. Carbon tetrachloride is a nonpolar molecule which has four polar covalent bonds. Explain why this statement is also true with respect to carbon tetrabromide.

Apply Your Understanding

1. The oceans contain enormous quantities of dissolved ionic compounds. Should the oceans conduct electricity? Explain.
2. Using any of the elements F, H, Cl, Mg, K, and referring to an electronegativity table, devise substances that demonstrate your understanding of ionic bonding, polar covalent bonding, and pure covalent bonding. Use electron dot formulas to show how electrons are shared or transferred.
3. Draw Lewis structures for the following compounds: **a)** SF_6; **b)** CBr_4; **c)** NCl_3; **d)** PCl_5; **e)** XeF_2. Which of these do not obey the octet rule?
4. Draw the Lewis structures for **a)** ICl_2^-; **b)** Cl_2CO.
5. Draw the Lewis structures for **a)** BH_4^-; **b)** $AlCl_4^-$.
6. Use your knowledge of bonding theory to explain why the formula for water is not H_3O.
7. The compound $BeCl_2$ is known to have polar Be—Cl bonds. However, the $BeCl_2$ molecule is not polar. Explain.
8. Extend your knowledge of VSEPR Theory to predict the bond angles in SF_6.
9. While phosphorus forms PCl_3 and PCl_5, nitrogen forms NCl_3 but does not form NCl_5. Suggest a possible explanation for this fact.
10. Two different compounds with the molecular formula C_2H_6O are known. Draw Lewis structures for these two compounds.
11. Diamond consists of carbon atoms each of which is bonded in a three-dimensional network to four other carbon atoms. How does this explain the hardness of diamond?

Investigations

1. Obtain a compound from your teacher. Perform experiments and make observations which will enable you to determine whether the substance has ionic or covalent bonding.

2. Write a research report on the preparation of the first noble gas compound. What other noble gas compounds have been prepared? What are some of their uses?

3. In 1962 Watson and Crick received the Nobel Prize for determining the structure of DNA, the molecule responsible for heredity. Their knowledge of the shape and bonding abilities of certain nitrogen-containing compounds led them to propose a double helix structure for DNA. Write a report on their work.

4. As a result of chemists' knowledge of molecular structure and chemical bonding, a technology known as genetic engineering has been developed. This method splices together DNA from different species. Read articles on this process and debate whether scientists should continue research in this area.

7

Nomenclature and Formula Writing

CHAPTER CONTENTS

Part One: **Oxidation Numbers**
    7.1   Assigning Oxidation Numbers
    7.2   Using Oxidation Numbers to Write Chemical Formulas

Part Two: **Chemical Nomenclature**
    7.3   Naming Chemical Compounds Containing Only Two Elements
    7.4   Naming Chemical Compounds Composed of More Than Two Elements
    7.5   Nomenclature of Acids
    7.6   Reading Chemical Equations

In earlier chapters we learned the symbols of the elements, and we discovered that atoms of one element combine with atoms of other elements in small whole-number ratios. In other words, the atoms of an element appear to have the ability to combine with the atoms of another element in only certain ratios to form compounds. Millions of compounds are known. Chemists use thousands of them routinely during their careers. How do chemists know what to call these compounds? Can we tell the formula of a compound from its name? Is there a standard system of writing the formulas for compounds and for naming compounds?

   Chemical nomenclature is the assignment of names to compounds. Students of chemistry must know the language of chemistry, that is, nomenclature and formula writing. In this chapter, we shall use the concept of oxidation numbers of atoms to predict the formulas for compounds such as the ones in the containers shown in Figure 7-1. Then we shall develop simple, systematic names that are related to the formulas of compounds. We shall conclude the chapter by using a knowledge of chemical nomenclature to translate chemical equations into words.

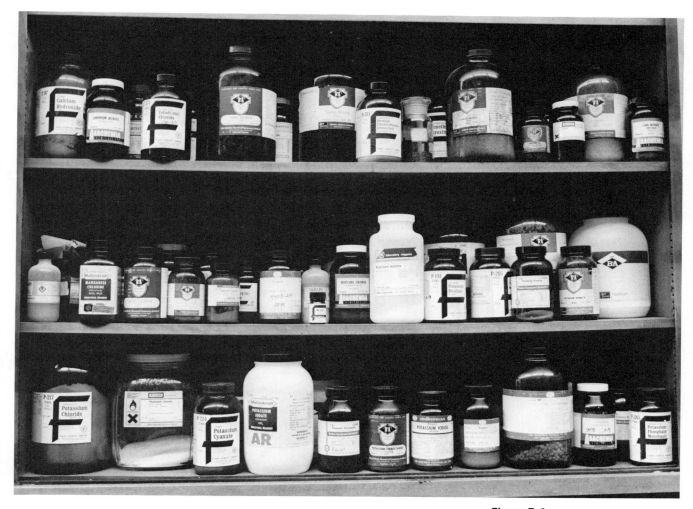

Figure 7-1
**A variety of compounds in a
chemical stockroom**

Key Objectives

When you have finished studying this chapter, you should be able to

1. write the formulas of the compounds formed by combinations of positive and negative ions, given the formulas of these ions.
2. write the name of a substance, given its formula.
3. write the formula of a substance, given its name.

Part One: Oxidation Numbers

7.1 Assigning Oxidation Numbers

As we learned in Chapter 6, when two or more atoms unite to form a compound, they form either ionic or covalent bonds. Atoms of some elements have a tendency to lose electrons and thereby acquire a positive charge. Atoms of

other elements have a tendency to gain electrons and thereby acquire a negative charge. In ionic compounds the atoms have full positive and negative
charges (for example, Na^+Cl^-). In contrast, in polar covalent compounds such
as water the atoms have only partial positive and negative charges:

Chemists use *oxidation numbers* to keep track of the *positive* or *negative*
character of atoms or ions. When electrons are removed completely or shifted
partially away from an atom during a chemical reaction, the atom is given a
more positive oxidation number. When electrons are gained or shifted toward
an atom during a reaction, the atom is given a more negative oxidation
number. The actual numerical value of the oxidation number depends on the
number of electrons shifted partially or transferred completely. In NaCl, for
example, the oxidation number of Na is +1, because one electron has been
transferred from it. The oxidation number of Cl is –1, because one electron has
been transferred to it from the sodium. In a covalent substance such as H—Cl
the hydrogen atom has an oxidation number of +1 due to a shift (but not a
transfer) of its valence electron toward the more electronegative chlorine atom.
The chlorine atom has an oxidation number of –1 due to a shift of one electron
toward it. In water the oxygen atom is more electronegative than a hydrogen
atom. The oxidation number of each hydrogen atom is +1 due to a shift of its
valence electron toward the oxygen atom. Therefore, the oxidation number of
the oxygen atom is –2 due to a shift of two electrons (one from each hydrogen
atom) toward it.

Oxidation numbers may be deduced from energy level populations, but it is
easier, by using the periodic table, to learn symbols of elements and formulas
for ions, and the oxidation numbers that go with them. Let's see how this is
done.

Free Elements

The atoms of an element are in the free state when they are bonded to other
atoms of the same element. The atoms of free elements are not combined with
the atoms of other elements. Thus, when they exist in the free state, elements
have not combined to form compounds. Consider the formation of a molecule
of H_2, O_2, or N_2, from individual atoms. We would write the equations as

$$H \cdot + \cdot H \longrightarrow H{:}H$$

$$:\ddot{O}\cdot + \cdot\ddot{O}: \longrightarrow :\ddot{O}::\ddot{O}:$$

$$:\dot{N}\cdot + \cdot\dot{N}: \longrightarrow :N{:}\!{:}N:$$

In each case, the molecule is made up of identical atoms (H, O, or N). Since
each atom in the pair has the same electronegativity, both have the same
tendency to attract the shared electrons. Thus, the electrons are shared equally
(pure covalent bonding). Both atoms of the molecule therefore have a charge
of zero, just as they do in the isolated state.

These results allow us to state a general rule: *The oxidation numbers of atoms of elements in the free state are always zero*. Thus, for He, Na, F_2, and As_4, the oxidation numbers are all zero.

Elements in Ionic Compounds

Table 7-1 lists the usual oxidation numbers of the ions of some of the common elements in the format of a periodic table. Note that ions of the metals (and of hydrogen) in Group 1 of the periodic table all have oxidation numbers of +1, since the atoms all have one valence electron which they can lose to form the ion. Similarly, in compounds, the elements of Group 2 have oxidation numbers of +2, and the elements of Group 13 usually have oxidation numbers of +3.

Table 7-1 Usual Oxidation Numbers of the Ions of Some Common Elements

1	2	Transition Metals										13	14	15	16	17	18
H(+1)																	
Li(+1)	Be(+2)											B(+3)		N(−3)	O(−2)	F(−1)	
Na(+1)	Mg(+2)											Al(+3)		P(−3)	S(−2)	Cl(−1)	
K(+1)	Ca(+2)				Cr(+2) Cr(+3) Cr(+6)		Fe(+2) Fe(+3)	Co(+2) Co(+3)	Ni(+2) Ni(+3)	Cu(+1) Cu(+2)	Zn(+2)	Ga(+3)		As(+3) As(+5)		Br(−1)	
Rb(+1)	Sr(+2)									Ag(+1)	Cd(+2)		Sn(+2) Sn(+4)			I(−1)	
Cs(+1)	Ba(+2)									Au(+1) Au(+3)	Hg(+1) Hg(+2)		Pb(+2) Pb(+4)				

The ions of the halogens (Group 17 elements) usually have oxidation numbers of −1. Since these atoms all have seven valence electrons and can gain one more electron to complete a stable octet, they easily form a negative ion. Similarly, the ions of Group 16 elements usually have oxidation numbers of −2.

Some metal atoms can have several different oxidation numbers. For example, a tin atom can lose either four electrons to form a Sn^{4+} ion with an oxidation number of +4, or it can lose only two electrons to form a Sn^{2+} ion with an oxidation number of +2. Similarly, an arsenic atom can lose either three or five electrons to form ions with oxidation numbers of +3 or +5, respectively.

In summary, *the oxidation number of an ion is the same as its charge*. For example, the Mg^{2+} ion has an oxidation number of +2; for Al^{3+} the oxidation number is +3; and for S^{2-} it is −2. Note that the superscript 2+, which is part of the symbol Mg^{2+}, is itself a symbol which indicates the number of + charges on the ion, while the oxidation number is written as +2, because it is a number, not a symbol.

EXAMPLE PROBLEM 7-1

Using a periodic table instead of Table 7-1, give the oxidation number of each element in: **a)** BaO; **b)** $GaBr_3$.

SOLUTION

a) Since Ba is in Group 2, it loses two electrons to form Ba^{2+}. Therefore, the oxidation number of Ba is +2. Since O is in Group 16, it gains two electrons to form O^{2-}. Therefore, the oxidation number of O is −2.

b) Since Ga is in Group 3, it loses three electrons to form Ga^{3+}. Therefore, the oxidation number of Ga is +3. Since Br is in Group 17, it gains one electron to form Br^-. Therefore, the oxidation number of Br is −1.

PRACTICE PROBLEM 7-1

Using a periodic table instead of Table 7-1, give the oxidation number of each element in **a)** $SrBr_2$; **b)** RbF; **c)** Ga_2O_3.

Elements in Covalent Substances

In covalent substances, the more electronegative element is assigned a negative oxidation number, and the less electronegative element is assigned a positive oxidation number. In HCl, the H atom shares its one electron with the chlorine atom, and the chlorine shares one of its electrons with the hydrogen atom. Because the chlorine atom is more electronegative, there is a *shift* of the shared electrons towards the chlorine atom. There is not a complete transfer of an electron from the hydrogen atom to the chlorine atom, because the bond still retains some covalent character. However, for purposes of assigning oxidation numbers, the electron is counted *as if* it were completely transferred from the hydrogen atom to the chlorine atom. Thus, the chlorine atom is assigned an oxidation number of −1. The hydrogen is then assigned an oxidation number of +1. In ammonia (NH_3), each hydrogen atom shares its one electron with the nitrogen atom. Because hydrogen is less electronegative than nitrogen, the hydrogen electron is shifted towards the nitrogen atom. The hydrogen atoms each acquire an oxidation number of +1. Three electrons (one from each hydrogen atom) have been shifted toward the nitrogen atom, and its oxidation number is therefore −3.

Thus, whether a given compound is ionic or covalent, the same method is used to determine the oxidation numbers of its atoms. This similarity arises because electrons are counted *as if* they had transferred completely from one atom to another, even if they are only shifted towards an atom.

EXAMPLE PROBLEM 7-2

What are the oxidation numbers of each atom in the following covalent substances:
a) H_2S; b) PCl_3; c) P_4?

SOLUTION

a) In H_2S, each hydrogen atom shares one electron with the sulfur atom. Because hydrogen is less electronegative than sulfur, hydrogen acquires an oxidation number of +1. Two electrons (one from each hydrogen atom) have been shifted toward the sulfur atom, and its oxidation number is therefore −2.

b) In PCl_3 the phosphorus atom shares three of its electrons—one with each chlorine atom. Because phosphorus is less electronegative than chlorine, the oxidation number of phosphorus is therefore +3. One electron has been shifted towards each chlorine atom, and the oxidation number of each chlorine atom is therefore −1.

c) The oxidation number of P in P_4 must be zero, because all the atoms in the molecule are identical (P_4 is the formula of elemental phosphorus).

PRACTICE PROBLEM 7-2

What are the oxidation numbers of each atom in each of the following:
a) Ag; b) Cl_2; c) CH_4; d) PF_5; e) H_2Te; f) $SiBr_4$?

Polyatomic Ions

A *polyatomic ion* is an ion which contains two or more atoms. The most common positively charged polyatomic ion is NH_4^+. It is found in compounds such as ammonium nitrate (NH_4NO_3) and ammonium phosphate [$(NH_4)_3PO_4$], both of which are important ingredients in fertilizers. Common negatively charged polyatomic ions are OH^- and SO_4^{2-}. The hydroxide ion, OH^-, is found in compounds such as sodium hydroxide (NaOH), a component of household drain cleaners. The sulfate ion, SO_4^{2-}, is found in compounds such as calcium sulfate ($CaSO_4$). Both gypsum and plaster of Paris consist of calcium sulfate combined with different numbers of water molecules. Table 7-2 is a listing of common negatively charged polyatomic ions. Their oxidation numbers are the same as their charge.

Table 7-2 Oxidation Numbers and Names of Some Negative Polyatomic Ions

Oxidation Number = -1			
Ion	Name	Ion	Name
CN^-	cyanide	HSO_4^-	hydrogen sulfate
$C_2H_3O_2^-$	acetate	$H_2PO_3^-$	dihydrogen phosphite
ClO^-	hypochlorite	$H_2PO_4^-$	dihydrogen phosphate
ClO_2^-	chlorite	MnO_4^-	permanganate
ClO_3^-	chlorate	NO_2^-	nitrite
ClO_4^-	perchlorate	NO_3^-	nitrate
HCO_3^-	hydrogen carbonate	OCN^-	cyanate
$HC_2O_4^-$	hydrogen oxalate	OH^-	hydroxide
HSO_3^-	hydrogen sulfite	SCN^-	thiocyanate

Oxidation Number = -2			
Ion	Name	Ion	Name
CO_3^{2-}	carbonate	HPO_4^{2-}	hydrogen phosphate
$C_2O_4^{2-}$	oxalate	O_2^{2-}	peroxide
CrO_4^{2-}	chromate	SO_3^{2-}	sulfite
$Cr_2O_7^{2-}$	dichromate	SO_4^{2-}	sulfate
HPO_3^{2-}	hydrogen phosphite	$S_2O_3^{2-}$	thiosulfate

Oxidation Number = -3			
Ion	Name	Ion	Name
AsO_3^{3-}	arsenite		
AsO_4^{3-}	arsenate	PO_4^{3-}	phosphate

The presence of a charge on a polyatomic ion means that electrons have been gained or lost. A superscript +, as in NH_4^+, indicates that the NH_4 unit has one electron less than the total contributed by the one nitrogen and four hydrogen atoms. The charge on the whole unit is +1. Similarly, a superscript 2–, as in SO_4^{2-}, means that the SO_4 unit has gained two extra electrons, and the charge on the whole unit is −2.

7.2 Using Oxidation Numbers to Write Chemical Formulas

In a **chemical formula**, the symbols of the elements are used to represent the composition of a compound. The oxidation numbers of ions and the zero sum rule are used to write the chemical formulas of compounds. The *zero sum rule* states that the sum of the positive oxidation numbers and the negative oxidation numbers of the elements in a compound must be zero. Thus, in the compound, AlF_3, aluminum has an oxidation number of +3, and each fluorine has an oxidation number of −1. The sum of the oxidation numbers is zero. In

PbO_2, each oxygen has an oxidation number of -2, and lead has an oxidation number of $+4$. Again, the sum of the oxidation numbers is zero. Notice that the ion which has a positive oxidation number is usually written first in a chemical formula.

EXAMPLE PROBLEM 7-3

Write the formulas of the compounds of bromide ion (Br^-) with **a)** potassium ion (K^+); **b)** magnesium ion (Mg^{2+}); and **c)** aluminum ion (Al^{3+}).

SOLUTION

a) Since potassium ion has an oxidation number of $+1$ and bromide ion has an oxidation number of -1, the formula of this compound is KBr.

$$KBr \quad (+1) + (-1) = 0$$

b) However, magnesium ion has an oxidation number of $+2$, and two bromide ions will be required since each has an oxidation number of -1. The formula of this compound is $MgBr_2$, and the subscript 2 indicates that there are two bromide ions for every magnesium ion in the compound.

$$MgBr_2 \quad (+2) + 2(-1) = 0$$

c) Finally, aluminum ion has an oxidation number of $+3$ and three bromide ions will be required for every aluminum ion. The formula of this compound is $AlBr_3$.

$$AlBr_3 \quad (+3) + 3(-1) = 0$$

PRACTICE PROBLEM 7-3

Write the formulas of the compounds of iodide ion (I^-) with **a)** K^+; **b)** Mg^{2+}; **c)** Al^{3+}.

EXAMPLE PROBLEM 7-4

Write the formulas of the compounds of sulfide ion (S^{2-}) with **a)** K^+; **b)** Mg^{2+}; **c)** Al^{3+}.

SOLUTION

a) Potassium ion has an oxidation number of $+1$, but sulfide ion has an oxidation number of -2. Thus, two potassium ions will be required to balance one sulfide ion. The formula of this compound is K_2S. The subscript 2 indicates that there are two potassium ions for every sulfide ion in the compound.

$$K_2S \quad 2(+1) + (-2) = 0$$

b) However, magnesium ion has an oxidation number (+2) which balances the oxidation number of the sulfide ion (−2). Magnesium sulfide is simply MgS.

$$\text{MgS} \ (+2) + (-2) = 0$$

c) Finally, aluminum ion has an oxidation number of +3, and sulfide ion has an oxidation number of −2. The smallest whole number into which both 3 and 2 divide evenly is 6. Two aluminum ions have a combined oxidation number of +6, and three sulfide ions have a combined oxidation number of −6. In the case of aluminum sulfide, the simplest way to ensure that the sum of the positive oxidation numbers will equal the sum of the negative oxidation numbers is to use the formula Al_2S_3. There must be two aluminum ions for three sulfide ions.

$$Al_2S_3 \ 2(+3) + 3(-2) = 0$$

PRACTICE PROBLEM 7-4

Write the formulas of the compounds of selenide ion (Se^{2-}) with **a)** K^+; **b)** Mg^{2+}; **c)** Al^{3+}.

Table 7-1 lists 38 positive monatomic ions. In addition, we know the formula for one positive polyatomic ion (NH_4^+). Table 7-1 also lists eight negative monatomic ions, and 31 negative polyatomic ions are listed in Table 7-2. The 39 positive ions can (in theory, at least) react with each of the 39 negative ions to form a total of $39 \times 39 = 1521$ possible compounds. Most of these are stable compounds, and you now have enough information to write the formulas resulting from each of the 1521 combinations.

EXAMPLE PROBLEM 7-5

Write the formulas of the compounds of **a)** NH_4^+ and Cl^-; **b)** Na^+ and CO_3^{2-}; **c)** Ba^{2+} and SO_4^{2-}.

SOLUTION
a) Ammonium ion has an oxidation number of +1 and chloride ion has an oxidation number of −1. The formula of this compound is NH_4Cl.

$$NH_4Cl \ (+1) + (-1) = 0$$

b) Sodium ion has an oxidation number of +1, but carbonate ion has an oxidation number of −2. Thus there must be two sodium ions for every carbonate ion and this is indicated by a subscript 2 following the Na. The formula is Na_2CO_3.

$$Na_2CO_3 \ 2(+1) + (-2) = 0$$

c) Barium ion has oxidation number of +2 and sulfate ion has an oxidation number of -2. The formula of this compound is $BaSO_4$.

$$BaSO_4 \quad (+2) + (-2) = 0$$

PRACTICE PROBLEM 7-5

Write the formulas of the compounds of **a)** NH_4^+ and NO_3^-; **b)** Ag^+ and PO_4^{3-}; **c)** Al^{3+} and PO_4^{3-}.

In writing the formulas of compounds which include more than one unit of a given polyatomic ion, the formula of that polyatomic ion is enclosed in parentheses, and the number of units of the polyatomic ion is indicated with a subscript. Example Problem 7-6 illustrates this point.

EXAMPLE PROBLEM 7-6

Write the formulas of the compounds of: **a)** Ca^{2+} and NO_3^-; **b)** NH_4^+ and PO_4^{3-}; **c)** Al^{3+} and SO_4^{2-}.

SOLUTION

a) Two nitrate ions (each with oxidation number of -1) are required for each calcium ion ($+2$). Thus, the formula for nitrate ion is enclosed in parentheses, and a subscript 2 is used. The formula is $Ca(NO_3)_2$.

$$Ca(NO_3)_2 \quad (+2) + 2(-1) = 0$$

$Ca(NO_3)_2$ may read "Ca, NO_3 taken twice." The word *taken* indicates that there is more than one unit of the polyatomic ion which precedes it.

b) Three ammonium ions (each $+1$) are required for each phosphate ion (-3). The formula for ammonium ion is enclosed in parentheses, and a subscript 3 is used. The formula is $(NH_4)_3PO_4$. It is read "NH_4 taken three times, PO_4."

$$(NH_4)_3PO_4 \quad 3(+1) + (-3) = 0$$

c) Two aluminum ions (each $+3$) are required for three sulfate ions (each -2). The formula for sulfate ion is enclosed in parentheses and a subscript 3 is used. Furthermore, a subscript 2 follows the aluminum. The formula is $Al_2(SO_4)_3$. It is read "Al two, SO_4 taken three times."

$$Al_2(SO_4)_3 \quad 2(+3) + 3(-2) = 0$$

PRACTICE PROBLEM 7-6

Write the formulas of the compounds of **a)** Al^{3+} and NO_3^-; **b)** NH_4^+ and SO_3^{2-}; **c)** Al^{3+} and SO_3^{2-}.

Part One Review Questions

1. Explain what is meant by the term *oxidation number*.
2. What use is made of oxidation numbers?
3. What is a free element?
4. Using the periodic table, give the oxidation number of each element in the ionic substances:
   a) Na_2S; **b)** CaF_2; **c)** MgO; **d)** $CsCl$; **e)** AlF_3; **f)** KCl; **g)** MgF_2; **h)** Al_2S_3.
5. Give the oxidation number of each atom in the covalent substances:
   a) NH_3; **b)** SiS_2; **c)** PH_3; **d)** I_2; **e)** H_2Se; **f)** CO_2; **g)** $AsBr_3$; **h)** CCl_4.
6. What is stated by the zero sum rule?
7. When is a polyatomic ion enclosed in parentheses in a chemical formula? Why is this convention followed?
8. Write the formulas of the compounds of: **a)** Al^{3+} and CO_3^{2-}; **b)** NH_4^+ and SO_4^{2-}; **c)** K^+ and CO_3^{2-}; **d)** Ba^{2+} and O^{2-}; **e)** Al^{3+} and F^-; **f)** Mg^{2+} and Cl^-.

Part Two: Chemical Nomenclature

7.3 Naming Chemical Compounds Containing Only Two Elements

In the last section, we learned how to use oxidation numbers to write chemical formulas. A compound is identified by its chemical formula. For this reason, chemists use formulas to refer to various substances. For example, H_2O is the formula which refers to a specific compound. But each compound also has a name to distinguish it from other compounds. Thus, the compound represented by the formula H_2O is called water. Such a name is called a *trivial* name. There is no way that such a name could be figured out from the formula. Other compounds with trival names are NH_3 (ammonia) and $C_{12}H_{22}O_{11}$ (sugar). These names arose in the early days of chemistry when names were invented on the spot as compounds were discovered. As the number of known compounds began to increase, it quickly became apparent that there must be a *systematic* method of nomenclature which uniquely describes each compound. Thus, although the trivial name of NaCl is "salt," its systematic name is "sodium chloride." The latter name certainly gives a better indication of the make-up of the compound. In this section we shall describe the method of deriving systematic names for compounds. Since many compounds are composed of ions, or can be considered as if they were composed of ions, we shall first examine the means of naming monatomic ions and the compounds that are formed by combinations of these ions.

Positive Monatomic Ions

The name of a positive monatomic ion is the same as the name of the element. Thus, Li^+ is lithium (ion), Mg^{2+} is magnesium (ion), and Al^{3+} is aluminum (ion).

The atoms of some metals form more than one stable ion. For example, tin forms the ions Sn^{2+} and Sn^{4+}; iron forms the ions Fe^{2+} and Fe^{3+}; and copper forms the ions Cu^+ and Cu^{2+} (Table 7-1). In such cases a Roman numeral in parentheses is added immediately after the name of the element to indicate the oxidation number of the element. Thus, Sn^{2+} is tin(II) and Sn^{4+} is tin(IV); Fe^{2+} is iron(II) and Fe^{3+} is iron(III); Cu^+ is copper(I) and Cu^{2+} is copper(II). This system of nomenclature is called the *Stock system*, after the German chemist Alfred Stock, who first proposed it. Roman numerals are not used if the positive ion has only one possible oxidation number, for example, ions of elements of Groups 1, 2, and 13.

An older method of naming such ions was to use the suffixes *-ous* and *-ic*. For the ion of lower oxidation number, the suffix *-ous* was added to the stem of the name of the element. For the ion of higher oxidation number, the suffix *-ic* was used. Thus, Sn^{2+} was stann*ous* ion and Sn^{4+} was stann*ic* ion; Fe^{2+} was ferr*ous* ion and Fe^{3+} was ferr*ic* ion; while Cu^+ was cupr*ous* ion and Cu^{2+} was cupr*ic* ion. The system had the disadvantage that a given suffix does not consistently represent the same oxidation number. For example, the suffix *-ic* stands for an oxidation number of +4 in stannic ion, +3 in ferric ion, and +2 in cupric ion. Furthermore, the system does not work at all for ions which may have more than two oxidation numbers (chromium, for example, may have oxidation numbers of +2, +3, and +6). Because this system is still used by chemists, it is (unfortunately) necessary for students to learn it. Both the Stock system and this older system are illustrated in Table 7-3.

Table 7-3   Names of Some Positive Ions that use the Suffixes *-ous* and *-ic*

Element	Oxidation Number	Old System	Stock System
Au	+1	aur*ous*	gold(I)
Au	+3	aur*ic*	gold(III)
Co	+2	cobalt*ous*	cobalt(II)
Co	+3	cobalt*ic*	cobalt(III)
Cu	+1	cupr*ous*	copper(I)
Cu	+2	cupr*ic*	copper(II)
Fe	+2	ferr*ous*	iron(II)
Fe	+3	ferr*ic*	iron(III)
Hg	+1	mercur*ous*	mercury(I)
Hg	+2	mercur*ic*	mercury(II)
Sn	+2	stann*ous*	tin(II)
Sn	+4	stann*ic*	tin(IV)

Table 7-4   Stems (boldface) of Some Common Elements

> **hydr**ogen
> **carb**on
> **nitr**ogen
> **ox**ygen
> **fluor**ine
> **phosph**orus
> **sulf**ur
> **chlor**ine
> **selen**ium
> **brom**ine
> **iod**ine

Negative Monatomic Ions

The name of a negative monatomic ion is formed by adding the suffix *-ide* to the stem (the first part) of the name of the element. The stems of some common elements are listed in Table 7-4.

Binary Compounds Containing a Metal and a Nonmetal

A *binary compound* is one which contains only two elements. The systematic name of a binary compound consists of two words, the name of the positive ion and the name of the negative ion. Usually the more positive, more metallic element is written first (NaCl, not ClNa). Thus, NaCl is sodium chloride; CaO is calcium oxide; FeS is iron(II) sulfide or ferrous sulfide; and $FeCl_3$ is iron(III) chloride or ferric chloride.

EXAMPLE PROBLEM 7-7

Name each of the following compounds: **a)** $BaCl_2$; **b)** $FeBr_3$; **c)** SrI_2.

SOLUTION
The first part of the name is the name of the first element in the formula. The second part of the name is the stem of the element followed by the suffix *-ide*: **a)** Therefore the name of $BaCl_2$ is barium chloride; **b)** for $FeBr_3$ it is iron(III) bromide or ferric bromide; **c)** for SrI_2 the name is strontium iodide.

PRACTICE PROBLEM 7-7

Name each of the following compounds: **a)** MgF_2; **b)** $CuCl_2$; **c)** K_2S; **d)** Li_2O; **e)** BeF_2; **f)** MgO; **g)** Ca_3N_2; **h)** Na_2O; **i)** Ba_3N_2; **j)** RbBr.

Binary Compounds Containing Hydrogen and Another Element

Compounds of hydrogen and a nonmetal from Groups 16 and 17 are named by writing the word *hydrogen* followed by the stem of the name of the nonmetal to which the ending *-ide* has been added. Thus, HCl is hydrogen chloride and H_2S is hydrogen sulfide. Of course, H_2O has the trivial name water.

In compounds of hydrogen and a metal, the hydrogen is usually the more electronegative element. Therefore, the symbol for the metal is written first in the formula (for example, NaH). The names are formed according to the usual rules for binary compounds of a metal and a nonmetal (for example, sodium hydride).

The Group 14 element carbon forms many compounds with hydrogen. These are called *hydrocarbons* and they are discussed in Chapter 15. Nonsystematic or trivial names are normally used for compounds containing hydrogen and a Group 15 element. For example, NH_3 is called ammonia, PH_3 is called phosphine, AsH_3 is called arsine, and SbH_3 is called stibine.

Binary Compounds Containing Two Nonmetals

Binary compounds containing two nonmetals are named by using yet another naming system. In this system the name of each element is preceded by a prefix which indicates the number of atoms of the element. The prefixes are listed in Table 7-5. Normally the prefix *mono-* is omitted from the name of the first element. The absence of a prefix indicates the number one. Since the formula of a binary compound is written with the less electronegative atom first, the name of a binary compound of two nonmetals is formed according to the pattern

prefix + element name   prefix + element stem + -ide.

Thus, As_2S_3 is called diarsenic trisulfide; NO_2 is nitrogen dioxide (not mononitrogen dioxide); and CCl_4 is carbon tetrachloride. The final *a* or *o* of a prefix is omitted if the name of the atom or ion being counted begins with either an *a* or an *o*. Thus, N_2O_5 is dinitrogen pentoxide (not dinitrogen pentaoxide).

In most cases the Stock system can also be used to name binary compounds containing two nonmetals. Thus, As_2S_3 can also be called arsenic(III) sulfide, and N_2O_5 can be called nitrogen(V) oxide. However, problems can occasionally arise when the Stock system is used to name binary compounds containing two nonmetals. For example, both NO_2 and N_2O_4 exist. Using the Stock system, they would both have to be called nitrogen(IV) oxide. The problem is solved if the former is called nitrogen dioxide and the latter is called dinitrogen tetroxide.

Table 7-5   Commonly Used Numerical Prefixes

Number	Prefix
1	mono-
2	di-
3	tri-
4	tetra-
5	penta-
6	hexa-
7	hepta-
8	octa-
9	nona-
10	deca-

EXAMPLE PROBLEM 7-8

Name the compounds represented by the formulas:
a) NO; **b)** P_2S_5; **c)** SCl_2; **d)** PCl_3; **e)** HI.

SOLUTION

a) Both atoms are nonmetals. There is one atom of each element. The appropriate prefix is therefore *mono-*. But the prefix is omitted before the *nitrogen*, and its final *o* is omitted before *oxide*. The correct name is therefore nitrogen monoxide.

b) Both atoms are nonmetals. The numbers are 2 and 5, so the prefixes are *di-* and *penta-*. The name is diphosphorus pentasulfide.

c) Both atoms are nonmetals. The numbers are 1 and 2, so the prefixes are *mono-* (omitted) and *di-*. The name is sulfur dichloride.

d) Both elements are nonmetals. The numbers are 1 and 3, so the prefixes are *mono-* (omitted) and *tri-*. The name is phosphorus trichloride.

e) Compounds of hydrogen and a nonmetal are named in the same manner as are compounds of metals and nonmetals. The name is therefore hydrogen iodide.

PRACTICE PROBLEM 7-8

Name the compounds represented by the following formulas: **a)** CF_4; **b)** $AsCl_5$; **c)** PBr_3; **d)** HBr; **e)** SO_2; **f)** P_2O_5; **g)** $SbCl_3$; **h)** CBr_4.

7.4 Naming Chemical Compounds Composed of More Than Two Elements

Figure 7-2
Compounds in the laboratory are labelled with systematic names.

Many compounds consist of a polyatomic ion and a metal ion, another polyatomic ion, or hydrogen. As you might expect, the names of such compounds are formed by writing first the name of the positive portion of the compound, followed by the name of the negative ion. Whereas the names of all monatomic negative ions end in *-ide* (for example, chloride, oxide), the names of negative polyatomic ions generally end in *-ate* or *-ite*. The names and formulas of some common negative polyatomic ions are given in Table 7-2. Only one common positive polyatomic ion is found in compounds, and that is NH_4^+, ammonium ion.

Of the ions listed in Table 7-2 (page 216), only three do not have either the *-ate* or *-ite* ending: cyanide (CN^-), hydroxide (OH^-), and peroxide (O_2^{2-}). Only two do not contain oxygen: cyanide (CN^-), and thiocyanate (SCN^-). The thiocyanate ion can be considered as being formed from a cyanate ion (OCN^-) by replacing the oxygen atom with a sulfur atom. The prefix *thio-* is used to indicate the sulfur atom. In a similar manner, the thiosulfate ion ($S_2O_3^{2-}$) can be considered as being formed from a sulfate ion (SO_4^{2-}) by replacing one oxygen atom with a sulfur atom. The extra sulfur atom is indicated by the prefix *thio-*.

Eight of the negative polyatomic ions listed in Table 7-2 are formed by the combination of H^+ with a negative ion. For example, HSO_4^- is formed by the combination of H^+ with SO_4^{2-} according to the reaction

$$H^+ + SO_4^{2-} \longrightarrow HSO_4^-$$

These ions are named by placing the word *hydrogen* before the name of the negative ion. Thus, HSO_4^- is called the hydrogen sulfate ion; HCO_3^- is the hydrogen carbonate ion; and HPO_4^{2-} is the hydrogen phosphate ion. The presence of two hydrogen atoms is indicated by the prefix *di-*, as in $H_2PO_4^-$, dihydrogen phosphate ion.

Two simple principles will help you to remember the names of the oxygen-containing ions. First, if there are only two oxygen-containing ions of an element (for example, SO_4^{2-} and SO_3^{2-}), the ion with the greater number of oxygen atoms has a name ending in *-ate*, while the ion with the smaller number of oxygen atoms has a name ending in *-ite* (for example, sulfate and sulfite). Second, if there are four oxygen-containing ions of an element (for example, ClO^-, ClO_2^-, ClO_3^-, ClO_4^-), the two ions with the fewest oxygen atoms have names ending in *-ite*, and the two with the greatest number of oxygen atoms have names ending in *-ate*. The ion with the most oxygens is distinguished by the prefix *per-*, and the one with the fewest oxygen atoms is given the prefix *hypo-* (for example, hypochlorite, chlorite, chlorate, and perchlorate). Table 7-6 summarizes these principles.

Table 7-6   Nomenclature of Polyatomic Oxygen-Containing Ions

Nomenclature	Group 15	Group 16	Group 17	Examples
hypo_____ite			ClO^-	hypochlorite
_____ite	NO_2^-	SO_3^{2-}	ClO_2^-	nitrite, sulfite, chlorite
_____ate	NO_3^-	SO_4^{2-}	ClO_3^-	nitrate, sulfate, chlorate
per_____ate			ClO_4^-	perchlorate

EXAMPLE PROBLEM 7-9

Name the compounds represented by the following formulas: **a)** $KClO_3$; **b)** NH_4I; **c)** $Ba(BrO_3)_2$; **d)** $CuSO_3$; **e)** Cu_2SO_4; **f)** $(NH_4)_2CO_3$; **g)** $Sn(SO_4)_2$.

SOLUTION

a) We know that ClO_3^- is an ion called chlorate. Therefore, this is potassium chlorate.

b) We know that NH_4^+ is an ion called ammonium. Therefore, this is ammonium iodide.

c) We know that ClO_3^- is an ion called chlorate. Since bromine is in the same group as chlorine, we would expect BrO_3^- to be called bromate. The compound is therefore barium bromate.

d) We know that SO_3^{2-} is an ion called sulfite. There is one sulfite ion with an oxidation number of −2. There is also one copper ion. Since the sum of the positive and negative oxidation numbers must equal zero, the oxidation number of the copper ion must be +2. Therefore, this is copper(II) sulfite. The name copper sulfite is incorrect, because it does not specify which of the two possible oxidation states of copper is present.

e) We know that SO_4^{2-} is an ion called sulfate. There is one sulfate ion with an oxidation number of −2. There are also two copper ions, and the zero sum rule requires that their total oxidation number equal +2. Each copper ion has an oxidation number of +1, and the compound is copper(I) sulfate (not copper sulfate!).

f) We know that NH_4^+ is the ammonium ion, and CO_3^{2-} is the carbonate ion. Therefore, this is ammonium carbonate.

g) We know that SO_4^{2-} is the sulfate ion. There are two sulfate ions, each with an oxidation number of −2, for a total of −4. There is one tin ion. The zero sum rule requires Sn to have an oxidation number of +4. The compound is tin(IV) sulfate.

PRACTICE PROBLEM 7-9

Name the compounds represented by the following formulas: **a)** K_3PO_4; **b)** $Ca(ClO_4)_2$; **c)** $Al(OH)_3$; **d)** $SrCO_3$; **e)** NH_4NO_3; **f)** $Fe(C_2H_3O_2)_3$.

APPLICATION

The Chemistry of Fireworks

When electrons absorb energy, they move to higher energy levels. This is an unstable condition for electrons, so they fall back to lower energy levels. In the process, the electrons emit their excess energy, which may be in the form of visible light. We see examples of this process whenever we watch fireworks.

The main ingredients of fireworks are a fuel, a substance to react with the fuel (an oxidizer), a material to hold all the chemicals together (a binder), and chemicals for special effects. The energy required to excite the electrons to higher energy levels comes from the reaction between the fuel and the oxidizer. Some of the fuels that are used in fireworks are aluminum, magnesium, sulfur, titanium, charcoal, antimony trisulfide, and antimony pentasulfide. Some chemicals that are used as oxidizers are potassium perchlorate, potassium nitrate, ammonium perchlorate, and potassium chlorate. A typical mixture of fuel and oxidizer might be aluminum and sulfur for the fuel, and potassium perchlorate for the oxidizer. The potassium perchlorate causes a chemical reaction with the fuel, and a large amount of energy is released. This reaction produces a bright flash from the aluminum and a loud bang from the rapidly expanding gases that are produced.

Various chemicals are used for their ability to create specific special effects. Strontium nitrate or strontium carbonate is used to produce a red colour; barium nitrate and barium chlorate both produce a green; copper(II) carbonate, copper(II) sulfate, and copper(I) chloride produce blue; and sodium oxalate produces yellow. White flame is produced by magnesium and aluminum metals, and gold sparks are produced by iron filings and charcoal. A whistle effect is produced by potassium benzoate or sodium salicylate, and white smoke is caused by a mixture of potassium nitrate and sulfur.

The chemicals producing the fireworks are launched from a steel cylinder buried in the ground. The propellant is a mixture of potassium nitrate, charcoal, and sulfur (known as black powder).

Although the setting off of fireworks may look like a simple matter, their composition is complex, and certain problems have to be avoided. For example, a fast-burning fuse is needed to ignite the propellant. Slow-burning fuses must be set to go off at different times to produce different effects. High flame temperatures produce white flashes so the desired colours tend to be washed out. For this reason it is sometimes necessary to use fuels that produce relatively low flame temperatures.

The most difficult colour to achieve is blue. Copper(I) chloride provides the best blue when used at low temperatures. Fireworks manufacturers are still trying to find a chemical that produces a deep blue colour. Finally, oxidizers such as potassium and ammonium perchlorate are extremely unstable and can ignite accidentally. The amount of perchlorates available to the fireworks industry is limited because there are few suppliers. Moreover, each launch of the Space Shuttle uses 680 t of ammonium perchlorate as the oxidizer in its solid-fuel booster rockets! The next time you observe fireworks, marvel not only at their beauty, but also at the skill involved in producing their special effects.

7.5 Nomenclature of Acids

An acid is a compound which has a sour taste. Acids turn blue litmus paper red. Normally, acid molecules contain at least one hydrogen atom that can be replaced by metals such as zinc and magnesium. The displaced hydrogen is given off as a gas. Acids are discussed more completely in Chapter 12.

Binary Acids

A compound of hydrogen and a nonmetal from Group 16 or 17 is named by writing the word *hydrogen* followed by the stem of the name of the nonmetal to which the ending -*ide* has been added. Solutions of these compounds dissolved in water (aqueous solutions) are acids. They are named by using the prefix *hydro*-, the stem of the name of the nonmetal, and the ending -*ic*, followed by the word *acid*. Thus, hydrogen chloride (HCl) is a gas, but a solution of HCl dissolved in water is called hydrochloric acid, a solution of HBr in water is called hydrobromic acid, and a solution of H_2S in water is called hydrosulfuric acid. The names of some common binary acids are listed in Table 7-7.

Table 7-7   Names of Some Common Binary Acids

Formula	Name of Acid
HF	hydrofluoric acid
HCl	hydrochloric acid
HBr	hydrobromic acid
HI	hydriodic* acid
H_2S	hydrosulfuric acid
H_2Se	hydroselenic acid

*The "o" of the prefix *hydro*- is dropped for ease of pronunciation.

Oxygen-Containing Acids

A compound composed of hydrogen and a negative polyatomic ion from Table 7-2 is named by writing the word *hydrogen* followed by the name of the polyatomic ion. Thus, H_2SO_4 is hydrogen sulfate, HNO_2 is hydrogen nitrite, and H_3PO_4 is hydrogen phosphate. However, aqueous solutions made by dissolving compounds of hydrogen and negative polyatomic ions in water are acids. Their solutions are given special names. The name of the aqueous solution is formed by dropping the word *hydrogen*, as well as the suffix from the name of the negative ion. The suffix is replaced with a new one, plus the word *acid*. The suffix -*ite* becomes -*ous acid*, and the suffix -*ate* becomes -*ic acid*. Thus, an aqueous solution of H_2SO_4 is sulfuric acid; an aqueous solution of $HClO_4$ is perchloric acid; and an aqueous solution of HNO_2 is nitrous acid. The only common exceptions are the acids containing phosphorus—for these the stem is not *phosp*-, but *phosphor*-, as in H_3PO_4, phosphoric acid. The nomenclature of aqueous solutions of oxygen-containing acids is summarized in Table 7-8.

Table 7-8   Nomenclature of Aqueous Solutions of Some Oxygen-Containing Acids

Nomenclature	Group 15	Group 16	Group 17	Examples
hypo_____ous acid			HClO	hypochlorous acid
_____ous acid	HNO_2			nitrous acid
		H_2SO_3		sulfurous acid
			$HClO_2$	chlorous acid
_____ic acid	HNO_3			nitric acid
		H_2SO_4		sulfuric acid
			$HClO_3$	chloric acid
per_____ic acid			$HClO_4$	perchloric acid

EXAMPLE PROBLEM 7-10

Name the following oxygen-containing acids of iodine, both in the pure state and in aqueous solution: **a)** HIO; **b)** HIO_2; **c)** HIO_3; **d)** HIO_4.

SOLUTION
a) The IO^- ion contains the smallest number of oxygen atoms. It is called hypoiodite ion. The pure compound is hydrogen hypoiodite. The aqueous solution is then called hypoiodous acid.

b) The IO_2^- ion contains the next smallest number of oxygen atoms. It is therefore the iodite ion. The pure compound is then called hydrogen iodite; its aqueous solution is iodous acid.

c) The IO_3^- ion contains the second largest number of oxygen atoms. Its name is iodate ion, so the pure compound is called hydrogen iodate, and its aqueous solution is iodic acid.

d) The IO_4^- ion contains the largest number of oxygen atoms. It is therefore the periodate ion. The pure compound is hydrogen periodate, and the aqueous solution is called periodic acid.

PRACTICE PROBLEM 7-10

Name the aqueous solutions of the following compounds:
a) $HC_2H_3O_2$; **b)** H_2CO_3; **c)** $HMnO_4$; **d)** $H_2C_2O_4$; **e)** H_3AsO_4.

In this chapter we have written formulas for and named a wide variety of chemical substances. Many of these chemicals are noteworthy either because they have important uses in industry or because they have a notable property. Some of these noteworthy chemicals are listed in Table 7-9.

Table 7-9   A Selection of Noteworthy Chemicals

Systematic Name	Chemical Formula	Alternative Name	Use or Notable Property
Aluminum oxide	Al_2O_3	alumina	abrasive materials
Ammonia	NH_3		in manufacture of fertilizers and in household cleaners
Ammonium carbonate	$(NH_4)_2CO_3$		smelling salts
Calcium carbonate	$CaCO_3$	limestone	agriculture
Calcium oxide	CaO	quicklime	mortar and plaster
Carbon monoxide	CO		poisonous gas in automobile exhaust
Hydrochloric acid	HCl	muriatic acid	cleaning metals
Hydrofluoric acid	HF		frosting or etching glass
Hydrogen peroxide	H_2O_2		topical antiseptic
Hydrogen sulfide	H_2S		poisonous gas with rotten-egg odour
Hypochlorous acid	$HClO$		sterlizing agent
Magnesium hydroxide	$Mg(OH)_2$	milk of magnesia	antacid
Dinitrogen monoxide	N_2O	laughing gas	inhalation anesthetic
Silicon dioxide	SiO_2	quartz, sand	manufacture of glass
Silver bromide	$AgBr$		in photographic film
Sodium carbonate	Na_2CO_3	washing soda	general cleaner
Sodium chloride	$NaCl$	salt	preserving food
Sodium hydrogen carbonate	$NaHCO_3$	baking soda	in baking powder
Sodium hydroxide	$NaOH$	soda lye	to neutralize acids
Sodium hypochlorite	$NaClO$		as a bleach (dissolved in water)
Sodium thiosulfate	$Na_2S_2O_3$	hypo	fixer in photography
Sulfur dioxide	SO_2		preserving fruits and vegetables
Sulfuric acid	H_2SO_4	oil of vitriol	manufacture of fertilizers, explosives, and other acids
Tin(II) fluoride	SnF_2	fluoristan	in toothpaste
Titanium dioxide	TiO_2		abrasive in toothpaste; in white house paint
Zinc oxide	ZnO	flowers of zinc	for treating certain skin conditions (zinc ointment)

7.6 Reading Chemical Equations

If chemical formulas are the words in the language of chemistry, chemical equations are the sentences. A **chemical equation** is a shorthand description of a chemical change, using symbols and formulas to indicate the substances involved in the change.

In order to read a chemical equation, we must be able to name all of the substances whose formulas and symbols appear in the equation. We must also be able to interpret the other shorthand devices that are used in an equation. Consider the following chemical equation:

$$NaOH(aq) + HCl(aq) \longrightarrow NaCl(aq) + H_2O(\ell)$$

The reactants, sodium hydroxide ($NaOH$) and hydrochloric acid (HCl), are written to the left of the arrow. The products, sodium chloride ($NaCl$) and water, are written to the right of the arrow. The arrow is read as *to yield*. A plus sign which separates reactants implies *reacts with*. A plus sign which separates products is read as *and*. The (aq) indicates an aqueous solution which means that the substance is dissolved in water. The (ℓ) indicates a pure liquid. This equation is read: An aqueous solution of sodium hydroxide reacts with hydrochloric acid (an aqueous solution of hydrogen chloride) to yield an aqueous solution of sodium chloride and liquid water.

Other shorthand devices which are used in chemical equations are (g), (s), and $\triangle$. A (g) indicates a gas; (s) indicates a solid; and a $\triangle$ placed above the yield arrow indicates that heat is required to cause the reactants to become products.

Often you will notice the presence of coefficients (numbers) in front of chemical formulas in equations. These coefficients are necessary to balance the chemical equation. In a *balanced chemical equation*, there are equal numbers of each kind of atom on both the left hand side and the right hand side. It is not sufficient to write the decomposition of hydrogen peroxide to water and oxygen as

$$H_2O_2(\ell) \longrightarrow H_2O(\ell) + O_2(g)$$

This equation has two oxygen atoms on the left hand side but three oxygen atoms on the right hand side. The equation is not balanced with respect to oxygen. Notice that the equation is balanced with respect to the hydrogen atoms. The balanced chemical equation for this reaction is

$$2H_2O_2(\ell) \longrightarrow 2H_2O(\ell) + O_2(g)$$

The equation is balanced with four oxygen atoms on both the left and right hand side. The same is true for the hydrogen atoms. The equation is read: Two molecules of hydrogen peroxide in the liquid state yield two molecules of water in the liquid state and one molecule of oxygen gas.

EXAMPLE PROBLEM 7-11

What is meant by the following chemical equation?

$$Na_2CO_3(s) + 2HCl(aq) \longrightarrow 2NaCl(aq) + H_2O(\ell) + CO_2(g)$$

SOLUTION

The equation means that one formula unit of sodium carbonate in the solid state reacts with two molecules of hydrochloric acid (an aqueous solution of hydrogen chloride) to yield two formula units of sodium chloride dissolved in water, one molecule of water in the liquid state, and one molecule of gaseous carbon dioxide. Notice that we do not refer to molecules of ionically bonded compounds. In Chapter 6, we stated that ionic substances do not consist of molecules. Thus, we use the term *formula unit*.

PRACTICE PROBLEM 7-11

What is meant by the following chemical equation?

$$3Na_2CO_3(s) + 2H_3PO_4(aq) \longrightarrow 2Na_3PO_4(aq) + 3H_2O(\ell) + 3CO_2(g)$$

CAREER

The Chemical Sales Representative

Figure 7-3
A representative of a pharmaceutical company presents data on one of his products.

Over 60% of all chemists work in the industries which produce products and technologies that shape our daily lives—textiles, rubber, glass, polymers, pharmaceuticals, electronics, paper, packaging, machinery, and food. Not all of these chemists, however, work directly to produce these products. Many of them are employed in technical service, advertising, marketing, management, and sales, all areas that are vital to industry.

Chemical sales representatives generally solicit business actively, trying to find markets for their companies' products. To do so, they must be able to explain chemistry to non-chemists. They must know their products in sufficient detail to be able to explain the advantages and limitations of these products to customers. This means that they must be able to combine their chemical knowledge with their sales skills. They

need to be articulate and persuasive and to enjoy people; they must also have a sound understanding of their companies' products and those of their competitors.

Increasingly, chemical sales personnel are required to be technical experts in their field. They should therefore have taken many courses in chemistry. Many have also benefited from introductory business courses in areas such as accounting, organizational behaviour, marketing, and the overall business environment. Because the work involves dealing with people to such a great extent, students who plan to become sales representatives should participate in campus activities, in debating, in public speaking, or in campus government. The experience gained in presenting ideas to a critical audience will be extremely valuable in a career in chemical sales.

Part Two Review Questions

1. When are Roman numerals omitted during the naming of compounds by the Stock system?
2. What is a binary compound?
3. Name each of the following: **a)** MgI_2; **b)** FeI_3; **c)** H_2Se; **d)** SO_3.
4. Draw a large table using Table 7-10 as a guide. Fill in each blank with the formula and an acceptable name of the compound that results when the positive ion at the left is combined with the negative ion at the top.

Table 7-10   Names and Formulas of Compounds

	OH^-	F^-	$C_2H_3O_2^-$	O^{2-}	$C_2O_4^{2-}$	$S_2O_3^{2-}$	PO_4^{3-}
H^+							
Na^+							
NH_4^+							
Ag^+							
Cu^{2+}							
Ca^{2+}							
Ni^{2+}							
Zn^{2+}							
Hg^{2+}							
Al^{3+}							

5. Give formulas for the following compounds: **a)** hydrogen bromide; **b)** hydriodic acid; **c)** carbon tetrabromide; **d)** sulfur trioxide; **e)** potassium oxalate; **f)** iron(III) nitrate; **g)** iron(III) sulfite; **h)** cobalt(II) chlorate; **i)** cobalt(II) oxalate; **j)** tin(IV) nitrate; **k)** tin(IV) sulfite.
6. What is the definition of a chemical equation?
7. What is meant by these chemical equations?
   a) $CaO(s) + H_2O(\ell) \longrightarrow Ca(OH)_2(aq)$
   b) $NaHCO_3(s) + HCl(aq) \longrightarrow NaCl(aq) + H_2O(\ell) + CO_2(g)$

Chapter Summary

- Chemical nomenclature is the assignment of names to compounds.
- Oxidation numbers are used to keep track of the positive or negative character of atoms or ions.
- The oxidation numbers of atoms of elements in the free state are always zero.
- The oxidation number of an ion is the same as its charge.
- In assigning oxidation numbers to atoms involved in polar covalent bond-

ing, electrons are counted as if they had transferred completely from one atom to another, even if they are only shifted towards an atom.

- A polyatomic ion is an ion which contains two or more atoms.
- The symbols of the elements are used in a chemical formula to represent the composition of a compound.
- The zero sum rule states that the sum of all the positive oxidation numbers and all the negative oxidation numbers of the elements in a compound must be zero.
- The Stock system must be used when naming ions that have more than one possible oxidation number.
- In the older method of naming positive ions which have more than one oxidation number, the suffix *-ous* is added to the stem of the element when referring to the lower oxidation number and the suffix *-ic* is added to the stem of the element when referring to the higher oxidation number.
- The name of a negative monatomic ion is formed by adding the suffix *-ide* to the stem (the first part) of the name of the element.
- A binary compound is one which contains only two elements.
- The name of a binary compound consisting of a metal and a nonmetal consists of the name of the positive ion, followed by the name of the negative ion.
- Compounds of hydrogen and a nonmetal from Groups 16 and 17 are named by writing the word *hydrogen* followed by the stem of the name of the nonmetal to which the ending *-ide* has been added.
- The name of a binary compound of two nonmetals is formed according to the pattern: prefix + element name, followed by prefix + element stem + *-ide*. Normally the prefix *mono-* is omitted from the name of the first element.
- The name of a compound consisting of a metal and a polyatomic ion consists of the name of the metal, followed by the name of the polyatomic ion.
- Solutions of compounds of hydrogen and a nonmetal from Groups 16 and 17 dissolved in water (aqueous solutions) are acids. They are named by using the prefix *hydro-*, the nonmetal stem, and the suffix *-ic*, followed by the word *acid*.
- Aqueous solutions containing compounds of hydrogen and negative polyatomic ions are acids. They are named in the following manner: the word *hydrogen* as well as the suffix from the name of the negative ion are dropped; the suffix is replaced with a new one; the suffix *-ite* becomes *-ous acid*, and the suffix *-ate* becomes *-ic acid*.
- A chemical equation is a shorthand description of a chemical change, using symbols and formulas to indicate the substances involved in the change.
- There are equal numbers of each kind of atom on both the left hand side and the right hand side of a balanced chemical equation.

Key Terms

chemical nomenclature
chemical formula
chemical equation

Test Your Understanding

Each of these statements or questions is followed by four responses. Choose the correct response in each case. (Do not write in this book.)

1. What is the oxidation number of the underlined ion in $\underline{Al}_2O_3$?
   a) +3   b) −3   c) +2   d) −2

2. What is the oxidation number of the underlined ion in $Ca_3\underline{N}_2$?
   a) −3   b) +3   c) −2   d) +2

3. What is the oxidation number of the underlined ion in $Na_2\underline{S}_2O_6$?
   a) −1   b) +1   c) −2   d) +2

4. What is the oxidation number of the underlined ion in $Zn_3\underline{Te}O_6$?
   a) −1   b) −3   c) −6   d) +3

5. The formula of the compound formed from Ba^{2+} and Cl^- is
   a) $BaCl$.   b) Ba_2Cl.   c) $ClBa_2$.   d) $BaCl_2$.

6. The formula of the compound formed from Ga^{3+} and S^{2-} is
   a) GaS_3.   b) Ga_2S_3. c) Ga_3S_2. d) Ga_2S.

7. The formula of the compound formed from Al^{3+} and $C_2H_3O_2^-$ is
   a) $AlC_2H_3O_2$.   b) $Al(C_2H_3O_2)_3$.   c) $Al_3C_2H_3O_2$.   d) $Al(C_2H_3O_2)_2$.

8. The formula of the compound formed from Sr^{2+} and PO_4^{3-} is
   a) $Sr(PO_4)_3$.   b) $Sr(PO_4)_2$.   c) $Sr_2(PO_4)_3$.   d) $Sr_3(PO_4)_2$.

9. What is the formula for tin(IV) oxide?
   a) SnO_4 b) SnO_2 c) Sn_4O d) SnO

10. What is the formula for copper(II) acetate?
    a) $Cu(C_2H_3O_2)_2$ b) $CuC_2H_3O_2$ c) $Cu_2C_2H_3O_2$ d) $Cu_2(C_2H_3O_2)_2$

11. The formula for phosphorus pentachloride is
    a) PCl_3.   b) P_5Cl_5. c) P_5Cl.   d) PCl_5.

12. What is the formula for dinitrogen monoxide?
    a) N_2O b) NO_2 c) N_2O_2 d) N_2O_4

13. What is the name of MgS?
    a) magnesium(I) sulfide   b) magnesium sulfate
    c) magnesium sulfite   d) magnesium sulfide

14. The name of the binary compound CaC_2 is
    a) carbon calcite. b) calcium carbide. c) calcous carbide. d) calcium carl

15. The name of Na_3PO_4 is
    a) sodium phosphide.   b) sodium phosphite.
    c) sodium phosphate.   d) sodium perphosphate.

16. The name of $Zn(HCO_3)_2$ is
    a) zinc hydrocarbonate.   b) zinc(II) carbonate.
    c) zinc carbonate.   d) zinc hydrogen carbonate.

17. What is the name of $Cu(NO_3)_2$?
    a) copper dinitrite   b) copper(II) dinitrate
    c) copper(II) nitrate   d) copper(I) nitrate

18. What is the name of Hg_2SO_4?
    a) mercury(II) sulfate   b) mercury(I) sulfate
    c) mercury(I) sulfite   d) mercury(II) sulfide

19. Which of the following is named antimony trichloride?
    a) $SbCl_3$ b) $AnCl_3$ c) $SeCl_3$ d) An_3Cl

20. What is the name of IF_5?
    a) monoiodine pentafluoride   **b)** iodide pentafluoride
    c) iodine pentafluoride   **d)** monoiodide pentafluoride

21. What is the name of $HClO_4$?
    a) chlorous acid   **b)** chloric acid
    c) perchloric acid   **d)** hypochlorous acid

Review Your Understanding

1. Indicate the oxidation number of the underlined ion in each of the following substances: **a)** $\underline{Zn}^{2+}$; **b)** Ba$\underline{S}$; **c)** Ca$\underline{SO_3}$; **d)** Na$\underline{OH}$; **e)** H$\underline{NO_3}$.

2. Indicate the oxidation number of the underlined ion in each of the following substances: **a)** $\underline{BrO_3^-}$; **b)** Na$_2\underline{S_2O_3}$; **c)** (NH$_4$)$_2\underline{Cr_2O_7}$; **d)** K$\underline{C_2H_3O_2}$.

3. What is the oxidation number of the polyatomic ion in each of the following compounds: **a)** Na_2MoO_4; **b)** $Mg(IO_3)_2$; **c)** $CaSiO_3$; **d)** $Al(C_2H_5O)_3$; **e)** $Li_2B_4O_7$?

4. What is the oxidation number of the metal in each of the following compounds: **a)** $SrCl_2$; **b)** YbI_3; **c)** $TlOCN$; **d)** V_2O_5; **e)** $Y(OH)_3$?

5. Give the oxidation number of each element in the following: **a)** P_4O_{10}; **b)** AsH_3; **c)** O_3; **d)** $GeCl_4$; **e)** SF_6.

6. Give the oxidation number of each element in the following: **a)** BrF_5; **b)** S_8; **c)** $SiCl_4$; **d)** SCl_2; **e)** $SbCl_5$.

7. Write formulas for the compounds formed by all possible combinations of Cs^+, Ba^{2+}, and Ga^{3+} with I^-, S^{2-}, and P^{3-} (nine possibilities).

8. Write formulas for all possible combinations of K^+, Mg^{2+}, and B^{3+} with $C_2H_3O_2^-$, $C_2O_4^{2-}$, and PO_3^{3-} (nine possibilities).

9. Write formulas for all possible combinations of NH_4^+, Ca^{2+}, Al^{3+}, and Sn^{4+} with CN^-, CO_3^{2-}, and PO_4^{3-} (12 possibilities). (Not all of the compounds represented will actually exist.)

10. Write formulas for all possible combinations of Rb^+, Sr^{2+}, Ga^{3+}, Pb^{4+}, and Bi^{5+} with Br^-, S^{2-}, and N^{3-}. (There are 15 possibilities, but not all of them actually exist.)

11. What is the formula for **a)** iron(III) chloride; **b)** iron(II) oxide; **c)** cobalt(II) bromide; **d)** tin(IV) chloride?

12. What is the formula for **a)** lithium sulfide; **b)** potassium iodide; **c)** magnesium bromide; **d)** magnesium sulfide; **e)** aluminum sulfide.

13. What is the formula for **a)** magnesium hydroxide; **b)** iron(III) carbonate; **c)** carbon tetrabromide; **d)** diphosphorus pentasulfide; **e)** nitrogen(V) sulfide?

14. Write formulas for the following compounds: **a)** stannous chloride; **b)** ferric sulfate; **c)** vanadium(II) bromide; **d)** tin(IV) oxide; **e)** gold(III) cyanide.

15. Write formulas for the following compounds: **a)** aluminum acetate; **b)** sodium oxide; **c)** potassium permanganate; **d)** barium sulfite; **e)** silicon tetrachloride; **f)** calcium hydrogen carbonate; **g)** silver phosphate; **h)** ferrous arsenate; **i)** ferric sulfate; **j)** iron(III) phosphite.

16. Name each of the following compounds: **a)** LiH; **b)** NH_3; **c)** FeS; **d)** $SnBr_4$; **e)** Hg_2Cl_2; **f)** N_2S_5; **g)** N_2O; **h)** Al_2O_3.

17. Name each of the following compounds: **a)** NaCl; **b)** Li_2S; **c)** PH_3; **d)** BaF_2; **e)** BeO; **f)** Mg_3N_2; **g)** GaI_3; **h)** Al_2S_3; **i)** BN.

18. Use the Stock system to rename the following compounds: **a)** cobaltous nitrate; **b)** mercuric selenide; **c)** stannous fluoride; **d)** ferric chloride.

19. Use the Stock system to rename the following compounds: **a)** ferrous sulfate; **b)** cobaltic sulfide; **c)** mercurous iodate; **d)** stannic iodide.

20. Name each of the following compounds: **a)** $(NH_4)_2Cr_2O_7$; **b)** $Cu(NO_3)_2$; **c)** $Ba(HCO_3)_2$; **d)** CaC_2O_4; **e)** $Pb(C_2H_3O_2)_2$; **f)** $Al_2(S_2O_3)_3$.

21. Name the following compounds according to the Stock system of nomenclature: **a)** $CuClO_2$; **b)** $Hg_2Cr_2O_7$; **c)** Tl_3PO_4; **d)** $Cu(HCO_3)_2$; **e)** $Fe(ClO_3)_2$; **f)** $Hg_3(PO_3)_2$; **g)** $Au(ClO_4)_3$; **h)** $Co_2(Cr_2O_7)_3$; **i)** $FeAsO_4$.

22. Give formulas and Stock system names for all possible combinations of Au^+, Co^{2+}, Fe^{3+}, and Pb^{4+} with ClO^-, CrO_4^{2-}, and AsO_4^{3-} (12 possibilities). (Not all of the represented compounds actually exist.)

23. Write the name of the aqueous solutions of each of the following: **a)** H_2SO_4; **b)** HCl; **c)** HNO_3; **d)** $HC_2H_3O_2$; **e)** H_3PO_4, **f)** $HClO_4$.

24. What is the formula for **a)** permanganic acid; **b)** oxalic acid; **c)** ammonium chromate; **d)** sodium perchlorate; **e)** potassium thiosulfate?

25. What is meant by the following equations:
    a) $2Na(s) + 2H_2O(\ell) \longrightarrow 2NaOH(aq) + H_2(g)$
    b) $Mg(s) + 2H_2O(\ell) \xrightarrow{\triangle} Mg(OH)_2(aq) + H_2(g)$
    c) $BaO(s) + H_2O(\ell) \longrightarrow Ba(OH)_2(s)$
    d) $SO_3(g) + H_2O(\ell) \longrightarrow H_2SO_4(aq)$

Apply Your Understanding

1. What are the advantages of the Stock system over the older method of using the suffixes *-ous* and *-ic*?

2. Draw the Lewis structures of the hypochlorite ion, the chlorite ion, the chlorate ion, and the perchlorate ion. Show why each of these ions has an oxidation number of −1.

3. The oxidation number of hydrogen in hydrogen chloride is +1. However, the oxidation number of hydrogen in potassium hydride is −1. What is the reason for the difference in hydrogen's oxidation number?

Investigations

1. It has been said that a country's consumption of sulfuric acid is a measure of its industrial development. Research the uses of sulfuric acid in order to determine the validity of this statement.

2. Write brief reports on the uses and any notable properties of five of the chemicals found in Table 7-9.

3. Write a research report on a chemical that has been in the news during the past six months.

4. Survey the ingredients of a representative sample of ten household products. Using the rules discussed in this chapter, write the formulas of those ingredients which are named.

5. Debate the following: Manufacturers should be forced to give the systematic names of *all* of the chemicals used in their products.

The Liquid Phase

8

CHAPTER CONTENTS

Part One: **The Liquid Phase**
    8.1 The Structure of Liquids
    8.2 Attractive Forces in the Liquid Phase
    8.3 Physical Properties of Liquids
    8.4 Freezing
    8.5 Evaporation and Boiling
    8.6 Distillation

Part Two: **Water**
    8.7 Occurrence of Water
    8.8 Physical Properties of Water
    8.9 Chemical Properties of Water
    8.10 Heavy Water
    8.11 Hard Water

Part Three: **Water Pollution**
    8.12 Causes of Water Pollution
    8.13 Purifying Polluted Water
    8.14 Water Pollution in Canada

Why would a mountain climber find it difficult to cook an egg in boiling water at the top of Mount Robson (Fig. 8-1), at 3.59 km, the highest peak in the Canadian Rockies? What would life on earth be like if water had different melting and boiling temperatures? What determines whether a substance is a solid, liquid, or gas? What is heavy water? What do making freeze-dried coffee and salvaging water-damaged books have in common? You will find the answers to these questions as you study this chapter.

Figure 8-1
Mount Robson

In this chapter we begin with a discussion of the liquid phase. A model is developed to explain the observed properties of liquids. This model is then used in a discussion of the properties of water—the most important of all liquids. We shall then describe the types and causes of water pollution and the purification of polluted water. The chapter concludes with a discussion of water pollution in Canada.

Key Objectives

When you have finished studying this chapter, you should be able to

1. explain the physical properties of liquids in terms of their structure and forces of attraction.
2. explain what is meant by a dynamic equilibrium, using a liquid-vapour equilibrium as an example.
3. describe the occurrence of water.
4. describe the physical properties of water.
5. describe the chemical properties of water.
6. list the uses of water.
7. name and describe the types of water pollution.

Part One: The Liquid Phase

8.1 The Structure of Liquids

The structure of liquids is not as well known as that of either solids or gases. However, it is generally assumed that the particles in a liquid are not as close to each other as those of a solid, but they are much closer together than the particles of a gas.

As you can see in Figure 8-2, the molecules in liquids are close together but are packed with no apparent order. There appear to be empty spaces throughout the liquid. What you cannot see in Figure 8-2 is that these empty spaces are not fixed in the same locations all the time, because the molecules are moving about continuously in a random manner.

Figure 8-2
Comparison of the structures of the states of matter

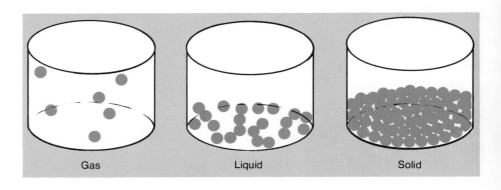

Gas   Liquid   Solid

8.2 Attractive Forces in the Liquid Phase

In the liquid state a number of attractive forces tend to keep the particles close together. Some of these forces of attraction were first characterized by a German physicist, Fritz London. They are known as **London dispersion forces**. When electrons move about in an atom, their motion is somewhat random. In any atom, there is a chance that at a given instant there will be more electrons on one side of the atom than the other. Thus, one side of the atom will be slightly negative, and the other side will be slightly positive. The atom will contain an instantaneous dipole (two poles of electrical charge). This instantaneous dipole in the atom can induce a dipole in a neighbouring atom. That is, as the negative end of an instantaneous dipole begins to form in one atom, it pushes electrons away in the atom alongside. Then, since the negative end of one dipole is close to the positive end of the other dipole, there is an attraction between them. This attraction helps hold the particles together in the liquid.

London dispersion forces are quite weak because the instantaneous dipoles exist only for a moment and then only when the particles are very close to each other. That is why the noble gases, for example, have to be cooled to low temperatures in order to form liquids. At higher temperatures the kinetic energies of the atoms are too great for the attractive forces to hold them together in the liquid state. The London dispersion forces increase as the number of electrons in the atom increases. The boiling temperatures of He, Ne, Ar, Kr, Xe, and Rn are −269°C, −246°C, −186°C, −152°C, −107°C, and −62°C, respectively. As we move from He to Rn, both the number of electrons and the size of the atoms increases. The London dispersion forces become stronger, and more thermal energy (higher temperatures) must be used to overcome the forces of attraction and boil the liquid. The forces of attraction between nonpolar molecules such as carbon tetrachloride are the same as between atoms: London dispersion forces.

Polar molecules such as hydrogen chloride are also attracted to one another by London dispersion forces. However, they feel an additional attraction to one another which is caused by their permanent dipoles. The slightly negative end of one molecule attracts the slightly positive end of the neighbouring molecule. This force is called the **dipole-dipole force** of attraction. Thus, the forces of attraction between polar molecules are the sum of London dispersion and dipole-dipole forces. These forces of attraction as a group are called **van der Waals forces** of attraction in honour of the Dutch physicist, Johannes van der Waals, who studied them.

Hydrogen Bonding

Within a group of related molecules, the van der Waals forces will increase as the number of electrons in the molecules increases. For example, the boiling points of the group HCl, HBr, and HI are −85°C, −67°C, and −35°C. As the molecules in this series increase in size, higher temperatures must be used to overcome the forces of attraction and boil the liquids. Furthermore, we would expect that HF would boil at a temperature which is lower than −85°C. In fact, however, it boils at +20°C, which is 105°C higher than −85°C. That is, in this group of molecules, the member having the fewest electrons does not have the

Figure 8-3
Boiling points of hydrogen compounds of Group 16 elements

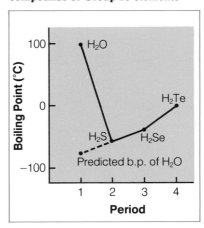

Table 8-1   Boiling Points of Hydrogen Compounds of Group 16 Elements

Compound	Boiling Point (°C)
H_2Te	−2.3
H_2Se	−41.3
H_2S	−60.3
H_2O	100.0

lowest boiling point. In the same way, if we consider the boiling points of hydrogen compounds of nonmetals in Group 16 (Table 8-1, Fig. 8-3), the trend would lead us to predict that water should boil near −80°C. Instead, we find that water does not boil until a temperature of 100°C is reached.

Why do HF and H_2O boil at temperatures that are higher than we would expect from the boiling points of related molecules? Liquids such as HF and H_2O must be held together by forces of attraction which are greater than van der Waals forces. What is the nature of these larger forces of attraction?

The H-F bond is polar covalent. In hydrogen fluoride, the slightly positive hydrogen atom of one molecule is chemically bonded to a fluorine atom, but it is also attracted to the slightly negative fluorine atom of another molecule. This is a dipole-dipole attraction, but it is a large one because the bond in the hydrogen fluoride molecule is very polar. This attraction is called a **hydrogen bond**. It is about one-tenth as strong as a covalent bond, but it is about ten times stronger than normal van der Waals forces. Hydrogen bonds occur only when hydrogen atoms are bonded to extremely electronegative elements such as fluorine, oxygen, or nitrogen. These three electronegative elements form covalent bonds with hydrogen atoms which are of such an extremely polar nature that the attractive forces between neighbouring molecules are too large to be considered normal van der Waals forces. A hydrogen bond has been described as the sharing of a hydrogen atom between two strongly electronegative atoms. However, a hydrogen bond (like any bond) is a force of attraction—the attraction between the hydrogen atom of one molecule and a fluorine, oxygen, or nitrogen atom of a second molecule.

Of the three types of forces of attraction mentioned above, London dispersion forces and dipole-dipole attractions are weaker than hydrogen bonding. In addition, each of these is much weaker than the ionic attractions which would be found in molten sodium chloride or the covalent bonds which hold atoms together in molecules.

8.3 Physical Properties of Liquids

We can explain some of the physical properties of liquids by considering the structure of, and attractive forces within, liquids.

The densities of liquids are ordinarily much greater than those of gases. For example, the density of liquid water at 100°C and standard atmospheric pressure is 8.58×10^2 kg/m^3. Water vapour under the same conditions has a predicted density of 5.88×10^{-1} kg/m^3. Thus, the density of this liquid is 1630 times greater than that of the gas. This implies that the molecules in the liquid state must be much closer together.

Empty spaces in the liquid phase give molecules some freedom to move past one another. This explains why liquids flow readily and take the shapes of their containers. The empty spaces in liquids allow the molecules in this phase to diffuse (mix together), but not as readily as do the molecules in the gas phase. Gases have much more empty space; this enables them to mix uniformly almost instantly. Since there are relatively few open spaces between molecules in a liquid, it is much more difficult to compress a liquid than to compress a gas. Liquids expand much less when heated than do gases. This is because there are stronger forces of attraction to overcome in liquids when the molecules are separated from each other by heating.

The volume resulting from mixing two different liquids together is not necessarily the sum of the original volumes. For example, ethanol (ethyl alcohol) does not react with water. However, when 50 mL of ethanol is mixed with 50 mL of water the volume of the resulting solution is 96 mL (Fig. 8-4). This decrease in volume upon mixing is mainly due to strong intermolecular forces (hydrogen bonds) between the two different liquids which draw the molecules closer together.

Figure 8-4
Mixing ethanol and water

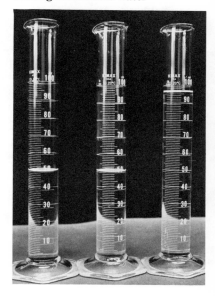

8.4 Freezing

If a liquid is cooled, its molecules will slow down and lose kinetic energy. With further cooling, the molecules will move even more slowly. Since they are moving more slowly, they will come into closer contact and be better able to attract one another. Eventually, a temperature will be reached at which the liquid will change phase to become a solid. The process of changing phase from liquid to solid is called **freezing**, and the temperature at which this occurs is known as the **freezing point** of the liquid.

The freezing point of any liquid is characteristic of that particular liquid. A pure liquid will freeze at a specific temperature. If an impurity is present, the liquid will not start to freeze until a lower temperature is reached. It will continue to freeze only if the temperature is lowered further. Thus, an impure liquid does not have a specific freezing point. It freezes over a range of temperatures. For example, a mixture of salt and water freezes over a range of temperatures below 0°C (the freezing point of pure water). Salt can therefore be spread on icy roads in winter, because the ice-salt mixture will melt at temperatures below 0°C. This effect is also used when ethylene glycol is added to the water in a car radiator to prevent damage to it during cold weather. The ethylene glycol impurity lowers the freezing point of the water. Enough ethylene glycol is used to ensure that the water will not freeze, even during the coldest expected weather. The radiator damage that could be caused by the freezing of the water is avoided.

8.5 Evaporation and Boiling

If a liquid in an open container is allowed to stand for a few days, the level of the liquid in the container decreases. For example, people who complain of dry air caused by central heating often use dishes of water as humidifiers. The evaporation of the water adds moisture to the air and increases the humidity. **Evaporation** of a liquid is the escape of molecules from the surface of the liquid to enter the gas phase above the liquid. The process of evaporation can be made to occur faster by heating the liquid. This causes an increase in the kinetic energy of the molecules. Thus, since the molecules are moving faster, more molecules possess enough energy to escape into the vapour phase. (The

Figure 8-5
A boiling liquid

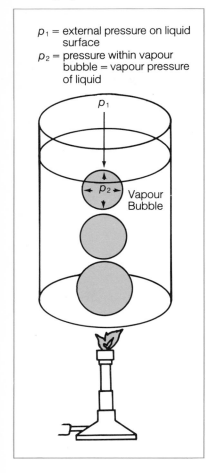

p_1 = external pressure on liquid surface
p_2 = pressure within vapour bubble = vapour pressure of liquid

p_1

p_2

Vapour Bubble

Figure 8-6
Liquid-vapour equilibrium

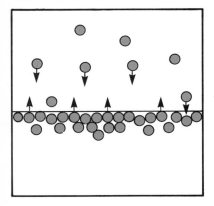

term *vapour* is generally used instead of *gas* to refer to the gaseous state of substances which are normally liquids.)

If heating is continued, eventually the molecules in the liquid will have sufficient kinetic energy to overcome the attractive forces of their neighbours. They will enter the vapour phase. If this transformation happens within the body of the liquid, tiny bubbles will form and rise to the surface of the liquid (Fig. 8-5). The external pressure exerted by the atmosphere on the surface of the liquid is transmitted undiminished and equally in all directions throughout the liquid. Thus, the pressure in all parts of the liquid equals atmospheric pressure. When the temperature of the liquid is high enough so that the pressure inside the bubble equals the atmospheric pressure, the bubble is able to rise to the surface. That is, the pressure inside the bubble is large enough to prevent the outside pressure from collapsing the bubble. When the vapour bubbles reach the surface of the liquid they break and allow the vapour to escape into the air. This process is called **boiling**. The temperature at which boiling occurs, that is, the temperature at which the pressure exerted by the vapour equals the external pressure exerted by the atmosphere, is called the **boiling point** of the liquid. If the atmospheric pressure is *standard pressure* (101.3 kPa), the temperature at which a liquid boils is called its **normal boiling point**.

The liquid does not boil away all at once. Rather, the rate at which a liquid is converted into the vapour phase depends on the rate at which heat energy is added to the system. Also, the temperature of a boiling liquid will remain constant as long as the pressure on the surface of the liquid stays the same. The temperature at which a liquid boils therefore depends upon the pressure on the surface of the liquid. Water has a boiling point of 100°C when the pressure on its surface is standard atmospheric pressure (101.3 kPa), which is the approximate average atmospheric pressure at sea level. Atmospheric pressure decreases as the height above sea level increases. Since atmospheric pressure decreases, the pressure inside the bubbles of vapour will equal the atmospheric pressure at a lower temperature. This explains why it takes longer to boil an egg on a mountain top than at sea level. The water boils at a lower temperature.

Are there differences between a liquid in an open container and a liquid in a closed container? One obvious difference is that the liquid in the closed container may not completely evaporate because the vapour is confined. Evaporation will begin to occur as it did in the open container, but in the closed container, the vapour will have to remain above the liquid surface. As more liquid evaporates to become a gas, more of the vapour is confined above the surface of the liquid. Like all molecules in the gaseous state, the molecules of a vapour are in constant, random motion, colliding with each other, with the walls of the container, and with the surface of the liquid. As the number of molecules in the vapour state increases, it becomes more likely that some of them will collide with molecules on the surface of the liquid and enter the liquid phase once again. This change of state from a gas to a liquid is known as **condensation**. Eventually, the rate at which molecules are leaving the liquid phase to enter the gas phase is equal to the rate at which molecules are leaving the gas phase to enter the liquid phase once again (Fig. 8-6). This condition is known as a *physical equilibrium*, because the physical changes of evaporation and condensation are occurring at *equal rates*. This example is called liquid-vapour equilibrium.

A liquid-vapour equilibrium is *dynamic* because evaporation and condensation continue to occur at the same time. A **dynamic equilibrium** is a state in which two opposing processes continue to occur, but at equal rates. Since evaporation and condensation are occurring at equal rates, as many molecules enter the vapour phase by a evaporation as leave it by condensation. Thus, the number of molecules in the vapour phase stops increasing and remains unchanged.

If the temperature of the liquid were raised, the number of molecules that entered the gas phase would increase. More molecules would leave the liquid at the higher temperature because more molecules would possess enough energy to escape from the surface of the liquid.

Another aspect of the liquid-vapour equilibrium is that the molecules in the vapour phase are colliding with the walls of the container and are exerting pressure on those walls. This pressure exerted by a vapour in equilibrium with its liquid is called the **vapour pressure** of the liquid. The vapour pressure depends upon the number of molecules colliding with the walls of the container and on the average force of the collisions. At constant temperature the average force of the collisions is constant. Therefore the vapour pressure is a measure of the tendency for molecules to leave the liquid and enter the vapour phase. If two different liquids have different vapour pressures at the same temperature, the molecules of the liquid with the higher vapour pressure have a greater tendency to escape from the liquid phase.

Vapour pressure is a physical property of a particular liquid and is related to the intermolecular forces of attraction in the liquid. At 20°C, for instance, the vapour pressure of ether is about 25 times as great as the vapour pressure of water. Ether molecules have a much greater tendency than water molecules to escape the liquid phase, because the intermolecular forces of attraction are London dispersion forces (weak) in ether and hydrogen bonding (strong) in water.

As the temperature of a liquid increases, its vapour pressure also increases (Fig. 8-7). This is because more molecules have enough energy to enter the vapour phase. These molecules move faster because of the higher temperature. Their collisions with the walls of the container are more energetic and more frequent, causing a higher vapour pressure.

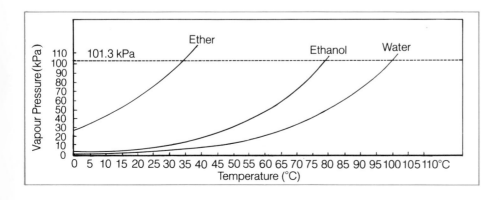

Figure 8-7
Effect of temperature on the vapour pressure of liquids

The operation of a pressure cooker (Fig. 8-8) depends on some of the concepts you have studied in this section. In an open saucepan, the pressure on the surface of the water is atmospheric pressure (about 100 kPa), and the water boils (at 100°C) when its vapour pressure reaches atmospheric pressure. In a pressure cooker, the pressure relief valve does not operate until the pressure inside the cooker is greater than atmospheric pressure. Thus, the water must be heated to a temperature greater than 100°C in order to generate enough vapour pressure to operate the relief valve. The water will then be boiling, because the vapour pressure of the water equals the operating pressure of the cooker. Since the temperature of the boiling water in the pressure cooker is greater than 100°C, the food will cook faster than it does in an open saucepan.

Figure 8-8
A pressure cooker

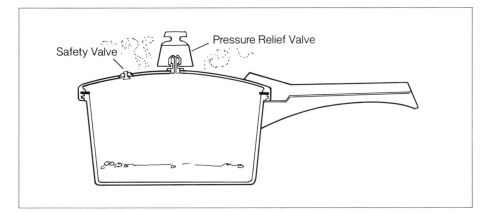

8.6 Distillation

Liquids may contain impurities such as dissolved solids or other liquids. An apparatus like the one shown in Figure 8-9 may be used to separate water or other liquids from nonvolatile impurities (substances with very low vapour pressures and little tendency to evaporate). When the distilling flask is heated, the liquid boils. The vapour travels down the inner glass tube of the condenser, which is cooled by cold water flowing along the outside of the tube. The vapour condenses on the cold surface to form a liquid which runs down the tube and is collected. This purified liquid is referred to as the *distillate*. The process is known as **distillation**. Beverages such as rum, vodka, and whisky are prepared by distillation. The process of distillation is also used by the oil industry to separate the valuable components of crude oil.

Another type of distillation, called *vacuum distillation*, is used by chemists to purify substances which may be unstable and which decompose at the higher temperatures of a regular distillation. Air is removed from above the liquid in the distilling flask to reduce the pressure. Since the pressure above the liquid is lowered, the compound in question boils at a lower, safer temperature at which it does not decompose. As in a simple distillation, the vapour is condensed on the inner cold surfaces of the condenser, and the distillate is collected in a suitable receiving flask.

A related process called *freeze drying* is used in the food industry to prepare dehydrated foods such as instant coffee. Coffee is first brewed. It is then

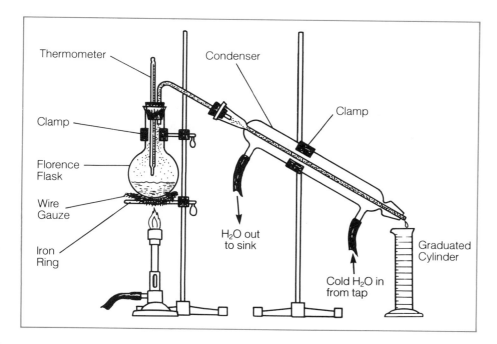

Figure 8-9
Apparatus for distillation

Labels in figure: Thermometer, Condenser, Clamp, Clamp, Florence Flask, Wire Gauze, Iron Ring, H₂O out to sink, Cold H₂O in from tap, Graduated Cylinder

immediately frozen and the water is removed under vacuum. This process is used because removing water by ordinary means at higher temperatures would spoil the taste of the coffee. By using freeze drying, water can be removed at much lower temperatures with little loss of flavour.

Freeze drying has also been used to save books which have been soaked by water. The wet books are frozen as soon as possible. They are then warmed slightly in a vacuum. The molecules of water are able to escape directly from the solid phase (ice) into the vapour phase. The change of a solid directly to its vapour is called **sublimation**. When all of the water has been removed by sublimation, the books are then placed in a high-humidity environment where they can absorb the proper amount of moisture, since they are dry and brittle after the freeze drying process.

In Part One of this chapter, we have discussed the structure of pure liquids and the forces that hold the molecules together in the liquid state. We have used this information to explain the properties of pure liquids and the purification of impure liquids by the technique of distillation.

Part One Review Questions

1. On the molecular level, how does a liquid differ from a gas? How does it differ from a solid?
2. State the three major forces that can exist between molecules in the liquid phase. Which of these is the strongest force?
3. Define the term *freezing point*.
4. Describe the processes which occur on a molecular level as a liquid is cooled to its freezing point.
5. What effect does an impurity have on the freezing point of a pure liquid?
6. Describe vapour pressure in molecular terms. What do we mean by saying that it involves a dynamic equilibrium?

7. How does the vapour pressure of a liquid depend on the intermolecular forces?

8. Describe in molecular terms the behaviour of a liquid and its vapour in an open container as the temperature is increased.

9. What is the difference between the terms *boiling point* and *normal boiling point*?

10. Why does water boil at a lower temperature on a mountain top?

11. What is the principle behind the operation of a pressure cooker?

12. Describe the processes which occur when a liquid is separated from a nonvolatile impurity by distillation.

CAREER   The Science Writer

Figure 8-10
Science writers communicate the results of scientific research to the general public.

Chemists must communicate the results of their experiments to other chemists, but it is also important that the general public be informed of their findings. It is generally the job of the science writer to communicate this information in language the public can understand—to write a science story so well that people will want to read it.

The most important qualifica-tion of a science writer is the ability to take complex scientific topics and invent analogies that will make them understandable. This requires considerable imagi-nation as well as a wide under-standing of scientific concepts. A typical science writer has a univer-sity degree in a science discipline such as chemistry, but is more in-terested in interacting with people than with chemicals. Prime re-quirements are, naturally, a love of writing and the ability to do it well.

Many science writers research and write articles using story ideas that they have developed them-selves. Therefore, they must be able to decide what topics and issues in science are important, interesting, and worth commu-nicating.

Science writers frequently work as science reporters for newspa-pers, magazines, or radio and tele-vision stations. Not only must they be able to communicate re-search results to the public, but they must also be capable of inter-preting government policy in sci-ence, medicine, and technology and explaining the impact of re-search and technology on society. Research scientists and govern-ment officials should not be al-lowed to answer by themselves the questions posed by science and technology; the public should be involved in the decision-making process. Giving people the infor-mation they need to make in-formed decisions is a major function of the science writer.

Part Two: Water

8.7 Occurrence of Water

Water is the only chemical which is present on earth abundantly and simultaneously as a gas (vapour), a liquid, and a solid. Large quantities of water vapour are present in the atmosphere. As a liquid, water forms oceans, lakes, and streams, covering nearly three-fourths of the earth's surface. In the solid state, it is present as ice in the polar regions and mountainous areas of the earth (Fig. 8-11). Water is present in the earth's crust and is a component of many minerals. It is also the principal constituent of living matter. Most plants and animals contain 60% to 90% water by mass. For example, the water content of blood is about 80%, while bones contain 12% to 15% water.

Figure 8-11
Water is present in this scene in the Canadian Rockies in all its forms: in the clouds (vapour), in the ice (solid), and in the tissues of the mountain climbers (liquid).

8.8 Physical Properties of Water

Water was thought to be an element as late as the latter part of the eighteenth century. But Henry Cavendish, in 1781, showed that water is formed when hydrogen burns in air. Lavoisier determined the percentage composition of water a few years later. Since then, the relative masses of hydrogen and oxygen in water have been determined to be 1.0076 parts hydrogen to 8.0000 parts oxygen. Water is a compound with chemical formula H_2O.

A molecule of water is polar and has the following structure:

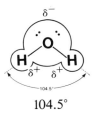

104.5°

Water is therefore capable of forming hydrogen bonds to other water molecules:

$$H \atop H \Big\rangle O^{\delta-} \cdots H^{\delta+} - O \big\backslash_H$$

At ordinary temperatures, pure water is a transparent, colourless, tasteless, and odourless liquid. At normal atmospheric pressure (101.3 kPa), water has a melting point of 0°C and a boiling point of 100°C. As was previously mentioned (Section 8.2), these temperatures are much higher than would be predicted, because of hydrogen bonding.

Water expands during freezing because ice has a hydrogen-bonded crystal structure as in Figure 8-12. The small spheres represent the hydrogen atoms; the larger spheres represent oxygen atoms. Each water molecule is hydrogen bonded to its four neighbours. The result is that the solid is a three-dimensional array with an open structure which includes much empty space. The expansion of water as it freezes is observed when a bottle of soft drink is left too long in a freezer. The result is a broken bottle.

Figure 8-12
The open hydrogen-bonded structure of ice

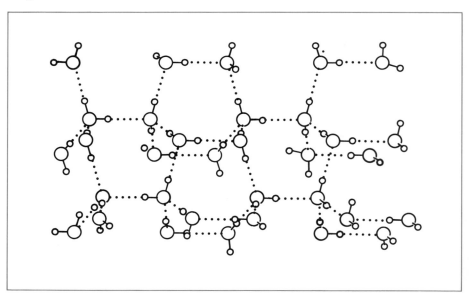

Since water expands on freezing, the density of ice is less than that of water. This property of water is extremely unusual. The solid forms of most other liquids are denser than their liquid forms. Thus, the solid forms of most liquids will sink in their liquids.

When ice melts, many of the hydrogen bonds are broken, and the rigid three-dimensional structure collapses. The molecules in liquid water are therefore closer together than in ice. Because there are fewer molecules in a given volume of ice, the density of ice is less than that of water. This explains why ice floats on liquid water.

Even in liquid water there is a large amount of hydrogen bonding, because not all of the hydrogen bonds are broken when the ice melts at 0°C. The molecules can move past each other in the liquid, so that hydrogen bonds are forming between some molecules as others are being broken. Thus, at 0°C the liquid can be viewed as consisting of clusters of four to eight water molecules whose partners are continuously changing.

As the temperature rises above 0°C, the extra thermal kinetic energy breaks more of the hydrogen bonds and the molecules move closer to each other. There is, however, an opposing effect. When the molecules of any liquid move faster, they occupy more space. This causes the density of the liquid to decrease. These two opposing processes balance each other at 4°C. Above 4°C the extra space required by the rapidly moving molecules causes the density of water to decrease. Thus, water has a maximum density at 4°C (Fig. 8-13).

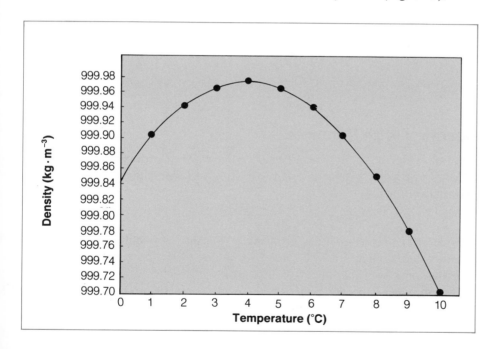

Figure 8-13
Variation of density of water

8.9 Chemical Properties of Water

The compound water, unlike its elements hydrogen and oxygen, reacts with many substances at room temperature. In some reactions in which water is a reactant, one of the covalent oxygen-hydrogen bonds is broken. The shared electron pair then remains with the oxygen atom while the hydrogen nucleus (proton) leaves to join some other molecule or ion.

Dissociation of Water

Pure water contains a very small concentration of ions. These ions are formed from neutral water molecules by a process called dissociation. On collision of two water molecules, it is possible for a hydrogen ion of one molecule to transfer to the second water molecule:

$$H_2O(\ell) + H_2O(\ell) \longrightarrow H_3O^+(aq) + OH^-(aq)$$

A hydronium ion (H_3O^+) and a hydroxide ion (OH^-) are formed. Both ions are *hydrated*; that is, they exist as a cluster consisting of the ion and one or more water molecules. Only about one of every 300 million water molecules is involved in the dissociation process.

The dissociation of water may appear to be an unimportant process since such a small proportion of the water molecules actually dissociates. However, the ability to dissociate is one of water's most important properties. The quantities of hydronium ions and hydroxide ions present, small as they may seem, can have a great effect on many of the metabolic reactions in living cells.

Reaction With Metals

Water reacts with many metals to produce hydrogen gas:

$$2Na(s) + 2H_2O(\ell) \longrightarrow 2NaOH(aq) + H_2(g)$$

$$Ba(s) + 2H_2O(\ell) \longrightarrow Ba(OH)_2(aq) + H_2(g)$$

$$Mg(s) + 2H_2O(\ell) \xrightarrow{\Delta} Mg(OH)_2(aq) + H_2(g)$$

$$3Fe(s) + 4H_2O(\ell) \xrightarrow{\Delta} Fe_3O_4(s) + 4H_2(g)$$

Reaction With Nonmetals

Fluorine, chlorine, and bromine react with water at ordinary temperatures. Chlorine reacts with water to form a mixture of hydrochloric acid (HCl) and hypochlorous acid (HClO):

$$Cl_2(g) + H_2O(\ell) \longrightarrow HCl(aq) + HClO(aq)$$

This reaction has a practical application because hypochlorous acid is an effective sterilizing agent. Thus, chlorine is widely used for the purification of drinking water.

Carbon, when strongly heated, reacts with steam to produce a mixture of carbon monoxide and hydrogen:

$$C(s) + H_2O(g) \xrightarrow{\Delta} CO(g) + H_2(g)$$

Both carbon monoxide and hydrogen burn in air to produce heat. This mixture, known as *water gas*, is used as an industrial fuel.

Reaction With Oxides of Metals and Nonmetals

Water reacts with certain metallic oxides to form metallic hydroxides (bases):

$$Na_2O(s) + H_2O(\ell) \longrightarrow 2NaOH(aq)$$

$$K_2O(s) + H_2O(\ell) \longrightarrow 2KOH(aq)$$

$$CaO(s) + H_2O(\ell) \longrightarrow Ca(OH)_2(s)$$

The oxides of certain nonmetals react with water to form acids:

$$CO_2(g) + H_2O(\ell) \longrightarrow H_2CO_3(aq)$$

$$SO_2(g) + H_2O(\ell) \longrightarrow H_2SO_3(aq)$$

$$SO_3(g) + H_2O(\ell) \longrightarrow H_2SO_4(aq)$$

$$P_4O_{10}(s) + 6H_2O(\ell) \longrightarrow 4H_3PO_4(aq)$$

Acids and bases are extremely important types of compounds. Their chemistry is discussed in Chapter 12.

Hydrates

When aqueous solutions of many of the soluble salts are evaporated, a precise number of water molecules may be retained as the ions form crystals. The water that becomes a part of the crystal is called **water of hydration**, and the compound is known as a **hydrate**. Examples of hydrates are Epsom salts, $MgSO_4 \cdot 7H_2O$; copper(II) sulfate pentahydrate, $CuSO_4 \cdot 5H_2O$; and alum, $KAl(SO_4)_2 \cdot 12H_2O$. The raised dot placed before the number of water molecules indicates that the water molecules are part of the formula unit. Another hydrate, cobalt(II) chloride ($CoCl_2 \cdot 6H_2O$), can be used to tell how much moisture the air contains. Hydrated cobalt(II) chloride is red, but it can lose its water of hydration, forming a blue substance. Paper covered with cobalt(II) chloride is pink in damp weather. As the air becomes less moist, the cobalt(II) chloride loses its water of hydration, and the paper turns blue.

If gypsum ($CaSO_4 \cdot 2H_2O$) is heated gently, it loses some of its water of hydration, forming a white powder called plaster of Paris:

$$2CaSO_4 \cdot 2H_2O(s) \longrightarrow (CaSO_4)_2 \cdot H_2O(s) + 3H_2O(g)$$

gypsum   *plaster of Paris*

When plaster of Paris is mixed with water, the process is reversed. The plaster of Paris forms a paste which hardens rapidly. Plaster of Paris is used to make moulded plaster statues and to make casts for broken limbs.

8.10 Heavy Water

In Section 4.6 you learned that many elements are mixtures of isotopes. Hydrogen is a mixture consisting of 98.985% ^1_1H and 0.015% ^2_1H. The ^2_1H isotope is known as *deuterium*(D). It is not radioactive. Thus, ordinary water is a mixture of H_2O, HDO, and D_2O. The D_2O is called deuterium oxide or *heavy water*. It is present in ordinary water in the ratio of one part heavy water to 7000 parts ordinary water.

One method of separating heavy water from H_2O involves passing an electric current through the water. This procedure decomposes the water into hydrogen and oxygen. The D_2O decomposes more slowly than the H_2O and is therefore concentrated by the process.

Although ordinary water and heavy water have the same appearance, they have slightly different physical properties. Ordinary water melts at 0.00°C, boils at 100.00°C, and has a density of $8.97 \times 10^2 \, kg/m^3$ at 20°C. The corresponding values for heavy water are 3.79°C, 101.41°C, and $1.108 \times 10^3 \, kg/m^3$. The rates of chemical reactions involving D_2O are slower than similar reactions involving H_2O. Heavy water is not toxic, but it stunts the growth of small mammals when it is the only water they are allowed to drink regularly. Scientists presume that stunting occurs because the heavy water reacts more slowly than ordinary water in fundamental life processes. Heavy water is used in CANDU nuclear reactors (Chapter 17) both as a coolant and as a moderator. A moderator slows neutrons enough to ensure they can be captured more readily by the reactor fuel, causing further nuclear reactions.

8.11 Hard Water

Water containing dissolved magnesium and calcium salts is known as **hard water**. Magnesium and calcium ions react with soluble soaps such as sodium stearate, $C_{17}H_{35}COONa$, to form insoluble compounds:

$$2C_{17}H_{35}COONa(aq) + CaCl_2(aq) \longrightarrow (C_{17}H_{35}COO)_2Ca(s) + 2NaCl(aq)$$

These insoluble materials can ruin the appearance of clothing. They also are responsible for producing the "ring" in bathtubs. Such insoluble compounds have no cleaning power and prevent the formation of a soapy lather. Hard water is also responsible for the formation of scale in boilers. At high temperatures the calcium and magnesium ions dissolved in water are precipitated as scale in the form of insoluble calcium carbonate, magnesium carbonate, and calcium sulfate. This scale impedes the transfer of heat in boilers.

To "soften" hard water, Ca^{2+} and Mg^{2+} ions must be removed. The nature of the hardness will determine the method which is to be used. Water which has *temporary hardness* contains these metallic ions along with HCO_3^- (hydrogen carbonate) ions. Since the HCO_3^- ions are unstable with respect to heat, the water may be softened by boiling. The HCO_3^- ions are converted to CO_3^{2-} ions, and the Ca^{2+} and Mg^{2+} ions are removed from the water as carbonate precipitates:

$$2HCO_3^- (aq) \longrightarrow CO_3^{2-} (aq) + CO_2(g) + H_2O(\ell)$$

$$Ca^{2+}(aq) + CO_3^{2-}(aq) \longrightarrow CaCO_3(s)$$

The HCO_3^- ions are also unstable with respect to OH^- ions. Slaked lime, $Ca(OH)_2$, provides OH^- ions, which convert the HCO_3^- ions to CO_3^{2-} ions, and again carbonate precipitates form:

$$Ca(OH)_2(s) \longrightarrow Ca^{2+}(aq) + 2OH^-(aq)$$

$$2HCO_3^- (aq) + 2OH^- (aq) \xrightarrow{\triangle} 2CO_3^{2-}(aq) + 2H_2O(\ell)$$

$$Ca^{2+}(aq) + CO_3^{2-}(aq) \longrightarrow CaCO_3(s)$$

Water which has *permanent hardness* contains Ca^{2+} and Mg^{2+} ions along with SO_4^{2-} ions. The SO_4^{2-} ions are stable with respect to both heat and OH^- ions. However, sodium carbonate (Na_2CO_3) can be used to soften the water.

The CO_3^{2-} ions from the sodium carbonate will remove the Ca^{2+} and Mg^{2+} ions by precipitation:

$$Ca^{2+}(aq) + CO_3^{2-}(aq) \longrightarrow CaCO_3(s)$$

$$Mg^{2+}(aq) + CO_3^{2-}(aq) \longrightarrow MgCO_3(s)$$

In treating hard water with slaked lime or sodium carbonate, just enough chemical must be added to soften the water, and therefore the amounts added must be carefully controlled. Also, the carbonate precipitates formed give the water a cloudy appearance; they must be removed by settling or filtration.

Another method of softening water, known as the ion-exchange method, does not have these inconveniences and may be used in both small and large scale water-softening operations. Certain natural materials, known as zeolites, or synthetic organic materials called resins may be used in this method. Zeolites may have rather complicated empirical formulas such as $NaAlSi_2O_6$ and $NaAlSi_3O_8$. Atoms of silicon, aluminum, and oxygen bond together to form a network of negative ions. The positive ions, Na^+, are trapped inside the network in empty, cage-like spaces. When hard water is passed through a column of zeolite, the calcium and magnesium ions in the water are continually exchanging places with the sodium ions in the zeolite. The sodium ions are held less strongly by the zeolite and flow out in the solution from the bottom of the column. If $NaZ(s)$ represents a zeolite, the exchange can be represented by the following equation:

$$Ca^{2+}(aq) + 2NaZ(s) \longrightarrow CaZ_2(s) + 2Na^+(aq)$$

With this process there is no need to filter precipitates since the Ca^{2+} ions are merely trapped in the zeolite. The zeolite reaction is reversible. When the zeolite can no longer accept Ca^{2+} and Mg^{2+} ions, it is regenerated by passing a concentrated solution of NaCl through it. The high concentration of sodium ions from the NaCl releases the calcium and mangesium ions from the zeolite:

$$2Na^+(aq) + CaZ_2(s) \longrightarrow 2NaZ(s) + Ca^{2+}(aq)$$

The liquid flowing through the column flushes these ions from the system and the column is then ready to be used again. Synthetic ion exchange resins are widely used in home water softeners. They are also available for a variety of purposes such as removing ionic substances from solution and for separating different ions from one another.

Part Two Review Questions

1. State two locations where water occurs naturally in **a)** the solid phase; **b)** the liquid phase; **c)** the gas phase.
2. What is the percentage by mass of water in most plants and animals?
3. How was it first shown that water is a compound and not an element?
4. Explain why water expands during freezing.
5. Describe the two opposing processes which determine whether liquid water will expand or contract on heating. How do these two processes explain the fact that water has a maximum density at 4°C?

6. What gas is formed when water reacts with metals such as sodium and calcium?

7. What is water gas? How is it formed? For what is it used?

8. What types of compounds are produced when water reacts with the oxides of metals? with the oxides of nonmetals?

9. What is a hydrate? Give two examples of hydrates.

10. In what ways is heavy water similar to ordinary water? In what ways does it differ?

11. What is hard water? Why is it undesirable?

12. Describe two different methods of removing calcium and magnesium ions from hard water.

Part Three: Water Pollution

8.12 Causes of Water Pollution

The cause of water pollution is negligence rather than ignorance. A community which dumps its raw sewage into a body of water is not ignorant of the effect of this action. Those who are unwilling to pay the cost of sewage treatment facilities can no longer excuse themselves by claiming that the body of water is big enough to handle raw sewage. Even huge lakes such as Lake Erie cannot absorb an unending flow of household and industrial waste. Dilution is no solution for pollution.

Every manufacturing plant administrator must surely know that the liquid which carries off one plant's wastes, when joined by the effluents of hundreds of plants, will overtax the natural processes by which even the largest rivers are cleaned of pollutants.

The technology existing today could be applied to eliminate most water pollution if everyone were willing to pay the costs. The initial attack on the problem would be the removal of pollutants at the source.

We must also be ready to pay for further research. We must be able to analyze for insidious pollutants in minute concentrations. We must learn more about the methods nature uses for slowly cleaning polluted water. We must learn methods for purifying water efficiently.

Bacteriological Contamination

Bacteriological contamination is the form of water pollution which is most dangerous to human health. Harmful bacteria in drinking water can cause diseases such as cholera, typhoid fever, and hepatitis. In the past, outbreaks of cholera and typhoid fever killed thousands of people. Most municipalities in North America, Europe, and Japan now use chemicals such as chlorine to destroy harmful bacteria in drinking water. However, bacteriological contamination is still a problem in areas of the world where water treatment facilities are lacking.

Industrial Wastes

Industrial wastes tend to affect aquatic life rather than cause human illness. Compounds of lead and copper, certain phenol derivatives, and hydrogen sulfide can destroy life in streams where these substances are dumped. Acids from acid rain, coal mines, and chemical plants may greatly increase the acidity of rivers and lakes and make it impossible for plants or fish to live. Acid rain is discussed in Chapter 12.

Mercury pollution is a dangerous form of water pollution. The mercury is eventually absorbed by fish which may be eaten by humans. In both fish and humans, mercury poisoning can cause damage to the nervous system and death in extreme cases. Several lakes in northern Canada have been environmentally damaged by mercury pollution, and the accumulation of mercury in the fish has destroyed the traditional fishing life-style of the local native populations.

The problem of water pollution by industrial waste is not a new one. Sawdust was a pollutant in the Ottawa River and the Rideau Canal from 1866 to 1894. Water pollution in Canada has been a problem for a long time.

Thermal Pollution

Thermal pollution results when water is used for cooling purposes, heated in the process, and returned to the river or lake from which it was taken. Water is an ideal cooling fluid for thermal generating stations, nuclear generators, and industrial plants. The Thames River in England warmed up by four degrees in 20 years—from an average temperature of 12°C in 1930 to 16°C in 1950.

The increased temperature means that less oxygen can dissolve in the water, since gases dissolve more readily in cold liquids. The problem of oxygen depletion will be discussed more fully in the next section. It is enough to note here that thermal pollution aggravates the problems of the oxygen supply in water. A significant rise in the temperature of a body of water can destroy entire biological populations.

The problem of thermal pollution can be solved by allowing the warm water to cool in storage lakes or in evaporation towers before returning it to rivers or lakes. The problem can also be reduced by building industrial plants or power stations beside a large body of relatively cold water such as Lake Huron. Even a number of large plants requiring a great deal of cooling water will probably not warm up such a vast body of cold water.

Oxygen Depletion

When waste material from paper mills, coal-tar residues from dye factories, or certain other organic materials (for example, sewage) are dumped into streams, they react with the oxygen dissolved in water. Most of the degradation (breaking down) of organic molecules is accomplished by the action of microorganisms which use organic molecules as food by metabolizing them with oxygen dissolved in the water. This dissolved oxygen is breathed by fish and is as vital to them as atmospheric oxygen is to us. If large amounts of the dissolved oxygen are used up by the oxidation of wastes, the fish will not survive. The term **biochemical oxygen demand** (BOD) is used to describe the oxygen take-up by the microorganisms that decompose organic materials in the water. Water with a BOD of less than 1 mg of O_2 per litre of water (1 ppm) is considered to be pure. If the BOD exceeds 20 mg/L, the water is considered to be polluted. Raw sewage has a BOD of 100 to 400 mg/L.

Household Detergents

The enormous quantities of detergents used in homes for laundry and dishwashing operations have presented another pollution problem. Flocks of ducks have drowned after landing on detergent-filled ponds (Fig. 8-14). The detergents washed off the oil on their feathers which allows ducks to remain afloat. When this oil was removed the ducks sank.

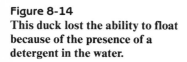

Figure 8-14
This duck lost the ability to float because of the presence of a detergent in the water.

The basic cause of detergent pollution lies in the fact that many commercial detergents have had chemical compositions which resist degradation by microorganisms. Municipal systems traditionally deal with sewage by relying on microorganisms for its degradation. The sewage is normally concentrated into a sludge which is then activated with microorganisms to hasten decay. When detergents were first introduced as replacements for soaps, many sewage treatment plants found themselves overwhelmed with foam. In recent years, detergents with structures that are easily attacked by microorganisms have been developed. These are called *biodegradable detergents*.

Eutrophication

During the normal life cycle of a lake, there is a gradual accumulation of sediment. As a lake grows rich in nutrients, algae, bacteria, and aquatic plants thrive. When they die, their remains begin to decompose and settle on the lake bottom. The process is called eutrophication (from Greek *eutrophein*, "to thrive"). Over a period of hundreds or even thousands of years, the processes associated with eutrophication cause the lake to become shallower. Eventually the lake becomes a marsh and, finally, dry land.

When critical nutrients such as nitrogen or phosphorus compounds enter a lake or a pond in high concentrations, eutrophication is tremendously accelerated. Algae and plants grow profusely. While growing they photosynthesize plant tissue and generate some oxygen, most of which goes into the air. When plants die, the carbohydrate material of which they were made decays or is degraded by microorganisms. This bacterial decomposition consumes oxygen. When the quantity of decaying matter becomes large enough, the sequence of processes uses up so much dissolved oxygen that aquatic life cannot survive.

In many eutrophication studies, compounds of phosphorus seem to be the most important nutrient pollutants. The evidence in the case of Lake Erie pointed to domestic sewage as the greatest source of phosphorus compounds. Another source of nutrients, especially nitrates, is fertilizer runoff from farms. Most governments make a great effort to analyze water for the presence of nitrates, phosphates, and other pollutants (Fig. 8-15).

Some studies indicate that carbon dioxide rather than phosphates or nitrates is the controlling nutrient in eutrophication. In waters where there are large amounts of degradable organic pollutants, algae seem to grow, even with little phosphate present. Microorganisms metabolize the organic molecules and supply large amounts of CO_2. This CO_2 is used by the algae for photosynthesis. The rapid growth of the algae speeds up the process of eutrophication, resulting in a prematurely aged water system which is unable to support aquatic life.

Figure 8-15
Cropping weeds to prevent eutrophication

8.13 Purifying Polluted Water

Municipal water supplies require treatment before the water can be used for human consumption. Sediment and suspended matter must be removed. Bacteria must be destroyed and industrial wastes must be neutralized or destroyed in water treatment plants such as the one shown in Figure 8-16.

Much of the water which comes from lakes or streams contains particles of clay, sand, or organic matter. The larger particles are allowed to settle out of their own accord in settling tanks (Fig. 8-17). The smaller particles are too small to settle out on their own. The water is treated with aluminum sulfate and calcium hydrogen carbonate. The two compounds form aluminum hydroxide which comes out of solution as a white gelatinous precipitate:

$$Al_2(SO_4)_3(aq) + 3Ca(HCO_3)_2(aq) \longrightarrow 2Al(OH)_3(s) + 6CO_2(g) + 3CaSO_4(s)$$

The aluminum hydroxide traps both the smaller particles and the solid $CaSO_4$ formed in the above reaction and carries them to the bottom of the tank. The carbon dioxide is a harmless gas. Water is drawn off from the top of the tank, and any particles of aluminum hydroxide remaining in the water are removed by filtration through a bed of sand.

Organic compounds which cause objectionable tastes or odours are removed by passage of the water through a bed of finely divided charcoal. The organic molecules are removed by attraction to the surfaces of the charcoal particles (adsorption). The water is then sprayed into the air in the aerator which removes some odours. It also increases the amount of dissolved air in the water because the surface area of the spray is larger. The increased amount of dissolved air improves the taste of the water. Water with little dissolved air tastes flat.

Figure 8-16
Sewage treatment plant on the Grand River, Ontario

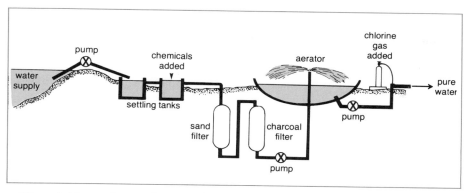

Figure 8-17
Schematic drawing of a water treatment plant

Harmful bacteria are destroyed when a chemical disinfectant such as chlorine is added. We saw in Section 8.9 that chlorine reacts with water to form hypochlorous acid:

$$Cl_2(g) + H_2O(\ell) \longrightarrow HClO(aq) + HCl(aq)$$

Although chlorine is a poisonous gas, its concentration in municipal water supplies is so low that it is not dangerous to humans, but it gives the water a noticeable taste. Both the dissolved chlorine and the hypochlorous acid are highly toxic to bacteria, however, even in the low concentrations used. Ozone and bromine monochloride can be used in place of chlorine. One advantage of chlorine is that it decomposes in the water much less rapidly than the ozone. The chlorine is therefore available in the water to kill any bacteria that enter the water after it has left the treatment plant. The chlorine may, however, react with organic substances present in the water to form harmful substances such as chloroform, which is now known to be a carcinogen (cancer-causing agent).

Municipal water-treatment plants often add fluoride compounds such as sodium fluoride to help prevent tooth decay. This fluoridated water reacts with compounds in tooth enamel, converting it to a substance which is harder, less soluble, and more resistant to acids. High concentrations of fluoride are toxic. Severe symptoms result from the ingestion of less than one gram, and death results from the consumption of five to ten grams of fluoride. Hence there is a debate over the use of fluoride compounds in drinking water. On the other hand, there is evidence that fluoridation of drinking water may reduce cavities by as much as 50%.

8.14 Water Pollution in Canada

It has been stated that the world is heading toward a water crisis caused by a rapidly-increasing world population. Society is beginning to realize that water is important and that it does not exist in unlimited quantities. Canada is fortunate in that it contains approximately one-quarter of the entire global supply of fresh water.

At present, Canada has more than enough fresh water to satisfy its own needs. The United States, however, is rapidly using up its supply of fresh water. Hence many Canadians recommend selling Canada's excess water to its neighbour. Some even recommend diverting Canadian rivers so that they flow into the United States. Many other Canadians, however, oppose the sale of Canadian water to other countries, for fear of losing control of a valuable natural resource.

It is estimated that one hundred and twenty billion litres of water are used in Canada each day. Some of this water is contaminated with toxic chemicals. Environment Canada defines toxic chemicals as substances which, when released, threaten the environment and human health. These chemicals often cause biological damage at low concentrations. They often persist for long periods and increase in concentration as they pass through the food chain. A number of toxic chemicals are thought to cause cancer and birth defects.

Chemists are studying the detection and movement of minute amounts of toxic chemicals in the environment at such places as the Canada Centre for Inland Waters at Burlington, Ontario. With advances in sophisticated electronic equipment, chemists are now able to detect some toxic chemicals in concentrations as low as ten picograms per litre of solution.

The toxicity of a substance is determined by testing it on animals and then extrapolating the results to humans on the basis of body mass. Toxicity is measured as the dosage that kills 50% of the test animals and is expressed as milligrams of substance per kilogram of body mass. This dosage is called the LD_{50} (for *l*ethal *d*osage, 50% die). Thus, if a dosage of 1 mg/kg kills 50% of a population of rats, the LD_{50} for this poison is 1 mg/kg. Formaldehyde has an LD_{50} of 70 mg/kg. The LD_{50} for sodium chloride is 4000 mg/kg, and the LD_{50} for water is 500 000 mg/kg. We see that even substances which are necessary for life (such as sodium chloride) are toxic when ingested in large amounts. Among the materials which are toxic when ingested by humans are fuels, lubricants, solvents, pesticides, radioactive chemicals, acids, bases, paints, and compounds containing mercury and lead. The toxicities of several substances are listed in Table 8-2.

What are the sources of toxic chemicals? Toxic chemicals come from the spraying of forests with pesticides, from industrial wastes or emissions, and also from leakage from dumps and landfills. Many household chemicals used by ordinary citizens also contribute to the level of toxic chemicals in water.

Two of the most toxic groups of chemicals in our environment are dioxins and polychlorinated bipheyls (PCBs). An example of the first group is TCDD (2,3,7,8-tetrachlorodibenzo-*p*-dioxin) which is the most toxic synthetic chemical. This form of dioxin is a contaminant in a number of disinfectants, herbicides, and wood preservatives. TCDD cannot be decomposed by heat, bases, or acids.

During the Vietnam War, the herbicides 2,4-D and 2,4,5-T were used as components of the defoliant called Agent Orange (from the colour of the drums in which it was stored). Defoliants strip plants of their leaves. U.S. Forces sprayed an estimated forty million litres on Vietnamese forests to deprive the Vietcong of cover. However, Agent Orange became notorious because it was later found to contain TCDD as an impurity in concentrations of a few milligrams per kilogram of defoliant.

Many Canadian power companies, railways, and provincial highways departments also used Agent Orange, or mixtures very similar to it, as defoliants to clear their rights of way. These defoliants were also found to contain TCDD as an impurity.

Table 8-2 Toxicities of Some Substances

Compound	LD_{50} (mg · kg^{-1})
TCDD (a dioxin)	0.0006
HCN	4
Strychnine	5
Arsenic acid	8
Nicotine	50
Formaldehyde	70
Caffeine	192
2,4,5-T	100–300
DDT	90–500
Aspirin	1750
Methanol	500–5000
NaCl	4000
Ethanol	5000–15 000
Water	500 000

TCDD and other dioxins are teratogenic (from Greek *teras*, "monster" + *genes*, "become"); that is, they cause birth defects. They are also suspected carcinogens. In sublethal quantities they cause severe skin rashes and abnormal changes in the structure of the skin. Dioxins have been blamed for many of the medical problems experienced by Vietnam War veterans and by the workers who sprayed the suspect defoliants in the Canadian countryside.

PCBs are also highly toxic. When biphenyl undergoes chlorination, a mixture of polychlorinated biphenyls is formed. These compounds contain three to five chlorine atoms per molecule. PCBs are found in electrical transformers as coolants. They have also been used as plasticizers in synthetic resins and as flame retardants. The government of Canada stopped all use of PCBs in 1980.

Figure 8-18
(a) These disused transformers contain PCBs.
(b) Extreme care is vital when handling equipment which contains PCBs.

Highly chlorinated compounds such as dioxins and PCBs are extremely stable. They do not react readily with air or water and cannot be broken down (degraded) by bacterial action. Therefore, they cause enormous environmental problems because they accumulate in the environment and remain for long periods of time. The only sure way of destroying these compounds is by burning them in incinerators at temperatures of 1000 to 1600°C. The products of the combustion are carbon dioxide, water and hydrogen chloride. Since carbon dioxide and water are harmless and the hydrogen chloride is easily trapped, the toxic substance is rendered harmless. Even here, though, considerable care is required. If the incinerator temperature is much below 1000°C, the chlorinated compound does not burn. It is simply vapourized and dispersed into the atmosphere to spread the pollution over an even larger portion of the environment.

An international commission, consisting of three Americans and three Canadians appointed by their respective governments, monitors the water quality in the Great Lakes. This committee was responsible for a nine billion dollar cleanup program which greatly reduced the pollution of Lake Erie in the 1970s. The Great Lakes are still polluted with many chemicals. It is believed that 90% of these chemicals come from the United States. For instance, pollutants from industries and leaks from toxic storage dumps in Niagara Falls, New York are entering the Niagara River.

The worst areas of toxic chemical pollution in Canada are the St. Lawrence River near Montreal and the St. Clair River near Sarnia (Canada's "Chemical Valley"). Why do these and other pollution problems exist in Canada? There are a number of reasons. We suffer from ignorance and complacency. Fines for polluting are relatively small. Companies often do not want the added expense required to avoid or lessen the amount of pollution from their plants. The cost of analysis for various pollutants is expensive. Some companies refuse court orders to clean up. Governments are sometimes unwilling to impose expensive cleanup programs because of the poor rapport with industry that might develop. Legal and political disagreements often lead to the delay of cleanup programs in certain areas. Canada has not fully adopted the latest technology to deal with toxic chemicals. Only Alberta and Ontario have plans to set up modern disposal and treatment plants. Finally, with hundreds of different toxic chemicals present in a body of water, chemists do not yet fully understand all of the complex chemical reactions taking place.

It is easy to argue that it is industry's responsibility to avoid polluting the environment. Nevertheless, there is little evidence that industry will face this responsibility on its own as long as it is cheaper to dump wastes than it is to prevent their release into the environment. It appears, therefore, that legal and political action are more effective means for constructive change. Society must preserve the environment for use by future generations. Governments must enact effective environmental laws, and these laws must be enforced. Only when this system is established will there be a reversal in the steady decline of the quality of the environment. We have the technology to bring this reversal about. What we need is the will to use our technology.

APPLICATION

Managing Hazardous Waste

A hazardous waste is one which can cause death or illness in humans or damage to the environment if not managed properly. Many hazardous wastes are organic compounds which contain carbon along with other elements such as hydrogen, nitrogen, oxygen, sulfur, or the halogens.

Hazardous wastes can be divided into four types: flammable, toxic, reactive, and corrosive. Flammable wastes burn readily when ignited, thus presenting a fire hazard. Toxic wastes contain or release toxic substances which endanger our health or the environment. Reactive wastes tend to react vigorously with water or air. They can explode or produce toxic products. Corrosive wastes, such as acids and bases, destroy ordinary containers and so require special ones.

Hazardous wastes are often the by-products of industrial processes which produce plastics, pesticides, medicines, paints, petroleum products, metals, leather, and textiles. If these wastes are not managed properly, they can contaminate our food, air, and water. Proper management of hazardous wastes means either destroying them or storing them safely.

Currently, governments are examining new technologies to address the problems of hazardous wastes. The best technologies are those that can be used on-site to destroy wastes and bury detoxified soil and debris. Some of the newer proposed methods of dealing with hazardous wastes include pyrolysis, encapsulation, and destruction by microorganisms.

Pyrolysis is the decomposition of organic compounds by heating them at high temperatures in contact with a limited amount of air. This method may be the best treatment for organic wastes. In some new pyrolysis techniques, the products of the pyrolysis are broken down to produce carbon dioxide, water, and harmless metal salts. A disadvantage of this technique is that, if temperatures are not high enough, even more dangerous substances may be produced. An advantage of this technique is that the volume of material remaining is small.

Encapsulation consists of enclosing the hazardous waste in a nonreactive concrete or glass-like material. This method of storage appears to be best suited for inorganic compounds, which are compounds of elements other than carbon. A disadvantage of this method is that the volume of the waste is increased as a result of adding the concrete or glass. Only a long period of time will tell whether this is an effective method of storing hazardous wastes safely.

Another promising method of destroying hazardous waste is the degradation (decomposition) of organic matter by microorganisms. The advantage of this method is that it is thorough and requires less energy than other methods. The disadvantage is that microorganisms may take months to break down waste materials. Another potential problem is that the effect of a mircroorganism may be different in the laboratory from what it is in the natural environment, and so its effectiveness at decomposing hazardous waste cannot be accurately predicted before it is used.

Part Three Review Questions

1. List five different ways in which water may become polluted. Give a one-sentence explanation of each type of pollution that you list.
2. What changes can occur in a body of water as its temperature is increased?

3. What is meant by the term *biochemical oxygen demand* (BOD)?
4. What does a high value of BOD indicate?
5. What effect does eutrophication have on the amount of oxygen dissolved in a body of water? Why?
6. Name three operations that must be performed before polluted water is suitable for drinking. How are they accomplished?
7. Describe the characteristics of dioxins and PCBs.

Chapter Summary

- There are more empty spaces in liquids than in solids. This accounts for some of the physical properties of liquids.
- The forces of attraction holding molecules together in a liquid are London dispersion forces, dipole-dipole forces and, in some cases, hydrogen bonding.
- Water has unexpectedly high melting and boiling points, due to hydrogen bonding.
- Liquids flow readily, take the shape of their container, and are relatively incompressible.
- The freezing point is the temperature at which a liquid changes phase to become a solid.
- The freezing point of a liquid is lowered by an impurity.
- Evaporation is the escape of molecules from the surface of a liquid into the gas phase above the liquid.
- A vapour is the gaseous state of a substance which is a liquid under ordinary conditions.
- The boiling point of a liquid is the temperature at which its vapour pressure equals the atmospheric pressure.
- The normal boiling point of a liquid is its boiling point at standard pressure (101.3 kPa).
- Condensation is the change of state from a gas to a liquid.
- A dynamic equilibrium is a state in which two opposing processes occur at equal rates.
- A liquid in a closed container eventually reaches a state of liquid-vapour equilibrium.
- The vapour pressure of a liquid is a measure of its ease of evaporation.
- The process of distillation is used to separate a liquid from a dissolved solid or another liquid.
- Freeze drying is the process of removing water from a substance by subjecting the frozen material to a vacuum.
- Sublimation is the change of a solid directly to its vapour.
- Water is the only chemical which is present on earth abundantly and simultaneously as a gas, a liquid, and a solid.
- A water molecule is polar.

- Water has a maximum density at 4°C because of hydrogen bonding effects.
- Water dissociates into ions.
- Water reacts with many metals to produce hydrogen gas and a metal hydroxide.
- Water reacts with the oxides of metals to form bases.
- Water reacts with the oxides of nonmetals to form acids.
- Water of hydration is water that is associated with certain ions when salts form crystals in an aqueous solution.
- Compounds which contain water of hydration are called hydrates.
- Deuterium is an isotope of hydrogen ($_1^2H$ or D).
- Heavy water is deuterium oxide, D_2O.
- Water containing dissolved magnesium and calcium salts is known as hard water.
- Water with temporary hardness can be softened by boiling.
- Water with permanent hardness must be softened by treatment with chemicals.
- Water may become polluted by bacteriological contamination, industrial wastes, thermal pollution, oxygen depletion, and household detergents.
- Eutrophication is the natural aging process of a lake or pond in which aquatic organisms live, die, and are decomposed by microorganisms in the water.
- Water treatment plants purify the water by a combination of filtration, aeration, and chlorination.
- Water is one of Canada's most important resources.
- PCBs and dioxins are two of the most toxic chemicals in the environment.

Key Terms

London dispersion force
dipole-dipole force
van der Waals force
hydrogen bond
freezing
freezing point
evaporation
boiling
boiling point
normal boiling point

condensation
dynamic equilibrium
vapour pressure
distillation
sublimation
water of hydration
hydrate
hard water
biochemical oxygen demand
eutrophication

Test Your Understanding

Each of these statements or questions is followed by four responses. Choose the correct response in each case. (Do not write in this book.)

1. Which of the following substances has the lowest boiling point?
   a) HCl   **b)** Ar   **c)** H_2 **d)** NaCl

2. The normal boiling points of ether, water, and mercury are 34°C, 100°C, and 357°C respectively. Which of these liquids has the weakest intermolecular forces?
   a) ether   **b)** water   **c)** mercury   **d)** impossible to identify

3. Which of the following substances has the lowest boiling point?
   a) H_2O　　　　b) H_2S　　　　c) H_2Se　　　　d) H_2Te

4. The conversion of a liquid to a solid is called
   a) sublimation.　b) condensation.　c) evaporation.　d) freezing.

5. The conversion of a liquid to a vapour is called
   a) sublimation.　b) condensation.　c) evaporation.　d) freezing.

6. The boiling point of a liquid is
   a) the temperature at which the molecules leave the liquid at the same rate as they return to the liquid.
   b) the temperature at which the vapour pressure equals the atmospheric pressure.
   c) 100°C.
   d) the temperature at which no molecules can return to the liquid.

7. The conversion of a vapour to a liquid is called
   a) sublimation.　b) condensation.　c) evaporation.　d) freezing.

8. A liquid is placed in a closed container at a constant temperature. When the vapour pressure becomes constant, the rate of evaporation
   a) is zero.
   b) is less than the rate of condensation.
   c) is greater than the rate of condensation.
   d) equals the rate of condensation.

9. The normal boiling points of water and ethanol are 100°C and 78.5°C, respectively. The vapour pressure of water at its normal boiling point
   a) equals that of ethanol at its normal boiling point.
   b) is greater than that of ethanol at its normal boiling point.
   c) is less than that of ethanol at its normal boiling point.
   d) cannot be compared with that of ethanol because of insufficient information.

10. Wet clothing hung outdoors at −10°C freezes solid at first, but then becomes soft and dry. This is an example of
    a) sublimation.　b) condensation.　c) evaporation.　d) freezing.

11. The number of hydrogen atoms surrounding an oxygen atom in ice is
    a) 8.　　　　b) 6.　　　　c) 4.　　　　d) 2.

12. The density of water is greater than that of ice because
    a) there are no dipole-dipole attractions in liquid water.
    b) there is more hydrogen bonding in liquid water than in ice.
    c) the molecules in liquid water are further apart than in ice.
    d) the hydrogen bonded structure of ice is partially broken when ice melts.

13. Some metals react with water to produce
    a) hydrogen.　　b) oxygen.　　c) metallic oxides.　　d) acids.

14. An acid is produced when water reacts with
    a) Na.　　　　b) K_2O.　　　　c) SO_3.　　　　d) Mg.

15. A base is produced when water reacts with
    a) CO_2.　　　　b) K_2O.　　　　c) SO_3.　　　　d) Cl_2.

Review Your Understanding

1. Why does ice float on water?
2. List five physical properties of liquids. Use the theory of the structure of liquids to explain each of these properties.
3. Explain the following:
   a) With enough cooling, gases liquefy.
   b) Water will sometimes boil at temperatures less than 100°C.
   c) Liquids are generally not as dense as solids.
4. Describe a relatively quick method of separating a solid from a liquid in which it is dissolved.
5. What determines whether a substance is usually found as a solid, liquid, or gas?
6. How does antifreeze protect the radiator of a car?
7. What rates are equal in a liquid-vapour equilibrium?
8. Why is no liquid-vapour equilibrium established in an open container of a liquid?
9. What is the percentage composition of water?
10. Why is water called a polar molecule?
11. Why is deuterium oxide known as heavy water?
12. What properties of water make it useful as a coolant in automobiles and in industry?
13. Why will the temperature of boiling water not rise as the container is heated further?
14. What information would you need to predict which of two liquids would evaporate faster under the same conditions? Explain your answer.
15. Explain why energy must be absorbed before a liquid can evaporate.
16. Describe the physical properties of water.
17. Describe the structure of a water molecule by drawing an electron dot formula and indicating the type of bonding.
18. Why is water considered so important?
19. Contrast the physical properties of ordinary water and heavy water.
20. Why are toxic chemicals often allowed to accumulate in the environment?

Apply Your Understanding

1. On separate graphs plot the boiling points of the hydrogen compounds of the Group 16 and the Group 17 elements. Use Figure 8-3 as a guide. Use these graphs to estimate what the boiling points of H_2O and HF would be if there were no hydrogen bonding in these compounds.
2. A cold bottle of soft drink often becomes covered with droplets of water when it is removed from the refrigerator on a humid summer day. Propose a molecular explanation.

3. Perspiration is one of the mechanisms used by the human body on a hot day to prevent the body temperature from rising above its normal value of 37°C. What do you think is the molecular basis for this cooling mechanism?
4. Why does solid lead sink in molten lead but ice float on liquid water?
5. What do you think the world would be like if water did not exhibit hydrogen bonding?
6. Is it possible for water to boil at 20°C? Explain your answer.
7. Attempts to establish industrial waste disposal sites frequently meet with public protests of the type, "Not in my back yard!" Discuss the issues involved.

Investigations

1. Investigate the main sources of water pollution in your area. Write a report on your findings.
2. What are the effects of long-term exposure to **a)** PCBs; **b)** TCDD?
3. Insecticides have increased crop yields, but they can also enter the water system as yet another toxic chemical. Do the benefits of insecticides outweigh the risks? Outline the arguments on each side of the question.
4. Write an essay on the purification methods used by your local water treatment plant.
5. Debate the following: Since Canada has more than enough fresh water to satisfy its own needs, it should sell excess water to the United States.
6. All municipal water supplies should be fluoridated. Debate this proposition.
7. Obtain a pure unknown liquid from your teacher and investigate its physical and chemical properties.

Chemical Reactions

9

CHAPTER CONTENTS

Part One: **The Nature of Chemical Reactions**
  9.1   Introduction to Chemical Reactions
  9.2   Chemical Reactions—The Rearrangement of Atoms
  9.3   Using the Energy of Chemical Reactions
  9.4   Factors Affecting the Rate of a Chemical Reaction
  9.5   Recognizing a Chemical Reaction

Part Two: **Chemical Reactions and Chemical Equations**
  9.6   Types of Reactions
  9.7   Balancing Chemical Equations
  9.8   Information Obtained from a Balanced Chemical Equation

Part Three: **Reactions Involving Electron Transfer**
  9.9   Identifying Electron Transfer Reactions
  9.10   The Activity Series of Metals

What happens during a chemical reaction? What changes occur when iron rusts, nitroglycerin explodes (Fig. 9-1), or gasoline burns? What are the chemical processes involved in the manufacture of steel, the making of silicon for computer chips, or the production of sulfuric acid and ammonia? You will find the answers to these questions as you study this chapter.

Figure 9-1
Blasting with dynamite

Our daily lives depend upon chemical reactions. The digestion of foods and the production of polyester clothing, fuels, food, and medicines all involve chemical reactions.

In this chapter we describe the nature of chemical reactions and discuss the evidence for their occurrence. Many chemical reactions are known. We classify them into four major categories and balance equations for simple examples of each type. Reactions involving the transfer of electrons from one substance to another are discussed in some detail.

Key Objectives

When you have finished studying this chapter, you should be able to
1. state the three quantities that are conserved in a chemical reaction.
2. describe the four types of chemical reactions and write equations for examples of each.
3. classify a given reaction into one of four categories, given the chemical equation for the reaction.
4. balance simple chemical equations by inspection.
5. given a list of chemical equations, identify those which involve transfer of electrons.
6. given a list of chemical equations, identify the oxidizing and reducing agents.
7. given a list of chemical equations, state which atoms are oxidized and reduced.
8. given a list of chemical equations, diagram the gain and loss of electrons.

Part One: The Nature of Chemical Reactions

9.1 Introduction to Chemical Reactions

We will begin a discussion of the terms used in describing chemical reactions by using as an example the formation of liquid water:

$$2H_2(g) + O_2(g) \longrightarrow 2H_2O(\ell) + \text{heat energy}$$

This reaction consists of hydrogen molecules and oxygen molecules reacting to produce water molecules plus energy in the form of heat. A chemical change has taken place, and a new substance, water, has been formed. This new substance has properties which differ from those of either hydrogen or oxygen. Hydrogen and oxygen are referred to as the **reactants** or starting materials, and water is the **product** or new substance formed as a result of the reaction.

We can explain a chemical reaction such as this by using atomic theory and a knowledge of chemical bonding. From atomic theory we know that matter is made up of tiny particles called atoms. From bonding theory we know that only the outer or valence electrons are involved in bonding and that there is a transfer or sharing of electrons between atoms. The result is a new molecule with a unique arrangement of bonds.

Conservation of Mass and Charge

No atoms are lost, destroyed, or created during the reaction that creates water. There are the same number of oxygen and hydrogen atoms after the reaction as before. Chemical reactions obey the Law of Conservation of Mass, as we learned in Chapter 3. The total mass of the reactants equals the total mass of the products.

The total number of electrons and protons in all the atoms after this chemical reaction is the same as before the reaction. Thus, the number of positive and negative charges remains constant. That is, charge is conserved.

Conservation of Energy

Energy is also conserved during a chemical reaction. The total amount of energy (kinetic plus potential) after the chemical reaction equals the total amount of energy before the reaction if energy is not allowed to enter or to leave the system. In the equation

$$2H_2(g) + O_2(g) \longrightarrow 2H_2O(\ell) + \text{heat energy}$$

we see that the energy term appears on the product side of the equation. This means that heat is released during the reaction. A reaction which involves the evolution of heat energy is said to be **exothermic** (from Greek *exo*, "out" + *therme*, "heat"). When water is decomposed into hydrogen and oxygen by the application of an electric current, the energy term appears on the left-hand side of the equation:

$$2H_2O(\ell) + \text{electrical energy} \longrightarrow 2H_2(g) + O_2(g)$$

This means that energy is absorbed during the reaction. A reaction which involves the input of energy from the surroundings is said to be **endothermic** (from Greek *endon*, "in").

9.2 Chemical Reactions—The Rearrangement of Atoms

Energy Required For Chemical Reactions

If there is no change in the total amount of mass, charge or energy during a chemical reaction, then what exactly *does* happen? To answer this question, let us take a closer look at the chemical reaction under discussion:

Hydrogen molecules

react with

oxygen molecules

to produce

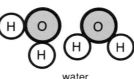

water molecules

The reactant molecules are undergoing translational, vibrational, and rotational motion. These types of motion were discussed in Chapter 3. Kinetic energy is associated with all of these motions. The reactant molecules also possess potential energy stored in the chemical bonds. This energy is a result of the attractive forces between the bonded atoms. The potential energy stored in chemical bonds is referred to as **chemical energy**. Thus, when we say that a substance has chemical energy, we mean that it has stored, potential energy which is related to the arrangement of atoms within its molecules or ions.

When hydrogen is mixed with oxygen, nothing happens. If a spark, a flame, or another source of energy is added to the mixture, however, the reaction may occur so rapidly that it is explosive. Chemical reactions often need an initial "energy kick" to get them going. In Figures 9-2 and 9-3 this energy kick is shown as a hump in which the potential energy of the reactants is raised to a higher level, after which the formation of products is energetically down-hill. In Figure 9-2, the potential energy of the products is greater than that of the reactants. Thus, the reaction is endothermic, and the extra energy has been supplied from the surroundings. For example, in the endothermic reaction

$$CaCO_3(s) \longrightarrow CaO(s) + CO_2(g)$$

the heat energy is provided from a source such as a Bunsen burner.

In Figure 9-3, the potential energy of the products is less than that of the reactants. Energy has been released from the starting materials, and the reaction is therefore exothermic. In such a reaction, no further energy is required after the initial energy kick. For example, as water molecules are formed from hydrogen and oxygen, energy is released. Some of this energy is used to keep the reaction going, and the rest is released to the surroundings. Why is energy needed to start the reaction and to keep it going? Energy is required to break chemical bonds—the hydrogen-hydrogen bonds and the oxygen-oxygen bonds.

Overall Energy Effects During Chemical Reactions

We are now able to see why energy is released or absorbed during a given chemical reaction. Whether energy is released or absorbed depends on the

Figure 9-2
Potential energy diagram for an endothermic reaction

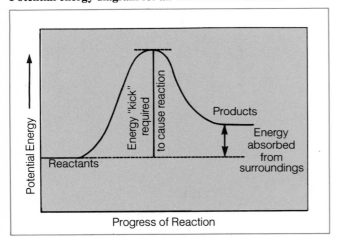

Figure 9-3
Potential energy diagram for an exothermic reaction

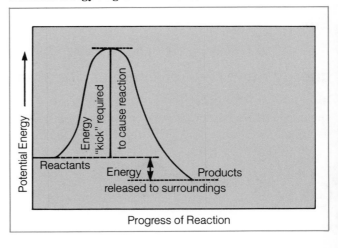

rearrangement of atoms during the reaction—that is, during the breaking of existing chemical bonds and the formation of new chemical bonds. Energy is required to break the old bonds, and energy is released when the new bonds are formed. Whether a reaction will be exothermic or endothermic depends upon the relative energies of the bond-breaking and bond-making events.

In the reaction between hydrogen and oxygen to form water, more energy is released when oxygen-hydrogen bonds are formed in the product than is needed to break the hydrogen-hydrogen and oxygen-oxygen bonds in the reactants. The overall reaction is therefore exothermic.

We have already seen that energy is conserved in a chemical reaction:

$$\text{Total energy of reactants} = \text{Total energy of products}$$

Since the total energy is the sum of the kinetic and potential energies, we can write:

$$\left\{ \begin{array}{c} \text{Potential energy of reactants} \\ + \\ \text{Kinetic energy of reactants} \end{array} \right\} = \left\{ \begin{array}{c} \text{Potential energy of products} \\ + \\ \text{Kinetic energy of products} \end{array} \right\}$$

The kinetic energies of molecules depend on the temperature of a system. The higher the temperature, the greater will be the average kinetic energy of the molecules. Conversely, the lower the temperature, the lower will be the average kinetic energy. Since heat energy is released during the formation of water, the kinetic energy of the products is greater than that of the reactants. Therefore, the potential energy of the products must be less than that of the reactants. Since the product molecules have less potential energy, they are more stable (that is, less likely to react).

The temperature will increase during this reaction, since temperature is a measure of the kinetic energy of the molecules. For this reason, temperature will rise during an exothermic reaction and fall during an endothermic reaction.

The energy released in chemical reactions may be in the form of heat energy, as in the formation of water. Heat energy is also released when coal or oil is burned. However, the energy released may also be electrical, as in a battery, or in the form of light, as in fireflies. Sometimes more than one form of energy is released during a reaction. For example, both heat and light may be given off during a combustion, and in an explosion some of the chemical energy is converted into sound energy.

In summary, chemical reactions and the energies involved in them are the result of rearranging the reactant atoms to form new substances as products.

9.3 Using the Energy of Chemical Reactions

People have used chemical reactions from earliest times. Fire was probably the first and most useful controlled chemical reaction. It provided heat and light and was useful in cooking foods. Another important early use of fire was in the separation of metals such as copper and tin from their ores to make jewelry, tools, and weapons. In addition, the combining of these two metals in their molten states produced the alloy, bronze, which was tougher and made superior tools and weapons.

In modern times many useful chemical reactions are used to provide energy in different forms such as heat, light, electrical, and mechanical energy. The high temperatures produced during some exothermic reactions are used for welding. For example, an oxyacetylene torch produces temperatures of 2400°C by using the reaction:

$$2C_2H_2(g) + 5O_2(g) \longrightarrow 4CO_2(g) + 2H_2O(g) + heat$$

Another reaction, formerly used to weld railroad rails, is the thermite reaction:

$$2Al(s) + Fe_2O_3(s) \longrightarrow 2Fe(\ell) + Al_2O_3(s) + heat$$

So much heat is produced in this reaction that the iron produced is in the liquid state.

The atomic hydrogen torch is also used for welding high-melting-point metals such as tantalum and tungsten. This torch can produce temperatures of up to 4000°C. In an atomic hydrogen torch, H atoms from hydrogen gas are produced in an electric arc. These atoms are then allowed to react with O_2 molecules to form H_2O molecules. Energy is released as the hydrogen-oxygen bonds are formed.

Sometimes useful energy in the form of light is obtained from chemical reactions. For instance, the light from the reaction

$$2Mg(s) + O_2(g) \longrightarrow 2MgO(s) + energy$$

is used in flares and flash bulbs. Candles provide light as a result of the combustion of hydrocarbons such as $C_{20}H_{42}$. There are plastic light sticks which contain chemicals in separate compartments: when the stick is bent, the chemicals mix and a reaction occurs which produces a greenish glow that lasts for hours. These light sticks can be used during emergencies. They can also be carried by children at Halloween. The light of fireflies is the result of biochemical reactions.

Major uses of energy from chemical reactions are the heating of buildings, the powering of machinery, and the cooking of foods. The chemicals used for these purposes are called fuels. Common fuels include coal, wood, oil, propane, and natural gas. Coal, oil, and gas are also burned to provide heat energy for the production of electricity. The most common automobile fuel is gasoline, which is a mixture of hydrocarbons usually produced from petroleum. One of the components in gasoline is isooctane. The combustion of isooctane could be written as

$$2C_8H_{18}(\ell) + 25O_2(g) \longrightarrow 16CO_2(g) + 18H_2O(g) + energy$$

The explosions taking place in the cylinders of a car push its pistons, and this mechanical motion eventually leads to the car's movement. Thus, the chemical energy released by this reaction is converted to the kinetic energy which is associated with the movement of the car.

Some of the chemicals that have been used to power spacecraft are hydrogen and oxygen; kerosene and oxygen; and hydrazine and hydrogen peroxide. Each of these pairs undergoes a vigorous reaction leading to the evolution of large amounts of energy when the components are mixed. Hydrazine and hydrogen peroxide, for example, undergo a highly exothermic reaction:

$$N_2H_4(\ell) + 2H_2O_2(\ell) \longrightarrow N_2(g) + 4H_2O(g) + energy$$

Chemical explosives also convert the chemical energy stored in their molecules into a useable form. Examples of explosives are nitroglycerin and trinitrotoluene (TNT). They react very rapidly according to the following equations:

$$4C_3H_5(NO_3)_3(\ell) \longrightarrow 12CO_2(g) + 10H_2O(g) + 6N_2(g) + O_2(g) + \text{energy}$$
nitroglycerin

$$2C_7H_5(NO_2)_3(s) \longrightarrow 12CO(g) + 2C(s) + 5H_2(g) + 3N_2(g) + \text{energy}$$
trinitrotoluene

In the chemical process, the relatively compact solid or liquid explosive is quickly converted into large volumes of gases. The simultaneous release of large amounts of heat causes these gases to expand very rapidly, causing a shock wave with a powerful shattering effect. Thus, the important characteristic of chemicals used as explosives is not the large release of heat energy, as with fuels, but rather the very rapid chemical reaction which produces large volumes of hot gases.

A very important fuel that we have not yet mentioned is food. Food is oxidized (digested) in our bodies to enable us to maintain body temperature, to move about, and to perform other functions. Nutritionists refer to the energy content of foods. What they mean is that a given amount of energy (kilojoules) would be released if the food underwent the chemical reactions involved in the digestive process.

The energy from chemical reactions is used for other purposes. The heat produced from the reaction between aluminum and sodium hydroxide in drain cleaners is important in clearing blocked outlets. Methane and propane are common fuels for laboratory burners. Butane is the fuel used in disposable cigarette lighters. Matches that can be ignited by striking any rough surface depend on friction to cause P_4S_3 to ignite in air. The heat given off causes other chemicals in the match head to react and produce a flame. The large amount of heat given off in a blast furnace during steelmaking decomposes limestone. The resulting calcium oxide combines with sand and gravel impurities to form a molten slag which can then be removed from the molten iron. In addition, many exothermic reactions are used to separate metals from their ores.

All of these reactions are useful because the stored potential chemical energy of the reactants is greater than the stored potential chemical energy of the products. This extra chemical potential energy of the reactants is released as the products are formed.

APPLICATION

Chemical Explosives

A chemical explosive is a substance that undergoes a rapid chemical reaction which produces large amounts of heat and gaseous products. The rapid production of large volumes of gases causes the shattering effect of an explosion. Explosives are usually solids or liquids. Typical explosives produce shock waves with velocities of 9000 m/s and temperatures of 3000 to 4000°C. Chemical explosions can be started by high temperatures, a sudden increase in pressure, or a physical shock.

The most important feature of a chemical explosion is the very rapid reaction rate. The large amount of heat released is not as important. For example, when 1 g of gasoline is burned in an automobile engine, 46 kJ of thermal energy is produced in 10 ms. However, when 1 g of gunpowder is used to propel an artillery shell, only 3.3 kJ of energy is released, but this reaction takes only 0.5 ms.

Explosives are divided into two general classes, high explosives and low explosives, based on the rate of the explosive reaction. High explosives undergo "detonation." The chemical reaction moves very rapidly through the body of the explosive. Low explosives "burn" by a much slower chemical reaction of material close to the surface of the explosive. Neither type of explosive reacts with the oxygen in the air. Rather, they use the oxygen which is part of the explosive material. Examples of high explosives include mercury fulminate, TNT (trinitrotoluene), nitroglycerin, and dynamite. The equation for the explosion of mercury fulminate is

$$Hg(ONC)_2(s) \longrightarrow Hg(g) + 2CO(g) + N_2(g)$$

The velocity of the detonation wave is 3920 m/s, and the explosion temperature is 4105°C. For the explosion of nitroglycerin, the velocity of the detonation wave is 8500 m/s and the explosion temperature is 3360°C. Gunpowder, the oldest explosive known, is a low explosive. It is a mixture of about 75% potassium nitrate, 13% carbon, and 12% sulfur.

Why does a large release of heat energy accompany an explosion? The release of heat energy during an explosion is the net result of the making and breaking of chemical bonds. The bonds in the products are much more stable than those in the reactants. Therefore, the net result is a large release of energy as the bonds in the products are formed.

Not every chemical that is capable of exploding is suitable for use as an explosive. In order to be used as an explosive, the compound must be able to be handled safely and to explode under controlled conditions. Although the chemistry of explosives has been studied for hundreds of years, we are only now beginning to understand it well enough to build theoretical models to explain explosive behaviour. These models will help in synthesizing new explosives which are even more powerful and even safer to handle.

9.4 Factors Affecting the Rate of a Chemical Reaction

A chemical reaction requires a collision between reactant particles. These particles must have sufficient energy and the proper orientation at the time of collision in order to break the chemical bonds of the reactants and form the chemical bonds of the products. The rate of the reaction is determined by measuring the amount of products formed in a given amount of time. Some chemical reactions proceed slowly, taking days or weeks. One example is the production of alcohol by fermentation. Other chemical reactions, such as acid-base neutralizations and explosions, take place extremely quickly.

Effect of Temperature

Increasing the temperature of the reactants is one way of increasing the rate of a chemical reaction. Thus, we use high temperatures to reduce the time it takes to cook food. In storing food, we wish to slow down the chemical reactions leading to food spoilage, so we keep it at low temperatures.

Effect of Catalysts

Another method of increasing the rate of a chemical reaction is to use a catalyst. A **catalyst** is a chemical which speeds up a reaction and can be recovered unchanged when the reaction is complete. When oxygen is prepared in the laboratory by heating potassium chlorate, manganese dioxide is added to speed up the reaction. A catalyst works by lowering the initial energy kick required to start a reaction (Fig. 9-4). Because less energy is required to get molecules over the energy barrier, more molecules will be able to cross over the barrier in a given time, and the reaction rate will increase. Catalysts are essential in many industrial processes such as petroleum refining and the production of sulfuric acid, ammonia, and plastics. Every automobile sold in Canada has a catalytic converter which converts carbon monoxide and unburned fuel in the exhaust into harmless carbon dioxide and water. Biological catalysts (enzymes) are essential for maintaining the life processes in all living organisms. An example of an enzyme is the amylase which is present in human saliva. Amylase catalyses the breakdown of starch into glucose molecules, beginning the process of digestion. If you hold a piece of bread in your mouth for a few minutes, it begins to taste sweet. This is sign that glucose is being produced.

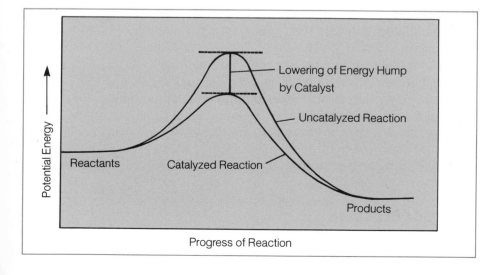

Figure 9-4
Potential energy diagram comparing a catalyzed reaction with an uncatalyzed reaction

Biography

Figure 9-5
Maud Menten (1879–1960)—
Investigated enzyme-catalyzed
reactions.

MAUD MENTEN was born in Lambton, Ontario, in 1879. While she was still young her family moved to Harrison Mills, British Columbia. Later, Maud returned to Ontario to study at the University of Toronto. She was one of the first women in Canada to graduate with a degree in medicine.

She won many scholarships, which enabled her to work in Europe. In 1913, at the age of 34, she was conducting research in biochemistry at the University of Berlin. It was there that she and Leonor Michaelis studied the behaviour of enzymes. Enzymes are catalysts which increase the rate of chemical reactions within biological organisms. The two researchers formulated what became known as the Michaelis-Menten Equation, which deals with the rate at which enzyme-catalyzed chemical reactions take place.

Upon returning to North America, Menten became a professor at the University of Pittsburg, where she worked chiefly on cancer research. She was immensely popular at the university, and many of her colleagues were happy to work with her, as co-authors, on over 70 important research papers. She left Pittsburg in her early seventies to continue her research in British Columbia. She died in 1960 and is buried in Chilliwack, British Columbia.

Effect of Concentration and Surface Area

The rate of a chemical reaction may also be increased by increasing either the concentration of the reactants or the surface area of reactants. Increasing the concentration of reactants increases the rate of the reaction because there are more collisions taking place between reactant molecules. For example, the propane in a Bunsen burner burns in air:

$$C_3H_8(g) + 5\,O_2(g) \longrightarrow 3CO_2(g) + 4H_2O(g)$$

If the concentrations of propane and oxygen are increased by compressing them into a much smaller volume (as in an automobile engine), the combustion occurs at explosive speed.

Increasing the surface area of a solid by dividing it into smaller particles exposes more molecules to react, and the reaction speeds up. The effect of increased surface area is sometimes shown quite dramatically in flour mills, grain elevators, and coal mines. The large amount of surface area provided by the finely divided dust particles can increase the rate of combustion to the extent that a spark can cause an explosion and fire. A simple example of this effect is that it is much easier to start a fire with very small pieces of wood than with a single large piece.

Effect of Pressure

The rate of a reaction depends on the rate at which molecules collide with each other. This rate, in turn, depends on the number of molecules in a given volume. A change in the pressure exerted on a liquid does not change the volume of the liquid, so it does not change the concentration of the molecules either. Thus, pressure has no effect on the rate of a reaction that is carried out in the liquid phase.

An increase in the pressure exerted on a gas will decrease the volume of the gas. Since the same number of molecules is now contained in a smaller volume, the concentration of the molecules and the rate of collisions between the molecules increases. Therefore, an increase in pressure *will* cause an increase in the rate of a gas-phase reaction. Similarly, a decrease in pressure will cause a decrease in the rate of a gas-phase reaction.

Effect of Reversibility

Not all chemical reactions proceed with the reactants being completely converted into products. Most chemical reactions are reversible. That is, the products can react to produce the original reactants. Eventually the rate at which reactants produce products becomes equal to the rate at which products produce the original reactants. When this occurs, a dynamic state of chemical equilibrium exists, and the chemical equation is written with arrows in both directions. For example, the Haber process for the industrial production of ammonia involves the reversible reaction

$$N_2(g) + 3H_2(g) \rightleftharpoons 2NH_3(g)$$

At equilibrium, the rate at which N_2 and H_2 react to produce NH_3 is equal to the rate at which NH_3 reacts to produce N_2 and H_2.

It is important to note that chemical reversibility does not influence the rate of a chemical reaction. Rather, as a result of chemical reversibility, not all reactants are converted to products. Once the state of chemical equilibrium has been reached, there is no further change in the amounts of reactants and products present under the given conditions of temperature and pressure. In other words, you have made as much product as you are going to make under those conditions.

9.5 Recognizing a Chemical Reaction

What type of evidence indicates that a chemical reaction is occurring? One or more of the following is usually observed:

1. *Change in temperature.* A chemical reaction may be accompanied by a decrease in temperature (absorption of heat) or by an increase in temperature (release of heat):

$$HCl(aq) + NaOH(aq) \longrightarrow NaCl(aq) + H_2O(\ell) + heat$$

2. *Evolution of a gas.* Some chemical reactions produce a gas as a product. If the reaction occurs in a liquid solution, tiny bubbles of the gas can be seen escaping from the liquid (Fig. 9-6). A seltzer tablet contains sodium hydrogen carbonate ($NaHCO_3$) and solid acids. When the tablet is dissolved in water, these acids react with the sodium hydrogen carbonate to release gaseous carbon dioxide. The equation for the reaction between $NaHCO_3$ and hydrochloric acid is

$$NaHCO_3(s) + HCl(aq) \longrightarrow NaCl(aq) + H_2O(\ell) + CO_2(g)$$

Figure 9-6
Seltzer tablet giving off gas in water

3. *Formation of a precipitate*. When some liquid solutions are mixed together, one of the products may be practically insoluble. This solid product will fall to the bottom of the solution and is referred to as a *precipitate*. For example, when a solution of silver nitrate is added to a solution of sodium chloride, a curdy white precipitate of insoluble silver chloride forms at once:

$$AgNO_3(aq) + NaCl(aq) \longrightarrow NaNO_3(aq) + AgCl(s)$$

Biography

Figure 9-7
John Polanyi (b. 1929) (left) — Winner of 1986 Nobel Prize for investigation of chemical reactions

JOHN C. POLANYI, son of a Hungarian scientist and philosopher, was brought up in England. Following his doctorate at Manchester University, he began working at the National Research Council's photochemistry and kinetics section in 1952. After spending some time at Princeton University, he moved in 1956 to the University of Toronto, where he has been ever since.

At Toronto, Polanyi has conducted research that is basic to a detailed understanding of chemical reactions. Prior to his work, chemists had only sketchy evidence on which to base their pictures of what the molecules are doing when chemicals react—the motions of the chemical products and the way in which the reagents move in order to cause a reaction. Chemical observations gave information only on the net effect of billions of molecules; the changes that occur to *individual* molecules had remained a mystery.

Polanyi developed the technique of infrared chemiluminescence for probing what happens when molecules collide and atoms rearrange themselves to form new molecules. The term *chemiluminescence* refers to the emission of photons of invisible infrared radiant energy from the vibrationally excited molecules that are formed in exothermic reactions. The excess energy is stored internally in the molecules as vibrational and rotational energy and is eventually emitted as infrared radiation. By measuring and analyzing this extremely weak radiation, chemists discover the way in which newly-born reaction products are vibrating, rotating, and hence 'translating' (moving around). Says Polanyi, "You can 'see' the dance of the molecules as old bonds break and new ones are formed."

Polanyi's data led to new insights regarding the specific forces operating when simple chemical reactions occur. In 1960 he suggested that it should be possible to construct an infrared laser in which the working material was a chemical reation—the so-called chemical laser. The first such laser was patented in 1964. Since 1965 chemical infrared lasers have been used in medicine, industry, and chemical research; what had been, in the basic scientist's laboratory, a barely detectible infrared emission became the most powerful source of infrared in existence.

For his work in providing "a much more detailed understanding of how chemical reactions take place," John Polanyi was awarded the 1986 Nobel Prize in Chemistry.

4. *Change in colour*. If the solvent is colourless, the intensity of the colour of the solution depends upon the concentration of the solute. If this solute takes part in a chemical reaction, there will be a change in the intensity of the colour of this solution as the reaction proceeds. For example, when metallic zinc is added to a solution of copper(II) sulfate, the solution gradually changes from being blue to being colourless as the blue copper(II) ions are removed from solution by the reaction:

$$\underset{\text{blue}}{Cu(NO_3)_2(aq)} + Zn(s) \longrightarrow \underset{\text{colourless}}{Zn(NO_3)_2(aq)} + Cu(s)$$

In other reactions a new colour may appear if one of the products is coloured. For example, the addition of aqueous ammonia to a solution of copper(II) nitrate results in the formation of a deep blue colour which is characteristic of the $Cu(NH_3)_4^{2+}$ ion:

$$\underset{\text{blue}}{Cu(NO_3)_2(aq)} + 4NH_3(aq) \longrightarrow \underset{\text{deep blue}}{Cu(NH_3)_4(NO_3)_2(aq)}$$

5. *Formation of an odour.* One of the products of a chemical reaction may have an odour which is noticeably different from that of the reactants. For example, iron(II) sulfide reacts with acids such as hydrochloric acid to produce hydrogen sulfide, a gas with the characteristic odour of rotten eggs:

$$FeS(s) + 2HCl(aq) \longrightarrow FeCl_2(aq) + H_2S(g)$$

CAUTION! You should *never* smell or taste a chemical unless you have been given instructions to do so by your teacher. Many chemicals are extremely poisonous, corrosive to living tissue, or highly irritating to the respiratory tract. Thus, even when you have received instructions to smell a substance, you should do so carefully, by gently wafting a small portion of its vapours toward your nose. Ammonia, for example, has a sharp, penetrating odour. In high concentration, it is highly irritating to the nose. Hydrogen sulfide is an extremely hazardous gas. Collapse, coma, and death from respiratory failure can occur within a few seconds after one or two inhalations of the gas. It is an insidious poison, since it anesthetizes the sense of smell. Thus, the victim becomes unaware of high concentrations of the gas. Treat this gas with the utmost respect.

Part One Review Questions

1. Explain the meaning of the terms *reactant* and *product*.
2. List the three quantities that are conserved in a chemical reaction.
3. Define the terms *exothermic* and *endothermic*. Write a chemical equation for an example of each of these types of reaction.
4. Energy is usually released or absorbed during a chemical reaction. Why does this happen?
5. What do we mean when we talk about the rate of a chemical reaction?
6. List three ways in which the rate of a chemical reaction may be increased.
7. What is a catalyst? How does it work?

8. For each of the following reactions indicate all evidence that a chemical reaction is taking place:

a) $Zn(s) + H_2SO_4(aq) \longrightarrow ZnSO_4(aq) + H_2(g) + heat$

b) $Pb(NO_3)_2(aq) + 2KI(aq) \longrightarrow PbI_2(s) + 2KNO_3(aq)$
   colourless   colourless   yellow   colourless

c) $Cu(s) + 2Ag^+(aq) \longrightarrow Cu^{2+}(aq) + 2Ag(s)$
   colourless   blue

d) $HNO_3(aq) + KOH(aq) \longrightarrow KNO_3(aq) + H_2O(\ell) + heat$

Part Two: Chemical Reactions and Chemical Equations

9.6 Types of Reactions

It is not easy to classify all chemical reactions precisely. Nevertheless, most reactions can be classified under one of four major categories. In each case the chemical reaction involves the rearrangement of atoms.

Combination

A **combination reaction**, also called an addition reaction or a synthesis, is a reaction in which atoms and molecules join together directly to produce larger molecules. Equations for combination reactions are usually of the type

$$A + B \longrightarrow AB$$

Some examples of combination reactions are

a) formation of sulfur dioxide by combustion of sulfur (one of the reactions in the formation of acid rain):

$$S(s) + O_2(g) \longrightarrow SO_2(g)$$

b) formation of ammonia (one of the most important industrially-produced chemicals):

$$3H_2(g) + N_2(g) \longrightarrow 2NH_3(g)$$

c) formation of calcium hydroxide (slaked lime in plaster and mortar) when calcium oxide reacts with water:

$$CaO(s) + H_2O(\ell) \longrightarrow Ca(OH)_2(aq)$$

Decomposition

Decomposition reactions are just the opposite of addition reactions. Equations for decomposition reactions are usually of the type

$$AB \longrightarrow A + B$$

Some examples are

a) decomposition of carbonic acid:

$$H_2CO_3(aq) \xrightarrow{\triangle} H_2O(\ell) + CO_2(g)$$

b) formation of quicklime, CaO (one of the reactions taking place in a blast furnace during steelmaking):

$$CaCO_3(s) \longrightarrow CaO(s) + CO_2(g)$$

c) explosion of TNT:

$$2C_7H_5(NO_2)_3(s) \longrightarrow 12CO(g) + 2C(s) + 5H_2(g) + 3N_2(g)$$

Displacement

Displacement or substitution reactions involve a change of partners. In these reactions one atom or group of atoms in a molecule is replaced by another atom or group of atoms, according to the general equation

$$A + BC \longrightarrow AC + B$$

Note than an atom or group of atoms in one compound (in this case, C) acquires a new partner.
Some examples are

a) the thermite reaction, once used in welding:

$$2Al(s) + Fe_2O_3(s) \longrightarrow 2Fe(\ell) + Al_2O_3(s)$$

b) a step in one method of producing gold:

$$2Au(CN)_2^-(aq) + Zn(s) \longrightarrow Zn(CN)_4^{2-}(aq) + 2Au(s)$$

c) the reduction of many metallic ores by heating with coke:

$$3C(s) + 2Fe_2O_3(s) \longrightarrow 4Fe(s) + 3CO_2(g)$$

Double displacement

Double displacement or metathetic reactions involve a joint exchange of partners, according to the general equation

$$AB + CD \longrightarrow AD + CB$$

Note that an atom or group of atoms in each of *two* compounds has a new partner. In this case, B and D have new partners.
Some examples are

a) neutralization of sodium hydroxide with hydrochloric acid:

$$NaOH(aq) + HCl(aq) \longrightarrow NaCl(aq) + H_2O(\ell)$$

b) formation of hydrogen sulfide by the action of hydrochloric acid on iron(II) sulfide:

$$FeS(s) + 2HCl(aq) \longrightarrow H_2S(g) + FeCl_2(aq)$$

c) precipitation of barium sulfate when solutions of barium chloride and ammonium sulfate are mixed (used for analysis of soluble sulfate):

$$BaCl_2(aq) + (NH_4)_2SO_4(aq) \longrightarrow BaSO_4(s) + 2NH_4Cl(aq)$$

Some common chemical reactions actually belong to one of the four major reaction types but are referred to by special names. For example, in the discussion of double displacement reactions, example *a* is known as an acid-base neutralization while example *c* is called a precipitation reaction. In a **precipitation reaction**, two aqueous solutions are mixed, and one of the products is an insoluble solid.

Figure 9-8 summarizes the four major types of chemical reactions by showing the rearrangement of atoms.

Figure 9-8
The four major types of chemical reactions

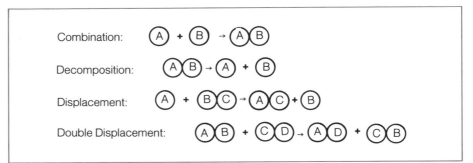

9.7 Balancing Chemical Equations

We learned in Chapter 7 to write chemical formulas. Now we can use our knowledge to write chemical equations. A **chemical equation** is simply a shorthand method for describing a chemical change. The symbols and formulas are used to represent the substances involved in the change.

The Law of Conservation of Mass states that matter is neither created nor destroyed in a chemical reaction. Therefore, in order to be correct, a chemical equation must be balanced. That is, there must be exactly the same number of each atom on each side of the equation.

In order to balance an equation, we must know what substances react and what substances are formed, and we must know the formulas of all substances involved in the reaction. We begin by writing an unbalanced equation. We write the correct formulas for the reactants and the products, with the reactants on the left and the products on the right separated by the yields symbol ($\longrightarrow$). Each reactant and each product is separated from the other by a + sign. Once the correct formulas have been written, *they must not be changed* during the balancing operations.

For example, aluminum reacts with chlorine to form aluminum chloride. The unbalanced equation is

$$Al + Cl_2 \longrightarrow AlCl_3$$

This equation does not obey the Law of Conservation of Mass. It shows two chlorine atoms on the left and three chlorine atoms on the right of the equation.

To balance the Cl, we put a 3 before the Cl_2 and a 2 before the $AlCl_3$ (that is, we change the coefficients of these formulas). This gives us 6 Cl on each side of the equation, which now becomes

$$Al + 3Cl_2 \longrightarrow 2AlCl_3$$

However, this equation is still unbalanced, because it shows 1 Al atom on the left and 2 Al atoms on the right. This is easily remedied by putting a coefficient of 2 in front of the first Al to give

$$2Al + 3Cl_2 \longrightarrow 2AlCl_3$$

The equation is now balanced, since there are 2 Al atoms and 6 Cl atoms on each side of the equation.

Usually the procedure that works best in balancing chemical equations is
1. Balance all atoms other than oxygen and hydrogen.
2. Balance oxygen.
3. Balance hydrogen.
4. Check to be sure that all atoms *are* balanced.
Let us put the principles into operation by balancing the equation

$$SF_4 + H_2O \longrightarrow SO_2 + HF$$

Step 1. Balance atoms other than oxygen and hydrogen. Since sulfur is balanced at the moment, we shall start by balancing fluorine. With four fluorine atoms on the left, there must be four fluorine atoms, or 4HF, on the right.

$$\underline{1}\,SF_4 + H_2O \longrightarrow SO_2 + \underline{4}\,HF$$

Note that, in balancing the number of fluorine atoms, the coefficient goes in front of the formula (for example, 4HF), as will always be the practice. This indicates four molecules of hydrogen fluoride, each containing one fluorine atom. If the coefficient were placed in error after the fluorine as a subscript in the formula, we would have HF_4. This does not represent hydrogen fluoride; rather, it represents a nonexistent compound! Thus, by putting the coefficients in front of the formula, we make sure that no new reactants or products are falsely indicated as a result of balancing the chemical equation. We shall temporarily underline the coefficients to remind ourselves that they are temporarily fixed and should not be changed unless absolutely necessary. The 1 in front of the SF_4 is not absolutely necessary, but it is a convenience during the balancing process.

Next, balance sulfur. Now that we have fixed one sulfur atom on the left, we need one sulfur atom, or 1 SO_2, on the right:

$$\underline{1}\,SF_4 + H_2O \longrightarrow \underline{1}\,SO_2 + \underline{4}\,HF$$

Step 2. Balance oxygen. With two oxygen atoms fixed on the right, we therefore need two oxygen atoms, or 2 H_2O, on the left:

$$\underline{1}\,SF_4 + \underline{2}\,H_2O \longrightarrow \underline{1}\,SO_2 + \underline{4}\,HF$$

Step 3. Balance hydrogen. Since all coefficients are fixed, hydrogen should be balanced.

Step 4. Check that all atoms are balanced. A check shows four hydrogen atoms (2H_2O) on the left and four hydrogen atoms (4HF) on the right. Similarly, there is one sulfur atom (1SF_4) on the left and one sulfur atom (1SO_2) on the right. There are four fluorine atoms ($1SF_4$) on the left and four fluorine atoms on the right (4HF). The balanced equation is therefore

$$SF_4 + 2\,H_2O \longrightarrow SO_2 + 4\,HF$$

Notice that coefficients of 1 are omitted from the balanced equation. If there is no coefficient in front of a formula, the coefficient is assumed to be 1.

It may be easier to understand the meaning of the coefficients if we write the equation another way:

$$SF_4 + \begin{matrix}H_2O\\H_2O\end{matrix} \longrightarrow SO_2 + \begin{matrix}HF\\HF\\HF\\HF\end{matrix}$$

It can be readily seen that the numbers of atoms are balanced.

EXAMPLE PROBLEM 9-1

Balance the equation $Al + H_2SO_4 \longrightarrow Al_2(SO_4)_3 + H_2$

SOLUTION

Step 1. Balance atoms other than oxygen and hydrogen. Since aluminum appears first in the equation, start with aluminum:

$$\underline{2}\,Al + H_2SO_4 \longrightarrow \underline{1}\,Al_2(SO_4)_3 + H_2$$

Next, balance sulfur:

$$\underline{2}\,Al + \underline{3}\,H_2SO_4 \longrightarrow \underline{1}\,Al_2(SO_4)_3 + H_2$$

Step 2. Balance oxygen. Oxygen is already balanced with 12 atoms on each side of the equation.

Step 3. Balance hydrogen:

$$\underline{2}\,Al + \underline{3}\,H_2SO_4 \longrightarrow \underline{1}\,Al_2(SO_4)_3 + \underline{3}\,H_2$$

Step 4. Check. On each side of the equation there are 2 atoms of Al, 6 of H, 3 of S, and 12 of O. All atoms are now balanced. The final equation is

$$2Al + 3H_2SO_4 \longrightarrow Al_2(SO_4)_3 + 3H_2$$

EXAMPLE PROBLEM 9-2

Balance the equation $Ca(OH)_2 + H_3PO_4 \longrightarrow Ca_3(PO_4)_2 + H_2O$

SOLUTION

Step 1. Balance atoms other than oxygen and hydrogen. Start with calcium:

$$\underline{3}\,Ca(OH)_2 + H_3PO_4 \longrightarrow \underline{1}\,Ca_3(PO_4)_2 + H_2O$$

Next, balance phosphorus:

$$\underline{3}\,Ca(OH)_2 + \underline{2}\,H_3PO_4 \longrightarrow \underline{1}\,Ca_3(PO_4)_2 + H_2O$$

Step 2. Balance oxygen. There are $6 + 8 = 14$ oxygen atoms on the left and only 8 [in the $Ca_3(PO_4)_2$] have been fixed on the right as a result of balancing calcium and phosphorus. The 6 remaining oxygen atoms must be present as 6 H_2O:

$$\underline{3}\ Ca(OH)_2 + \underline{2}\ H_3PO_4 \longrightarrow \underline{1}\ Ca_3(PO_4)_2 + \underline{6}\ H_2O$$

Step 3. Balance hydrogen. Hydrogen atoms are already balanced with 12 atoms on each side.

Step 4. Check. On each side of the equation there are 3 atoms of Ca, 14 of O, 12 of H, and 2 of P. The balanced equation is therefore

$$3Ca(OH)_2 + 2H_3PO_4 \longrightarrow Ca_3(PO_4)_2 + 6H_2O$$

EXAMPLE PROBLEM 9-3

Balance the equation $C_2H_6 + O_2 \longrightarrow CO_2 + H_2O$

SOLUTION

Step 1. Balance atoms other than oxygen and hydrogen. In this case, balance carbon:

$$\underline{1}\ C_2H_6 + O_2 \longrightarrow \underline{2}\ CO_2 + H_2O$$

Step 2. Balance oxygen. We cannot balance oxygen at this point, because we have not yet fixed the number of oxygen atoms on either side.

Step 3. Balance hydrogen atoms. In balancing carbon atoms, we have fixed the number of hydrogen atoms on the left. Therefore, we can balance hydrogen:

$$\underline{1}\ C_2H_6 + O_2 \longrightarrow \underline{2}\ CO_2 + \underline{3}\ H_2O$$

Now balance oxygen. To balance oxygen we must use a fractional coefficient:

$$\underline{1}\ C_2H_6 + \tfrac{7}{2} O_2 \longrightarrow \underline{2}\ CO_2 + \underline{3}\ H_2O$$

Balanced chemical equations are usually written using the smallest whole-number coefficients. To get rid of the fraction in this case, multiply the coefficients on both sides by 2:

$$2C_2H_6 + 7\ O_2 \longrightarrow 4CO_2 + 6H_2O$$

Step 4. Check. On each side of the equation there are 4 atoms of C, 12 of H, and 14 of O. The equation is therefore balanced.

This problem illustrates the fact that it is possible to have fractional coefficients during the balancing process. Both sides of the equation can then be multiplied by an appropriate factor to get the final balanced chemical equation with whole-number coefficients.

In the method that we have been using to balance equations, the unbalanced chemical equation has been rewritten each time we balanced a different atom.

You could do the balancing by writing the unbalanced equation only once in pencil. Leave enough space for the coefficients before each chemical formula. Now, you can enter the coefficients in pencil, and make changes readily.

PRACTICE PROBLEM 9-1

Write a balanced equation to represent the combustion of acetylene gas, C_2H_2, to form carbon dioxide and water.

PRACTICE PROBLEM 9-2

Write a balanced equation to represent the removal of carbon dioxide from the air in a spacecraft by its reaction with lithium hydroxide according to the unbalanced equation

$$CO_2 + LiOH \longrightarrow Li_2CO_3 + H_2O$$

PRACTICE PROBLEM 9-3

Write a balanced equation to represent the preparation of carbon dioxide gas by the action of sulfuric acid on sodium hydrogen carbonate according to the unbalanced equation

$$NaHCO_3 + H_2SO_4 \longrightarrow CO_2 + Na_2SO_4 + H_2O$$

9.8 Information Obtained from a Balanced Chemical Equation

We have learned how to balance chemical equations. How do we interpret them? What do they mean? For instance, what is meant by the following chemical equation?

$$2KClO_3(s) \xrightarrow{\triangle} 2KCl(s) + 3\ O_2(g)$$

A chemical equation affords more information than simply which substances are reactants and which are products. An equation normally indicates the states of the reactants and products. That is, it indicates whether they are solids, liquids, gases, or in aqueous solution. In addition, any special conditions required for the reaction are written above and/or below the reaction arrow. As we learned in Section 7.6, (s) indicates a solid, (ℓ) a liquid, (g) a gas, and (aq) an aqueous solution. The symbol $\triangle$ indicates heat. These symbols are summarized in Table 9-1.

Thus, the above equation provides the following information: Two formula units of solid potassium chlorate yield (when heated) two formula units of solid potassium chloride and three molecules of gaseous oxygen. As was stated in

Chapter 6, the word *molecule* does not have much meaning in the case of ionic solids such as potassium chlorate and potassium chloride (or any compound which contains the ammonium ion or a metal ion). Therefore, we refer to potassium chlorate and potassium chloride formula units rather than molecules. For example, a formula unit of barium hydroxide dissolved in water reacts with a formula unit of sulfuric acid to yield a formula unit of solid barium sulfate and two molecules of liquid water. This information can be conveyed by the equation

$$Ba(OH)_2(aq) + H_2SO_4(aq) \longrightarrow BaSO_4(s) + 2H_2O(\ell)$$

Table 9-1   Common Symbols Used in Equations

Symbol	Meaning
$\longrightarrow$	A reaction arrow, which is read as *yields* or *produces*. The arrow points in the direction of change.
(s)	Read as *solid*. Indicates that the substance is in the solid state.
(ℓ)	Read as *liquid*. Indicates that the substance is in the liquid state.
(g)	Read as *gas*. Indicates that the substance is in the gaseous state.
(aq)	Read as *aqueous*. Indicates that the substance is in aqueous solution (dissolved in water).
$\triangle$	Denotes heat energy. Indicates that heat is required for the reaction.
hv	Denotes light energy. Indicates that light energy is required for the reaction.

EXAMPLE PROBLEM 9-4

What information is conveyed by the following chemical equation?

$$CaCO_3(s) \xrightarrow{\triangle} CaO(s) + CO_2(g)$$

SOLUTION

This equation means that one formula unit of solid calcium carbonate yields (when heated) one formula unit of solid calcium oxide and one molecule of gaseous carbon dioxide.

EXAMPLE PROBLEM 9-5

If CH_4 is methane, what information is conveyed by the following chemical equation?

$$CH_4(g) + 2O_2(g) \longrightarrow CO_2(g) + 2H_2O(g)$$

SOLUTION

This equation means that one molecule of gaseous methane reacts with two molecules of gaseous oxygen to yield one molecule of gaseous carbon dioxide and two molecules of water vapour.

EXAMPLE PROBLEM 9-6

What information is conveyed by the following chemical equation?

$$Zn(s) + CuSO_4(aq) \longrightarrow ZnSO_4(aq) + Cu(s)$$

SOLUTION

This equation means that one atom of solid zinc reacts with one formula unit of copper(II) sulfate dissolved in water to yield one formula unit of zinc sulfate dissolved in water plus one atom of solid copper.

Part Two Review Questions

1. Explain what is meant by the following terms:
   a) combination reaction;
   b) decomposition reaction;
   c) displacement reaction;
   d) double displacement reaction.
   Give an example of each type.
2. Classify each of the following reactions as combination, decomposition, displacement, or double displacement reactions:
   a) $CaCO_3 \longrightarrow CaO + CO_2$ b) $K_2SO_4 + Ba(NO_3)_2 \longrightarrow BaSO_4 + 2KNO_3$
   c) $2H_2 + O_2 \longrightarrow 2H_2O$ d) $CaC_2 + 2H_2O \longrightarrow C_2H_2 + Ca(OH)_2$
   e) $2Al + 6\ HCl \longrightarrow 2AlCl_3 + 3H_2$
3. Balance the following equations:
   a) $Fe_3O_4 + H_2 \longrightarrow Fe + H_2O$ b) $Ga + H_2SO_4 \longrightarrow H_2 + Ga_2(SO_4)_3$
   c) $Fe_2O_3 + C \longrightarrow Fe + CO$ d) $Ca(OH)_2 + HCl \longrightarrow CaCl_2 + H_2O$
   e) $P_4 + O_2 \longrightarrow P_2O_5$
4. One reaction involved in the production of sulfuric acid is

$$SO_2(g) + O_2(g) \xrightarrow{\text{catalyst}} SO_3(g)$$

   Identify the type of reaction and balance the equation.
5. The chemical equation for the corrosion of iron is thought to be

$$Fe(s) + O_2(g) \xrightarrow{H_2O(\ell)} \underset{\text{rust}}{Fe_2O_3(s)}$$

   Identify the type of reaction and balance the equation.
6. What information is conveyed by each of the following equations?
   a) $CaO(s) + H_2O(\ell) \longrightarrow Ca(OH)_2(s)$
   b) $C_3H_8(g) + 5\ O_2(g) \longrightarrow 3CO_2(g) + 4H_2O(g)$ (C_3H_8 is propane)
   c) $AgNO_3(aq) + NaCl(aq) \longrightarrow AgCl(s) + NaNO_3(aq)$
   d) $Na_2SO_4(aq) + BaCl_2(aq) \longrightarrow BaSO_4(s) + 2NaCl(aq)$

Part Three: Reactions Involving Electron Transfer

9.9 Identifying Electron Transfer Reactions

Chemical reactions can be classified into two major types according to another criterion:

1. Reactions which do not involve transfer of electrons from one atom or ion to another.
2. Reactions which do involve transfer of electrons from one atom or ion to another.

Double displacement reactions do not generally involve electron transfers. Combination, decomposition, and displacement reactions, however, frequently do involve transfer of electrons between atoms.

The following reactions, for example, all involve electron transfers:

$$2Mg + O_2 \longrightarrow 2MgO \text{ (combination)}$$

$$2HgO \longrightarrow 2Hg + O_2 \text{ (decomposition)}$$

$$2Na + 2H_2O \longrightarrow 2NaOH + H_2 \text{ (displacement)}$$

$$Cl_2 + 2NaBr \longrightarrow Br_2 + 2NaCl \text{ (displacement)}$$

How do we know that electrons are being transferred between atoms? We calculate the oxidation numbers of the atoms involved (Section 7.1). When an atom loses electrons during a reaction, its oxidation number becomes more positive. When an atom gains electrons, its oxidation number becomes less positive (more negative). Thus, a change in oxidation number indicates that electrons are being transferred in a chemical reaction.

Consider, for example, a reaction such as

$$N_2 + O_2 \longrightarrow 2NO$$

Does this reaction involve a transfer of electrons? When we assign oxidation numbers, we recall that the oxidation numbers of atoms of elements in the free state are always zero. Therefore the oxidation numbers of the atoms in N_2 and O_2 are zero. The usual oxidation number of oxygen in a compound (such as NO) is -2, so the oxidation number of nitrogen in NO is $+2$. If we write these oxidation numbers above the atomic symbols, the equation becomes

$$\overset{0}{N_2} + \overset{0}{O_2} \longrightarrow 2\overset{+2 \; -2}{NO}$$

We see that the nitrogen atoms have changed oxidation numbers from 0 to $+2$, and the oxidation numbers of the oxygen atoms have changed from 0 to -2. Since the oxidation numbers have changed, electrons must have been transferred.

EXAMPLE PROBLEM 9-7

Is electron transfer involved in the following reaction?

$$H_2 + Br_2 \longrightarrow 2HBr$$

SOLUTION

The oxidation numbers are indicated above each atom in the equation

$$\overset{0}{H_2} + \overset{0}{Br_2} \longrightarrow 2\, \overset{+1\ -1}{HBr}$$

Both H and Br change oxidation state, so electron transfer must be involved.

EXAMPLE PROBLEM 9-8

Is electron transfer involved in the following reaction?

$$2Na_3PO_4 + 3BaCl_2 \longrightarrow Ba_3(PO_4)_2 + 6NaCl$$

SOLUTION

The oxidation numbers are indicated above each atom or ion:

$$2\, \overset{+1\ \ -3}{Na_3PO_4} + 3\, \overset{+2-1}{BaCl_2} \longrightarrow \overset{+2\ \ -3}{Ba_3(PO_4)_2} + 6\, \overset{+1\ -1}{NaCl}$$

No atoms change oxidation state, so electron transfer is not involved. This is a double displacement reaction.

EXAMPLE PROBLEM 9-9

Is electron transfer involved in the following reaction?

$$Zn + 2HCl \longrightarrow ZnCl_2 + H_2$$

SOLUTION

$$\overset{0}{Zn} + 2\overset{+1\ -1}{HCl} \longrightarrow \overset{+2\ -1}{ZnCl_2} + \overset{0}{H_2}$$

Both Zn and H change oxidation state, so electron transfer is involved.

EXAMPLE PROBLEM 9-10

Is electron transfer involved in the following reaction?

$$C + 2H_2 \longrightarrow CH_4$$

SOLUTION

$$\overset{0}{C} + 2\overset{0}{H_2} \longrightarrow \overset{-4\ +1}{CH_4}$$

Both C and H change oxidation state, so electron transfer is involved.

PRACTICE PROBLEM 9-4

Is electron transfer involved in the following reaction?

$$P_4 + 5O_2 \longrightarrow P_4O_{10}$$

(*Answer*: Yes.)

PRACTICE PROBLEM 9-5

Is electron transfer involved in the following reaction?

$$Ba + 2H_2O \longrightarrow Ba(OH)_2 + H_2$$

(*Answer*: Yes.)

PRACTICE PROBLEM 9-6

Is electron transfer involved in the following reaction?

$$(NH_4)_2SO_4 + Pb(NO_3)_2 \longrightarrow 2NH_4NO_3 + PbSO_4$$

(*Answer*: No.)

The loss of electrons from an atom, with a resulting increase in oxidation number, is called **oxidation**. The substance which loses the electrons is said to be **oxidized**, and the substance which removes them is called the **oxidizing agent**.

The gain of electrons by an atom, with a resulting decrease in oxidation number, is called **reduction**. The substance which gains the electrons is said to be **reduced**, and the substance which supplies them is called the **reducing agent**.

Oxidation and reduction always occur together. If one atom is oxidized during a chemical reaction, some other atom must be reduced. If an atom loses electrons, those electrons must be accepted by another atom. Thus, the oxidizing agent is reduced, and the reducing agent is oxidized. Hence reactions which involve electron transfer are usually called **oxidation-reduction** reactions or simply **redox** reactions.

The definitions of oxidation and reduction are frequently remembered by the sentence, "LEO the lion says GER." The capitalized letters stand for "Loss of Electrons is Oxidation" and "Gain of Electrons is Reduction." Their relationship to oxidation numbers may be remembered by a simple diagram as shown in the margin.

If the oxidation number of an atom goes up on the scale, the atom is oxidized. If the oxidation number goes down the scale, the atom is reduced. For example, in the reaction between magnesium and oxygen, the oxidation number of magnesium increases from 0 to +2. Magnesium, with only two valence electrons, tends to donate them to some other atom in order to achieve a stable octet. In losing these electrons, the magnesium atoms are oxidized. Each oxygen atom, having six valence electrons, will accept the two electrons to achieve an octet. The oxidation number of oxygen is decreased from 0 to −2. Therefore, oxygen is reduced in this process. These statements can be summarized as

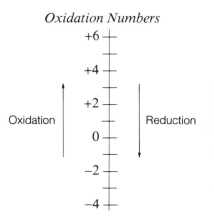

Oxidation Numbers

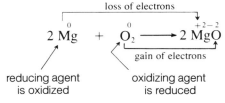

Because matter can neither be destroyed nor created, there must be as many electrons gained as are lost. In this reaction, for example, each of the two magnesium atoms loses two electrons, for a total of four electrons lost. Each of the two oxygen atoms gains two electrons, for a total of four electrons gained. Since four electrons are lost and four electrons are gained, the Law of Conservation of Mass is obeyed and the charge is conserved.

EXAMPLE PROBLEM 9-11

For the reaction $2HgO \longrightarrow 2Hg + O_2$, indicate the oxidizing agent, the reducing agent, the substance oxidized, and the substance reduced. Diagram the gain and loss of electrons.

SOLUTION

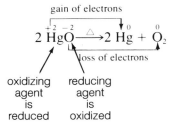

EXAMPLE PROBLEM 9-12

For the reaction $2Na + 2H_2O \longrightarrow 2NaOH + H_2$, indicate the oxidizing agent, the reducing agent, the substance oxidized, and the substance reduced. Diagram the gain and loss of electrons.

SOLUTION

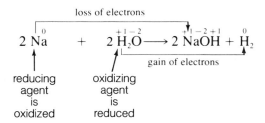

EXAMPLE PROBLEM 9-13

For the reaction $Cl_2 + 2NaBr \longrightarrow Br_2 + 2NaCl$, indicate the oxidizing agent, the reducing agent, the substance oxidized, and the substance reduced. Diagram the gain and loss of electrons.

SOLUTION

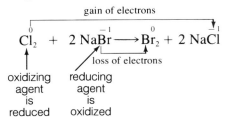

EXAMPLE PROBLEM 9-14

For the reaction $4NH_3 + 7 O_2 \longrightarrow 4NO_2 + 6H_2O$, indicate the oxidizing agent, the reducing agent, the substance oxidized, and the substance reduced. Diagram the gain and loss of electrons.

SOLUTION

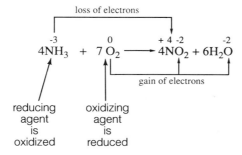

loss of electrons

$$\overset{-3}{4NH_3} + \overset{0}{7 O_2} \longrightarrow \overset{+4\ -2}{4NO_2} + \overset{-2}{6H_2O}$$

gain of electrons

reducing
agent
is
oxidized

oxidizing
agent
is
reduced

EXAMPLE PROBLEM 9-15

For the reaction $5CO + I_2O_5 \longrightarrow 5CO_2 + I_2$, indicate the oxidizing agent, the reducing agent, the substance oxidized, and the substance reduced. Diagram the gain and loss of electrons.

SOLUTION

loss of electrons

$$\overset{+2}{5CO} + \overset{+5}{I_2O_5} \longrightarrow \overset{+4}{5CO_2} + \overset{0}{I_2}$$

gain of electrons

reducing
agent
is
oxidized

oxidizing
agent
is
reduced

PRACTICE PROBLEM 9-7

For the reaction $Ni + HgCl_2 \longrightarrow Hg + NiCl_2$, indicate the oxidizing agent, the reducing agent, the substance oxidized, and the substance reduced. Diagram the gain and loss of electrons.

PRACTICE PROBLEM 9-8

For the reaction $Mg + 2HCl \longrightarrow MgCl_2 + H_2$, indicate the oxidizing agent, the reducing agent, the substance oxidized, and the substance reduced. Diagram the gain and loss of electrons.

9.10 The Activity Series of Metals

The elements and their compounds undergo many oxidation-reduction reactions. In this section we shall take a look at some reactions that are typical of metals. Metals have low ionization energies and small electron affinities. As a result, they lose electrons easily and have little tendency to gain them. When a metal reacts with another substance, therefore, the metal atoms tend to form positive metal ions by losing electrons. That is, the metal atoms are oxidized. The second substance accepts the electrons and is reduced. In this way, metals participate in oxidation-reduction reactions.

We know that metals differ in their reactivities. Gold, for example, is prized for its use in jewelry because it is unreactive to chemicals in the environment—gold does not tarnish. An iron nail, however, quickly becomes coated with rust when left in a moist environment. We can arrange the metals in the order of their chemical activities. Certain metals, such as sodium and potassium, react readily with water to displace hydrogen and form metallic hydroxides. Other metals, such as magnesium and iron, react with water in a similar reaction only when heated. Similarly, sodium and potassium react almost explosively with acids, and magnesium reacts vigorously with acids, but iron reacts relatively slowly. A brief form of an activity series is given in Figure 9-9. It is arranged so that the more reactive metals are at the top of the activity series and the least reactive metals are at the bottom of the series. As Column 3 shows explicitly, activity increases when going from the bottom to the top of the table.

We see in Column 1 that metals which are above hydrogen in the activity series are so reactive that they are found in nature only as compounds. The metals at the very bottom of the table (Pt, Au) are so unreactive that they occur in nature only as free metals. Between these two extremes are the metals with a moderate activity. They are found both as free metals and in combined forms as compounds.

A reactive metal has a strong tendency to form compounds. Conversely, somewhat drastic conditions are required to reverse this process and release the metals from their compounds. For the most active metals, chemists must resort to the rather drastic process of passing a current of electricity through a molten salt of the metal (electrolysis). The least active metals, on the other hand, can be released from their compounds by the application of heat. For example, mercury(II) oxide is decomposed to liquid mercury and gaseous oxygen by gentle heating:

$$2HgO(s) \xrightarrow{\triangle} 2Hg(\ell) + O_2(g)$$

The most active metals react with oxygen to form peroxides (for example, Li_2O_2, BaO_2), while the least reactive metals do not react with oxygen at all (Column 5). Most of the metals have a moderate activity and react with oxygen to form oxides.

As you might predict, the least active metals do not react with acids (these are located below hydrogen in the series). The metals located above hydrogen will react with acids at room temperatures. The metals from cadmium up are sufficiently active to react with water (a very weak acid) in the form of steam.

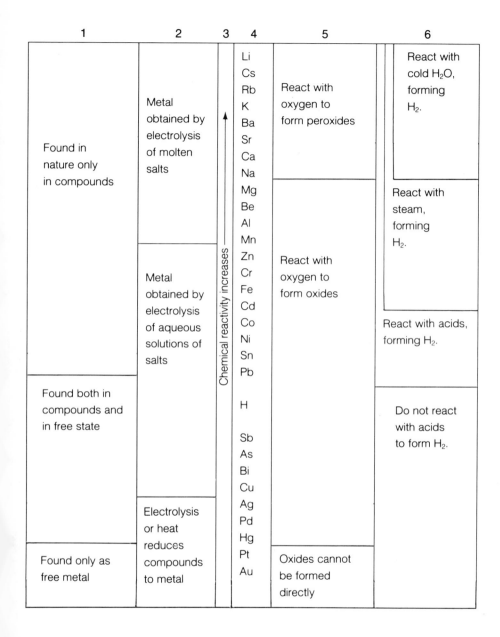

Figure 9-9
Activity series of metals

From sodium up, the metals are so active that they react even with cold water (Column 6).

Figure 9-9 shows that gold is the least reactive metal. As we go up the chart, each succeeding metal is more reactive than the one below it. In general, the Group 1 metals are more reactive than the Group 2 metals.

If a metal is unreactive towards one element (for example, oxygen), it is generally unreactive towards other elements (for example, hydrogen, sulfur, or the halogens). Those compounds of metals which are most likely to form will be least likely to decompose to give back the metals. Thus, the activity series enables us to predict the likelihood of the reaction of a given metal with oxygen, acids, water, sulfur, or the halogens.

An important characteristic of the activity series is that a metal will displace from a compound any other metal which is located *below* it in the series. For example, zinc is above copper in the activity series (or, copper is *below* zinc). Therefore, zinc will displace the copper from copper(II) sulfate:

$$Zn(s) + CuSO_4(aq) \longrightarrow ZnSO_4(aq) + Cu(s)$$

However, zinc cannot displace the magnesium from magnesium sulfate, because magnesium is above zinc in the activity series.

We can use the activity series of metals to help us predict whether a given reaction will occur. For example, will there be a reaction if we place metallic copper in a solution of silver nitrate? The question can be reworded: Will copper displace silver from the silver nitrate solution according to the following equation?

$$Cu(s) + 2AgNO_3(aq) \longrightarrow Cu(NO_3)_2(aq) + 2Ag(s)$$

We see that copper is above silver (or, silver is below copper) in the activity series. Therefore, copper is more reactive and will displace silver from silver nitrate.

EXAMPLE PROBLEM 9-16

Predict whether the following reaction takes place:

$$Cu(s) + Zn(NO_3)_2(aq) \longrightarrow Zn(s) + Cu(NO_3)_2(aq)$$

SOLUTION
Copper metal is below zinc in the activity series. It is not as reactive as zinc. Therefore, copper metal will not displace zinc from zinc nitrate. The reaction will not take place.

EXAMPLE PROBLEM 9-17

Predict whether the following reaction will take place:

$$Zn(s) + H_2SO_4(aq) \longrightarrow ZnSO_4(aq) + H_2(g)$$

SOLUTION
The activity series indicates that any metal above H in the series will react with an acid to displace H from the acid and form H_2 and a salt. Zinc is above H in the series and therefore will react with sulfuric acid to form H_2 and the salt, $ZnSO_4$.

EXAMPLE PROBLEM 9-18

Predict whether the following reaction will take place:

$$Zn(s) + 2H_2O(\ell) \underset{cold}{\longrightarrow} H_2(g) + Zn(OH)_2(s)$$

SOLUTION

Zinc is not in the list of metals which will react with cold water to form H_2 and a hydroxide. Therefore, this reaction will not take place.

PRACTICE PROBLEM 9-9

Will the reaction $Zn(s) + K_2SO_4(aq) \longrightarrow 2K(s) + ZnSO_4(aq)$ occur? State your reasons.

(*Answer*: No. Since Zn is below K in the activity series, Zn will not displace K from K_2SO_4.)

PRACTICE PROBLEM 9-10

Will the reaction $Mg(s) + 2AgNO_3(aq) \longrightarrow 2Ag(s) + Mg(NO_3)_2(aq)$ occur? State your reasons.

PRACTICE PROBLEM 9-11

Will the reaction $2Al(s) + 6HCl(aq) \longrightarrow 3H_2(g) + 2AlCl_3(aq)$ occur? State your reasons.

PRACTICE PROBLEM 9-12

Will the reaction $Ni(s) + 2\underset{\text{steam}}{H_2O(g)} \longrightarrow H_2(g) + Ni(OH)_2(s)$ occur? State your reasons.

(*Answer*: No. Figure 9-9 shows that, although Ni is above H in the activity series, Ni is not reactive enough to displace H from H_2O.)

Part Three Review Questions

1. In which of the following reactions is electron transfer involved?
   a) $2Ca + O_2 \longrightarrow 2CaO$
   b) $CaO + H_2O \longrightarrow Ca(OH)_2$
   c) $Ca + 2H_2O \longrightarrow Ca(OH)_2 + H_2$
2. What is an oxidizing agent? What is a reducing agent?
3. In each of the following oxidation-reduction reactions, indicate the oxidizing agent, the reducing agent, the substance oxidized, the substance reduced, and the number of electrons transferred by one atom of the reducing agent:
   a) $2Fe + O_2 \longrightarrow 2FeO$
   b) $2Fe + 3Br_2 \longrightarrow 2FeBr_3$
   c) $Fe + 2HCl \longrightarrow FeCl_2 + H_2$

4. In each of the following cases, write a balanced chemical equation for any reaction that occurs. If there is no reaction, simply write N.R. (no reaction).
   a) A piece of chromium is dipped into a solution of silver nitrate.
   b) A piece of lead is dipped into a solution of aluminum sulfate.
   c) A piece of aluminum is dipped into a solution of zinc chloride.
   d) A piece of tin is dipped into a solution of copper(II) bromide.

5. Predict whether each of the following reactions will occur. Give your reasons.
   a) $2Li(s) + 2H_2O(\ell) \longrightarrow H_2(g) + 2LiOH(aq)$
   b) $2Ag(s) + 2HCl(aq) \longrightarrow H_2(g) + 2AgCl(s)$
   c) $3Mg(s) + 2Al(NO_3)_3(aq) \longrightarrow 3Mg(NO_3)_2(aq) + 2Al(s)$
   d) $Ni(s) + Zn(NO_3)_2(aq) \longrightarrow Ni(NO_3)_2(aq) + Zn(s)$

Chapter Summary

- Mass, charge, and energy are conserved in a chemical reaction.
- Chemical reactions involve a rearrangement of atoms.
- Chemical reactions may be exothermic (release energy) or endothermic (absorb energy).
- The release or absorption of energy during a chemical reaction is the net result of bond making and bond breaking.
- Energy may be released during a chemical reaction in the form of heat, light, electrical, or mechanical energy.
- Energy is usually required to start a chemical reaction.
- Temperature, catalysts, concentration, surface area, and pressure affect the rate of a chemical reaction.
- The indications that a chemical reaction is occurring include change in temperature or colour; formation of a precipitate or odour; and evolution of a gas.
- There are four major types of reactions: combination, decomposition, displacement, and double displacement.
- Chemical equations can be balanced only by placing the proper coefficients in front of chemical formulas.
- Redox reactions involve the transfer of electrons and are accompanied by changes in oxidation numbers.
- Oxidation is the loss of electrons. Reduction is the gain of electrons.
- In oxidation, the oxidizing agent is reduced; the reducing agent is oxidized.
- The activity series of metals helps us to predict the products of simple chemical reactions.
- A metal will displace from a compound any other metal which is located below it in the activity series.

Key Terms

reactant

product

exothermic

endothermic

chemical energy

catalyst

combination reaction

decomposition reaction

displacement reaction

double displacement reaction

precipitation reaction

chemical equation

oxidation

oxidized

oxidizing agent

reduction

reduced

reducing agent

oxidation-reduction (redox) reaction

Test Your Understanding

Each of these statements or questions is followed by four responses. Choose the correct response in each case. (Do not write in this book.)

1. The rate of a reaction carried out in solution is not affected by
   a) concentration.
   b) temperature.
   c) pressure.
   d) a catalyst.

2. The reaction $2Na(s) + 2H_2O(\ell) \longrightarrow 2NaOH(aq) + H_2(g)$ is an example of a
   a) combination reaction.
   b) decomposition reaction.
   c) displacement reaction.
   d) double displacement reaction.

3. The reaction $Ba(OH)_2(aq) + 2HNO_3(aq) \longrightarrow Ba(NO_3)_2(aq) + 2H_2O(\ell)$ is an example of a
   a) combination reaction.
   b) decomposition reaction.
   c) displacement reaction.
   d) double displacement reaction.

4. The reaction $Li_2O(s) + CO_2(g) \longrightarrow Li_2CO_3(s)$ is classified as a
   a) combination reaction.
   b) decomposition reaction.
   c) displacement reaction.
   d) double displacement reaction.

5. The reaction $TiI_4(g) \longrightarrow Ti(s) + 2I_2(g)$ is classified as a
   a) combination reaction.
   b) decomposition reaction.
   c) displacement reaction.
   d) double displacement reaction.

6. The reaction $2AgNO_3(aq) + CaCl_2(aq) \longrightarrow 2AgCl(s) + Ca(NO_3)_2(aq)$ is an example of a
   a) precipitation reaction.
   b) combination reaction.
   c) decomposition reaction.
   d) displacement reaction.

7. The reaction of solutions of ammonium phosphate and barium nitrate gives a precipitate of barium phosphate. The equation which best represents this statement is
   a) $2(NH_4)_3PO_4(s) + 3Ba(NO_3)_2(aq) \longrightarrow Ba_3(PO_4)_2(aq) + 6NH_4NO_3(s)$.
   b) $2(NH_4)_3PO_4(aq) + 3Ba(NO_3)_2(aq) \longrightarrow Ba_3(PO_4)_2(s) + 6NH_4NO_3(aq)$.
   c) $2(NH_4)_3PO_4(aq) + 3Ba(NO_3)_2(s) \longrightarrow Ba_3(PO_4)_2(s) + 6NH_4NO_3(aq)$.
   d) $2(NH_4)_3PO_4(aq) + 3Ba(NO_3)_2(aq) \longrightarrow Ba_3(PO_4)_2(aq) + 6NH_4NO_3(aq)$.

8. Of the following equations, the only one which is balanced is
   a) $H_2 + O_2 \longrightarrow H_2O$.
   b) $Cu + Cl_2 \longrightarrow CuCl_2$.
   c) $Ag + Cl_2 \longrightarrow AgCl$.
   d) $Na + H_2O \longrightarrow NaOH + H_2$.

9. In the balanced equation for the combustion of ethane (C_2H_6),

$$2C_2H_6 + 7\,O_2 \longrightarrow CO_2 + 6H_2O$$

the coefficient of CO_2 is

a) 1.   b) 2.   c) 4.   d) 6.

10. In the incomplete equation $2NaHCO_3 \longrightarrow + 2CO_2 + H_2O$ the missing product has the formula:

a) Na_2O.   b) Na_2O_2.   c) NaO_2.   d) NaO.

11. In the incomplete equation $K_2SO_4 + 2AgClO_3 \longrightarrow 2KClO_3 + $ the missing product has the formula:

a) Ag_2SO_3.   b) $AgClO_4$.   c) $AgSO_4$.   d) Ag_2SO_4.

12. Lithium reacts with water to form hydrogen gas. In the balanced equation for this reaction, the other product with its coefficient is:

a) Li_2O.   b) $LiOH$.   c) $2LiOH$.   d) $2Li_2O$.

13. In an oxidation-reduction reaction, the reducing agent

a) is oxidized by the oxidizing agent.
b) takes electrons from the oxidizing agent.
c) is always reduced.
d) acts as a catalyst.

Answer questions 14 to 17 for the equation:

$$MnO_2(s) + 4HCl(aq) \longrightarrow MnCl_2(aq) + Cl_2(g) + 2H_2O(\ell)$$

14. The oxidizing agent is:

a) $MnO_2(s)$.   b) $HCl(aq)$.   c) $MnCl_2(aq)$.   d) $Cl_2(g)$.

15. The reducing agent is:

a) $MnO_2(s)$.   b) $HCl(aq)$.   c) $MnCl_2(aq)$.   d) $Cl_2(g)$.

16. The substance oxidized is:

a) $MnO_2(s)$.   b) $HCl(aq)$.   c) $MnCl_2(aq)$.   d) $Cl_2(g)$.

17. The substance reduced is:

a) $MnO_2(s)$.   b) $HCl(aq)$.   c) $MnCl_2(aq)$.   d) $Cl_2(g)$.

18. When the metals Cu, Au, Fe, and Ca are arranged in order of decreasing activity (most active metal first), the order is:

a) Ca, Cu, Au, Fe.   b) Cu, Au, Fe, Ca.
c) Fe, Au, Cu, Ca.   d) Ca, Fe, Cu, Au.

19. When a piece of Al is immersed in a solution of $CuCl_2$, the result is:

a) $AlCl_3(aq)$ and $CuCl_2(aq)$.   b) $Al(s)$, $Cu(s)$, and $Cl_2(g)$.
c) $AlCl_3(aq)$ and $Cu(s)$.   d) no reaction.

20. Which of the following reactions does not occur?

a) $2Li(\ell) + PbCl_2(\ell) \longrightarrow 2LiCl(s) + Pb(\ell)$
b) $Mg(s) + PbCl_2(aq) \longrightarrow MgCl_2(aq) + Pb(s)$
c) $2Al(s) + 3PbCl_2(aq) \longrightarrow 2AlCl_3(aq) + 3Pb(s)$
d) $Cu(s) + PbCl_2(aq) \longrightarrow CuCl_2(aq) + Pb(s)$

Review your Understanding

1. What is meant by chemical energy? Why is it released during an exothermic reaction?

2. What determines whether a chemical reaction will be exothermic or endothermic?

3. Why must an initial source of energy be provided even if a reaction is exothermic?

4. Why does one substance have a different amount of chemical energy from another substance?

5. Explain the following:
   a) When an iron nail is placed in a blue solution of copper(II) sulfate, the colour of the solution becomes less intense.
   b) A bonfire will burn faster when there is a strong wind.

6. A good rule of thumb for many reactions is that the rate of a reaction doubles for every 10°C rise in temperature. If a reaction takes 80 min at 20°C, how long will it take at 50°C?

7. What evidence is there that a chemical reaction has taken place in the following equation?

$$Mg(s) + 2HCl(aq) \longrightarrow MgCl_2(aq) + H_2(g) + heat$$

8. For each of the following reactions indicate all evidence that a chemical reaction is taking place:
   a) $S(s) + O_2(g) \longrightarrow SO_2(g) + heat$
   b) $2HgO(s) + heat \longrightarrow 2Hg(\ell) + O_2(g)$
   c) $Ni(NO_3)_2(aq) + Zn(s) \longrightarrow Ni(s) + Zn(NO_3)_2(aq)$
      green   colourless ıu
   d) $(NH_4)_2CO_3(aq) + Ca(NO_3)_2(aq) \longrightarrow CaCO_3(s) + 2NH_4NO_3(aq)$

9. Is the following one of the four major reaction types that we studied? If so, which type is it?

$$3NO_2 + H_2O \longrightarrow 2HNO_3 + NO$$

10. Balance the following equations and indicate the type of chemical reaction taking place:
    a) $Zn + HCl \longrightarrow ZnCl_2 + H_2$
    b) $Mg + CO_2 \longrightarrow MgO + C$
    c) $Al_2O_3 + H_2 \longrightarrow Al + H_2O$
    d) $Cu(NO_3)_2 + H_2S \longrightarrow CuS + HNO_3$
    e) $CaO + HNO_3 \longrightarrow Ca(NO_3)_2 + H_2O$

11. Balance the following equations:
    a) $Ca(AlO_2)_2 + HCl \longrightarrow AlCl_3 + CaCl_2 + H_2O$
    b) $O_2 + Sb_2S_3 \longrightarrow Sb_2O_3 + SO_2$
    c) $Cr + O_2 \longrightarrow Cr_2O_3$
    d) $C_2H_6 + O_2 \longrightarrow CO_2 + H_2O$
    e) $F_2 + NaOH \longrightarrow O_2 + NaF + H_2O$
    f) $NH_3 + O_2 \longrightarrow H_2O + NO$
    g) $Cl_2 + CrBr_3 \longrightarrow Br_2 + CrCl_3$
    h) $Sr(IO_3)_2 \longrightarrow SrI_2 + O_2$
    i) $Al_2(SO_4)_3 + NH_3 + H_2O \longrightarrow Al(OH)_3 + (NH_4)_2SO_4$
    j) $SiO_2 + Al \longrightarrow Si + Al_2O_3$

12. Balance the equation

$$C_7H_5N_3O_6(s) + O_2(g) \longrightarrow CO_2(g) + N_2(g) + H_2O(g)$$

13. What information is conveyed by the following equations?
    a) $2CO(g) + O_2(g) \longrightarrow 2CO_2(g)$
    b) $PCl_3(\ell) + Cl_2(g) \longrightarrow PCl_5(g)$
    c) $AgNO_3(aq) + HCl(aq) \longrightarrow AgCl(s) + HNO_3(aq)$
    d) $2NaHCO_3(aq) + H_2SO_4(aq) \longrightarrow Na_2SO_4(aq) + 2H_2O(\ell) + 2CO_2(g)$

14. Predict whether each of the following reactions will occur. Give your reasons in each case:
   a) $Co(s) + 2HCl(aq) \longrightarrow H_2(g) + CoCl_2(aq)$
   b) $Mg(s) + 2H_2O(\ell)_{cold} \longrightarrow H_2(g) + Mg(OH)_2(s)$
   c) $Ni(s) + Mg(NO_3)_2(aq) \longrightarrow Ni(NO_3)_2(aq) + Mg(s)$
   d) $Zn(s) + Ni(NO_3)_2(aq) \longrightarrow Zn(NO_3)_2(aq) + Ni(s)$
   e) $Zn(s) + 2H_2O(g)_{steam} \longrightarrow H_2(g) + Zn(OH)_2(s)$
   f) $Sr(s) + 2H_2O(\ell) \longrightarrow H_2(g) + Sr(OH)_2(aq)$
   g) $2Al(s) + 3ZnSO_4(aq) \longrightarrow Al_2(SO_4)_3(aq) + 3Zn(s)$
   h) $3Cu(s) + Al_2(SO_4)_3(aq) \longrightarrow 3CuSO_4(aq) + 2Al(s)$
   i) $Fe(s) + 2HCl(aq) \longrightarrow H_2(g) + FeCl_2(aq)$
   j) $2Ag(s) + 2HCl(aq) \longrightarrow H_2(g) + 2AgCl(s)$

15. Predict the products and write a balanced chemical equation for each reaction. If there is no reaction, write N.R.
   a) $K(s) + HCl(aq) \longrightarrow$
   b) $Na(s) + H_2O(\ell)_{cold} \longrightarrow$
   c) $Mg(s) + O_2(g) \longrightarrow$
   d) $Mg(s) + CuSO_4(aq) \longrightarrow$
   e) $Ni(s) + Pb(NO_3)_2(aq) \longrightarrow$
   f) $Pb(s) + H_2SO_4(aq) \longrightarrow$
   g) $Fe(s) + SnCl_2(aq) \longrightarrow$
   h) $Ca(s) + H_2SO_4(aq) \longrightarrow$
   i) $Na(\ell) + AlCl_3(s) \longrightarrow$

16. In which chemical family would you expect to find a) the most readily oxidized elements; b) the most readily reduced elements? Explain your answers.

17. In which of the following reactions is electron transfer involved?
   a) $Mg + 2HCl \longrightarrow H_2 + MgCl_2$
   b) $2Mg + O_2 \longrightarrow 2MgO$
   c) $AgNO_3 + NaCl \longrightarrow NaNO_3 + AgCl$
   d) $3Cl_2 + 2CrBr_3 \longrightarrow 3Br_2 + 2CrCl_3$
   e) $HCl + NaOH \longrightarrow NaCl + H_2O$
   f) $CaCO_3 \longrightarrow CaO + CO_2$
   g) $Zn + 2AgNO_3 \longrightarrow Zn(NO_3)_2 + 2Ag$

18. Indicate the oxidizing agent, the reducing agent, the substance oxidized, and the substance reduced for the following reaction:

$$KClO_3 + 3Na_2SnO_2 \longrightarrow KCl + 3Na_2SnO_3$$

19. In each of the following oxidation-reduction reactions indicate the oxidizing agent, the reducing agent, the substance oxidized, and the substance reduced:
   a) $NaNO_3 + Pb \longrightarrow NaNO_2 + PbO$
   b) $Na_2SO_4 + 4C \longrightarrow Na_2S + 4CO$
   c) $4Al + 3O_2 \longrightarrow 2Al_2O_3$
   d) $Mg + 2HBr \longrightarrow MgBr_2 + H_2$

Apply Your Understanding

1. A chlorine atom has a much higher ionization energy than a sodium atom. What does this mean in terms of the chemical behaviour of the elements?
2. How would you expect electronegativity to be related to oxidizing and reducing power?

3. Many biological processes involve oxidation and reduction. For example, ethanol is metabolized in the body in a series of steps:

$$CH_3CH_2OH \longrightarrow CH_3CHO \longrightarrow CH_3CO_2H \longrightarrow CO_2$$
$$\underset{\text{ethanol}}{} \quad \underset{\text{acetaldehyde}}{} \quad \underset{\text{acetic acid}}{}$$

In each of these compounds, what is the *average* oxidation number of the carbon atoms? Are the carbon atoms in each step being oxidized or reduced?

4. Why do metals nearly always behave as reducing agents in chemical reactions?

5. Where in the periodic table would you expect to find the most easily oxidized metals? Where would you expect to find the least easily oxidized metals?

6. Which metal in each of the following pairs should be more easily oxidized? Give reasons for your answers.

   a) Cs or Ba   b) K or Ga   c) K or Cs   d) Ga or Sr

7. Why aren't we permitted to change the subscripts in formulas when we balance a chemical equation? Why are we permitted to change the coefficients of formulas when we balance chemical equations?

8. Oxidation numbers change during a redox reaction. Are oxidation numbers conserved; that is, does the sum of the oxidation numbers on the left hand side of a balanced equation necessarily equal the sum of the oxidation numbers on the right hand side of the equation? Explain your answer.

9. Why must oxidation and reduction occur together in a reaction?

Investigations

1. Hydrogen peroxide is unusual in that sometimes it acts as a reducing agent and at other times as an oxidizing agent. Find two reactions in which it behaves as an oxidizing agent and two in which it behaves as a reducing agent. Write the balanced equations for these reactions.

2. Devise and conduct an experiment to determine whether a reaction assigned by your teacher is exothermic or endothermic.

3. Marble chips ($CaCO_3$) dissolve in dilute hydrochloric acid to produce gaseous carbon dioxide. Devise an experiment to determine the effect of temperature on the rate of this reaction.

4. The corrosion of iron by oxidation (rusting) has economic consequences for any industrialized society. Write a research report on the corrosion of iron. Your report should include the factors that affect the rate of corrosion of iron and methods used to minimize the rusting of iron.

5. Debate the following resolution: Because fossil fuels are convenient to obtain, they should be burned as our prime sources of energy in thermal electric generating stations.

10 Solutions

CHAPTER CONTENTS

Part One: **Characteristics of Solutions**
    10.1 General Features of a Solution
    10.2 Types of Solutions
    10.3 Applications of Solutions

Part Two: **Solubility**
    10.4 The Solution Process
    10.5 Factors Affecting Solubility

Part Three: **Two Classes of Solutes—Electrolytes and Nonelectrolytes**
    10.6 Electrolytes and Nonelectrolytes
    10.7 Dissolving Ionic Substances in Water
    10.8 Dissolving Polar Molecules in Water
    10.9 Like Dissolves Like

Part Four: **Reactions of Ions in Solution**
    10.10 Precipitation
    10.11 Writing Ionic and Net Ionic Equations
    10.12 Applications of Precipitation Reactions
    10.13 Qualitative Analysis of Aqueous Solutions

Why can't vitamin C be stored to any great extent in our bodies? How is soap able to remove a grease spot from clothing? Why doesn't the blood in fish native to the cold waters of Antarctica freeze? Why doesn't gasoline dissolve in water? How does the ability of ammonium nitrate to dissolve in water make it useful for cold packs? What causes the diving condition known as the "bends"? You will find the answers to these questions as you study this chapter.

We begin with a discussion of the general features of solutions and the reasons for their importance. Various types of solutions and the factors that affect solubility are considered. We examine the formation of solutions and describe solutions of electrolytes and nonelectrolytes. We discuss the process of precipitation, including the prediction of whether a precipitate will form when two solutions are mixed. The chapter concludes with an examination of the methods used for detecting the presence of various ions in aqueous solution

Figure 10-1
A solution may be liquid (as shown here), gaseous, or solid.

Key Objectives

When you have finished studying this chapter, you should be able to
1. describe the general features of a solution.
2. distinguish among unsaturated, saturated, and supersaturated solutions.
3. draw and interpret solubility curves.
4. explain the effects of molecular polarity on the solubility of a solute in a solvent.
5. write ionic equations for the dissociation of ionic compounds and polar molecules in water.
6. write ionic and net ionic equations for precipitation reactions.
7. describe how to perform a qualitative analysis of an unknown aqueous solution containing two ions.

Part One: Characteristics of Solutions

10.1 General Features of a Solution

When a lump of sugar is placed in a cup of tea, the sugar disappears, but it remains chemically unchanged. We know that the sugar is unchanged when it mixes with the water of the tea because we can still taste it. This is an example of a solution. The sugar has dissolved in the hot water of the tea.

A **solution** is a homogeneous (the same throughout) mixture of two or more substances. The particles in a solution are of molecular size. These particles are scattered randomly throughout the solution and are in continuous motion. The composition of a solution can vary within certain limits. You can dissolve one lump, two lumps, or three lumps of sugar in a cup of tea, but you cannot dissolve a box of sugar cubes in one cup of tea. When the sugar dissolves in the tea, the sugar molecules are distributed randomly and evenly throughout the cup of tea. Once the tea has been stirred, the uniform sweetness of the tea tells us that the solution is homogeneous.

In some ways, the tea example is confusing. Tea itself is a solution of many substances dissolved in water. Therefore, let us think of sugar dissolved in pure water. Two words that we should define are solvent and solute. Generally we refer to the substance that is present in larger quantity as the **solvent**. The **solute** is the substance present in smaller quantity which is dissolved in the solvent. In our example, water is the solvent and sugar is the solute. The sugar dissolves in the water. However, we are allowed to interchange the terms whenever it is convenient. In solutions where one substance is a liquid and the other substance is a gas or a solid, the liquid is usually called the solvent.

Solutions are often referred to as dilute or concentrated. A **dilute** solution is one which contains a relatively small amount of solute compared with the amount of solvent. A **concentrated** solution contains a relatively large amount of solute. These terms are used qualitatively. We learn how to express the concentration of a solute quantitatively (that is, using numbered quantities) in Chapter 11.

10.2 Types of Solutions

If we consider the state of the solvent in a solution, there are three types of solutions: liquid solutions, gaseous solutions, and solid solutions.

Liquid Solutions

Liquid solutions are made by dissolving solids, liquids, or gases in liquids. Iodine in alcohol (tincture of iodine) is an example of a solution of a solid dissolved in a liquid. Other common examples of a solution of a solid dissolved in a liquid include sugar in water (Fig. 10-2) and salt water. Solutions made by

Figure 10-2
A liquid solution (for example, sugar in water)

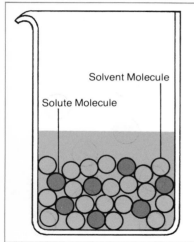

Solvent Molecule

Solute Molecule

dissolving solutes in water are called **aqueous solutions**. Much of our discussion of solutions will deal with aqueous solutions. Other aqueous solutions include alcohol dissolved in water (liquid in a liquid) and oxygen dissolved in water (gas in a liquid).

A solution of alcohol in water is an example of a liquid dissolved in another liquid. Alcohol and water mix in any proportion. They are said to be miscible (easily mixed in all proportions). However, some combinations of liquids do not mix to any great extent. Gasoline floats on top of water, but it will not dissolve very well in the water. Gasoline and water are practically immiscible and therefore do not form a solution. Gasoline itself is a liquid solution consisting of a mixture of organic compounds (hydrocarbons). Another example of a liquid solution is seawater (sodium chloride and other ionic solids dissolved in water). Examples of liquid solutions found in the household include juice, soda pop, nail polish remover, mouthwash, shampoo, bleach, insecticides, antifreeze, and perfume.

Gaseous Solutions

Gaseous solutions are made by dissolving a gas in another gas (Fig. 10-3). All gases mix in all proportions to produce homogeneous solutions. Air itself is a solution of oxygen and many other gases dissolved in nitrogen.

Solid Solutions

The most common type of solid solution is formed when one solid substance is dissolved in another to produce a homogeneous mixture (Fig. 10-4). Many alloys are solid solutions. For example, brass is a solid solution in which zinc atoms have been mixed homogeneously into the solid crystal of copper atoms.

Table 10-1 contains other examples of the various types of solutions with their uses. There are seven possible solute-solvent combinations.

Table 10-1   Examples of Solutions

Solvent	Solute	Example	Use
liquid	liquid	ethylene glycol in water	antifreeze
liquid	solid	ammonium nitrate in water	cold pack
liquid	gas	carbon dioxide in water	carbonated beverages
gas	gas	oxygen in helium	scuba tank
solid	solid	copper in gold	jewelry
solid	liquid	mercury in dental alloy	dental amalgams
solid	gas	hydrogen in palladium	gas stove lighter

10.3 Applications of Solutions

Solutions are important for many reasons. Many useful products of the chemical industry involve chemical reactions in which the reacting substances are dissolved in various solvents. In fact, many chemical reactions take place at an

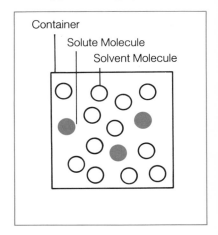

Figure 10-3
A gaseous solution (for example, air—oxygen in nitrogen)

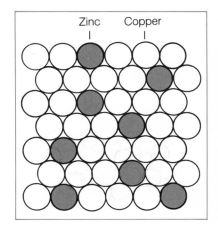

Figure 10-4
A solid solution (for example, brass—zinc in copper)

acceptable rate only when the reacting substances are in solution. Medicines such as cough syrups and antacids are frequently taken as solutions.

Chemists work mainly with liquid solutions rather than with gaseous or solid solutions. In particular, the liquid solvent most often used is water. Water is an excellent solvent.

Water is one of our most abundant chemicals. Two-thirds of our body mass is water, and our life processes depend on reactions in water. Important chemicals such as hormones and nutrients are transported throughout the human body dissolved in the bloodstream, which is composed mainly of water.

One of the characteristics of solutions that make them useful is that they freeze at lower temperatures and boil at higher temperatures than pure solvents.

Part One Review Questions

1. Give three characteristics of a solution.
2. What are the meanings of the terms *dilute* and *concentrated solution*?
3. How many types of solutions are possible? Name and describe each type.
4. What term is used to describe two liquids that will mix in any proportion?
5. What are the states of the solute and solvent in the most common type of solution?
6. List three uses of solutions.

Part Two: Solubility

Figure 10-5
Water molecules are attracted to OH groups on the sugar molecule. In this figure only one of the eight OH groups of sugar is shown.

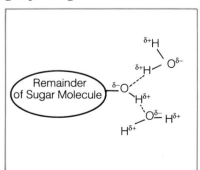

10.4 The Solution Process

Consider the dissolving of sugar crystals in water. Once the sugar crystals are placed in the water, sugar molecules from the surface of the crystals are attracted by water molecules, become dislodged, and move into spaces between other water molecules. This process is called *dissolving*.

Sugar molecules contain polar oxygen-hydrogen bonds. These attract polar water molecules (Fig. 10-5). The slightly positive hydrogen atoms of the sugar's OH groups attract the slightly negative oxygen atoms in water molecules. The slightly negative oxygen atoms of the sugar's OH groups attract the slightly positive hydrogen atoms in water molecules. As the sugar molecules leave the surface of the sugar crystals, the crystals become smaller and the amount of sugar dissolved in water becomes larger.

However, an opposing process soon begins. The positive ends of polar oxygen-hydrogen bonds in sugar molecules (the hydrogen atoms) are attracted to the negative ends of oxygen-hydrogen bonds in other sugar molecules (the oxygen atoms). Thus when sugar molecules in the solution collide with undissolved sugar crystals, they are attracted and stick to the surface of the crystals. This process is called *crystallization*. The dissolved sugar molecules are crystallizing.

As more and more sugar molecules dissolve, more sugar molecules crystallize. Finally a state of balance or equilibrium occurs:

$$\text{solute + solvent} \underset{\text{crystallizing}}{\overset{\text{dissolving}}{\rightleftharpoons}} \text{solution}$$

The rate at which sugar molecules dissolve equals the rate at which sugar molecules crystallize. This is called a state of **dynamic equilibrium** because two opposing processes (dissolving and crystallizing) occur at equal rates.

A simple model can help us understand a dynamic equilibrium. Suppose two rooms are connected by an open door. Ten students are in room *A* and 18 students are in room *B*. Each student is allowed to go from one room to the other as long as a student in the other room will trade places. At all times there are ten students in room *A* and eighteen students in room *B*. However, they are not always the same students. Some students are going from one room to the other, and other students are going in the opposite direction. In the same way, some sugar molecules are dissolving while other sugar molecules are crystallizing. However, there is always the same amount of sugar in the solution. At the point of dynamic equilibrium, the solution is said to be *saturated* with sugar.

A **saturated solution** contains the maximum amount of solute that can remain dissolved in a given amount of solvent at a particular temperature. In a saturated solution, the rate at which the solute molecules dissolve equals the rate at which the solute molecules crystallize (Fig. 10-6). A solution equilibrium is reached. Usually a solution is saturated if undissolved solute remains visible no matter how hard the solution is stirred or how long the solution sits.

Sometimes a saturated solution may be heated to dissolve extra solid and then carefully cooled to the original temperature without reappearance of the extra solid. This solution is not stable because a small crystal or agitation of the solution may cause crystallization to occur. A solution which contains more dissolved solute than it would if it were saturated is said to be a **supersaturated solution**.

If the solution is *unsaturated*, all of the solute molecules will be dissolved. There will be no solution equilibrium because there will be no opposing processes of dissolving and crystallizing. An **unsaturated solution** contains less than the maximum amount of solute that can dissolve in a given amount of solvent at a particular temperature.

During the solution process (Fig. 10-7) there are really three processes occurring simultaneously. The solute particles must separate from each other. This process is *endothermic* (heat energy is absorbed). The solvent molecules must separate from each other, and this process is also endothermic. Solute particles attract solvent molecules. Heat energy is released, and this process is therefore *exothermic*. An overall solution process could be exothermic, with heat released, and an accompanying rise in temperature of the solution. However, a solution process could also be endothermic, with heat absorbed, and an accompanying decrease in the temperature of the solution. Whether the solution process is exothermic or endothermic depends upon the magnitude of the

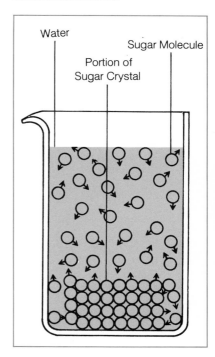

Figure 10-6
A saturated solution

three heat terms involved. In general, when a solid dissolves in a liquid the overall process is endothermic:

$$solute + solvent + heat\ energy \longrightarrow solution$$

Figure 10-7
The formation of a solution
(a) Solvent molecules, O, attract solute particles, ⊗, on the surface.
(b) Solute particles removed from the surface become surrounded by solvent molecules.

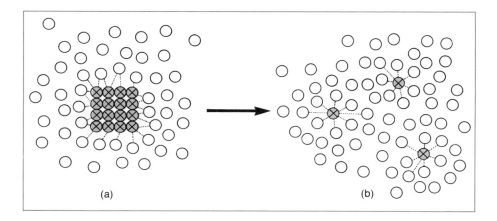

(a)   (b)

We can make practical use of the fact that the dissolving of most solids in a liquid is endothermic. Ammonium nitrate, NH_4NO_3, absorbs heat when it dissolves in water. It is used to make cold packs. The solid ammonium nitrate is placed in a thin-walled plastic bag. This bag is sealed inside a thicker plastic bag containing water. When the cold pack is to be used, the bag containing the ammonium nitrate is broken. The solid dissolves in the water, and the resulting solution becomes very cold. The cold pack can then be used to reduce swelling in the affected area of the body. Thirty grams of NH_4NO_3 dissolved in 100 mL of water at 20°C lowers the temperature of the water to 0°C.

There is also an application for a solution process that is exothermic. Hot packs usually use solid calcium chloride or solid magnesium sulfate. When the solid is allowed to dissolve in water, heat is released. Forty grams of $CaCl_2$ dissolved in 100 mL of water at 20°C raises the temperature of the water to 90°C.

10.5 Factors Affecting Solubility

The term **solubility** refers to the maximum amount of a solute that can be dissolved in a given amount of solvent at a particular temperature and pressure. Solubility is the amount of solute required to form a saturated solution.

Pressure

Of the various types of solutions, only gases dissolved in liquids (or solids) are affected very much by pressure. Henry's Law states that the solubility of gases in liquids is directly proportional to the pressure of the gas above the liquid. Figure 10-8 shows that under higher pressure the gas molecules are more likely to come in contact with and dissolve in the liquid layer. This principle is used in the soft drink industry, where carbon dioxide gas is dissolved in a water-based syrup under pressure. When the container is opened and the liquid is exposed to the air, which has less pressure, some of the carbon dioxide bubbles out of the bottle.

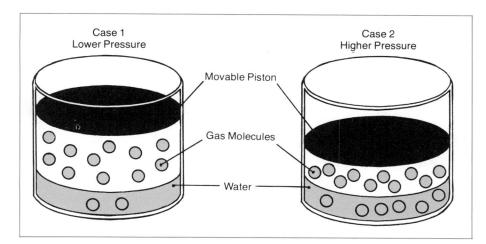

Figure 10-8
Illustration of Henry's Law

Sea divers rely on compressed air for their supply of oxygen. As a diver descends below the surface of water, the pressure exerted by the water increases. Because of this increased pressure, at a depth in excess of 15 m, a greater amount of nitrogen in the compressed air will dissolve in the blood and other body fluids, in accordance with Henry's Law. If a diver comes to the surface too rapidly, the decrease in pressure causes excess nitrogen to leave the body fluids in the form of tiny bubbles which affect circulation and nerve impulses. This condition is known as the "bends" and can be very painful and even fatal. It is called the "bends" because pain causes the diver to double up. During deep sea dives, helium gas is sometimes substituted for nitrogen because it is less soluble in body fluids. Thus, during deep dives, a mixture of helium and oxygen is less apt to lead to the bends than is compressed air.

Pressurized products that are used in the household, such as shaving cream, insecticides, and cleaners, also make use of the relationship between pressure and solubility of a gas. When the spray valve is opened, the gas pressure in the can drops. Because the gas pressure above the liquid is lower, dissolved gas comes out of the solution (Henry's Law), carrying the liquid with it as a spray or foam.

Temperature

The effect of pressure on any solution of a gas dissolved in a liquid is always the same. However, the effect of a temperature change on a solution depends upon the type of solution.

In the case of gases dissolved in liquids, rising temperatures cause the gas molecules to leave the liquid. Thus, the solubility of gases in liquids decreases as the temperature of the solution is increased. More oxygen gas is able to dissolve in water at lower temperatures. That is why most fish prefer cooler water. When water is heated, tiny bubbles form because the dissolved air is less soluble at higher temperatures.

Energy is usually required for solids to dissolve in liquids. Thus, solids are

Table 10-2   Solubilities of Various Solid Solutes in Water at 0°C and 100°C

Solute	Solubility in g solute/100 g of H_2O	
	0°C	100°C
$Ca(OH)_2$	0.18	0.07
$PbCl_2$	0.67	3.33
KNO_3	13	248
KCl	28	58
NaCl	36	40
NaOH	42	347
NH_4NO_3	118	871
sugar	179	487

usually more soluble in warmer than in cooler liquids. The required heat energy is used mainly to enable the solute particles to break away from the crystals. Some solids, however, are more soluble in cooler liquids. There is quite a difference in solubility among different solutes. For instance, approximately 3200 g of $LiClO_3$ can be dissolved in 1 L of water at 25°C, while the calculated solubility of HgS is only about 2.3×10^{-24} g/L at 25°C. The difference in solubility among different solid solutes can be seen in Table 10-2. This table shows that some solutes are very soluble while others have a very low solubility. The table also shows that the solubility of some solutes increases dramatically from 0°C to 100°C, while other solutes show little change in solubility with increasing temperature.

If the solubility of a solute is known at different temperatures, it is possible to plot a graph of solubility (*y* axis) versus temperature (*x* axis). Such a graph is called a **solubility curve** (Fig. 10-9).

The solubility curves in Figure 10-9 show how differently the solubility of each solute is affected by a change in temperature. If we wish to determine the solubility at a given temperature, we merely locate the temperature on the solubility curve and read the solubility. For example, the solubility of KNO_3 at 40°C is approximately 65 g/100 g H_2O. This means that the maximum amount of KNO_3 that can be dissolved in 100 g of water at 40°C is 65 g. If we added less than 65 g of KNO_3 to 100 g of water at 40°C it would all dissolve. If we added more than 65 g of KNO_3 to 100 g of water at 40°C, only 65 g of the KNO_3 would dissolve. The rest of the KNO_3 would remain as undissolved solid.

The ability to interpret solubility curves can be useful in purifying compounds. Consider a sample of 100 g of solid KNO_3 contaminated with 20 g of solid NaCl. From the solubility curves (Fig. 10-9) we see that, if the mixture is added to 100 mL of water at 60°C, all of the solid mixture will dissolve. Let us

**Figure 10-9
Solubility curves**

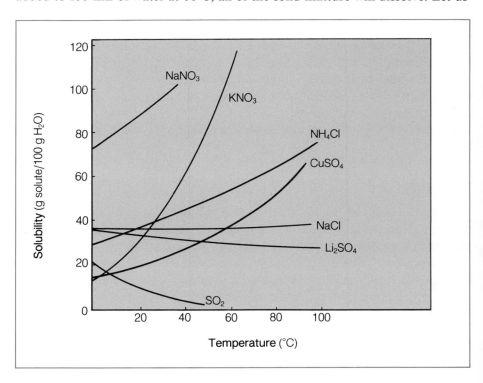

assume we slowly cool the solution to 0°C. At 0°C the solubility of KNO_3 is 13 g/100 g H_2O, and for NaCl it is 36 g/100 g H_2O. Thus at 0°C all of the NaCl remains in solution, but only 13 g of KNO_3 remains dissolved. The rest of the KNO_3, 87 g (100 g – 13 g), is present as a solid and can now be separated from the solution by filtration. About 87% of the KNO_3 in the original contaminated mixture has been recovered in pure form. This method of purifying compounds is used frequently by the chemical industry and is known as *recrystallization*.

Less energy is required to cause mixing of liquids in liquids, because the forces between molecules in the liquid state are not as strong as those holding solids together. The solubility of a liquid in a liquid seldom depends on the temperature of the mixture.

APPLICATION

Artificial Blood

Recently, scientists have been investigating the use of artificial blood for purposes such as transfusion. A colloidal dispersion of perfluorobutyltetrahydrofuran in water has been used as a blood substitute. A colloidal dispersion is a mixture which contains suspended particles that are large enough to be visible under high magnification but too small to settle. In addition, the components of a colloidal dispersion cannot be separated by filtration. Perfluorobutyltetrahydrofuran is a fluorocarbon. It has the following structure:

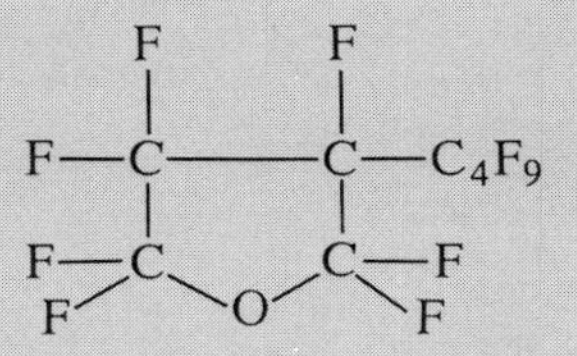

Other fluorocarbons have also been used as blood substitutes. These fluorocarbons have a number of important properties. They are able to transport oxygen just as natural blood does, but they can dissolve almost three times the quantity of oxygen per unit volume as natural whole blood. In addition, they are so inert that they do not harm the body.

"Bloodless" rats have survived with artificial blood replacing natural blood in their veins. Humans have survived transfusions of artificial blood. In fact, artificial blood has some advantages over natural blood. Perhaps the most important of these advantages is that blood typing is not required before transfusion. Also, artificial blood is not susceptible to certain poisons. For example, carbon monoxide prevents natural blood from transporting oxygen but has no such effect on artificial blood.

Part Two Review Questions

1. Explain the processes of dissolving and crystallization.
2. Describe the interaction between sugar molecules and water molecules that enables sugar to dissolve in water.
3. How does a supersaturated solution differ from a saturated solution?
4. What three actions occur during the dissolving process?
5. What is meant by exothermic and endothermic processes?
6. What is meant by the term *solubility*?
7. What factors affect solubility of **a)** a gas in a liquid; and **b)** a solid in a liquid?
8. What is a solubility curve? What information does it provide?
9. Explain the technique known as recrystallization. Why is it useful?
10. Given the following solubility data, plot a solubility curve for NH_4ClO_4.

Temperature (°C)	0	20	40	60	80	100
Solubility (g/100 g H_2O)	12.9	23.4	36.7	51.5	67.8	87.1

   a) How many grams of NH_4ClO_4 do you predict can dissolve at 30°C?
   b) If you tried to dissolve 50 g of NH_4ClO_4 in 100 g of water at 20°C, what would happen?
   c) At what temperature would you be able to dissolve 75 g of the NH_4ClO_4 in 100 g of H_2O?

Part Three: Two Classes of Solutes—Electrolytes and Nonelectrolytes

10.6 Electrolytes and Nonelectrolytes

An electric current will pass through some solutions. The solutes in these solutions are called **electrolytes**. However, an electric current will not pass through other solutions, and so the solutes in these solutions are called **nonelectrolytes**. Sugar is a nonelectrolyte, but sodium chloride is an electrolyte.

   Solutions of electrolytes are important within the body. Blood plasma is a solution containing the ions Mg^{2+}, K^+, Cl^-, Na^+, Ca^{2+}, HPO_4^{2-}, and HCO_3^-. The concentrations of these ions in blood are kept constant under normal conditions, and small variations may be fatal.

   Both electrolytes and nonelectrolytes affect the properties of solvents. They cause the solvent to melt (or freeze) at a lower temperature and to boil at a higher temperature than normal. For example, a saturated solution of NaCl in water at 0°C will lower the freezing temperature of water to −18°C. Thus, when salt is spread on roads, it dissolves in any water present to lower the freezing point so that much colder temperatures are required to cause freezing. If salt is added to ice, the salt melts the ice, forming a saturated solution that once again has a lower freezing point. This principle is used when salt is added to ice to make homemade ice cream. The lower temperature of the ice-salt mixture is cold enough to solidify the cream.

The freezing point of sea water is approximately −1.85°C because of the presence of solutes. The blood of fish in the very cold waters of Antarctica does not freeze because it contains dissolved salts and proteins.

Water expands about 9% by volume on freezing and generates a tremendous force in the process. Such a force could easily crack an automobile's radiator. Liquid antifreeze is dissolved in the cooling system of the car, causing the water in it to freeze at a lower temperature. For example, a 50% by volume solution of ethylene glycol and water causes the water in the cooling system to freeze at −36°C instead of 0°C. This same fluid in the radiator will cause the water to boil at a higher temperature, helping to prevent the coolant from boiling away in the summer.

The ability of a solution to conduct electricity can be determined with the simple apparatus shown in Figure 10-10. A pair of electrodes in series with a light bulb is connected to a source of electricity. No current can flow through the circuit as long as the electrodes are separated. Since pure water is essentially a nonconductor of electricity, no current can flow through the circuit when the electrodes are immersed in water. However, when an electrolyte is dissolved in the water and the electrodes are immersed in the solution, a current will flow and the light bulb will glow.

In order for a solution to conduct electricity, charged particles (ions) must exist in the solution. In fact, up to a certain limit, the more charged particles there are, the better the solution will conduct electricity. It is the movement of the charged particles that allows the current to pass through a solution of an electrolyte (Fig. 10-11). This movement can take place because of empty spaces in liquid water. If a solid ionic compound such as sodium chloride is placed between the electrodes, the light bulb will not glow because, although ions are present, they are not free to move. If a molten ionic compound is placed between the electrodes, the light bulb will glow because of the movement of ions in the liquid state. If an aqueous solution of an ionic compound is used, the bulb will again glow because the ions have even more freedom of movement. Thus, solutions of ionic compounds conduct electricity. Ionic compounds are electrolytes.

The English scientist Michael Faraday assumed that the formation of charged particles in the solution of electrolytes was caused by the presence of the charged electrodes of the conductivity apparatus. However, further experiments, especially the important work of the Swedish chemist Svante Arrhenius, showed Faraday's assumption to be erroneous and led to the development of the modern theory of electrolytes. The principles of this theory are:

1. Electrolytes in solution exist in the form of ions.
2. An ion is an atom or group of atoms which has an electric charge.
3. In electrolyte solutions, the total positive ionic charge must equal the total negative ionic charge.

Figure 10-10
Apparatus for determining the conductivity of a solution

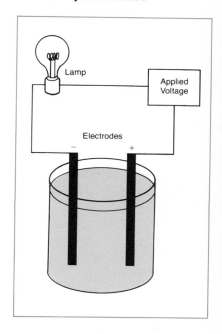

Figure 10-11
Conductivity test with an electrolyte (NaCl) solution

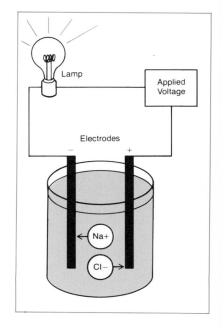

Electrolytes may be classified as either strong or weak. A **strong electrolyte** readily furnishes ions in solution. A **weak electrolyte** does not readily furnish ions in solution. A comparison of a strong and a weak electrolyte of equal concentrations is shown in Table 10-3.

Figure 10-12
Conductivity test with a nonelectrolyte (sugar) solution

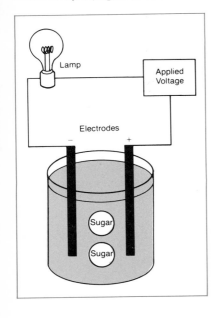

Table 10-3   A Comparison of Conductivities of Aqueous Solutions

Solute	Percentage of Solute Existing as Ions	Conductivity Test
Strong electrolyte	high	bright light
Weak electrolyte	low	dim light
Nonelectrolyte	zero	no light

When ionic compounds dissolve, they are essentially 100% dissociated into ions. Therefore, ionic compounds are strong electrolytes. **Dissociation** is the separation or breaking apart of solute particles, resulting in ions in solution. Strong electrolytes could also be polar covalent molecules that are essentially 100% dissociated into ions in aqueous solution. Weak electrolytes are polar covalent molecules of which a very small percentage break apart into ions in solution.

In the case of a nonelectrolyte—such as sugar—dissolved in water, no ions are present. When sugar molecules dissolve in water, they become surrounded by a layer of water molecules, but their structure is unchanged and they remain uncharged. When charged electrodes are placed in a solution containing hydrated sugar molecules (sugar molecules surrounded by a layer of water molecules), the neutral sugar molecules do not migrate towards one electrode or the other. Thus, the nonelectrolyte sugar will not conduct electricity in solution (Fig. 10-12).

APPLICATION

Antifreeze and Arctic Insects

Dissolved substances lower the freezing temperature of water. This is the reason for adding antifreeze to automobiles in cold weather. This is also the reason a remarkable insect is able to survive.

The Arctic wooly-bear caterpillar lives on Canada's Ellesmere and Devon Islands, far north of the Arctic Circle. It is able to survive temperatures of −70°C or lower. The wooly-bear insect spends more than 90% of its 14 year life cycle as a larva. It is able to tolerate the extreme conditions of cold because it produces glycerin which acts as antifreeze.

In order to investigate the action of glycerin in the insect, scientists used an isotope of carbon (carbon-13) to replace one of the normal carbon-12 atoms in glucose. Then they injected this labelled glucose into the larva. They followed the conversion of the labelled glucose to glycerin and monitored the disappearance of glycerin as a function of temperature. They found that the larva produced glycerin between −30°C and 30°C but that the glycerin accumulated only at temperatures below 5°C. One explanation for the failure of glycerin to accumulate above 5°C is that the larva takes in more oxygen at temperatures above 5°C. Thus, it produces a faster metabolic reaction of the glycerin with oxygen, preventing the buildup of unnecessary glycerin in warmer temperatures.

10.7 Dissolving Ionic Substances in Water

Sodium chloride is a typical ionic compound. Sodium is in Group 1. A sodium atom has one outer electron which it can lose. Chlorine is in Group 17 of the periodic table. A chlorine atom has seven outer electrons and can gain one electron to complete its octet:

$$Na\cdot \longrightarrow Na^+ + 1\,e^-$$

$$:\ddot{C}l\cdot + 1\,e^- \longrightarrow :\ddot{C}l:^-$$

When crystals of sodium chloride form, the positive sodium ions are attracted to the negative chloride ions because unlike charges attract. A portion of a sodium chloride crystal is shown in Figure 10-13.

A solution of sodium chloride in water conducts electricity and therefore must contain ions. The question is, what process is able to overcome the attraction between the sodium ions and the chloride ions? Why do the sodium ions separate from the chloride ions as they dissolve?

To answer this question, we must look at the structure of the water molecule. Water is a bent molecule in which the two oxygen-to-hydrogen bonds are covalent bonds:

$$\underset{H \quad H}{:\ddot{O}:} \quad \textbf{or} \quad \underset{H \qquad H}{O}$$

Because oxygen is a more electronegative element than hydrogen, the oxygen atom attracts the two shared pairs of electrons more effectively than do the two hydrogen atoms. The oxygen end of the molecule is slightly negative and the hydrogen end of the molecule is slightly positive.

$$\underset{H \quad \delta+ \quad H}{\overset{\delta-}{O}}$$

A molecule which has one slightly positive end and one slightly negative end is a polar molecule. Thus, water is a polar molecule which can be represented schematically as

When a crystal of sodium chloride is placed in water, the negative oxygen ends of the water molecules are attracted to the positive sodium ions, and the

Figure 10-13
One layer of a sodium chloride crystal

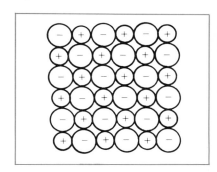

Figure 10-14
A crystal of NaCl in a beaker of water

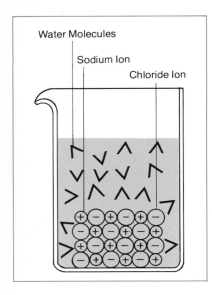

Figure 10-15
A schematic representation of the dissolving of NaCl

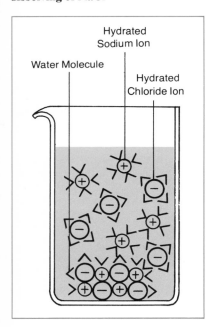

positive hydrogen ends of the water molecules are attracted to the negative chloride ions as illustrated in Figure 10-14. These mutual forces of attraction between the sodium ions and water, and between the chloride ions and water, are strong enough to overcome the attractive forces between sodium ions and the chloride ions.

The water-surrounded sodium ions move off into the solution. They are called *hydrated* or *aqueous* sodium ions, $Na^+(aq)$. The water-surrounded chloride ions, called hydrated or aqueous chloride ions, $Cl^-(aq)$, also move into solution as shown in Figure 10-15. This process is represented by the equation

$$NaCl(s) \longrightarrow Na^+(aq) + Cl^-(aq)$$

The (aq) shows that each ion is surrounded by a number of water molecules. However, we have not indicated how many water molecules are involved, since different ions will be surrounded by different numbers of water molecules. It is the layer of water molecule around each ion which diminishes the attraction of the sodium ions for the chloride ions. Ions are already present in the sodium chloride crystal. In solution, the ions have separated or dissociated. Other ionic substances dissociate in water much as sodium chloride does. However, some ionic substances dissolve more readily in water than does sodium chloride, and some dissolve less readily in water. The following equations represent the dissociation of three other ionic solids in water:

$$Na_2SO_4(s) \longrightarrow 2\ Na^+(aq) + SO_4^{2-}(aq)$$

$$(NH_4)_3PO_4(s) \longrightarrow 3\ NH_4^+(aq) + PO_4^{3-}(aq)$$

$$NaOH(s) \longrightarrow Na^+(aq) + OH^-(aq)$$

10.8 Dissolving Polar Molecules in Water

Certain covalent molecules form ions when they dissolve in water. A good example is the hydrogen chloride molecule (HCl). Normally, hydrogen chloride is a gas. However, it can be liquefied and solidified by cooling. Whether it is a gas, solid, or liquid, hydrogen chloride will not conduct electricity. No ions are present in pure hydrogen chloride.

When hydrogen chloride molecules are dissolved in water, the solution (called hydrochloric acid) will conduct electricity. Ions *are* present. The hydrogen chloride molecules must have *reacted* with the water molecules to *produce* ions.

Hydrogen chloride, like water, is a polar molecule:

$$^{\delta+}H:\ddot{\underset{..}{Cl}}:^{\delta-}$$

The chlorine end of the molecule is slightly negative and the hydrogen end of the molecule is slightly positive. One would expect that a hydrogen chloride molecule would be attracted to a water molecule:

$$^{\delta+}H\!:\!\ddot{O}\!:^{\delta-}\!-\!-\!-\!-\!^{\delta+}H\!:\!\ddot{\underset{..}{Cl}}\!:^{\delta-}$$
$$\underset{\delta+}{\overset{|}{H}}$$

During a collision between a hydrogen chloride molecule and a water molecule, a hydrogen ion (i.e., a proton—H^+) is transferred from the hydrogen chloride to the water:

$$H\!:\!\underset{H}{\ddot{O}}\!: + H\!:\!\ddot{\underset{..}{Cl}}\!: \longrightarrow \left[H\!:\!\underset{H}{\ddot{O}}\!:\!H\right]^+ + :\ddot{\underset{..}{Cl}}\!:^-$$

$$H_2O + HCl \longrightarrow H_3O^+ + Cl^-$$

This proton (H^+) forms a coordinate bond with the oxygen in water since the oxygen atom provides both electrons to be shared in the covalent bond. When the hydrogen ion leaves the HCl molecule, it leaves its electron behind, forming a chloride ion.

The H_3O^+ is called the hydronium ion. Both the hydronium ion and the chloride ion are surrounded by a layer of water molecules (they are hydrated):

$$H_2O + HCl \longrightarrow H_3O^+(aq) + Cl^-(aq)$$

Further examples of dissociation of polar molecules in water are:

$$HNO_3(\ell) + H_2O(\ell) \longrightarrow H_3O^+(aq) + NO_3^-(aq)$$

$$HClO_4(\ell) + H_2O(\ell) \longrightarrow H_3O^+(aq) + ClO_4^-(aq)$$

Polar molecules that are strong electrolytes dissociate essentially 100%. Polar molecules that are weak electrolytes dissociate only slightly.

10.9 Like Dissolves Like

Ionic compounds are able to dissolve in water (a polar solvent) because there are mutual forces of attraction. The positive ions of the ionic solid and the negative ends of the water molecules attract one another. The negative ions of the ionic solid and the positive ends of the water molecules also attract one another. These mutual forces of attraction are enough to separate the positive and negative ions from each other and to separate the water molecules from each other. A solution is thus formed.

Polar compounds are able to dissolve in water because there are mutual forces of attraction which enable solute molecules to become separated from one another and solvent molecules to become separated from one another. The negative ends of polar solute molecules and the positive ends of the polar water molecules attract each other. The positive ends of polar solute molecules and the negative ends of the polar water molecules also attract each other. The result is the formation of a solution.

Nonpolar solutes are able to dissolve in nonpolar solvents because only weak van der Waals forces attract solute molecules to one another. Similarly, only van der Waals forces attract solvent molecules to one another. These weak forces are easily overcome by the continuous random motion of the molecules, allowing them in this way to mix and form a solution. Carbon tetrachloride is a nonpolar molecule, since it does not have a definite negative end or a definite positive end. When a crystal of sodium chloride is placed in a beaker of carbon tetrachloride, the carbon tetrachloride molecules are not appreciably attracted to the positive sodium ions or to the negative chloride ions. There is no force acting to separate these ions and pull them into solution; therefore the sodium chloride is not as soluble in nonpolar carbon tetrachloride as it is in polar water.

Why is gasoline, a mixture of nonpolar molecules, not as soluble in polar water as it is in nonpolar carbon tetrachloride? In water, all of the polar water molecules are attracted to one another. Thus, the attractive forces between the water molecules have to be overcome before gasoline molecules can move between the water molecules. Since the water molecules have no attraction for those of the gasoline, they tend not to separate from one another to make room for the gasoline. On the other hand, carbon tetrachloride molecules are not polar and are not strongly attracted to one another. They are also not strongly attracted to the molecules of the gasoline. However, there is no great force to be overcome in order to get the carbon tetrachloride molecules to separate from one another to make room for the gasoline. The weak van der Waals forces are overcome by the continuous random motion of the molecules. Thus, nonpolar gasoline dissolves much more readily in nonpolar carbon tetrachloride than it does in polar water.

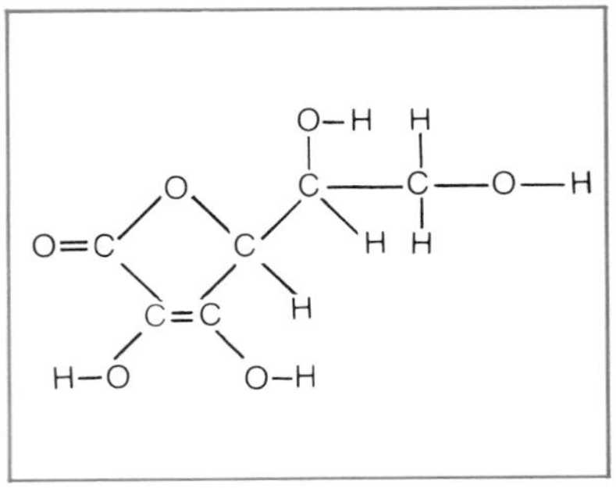

Polar or ionic substances are more soluble in polar solvents than in nonpolar solvents. Nonpolar substances are more soluble in nonpolar solvents than in polar solvents. Thus, "like dissolves like."

This explains why vitamins B and C are water soluble while vitamins A, D, E, and K are fat soluble. By fat soluble we mean they are stored in the nonpolar fatty tissue of the body.

Let us examine the structure of vitamin C (Fig. 10-16). There are a number of polar −OH groups present. This polar molecule will readily dissolve in polar water and cannot be stored to any great extent in the body, since it is eliminated. Therefore daily dosages must be ingested as part of a healthy diet.

On the other hand, vitamin A (Fig. 10-17) has only one polar −OH group. Since the molecule is so large, it can be considered to be a nonpolar molecule. It therefore dissolves in the nonpolar fatty tissues of the body and is stored there. A temporary absence of these fat soluble vitamins in one's diet will cause no harm, because the deficiency will be overcome by the vitamins stored in the fatty tissue.

Figure 10-16
Vitamin C

Figure 10-17
Vitamin A

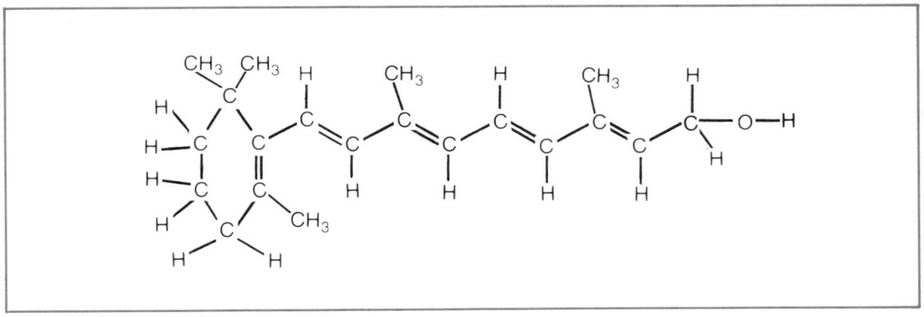

The rule of like dissolves like has other practical applications. A molecule of a typical soap has a long, nonpolar "tail" and a polar, water-soluble "head." The tail should be easily soluble in nonpolar materials, and the head should be quite soluble in water:

$$CH_3(CH_2)_{16}\!-\!\overset{\overset{\displaystyle O}{\|}}{C}\!-\!O^-Na^+$$

nonpolar end   *polar end*
(covalent bonds)   *(ionic bond)*

When soapy water comes in contact with a soiled object such as a grease spot on a sweater, the tails of the soap molecules tend to dissolve in the grease layer. The polar ends will remain in the polar water solution. A little agitation will loosen the film of grease and cause it to break up into tiny globules which remain suspended in the solution (Fig. 10-18).

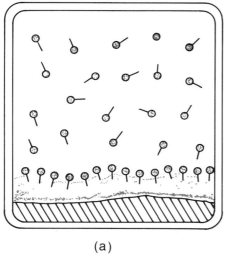 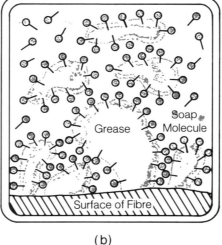

(a)   (b)

Figure 10-18
Detergent action
(a) Nonpolar tails of soap molecules become embedded in the grease layer.
(b) Polar heads of soap molecules assist in breaking up the grease layer, removing it from the surface of the fibre, and keeping the droplets suspended in the solution.

Sickle cell anemia is a disease in which the red blood cells become elongated (sickle-shaped). Abnormal hemoglobin, found in the red blood cells of people with this disease, differs from normal hemoglobin in only one small part of the molecule. The different group of atoms in the abnormal hemoglobin is nonpolar, whereas in normal hemoglobin the group is polar. Because the abnormal hemoglobin has the nonpolar group, it is not soluble in water. It forms a solid in the red blood cells, deforming them into elongated, sickle shapes. These deformed red blood cells can no longer carry oxygen. Sickled cells can also clog capillaries and interfere with the transport of blood.

Part Three Review Questions

1. Explain what is meant by the terms *electrolyte* and *nonelectrolyte*, and give an example of each.
2. Why is salt added to the roads in winter?
3. Why does sea water freeze below 0°C?

4. How would you test a substance to determine whether it is an electrolyte?
5. What is the difference between a weak electrolyte and a strong electrolyte?
6. What do solutions of all electrolytes contain that distinguishes them from nonelectrolytes?
7. What is meant by dissociation?
8. Write an equation for the dissociation of **a)** $BaCl_2$ in water; **b)** HBr in water.
9. What types of substances are usually considered electrolytes?
10. Explain why gasoline will not dissolve in water.

Part Four: Reactions of Ions in Solution

10.10 Precipitation

As we learned earlier in the chapter, different compounds have different degrees of solubility in water.

We will divide solutes into three arbitrary categories: soluble, sparingly or slightly soluble, and insoluble. We will consider a substance to be soluble if its solubility is greater than 10 g/L, as sparingly soluble if its solubility is between 1 g/L and 10 g/L, and as insoluble if its solubility is less than 1 g/L. Remember that these are arbitrary designations, and that nothing is completely insoluble. We can use a table of solubilities, Table 10-4, to predict which ionic compounds will tend to be soluble, sparingly soluble, or insoluble in water. From this solubility table you can see that carbonate ions (CO_3^{2-}), phosphate ions (PO_4^{3-}) and sulfide ions (S^{2-}) combine with practically all metal ions to form insoluble compounds.

Suppose we add an aqueous solution of $AgNO_3$ to an aqueous solution of NaCl. What will happen? We can put this information in the form of a chemical equation:

$$AgNO_3(aq) + NaCl(aq) \longrightarrow ?$$

Since we have solutions of these ionic compounds, we know that the ionic solids $AgNO_3$ and NaCl dissociated in water to form aqueous solutions of the ions. We may then write the following:

$$Ag^+(aq) + NO_3^-(aq) + Na^+(aq) + Cl^-(aq) \longrightarrow ?$$

Aqueous solutions of ionic compounds which react when mixed together undergo a double displacement reaction. The positive ions of one compound join with the negative ions of the other compound. Thus Ag^+ joins with Cl^- to produce AgCl, and Na^+ joins with NO_3^- to produce $NaNO_3$:

$$AgNO_3(aq) + NaCl(aq) \longrightarrow AgCl + NaNO_3$$

Table 10-4   Solubility Table for Ionic Compounds in Water

	NH_4^+	Na^+	K^+	Mg^{2+}	Ca^{2+}	Sr^{2+}	Ba^{2+}	Cr^{3+}	Mn^{2+}	Fe^{2+}	Fe^{3+}	Co^{2+}	Ni^{2+}	Cu^{2+}	Ag^+	Zn^{2+}	Cd^{2+}	Hg_2^{2+}	Hg^{2+}	Al^{3+}	Sn^{2+}	Sn^{4+}	Pb^{2+}
F^-	S	S	S	I	I	I	*	I	*	*	*	S	S	S	S	S	S	—	—	*	S	S	I
Cl^-	S	S	S	S	S	S	S	S	S	S	S	S	S	S	I	S	S	I	S	S	S	S	*
Br^-	S	S	S	S	S	S	S	S	S	S	S	S	S	S	I	S	S	I	*	S	S	S	*
I^-	S	S	S	S	S	S	S	—	S	S	—	S	S	—	I	S	S	I	I	S	S	S	I
NO_3^-	S	S	S	S	S	S	S	S	S	S	S	S	S	S	S	S	S	—	S	S	—	—	S
ClO_3^-	S	S	S	S	S	S	S	—	—	—	—	S	S	S	S	S	S	S	S	S	—	—	S
$C_2H_3O_2^-$	S	S	S	S	S	S	S	S	S	S	—	S	—	S	S	S	S	*	S	—	—	—	S
OH^-	S	S	S	I	*	*	S	—	I	I	—	I	I	I	—	I	I	—	—	I	—	—	I
S^{2-}	S	S	S	—	I	*	*	I	I	I	I	I	I	I	I	I	I	I	I	I	—	I	I
SO_4^{2-}	S	S	S	S	*	I	I	S	S	*	*	S	S	S	*	S	S	I	—	S	S	S	I
CO_3^{2-}	S	S	S	I	I	I	I	—	I	I	—	I	I	—	I	I	I	I	—	—	—	—	I
PO_4^{3-}	S	S	S	I	I	I	I	—	—	I	I	I	I	I	I	I	I	I	—	—	I	I	I

S = soluble   * = sparingly soluble   I = insoluble   — = solubility data not available

If we consult the solubility table, we find that $NaNO_3$ is soluble and that $AgCl$ is insoluble. Since $NaNO_3$ is soluble, it forms an aqueous solution which we indicate by writing $NaNO_3(aq)$. Since $AgCl$ is insoluble, most of it is present as a solid. We indicate this by writing $AgCl(s)$. The balanced chemical equation for the reaction can now be written:

$$AgNO_3(aq) + NaCl(aq) \longrightarrow AgCl(s) + NaNO_3(aq)$$

A chemical reaction in which one or more solids is produced as the result of mixing two aqueous solutions of ionic compounds is known as a **precipitation reaction**. The process of forming this solid is called **precipitation**. The solid that is formed is called a **precipitate**. Thus, in our example, $AgCl(s)$ is the precipitate.

We will use Table 10-4 to predict whether a precipitate is formed, even if we do not know actual concentrations of reactants or products. Compounds that are listed as insoluble or sparingly soluble will be considered as precipitates.

10.11 Writing Ionic and Net Ionic Equations

Let us examine the chemical equation for the precipitation reaction between aqueous solutions of $AgNO_3$ and $NaCl$:

$$AgNO_3(aq) + NaCl(aq) \longrightarrow AgCl(s) + NaNO_3(aq)$$

This represents the balanced chemical equation for the reaction. However, the solution of $AgNO_3$ does not contain $AgNO_3$ units but rather it has dissociated to form $Ag^+(aq)$ and $NO_3^-(aq)$. A similar argument can be made for the other aqueous solutions in the equation. These aqueous solutions contain separate hydrated ions. Therefore, it is appropriate to write the equation as

$$Ag^+(aq) + NO_3^-(aq) + Na^+(aq) + Cl^-(aq) \longrightarrow AgCl(s) + Na^+(aq) + NO_3^-(aq)$$

Note that we have written the silver chloride as $AgCl(s)$ and not $Ag^+(aq)$ and $Cl^-(aq)$. This is because most of the silver ions and the chloride ions remain joined as undissolved $AgCl(s)$ and are not separate ions. Equations in which soluble compounds that dissociate are written as ions are called **ionic equations**.

We notice that $Na^+(aq)$ and $NO_3^-(aq)$ ions appear as separate, hydrated ions on both sides of the equation. They have not taken part in the reaction at all. As many sodium ions are present at the end of the reaction as at the start. Similarly, as many nitrate ions are present at the end of the reaction as at the start. Since they do not participate directly in the reaction, they are called **spectator ions** and can be cancelled from each side of the equation:

$$Ag^+(aq) + \cancel{NO_3^-}(aq) + \cancel{Na^+}(aq) + Cl^-(aq) \longrightarrow AgCl(s) + \cancel{Na^+}(aq) + \cancel{NO_3^-}(aq)$$

After removing the spectator ions, we are left with the **net ionic equation**:

$$Ag^+(aq) + Cl^-(aq) \longrightarrow AgCl(s)$$

The net ionic equation for a reaction shows only those species which take part in the reaction. To summarize, we have written the balanced chemical equation:

$$AgNO_3(aq) + NaCl(aq) \longrightarrow AgCl(s) + NaNO_3(aq)$$

the ionic equation:

$$Ag^+(aq) + NO_3^-(aq) + Na^+(aq) + Cl^-(aq) \longrightarrow AgCl(s) + Na^+(aq) + NO_3^-(aq)$$

and the net ionic equation:

$$Ag^+(aq) + Cl^-(aq) \longrightarrow AgCl(s)$$

EXAMPLE PROBLEM 10-1

What would happen if we mixed aqueous solutions of KNO_3 and $NaCl$?

SOLUTION

$$KNO_3(aq) + NaCl(aq) \longrightarrow ?$$

The predicted products are KCl and $NaNO_3$:

$$KNO_3(aq) + NaCl(aq) \longrightarrow KCl + NaNO_3$$

Both KCl and $NaNO_3$ are soluble. Therefore, the balanced chemical equation would be:

$$KNO_3(aq) + NaCl(aq) \longrightarrow KCl(aq) + NaNO_3(aq)$$

The ionic equation is:

$$K^+(aq) + NO_3^-(aq) + Na^+(aq) + Cl^-(aq) \longrightarrow$$
$$K^+(aq) + Cl^-(aq) + Na^+(aq) + NO_3^-(aq)$$

The ionic equation indicates that all the ions present in the equation are spectator ions. No ions have been involved in a chemical reaction. Therefore, no chemical reaction has occurred. There is no net ionic equation, and we write:

$$KNO_3(aq) + NaCl(aq) \longrightarrow NO\ REACTION.$$

EXAMPLE PROBLEM 10-2

What would happen if we mixed aqueous solutions of $CaBr_2$ and $(NH_4)_2CO_3$?

SOLUTION

$$CaBr_2(aq) + (NH_4)_2CO_3(aq) \longrightarrow ?$$

Interchanging positive and negative ions and writing the correct formulas for the products, we obtain:

$$CaBr_2(aq) + (NH_4)_2CO_3(aq) \longrightarrow CaCO_3 + NH_4Br$$

This time the equation is not balanced. The balanced equation is:

$$CaBr_2(aq) + (NH_4)_2CO_3(aq) \longrightarrow CaCO_3 + 2NH_4Br$$

According to the solubility table, NH_4Br is soluble, and $CaCO_3$ is sparingly soluble. We will, therefore, consider $CaCO_3$ to be a precipitate. The balanced chemical equation is:

$$CaBr_2(aq) + (NH_4)_2CO_3(aq) \longrightarrow CaCO_3(s) + 2NH_4Br(aq)$$

The ionic equation is:

$$Ca^{2+}(aq) + 2Br^-(aq) + 2NH_4^+(aq) + CO_3^{2-}(aq) \longrightarrow$$
$$CaCO_3(s) + 2NH_4^+(aq) + 2Br^-(aq)$$

The net ionic equation is:

$$Ca^{2+}(aq) + CO_3^{2-}(aq) \longrightarrow CaCO_3(s)$$

In summary, when you are asked what happens when two aqueous solutions are mixed, assume a double displacement. Then use the following rules:

1. Switch positive ions in the compounds, and write the correct formulas for the products.
2. Write the chemical equation and balance it if necessary.
3. Determine the solubilities of the products from the solubility table. If the substance is soluble write (aq) after its chemical formula. If it is insoluble or sparingly soluble write (s) after its chemical formula.
4. Write the correct states for all reactants and products.
5. Write the ionic equation if necessary.
6. Write the net ionic equation if necessary.

PRACTICE PROBLEM 10-1

What happens when aqueous solutions of $Al(NO_3)_3$ and $BaCl_2$ are mixed?

PRACTICE PROBLEM 10-2

What happens when aqueous solutions of $(NH_4)_2S$ and $Cu(NO_3)_2$ are mixed?

10.12 Applications of Precipitation Reactions

Many applications of precipitation reactions are used by industry. For example, in one of the early steps in the production of aluminum, a precipitate of $Al(OH)_3$ is produced. In the production of nickel, iron impurities are precipitated as hydrated iron(III) oxide and are filtered off. The first step in the extraction of magnesium from sea water is the precipitation of Mg^{2+} as $Mg(OH)_2$.

Water is purified by treatment with aluminum sulfate and calcium hydrogen carbonate. The two compounds form aluminum hydroxide, which comes out of solution as a white gelatinous precipitate:

$$Al_2(SO_4)_3(aq) + 3Ca(HCO_3)_2(aq) \longrightarrow 2Al(OH)_3(s) + 6CO_2(g) + 3CaSO_4(s)$$

The aluminum hydroxide traps small particles which are impurities and carries them to the bottom of the tank of water.

Hard water contains Ca^{2+} and Mg^{2+} ions. Sodium carbonate can be used to soften the water. The CO_3^{2-} ions from the sodium carbonate will remove the Ca^{2+} and Mg^{2+} ions by precipitation:

$$Ca^{2+}(aq) + CO_3^{2-}(aq) \longrightarrow CaCO_3(s)$$

$$Mg^{2+}(aq) + CO_3^{2-}(aq) \longrightarrow MgCO_3(s)$$

10.13 Qualitative Analysis of Aqueous Solutions

Qualitative analysis of aqueous solutions is the identification of the chemicals present. Through use of the solubility rules and the formation of precipitates it is possible to separate and identify various ions in an aqueous solution. Although chemists would also be concerned with the presence of other chemicals in a solution, we will limit our discussion to the analysis of ions.

Why is it important to know what is present in an aqueous solution? In the manufacture of medicines, foods, and beverages it is crucial to monitor solutions produced during various stages as well as the final product to ensure that no harmful impurities are present. The producers of chemicals are concerned with the purity of their products. The purity of drinking water is always being checked. Government agencies monitor the environment for possible water pollution. Blood and urine are analyzed at hospitals. Solutions are analyzed at crime detection laboratories. Clearly, many people are concerned with identifying the chemicals in solutions.

In this section we will discuss some simple examples of qualitative analysis. For example, knowing the ions that might be present in an aqueous solution, we will use simple techniques to perform the analysis. Chemists and chemical technicians must analyze solutions which could contain a wide variety of substances. Often the analysis must be done on very small amounts, using complex and costly electronic equipment. However, some analysis is still done using the simple techniques which we will describe.

Suppose we wish to analyze a solution that might contain either or both Ca^{2+} and Ag^+ ions. One possible method of analysis would be to begin by adding hydrochloric acid to provide Cl^- ions. If both ions are present, possible products would be AgCl and $CaCl_2$. A look at Table 10-4 tells us that $CaCl_2$ is soluble. However, AgCl(s) is insoluble. Thus, if Ag^+ is present, a precipitate of AgCl forms. If Ag^+ is not present, there will be no precipitate. If no precipitate forms, we could then add Na_2CO_3(aq) to the unknown solution. If Ca^{2+} ions are present, a precipitate of $CaCO_3$ is formed. No precipitate indicates the absence of Ca^{2+} ions. If, in the first step, the addition of HCl(aq) forms a precipitate (indicating the presence of Ag^+ ions), we allow the precipitate to settle and pour off the liquid. We can now use this liquid to test for the presence of Ca^{2+} ions.

In qualitative analysis a few important rules should be followed. The separation of ions in the unknown solution should be as selective and as complete as possible. For instance, if we had added Na_2CO_3(aq) to the unknown solution as a first step, we would not know for sure the identity of the ions because there could be precipitates of both Ag_2CO_3 and $CaCO_3$. We would not have selectively separated one of the ions. If, as a first step, we had not added enough HCl(aq), there would not be a complete removal of all Ag^+ ions. This would mean that when we tested the liquid above the AgCl precipitate with Na_2CO_3(aq), any remaining Ag^+ ions would have precipitated as Ag_2CO_3(s). We would then erroneously have concluded that Ca^{2+} ions are present. The order in which chemicals are added during the analysis is important.

It is useful before actually doing the qualitative analysis to have a plan of attack written down. This plan is usually called a qualitative analysis scheme or flowchart. Such a scheme for the analysis of Ag^+ and Ca^{2+} ions is shown in Figure 10-19.

Figure 10-19
Qualitative analysis flowchart for a solution that could contain Ag^+ and Ca^{2+} ions

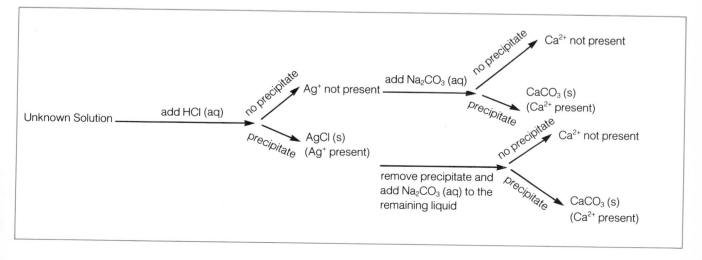

Table 10-5   Flame Colours of Various Metallic Ions

Metallic Ion	Flame Colour
Li^+	crimson
Na^+	yellow
K^+	violet
Ca^{2+}	orange-red
Ba^{2+}	yellow-green
Sr^{2+}	red
Cu^{2+}	green

Table 10-6   Colours of Aqueous Solutions of Various Ions

Ion	Colour of Aqueous Solution
Cu^{2+}	blue
Co^{2+}	pink
MnO_4^-	purple
Fe^{2+}	pale green
CrO_4^{2-}	yellow
$Cr_2O_7^{2-}$	orange
Ni^{2+}	green

Figure 10-20
Flowchart for the analysis of a solution that could contain Ba^{2+} and K^+ ions.

Another useful technique in qualitative analysis is the flame test. Certain metallic elements can be identified by the characteristic colour that aqueous solutions of their ions produce in a flame. The usual procedure is to dip the end of a platinum wire or nichrome wire into the solution containing the ion, and then to place the wire immediately into the tip of the inner flame of a burner. Table 10-5 indicates the flame colour for a given metallic ion.

Another aid in determining the presence of a particular ion is the fact that certain ions impart characteristic colours to their aqueous solutions. Thus, we might suspect the presence of an ion merely by noting the colour of its aqueous solution. Table 10-6 shows the colours of some ions in aqueous solution. These colours should not be taken as definite confirmation that the particular ions are present. Further analysis is required for confirmation of their presence.

Thus, in addition to forming precipitates, we can use flame tests and colours of solutions in our analysis. Flame tests should be used only when we are sure that the solution contains the ions of no more than one of the metals. If the ions of more than one metal are present, the flame colour of one ion could obscure that of the other.

EXAMPLE PROBLEM 10-3

How would we analyze a solution that might contain either or both of Ba^{2+} and K^+ ions?

SOLUTION

We could add $(NH_4)_2CO_3$ to the solution. If a precipitate formed, it could be only $BaCO_3$, since K_2CO_3 is soluble. If no precipitate formed, Ba^{2+} is not present. If there is a precipitate, we could use the solution above it to carry out a flame test to detect K^+. If there is no precipitate, the unknown solution itself could be used for the flame test. A violet flame constitutes a positive test for K^+. The absence of a violet flame indicates the absence of K^+. We chose to use $(NH_4)_2CO_3$ instead of Na_2CO_3 because the Na^+ would have interfered with the flame test. The qualitative analysis scheme is shown in Figure 10-20.

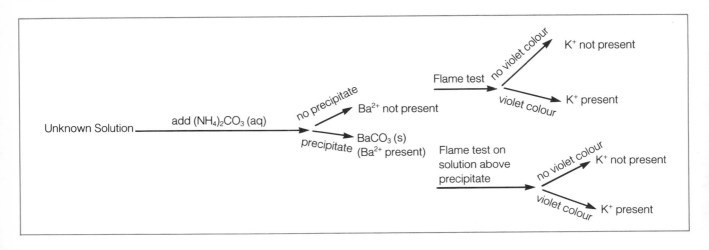

The following are some of the solutions that are commonly used to generate precipitates in qualitative analysis schemes:

$(NH_4)_2CO_3(aq)$ to produce CO_3^{2-} precipitates

$HCl(aq)$ to produce Cl^- precipitates

$H_2S(g)$ to produce S^{2-} precipitates

$NH_3(aq)$ (provides NH_4^+, OH^- in water) to produce OH^- precipitates.

Part Four Review Questions

1. Explain the meaning of the following terms: *precipitation reaction*, *precipitate*, *spectator ion*, and *qualitative analysis*.
2. A solution of NaBr is added to a solution of $AgNO_3$. **a)** Write the balanced chemical equation. **b)** Write the ionic equation. **c)** Write the net ionic equation. **d)** Identify any spectator ions.
3. A solution of K_2SO_4 is added to a solution of $Ca(NO_3)_2$. **a)** Write the balanced chemical equation. **b)** Write the ionic equation. **c)** Write the net ionic equation. **d)** Identify any spectator ions.
4. Why is the qualitative analysis of aqueous solutions important?
5. How would you analyze a solution for the presence of Na^+ ions and/or Ag^+ ions? Draw a qualitative analysis flow chart.

Chapter Summary

- A solution is a homogeneous mixture.
- The solute dissolves in the solvent of a solution.
- *Dilute* and *concentrated* refer to the relative amount of solute in a solution.
- There are seven possible types of solution.
- An aqueous solution contains water as the solvent.
- A state of dynamic equilibrium has two opposing processes occurring at equal rates.
- Solution equilibria are dynamic.
- A saturated solution contains the maximum amount of solute that can be dissolved.
- A saturated solution has reached a state of solution equilibrium.
- An unsaturated solution contains less than the maximum amount of solute that can be dissolved.
- A supersaturated solution contains more dissolved solute than it would if it were saturated.
- The formation of a solution may be endothermic or exothermic.
- Solubility refers to the maximum amount of solute that can be dissolved under specified conditions of temperature and pressure.
- Henry's Law states that the solubility of a gas in a liquid is directly proportional to the pressure of the gas above the liquid.

- The solubility of a gas in a liquid decreases with increasing temperature.
- The solubility of a solid in a liquid generally increases with increasing temperature.
- Solubility curves show solubility of a solute over a wide temperature range.
- Aqueous solutions of electrolytes conduct electricity.
- Aqueous solutions of nonelectrolytes do not conduct electricity.
- The behaviour of electrolytes is explained by assuming that electrolytes exist in solution as charged particles called ions.
- Dissociation is the separation or breaking apart of solute particles, resulting in ions in solution.
- Strong electrolytes dissociate almost 100%.
- Weak electrolytes dissociate to a small extent.
- "Like dissolves like."
- A precipitate is an insoluble product of a chemical reaction which is being carried out in solution.
- Ionic equations are equations in which soluble compounds that dissociate are written as ions.
- Net ionic equations show only those ions which have reacted.
- Qualitative analysis of aqueous solutions determines the presence or absence of certain ions.

Key Terms

solution	supersaturated solution	dissociation
solvent	unsaturated solution	precipitation reaction
solute	solubility	precipitation
dilute	solubility curve	precipitate
concentrated	electrolyte	ionic equation
aqueous solution	nonelectrolyte	spectator ion
dynamic equilibrium	strong electrolyte	net ionic equation
saturated solution	weak electrolyte	qualitative analysis

Test Your Understanding

Each of these statements or questions is followed by four responses. Choose the correct response in each case. (Do not write in this book.)

1. Which of the following is not a property of solutions?
   a) Solutions are homogeneous.
   b) Solutions contain more than one phase.
   c) Solutions are mixtures.
   d) Solutions can be separated by physical means.
2. Which of the following is true about solutions?
   a) They are pure substances.
   b) They are heterogeneous.
   c) They obey the Law of Definite Composition.
   d) They are mixtures.

3. Which of the following can form a solution?
   a) sawdust and water   b) oil and water
   c) gasoline and water   d) sugar and water

4. Which of the following pairs could form a solution?
   a) element, element
   b) element, compound
   c) compound, compound
   d) all of these

5. Which of the following is true for an aqueous solution of sugar at equilibrium?
   a) It is a saturated solution.
   b) The rate of dissolving is greater than the rate of crystallizing.
   c) The rate of crystallizing is greater than the rate of dissolving.
   d) If more sugar is added, it will dissolve.

6. You are given a solution in which no solid is visible. Which of the following statements is true?
   a) The solution is saturated.
   b) The solution is supersaturated.
   c) The solution is unsaturated.
   d) There is not enough information to decide.

7. Which of the following processes, during the formation of a solution, is exothermic?
   a) the separation of solute particles
   b) the separation of solvent particles
   c) the attraction of solute particles to solvent particles
   d) all of these

8. For which of the following is the solubility most affected by pressure?
   a) a gas dissolving in a liquid
   b) a gas dissolving in a gas
   c) a solid dissolving in a liquid
   d) a liquid dissolving in a liquid

9. For which of the following combinations does an increase in temperature usually cause an increase in solubility?
   a) a gas dissolving in a solid
   b) a solid dissolving in a liquid
   c) a gas dissolving in a liquid
   d) a gas dissolving in a gas

10. To increase the number of sugar molecules in a saturated solution which is in equilibrium with some undissolved sugar, you would
    a) cool the solution.
    b) add more solvent.
    c) shake the solution.
    d) evaporate some solvent from the solution.

11. Which of the following changes the solubility of a solid in a saturated solution?
    a) stirring the solution   b) heating the solution
    c) adding more solute   d) grinding the solid

12. Which of the following is a property of both electrolytes and nonelectrolytes?
   a) They lower the freezing point and raise the boiling point of the solvent.
   b) They lower both the freezing point and the boiling point of the solvent.
   c) They raise both the freezing point and the boiling point of the solvent.
   d) They raise the freezing point and lower the boiling point of the solvent.

13. A concentrated solution of Compound X is tested with a conductivity apparatus. The light bulb glows dimly. What may we conclude?
   a) Compound X is a weak electrolyte.
   b) Compound X is a strong electrolyte.
   c) Compound X is a nonelectrolyte.
   d) Compound X does not dissociate.

14. Which of the following is a nonelectrolyte?
   a) carbon tetrachloride   b) hydrogen chloride
   c) sodium chloride   d) sodium hydroxide

15. Which of the following pairs of aqueous solutions is least likely to produce a precipitate when mixed?
   a) $AgNO_3$ and KCl b) Na_3PO_4 and $MgBr_2$
   c) Na_2CO_3 and NH_4Cl d) $CaCl_2$ and KF

16. Which of the following pairs of aqueous solutions would be most likely to produce a precipitate when mixed?
   a) KCl and $Mg(NO_3)_2$ b) Na_2S and $Cr(NO_3)_3$
   c) $MgCl_2$ and Na_2SO_4 d) KF and $NiBr_2$

17. What are the spectator ions in the reaction

$$Ba^{2+}(aq) + 2NO_3^-(aq) + 2Na^+(aq) + SO_4^{2-}(aq) \longrightarrow$$
$$BaSO_4(s) + 2Na^+(aq) + 2NO_3^-(aq)?$$

   a) Ba^{2+} and NO_3^- b) Na^+ and SO_4^{2-}
   c) Na^+ and NO_3^- d) Ba^{2+} and SO_4^{2-}

18. Which of the following is a net ionic equation?
   a) $Ba(NO_3)_2 + K_2SO_4 \longrightarrow BaSO_4 + 2KNO_3$
   b) $Ba^{2+}(aq) + SO_4^{2-}(aq) \longrightarrow BaSO_4(s)$
   c) $Ba^{2+}(aq) + 2NO_3^-(aq) + 2K^+(aq) + SO_4^{2-}(aq) \longrightarrow$
      $BaSO_4(s) + 2K^+(aq) + 2NO_3^-(aq)$
   d) $Ba(NO_3)_2(aq) + K_2SO_4(aq) \longrightarrow BaSO_4(s) + 2KNO_3(aq)$

19. For which of the following solutions could the metal ions be separated from one another by adding a solution of sodium hydroxide?
   a) A solution containing $Co(NO_3)_2$ and $Al(NO_3)_3$
   b) A solution containing $Ba(NO_3)_2$ and $KClO_3$
   c) A solution containing $Mg(NO_3)_2$ and $Ni(NO_3)_2$
   d) A solution containing KNO_3 and $Fe(NO_3)_2$

20. For which of the following solutions could the metal ions be separated from one another by adding a solution of hydrochloric acid?
   a) a solution containing $NaNO_3$ and $Hg_2(NO_3)_2$
   b) a solution containing $Ca(NO_3)_2$ and $Sr(NO_3)_2$
   c) a solution containing $AgNO_3$ and $Hg_2(NO_3)_2$
   d) a solution containing $Fe(NO_3)_2$ and $Fe(NO_3)_3$

Review your Understanding

1. In what ways does a solution differ from a compound?
2. Which actions in the dissolving process absorb energy?
3. The dissolving of most solids in liquids is endothermic. Explain why the temperature decreases as the solution forms.
4. Why isn't a mixture of sand in water considered a solution?
5. Describe the concept of dynamic equilibrium as it applies to a saturated solution.
6. a) Eighty grams of KNO_3 are added to 100 mL of water at 40°C. Approximately how many grams of KNO_3 dissolve?
   b) How could you get the rest of the KNO_3 to dissolve?
7. Consider the following solubility data:

Temperature (°C)	0	20	40	60	70	80	100
Solubility (g $(NH_4)_2SO_4$/100 g H_2O)	70.7	75.7	81.2		91.7	95.6	103.3
Solubility (g NH_4Br/100 g H_2O)	59.7	75.5	91.8	107.8		126.0	145.9
Solubility (g $NaClO_3$/100 g H_2O)	79.9	98.0	118.2	142.7		171.5	206.6

   a) Draw solubility curves for $(NH_4)_2SO_4$, NH_4Br, and $NaClO_3$.
   b) What is the predicted solubility of each at 50°C?
   c) What is the maximum number of grams of each that could be dissolved in 100 g of water at 15°C?
   d) What would happen to a saturated solution of each at 40°C if it were cooled to 30°C?
   e) How many grams of each could be dissolved in 250 g of water at 80°C?
   f) What minimum temperature would be required to dissolve 100 g of each in 100 g of water?
8. How would you experimentally distinguish a strong electrolyte from a weak electrolyte?
9. There is much public concern about oil spills. Why doesn't most of the oil just dissolve in the ocean?
10. Write an equation for the dissociation of HI in water.
11. Why will a nonelectrolyte not conduct electricity when it is dissolved in water?
12. Describe the dissolving of the ionic substance, KBr, in water to produce a solution which conducts electricity. What is the role of the water molecules in this process?
13. What is meant by the equation $CaCl_2(s) \longrightarrow Ca^{2+}(aq) + 2Cl^-(aq)$?
14. Why does a nonpolar solute not dissolve in a polar solvent?
15. Why does an ionic substance not dissolve in a nonpolar solvent?
16. Why was carbon tetrachloride used in the dry cleaning industry? Why is it no longer used in this industry?

17. Write equations for the dissociation of the following in water: **a)** $MgCl_2$; **b)** KCl; **c)** $Ca(NO_3)_2$; **d)** $Ba(OH)_2$; **e)** $Al_2(SO_4)_3$.

18. Solid naphthalene will dissolve in nonpolar octane. What can we conclude about the nature of naphthalene?

19. A compound, X, dissolves readily in water. Can we conclude that X is an electrolyte? Explain.

20. Two polar substances of equal concentration were tested with a conductivity apparatus. Compound A caused the bulb to glow brightly. Compound B caused the bulb to glow dimly. Comment on the nature of these two compounds.

21. For each of the following mixtures of two aqueous solutions, **a)** write the balanced chemical equation; **b)** write the ionic equation; **c)** write the net ionic equation; and **d)** indicate any spectator ions:

    $BaCl_2$ and Na_2SO_4
    $AgNO_3$ and KI
    $AlBr_3$ and LiOH
    $Fe(NO_3)_2$ and $(NH_4)_2CO_3$
    KBr and $MgCl_2$
    Na_3PO_4 and $CaCl_2$
    $Ca(NO_3)_2$ and LiF
    $Pb(NO_3)_2$ and NaI

22. Using a flow chart, explain how you would perform a qualitative analysis on the aqueous solutions indicated: **a)** Ag^+ and/or Li^+; **b)** Mg^{2+} and/or Ba^{2+}; **c)** Sr^{2+} and/or Ca^{2+}; **d)** Cu^{2+} and/or Ba^{2+}; **e)** Cu^{2+} and/or K^+.

23. Draw a qualitative analysis scheme for the following aqueous solutions: **a)** Mg^{2+} and/or Na^+; **b)** Ca^{2+} and/or Cu^{2+}.

24. The solution in each part of this question contains only one metal ion. The colour of the solution or the colour for a flame test is provided. Identify the most likely ion that is present in the solution.

	Colour of solution	Colour of flame
a)	blue	
b)	green	
c)		crimson
d)		violet
e)	pink	

25. **a)** Given $H_2S(g)$ and $(NH_4)_2CO_3(aq)$, describe the analysis of a solution of Ca^{2+} and/or Li^+.

    b) Given HCl(aq), describe the analysis of a solution of Cr^{3+} and/or Mg^{2+}.

26. Use the solubility rules to predict whether each of the following compounds is soluble, insoluble, or sparingly soluble in water: **a)** $PbSO_4$; **b)** K_2CO_3;   **c)** $Cu(NO_3)_2$;   **d)** KOH;   **e)** Ag_2S;   **f)** SrS;   **g)** $FePO_4$; **h)** $Ba(OH)_2$;   **i)** $PbCl_2$;   **j)** MgF_2.

27. Describe a test that would enable you to distinguish between $AgNO_3(s)$ and $Cu(NO_3)_2(s)$.

28. Both NH_4Cl and KCl are white solids. How would you distinguish between them?

Apply Your Understanding

1. Why do you think there are only seven solvent-solute combinations listed in Table 10-1?
2. Is it possible for a solution to be both saturated and dilute? Explain.
3. Four solutions of KNO_3, each containing 100 mL of water at 30°C, were analyzed. Shown below are the number of grams of dissolved KNO_3 found in each. Determine what type of solution (that is, unsaturated, saturated, and supersaturated) is present in each case and give your reason.

Solution A	70 g	no solid visible
Solution B	20 g	no solid visible
Solution C	65 g	solid visible
Solution D	55 g	no solid visible

4. What test could you perform to determine whether a solution is unsaturated, saturated, or supersaturated? Explain your answer.
5. Why is there more dissolved nitrogen than dissolved oxygen in any pond, lake, or river?
6. Why is it pointless to add NaCl to icy roads when the temperature is below −20°C? Explain.
7. Why is it dangerous to operate an electrical appliance in a wet area?
8. Originally Arrhenius believed that there must be the same number of positive and negative ions in a solution. However, the modern theory of electrolytes states that the total positive ionic charge must equal the total negative ionic charge. Is there a difference between these two statements? If so, why did Arrhenius' original theory have to be modified?
9. Will we ever find a chemical that will dissolve all substances (a universal solvent)? Give reasons for your answer.
10. A suggested treatment for poisoning caused by ingestion of soluble lead compounds is to give $Na_2SO_4(aq)$ or $MgSO_4(aq)$ immediately. Why should this procedure work?
11. Barium is used in X-ray examinations to outline the digestive system because it blocks X-rays so readily. The barium that is used is not in the form of $Ba^{2+}(aq)$, however, since $Ba^{2+}(aq)$ ions are very poisonous. Instead, $BaSO_4(s)$ is used. Why is this a safe procedure?

Investigations

1. Search your home for substances that could be classified as solutions. Write a report on your findings.
2. Is blood a solution? Write a research report on the composition of blood, including current research into the synthesis of artificial blood.
3. Write a research report on the methods used by industry and government to deal with chemical waste management.
4. Devise and perform experiments to determine the factors that affect the rate at which a solid dissolves in a liquid.
5. Both sodium chloride and calcium chloride are used as road salt. Devise an experiment to determine which is more effective in preventing the formation of ice.
6. Debate the proposition, "Dilution is no solution for pollution."

11

The Mole and Its Use

CHAPTER CONTENTS

Part One: Introducing the Mole
   11.1   Avogadro's Number and the Mole
   11.2   Determining Avogadro's Number
   11.3   The Mole Triangle
   11.4   Molecular Mass and Formula Mass
   11.5   Percentage Composition

Part Two: The Mole and Chemical Formulas
   11.6   Expanding the Mole Concept
   11.7   Calculating Empirical Formulas
   11.8   Molecular Formula Determination

Part Three: The Mole and Chemical Equations
   11.9   Relating the Mole to Chemical Equations
   11.10 Mass Calculations Involving Chemical Equations
   11.11 Limiting Reagents in Mass Calculations
   11.12 The Yield of a Chemical Reaction
   11.13 The Mole and Industrial Chemical Reactions

Part Four: Calculations Involving Solutions
   11.14 Concentration Measured in Moles Per Litre
   11.15 Concentration Measured in Percentage by Mass

Chemists spend much of their time studying compounds and chemical reactions. When they discover a new compound, one of the questions they ask is, "What is the formula of this compound?" When they study a chemical reaction, one of the questions they need to ask is, "How much of each substance is required for this reaction?"

Why can we say that the mole is the chemist's best friend? In other words, why is the concept of the mole one of the most important in chemistry? How can we use the mole concept to determine mass relationships implied in chemical formulas and chemical equations?

Suppose that you were the manager of a thermal electric generating station (Fig. 11-1). If sulfur-containing coal is burned at your plant, sulfur dioxide (a major cause of acid rain) is one of the exhaust gases. The sulfur dioxide can be removed by reacting it with calcium oxide. As the plant manager, you would have to calculate how much calcium oxide would be required each year. In order to do this calculation, you would have to use the mole concept.

Figure 11-1
This coal-fired thermal electric generating station at Lingan, Nova Scotia produces 600 MW of power.

Key Objectives

When you have finished studying this chapter, you should be able
1. to do calculations involving moles, grams of substance, numbers of atoms (or molecules or formula units), and molecular (or formula) masses.
2. to calculate the molecular mass (or formula mass), molar mass, and percentage composition of a substance, given its formula.
3. to use the molar mass of a substance to convert from grams to moles or from moles to grams.
4. to calculate the empirical formula of a compound, given its percentage composition.
5. to calculate the molecular formula of a compound, given its molecular mass and either its empirical formula or its percentage composition.
6. to calculate the corresponding number of grams or moles of any other substance appearing in the equation for a reaction, given a balanced equation and the number of grams or moles of one or two substances involved in the reaction.
7. to identify the limiting reagent and calculate the amount of excess reagent, the theoretical yield, and the percent yield in a chemical reaction.
8. to calculate the third quantity, given any two of mass of solute, volume of solution, and concentration of solution in moles per litre.

Part One: Introducing the Mole

11.1 Avogadro's Number and the Mole

One of the most important concepts to follow from the theory of the atom is that of atomic mass. Actually, the mass of an atom is extremely small. For example, the mass of an atom of the most common isotope of oxygen is 2.66×10^{-23} g. The size of this number indicates that the gram is too large a unit to be used for determining the mass of an oxygen atom. In fact, the gram is much too large a mass unit to be used for determining the mass of any atom.

As we have learned, the atomic mass unit (u) is the unit of mass we use when obtaining the mass of any atom. For many years, oxygen was used as the standard reference element for atomic masses. This was a logical choice, because oxygen combines readily with most other elements. Thus, it was easy to compare the masses of atoms of most elements with the mass of an oxygen atom.

There was a problem, however, because oxygen consists of a mixture of three isotopes (99.762% $^{16}_{8}O$, 0.038% $^{17}_{8}O$, and 0.200% $^{18}_{8}O$). In developing a scale of atomic masses, physicists assigned a mass of 16.0000 u to the $^{16}_{8}O$ isotope, while chemists assigned a mass of 16.0000 u to the naturally occurring mixture of isotopes. The result was two different scales of atomic masses with slightly different values for all the elements.

In 1961 a new standard reference element was chosen to eliminate the dual scales. The carbon-12 atom was assigned a mass of exactly 12 u. Therefore, one atomic mass unit is 1/12 the mass of a carbon-12 atom.

The mass of the most common oxygen isotope is about $^4/_3$ as great as the mass of carbon-12. Therefore, the mass of an atom of oxygen is about 16 u ($12 \times \frac{4}{3} = 16$). A more accurate value is 15.994 91 u. This is an extremely small amount of mass. No balance is capable of measuring such a small quantity of matter. For this reason, it is not possible to measure the masses of individual atoms. However, an important aspect of chemistry is calculating mass relationships involved in chemical reactions. Chemists frequently carry out chemical reactions between measured quantities of substances. They must be able to calculate the amount of one substance needed to react with or to produce a required quantity of another substance.

Let us consider a simple chemical reaction. Iron and sulfur react to form iron(II) sulfide. If we wanted to use just enough sulfur to react with a measured amount of iron, it would be useful to know the relationship between the mass of iron and the mass of sulfur involved in the reaction. That is, how much iron reacts with how much sulfur? The chemical equation for this reaction is

$$Fe(s) + S(s) \longrightarrow FeS(s)$$

Since the iron atoms and the sulfur atoms react on a one-to-one basis to form iron(II) sulfide, a mass relationship in this reaction can be determined by looking up the atomic masses of iron and sulfur in a set of tables. Thus, 56 u of iron reacts with 32 u of sulfur to produce iron(II) sulfide. However, it is not possible to measure 56 u of iron or 32 u of sulfur, so this mass relationship is not particularly useful.

Since it is impossible to react only one iron atom with one sulfur atom, a large number of iron atoms must be reacted with an *equal* number of sulfur atoms. The large number which is convenient to use is called Avogadro's number, in honour of the Italian physicist Amedeo Avogadro (Fig. 11-2). It is the number of carbon-12 atoms needed to have a mass of 12.0000 g. Avogadro's number of carbon-12 atoms has a mass in grams that is numerically equal to the atomic mass of carbon-12 in atomic mass units. In fact, Avogadro's number of atoms of *any* element has a mass in grams numerically equal to the atomic mass of the element in atomic mass units. Thus, if the atomic mass of iron is 56 u, the mass of Avogadro's number of iron atoms is 56 g. If the atomic mass of sulfur is 32 u, the mass of Avogadro's number of sulfur atoms is 32 g. As a result, a useful mass relationship for the reaction of iron and sulfur is that 56 g of iron (Avogadro's number of iron atoms) reacts with 32 g of sulfur (Avogadro's number of sulfur atoms) to produce iron(II) sulfide.

Biography

AMEDEO AVOGADRO was born in Turin, Italy, where he obtained degrees in philosophy at the age of 13, in jurisprudence at 16, and in church law four years later. He practised law for about three years, but then tired of the petty squabbles in the courts and began a serious study of mathematics and physics.

In 1809 Avogadro became professor of physics at the Royal College at Vercelli. Two years later he published the famous hypothesis which bears his name. Not a single scientist in the world commented on it! It was partly because of his quiet, unassuming nature that his hypothesis was not immediately accepted and adopted.

Despite the lack of attention paid to his hypothesis, Avogadro continued to lead a busy life. He persisted with his studies in physics, especially of capillary action and thermal expansion of liquids. He filled many public offices, especially those connected with public education, national statistics, meteorology, and weights and measures. He did not care for prominent positions or public honours. Rather, he led the life of a philosopher, wholly occupied in his studies but not forgetting his duties as a citizen and as a father.

In 1820 Avogadro was appointed professor of mathematical physics at the University of Turin, where he worked quietly until his retirement. He died peacefully in his 80th year.

Figure 11-2
Amedeo Avogadro (1776-1856)—
Now famous for Avogadro's number
and Avogadro's Hypothesis

The Mole Concept

The quantity of a substance which is useful for determining mass relationships in chemical reactions is called the mole. The definition of the atomic mass unit is based on the carbon-12 atom, and so is the definition of the mole. A **mole** of a substance is the quantity of that substance which contains the same number of chemical units (atoms, molecules, formula units, or ions) as there are atoms in exactly 12 g of carbon-12. Of course, 12 g is the mass of carbon which is numerically equal to the atomic mass of carbon, and therefore 12 g of carbon contains Avogadro's number of carbon atoms. A mole can thus be considered to be the mass of any substance which contains Avogadro's number of chemical units of the substance.

The **molar mass** of a substance is the mass of one mole of the substance. In the case of an element, the molar mass is the atomic mass of the element expressed in grams. The molar mass of carbon is 12 g, and this mass of carbon consists of Avogadro's number of carbon atoms. The molar mass of iron is 56 g, and this mass of iron consists of Avogadro's number of iron atoms. The molar mass of sulfur is 32 g, and this mass of sulfur consists of Avogadro's number of sulfur atoms.

11.2 Determining Avogadro's Number

How large is Avogadro's number? To determine this, we must devise an experiment to count the number of particles in one mole of a substance. Several methods are available, but a simple method involves the use of radioactive substances which emit alpha particles. In such cases, we can use a Geiger counter to count the alpha particles that are emitted in a given time. Since alpha particles are helium nuclei, we can design the experiment so that the charges on the alpha particles are neutralized and helium atoms are produced.

$$He^{2+} + 2e^- \longrightarrow He$$

We then measure the volume of helium gas produced at standard temperature and pressure (0°C and 101.3 kPa). In Section 14.4 we shall learn that one mole of any gas will occupy 22.4 L at standard temperature and pressure. We now have enough information to calculate Avogadro's number:

$$N \text{ particles/mol} = \frac{\text{alpha particles counted}}{\text{litres of He produced}} \times \frac{22.4 \text{ L}}{1 \text{ mol}}$$

Experiments like this have shown that Avogadro's number is 6.02×10^{23}. Therefore, if the atomic mass of iron is 55.8 u, then the mass of 6.02×10^{23} iron atoms is 55.8 g. If the atomic mass of sulfur is 32.1 u, the mass of 6.02×10^{23} sulfur atoms is 32.1 g. If the atomic mass of any element is x u, the mass of 6.02×10^{23} atoms of that element is x g. One mole of an element consists of 6.02×10^{23} atoms of the element. The mass of one mole of the element is the atomic mass of the element expressed in grams.

PRACTICE PROBLEM 11-1

In an actual experiment, 1.82×10^{17} alpha particles were counted. The helium gas which was produced occupied a volume of 0.006 73 mL at 0°C and 101.3 kPa. Show that this experiment gives a value of 6.06×10^{23} for Avogadro's number.

11.3 The Mole Triangle

The mole is the mass of Avogadro's number of atoms or molecules. Avogadro's number is very large: 6.022×10^{23}. In addition, the mass in grams of Avogadro's number of atoms of any element is numerically equal to the atomic mass of an atom of the element in atomic mass units. Similarly, the mass in grams of Avogadro's number of molecules of any compound is numerically equal to the mass of a molecule of the compound in atomic mass units.

These relationships can be visualized with the help of the mole triangle (Fig. 11-3). The two quantities at the base of the triangle can be converted into moles or vice versa by multiplying by the conversion factors shown on the sides. A conversion from one side of the base to the other must be by way of the apex of the triangle.

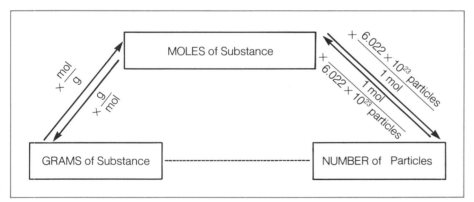

Figure 11-3
The mole triangle

For example, what is the mass of 3 mol of atoms of an element with an atomic mass of 9.0 u? We wish to convert moles of substances to grams of substance. In Figure 11-3, this conversion involves going from the apex of the triangle to the left end of its base. We see that in doing so, we must multiply by the conversion factor "g/mol." Our argument would therefore be:

1 atom has a mass of 9.0 u
∴ 1 mol of atoms has a mass of 9.0 g
∴ The conversion factor is 9.0 g/mol

$$3 \ \text{mol} \times \frac{9.0 \ \text{g}}{1 \ \text{mol}} = 27 \ \text{g}$$

∴ The mass of 3 mol of atoms is 27 g.

EXAMPLE PROBLEM 11-1

If a certain element has an atomic mass of 9.0 u, what is the mass in grams of one-third mole of atoms of the element?

SOLUTION

We want to convert moles to grams, so we multiply by the conversion factor.

1 atom has a mass of 9.0 u
∴ 1 mol of atoms has a mass of 9.0 g
∴ The conversion factor is 9.0 g/mol

$$\frac{1}{3} \, \text{mol} \times \frac{9.0 \, \text{g}}{1 \, \text{mol}} = 3.0 \, \text{g}$$

EXAMPLE PROBLEM 11-2

A certain element has an atomic mass of 20.2 u. How many moles of atoms are present in 50.5 g of the element? How many in 3.00 g of the element?

SOLUTION

In this case we wish to convert from grams to moles, so we multiply by the conversion factor, 1 mol/20.2 g.

$$50.5 \, \text{g} \times \frac{1 \, \text{mol}}{20.2 \, \text{g}} = 2.50 \, \text{mol}$$

$$3.00 \, \text{g} \times \frac{1 \, \text{mol}}{20.2 \, \text{g}} = 0.149 \, \text{mol}$$

There are 2.50 mol of atoms in 50.5 g of the element and 0.149 mol of atoms in 3.00 g of the element.

EXAMPLE PROBLEM 11-3

How many zinc atoms are present in 20.0 g of zinc?

SOLUTION

From the table of relative atomic masses on the inside front cover of this text, we find that the atomic mass of zinc is 65.4 u. Therefore the mass of one mole of zinc is 65.4 g. According to the instructions for using the mole triangle, we must go through the apex of the triangle when converting from one side of the base to the other. We will therefore first convert grams of zinc to moles of zinc, and then moles of zinc to atoms of zinc.

$$20.0 \, \text{g Zn} \times \frac{1 \, \text{mol Zn}}{65.4 \, \text{g Zn}} = 0.306 \, \text{mol Zn}$$

$$0.306 \, \text{mol Zn} \times \frac{6.02 \times 10^{23} \, \text{atoms Zn}}{1 \, \text{mol Zn}} = 1.84 \times 10^{23} \, \text{atoms Zn}$$

Note: Dimensional analysis allows solution of this problem in one step by simply ensuring that all units except the desired ones cancel.

$$20.0 \text{ g Zn} \times \frac{1 \text{ mol Zn}}{65.4 \text{ g Zn}} \times \frac{6.02 \times 10^{23} \text{ atoms Zn}}{1 \text{ mol Zn}} = 1.84 \times 10^{23} \text{ atoms Zn}$$

Whichever method we use, we conclude that there are 1.84×10^{23} atoms of zinc in 20.0 g of the element.

PRACTICE PROBLEM 11-2

If 2.50 mol of atoms of an element have a mass of 60.8 g, what is the average mass of a single atom of that element? (*Answer*: 24.3 u)

PRACTICE PROBLEM 11-3

How many moles of atoms are present in 100 g of an element of atomic mass 39.1 u? (*Answer*: 2.56 mol)

PRACTICE PROBLEM 11-4

How many atoms of carbon are there in 4.00 mol of carbon? (*Answer*: 2.41×10^{24} atoms)

PRACTICE PROBLEM 11-5

How many atoms of sodium are there in 46.0 g of sodium? (*Answer*: 1.20×10^{24} atoms)

11.4 Molecular Mass and Formula Mass

When atoms combine to form molecules, mass is neither created nor destroyed. Thus the mass of a molecule such as carbon dioxide (CO_2) can be obtained by adding the masses of each atom in the molecule. In the case of carbon dioxide there is one carbon atom (12.0 u) and two oxygen atoms (16.0 u each) in every molecule. Therefore the *molecular mass* of carbon dioxide is $12.0 + 16.0 + 16.0 = 44.0$ u.

In the case of a crystal of ionically-bonded material like $NaCl$, CaF_2, or AlF_3 the formulas do not describe molecules. The formula $NaCl$ should suggest to you that a crystal of sodium chloride is made up of positive sodium ions attracted to negative chloride ions and that there is one sodium ion for every one chloride ion in the crystal. The formula CaF_2 should suggest that there are two fluoride ions for every one calcium ion in a crystal of calcium fluoride; and the formula $Al(NO_3)_3$ should suggest that there are three nitrate ions for every one aluminum ion in a crystal of aluminum nitrate. Thus, in solid ionic materials like sodium choride there are no molecules—only arrangements of ions. Therefore, we do not speak of the molecular mass of NaCl. We speak of the *formula mass*. However, there is no problem in obtaining the formula mass of sodium chloride. One merely adds the mass of sodium (23.0 u) to the mass of chlorine (35.5 u) and obtains a formula mass of 58.5 u for NaCl.

EXAMPLE PROBLEM 11-4

Alanine, $C_3H_7NO_2$, is a compound which is one of the building blocks of protein. What is the molecular mass of alanine?

SOLUTION

$$3\,C = 3 \times 12.0\ u = 36.0\ u$$
$$7\,H = 7 \times 1.0\ u = 7.0\ u$$
$$1\,N = 1 \times 14.0\ u = 14.0\ u$$
$$2\,O = 2 \times 16.0\ u = 32.0\ u$$
$$\text{The molecular mass of alanine} = 89.0\ u$$

EXAMPLE PROBLEM 11-5

What is the formula mass of aluminum nitrate, $Al(NO_3)_3$?

SOLUTION

$$1\,Al = 1 \times 27.0\ u = 27.0\ u$$
$$3\,N = 3 \times 14.0\ u = 42.0\ u$$
$$9\,O = 9 \times 16.0\ u = 144.0\ u$$
$$\text{The formula mass of aluminum nitrate} = 213.0\ u$$

PRACTICE PROBLEM 11-6

What is the molecular mass of H_2SO_4? What is the formula mass of $Ca(HSO_4)_2$? (*Answers*: 98.1 u; 234.3 u)

11.5 Percentage Composition

In Example Problem 11-4 the molecular mass of alanine was calculated to be 89.0 u. How much of this mass was contributed respectively by the 3 carbon atoms, by the 7 hydrogen atoms, by the nitrogen atom, and by the 2 oxygen atoms? In other words, what is the *percentage composition* of alanine? The percentage composition by mass of a substance can be calculated as shown in Example Problems 11-6, 11-7, and 11-8.

EXAMPLE PROBLEM 11-6

What is the percentage composition of hydrogen peroxide, H_2O_2?

SOLUTION

The total atomic mass of hydrogen in H_2O_2 is 2×1.0 u $=$ 2.0 u
The total atomic mass of oxygen in H_2O_2 is 2×16.0 u $=$ 32.0 u

The total molecular mass of $H_2O_2 = \overline{34.0}$ u

$$\% \text{ H} = \frac{\text{mass of H in 1 molecule}}{\text{mass of molecule}} \times 100\%$$

$$= \frac{2.0 \text{ u}}{34.0 \text{ u}} \times 100\% = 5.9\%$$

$$\% \text{ O} = \frac{\text{mass of O in 1 molecule}}{\text{mass of molecule}} \times 100\%$$

$$= \frac{32.0 \text{ u}}{34.0 \text{ u}} \times 100\% = 94.1\%$$

The sum of the two percentages adds up to 100.0%. This gives us confidence that we have made no mathematical errors. Hydrogen peroxide is therefore 5.9% H and 94.1% O.

EXAMPLE PROBLEM 11-7

What is the percentage composition of nitric acid, HNO_3?

SOLUTION

The total atomic mass of hydrogen in HNO_3 is 1×1.0 u $=$ 1.0 u
The total atomic mass of nitrogen in HNO_3 is 1×14.0 u $=$ 14.0 u
The total atomic mass of oxygen in HNO_3 is 3×16.0 u $=$ $\overline{48.0}$ u

The total molecular mass of $HNO_3 = \overline{63.0}$ u

$$\% \text{ H} = \frac{\text{mass of H in 1 molecule}}{\text{mass of molecule}} \times 100\%$$

$$= \frac{1.0 \text{ u}}{63.0 \text{ u}} \times 100\% = 1.6\%$$

$$\% \text{ N} = \frac{\text{mass of N in 1 molecule}}{\text{mass of molecule}} \times 100\%$$

$$= \frac{14.0 \text{ u}}{63.0 \text{ u}} \times 100\% = 22.2\%$$

$$\% \text{ O} = \frac{\text{mass of O in 1 molecule}}{\text{mass of molecule}} \times 100\%$$

$$= \frac{48.0 \text{ u}}{63.0 \text{ u}} \times 100\% = 76.2\%$$

EXAMPLE PROBLEM 11-8

What is the percentage composition of calcium hydrogen carbonate, $Ca(HCO_3)_2$?

SOLUTION

The total atomic mass of calcium in $Ca(HCO_3)_2$ is 1×40.1 u $= 40.1$ u
The total atomic mass of hydrogen in $Ca(HCO_3)_2$ is 2×1.0 u $= 2.0$ u
The total atomic mass of carbon in $Ca(HCO_3)_2$ is 2×12.0 u $= 24.0$ u
The total atomic mass of oxygen in $Ca(HCO_3)_2$ is 6×16.0 u $= 96.0$ u

The total formula mass of $Ca(HCO_3)_2$ $= 162.1$ u

$$\% \, Ca = \frac{\text{mass of Ca in 1 formula unit}}{\text{mass of formula unit}} \times 100\%$$

$$= \frac{40.1 \text{ u}}{162.1 \text{ u}} \times 100\% = 24.7\%$$

$$\% \, H = \frac{\text{mass of H in 1 formula unit}}{\text{mass of formula unit}} \times 100\%$$

$$= \frac{2.0 \text{ u}}{162.1 \text{ u}} \times 100\% = 1.2\%$$

$$\% \, C = \frac{\text{mass of C in 1 formula unit}}{\text{mass of formula unit}} \times 100\%$$

$$= \frac{24.0 \text{ u}}{162.1 \text{ u}} \times 100\% = 14.8\%$$

$$\% \, O = \frac{\text{mass of O in 1 formula unit}}{\text{mass of formula unit}} \times 100\%$$

$$= \frac{96.0 \text{ u}}{162.1 \text{ u}} \times 100\% = 59.2\%$$

The total is 99.9% due to the accumulated errors caused by rounding off the individual percentages to one place after the decimal point. However, this total is equal to 100.0% within experimental uncertainty. Therefore the composition of $Ca(HCO_3)_2$ is 24.7% Ca, 1.2% H, 14.8% C, and 59.2% O.

PRACTICE PROBLEM 11-7

What is the percentage composition of barium peroxide, BaO_2? (*Answer*: 81.1% Ba; 18.9% O)

PRACTICE PROBLEM 11-8

What is the percentage composition of sulfuric acid, H_2SO_4? (*Answer*: 2.0% H; 32.7% S; 65.2% O)

PRACTICE PROBLEM 11-9

What is the percentage composition of aluminum carbonate, $Al_2(CO_3)_3$? (*Answer*: 23.1% Al, 15.4% C, 61.5% O)

Part One Review Questions

1. If the mass of one atom of an element is 84 u, what is the mass of Avogadro's number of atoms of the element?
2. What is a mole of a substance?
3. What is the molar mass of a substance?
4. What is the numerical value of Avogadro's number?
5. If a certain element has an atomic mass of 197 u, what is the mass of 0.65 mol of the element?
6. How many moles are present in 25.0 g of an element that has an atomic mass of 89 u?
7. How many argon atoms are present in a 10.0 g sample of argon?
8. How many iron atoms are present in a 167 g sample of iron?
9. Why do we refer to the *molecular mass* of carbon dioxide but to the *formula mass* of sodium chloride?
10. What is the formula mass of sodium carbonate, Na_2CO_3?
11. What is the molecular mass of sulfanilamide (the first of the sulfa drugs), $C_6H_8N_2O_2S$?
12. What is the percentage composition of Na_2CO_3? $C_6H_8N_2O_2S$?

Part Two: The Mole and Chemical Formulas

11.6 Expanding the Mole Concept

We have learned that Avogadro's number of atoms of any element has a mass in grams numerically equal to the atomic mass of the element. We have also learned that a mole is the quantity of any substance which contains Avogadro's number of chemical units of the substance. We can expand the mole concept to include compounds.

A sample of formaldehyde consists of CH_2O molecules:

$$1C = 1 \times 12.0 \text{ u} = 12.0 \text{ u}$$
$$2H = 2 \times 1.0 \text{ u} = 2.0 \text{ u}$$
$$1O = 1 \times 16.0 \text{ u} = 16.0 \text{ u}$$
$$\text{molecular mass} = \overline{30.0 \text{ u}}$$

One mole of formaldehyde is the quantity which contains Avogadro's number (N) of formaldehyde molecules. Each formaldehyde molecule contains one carbon atom, two hydrogen atoms, and one oxygen atom. Therefore, Avogadro's number of formaldehyde molecules contains N carbon atoms, $2N$ hydrogen atoms, and N oxygen atoms:

$$1N \text{ C} = 1 \times 12.0\text{g} = 12.0\text{g}$$
$$2N \text{ H} = 2 \times 1.0\text{g} = 2.0 \text{ g}$$
$$1N \text{ O} = 1 \times 16.0\text{g} = 16.0 \text{ g}$$
$$\text{molar mass} = \overline{30.0 \text{ g}}$$

The molar mass of formaldehyde is its molecular mass expressed in grams.

The molar mass of a molecular compound is its molecular mass expressed in grams. Similarly, the molar mass of an ionic compound is its formula mass expressed in grams, and the molar mass of an ion is its ionic mass expressed in grams. For example, the molar mass of NaCl is 58.5 g, and the molar mass of OH^- is 17.0 g.

PRACTICE PROBLEM 11-10

What is the molar mass of each of the following: **a)** $C_3H_7NO_2$; **b)** $Al(NO_3)_3$; **c)** H_2SO_4; **d)** $Ca(HSO_4)_2$; and **e)** PO_4^{3-}? (*Answers*: **a)** 89.0 g; **b)** 213.0 g; **c)** 98.1 g; **d)** 234.3 g; **e)** 95.0 g)

EXAMPLE PROBLEM 11-9

a) How many molecules are present in 5.00 mol of CH_2O?
b) How many atoms are present in 5.00 mol of CH_2O?

SOLUTION
a) 1 mol of CH_2O contains 6.02×10^{23} molecules of CH_2O

$$5.00 \text{ mol } CH_2O \times \frac{6.02 \times 10^{23} \text{ molecules } CH_2O}{1 \text{ mol } CH_2O}$$

$$= 3.01 \times 10^{24} \text{ molecules } CH_2O$$

b) 1 molecule of CH_2O contains 4 atoms

$$3.01 \times 10^{24} \text{ molecules } CH_2O \times \frac{4 \text{ atoms}}{1 \text{ molecule } CH_2O}$$

$$= 1.20 \times 10^{25} \text{ atoms}$$

PRACTICE PROBLEM 11-11

How many molecules are present in 7.50 mol of H_2SO_4? How many atoms are present in 0.45 mol of H_2SO_4? (*Answers*: 4.52×10^{24} molecules; 1.9×10^{24} atoms)

EXAMPLE PROBLEM 11-10

How many moles of $Al(NO_3)_3$ are present in 23.1 g of $Al(NO_3)_3$?

SOLUTION
The molar mass of $Al(NO_3)_3$ is 213 g.

$$23.1 \text{ g } Al(NO_3)_3 \times \frac{1.00 \text{ mol } Al(NO_3)_3}{213 \text{ g } Al(NO_3)_3}$$

$$= 0.108 \text{ mol } Al(NO_3)_3$$

PRACTICE PROBLEM 11-12

How many moles are present in 139 g of $HC_2H_3O_2$? (*Answer*: 2.32 mol)

EXAMPLE PROBLEM 11-11

What is the mass of 0.763 mol of HNO_3?

SOLUTION

The molar mass of HNO_3 is 63.0 g.

$$0.763 \text{ mol HNO}_3 \times \frac{63.0 \text{ g HNO}_3}{1 \text{ mol HNO}_3}$$

$$= 48.1 \text{ g HNO}_3$$

PRACTICE PROBLEM 11-13

What is the mass of 2.40 mol of NaCl? (*Answer*: 140 g)

EXAMPLE PROBLEM 11-12

How many atoms are present in 15.0 g of CH_2O?

SOLUTION

The molar mass of CH_2O is 30.0 g, and 1 molecule of CH_2O contains 4 atoms.

$$15.0 \text{ g CH}_2\text{O} \times \frac{1 \text{ mol CH}_2\text{O}}{30.0 \text{ g CH}_2\text{O}} \times \frac{6.02 \times 10^{23} \text{ molecules CH}_2\text{O}}{1 \text{ mol CH}_2\text{O}} \times$$

$$\frac{4 \text{ atoms}}{1 \text{ molecule CH}_2\text{O}}$$

$$= 1.20 \times 10^{24} \text{ atoms}$$

PRACTICE PROBLEM 11-14

How many atoms are present in **a)** 27 g of H_2SO_4; **b)** 16 g of Br_2? (*Answers*: **a)** 1.2×10^{24} atoms; **b)** 1.2×10^{23} atoms)

EXAMPLE PROBLEM 11-13

What is the mass of 1.71×10^{24} molecules of H_2SO_4?

SOLUTION

The molar mass of H_2SO_4 is 98.1 g.

$$1.71 \times 10^{24} \text{ molecules H}_2\text{SO}_4 \times \frac{1 \text{ mol H}_2\text{SO}_4}{6.02 \times 10^{23} \text{ molecules H}_2\text{SO}_4} \times$$

$$\frac{98.1 \text{ g H}_2\text{SO}_4}{1 \text{ mol H}_2\text{SO}_4}$$

$$= 279 \text{ g H}_2\text{SO}_4$$

PRACTICE PROBLEM 11-15

What is the mass of 2.50×10^{22} molecules of $C_3H_7NO_2$? (*Answer*: 3.70 g)

11.7 Calculating Empirical Formulas

When a new compound is prepared, chemists must determine its formula. They do so by analyzing the molecule to find its percentage composition. The percentage composition information is then used to determine the empirical formula.

The **empirical formula** of a substance is its simplest formula. It gives a bare minimum of information about the compound because it shows only the relative numbers of moles of each type of atom in the compound. In writing the empirical formula, we write the symbols of the elements with subscripts to designate the relative numbers of moles of these elements. The empirical formula CH_2O represents a compound in which there is 1 mol of carbon atoms and 2 mol of hydrogen atoms for every 1 mol of oxygen atoms. However, 1 mol of atoms of any element contains the same number (Avogadro's number) of atoms as 1 mol of atoms of any other element. Therefore we can also state that the empirical formula CH_2O represents a compound in which there is 1 atom of carbon and 2 atoms of hydrogen for every 1 atom of oxygen.

The empirical formula tells us the relative number of atoms in a compound. It does not tell us how many atoms of each type are in a molecule of the compound. For example, the actual formula of a compound with empirical formula CH_2O could be CH_2O, $C_2H_4O_2$, $C_3H_6O_3$, or some other multiple of the empirical formula.

Suppose that a certain compound contains 5.9% hydrogen and 94.1% oxygen. For the empirical formula we need only the relative number of moles of hydrogen and moles of oxygen in the compound. Since only relative numbers are involved, we can consider any amount of the compound. For example, 100 g of compound contains 5.9 g of hydrogen and 94.1 g of oxygen. We can use a table of atomic masses (inside front cover) to determine the molar mass of each element and use this information to determine the number of moles of each element in 100 g of compound:

$$5.9 \; \cancel{g \; H} \times \frac{1 \; mol \; H}{1.0 \; \cancel{g \; H}} = 5.9 \; mol \; H$$

$$94.1 \; \cancel{g \; O} \times \frac{1 \; mol \; O}{16.0 \; \cancel{g \; O}} = 5.9 \; mol \; O$$

The ratio of moles of hydrogen atoms to moles of oxygen atoms is 5.9 to 5.9 or 1 to 1. Thus, the ratio of hydrogen atoms to oxygen atoms is also 1 to 1. This gives an empirical formula of HO.

EXAMPLE PROBLEM 11-14

What is the empirical formula of a compound which is found by analysis to contain 92.3% carbon and 7.7% hydrogen?

SOLUTION

In 100 g of the compound there would be 92.3 g of carbon and 7.7 g of hydrogen.

$$C: 92.3 \text{ g C} \times \frac{1 \text{ mol C}}{12.0 \text{ g C}} = 7.69 \text{ mol C}$$

$$H: 7.7 \text{ g H} \times \frac{1 \text{ mol H}}{1.0 \text{ g H}} = 7.7 \text{ mol H}$$

The molar ratio is $C:H = 7.69 \text{ mol}:7.7 \text{ mol}$

To obtain the simplest molar ratio divide both quantities by the smaller number of moles present (7.69):

$$C:H = \frac{7.69 \text{ mol}}{7.69} : \frac{7.7 \text{ mol}}{7.69} = 1.00 \text{ mol}:1.0 \text{ mol}$$

Within experimental error the ratio is $1:1$. $\therefore$ The empirical formula of the substance is CH.

EXAMPLE PROBLEM 11-15

What is the empirical formula of a compound which on analysis shows 2.2% H, 26.7% C, and 71.1% O?

SOLUTION

In 100 g of the compound there would be 2.2 g H, 26.7 g C, and 71.1 g O.

$$H: 2.2 \text{ g H} \times \frac{1 \text{ mol H}}{1.0 \text{ g H}} = 2.2 \text{ mol H}$$

$$C: 26.7 \text{ g C} \times \frac{1 \text{ mol C}}{12.0 \text{ g C}} = 2.23 \text{ mol C}$$

$$O: 71.1 \text{ g O} \times \frac{1 \text{ mol O}}{16.0 \text{ g O}} = 4.44 \text{ mol O}$$

The molar ratios are $H:C:O = 2.2 \text{ mol}:2.23 \text{ mol}:4.44 \text{ mol}$.

To obtain the simplest molar ratios, divide all molar quantities by the smallest number of moles present (2.2):

$$H:C:O = \frac{2.2 \text{ mol}}{2.2} : \frac{2.23 \text{ mol}}{2.2} : \frac{4.44 \text{ mol}}{2.2} = 1.0 \text{ mol}:1.01 \text{ mol}:2.02 \text{ mol}$$

Within experimental error, $H:C:O = 1:1:2$. Therefore the empirical formula of the substance is HCO_2.

EXAMPLE PROBLEM 11-16

What is the empirical formula of a compound which on analysis shows 69.9% Fe and 30.1% O?

SOLUTION

In 100 g of the compound there would be 69.9 g Fe and 30.1 g O.

$$\text{Fe}: 69.9 \text{ g Fe} \times \frac{1 \text{ mol Fe}}{55.8 \text{ g Fe}} = 1.25 \text{ mol Fe}$$

$$\text{O}: 30.1 \text{ g O} \times \frac{1 \text{ mol O}}{16.0 \text{ g O}} = 1.88 \text{ mol O}$$

The mole ratio is $\text{Fe}:\text{O} = 1.25 \text{ mol}:1.88 \text{ mol}$.

To obtain the simplest molar ratio, divide both molar quantities by the smaller number of moles.

$$\text{Fe}:\text{O} = \frac{1.25 \text{ mol}}{1.25}:\frac{1.88 \text{ mol}}{1.25} = 1.00 \text{ mol}:1.50 \text{ mol}$$

This is not a ratio of small whole numbers. However, if we now multiply both terms in the ratio by 2, we get

$$\text{Fe}:\text{O} = 2.00 \text{ mol}:3.00 \text{ mol}$$

Within experimental error, the ratio of Fe atoms to O atoms is 2 to 3. Therefore the empirical formula is Fe_2O_3.

PRACTICE PROBLEM 11-16

What is the empirical formula of a compound which is found by analysis to contain 80.0% carbon and 20.0% hydrogen? (*Answer*: CH_3)

PRACTICE PROBLEM 11-17

What is the empirical formula of a compound which contains 35.9% aluminum and 64.1% sulfur? (*Answer*: Al_2S_3)

PRACTICE PROBLEM 11-18

What is the empirical formula of a compound which contains 26.6% K, 35.4% Cr, and 38.1% O? (*Answer*: $K_2Cr_2O_7$)

11.8 Molecular Formula Determination

If we know the molecular mass and the empirical formula we can determine the molecular formula. The **molecular formula** shows the actual number of atoms of each element in a molecule of the compound. This is the next step in the determination of the formula of a new compound. (The final step would be to show how the various atoms in the compound are bonded together.)

Suppose we find by experiment that one mole of the compound from Example Problem 11-15 has a mass of 90 g. Therefore one molecule of the compound has a mass of 90 u. The empirical formula was found to be HCO_2. The empirical formula mass may be calculated:

$$\begin{aligned}
\text{Mass of hydrogen is } 1 \times 1\,\text{u} &= 1\,\text{u} \\
\text{Mass of carbon is } 1 \times 12\,\text{u} &= 12\,\text{u} \\
\text{Mass of oxygen is } 2 \times 16\,\text{u} &= \underline{32\,\text{u}} \\
& 45\,\text{u}
\end{aligned}$$

Thus 1 molecule of the compound must contain 2 empirical formula units, $(HCO_2)_2$. This is usually written as $H_2C_2O_4$. The molecular formula is therefore $H_2C_2O_4$.

EXAMPLE PROBLEM 11-17

CH_2O is the empirical formula of a certain compound whose molecular mass is 180 u. What is the molecular formula of the compound?

SOLUTION

The empirical formula mass of CH_2O may be calculated:

$$\begin{aligned}
\text{Mass of carbon is } 1 \times 12\,\text{u} &= 12\,\text{u} \\
\text{Mass of hydrogen is } 2 \times 1\,\text{u} &= 2\,\text{u} \\
\text{Mass of oxygen is } 1 \times 16\,\text{u} &= \underline{16\,\text{u}} \\
& 30\,\text{u}
\end{aligned}$$

One molecule of the compound contains six empirical formula units (180 u/30 u = 6), and the molecular formula is $C_6H_{12}O_6$.

EXAMPLE PROBLEM 11-18

The analysis of a compound shows that it is made up of 21.9% Na, 45.7% C, 1.9% H, and 30.5% O. What is the molecular formula of the compound if its molecular mass is 210 u?

SOLUTION

There are two ways to solve this problem. The first method involves determining the empirical formula as shown in Example Problems 11-14 to 11-16. This is followed by using the empirical formula and molecular mass to determine the molecular formula as shown in Example Problem 11-17.

The second method uses a faster solution. Instead of working with 100 g of the compound, we use the molecular mass, 210 u. Thus, the actual mass of sodium in the compound is 21.9% of 210 u; the actual mass of carbon is 45.7% of 210 u; the actual mass of hydrogen is 1.9% of 210 u; and the actual mass of oxygen is 30.57% of 210 u. The resulting masses of each element can be converted to the number of atoms of each element in one molecule of the compound:

$$0.219 \times 210 \text{ u} = 46.0 \text{ u Na}$$
$$0.457 \times 210 \text{ u} = 96.0 \text{ u C}$$
$$0.019 \times 210 \text{ u} = 4.0 \text{ u H}$$
$$0.305 \times 210 \text{ u} = 64.0 \text{ u O}$$

$$\text{Na: } 46.0 \text{ u Na} \times \frac{1 \text{ atom Na}}{23.0 \text{ u Na}} = 2.00 \text{ atoms Na}$$

$$\text{C: } 96.0 \text{ u C} \times \frac{1 \text{ atom C}}{12.0 \text{ u C}} = 8.00 \text{ atoms C}$$

$$\text{H: } 4.0 \text{ u H} \times \frac{1 \text{ atom H}}{1.0 \text{ u H}} = 4.0 \text{ atoms H}$$

$$\text{O: } 64.0 \text{ u O} \times \frac{1 \text{ atom O}}{16.0 \text{ u O}} = 4.00 \text{ atoms O}$$

The formula must contain an integral number of each type of atom. The molecular formula of the compound is $Na_2C_8H_4O_4$.

PRACTICE PROBLEM 11-19

A certain compound was found to have empirical formula CH. Its molecular mass was found to be 52.0 u. What is its molecular formula? (*Answer*: C_4H_4)

PRACTICE PROBLEM 11-20

A unknown compound is found to have a molecular mass of 75.0 u and to contain 32.0% carbon, 6.7% hydrogen, and 18.7% nitrogen, with the rest of the molecule consisting of oxygen. What is the molecular formula of the compound? (*Answer*: $C_2H_5NO_2$)

APPLICATION

Computers in Chemistry

The declining cost of computer hardware and the availability of supporting accessories and programs have contributed to a growing use of computers by chemists. Chemists use computers to control instruments during experiments; to collect, store, analyze, and plot data; to search through large volumes of chemical information; and to simulate molecules and chemical reactions.

Many modern measuring instruments are directly linked to computers. The computer automates the measuring process and can even analyze the results. The computer may also suggest the possible identity of an unknown substance.

Computers are used to calculate the structure and stability of small molecules. The results of these calculations are very close to those obtained from actual measurements. Thus, a computer can be used to estimate the structure and stability of compounds that are not convenient to experiment with in the laboratory.

Computers are also used in theoretical studies of the atom. For example, chemists use computers to solve complex mathematical equations. They hope to show close agreement of computer predicted values with experimentally determined ones.

Another important use of computers in chemistry is keeping track of the large amounts of data accumulated on compounds that have been synthesized or tested over the years. It is inefficient and time consuming to search manually through written reports or notebooks to find specific information. Instead, a chemist who wants to prepare a new compound can have the computer search a database to see if someone else has already done the job. Some database programs even search for a compound that matches a structure drawn on the computer screen.

Computers are also used to help in the synthesis of chemical compounds. Using a technique called molecular modelling, chemists can display the structure of a molecule in three dimensions on the screen of a monitor. There, the molecule can be manipulated and rotated. In this way, chemists who are trying to synthesize a new drug can see if its molecules will fit the active receptor site. This can all be done before the potential drug is made. Thus, chemists use computer generated models to predict the physical properties of a compound and to determine which of its properties are responsible for biological activity. There are even computer programs which will conduct a search to determine the most efficient method to make a particular compound.

Chemical reactions can be studied in detail on computer. The computer can be used to predict the energies and chemicals produced during the course of a reaction. It can also calculate the forces on all the atoms as the bonds become reorganized when products are formed from reactants.

Part Two Review Questions

1. What is meant by the molar mass of **a)** a molecular compound; **b)** an ionic compound; **c)** an ion?
2. What is an empirical formula?
3. What information do we require to determine the empirical formula of a compound?
4. What does the empirical formula fail to tell us about a molecule of a compound?
5. What is a molecular formula?

6. What information do we require to determine the molecular formula of a compound?

7. When copper is heated in the presence of oxygen, an oxide of copper is formed. It was found in the laboratory that 8.24 g of copper react with 2.08 g of oxygen. What is the empirical formula of the oxide?

8. Acetone, a liquid often used as a nail polish remover, is found to contain 62.0% carbon, 10.4% hydrogen, and 27.5% oxygen. Its molecular mass is found to be 58.1 u. What is its molecular formula?

Part Three: The Mole and Chemical Equations

11.9 Relating the Mole to Chemical Equations

The chemical equation is a shorthand method for describing a chemical change. The symbols and formulas are used to indicate the substances involved in the change. In order to balance a chemical equation we must know what substances react, what substances are produced, and the correct symbols and formulas for all of the substances involved in the reaction. We must also obey the Law of Conservation of Mass. That is, there must be as many atoms of each element on the right side of the yields sign as there are on the left side.

Let us consider a simple equation:

$$C(s) + O_2(g) \longrightarrow CO_2(g)$$

This balanced equation states that 1 atom of carbon reacts with 1 molecule of oxygen to give 1 molecule of carbon dioxide. It also means that 12 u of carbon react with 32 u of oxygen to give 44 u of carbon dioxide. However, 10 carbon atoms (120 u) would also react with 10 oxygen molecules (320 u) to give 10 molecules of carbon dioxide (440 u). Furthermore, 6.02×10^{23} carbon atoms (12 g or 1 mol) would react with 6.02×10^{23} oxygen molecules (32 g or 1 mol) to give 6.02×10^{23} molecules of carbon dioxide (44 g or 1 mol). This information is summarized below:

C(s) +	O$_2$(g) $\longrightarrow$	CO$_2$(g)
1 atom	1 molecule	1 molecule
12 u	32 u	44 u
10 atoms	10 molecules	10 molecules
120 u	320 u	440 u
6.02×10^{23} atoms	6.02×10^{23} molecules	6.02×10^{23} molecules
12 g	32 g	44 g
1 mol	1 mol	1 mol

11.10 Mass Calculations Involving Chemical Equations

We can use our knowledge of chemical equations and moles to calculate mass relationships among the substances involved in a chemical reaction. Chemists usually calculate the mass relationships before experimenting with a chemical reaction. This enables them to avoid wasting expensive chemicals in reactions, thereby saving both money and the time needed to recover the unreacted chemicals. As a simple example, how many grams of oxygen would be needed to react with 60 g of carbon to produce carbon dioxide, and how many grams of carbon dioxide would result? We write the equation:

$$C(s) + O_2(g) \longrightarrow CO_2(g)$$

1 mol	1 mol	1 mol
12 g	32 g	44 g

Thus 1 mol (12 g) of carbon will react with 1 mol (32 g) of oxygen to produce 1 mol (44 g) of carbon dioxide. How many moles of carbon is 60 g?

$$60 \text{ g C} \times \frac{1 \text{ mol C}}{12 \text{ g C}} = 5.0 \text{ mol C}$$

According to the equation, 1 mol of carbon reacts with 1 mol of oxygen. Therefore 5.0 mol of carbon will react with 5.0 mol of oxygen. How many grams of oxygen are present in 5.0 mol?

$$5.0 \text{ mol } O_2 \times \frac{32 \text{ g } O_2}{1 \text{ mol } O_2} = 160 \text{ g } O_2$$

Also, 5.0 mol of carbon will produce 5.0 mol of carbon dioxide. How many grams of carbon dioxide are present in 5.0 mol?

$$5.0 \text{ mol } CO_2 \times \frac{44 \text{ g } CO_2}{1 \text{ mol } CO_2} = 220 \text{ g } CO_2$$

EXAMPLE PROBLEM 11-19

How many moles of oxygen are required to react with 9.7 g of magnesium to produce magnesium oxide? How many grams of oxygen are required?

SOLUTION

Unbalanced Equation: $Mg(s) + O_2(g) \longrightarrow MgO(s)$

Balanced Equation: $2 Mg(s) + O_2(g) \longrightarrow 2 MgO(s)$

2 mol	1 mol	2 mol

$$9.7 \text{ g Mg} \times \frac{1 \text{ mol Mg}}{24.3 \text{ g Mg}} = 0.40 \text{ mol Mg}$$

2 mol of Mg react with 1 mol O_2.

$$0.40 \text{ mol Mg} \times \frac{1 \text{ mol } O_2}{2 \text{ mol Mg}} = 0.20 \text{ mol } O_2$$

$$0.20 \text{ mol } O_2 \times \frac{32 \text{ g } O_2}{1 \text{ mol } O_2} = 6.4 \text{ g } O_2$$

There are a number of methods for solving this type of problem. Example Problem 11-20 illustrates two common methods.

EXAMPLE PROBLEM 11-20

a) How many grams of $KClO_3$ must be decomposed to give 0.96 g of oxygen?

b) How many moles of KCl will be produced during this same reaction?

SOLUTION

Unbalanced Equation: $KClO_3(s) \xrightarrow{\Delta} KCl(s) + O_2(g)$
 Balanced Equation: $2 KClO_3(s) \xrightarrow{\Delta} 2 KCl(s) + 3 O_2(g)$
                          2 mol   2 mol   3 mol

$$0.96 \text{ g } O_2 \times \frac{1.0 \text{ mol } O_2}{32 \text{ g } O_2} = 0.030 \text{ mol } O_2$$

a)   3 mol of O_2 are produced from 2 mol of $KClO_3$

$$0.030 \text{ mol } O_2 \times \frac{2 \text{ mol } KClO_3}{3 \text{ mol } O_2} = 0.020 \text{ mol } KClO_3$$

$$0.020 \text{ mol } KClO_3 \times \frac{122.6 \text{ g } KClO_3}{1 \text{ mol } KClO_3} = 2.5 \text{ g } KClO_3$$

b)   3 mol of O_2 are produced with 2 mol of KCl

$$0.030 \text{ mol } O_2 \times \frac{2 \text{ mol } KCl}{3 \text{ mol } O_2} = 0.020 \text{ mol } KCl$$

The second method makes use of the mole ratio and the mass ratio:

Unbalanced Equation: $KClO_3(s) \xrightarrow{\Delta} KCl(s) + O_2(g)$

			0.96 g
Balanced Equation: $2KClO_3(s) \xrightarrow{\Delta}$	$2KCl(s)$	+	$3 O_2(g)$
mole ratio:	2 mol	2 mol	3 mol
	2(122.6 g)	2(74.6 g)	3(32.0 g)
mass ratio:	245.2 g	149.2 g	96.0 g

Above the balanced equation, the given quantity is written. The mole ratio and the mass ratio, which are used to provide conversion factors, are written below the balanced equation. In each case, we merely convert from the given quantity (0.96 g O_2) to one of the unknown quantities by making use of the mole and mass ratios:

a)

$$0.96 \, \cancel{g \, O_2} \times \frac{245.2 \, g \, KClO_3}{96.0 \, \cancel{g \, O_2}}$$

$$= 2.5 \, g \, KClO_3$$

b) We note that 2 mol of KCl are formed together with 96.0 g of oxygen. Therefore we can write

$$0.96 \, g \, O_2 \times \frac{2 \, mol \, KCl}{96.0 \, g \, O_2}$$

$$= 0.020 \, mol \, KCl$$

PRACTICE PROBLEM 11-21

In the reaction Mg(s) + 2HCl(aq) $\longrightarrow$ MgCl$_2$(aq) + H$_2$(g), how many grams of magnesium are required to produce 5.00 g of hydrogen? (*Answer*: 60.8 g)

PRACTICE PROBLEM 11-22

How many moles of hydrogen are required to react with 15.4 g of nitrogen to produce ammonia, NH$_3$? How many grams of hydrogen are required? (*Answer*: 1.65 mol; 3.30 g)

11.11 Limiting Reagents in Mass Calculations

In most chemical reactions, only one of the reactants is completely consumed. This reactant is known as the **limiting reagent**. Any reactant which is not completely used up is present in excess.

Consider the chemical reaction between iron and sulfur to form iron(II) sulfide:

$$Fe(s) + S(s) \longrightarrow FeS(s)$$

Suppose that 14 g of Fe and 12 g of S are available for reaction. The balanced equation indicates that one mole of iron reacts with one mole of sulfur to produce one mole of iron(II) sulfide. We must first calculate how many moles of iron and of sulfur are actually available for the reaction:

$$14 \, \cancel{g \, Fe} \times \frac{1 \, mol \, Fe}{55.8 \, \cancel{g \, Fe}} = 0.25 \, mol \, Fe$$

$$12 \, \cancel{g \, S} \times \frac{1 \, mol \, S}{32.1 \, \cancel{g \, S}} = 0.37 \, mol \, S$$

The limiting reagent is the iron. The 0.25 mol Fe will be completey consumed by 0.25 mol S, and 0.12 mol S will be left over. The sulfur is present in excess. The reaction produces 0.25 mol of FeS.

EXAMPLE PROBLEM 11-21

How much H_2SO_4 can be prepared from 50.0 g of SO_2, 15.0 g of O_2, and an unlimited quantity of water? The balanced chemical equation is

$$2SO_2(g) + O_2(g) + 2H_2O(\ell) \longrightarrow 2H_2SO_4(\ell)$$

SOLUTION
To solve the problem, we must first determine which is the limiting reagent by calculating the number of moles of SO_2 and O_2:

$$50.0 \text{ g } SO_2 \times \frac{1 \text{ mol } SO_2}{64.1 \text{ g } SO_2} = 0.780 \text{ mol } SO_2$$

$$15.0 \text{ g } O_2 \times \frac{1 \text{ mol } O_2}{32.0 \text{ g } O_2} = 0.469 \text{ mol } O_2$$

According to the balanced chemical equation, two moles of SO_2 require one mole of O_2. Thus, 0.780 mol of SO_2 requires only

$$0.780 \text{ mol } SO_2 \times \frac{1 \text{ mol } O_2}{2 \text{ mol } SO_2} = 0.390 \text{ mol } O_2.$$

The O_2 is present in excess, and the SO_2 is the limiting reagent.

We use the quantity of the limiting reagent, SO_2, to calculate the quantity of H_2SO_4 prepared. According to the balanced chemical equation, two moles of SO_2 produce two moles of H_2SO_4. Therefore, one mole of SO_2 produces one mole of H_2SO_4, and 0.780 mol of SO_2 produces 0.780 mol of H_2SO_4:

$$0.780 \text{ mol } H_2SO_4 \times \frac{98.1 \text{ g } H_2SO_4}{1 \text{ mol } H_2SO_4} = 76.5 \text{ g } H_2SO_4$$

PRACTICE PROBLEM 11-23

How many grams of CO_2 will be produced when 8.50 g of CH_4 react with 15.9 g of O_2, according to the chemical equation

$$CH_4(g) + 2 O_2(g) \longrightarrow CO_2(g) + 2H_2O(g)?$$

(*Answer*: 10.9 g)

11.12 The Yield of a Chemical Reaction

The **theoretical yield** is the maximum amount of a product that can form in a chemical reaction. The theoretical yield is calculated by assuming that all of the limiting reagent has reacted to form the product. This assumption may not be correct for a number of reasons. First, many reactions do not go to completion. Instead, they appear to stop before all of the limiting reagent has reacted to form the product or products. The result is a mixture of reactants and products. Second, there may be other reactions, called *side reactions*, which occur in addition to the main reaction. The occurrence of one or more side reactions prevents the maximum amount of a product from forming. Third, some product may be lost during the process of separating it from the reaction mixture. The **actual yield** is the amount of a product that is actually obtained from a chemical reaction. For the reasons we have given, the actual yield is almost always less than the theoretical yield.

The theoretical yield is a calculated quantity, and the actual yield is an experimentally-determined quantity. The **percent yield** is the actual yield expressed as a percentage of the theoretical yield:

$$\text{percent yield} = \frac{\text{actual yield}}{\text{theoretical yield}} \times 100\%$$

EXAMPLE PROBLEM 11-22

Suppose 7.00 g of $AgNO_3$ is added to a solution which contains an excess of dissolved KBr. If 7.32 g of AgBr is obtained, what is the percent yield?

SOLUTION

The balanced chemical equation for this reaction is

$$AgNO_3(aq) + KBr(aq) \longrightarrow AgBr(s) + KNO_3(aq)$$

We can calculate the theoretical yield by using the molar masses of $AgNO_3$ and AgBr and the mole ratio of $AgNO_3$ to AgBr ($1:1$) from the balanced equation:

$$7.00 \text{ g AgNO}_3 \times \frac{1 \text{ mol AgNO}_3}{170 \text{ g AgNO}_3} \times \frac{1 \text{ mol AgBr}}{1 \text{ mol AgNO}_3} \times \frac{188 \text{ g AgBr}}{1 \text{ mol AgBr}}$$

$$= 7.74 \text{ g AgBr}$$

The theoretical yield of AgBr is 7.74 g and the actual yield is 7.32 g. The percent yield may be calculated:

$$\text{percent yield} = \frac{\text{actual yield}}{\text{theoretical yield}} \times 100\%$$

$$= \frac{7.32 \text{ g}}{7.74 \text{ g}} \times 100\%$$

$$= 94.6\%$$

PRACTICE PROBLEM 11-24

If 8.76 g of PbF_2 precipitate is obtained when 4.50 g of KF is added to a solution which contains an excess of dissolved $Pb(NO_3)_2$, what is the percent yield? (*Answer*: 92.3%)

EXAMPLE PROBLEM 11-23

A mass of 4.55 g of NaOH is dissolved in water and 3.20 g of gaseous H_2S is bubbled in. The following reaction takes place:

$$H_2S(g) + 2NaOH(aq) \longrightarrow Na_2S(aq) + 2H_2O(\ell)$$

If 3.92 g of Na_2S is recovered from the reaction mixture, what is the percent yield?

SOLUTION

We must first determine which is the limiting reagent by calculating the number of moles of H_2S and NaOH:

$$3.20 \text{ g } H_2S \times \frac{1 \text{ mol } H_2S}{34.1 \text{ g } H_2S} = 0.0938 \text{ mol } H_2S$$

$$4.55 \text{ g NaOH} \times \frac{1 \text{ mol NaOH}}{40.0 \text{ g NaOH}} = 0.114 \text{ mol NaOH}$$

According to the balanced chemical equation, two moles of NaOH require one mole of H_2S. Thus, 0.114 mol of NaOH requires only

$$0.114 \text{ mol NaOH} \times \frac{1 \text{ mol } H_2S}{2 \text{ mol NaOH}} = 0.0570 \text{ mol } H_2S$$

The H_2S is present in excess, and the NaOH is the limiting reagent which is used to calculate the theoretical yield of Na_2S:

$$0.114 \text{ mol NaOH} \times \frac{1 \text{ mol } Na_2S}{2 \text{ mol NaOH}} \times \frac{78.0 \text{ g } Na_2S}{1 \text{ mol } Na_2S}$$

$$= 4.45 \text{ g } Na_2S$$

The theoretical yield of Na_2S is 4.45 g, and the actual yield is 3.92 g. The percent yield may be calculated:

$$\text{percent yield} = \frac{\text{actual yield}}{\text{theoretical yield}} \times 100\%$$

$$= \frac{3.92 \text{ g}}{4.45 \text{ g}} \times 100\%$$

$$= 88.1\%$$

PRACTICE PROBLEM 11-25

Suppose a solution containing 5.00 g of Na_3PO_4 is mixed with a solution containing 10.90 g of $Ba(NO_3)_2$. If 7.69 g of $Ba_3(PO_4)_2$ precipitate is recovered from the reaction mixture, what is the percent yield? The chemical reaction is

$$2Na_3PO_4(aq) + 3Ba(NO_3)_2(aq) \longrightarrow Ba_3(PO_4)_2(s) + 6NaNO_3(aq)$$

(*Answer*: 91.8%)

11.13 The Mole and Industrial Chemical Reactions

Many chemical reactions are of great practical importance. For example, iron is produced from its ore, hematite, by reaction with carbon monoxide in a blast furnace:

$$Fe_2O_3 + 3CO \longrightarrow 2Fe + 3CO_2$$

EXAMPLE PROBLEM 11-24

How many kilograms of iron could be produced from 1.00 t of hematite?

SOLUTION

To solve this problem, we shall first convert the mass of hematite to moles:

$$1.00\ t\ Fe_2O_3 \times \frac{1.00 \times 10^6\ g\ Fe_2O_3}{1\ t\ Fe_2O_3} \times \frac{1\ mol\ Fe_2O_3}{160\ g\ Fe_2O_3} = 6250\ mol\ Fe_2O_3$$

Next, we use the mole ratio of Fe_2O_3 to Fe (1:2) to calculate the moles of Fe formed:

$$6250\ mol\ Fe_2O_3 \times \frac{2\ mol\ Fe}{1\ mol\ Fe_2O_3} = 12\ 500\ mol\ Fe$$

Finally, we convert the moles of Fe to kilograms:

$$12\ 500\ mol\ Fe \times \frac{55.8\ g\ Fe}{1\ mol\ Fe} \times \frac{1\ kg\ Fe}{1000\ g\ Fe} = 698\ kg\ Fe$$

Therefore 698 kg of iron can be produced from 1.00 t of hematite. Note that we could also have chained the conversion factors to solve the problem in one step:

$$1.00\ t\ Fe_2O_3 \times \frac{1.00 \times 10^6\ g\ Fe_2O_3}{1\ t\ Fe_2O_3} \times \frac{1\ mol\ Fe_2O_3}{160\ g\ Fe_2O_3} \times \frac{2\ mol\ Fe}{1\ mol\ Fe_2O_3} \times$$

$$\frac{55.8\ g\ Fe}{1\ mol\ Fe} \times \frac{1\ kg\ Fe}{1000\ g\ Fe}$$

$$= 698\ kg\ Fe$$

PRACTICE PROBLEM 11-26

The most common ore of arsenic, mispickel, can be heated to produce arsenic:

$$FeSAs(s) \longrightarrow FeS(s) + As(s)$$

How many kilograms of mispickel are required to produce one kilogram of arsenic? (*Answer*: 2.17 kg)

EXAMPLE PROBLEM 11-25

Many industrial reactions involve more than one step. For example, nitric acid is made from ammonia by the Ostwald process. This process consists of three steps:

$$4NH_3(g) + 5\ O_2(g) \longrightarrow 4NO(g) + 6H_2O(g)$$

$$2NO(g) + O_2(g) \longrightarrow 2NO_2(g)$$

$$3NO_2(g) + H_2O(\ell) \longrightarrow 2HNO_3(aq) + NO(g)$$

How many kilograms of nitric acid can be produced from 1.00 kg of ammonia?

SOLUTION

The steps involved in the solution would be:

a) Convert kilograms of NH_3 to moles of NH_3:

$$1.00 \text{ kg } NH_3 \times \frac{1000 \text{ g } NH_3}{1 \text{ kg } NH_3} \times \frac{1 \text{ mol } NH_3}{17.0 \text{ g } NH_3} = 58.8 \text{ mol } NH_3$$

b) Use the molar ratios obtained from the balanced equations to convert moles of NH_3 to moles of HNO_3:

From the first reaction, the mole ratio of NH_3 to NO is 4 to 4.
From the second reaction, the mole ratio of NO to NO_2 is 2 to 2.
From the third reaction, the mole ratio of NO_2 to HNO_3 is 3 to 2.

$$\therefore 58.8 \text{ mol } NH_3 \times \frac{4 \text{ mol NO}}{4 \text{ mol } NH_3} \times \frac{2 \text{ mol } NO_2}{2 \text{ mol NO}} \times \frac{2 \text{ mol } HNO_3}{3 \text{ mol } NO_2}$$

$$= 39.2 \text{ mol } HNO_3$$

c) Finally, convert moles of HNO_3 to kilograms of HNO_3:

$$39.2 \text{ mol } HNO_3 \times \frac{63.0 \text{ g } HNO_3}{1 \text{ mol } HNO_3} \times \frac{1 \text{ kg } HNO_3}{1000 \text{ g } HNO_3} = 2.47 \text{ kg } HNO_3$$

Therefore, 2.47 kg of HNO_3 can be produced from 1.00 kg of NH_3. Note that the problem could also have been solved by multiplying the conversion factors in one (long) step:

$$1.00 \text{ kg } NH_3 \times \frac{1000 \text{ g } NH_3}{1 \text{ kg } NH_3} \times \frac{1 \text{ mol } NH_3}{17.0 \text{ g } NH_3} \times \frac{4 \text{ mol NO}}{4 \text{ mol } NH_3} \times$$

$$\frac{2 \text{ mol } NO_2}{2 \text{ mol NO}} \times \frac{2 \text{ mol } HNO_3}{3 \text{ mol } NO_2} \times \frac{63.0 \text{ g } HNO_3}{1 \text{ mol } HNO_3} \times \frac{1 \text{ kg } HNO_3}{1000 \text{ g } HNO_3}$$

$$= 2.47 \text{ kg } HNO_3$$

PRACTICE PROBLEM 11-27

Zinc is produced from sphalerite (ZnS), an important zinc ore, by a two-step process. The ore is first roasted to convert the sulfide to the oxide:

$$2ZnS(s) + 3\ O_2(g) \longrightarrow 2ZnO(s) + 2SO_2(g)$$

The zinc oxide is then reduced by heating with coke:

$$ZnO(s) + C(s) \longrightarrow Zn(s) + CO(g)$$

How many kilograms of zinc can be produced from 200 kg of ZnS? (*Answer*: 134 kg)

PRACTICE PROBLEM 11-28

There are many instances in which it is important to mix reactants so that they completely consume one another. One type of rocket engine makes use of a reaction similar to the following:

$$N_2O_4(\ell) + 2N_2H_4(\ell) \longrightarrow 3N_2(g) + 4H_2O(g)$$

When the dinitrogen tetroxide and hydrazine are brought together in the rocket engine, they react explosively. The two product gases exit from the rocket engine and push the rocket forward. Because of the enormous costs involved in carrying material into space, it is important to avoid carrying an excess of either reactant. Suppose a rocket is powered by a dinitrogen tetroxide-hydrazine engine. If 25.0 t of dinitrogen tetroxide is carried in one tank, how much hydrazine should be carried in the other tank? (*Answer*: 17.4 t)

Part Three Review Questions

1. Consider the chemical reaction:

   $$MnO_2 + 4HCl \longrightarrow MnCl_2 + Cl_2 + 2H_2O$$

   a) How many molecules of HCl react with one formula unit of MnO_2?
   b) How many moles of Cl_2 are produced if four moles of HCl are consumed? How many are produced if two moles of HCl are consumed?
   c) How many molecules of HCl are consumed if 6.02×10^{23} molecules of H_2O are produced? How many are consumed if 3.01×10^{23} molecules of H_2O are produced? How many are consumed if 6.02×10^{23} formula units of $MnCl_2$ are produced?

2. Consider the chemical reaction:

   $$3Fe + 4H_2O \longrightarrow Fe_3O_4 + 4H_2$$

   a) What are the mole ratios for this reaction?
   b) What are the mass ratios for this reaction?
   c) If 20 g of iron are consumed in this reaction, how many grams of Fe_3O_4 will be produced? How many moles of H_2 will be produced?

3. What is a limiting reagent?

4. What is the limiting reagent if 25 g of magnesium reacts with 20 g of oxygen in the following reaction:

   $$2Mg(s) + O_2(g) \longrightarrow 2MgO(s)$$

5. What is meant by theoretical yield? How is it calculated?
6. What is meant by actual yield?
7. List three reasons why the actual yield might be less than the theoretical yield.
8. What is meant by percent yield?
9. A chemist makes nitroglycerin, $C_3H_5(NO_3)_3$, from glycerol, $C_3H_5(OH)_3$, and nitric acid:

$$C_3H_5(OH)_3 + 3HNO_3 \longrightarrow C_3H_5(NO_3)_3 + 3H_2O$$

In one experiment, 4.1 g of glycerol and 13.5 g of HNO_3 produce 8.8 g of nitroglycerin.
   a) What is the limiting reagent?
   b) What is the theoretical yield of nitroglycerin?
   c) What is the percent yield of nitroglycerin?

Part Four: Calculations Involving Solutions

11.14 Concentration Measured in Moles Per Litre

Figure 11-4
Dilute and concentrated solutions of the same coloured substance

Chemists frequently carry out chemical reactions using reactants which are in solution. In order to perform calculations involving solutions, it is necessary to know the concentrations of the solutions. The **concentration** of a solution is the quantity of solute dissolved in a certain amount of solution. *Dilute* and *concentrated* are terms which qualitatively describe the concentration of a solution (Fig. 11-4). A *dilute solution* has a relatively small concentration, and a *concentrated solution* has a relatively large concentration.

There are several quantitative methods of expressing concentration. The concentration expression that is used most frequently by chemists is called molarity, because moles are the most convenient units to manipulate in calculations involving chemical reactions. The **molarity** of a solution is defined as the number of moles of a solute dissolved per litre of solution. For example, the molarity of a solution which contains one mole of solute dissolved in enough solvent to give one litre of solution is 1 mol/L.

If you were asked to make up one litre of 1.00 mol/L NaCl, you would measure 1.00 mol (58.5 g) of dry NaCl into a one-litre volumetric flask. A *volumetric flask* is a flask which is calibrated to contain a certain quantity of liquid at a certain temperature (Fig. 11.5). Distilled water would be added to the flask, and the contents would be swirled until the NaCl dissolved completely. Then more distilled water would be added until the water level reached the one-litre mark on the neck of the flask, when the contents of the flask were mixed thoroughly. You would not know how much water you had added to the volumetric flask. You would know only that you had *one* litre of *solution* (NaCl + H₂O), containing one mole of NaCl, in the flask.

If you were asked to make up one litre of 2.00 mol/L NaCl, you would place 2.00 mol (117 g) of dry NaCl in the one-litre volumetric flask and dissolve it in enough distilled water to fill the flask to the one-litre mark. If you were asked to make up one litre of 0.500 mol/L NaCl, you would use 0.500 mol (29.2 g) of dry NaCl and dissolve it in one litre of solution.

Volumetric flasks come in many sizes (For example, 50 mL, 100 mL, 250 mL, 500 mL, 1000 mL, and 2000 mL). Therefore, it is possible to make up almost any quantity of a solution of any required concentration. For example, if you were asked to make up 500 mL of 1.00 mol/L NaCl, you would use half as much NaCl as you would if you were preparing one litre of 1.00 mol/L NaCl. Therefore, you would dissolve 0.500 mol (29.2 g) of NaCl in 500 mL of solution.

It is convenient to construct a simple equation from the definition of molarity:

$$\text{molarity (moles per litre)} = \frac{\text{moles of solute}}{\text{volume of solution in litres}}$$

Since the concentration in moles per litre, the moles of solute, and the volume of solution in litres can all vary, we can use x, y, and z to stand for the variable quantities:

$$x \text{ mol/L} = \frac{y \text{ mol}}{z \text{ L}}$$

The use of this equation in many different types of chemical problems is demonstrated in the following problems:

Figure 11-5
Volumetric flasks are available in a variety of sizes.

EXAMPLE PROBLEM 11-26

How many grams of sugar, $C_{12}H_{22}O_{11}$, are contained in 50 mL of a 0.40 mol/L solution of sugar in water?

SOLUTION

$$x \text{ mol/L} = \frac{y \text{ mol}}{z \text{ L}}$$

$$50 \text{ mL} = 0.050 \text{ L}$$

$$0.40 \text{ mol/L} = \frac{y \text{ mol}}{0.050 \text{ L}}$$

$$y \text{ mol} = 0.40 \text{ mol/L} \times 0.050 \text{ L} = 0.020 \text{ mol}$$

$$0.020 \text{ mol } C_{12}H_{22}O_{11} \times \frac{342 \text{ g } C_{12}H_{22}O_{11}}{1 \text{ mol } C_{12}H_{22}O_{11}}$$

$$= 6.8 \text{ g } C_{12}H_{22}O_{11}$$

EXAMPLE PROBLEM 11-27

What is the concentration in moles per litre of a solution that contains 49 g of sulfuric acid (H_2SO_4) in 3.0 L of solution?

SOLUTION

$$x \text{ mol/L} = \frac{y \text{ mol}}{z \text{ L}}$$

$$y \text{ mol } H_2SO_4 = 49 \text{ g } \cancel{H_2SO_4} \times \frac{1 \text{ mol } H_2SO_4}{98 \text{ g } \cancel{H_2SO_4}}$$

$$= 0.50 \text{ mol } H_2SO_4$$

$$x \text{ mol/L} = \frac{0.50 \text{ mol}}{3.0 \text{ L}} = 0.17 \text{ mol/L}$$

PRACTICE PROBLEM 11-29

How many grams of formaldehyde, CH_2O, are contained in 500 mL of a 13.0 mol/L aqueous solution of formaldehyde? (*Answer*: 195 g)

EXAMPLE PROBLEM 11-28

What volume of 3.0 mol/L NaOH would be required to make 250 mL of 0.15 mol/L NaOH?

SOLUTION

We will first determine the number of moles of NaOH needed to prepare the 0.15 mol/L solution:

$$x \text{ mol/L} = \frac{y \text{ mol}}{z \text{ L}}$$

$$0.15 \text{ mol/L} = \frac{y \text{ mol}}{0.250 \text{ L}}$$

$$y \text{ mol} = 0.15 \text{ mol/L} \times 0.250 \text{ L}$$

$$= 0.038 \text{ mol}$$

Then, we can determine the volume of 3.0 mol/L solution which contains the required 0.038 mol:

$$x \text{ mol/L} = \frac{y \text{ mol}}{z \text{ L}}$$

$$3.0 \text{ mol/L} = \frac{0.038 \text{ mol}}{z \text{ L}}$$

$$z \text{ L} = \frac{0.038 \text{ mol}}{3.0 \text{ mol/L}} = 0.013 \text{ L} = 13 \text{ mL}$$

Thus, 13 mL of the 3.0 mol/L NaOH can be measured into a 250 mL volumetric flask and diluted with water to the mark to form 250 mL of 0.15 mol/L NaOH.

EXAMPLE PROBLEM 11-29

If 25.0 mL of 6.00 mol/L HNO_3 is diluted to a final volume of 500 mL, what is the new concentration of the HNO_3?

SOLUTION

$$x \text{ mol/L} = \frac{y \text{ mol}}{z \text{ L}}$$

$$6.00 \text{ mol/L} = \frac{y \text{ mol}}{0.0250 \text{ L}}$$

$$y \text{ mol} = 6.00 \text{ mol/L} \times 0.0250 \text{ L}$$

$$= 0.150 \text{ mol}$$

Then, we calculate the new concentration:

$$x \text{ mol/L} = \frac{y \text{ mol}}{z \text{ L}}$$

$$x \text{ mol/L} = \frac{0.150 \text{ mol}}{0.500 \text{ L}} = 0.300 \text{ mol/L}$$

PRACTICE PROBLEM 11-30

What volume of 6.0 mol/L KBr would be required to make 3.0 L of 0.20 mol/L KBr? (*Answer*: 100 mL)

PRACTICE PROBLEM 11-31

If 25.0 mL of 6.0 mol/L KBr is diluted to a final volume of 2.0 L, what is the new concentration of the KBr? (*Answer*: 0.075 mol/L)

EXAMPLE PROBLEM 11-30

If 25.0 mL of 6.00 mol/L HCl reacts with excess Zn, how many grams of $ZnCl_2$ are produced? The balanced chemical equation is

$$Zn(s) + 2HCl(aq) \longrightarrow ZnCl_2(aq) + H_2(g)$$

SOLUTION

$$Zn(s) + 2HCl(aq) \longrightarrow ZnCl_2(aq) + H_2(g)$$
$$1 \text{ mol} \quad 2 \text{ mol} \qquad 1 \text{ mol} \qquad 1 \text{ mol}$$

First, we determine the moles of HCl:

$$x \text{ mol/L} = \frac{y \text{ mol}}{z \text{ L}}$$

$$6.00 \text{ mol/L} = \frac{y \text{ mol}}{0.0250 \text{ L}}$$

$$y \text{ mol} = 6.00 \text{ mol/L} \times 0.0250 \text{ L}$$

$$= 0.150 \text{ mol HCl}$$

The reaction indicates that 2 mol of HCl produce 1 mol of $ZnCl_2$.

$$0.150 \text{ mol HCl} \times \frac{1 \text{ mol } ZnCl_2}{2 \text{ mol HCl}} \times \frac{136.4 \text{ g } ZnCl_2}{1 \text{ mol } ZnCl_2} = 10.2 \text{ g } ZnCl_2$$

PRACTICE PROBLEM 11-32

If 17.4 mL of 3.50 mol/L NaOH reacts with excess H_2SO_4, how many grams of Na_2SO_4 are produced? The balanced chemical equation is:

$$2NaOH(aq) + H_2SO_4(aq) \longrightarrow Na_2SO_4(aq) + 2H_2O(\ell)$$

(*Answer*: 4.32 g Na_2SO_4)

11.15 Concentration Measured in Percentage by Mass

Although concentration is most frequently expressed in terms of moles of solute per litre of solution, there are several other quantitative expressions of concentration. One of the simplest is percentage by mass. The *percentage by mass* of a solute in a solution is the mass of the solute expressed as a percentage of the total mass of the solution. The equation is:

$$\text{Mass \% of solute} = \frac{\text{mass of solute}}{\text{mass of solution}} \times 100\%$$

For example, a 10% aqueous sugar solution contains 10 g of sugar dissolved in 90 g of water, that is, 10 g of sugar in 100 g of solution.

Generally, percentage concentrations are percentages by mass. Percentage by volume (the volume of a solute expressed as a percentage of the total volume of the solution) is seldom used. Thus, concentrations recorded as percentages should be considered to be based on mass unless you are told they are based on volume.

EXAMPLE PROBLEM 11-31

A solution contains 6.5 g of $NaHCO_3$ in 250.0 g of water. What is the percentage by mass of solute in this solution?

SOLUTION

First, we calculate the total mass of the solution.

$$\text{Total mass} = \text{mass of } NaHCO_3 + \text{mass of } H_2O$$

$$= 6.5 \text{ g} + 250.0 \text{ g} = 256.5 \text{ g}$$

$$\text{Mass \% of solute} = \frac{\text{mass of solute}}{\text{mass of solution}} \times 100\%$$

$$= \frac{6.5 \text{ g}}{256.5 \text{ g}} \times 100\% = 2.5\%$$

EXAMPLE PROBLEM 11-32

The label on a bottle of concentrated hydrochloric acid states "37.0% HCl, density = 1.18 g/mL." What is the molarity of this acid?

SOLUTION

Since we want to find the number of moles of HCl in one litre of solution, we shall use the following steps:

a) Calculate the mass of one litre of solution.

$$1.00 \text{ L solution} \times \frac{1000 \text{ mL solution}}{1 \text{ L solution}} \times \frac{1.18 \text{ g solution}}{1 \text{ mL solution}}$$

$$= 1180 \text{ g solution}$$

b) Calculate the mass of HCl in this solution.

$$1180 \text{ g solution} \times \frac{0.370 \text{ g HCl}}{1 \text{ g solution}} = 437 \text{ g HCl}$$

c) Convert grams of HCl to moles of HCl.

$$437 \text{ g HCl} \times \frac{1 \text{ mol HCl}}{36.5 \text{ g HCl}} = 12.0 \text{ mol HCl}$$

Since one litre of the concentrated solution contains 12.0 mol of HCl, the molarity of the acid is 12.0 mol/L. As usual, the conversion factors can be strung together to give the solution in one step:

$$1.00 \text{ L solution} \times \frac{1000 \text{ mL solution}}{1 \text{ L solution}} \times \frac{1.18 \text{ g solution}}{1 \text{ mL solution}} \times$$

$$\frac{0.370 \text{ g HCl}}{1 \text{ g solution}} \times \frac{1 \text{ mol HCl}}{36.5 \text{ g HCl}}$$

$$= 12.0 \text{ mol HCl}$$

PRACTICE PROBLEM 11-33

If concentrated phosphoric acid solution is 85.0% H_3PO_4 and has a density of 1.69 g/mL, what is the concentration of this acid in moles per litre? (*Answer*: 14.7 mol/L)

PRACTICE PROBLEM 11-34

A solution contains 15.0 g of $BaCl_2$ per 200.0 mL of water. What is the percentage by mass of solute in this solution? (*Answer*: 6.98% $BaCl_2$)

Part Four Review Questions

1. What is meant by the term *concentration*?
2. How does a dilute solution differ from a concentrated solution?
3. What concentration expression is used most frequently by chemists?
4. What is meant by percentage by mass of a solute in a solution?

CAREER

The Analytical Chemist

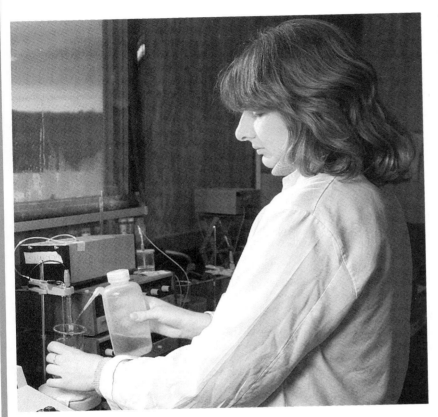

Figure 11-6
An analytical chemist tests water quality.

Analytical chemists perform a wide variety of tasks, usually for industry or government. The role of an analytical chemist is often to determine the substances present in given samples. In the past, analytical chemists relied on conventional laboratory equipment; more recently, electronic instruments have become an integral part of their work.

In industry, quality control chemists might analyze raw materials coming into a manufacturing plant and finished products leaving it, to ensure quality. At a water treatment plant, an analytical chemist would determine the dosage of chemicals required to treat the water and would monitor the levels of naturally-occurring chemicals. Some manufacturing plants use water to cool their equipment. Since this water is returned to the environment, the water treatment ponds must be analyzed to ensure that the water meets government specifications. Other analytical chemists prepare and analyze samples of ground water, waste water, oils, and soils to determine the levels of trace metals present.

Analytical chemists perform an equally wide variety of tasks for government agencies. Bioassay chemists determine the amount of radioactivity absorbed by workers who use radioactive materials. Food chemists analyze meat samples to determine the quantities of components such as protein, preservatives, food dyes, salt and fat. Because of the strict federal regulations governing the pharmaceutical industry, analytical chemists are required to analyze the major ingredients in pharmaceutical products. Forensic chemists analyze evidence such as drugs, blood stains, and broken glass for law enforcement officers. They submit official reports of the results and can be called to testify in court.

Analytical chemists may also be employed by hospitals, universities, and other research establishments. Most analytical chemists have an undergraduate degree in chemistry, and some have graduate level training in analytical chemistry.

Chapter Summary

- Avogadro's number is the number of atoms of carbon-12 needed to have a mass of 12.00 g.
- A mole of a substance is the quantity of that substance which contains the same number of chemical units (atoms, molecules, formula units, or ions) as there are atoms in exactly 12 g of carbon-12 (that is, Avogadro's number).
- A molar mass of a substance is the mass of one mole of the substance. The molar mass of an element is its atomic mass expressed in grams.
- Avogadro's number is 6.02×10^{23}.
- The molecular mass of a compound is the mass of one molecule of the compound.
- The formula mass of an ionic compound is the mass of one formula unit of the compound.
- The percentage composition of a compound is a measure of the percentage of the mass of the compound that is contributed by each element in the compound.
- The molar mass of a molecular compound is its molecular mass expressed in grams; the molar mass of an ionic compound is its formula mass expressed in grams; and the molar mass of an ion is its ionic mass expressed in grams.
- The empirical formula is the simplest formula and shows the relative numbers of atoms of each element in the compound.
- The molecular formula shows the actual number of atoms of each element in a molecule of the compound.
- A limiting reagent is the reactant which is completely consumed in a chemical reaction. It is the reactant upon which the chemical calculations must be based.
- The theoretical yield is the maximum amount of product that can form in a chemical reaction. It is calculated by assuming that all of the limiting reagent has reacted to form the product.
- The actual yield is the amount of product that is obtained from a chemical reaction.
- The percent yield is the actual yield expressed as a percentage of the theoretical yield.
- The concentration of a solution is the quantity of solute dissolved in a certain amount of solution.
- A dilute solution has a relatively small concentration, and a concentrated solution has a relatively large concentration.
- The molarity of a solution is the number of moles of solute contained in one litre of the solution.
- A volumetric flask is a flask which is calibrated to contain a certain quantity of a liquid at a certain temperature.
- The percentage by mass of a solute in a solution is the mass of the solute expressed as a percentage of the total mass of the solution.

Key Terms

mole	theoretical yield
molar mass	actual yield
empirical formula	percent yield
molecular formula	concentration
limiting reagent	molarity

Test Your Understanding

Each of these statements or questions is followed by four responses. Choose the correct response in each case. (Do not write in this book.)

1. The number of carbon atoms in 3.00 mol of glucose ($C_6H_{12}O_6$) is
   a) 1.08×10^{23} b) 6.02×10^{23} c) 18.0×10^{23} d) 108×10^{23}

2. The mass of 1.50 mol of calcium sulfate ($CaSO_4$) is
   a) 40.1 g b) 0.0110 g c) 204 g d) 136 g

3. The number of moles of $CaSO_4$ in 260 g of $CaSO_4$ is
   a) 1.91 b) 136 c) 0.191 d) 35 400

4. The number of sucrose molecules in 4.50 g of $C_{12}H_{22}O_{11}$ is
   a) 2.18×10^{-26} b) 342 c) 1.54×10^{21} d) 7.92×10^{21}

5. The mass of 1.50×10^{23} molecules of $C_{12}H_{22}O_{11}$ is
   a) 85.2 g b) 5.13×10^{25} g c) 342 g d) 1370 g

6. The number of carbon monoxide molecules contained in 14.0 g of carbon monoxide is
   a) 6.02×10^{23} b) 3.01×10^{23} c) 3.01×10^{22} d) 6.02×10^{22}

7. The number of atoms contained in 2.80 g of carbon monoxide is
   a) 6.02×10^{23} b) 6.02×10^{22} c) 1.20×10^{23} d) 1.20×10^{24}

8. The number of moles of H_2SO_4 present in 392 g of H_2SO_4 is
   a) 98.1 b) 4.00 c) 0.250 d) 38 500

9. The number of moles of oxygen atoms present in 392 g of H_2SO_4 is
   a) 4 b) 8 c) 12 d) 16

10. The percent of hydrogen in glucose ($C_6H_{12}O_6$) is
    a) 13.3% b) 12.0% c) 6.00% d) 6.67%

11. The empirical formula of a compound which contains 37.5% C, 12.5% H, and 50.0% O is
    a) $C_4H_{10}O_5$ b) C_3HO_4 c) CH_3O_4 d) CH_4O

12. In the reaction $CuS + H_2 \longrightarrow Cu + H_2S$, the mass of Cu that can be obtained from 9.56 g of CuS is
    a) 6.35 g b) 63.5 g c) 0.100 g d) 9.56 g

13. In the reaction $CuS + H_2 \longrightarrow Cu + H_2S$, the mass of H_2 required to convert 9.56 g of CuS completely to products is
    a) 2.00 g b) 0.10 g c) 0.20 g d) 6.35 g

14. In the reaction $CH_4 + 2O_2 \longrightarrow CO_2 + 2H_2O$, if 3 mol of CH_4 are reacted with 15 mol of O_2, the material remaining after the reaction has gone to completion will be
    a) 13 mol O_2 b) 9 mol O_2 c) 2 mol CH_4 and 13 mol O_2 d) 2 mol CH_4

15. In the reaction $2SO_2 + O_2 + 2H_2O \longrightarrow 2H_2SO_4$, the mass of H_2SO_4 that can be prepared from 15 mol of SO_2, 6 mol of O_2, and an unlimited amount of water is
    a) 2 mol   **b)** 6 mol   **c)** 12 mol   **d)** 15 mol
16. In the reaction $2Al + 3S \longrightarrow Al_2S_3$, the theoretical yield of Al_2S_3 expected from the reaction of 54.0 g of Al with 128 g of S is
    a) 150 g   **b)** 182 g   **c)** 32 g   **d)** 27 g
17. The amount of NaCl required to make 2.50 L of a 0.500 mol/L solution of NaCl is
    a) 0.200 mol   **b)** 0.500 mol   **c)** 1.25 mol   **d)** 5.00 mol
18. A 25 mL volume of 0.50 mol/L NaCl is diluted with water to make 2.5 L of solution. The molarity of the diluted solution is
    a) 5.0×10^{-3} mol/L   **b)** 0.50 mol/L **c)** 5.0×10^3 mol/L **d)** 0.20 mol/L
19. The amount of KBr in 20 mL of a 3.0 mol/L solution is
    a) 0.020 mol   **b)** 0.060 mol   **c)** 15 mol   **d)** 3.0 mol
20. The mass of KBr in 20 mL of a 3.0 mol/L solution is
    a) 2.4 g   **b)** 7.1 g   **c)** 71 g   **d)** 119 g

Review Your Understanding

1. What is the mass of one mole of **a)** mercury; **b)** sodium; **c)** argon?
2. How many atoms are present in one mole of any monatomic element?
3. In an experiment, 1.74×10^{17} alpha particles were counted. The helium gas which was produced occupied a volume of 0.006 49 mL at $0°$ C and 101.3 kPa. What was the experimental value of Avogadro's number?
4. What is the mass in grams of one atom of **a)** sulfur; **b)** iron; **c)** fluorine?
5. How many moles of Kr are present in a sample which is made up of 2.71×10^{21} atoms?
6. If a sample consists of 1.41×10^{23} molecules, how many moles are present?
7. How many atoms are present in **a)** 5.3 mol of Ar; **b)** 0.27 mol of C_2H_6?
8. How many sodium atoms are present in a 71.2 g sample of sodium?
9. How many neon atoms are present in a 20.0 g sample of neon?
10. What is the molecular mass of **a)** $C_{12}H_{22}O_{11}$; **b)** $HC_2H_3O_2$?
11. What is the formula mass of **a)** $(NH_4)_2SO_4$; **b)** Na_3PO_4?
12. What is the percentage composition of **a)** CaO; **b)** H_2S?
13. What is the percentage composition of **a)** Na_2SO_4; **b)** C_4H_9SH?
14. What is the molar mass of **a)** C_2H_5OH; **b)** $Al_2(SO_4)_3$?
15. What is the molar mass of each of the following: **a)** $H_2C_2O_4$; **b)** NH_4NO_3; **c)** Cu; and **d)** MnO_4^-?
16. How many moles are present in **a)** 14.2 g of $H_2C_2O_4$; **b)** 156 g of NH_4NO_3; **c)** 0.147 g of HNO_3?
17. How many grams are present in **a)** 0.422 mol of $SnCl_2$; **b)** 1.75 mol of C_3H_8; **c)** 4.66 mol of $HMnO_4$?

18. a) How many molecules are present in 7.10 mol of $HC_2H_3O_2$; **b)** How many atoms are present in 7.10 mol of $HC_2H_3O_2$; **c)** How many hydrogen atoms are present in 7.10 mol of $HC_2H_3O_2$?

19. What is the mass of **a)** 7.45×10^{22} molecules of $C_6H_{12}O_6$; **b)** 3.11×10^{24} molecules of C_4H_9SH?

20. How many atoms are present in **a)** 33.0 g of $Ca(HCO_3)_2$; **b)** 256 g of $Sr(NO_3)_2$?

21. What are the empirical formulas for compounds whose percentage compositions are given:
a) Fe = 63.53%, S = 36.47%
b) Na = 21.6%, Cl = 33.3%, O = 45.1%
c) Cr = 26.52%, S = 24.52%, O = 48.96%
d) C = 63.1%, H = 11.92%, F = 24.97%

22. When 2.435 g of antimony is heated with excess sulfur, a chemical reaction occurs. The excess sulfur is driven off, leaving only the compound. If 3.397 g of compound are produced, what is the empirical formula of the compound?

23. What is the molecular formula for a compound whose molecular mass is 116 u and whose empirical formula is CHO?

24. What is the molecular formula for a compound whose molecular mass is 90 u and whose empirical formula is CH_2O?

25. What is the molecular formula for a compound whose molecular mass is 198 u and whose percentage composition is C = 48.48%, H = 5.05%, N = 14.14%, O = 32.32%?

26. What is the molecular formula for a compound whose molecular mass is 50.5 u and whose percentage composition is C = 23.8%, H = 5.99%, and Cl = 70.2%?

27. What is the molecular formula for a compound whose molecular mass is 170.2 u and whose percentage composition is C = 49.38%, H = 3.55%, O = 9.40%, and S = 37.67%?

28. What is the formula for a compound whose formula mass is 238 u and whose percentage composition is Na = 19.3%, S = 26.9%, and O = 53.8%?

29. Summarize the information contained in the following balanced equation:

$$2VO(s) + 3Fe_2O_3(s) \longrightarrow 6FeO(s) + V_2O_5(s)$$

Your answer should include the ratio of moles of reactants and products.

30. What mass of NH_3 is formed when 4.84 g of Li_3N reacts with water according to the following equation?

$$Li_3N(s) + 3H_2O(\ell) \longrightarrow 3LiOH(s) + NH_3(g)$$

31. a) How many grams of $O_2(g)$ are needed to react with 6.4 g of methane (CH_4) to produce $CO_2(g)$ and $H_2O(g)$ according to the following equation?

$$CH_4(g) + 2\,O_2(g) \longrightarrow CO_2(g) + 2H_2O(g)$$

b) How many grams of $CO_2(g)$ will be formed?

32. a) How many grams of FeS are necessary to react with 7.81 g of oxygen according to the following equation?

$$4FeS(s) + 7\,O_2(g) \longrightarrow 2Fe_2O_3(s) + 4SO_2(g)$$

b) How many moles of O_2 are necessary to react with 6.79 mol of FeS?

33. a) How many moles of NaOH are necessary to produce 8.61 mol of Na_2SO_4 according to the following equation?

$$2NaOH(aq) + H_2SO_4(aq) \longrightarrow Na_2SO_4(aq) + 2H_2O(\ell)$$

b) How many grams of H_2SO_4 are necessary to produce 4.77 mol of Na_2SO_4?

34. How many grams of nitrogen dioxide (NO_2) will be produced when 128 g of $O_2(g)$ react completely with nitrogen monoxide (NO)? Write a balanced equation before solving the problem.

35. Sodium carbonate and hydrochloric acid react to give sodium chloride, carbon dioxide, and water. How many grams of sodium carbonate and hydrochloric acid would be required to produce 286 g of carbon dioxide?

36. How many grams of Fe_2O_3 are produced when 20.9 g of FeS react with 14.1 g of O_2 according to the following equation?

$$4FeS(s) + 7\,O_2(g) \longrightarrow 2Fe_2O_3(s) + 4SO_2(g)$$

37. How many grams of $CaCl_2$ can be prepared from 60.4 g of CaO and 69.0 g of HCl in the following reaction?

$$CaO(s) + 2HCl(aq) \longrightarrow CaCl_2(aq) + H_2O(\ell)$$

38. How many grams of $Al_2(SO_4)_3$ form when 6.71 g of Al react with 12.95 g of H_2SO_4 according to the following equation?

$$2Al(s) + 3H_2SO_4(aq) \longrightarrow Al_2(SO_4)_3(aq) + 3H_2(g)$$

39. Aspirin, $C_9H_8O_4$, is prepared by heating salicylic acid, $C_7H_6O_3$, with acetic anhydride, $C_4H_6O_3$:

$$C_7H_6O_3 + C_4H_6O_3 \xrightarrow{\triangle} C_9H_8O_4 + C_2H_4O_2$$

If 4.00 g of salicylic acid are heated with 8.00 g of acetic anhydride, what is the theoretical yield? If the actual yield of aspirin is 4.36 g, what is the percent yield?

40. Oil of wintergreen (methyl salicylate) is prepared by heating salicylic acid, $C_7H_6O_3$, with methanol, CH_3OH:

$$C_7H_6O_3 + CH_3OH \longrightarrow C_8H_8O_3 + H_2O$$

If 4.50 g of salicyclic acid is reacted with excess methanol and the yield of oil of wintergreen is 3.73 g, what is the percent yield?

41. In an experiment 10.80 g of butanoic acid, $C_4H_8O_2$, was heated with 3.40 g of ethanol, C_2H_6O, to produce ethyl butanoate, $C_6H_{12}O_2$, and water. What is the theoretical yield of the ethyl butanoate? If the actual yield was 5.57 g, what is the percent yield?

42. Phosphorus oxide is extracted from phosphate-containing rocks by reaction with silicon dioxide found in sand:

$$2Ca_3(PO_4)_2(s) + 6SiO_2(s) \longrightarrow 6CaSiO_3(s) + P_4O_{10}(g)$$

If 1.00 t of sand that is 60% SiO_2 is used, how many kilograms of P_4O_{10} could be produced?

43. A tank of impure CO(g) contains 6.8 g of oxygen impurity per 100 kg of gas. The O_2(g) is removed by reacting it with cobalt(II) oxide:

$$4CoO(s) + O_2(g) \longrightarrow 2Co_2O_3(s)$$

How many grams of impure gas can be purified by passing it through a column containing 47.9 g of CoO(s)?

44. Titanium is produced from the mineral rutile, TiO_2, by a two-step process:

$$TiO_2(s) + 2Cl_2(g) + 2C(s) \longrightarrow TiCl_4(g) + 2CO(g)$$

$$TiCl_4(g) + 2Mg(s) \longrightarrow Ti(s) + 2MgCl_2(s)$$

How many kilograms of titanium can be obtained from a tonne of rutile?

45. Antimony is usually found as the mineral stibnite, Sb_2S_3. Pure antimony can be obtained by a two-step process:

$$2Sb_2S_3(s) + 9\ O_2(g) \longrightarrow Sb_4O_6(s) + 6SO_2(g)$$

$$Sb_4O_6(s) + 6C(s) \longrightarrow 4Sb(s) + 6CO(g)$$

How many grams of antimony can be obtained from a kilogram of stibnite?

46. Coal from a certain mine contains 2.8% S. When the coal is burned at a power generating station, the sulfur is converted to SO_2. The SO_2 is then reacted at the plant with CaO to form $CaSO_3$. If 1200 t of coal is burned at the power plant each day, what is the daily output of $CaSO_3$?

47. What is the concentration in moles per litre of a solution that contains 39.2 g of H_3PO_4 in 500 mL of solution?

48. What is the concentration in moles per litre of a solution that contains 100 g of Na_2SO_4 in 10.0 L of solution?

49. How many grams of $C_6H_{12}O_6$ are contained in 250 mL of a 0.050 mol/L solution of $C_6H_{12}O_6$ in water?

50. How many grams of $CuSO_4 \cdot 5H_2O$ are required to prepare 2.0 L of a 3.0 mol/L copper sulfate solution?

51. Sodium phosphate, Na_3PO_4 (known commercially as TSP), is used for cleaning grease and oil spills. Describe precisely how you would prepare 250 mL of a 0.320 mol/L solution of Na_3PO_4.

52. What volume of 0.14 mol/L hydrochloric acid would contain 5.0 g of HCl?

53. What volume of 0.95 mol/L Na_2SO_4 would be required to prepare 200 mL of 0.15 mol/L Na_2SO_4?

54. If 55.0 mL of 0.55 mol/L Na_2SO_4 are diluted to a final volume of 250 mL, what is the new concentration of the Na_2SO_4 in moles per litre?

55. To what final volume would 50.0 mL of 1.50 mol/L HNO_3 have to be diluted to prepare 0.45 mol/L HNO_3?

56. How many grams of $Ca(NO_3)_2$ can be prepared by reacting 125 mL of 5.00 mol/L HNO_3 with an excess of $Ca(OH)_2$?

$$2HNO_3(aq) + Ca(OH)_2(s) \longrightarrow Ca(NO_3)_2(aq) + 2H_2O(\ell)$$

57. If 0.200 g of Na_2CO_3 completely reacts with 30.0 mL of HCl, what is the concentration of the HCl in moles per litre?

$$Na_2CO_3(aq) + 2HCl(aq) \longrightarrow 2NaCl(aq) + H_2O(\ell) + CO_2(g)$$

58. If 50.0 mL of H_2SO_4 yields 0.300 g of $BaSO_4$ when reacted with excess $BaCl_2$, what is the concentration of the H_2SO_4 in moles per litre?

$$BaCl_2(aq) + H_2SO_4(aq) \longrightarrow BaSO_4(s) + 2HCl(aq)$$

59. How many grams of Fe^{2+} are required to react with 30.1 mL of 0.0165 mol/L $K_2Cr_2O_7$ solution?

$$14H^+(aq) + 6Fe^{2+}(aq) + K_2Cr_2O_7(aq) \longrightarrow 6Fe^{3+}(aq) + 2Cr^{3+}(aq) + $$
$$2K^+(aq) + 7H_2O(\ell)$$

60. What is the maximum number of grams of NaCl that can be produced when 50.0 mL of 0.120 mol/L NaOH reacts with 39.4 mL of 0.165 mol/L HCl?

$$NaOH(aq) + HCl(aq) \longrightarrow NaCl(aq) + H_2O(\ell)$$

61. How would you prepare 500 g of an aqueous solution containing 2.50% by mass of sodium chloride?

62. If concentrated ammonia solution is 27.0% NH_3 and has a density of 0.90 g/mL, what is the concentration of this solution in moles per litre?

63. If a concentrated solution of acetic acid is 99.5% $HC_2H_3O_2$ and has a density of 1.05 g/mL, what is the concentration of this acid in moles per litre?

64. The anesthetic methoxyflurane contains two fluorine atoms per molecule. If the chemical analysis shows that methoxyflurane is 23.0% F, what is its molecular mass?

Apply Your Understanding

1. One method of analyzing for arsenic in a pesticide is to treat the sample chemically to convert the arsenic into soluble sodium arsenate (Na_3AsO_4). Then a solution of silver nitrate is added until a precipitate of Ag_3AsO_4 is no longer formed. If a 1.10 g sample of a pesticide required 23.7 mL of 0.0968 mol/L $AgNO_3$ in a given analysis, what was the percentage of arsenic present in the pesticide?

2. Aspirin ($C_9H_8O_4$) is produced commercially from salicylic acid ($C_7H_6O_3$) and acetic anhydride ($C_4H_6O_3$) according to the equation

$$C_7H_6O_3 + C_4H_6O_3 \longrightarrow C_9H_8O_4 + HC_2H_3O_2$$

   a) If all of the salicylic acid is converted to aspirin, how much salicylic acid is required to prepare 175 kg of aspirin?

b) If only 75.0% of the salicylic acid is converted to aspirin, how much salicylic acid would be required?

c) If salicylic acid costs \$11.00/kg and acetic anhydride costs \$13.00/kg, which compound would you choose as the limiting reagent in order to have the most economical process?

d) What is the theoretical yield of aspirin if 205 kg of salicylic acid are allowed to react with 140 kg of acetic anhydride?

e) If the actual yield of aspirin from part (d) is 202 kg, what is the percent yield?

f) What would you have to charge for a kilogram of aspirin to cover the cost of the raw materials? (Ignore the cost of labour, electricity, machinery, taxes, etc.)

3. Iron(III) chloride can be prepared by reacting iron metal with hydrochloric acid. The other product is hydrogen.

a) What is the balanced equation for this reaction?

b) How many grams of iron are required to make 2.25 L of an aqueous solution containing 8.00% iron(III) chloride? The density of the solution is 1.067 g/mL.

4. Polychlorinated biphenyls (PCBs) are environmental pollutants with the general molecular formula $C_{12}H_{10-n}Cl_n$, where n is an integer. What is the value of n for a PCB that is found by analysis to contain 62.8% Cl?

5. One method of analyzing gold ores is to convert the gold to soluble $AuCl_3$ and treat the solution with an excess of a solution of KI. The reaction that occurs is

$$AuCl_3(aq) + 3KI(aq) \longrightarrow AuI(s) + I_2(aq) + 3KCl(aq)$$

The liberated iodine is then reacted with a solution of sodium thiosulfate $(Na_2S_2O_3)$ until all the iodine has disappeared. The equation for this reaction is

$$2Na_2S_2O_3(aq) + I_2(aq) \longrightarrow Na_2S_4O_6(aq) + 2NaI(aq)$$

If 28.8 mL of 1.00×10^{-4} mol/L $Na_2S_2O_3$ are required to react with the iodine generated by a 0.945 g sample of gold ore, what is the percentage of gold in the ore?

6. Vinyl chloride is the raw material for the production of a commercial plastic called PVC. Vinyl chloride contains only carbon, hydrogen, and chlorine, and it burns in air to form CO_2, H_2O, and HCl. If the combustion of 2.152 g of vinyl chloride produces 3.029 g of CO_2 and 1.257 g of HCl, what is the empirical formula of vinyl chloride?

7. A white substance, suspected of being cocaine, was first purified and then analyzed by a forensic chemist. A 44.10 mg sample produced 130.0 mg of CO_2 and 39.93 mg of H_2O. The sample was also found to contain 9.40% N. Determine the empirical formula of the substance. Was the substance cocaine?

8. A sample of solid $Zn(OH)_2$ is added slowly to 228 mL of 0.609 mol/L HBr. The resulting solution contains unreacted HBr. A 0.450 mol/L solution of NaOH is added, and 155 mL is required to react with all of the remaining HBr. What is the mass of the sample of $Zn(OH)_2$?

9. A 0.656 g sample of a mixture of Na_2SO_4 and $MgSO_4$ is dissolved in water, and an excess of $BaCl_2$ is added to the solution. All of the sulfate ion in the mixture is precipitated as $BaSO_4$. A mass of 1.256 g of $BaSO_4$ is obtained. Determine the percent of $MgSO_4$ in the mixture.

10. A mixture of pure AgCl and pure AgBr contains 68.9% Ag. What is the percentage of Cl in the mixture?

Investigations

1. Drinking water is frequently analyzed to determine the concentrations of iron, arsenic, and uranium. Write a research report on the methods used by chemists to determine the concentrations of these three elements in drinking water.

2. What are the accepted safe limits of iron, arsenic, and uranium in drinking water? What problems could result if these limits are exceeded? Write a report that answers these questions.

3. In consultation with your teacher, select a simple, safe reaction that can be used to illustrate the concept of a limiting reagent. Devise and conduct an experiment using your reaction to illustrate this concept. Report your results.

4. Write a research report on the different methods that have been used to determine Avogadro's number. Where does our present value of Avogadro's number come from?

12 Acids, Bases, and Acid Rain

CHAPTER CONTENTS

Part One: **The Nature of Acids and Bases**
12.1 Properties of Acids and Bases
12.2 Hydronium and Hydroxide Ions
12.3 Neutralization and Salts
12.4 Acid-Base Titration
12.5 Sources and Uses of Acids and Bases

Part Two: **Acids and Bases—Additional Concepts**
12.6 The pH Scale
12.7 Determining pH with Indicators
12.8 The Brønsted-Lowry Theory: A More General Theory of Acids and Bases

Part Three: **Acids in the Atmosphere**
12.9 Sulfur Dioxide: London-Type Smog
12.10 Acid Rain

What is pH? How is pH related to acids and bases? Why is it important to control the acidity of the water in swimming pools? How are major spills of corrosive acids cleaned up? How is the acidity of blood kept relatively constant? How do antacids work? What causes the paper in many books to disintegrate gradually with time? What is acid rain? You will find the answers to these questions as you study this chapter.

We begin by examining the properties and developing definitions of acids and bases. We focus on one particular property of acids and bases, the ability of acids to cause bases to lose their characteristic properties and vice versa. We discuss a number of uses for acids and bases, and explain the meaning of pH and the use of indicators. Finally, we examine the problem of acid rain.

Figure 12-1
Liming a lake lowers its acidity, but provides only a short-term solution to problems of pollution.

Key Objectives

When you have completed studying this chapter, you should be able to:
1. state operational and conceptual definitions of acids and bases.
2. distinguish between a strong acid and a weak acid.
3. write balanced chemical equations, ionic equations, and net ionic equations for acid-base neutralization reactions.
4. describe the steps in an acid-base titration.
5. calculate the concentration of an acid or base from titration data.
6. explain the meaning of the term *pH*.
7. describe the origin and effects of acid rain.

Part One: The Nature of Acids and Bases

12.1 Properties of Acids and Bases

Acids have certain common properties. Acids are electrolytes since they dissolve in water to give solutions that conduct electricity. Acidic solutions have a sour taste; for example, the acetic acid in vinegar is sour, as is citric acid in lemon juice. Solutions of acids turn certain dyes, called indicators, from one colour to another. For example, blue litmus paper is turned red by acids; pink phenolphthalein is made colourless by acids; and bromthymol blue is changed to yellow by acids. Acids react with certain metals to generate hydrogen gas:

$$Mg(s) + 2HCl(aq) \longrightarrow MgCl_2(aq) + H_2(g)$$

$$Zn(s) + H_2SO_4(aq) \longrightarrow ZnSO_4(aq) + H_2(g)$$

$$Fe(s) + 2HCl(aq) \longrightarrow FeCl_2(aq) + H_2(g)$$

Acids react with metallic carbonates and metallic hydrogen carbonates to generate carbon dioxide gas:

$$CaCO_3(s) + 2HCl(aq) \longrightarrow CO_2(g) + H_2O(\ell) + CaCl_2(aq)$$

$$NaHCO_3(aq) + HCl(aq) \longrightarrow CO_2(g) + H_2O(\ell) + NaCl(aq)$$

The reaction of an acid with $NaHCO_3$, sodium hydrogen carbonate, to produce carbon dioxide has a number of applications. Baking powder contains $NaHCO_3$, which is also known as sodium bicarbonate, bicarbonate of soda, or baking soda. The HCO_3^- ion reacts with an acid to form carbonic acid, H_2CO_3, which is unstable and decomposes to give CO_2 and H_2O. These CO_2 bubbles cause dough to rise in baking. A similar reaction occurs in certain fire extinguishers called soda-acid extinguishers.

Bases also have certain common properties. Like acids, bases dissolve in water to form solutions that conduct electricity. Basic solutions have a bitter taste. Solutions of bases, like solutions of acids, affect indicators. Red litmus is turned blue by bases; phenolphthalein is changed from colourless to pink by bases; and bromthymol blue is changed from yellow to blue by bases. (Notice that bases and acids have opposite effects on these indicators.) Basic solutions feel slippery, like soap.

Acids and bases are related to one another. An acid loses its characteristic properties when it is reacted with a base. A base loses its characteristic properties when it is reacted with an acid. In other words, an acid is neutralized by a base, and a base is neutralized by an acid. One property which is common to both acids and bases is not lost during the neutralization process. When an acid and a base react with one another, the resulting solution will still conduct electricity.

We can use a characteristic property of acids to form an operational definition of acids. An operational definition is based on the observed experimental behaviour of a category or class. For example, here is an operational definition of acids: acids are substances which turn blue litmus red. According to this definition, any solution which turns blue litmus red is an acidic solution. Here is an operational definition of bases: bases are substances which turn red litmus blue. According to this definition, any solution which turns red litmus blue is a basic solution.

We can use operational definitions to classify solutions quickly as acidic or basic. However, there are other definitions, such as the Arrhenius definitions, which add to our understanding of acids and bases. The definitions proposed by Arrhenius are introduced in the next section.

12.2 Hydronium and Hydroxide Ions

One substance which we normally consider to be a nonelectrolyte is, in fact, a very weak electrolyte. It dissociates slightly and is a poor conductor of electricity. That substance is water. Although water is a poor conductor of electricity, it is dangerous to use electrical equipment in wet areas because water does conduct electricity to some extent. When two water molecules collide, it is possible for a hydrogen ion to move from one water molecule to the second water molecule:

$$H_2O(\ell) + H_2O(\ell) \longrightarrow H_3O^+(aq) + OH^-(aq)$$

A hydronium ion (H_3O^+) and a hydroxide ion (OH^-) are formed. Both ions are hydrated.

This is a dissociation process, but only one out of every 300 million water molecules is involved in such a process. The dissociation of water may appear to be an unimportant process since such a small proportion of the water molecules actually dissociate. However, the ability to dissociate is one of water's most important properties. The quantities of hydronium ions and hydroxide ions present can, for example, have a great effect on many of the metabolic reactions in living cells.

In pure water, the number of hydronium ions must equal the number of hydroxide ions since they are both formed together during the dissociation process. Whenever the quantity of hydronium ions equals the quantity of hydroxide ions in any aqueous solution, the solution is neutral. It is possible to add substances to water which will increase either the quantity of hydronium ions or the quantity of hydroxide ions.

In 1889, Svante Arrhenius (Fig. 12-2), who was the first scientist to recognize the presence of ions in aqueous solutions, provided us with one of the first conceptual definitions of acids and bases. A conceptual definition attempts to explain why a certain category or class has its properties. According to Arrhenius, an acid is a substance that produces hydrogen ions, $H^+(aq)$, in water. These hydrogen ions always combine with at least one water molecule and are frequently symbolized in chemical equations as H_3O^+, called the **hydronium ion**. On the other hand, a base, according to Arrhenius, is a substance that produces hydroxide ions, $OH^-(aq)$, in water. The hydronium ion accounts for the chemical properties of acids, and the hydroxide ion accounts for the chemical properties of bases.

We have already seen that hydrogen chloride dissolves in water and dissociates to form hydronium ions and chloride ions. When hydrogen chloride dissolves in water it increases the number of hydronium ions present in the water, and it is therefore called an acid—hydrochloric acid. Any substance which increases the number of hydronium ions present in an aqueous solution is called an **acid**.

Strengths of Acids

The relative strength of an acid (or a base) depends on the degree to which it dissociates in solution. Hydrochloric acid is called a **strong acid** because it is completely dissociated in solution. Essentially, 100% of the hydrogen chloride molecules dissociate to form hydronium ions and chloride ions when the hydrogen chloride dissolves in water. The dissociation process is shown in these equations:

$$H\!:\!\ddot{O}\!: + H\!:\!\ddot{\underset{..}{Cl}}\!: \longrightarrow \left[H\!:\!\ddot{O}\!:\!H \right]^{+} + :\!\ddot{\underset{..}{Cl}}\!:^{-}$$
$$\underset{H}{} \qquad\qquad \underset{H}{}$$

$$H_2O(\ell) + HCl(g) \longrightarrow H_3O^+(aq) + Cl^-(aq)$$

Nitric acid is another example of a strong acid. It has one removable hydrogen atom which can be transferred to a water molecule:

$$HNO_3(\ell) + H_2O(\ell) \longrightarrow H_3O^+(aq) + NO_3^-(aq)$$

Not all acids dissociate completely. Acetic acid is called a **weak acid** because it dissociates only partially when it is dissolved in water:

$$HC_2H_3O_2(\ell) + H_2O(\ell) \longrightarrow H_3O^+(aq) + C_2H_3O_2^-(aq)$$

Biography

Figure 12-2
Svante August Arrhenius (1859-1927)—Defined acids and bases.

SVANTE ARRHENIUS was born in Wijk, Sweden, in 1859. In 1876, Arrhenius entered the University of Uppsala to study mathematics, chemistry, and physics. He was not satisfied with the instruction he was receiving in physics, so he went to Stockholm to study at the Academy of Sciences.

While in Stockholm, Arrhenius tried to gain an understanding of what occurs in a solution when an electric current is passed through it. From the results of his experiments, Arrhenius decided that electrolytes become split into positive and negative ions when they dissolve in water. He recorded his theory of ionization in the doctoral thesis which he submitted to the University of Uppsala. The science faculty at Uppsala was not impressed with his theory, and he was just barely awarded his doctorate.

Arrhenius believed in the theory of ionization, and he eventually gained the support of Ostwald and van't Hoff, two respected scientists. The Battle of the Ions began. Ostwald led the army of Ionians; his lieutenants were Arrhenius and van't Hoff. Their opponents included Lord Kelvin. The enemies of the Ionians were able to postpone temporarily the appointment of Arrhenius as professor at the University of Stockholm.

The Ionians were eventually victorious, and Arrhenius was appointed professor and two years later president of the University of Stockholm. In 1903, Arrhenius was awarded the Nobel Prize in Chemistry. In 1905, the King of Sweden formed the Nobel Institute for Physical Research at Stockholm, and Arrhenius was made director. The stormy period of his career had ended. He had progressed from being a scientific outcast to being a scientific oracle.

The majority of the acetic acid molecules dissolve as molecules, just as sugar dissolves as molecules. Only a small percentage of the acetic acid molecules actually dissociate. Weak acids dissociate slightly when they are dissolved in water.

Sulfuric acid is an acid which can transfer two hydrogen ions to two water molecules. The first hydrogen ion leaves the sulfuric acid molecule easily. Therefore sulfuric acid is a strong acid:

$$H_2SO_4(\ell) + H_2O(\ell) \longrightarrow H_3O^+(aq) + HSO_4^-(aq)$$

The second hydrogen ion can also leave the HSO_4^- ion, but it does not leave as readily as the first hydrogen ion:

$$HSO_4^-(aq) + H_2O(\ell) \longrightarrow H_3O^+(aq) + SO_4^{2-}(aq)$$

This is definitely a *two-step* process; however, it is often written as an overall reaction:

$$H_2SO_4(\ell) + 2H_2O(\ell) \longrightarrow 2H_3O^+(aq) + SO_4^{2-}(aq)$$

The term *weak acid* has no relation to the reactivity of the acid. For example, hydrofluoric acid, HF, is a weak acid. However, hydrofluoric acid reacts vigorously with many substances, including glass. In fact, hydrofluoric acid is used to frost and etch glass.

Strengths of Bases

Any substance which increases the number of hydroxide ions present in an aqueous solution is called a **base**. Typical bases are sodium hydroxide and ammonia:

$$NaOH(s) \longrightarrow Na^+(aq) + OH^-(aq)$$

$$NH_3(g) + H_2O(\ell) \longrightarrow NH_4^+(aq) + OH^-(aq)$$

Bases can also be classified as strong and weak. Typical strong bases are sodium hydroxide and potassium hydroxide. Since these bases are ionic, they dissociate 100% in aqueous solution:

$$NaOH(s) \longrightarrow Na^+(aq) + OH^-(aq)$$

$$KOH(s) \longrightarrow K^+(aq) + OH^-(aq)$$

Ammonia acts as a weak base since, as it dissolves in water, only a few of the ammonia molecules accept hydrogen ions from water molecules to cause an increase in the hydroxide ion concentration:

$$NH_3(g) + H_2O(\ell) \longrightarrow NH_4^+(aq) + OH^-(aq)$$

Table 12-1 lists some strong and weak acids and bases, and Table 12-2 shows the structures of some of these acids and bases. All of the acids listed in Table 12-1 dissociate in water and transfer H^+ ions to water molecules, forming hydronium ions, H_3O^+. All of the bases listed in Table 12-1 dissociate in water, releasing hydroxide ions, OH^-, into the water. Each of the weak bases contains

The label on this bottle was made by etching the glass.

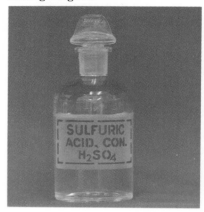

Table 12-1   Strong and Weak Acids and Bases

Strong Acids	
$HClO_4$	perchloric acid
HCl	hydrochloric acid
HNO_3	nitric acid
H_2SO_4	sulfuric acid
HBr	hydrobromic acid
HI	hydriodic acid
Weak Acids	
$HC_2H_3O_2$	acetic acid
HF	hydrofluoric acid
H_2CO_3	carbonic acid
HNO_2	nitrous acid
C_6H_5COOH	benzoic acid
Strong Bases	
NaOH	sodium hydroxide
KOH	potassium hydroxide
LiOH	lithium hydroxide
RbOH	rubidium hydroxide
$Ca(OH)_2$	calcium hydroxide
$Ba(OH)_2$	barium hydroxide
$Sr(OH)_2$	strontium hydroxide
Weak Bases	
NH_3	ammonia
$C_6H_5NH_2$	aniline
CH_3NH_2	methylamine
C_5H_5N	pyridine
N_2H_4CO	urea

a nitrogen atom with a lone pair of electrons. When these compounds are added to water, a hydrogen ion leaves the water molecule to form a coordinate covalent bond to the nitrogen and an OH⁻ ion remains.

Table 12-2   Structures of Some Acids and Bases

nitric acid	carbonic acid	methylamine
sulfuric acid	acetic acid	aniline
perchloric acid	ammonia	

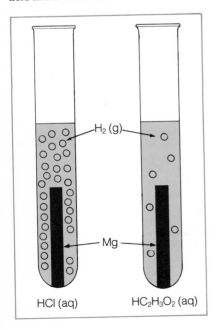

Figure 12-3
Reaction of magnesium with a strong acid and a weak acid

H₂ (g)

Mg

HCl (aq)   HC₂H₃O₂ (aq)

The strength of an acid or a base can affect the speed with which it will react with other chemicals. For instance, if we put equal amounts of magnesium into two test tubes and add hydrochloric acid to one of the tubes, we notice a vigorous reaction with many bubbles of hydrogen gas escaping. If we add acetic acid of the same volume and concentration to the other test tube, we see a much less vigorous reaction with fewer bubbles of hydrogen being given off (Fig. 12-3). The reason for the difference is that there are fewer hydronium ions present in the weak acetic acid to react with the magnesium, hence there is a slower reaction.

$$Mg(s) + 2H_3O^+(aq) \longrightarrow Mg^{2+}(aq) + H_2(g) + 2H_2O(\ell)$$

12.3 Neutralization and Salts

Water dissociates only slightly into hydronium ions and hydroxide ions. When hydronium ions and hydroxide ions are brought together from separate solutions, their natural tendency is to react to form water molecules:

$$H_3O^+(aq) + OH^-(aq) \longrightarrow 2H_2O(\ell)$$

This process is called **acid-base neutralization**. When hydronium ions from an acid solution react with an equal number of hydroxide ions from a basic solution, the original properties of these solutions no longer exist. This is because the hydronium ions account for the properties of an acid, and the hydroxide ions account for the properties of a base. In a neutral solution most of the hydronium ions and hydroxide ions have combined to form water. The few hydronium ions and hydroxide ions that remain are present in equal quantities.

Let us examine what happens when a solution of hydrochloric acid reacts with a solution of sodium hydroxide. The balanced chemical equation is

$$HCl(aq) + NaOH(aq) \longrightarrow NaCl(aq) + H_2O(\ell)$$

If we allow for the dissociation of the HCl,

$$HCl(g) + H_2O(\ell) \longrightarrow H_3O^+(aq) + Cl^-(aq)$$

the dissociation of the NaOH,

$$NaOH(s) \longrightarrow Na^+(aq) + OH^-(aq)$$

and the reacting of $H_3O^+(aq)$ and $OH^-(aq)$,

$$H_3O^+(aq) + OH^-(aq) \longrightarrow 2H_2O(\ell)$$

the result is the following ionic equation:

$$H_3O^+(aq) + Cl^-(aq) + Na^+(aq) + OH^-(aq) \longrightarrow 2H_2O(\ell) + Na^+(aq) + Cl^-(aq)$$

You may notice that the ionic equation has $2H_2O$ while the balanced equation has just H_2O. This is because in the dissociation of HCl, we have written $H_3O^+(aq)$ rather than $H^+(aq)$.

Notice that hydrated sodium ions and hydrated chloride ions occur both as reactants and as products in this equation. This means that they do not take part in the reaction at all. There are as many sodium ions present at the start of the reaction as at the end. Similarly, as many chloride ions are present at the start of the reaction as at the end. We have learned in Chapter 10 that ions which do not participate directly in the reaction are called spectator ions and can be cancelled from each side of the equation:

$$H_3O^+(aq) + \cancel{Cl^-(aq)} + \cancel{Na^+(aq)} + OH^-(aq) \longrightarrow 2H_2O(\ell) + \cancel{Na^+(aq)} + \cancel{Cl^-(aq)}$$

The actual net ionic equation which shows the acid-base neutralization process is therefore:

$$H_3O^+(aq) + OH^-(aq) \longrightarrow 2H_2O(\ell)$$

This net ionic equation describes the neutralization of any strong acid by any strong base and shows, once again, that hydronium ions react with hydroxide ions to form water. If the water is removed by evaporation, solid sodium chloride is formed:

$$Na^+(aq) + Cl^-(aq) \longrightarrow NaCl(s)$$

Sodium chloride is a member of a class of substances called salts. **Salts** consist of a positive ion of a base combined with the negative ion of an acid. An acid and a base neutralize one another to form water and ions which can combine to form a salt:

$$Acid + Base \longrightarrow Salt + Water$$

Another example of neutralization is the reaction of the base, potassium hydroxide, with nitric acid. Potassium hydroxide reacts with nitric acid to form water and a salt. Potassium ions and nitrate ions remain dissolved in the water. The balanced chemical equation is:

$$KOH(aq) + HNO_3(aq) \longrightarrow H_2O(\ell) + KNO_3(aq)$$

Writing the equations for dissociation, we obtain:

$$KOH(s) \longrightarrow K^+(aq) + OH^-(aq)$$

$$HNO_3(\ell) + H_2O(\ell) \longrightarrow H_3O^+(aq) + NO_3^-(aq)$$

Hydroxide and hydronium ions react to form water. The ionic equation is:

$$K^+(aq) + OH^-(aq) + H_3O^+(aq) + NO_3^-(aq) \longrightarrow 2H_2O(\ell) + K^+(aq) + NO_3^-(aq)$$

If we remove spectator ions, the resulting net ionic equation is

$$OH^-(aq) + H_3O^+(aq) \longrightarrow 2H_2O(\ell)$$

If the water is removed by evaporation, the salt, potassium nitrate, is collected as a solid:

$$K^+(aq) + NO_3^-(aq) \longrightarrow KNO_3(s)$$

EXAMPLE PROBLEM 12-1

Write **a)** the balanced chemical equation; **b)** the ionic equation; and **c)** the net ionic equation for the neutralization of perchloric acid, $HClO_4(aq)$, by lithium hydroxide, $LiOH(aq)$.

SOLUTION

a) The balanced chemical equation can be written if you remember that the negative ion from the acid joins with the positive ion from the base to form a salt. Hydrogen ions from the acid react with hydroxide ions from the base to form H_2O:

$$HClO_4(aq) + LiOH(aq) \longrightarrow LiClO_4(aq) + H_2O(\ell)$$

b) Before we write the ionic equation we should write the equations for the dissociation of $HClO_4$ and $NaOH$:

$$HClO_4(\ell) + H_2O(\ell) \longrightarrow H_3O^+(aq) + ClO_4^-(aq)$$

$$LiOH(s) \longrightarrow Li^+(aq) + OH^-(aq)$$

c) Once we are aware of these processes we are able to readily write the ionic equation:

$$H_3O^+(aq) + ClO_4^-(aq) + Li^+(aq) + OH^-(aq) \longrightarrow$$
$$Li^+(aq) + ClO_4^-(aq) + 2H_2O(\ell)$$

If we cancel and remove spectator ions we obtain the net ionic equation:

$$H_3O^+(aq) + OH^-(aq) \longrightarrow 2H_2O(\ell)$$

EXAMPLE PROBLEM 12-2

Write **a)** the balanced chemical equation; **b)** the ionic equation; and **c)** the net ionic equation for the neutralization of nitric acid, $HNO_3(aq)$, by calcium hydroxide, $Ca(OH)_2(aq)$.

SOLUTION

a) Joining the positive ion of the base to the negative ion from the acid to form a salt and reacting H^+ from the acid with OH^- from the base, we obtain the chemical equation:

$$HNO_3(aq) + Ca(OH)_2(aq) \longrightarrow Ca(NO_3)_2(aq) + H_2O(\ell)$$

This equation has the correct formulas for reactants and products, but it is not balanced. The balanced chemical equation is:

$$2HNO_3(aq) + Ca(OH)_2(aq) \longrightarrow Ca(NO_3)_2(aq) + 2H_2O(\ell)$$

b) Before writing the ionic equation, we write the equations for the dissociation of HNO_3 and of $Ca(OH)_2$.

$$2HNO_3(\ell) + 2H_2O(\ell) \longrightarrow 2H_3O^+(aq) + 2NO_3^-(aq)$$

$$Ca(OH)_2(s) \longrightarrow Ca^{2+}(aq) + 2OH^-(aq)$$

The ionic equation is therefore:

$$2H_3O^+(aq) + 2NO_3^-(aq) + Ca^{2+}(aq) + 2OH^-(aq) \longrightarrow Ca^{2+}(aq) + 2NO_3^-(aq) + 4H_2O(\ell)$$

Cancelling and removing spectator ions, we obtain

$$2H_3O^+(aq) + 2OH^-(aq) \longrightarrow 4H_2O(\ell)$$

c) Since balanced equations usually have the smallest whole number coefficients, we divide the coefficients in this equation by the highest common factor (in this case, 2) to give the net ionic equation:

$$H_3O^+(aq) + OH^-(aq) \longrightarrow 2H_2O(\ell)$$

PRACTICE PROBLEM 12-1

Write **a)** the balanced chemical equation; **b)** the ionic equation; and **c)** the net ionic equation for the neutralization of hydrochloric acid, $HCl(aq)$, by rubidium hydroxide, $RbOH(aq)$.

PRACTICE PROBLEM 12-2

Write **a)** the balanced chemical equation; **b)** the ionic equation; and **c)** the net ionic equation for the neutralization of sulfuric acid, $H_2SO_4(aq)$, by magnesium hydroxide, $Mg(OH)_2(aq)$.

Applications of the Neutralization Reaction

The human stomach contains hydrochloric acid which is necessary for the digestive process. Excessive secretion of stomach acid causes heartburn. Antacids are compounds that neutralize excess stomach acid. Magnesium hydroxide, $Mg(OH)_2$, known as milk of magnesia, is one of many antacids.

Soil that is too acidic can be reduced in acidity through acid-base neutralization reactions. In addition, lakes that have become acidified by acid rain are sometimes treated with a base to reduce their acidity.

In Denver, Colorado, in 1983, about 90 000 L of nitric acid spilled from a railway tank car. The spill was neutralized with sodium carbonate according to the equation

$$2HNO_3(aq) + Na_2CO_3(s) \longrightarrow 2NaNO_3(aq) + H_2O(\ell) + CO_2(g)$$

Smaller laboratory spills of acid are usually neutralized with $NaHCO_3(s)$.

Since the mid 1800s, book printers have employed a compound that prevents the ink from spreading on the paper. This compound, however, reacts with moisture in the book and in the air to produce sulfuric acid. The acid then attacks the cellulose fibres in the paper, with the result that the books turn yellow and eventually crumble into dust. Chemists have come to the rescue. Diethylzinc is a gas which can diffuse (spread itself) through the pages of a book to do two jobs. First, it neutralizes the acids present. Second, it leaves a residue of zinc oxycarbonate, which is basic, distributed evenly throughout the paper fibres. In this way, it protects the pages from future attack by acid. Modern printers use paper containing a base to neutralize acid attack.

12.4 Acid-Base Titration

The procedure of carefully measuring the volume of a solution required to react completely with a known amount of another chemical is known as a **titration**. Titrations are used in chemistry to analyze the compositions of mixtures. For example, the percentage of acetic acid in vinegar or the amount of vitamin C in apple juice can be determined by titration. The concentration of an acidic or basic solution can be determined by the technique of acid-base titration.

Suppose that we wish to determine the concentration of a known volume of hydrochloric acid. We could use two burets as in Figure 12-4. A buret is a glass tube with a device at the bottom which controls the flow of the solution out of the tube. It has graduations that measure any changes in the volume of solution at any point during the titration.

The hydrochloric acid of unknown concentration is titrated with a solution of sodium hydroxide of known concentration. The hydrochloric acid is placed in an Erlenmeyer flask, and phenolphthalein indicator is added. The sodium hydroxide solution is titrated from a buret into the flask containing the acid. The acid and the base react with one another according to the equation

$$HCl(aq) + NaOH(aq) \longrightarrow NaCl(aq) + H_2O(\ell)$$

During the first stages of the titration, the sodium hydroxide will be completely neutralized, and an excess of acid will remain. However, eventually there will

Figure 12-4
Two burets for acid-base titrations

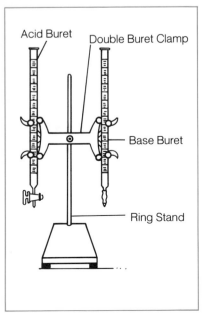

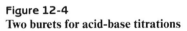

be a point, the theoretical endpoint, at which the acid and the base have neutralized one another exactly, and no more base should be added to the flask.

The phenolphthalein indicator is used to determine experimentally the point, called the experimental endpoint, at which the base has neutralized the acid. Phenophthalein is colourless in acid solution. It turns pink when the acid is completely neutralized and a slight excess of base is present. Because this experimental endpoint cannot be detected unless there is a slight excess of base, it differs slightly from the theoretical endpoint. In this titration, a successful endpoint is achieved if one drop of base turns the solution in the flask from colourless to pink. When this happens, the experimental endpoint differs from the theoretical endpoint by less than one drop of base.

Our goal is to determine the concentration of the hydrochloric acid in moles per litre. In Chapter 11, a simple equation was constructed from this concentration expression (moles per litre):

$$x \text{ mol/L} = \frac{y \text{ mol}}{z \text{ L}}$$

The original volume of the acid in the Erlenmeyer flask is already known. Therefore, to determine the concentration of HCl, we must first determine the number of moles of HCl present. How do we do this?

At the endpoint, the number of moles of hydrochloric acid used equals the number of moles of sodium hydroxide used. We know this because the balanced chemical equation for the reaction shows one mole of HCl reacting with one mole of NaOH. Since we know both the concentration in moles per litre and the volume in millilitres (which we can convert to litres) of the sodium hydroxide, we can calculate the number of moles of base used. The concentration of NaOH is expressed as x mol/L, where

$$x \text{ mol/L NaOH solution} = \frac{y \text{ mol NaOH}}{z \text{ L NaOH solution}}$$

We can cross-multiply this equation to determine the number of moles of NaOH:

$$y \text{ mol NaOH} = (x \text{ mol/L NaOH solution}) \times z \text{ L NaOH solution}$$

The number of moles of HCl in the measured volume of acid equals the number of moles of NaOH, and we can calculate the molarity of the acid.

Let us use some numbers and calculate the concentration of hydrochloric acid. Suppose we used a total of 25.00 mL of hydrochloric acid to neutralize 28.75 mL of 0.150 mol/L sodium hydroxide solution. We first calculate the number of moles of NaOH used in this titration:

$$y \text{ mol NaOH} = 0.150 \text{ mol/L NaOH solution} \times 0.028\ 75 \text{ L NaOH solution}$$

$$= 4.31 \times 10^{-3} \text{ mol NaOH}$$

Since the acid neutralized the base, equal moles of HCl and NaOH were used. Thus, 4.31×10^{-3} mol HCl was used, and the concentration of HCl is given by:

$$x \text{ mol/L HCl solution} = \frac{4.31 \times 10^{-3} \text{ mol HCl}}{0.025\ 00 \text{ L HCl solution}}$$

$$= 0.173 \text{ mol/L HCl solution}$$

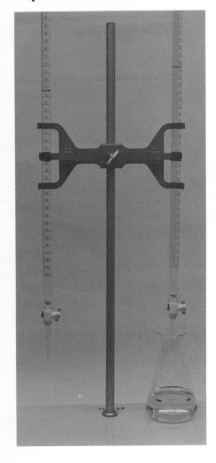

A pair of burets set up for an acid-base titration. The pink colour of the solution in the flask indicates the endpoint of the titration.

EXAMPLE PROBLEM 12-3

In an acid-base titration, 17.50 mL of hydrochloric acid were neutralized completely by 23.50 mL of 0.250 mol/L sodium hydroxide. Calculate the concentration of the hydrochloric acid.

SOLUTION

The first step is to write a balanced chemical equation for the acid-base neutralization reaction taking place:

$$HCl(aq) + NaOH(aq) \longrightarrow NaCl(aq) + H_2O(\ell)$$

Next, calculate the number of moles of solute in the solution whose concentration and volume are both known. In this case, we know the concentration and volume of the sodium hydroxide solution, and we can calculate the number of moles of NaOH used:

$$y \text{ mol} = 0.250 \text{ mol/L} \times 0.023\ 50 \text{ L}$$

$$= 5.88 \times 10^{-3} \text{ mol}$$

The balanced chemical equation for the reaction tells us that 1 mol of NaOH is required to neutralize 1 mol of HCl. Therefore, since 5.88×10^{-3} mol of NaOH were used, 5.88×10^{-3} mol of HCl were neutralized. We are now able to calculate the concentration of the acid. The concentration of hydrochloric acid is given by:

$$x \text{ mol/L} = \frac{5.88 \times 10^{-3} \text{ mol}}{0.017\ 50 \text{ L}} = 0.336 \text{ mol/L}$$

EXAMPLE PROBLEM 12-4

In an acid-base titration, 32.75 mL of sodium hydroxide were neutralized completely by 18.45 mL of 0.120 mol/L nitric acid. Calculate the concentration of the sodium hydroxide solution.

SOLUTION

The balanced chemical equation for the neutralization reaction is:

$$NaOH(aq) + HNO_3(aq) \longrightarrow NaNO_3(aq) + H_2O(\ell)$$

In this problem we are able to calculate the number of moles of HNO_3 used since we know the volume of the solution and the concentration of the HNO_3:

$$y \text{ mol} = 0.120 \text{ mol/L} \times 0.018\ 45 \text{ L}$$

$$= 2.21 \times 10^{-3} \text{ mol}$$

The balanced chemical equation for the reaction indicates that 1 mol of HNO_3 is required to neutralize 1 mol of NaOH. Therefore, since 2.21×10^{-3} mol of HNO_3 were used, 2.21×10^{-3} mol of NaOH were neutralized. We are now able to calculate the concentration of the sodium hydroxide. The concentration of the sodium hydroxide is given by:

$$x \text{ mol/L} = \frac{2.21 \times 10^{-3} \text{ mol}}{0.032\ 75 \text{ L}} = 0.0676 \text{ mol/L}$$

EXAMPLE PROBLEM 12-5

In an acid-base titration, 24.00 mL of sulfuric acid were neutralized completely by 37.50 mL of 0.250 mol/L potassium hydroxide. Calculate the concentration of the sulfuric acid.

SOLUTION

The balanced chemical equation for the neutralization reaction is:

$$H_2SO_4(aq) + 2KOH(aq) \longrightarrow K_2SO_4(aq) + 2H_2O(\ell)$$

In this problem we are able to calculate the number of moles of KOH used since we know the concentration and volume of its solution.

$$y \text{ mol} = 0.250 \text{ mol/L} \times 0.037\ 50 \text{ L}$$

$$= 9.38 \times 10^{-3} \text{ mol}$$

This problem differs from the preceding example problems because this is not a reaction in which 1 mol of acid reacts with 1 mol of base. The balanced chemical equation indicates that 2 mol of KOH react with 1 mol of H_2SO_4. Therefore the number of moles of sulfuric acid is given by:

$$9.38 \times 10^{-3} \text{ mol of KOH} \times \frac{1 \text{ mol of } H_2SO_4}{2 \text{ mol of KOH}} = 4.69 \times 10^{-3} \text{ mol of } H_2SO_4$$

The concentration of the sulfuric acid is given by:

$$x \text{ mol/L} = \frac{4.69 \times 10^{-3} \text{ mol}}{0.024\ 00 \text{ L}} = 0.195 \text{ mol/L}$$

PRACTICE PROBLEM 12-3

In an acid-base titration, 23.30 mL of hydrochloric acid were neutralized completely by 19.50 mL of 0.315 mol/L potassium hydroxide. Calculate the concentration of the hydrochloric acid.
(*Answer*: 0.264 mol/L HCl)

PRACTICE PROBLEM 12-4

In an acid-base titration, 15.65 mL of sodium hydroxide were neutralized completely by 22.50 mL of 0.250 mol/L hydrochloric acid. Calculate the concentration of the sodium hydroxide.
(*Answer*: 0.359 mol/L NaOH)

PRACTICE PROBLEM 12-5

In an acid-base titration, 15.00 mL of carbonic acid, $H_2CO_3(aq)$, were neutralized completely by 18.25 mL of 0.180 mol/L potassium hydroxide. Calculate the concentration of the carbonic acid.
(*Answer*: 0.110 mol/L H_2CO_3)

APPLICATION

Restoring the Challenger Tapes

In January 1986, the Space Shuttle Challenger exploded and disintegrated shortly after take-off, killing all seven astronauts on board. The tape of the astronauts' cockpit conversations was found by navy divers under 28 m of water. The tape had been exposed to seawater for about six weeks and had been damaged by the resulting chemical reactions.

When the container was opened, the tapes were "a foaming, concrete-like mess, all glued together." NASA tried to restore the tape through the usual procedures of washing in chilled, distilled water and drying. However, these techniques failed.

NASA turned the tape over to a 16-member team of chemists, physicists, mechanical engineers, and specialists in other fields. The main problem was that seawater had reacted with the magnesium in the tape reel to produce magnesium hydroxide. The magnesium hydroxide had then covered the tape layers and glued them together.

The team used nitric acid to wash the tape in a special bath designed to recirculate the liquid continuously. This bath was also pressurized to force the acid between the layers of tape. The nitric acid neutralized the basic magnesium hydroxide. The bath was filled alternately with dilute nitric acid and distilled water.

The next step in the restoration process was to use a bath of methanol to remove water from the tape. Once the tape was dry, it was lubricated with methylsilicone. It was then unwound by hand onto a new reel, a task that took over 12 hours. The data on the tape was rerecorded onto a fresh tape. In the final step of this tedious restoration process, technicians at NASA's Johnson Space Center converted the digital data on the new tape to the sound of the astronauts' voices.

12.5 Sources and Uses of Acids and Bases

Acids and bases have many important uses. Table 12-3 shows physical properties and uses for some important acids and bases. The chemicals in Table 12-3 include those chemicals produced in the largest amounts by the chemical industry. Of these, sulfuric acid is the one produced in the largest quantity.

Table 12-3   Physical Properties and Uses of Important Acids and Bases

Chemical	Properties	Uses
Sulfuric acid, $H_2SO_4(aq)$	clear, colourless, odourless, viscous, oily liquid. very corrosive.	fertilizers, drugs, paints, rayon fibres, cleaning of metals, automobile batteries, explosives, plastics, dyes, pigments, petroleum refining.
Hydrochloric acid, $HCl(aq)$	clear, colourless liquid. pungent odour.	cleaning metal surfaces, removing excess mortar from brick buildings, and in the petroleum and pharmaceutical industries.
Nitric acid, $HNO_3(aq)$	colourless liquid. fumes in moist air. choking odour.	production of fertilizers, explosives, plastics, drugs, dyes, engraving copper plates, and in metallurgy.
Phosphoric acid, $H_3PO_4(aq)$	clear, syrupy liquid.	fertilizers, detergents, metal-cleaning preparations, and in pharmaceutical and food industries (e.g., as flavouring in soft drinks such as root beer and cola).
Sodium hydroxide, $NaOH(s)$	white waxy solid. very corrosive.	paper industry, petroleum refining, in manufacture of cellophane, rayon, soap, textiles, drain cleaners, oven cleaners.
Aqueous ammonia, $NH_3(aq)$	clear, colourless solution. very pungent odour.	fertilizers, cleaning agents (e.g., window cleaners).

Acids and bases are found in foods and beverages. For example, carbonic acid, $H_2CO_3(aq)$, is present in soft drinks; ascorbic acid (vitamin C) is often added to many food products; tannic acid is found in tea; lactic acid is found in milk, sour cream, and other dairy products; and caffeine is a weak base found in coffee, tea, and colas.

Some medicines contain acids and bases. Aspirin, ASA, is acetylsalicylic acid. Antacids contain bases such as magnesium hydroxide. A number of toiletries contain acids or bases. Boric acid is found in some types of mouthwash.

Acids and bases are found in our bodies. Hydrochloric acid is found in the digestive fluid in the stomach. Lactic acid is formed during muscular exercise. It is the concentration of lactic acid, as it builds up, that produces pain if the muscle is used too strenuously and for too long a period of time. Amino acids are important building blocks of proteins in our bodies.

Acids and bases play a role in industry. The mining industry uses acids and bases to extract metals from ores. Aniline is a weak base which is used in many chemical reactions, in particular those involving the manufacture of synthetic dyes. Pyridine is another weak base which is used in the production of pharmaceuticals such as sulfa drugs and antihistamines. Perchloric acid is used to

Indicator sticks dipped into a solution will indicate a pH by matching a set of colours on the chart.

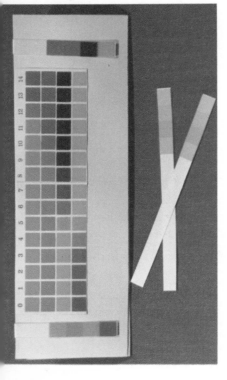

make fireworks, explosives, and matches. Hydrofluoric acid is used commercially to etch glass.

There are still other uses for acids and bases. Slaked lime, $Ca(OH)_2$, is used in mortar for building construction. Acids and bases are used in qualitative analysis of aqueous solutions. Water treatment plants use acids and bases. Oxalic acid is used in products that clean copper and brass, and removes rust. Formic acid is the irritative component in the sting of bees and the "bite" of many types of ants.

Part One Review Questions

1. One useful operational definition of an acid describes an acid as a substance which turns blue litmus red. What is another useful operational definition of an acid?
2. Acids cause certain indicators to change colour. Why is this statement, although true, not a useful operational definition of an acid?
3. What are the Arrhenius definitions of acids and bases?
4. Explain the meaning of the terms *strong acid* and *weak acid*. Give an example of each.
5. Write the balanced chemical equation, the ionic equation, and the net ionic equation for each of the following acid-base neutralization reactions:
   a) $NaOH(aq)$ and $H_2SO_4(aq)$
   b) $NaOH(aq)$ and $HNO_3(aq)$
6. What are the usual products of an acid-base neutralization reaction?
7. What is an acid-base titration?
8. Why is an indicator used during a titration?
9. List three uses for each of the following: a) sulfuric acid; b) nitric acid; c) sodium hydroxide.
10. Give three examples of acids and three examples of bases found in the household.

Oxalic acid is used in copper-cleaning products.

Part Two: Acids and Bases— Additional Concepts

12.6 The pH Scale

Chemists normally work with dilute solutions of acids and bases. For example, it is not uncommon for them to use solutions containing as little as 3.0×10^{-9} mol/L of hydronium ions. In 1909, the Danish biochemist S.P. Sørensen proposed a more suitable method of expressing the acidity of a solution. He proposed that the acidity of a solution be expressed in terms of a quantity known as pH (the potency of hydrogen). Recall that a hydrogen ion in water is really a hydronium ion, $H_3O^+(aq)$. Thus, the pH of a solution is related to the quantity of hydronium ions in a given amount of solution. The term pH has become widely used. Advertisers refer to the pH of grooming products; biologists refer to the pH of lake water; oceangraphers refer to the pH of seawater; and chemists refer to the pH of a solution.

The normal range of pH values is from 0 to 14. The pH scale is defined in such a way that the numerical pH value becomes lower as a solution becomes more acidic. For example, there are 10 times as many H_3O^+ ions in one litre of a solution of pH 3 as in one litre of a solution of pH 4.

The scale can be understood by looking at the mathematical definition of pH:

$$pH = -\log[H_3O^+]$$

The brackets which surround the H_3O^+ are used to indicate that the concentration of the ion is measured in moles per litre. Thus, this mathematical equation tells us that the **pH** of a solution is equal to the negative logarithm of the molar concentration of the hydronium ion. For each unit of pH increase, the concentration of hydronium ions decreases by a factor of ten. The relationship between hydronium ion concentration and pH can also be seen by writing the above equation as $[H_3O^+] = 10^{-pH}$. Thus, if the pH of a solution is 1, $[H_3O^+] = 10^{-1}$ mol/L; if the pH of a solution is 2, $[H_3O^+] = 10^{-2}$ mol/L; if the pH of a solution is 3, $[H_3O^+] = 10^{-3}$ mol/L; and so on.

In any neutral aqueous solution the quantity of H_3O^+ ions equals the quantity of OH^- ions, and the pH is 7. When the quantity of H_3O^+ ions exceeds the quantity of OH^- ions, the solution is acidic, and the pH is less than 7. On the other hand, if the quantity of OH^- ions exceeds the quantity of H_3O^+ ions, the solution is basic, and the pH is greater than 7. The following diagram shows how one can qualitatively use the pH scale to assess the acidity or basicity of a solution:

0 ◄————————————— 7 ————————————► 14

Very Acidic   Acidic   Neutral   Basic   Very Basic

The pH values of some common substances are given in Table 12-4.

Table 12-4 pH Values of Some Common Substances

Substance	pH
Gastric fluid	1.7
Lemon juice	2.6
Vinegar	2.8
Soft drinks	3.0
Apples	3.1
Grapefruit	3.1
Oranges	3.5
Tomatoes	4.5
Black coffee	5.1
Saliva	5.7-7.1
Milk	6.5
Pure water	7.0
Blood	7.4
Tears	7.4
Eggs	7.8
Sea water	7.8
Milk of magnesia	9
Detergents	10
Household ammonia	11
Lye	14

The pH of swimming pool water is controlled to prevent the growth of algae and bacteria. Acids such as hydrochloric acid and hypochlorous acid are added to lower the pH and kill bacteria.

Advertisers talk about pH-balanced shampoo. Hair is made up of protein and is slightly acidic. We would thus expect the acidity or basicity of a shampoo to affect hair. Therefore, most shampoos have a pH value between 5 and 8, so as not to differ too much from the hair's natural pH. Shampoos with a pH too low or too high would damage the hair as well as irritate the eyes.

The pH is different in different parts of the body. The pH of gastric juice is about 1.5. The pH of arterial blood is 7.4. The blood contains a mixture of chemicals called buffers which keeps the pH relatively constant. If there is a large enough change in pH, there is interference with the blood's capacity to carry oxygen and remove carbon dioxide. Thus, even relatively small changes in the pH of blood could result in death.

The pH of a solution is determined by using an electronic pH meter (Fig. 12-5). Another experimental, but less accurate, method of determining pH, is through the use of indicators.

**Figure 12-5
A pH meter**

12.7 Determining pH with Indicators

As we have seen, **indicators** are dyes which indicate the acidity or basicity of a solution. Some act as weak acids and others as weak bases in the presence of acidic and basic solutions. An indicator has one colour below a certain pH, and another colour above a second pH. It has an intermediate colour between these two pH values (Table 12-5)

Indicators can be used to give us an estimate of the pH or acidity of a solution. Suppose we are given an unknown solution and asked to find its pH. We would divide the solution into a number of parts and add a few drops of bromthymol blue indicator to one part. If this solution turns yellow we know that it is acidic and that the pH is between 0.0 and 6.0. There are many other indicators that can be used to test the remaining portions to determine the actual value of the pH.

Dyes that are common in natural substances are also indicators. Juices such as those of red cabbage, black cherries, and beets show colour changes in the presence of acids and bases. One of the most common acid-base indicators used in the school laboratory is litmus paper. Litmus paper contains a dye which is extracted from a species of lichen. Litmus is red in the presence of acids and blue in the presence of bases.

Table 12-5   pH Range of Indicators

Indicator	pH	Colour
Methyl red	0.0-4.2	red
	4.2-6.3	orange
	6.3-14.0	yellow
Litmus	0.0-5.0	red
	5.0-8.0	purple
	8.0-14.0	blue
Bromthymol blue	0.0-6.0	yellow
	6.0-7.6	green
	7.6-14.0	blue
Phenol-phthalein	0.0-8.0	colourless
	8.0-9.6	pink
	9.6-14.0	red

12.8 The Brønsted-Lowry Theory: A More General Theory of Acids and Bases

In Section 12.2 we presented the Arrhenius definition of acids and bases. This definition works quite well. However, most chemists use a more general definition of acids and bases, the Brønsted-Lowry definition. This definition is named after the two scientists who presented it. According to the **Brønsted-Lowry Theory**, an acid is a *proton* (that is, H^+ ion) *donor* and a base is a *proton acceptor*.

This definition is not limited to solutions where water is the solvent. It applies also to other systems. For example, gaseous hydrogen chloride donates a proton to gaseous ammonia to form an ammonium ion (NH_4^+) and a chloride ion (Cl^-). These ions immediately combine to form solid ammonium chloride:

$$HCl(g) + NH_3(g) \longrightarrow NH_4Cl(s)$$

$$Cl^{\delta-}\!\!-\!\!H^{\delta+} \ldots\ldots {}^{\delta-}\overset{\overset{\displaystyle H^{\delta+}}{\diagup}}{\underset{\underset{\displaystyle H}{\diagdown}{}^{\delta+}}{N}}\!\!-\!\!H^{\delta+} \rightarrow Cl^- + \left[\overset{\overset{\displaystyle H}{\diagup}}{\underset{\underset{\displaystyle H\,H}{|\diagdown}}{H\!\!-\!\!N}}\right]^+ \rightarrow NH_4Cl(s)$$

The hydrogen chloride acts as an acid, and the ammonia acts as a base.

When acetic acid dissolves in water, an acetic acid molecule donates a proton to a water molecule. The acetic acid is a Brønsted-Lowry acid:

$$HC_2H_3O_2(\ell) + H_2O(\ell) \rightleftharpoons H_3O^+(aq) + C_2H_3O_2^-(aq)$$

The double arrows in these two equations indicate that an equilibrium is established. The rate of the forward reaction equals the rate of the reverse reaction, and therefore there is no further change in the concentrations of reactants or products.

In the reaction between acetic acid and water, hydronium ions are formed. The H_3O^+ ion can donate a proton to a base. The substance formed by addition of a proton to a base is called the *conjugate acid* of that base. Thus, H_3O^+ is the conjugate acid of the base, H_2O. Similarly, the $C_2H_3O_2^-$ ion can accept a proton from an acid. The substance formed by the loss of a proton from an acid is called the *conjugate base* of that acid. Thus, $C_2H_3O_2^-$ is the conjugate base of the acid, $HC_2H_3O_2$.

The following series of equations will illustrate the use of the Brønsted-Lowry Theory:

$$\begin{array}{lllll}
HCl(g) & + \; H_2O(\ell) & \rightleftharpoons \; H_3O^+(aq) & + \; Cl^-(aq) \\
\textit{Acid} & \textit{Base} & \textit{Conjugate Acid} & \textit{Conjugate Base}
\end{array}$$

$$\begin{array}{lllll}
HNO_3(\ell) & + \; H_2O(\ell) & \rightleftharpoons \; H_3O^+(aq) & + \; NO_3^-(aq) \\
\textit{Acid} & \textit{Base} & \textit{Conjugate Acid} & \textit{Conjugate Base}
\end{array}$$

$$\begin{array}{lllll}
H_2O(\ell) & + \; NH_3(g) & \rightleftharpoons \; NH_4^+(aq) & + \; OH^-(aq) \\
\textit{Acid} & \textit{Base} & \textit{Conjugate Acid} & \textit{Conjugate Base}
\end{array}$$

$$\begin{array}{lllll}
H_2O(\ell) & + \; H_2O(\ell) & \rightleftharpoons \; H_3O^+(aq) & + \; OH^-(aq) \\
\textit{Acid} & \textit{Base} & \textit{Conjugate Acid} & \textit{Conjugate Base}
\end{array}$$

In two of the four acid-base reactions shown here, water acts only as a base. However, in one of these reactions, water is donating a proton to ammonia. In this case water is acting as an acid. In the fourth reaction, one water molecule donates a proton to a second water molecule. Compounds which are capable of acting as either acids or bases, depending upon the substance with which they are reacting, are called **amphoteric substances**. There are many amphoteric substances. One example is the hydrogen carbonate ion, HCO_3^-. This ion can lose a proton to a hydroxide ion, or it can gain a proton from a hydronium ion:

$$\begin{array}{llll}
HCO_3^-(aq) + OH^-(aq) & \longrightarrow & CO_3^{2-}(aq) & + \; H_2O(\ell) \\
\textit{Acid} \quad\quad \textit{Base} & & \textit{Conjugate Base} & \textit{Conjugate Acid}
\end{array}$$

$$\begin{array}{llll}
HCO_3^-(aq) + H_3O^+(aq) & \longrightarrow & H_2CO_3(aq) & + \; H_2O(\ell) \\
\textit{Base} \quad\quad \textit{Acid} & & \textit{Conjugate Acid} & \textit{Conjugate Base}
\end{array}$$

Table 12-6   Two Acid-Base Theories

	Arrhenius Theory	Brønsted-Lowry Theory
Acid	any substance that produces H^+ ions in water	any substance that is a proton donor
Base	any substance that produces OH^- ions in water	any substance that is a proton acceptor

The Brønsted-Lowry Theory of acids and bases includes more chemical systems than does the Arrhenius Theory of acids and bases. The Arrhenius Theory describes acids and bases in aqueous solutions. The Brønsted-Lowry Theory also describes acids and bases in aqueous solutions, but it is capable of describing acids in nonaqueous systems such as the gas phase or nonaqueous solvents. The two theories are compared in Table 12-6.

Part Two Review Questions

1. How is the pH of a solution mathematically related to the hydronium ion concentration in the solution?
2. Is a solution with a pH of 5 an acid or a base? Explain.
3. What are acid-base indicators?
4. Describe how you would determine the pH of a solution using indicators.
5. Methyl red turns yellow when it is mixed with a certain solution; however, phenolphthalein remains colourless when it is mixed with the same solution. Within what range is the pH of this solution?
6. According to the Brønsted-Lowry Theory, what is the definition of **a)** an acid; **b)** a base?
7. Identify the conjugate acid and the conjugate base in each of these reactions:
   a) $HC_2H_3O_2(aq) + F^-(aq) \rightleftharpoons C_2H_3O_2^-(aq) + HF(aq)$
   b) $HCO_3^-(aq) + NH_3(aq) \rightleftharpoons CO_3^{2-}(aq) + NH_4^+(aq)$
8. Write chemical equations that demonstrate that the hydrogen sulfide ion, HS^-, is an amphoteric substance.

Part Three: Acids in the Atmosphere

There is much concern about acid rain. In this part of the chapter we will begin with a few historical cases of acids in the atmosphere. We examine Copperhill, Tennessee and London, England. Then we discuss the present problems of acid rain, both internationally and in Canada.

12.9 Sulfur Dioxide: London–Type Smog

Copperhill, Tennessee, provides us with an early example of the effects of acids in the atmosphere. Air pollution damage near Copperhill was as dramatic in magnitude as the damage caused by an earthquake or a tidal wave. Copper ore has been mined and smelted there since 1847. In the smelting process copper ore containing copper(I) sulfide is converted to copper and sulfur dioxide:

$$Cu_2S(s) + O_2(g) \xrightarrow{\Delta} 2\ Cu(s) + SO_2(g)$$
copper(I) sulfide

The sulfur dioxide was discharged into the air until 1917. After 1917, the sulfur dioxide was used to make sulfuric acid. During the period from 1847 to 1917, the sulfur dioxide simply killed all vegetation in the area of Copperhill.

Another dramatic example of the effects of acids in the atmosphere is provided by **London-type smog**. The mixing of moist air from above the nearby warm Gulf Stream and air from above the frigid Arctic currents frequently causes fogs in London. However, the fog of December 1952 was unusual. It was especially thick and cold. Because it was so cold, Londoners fired their coal heaters hotter and hotter. The thick fog conditions did not let the smoke disappear into the upper atmosphere. The smoke mixed with the fog and became increasingly thicker. For five days, 12 million people filled their lungs with this smog (smoke + fog). Deaths from bronchitis jumped to nine times the usual rate; deaths from pneumonia increased fourfold. Over 4000 deaths were attributed to the effects of this smog. Those fatally affected were usually 45 years old or older, but the mortality rate of infants under one year also rose.

What was the killer, and how did it attack people? The killer was sulfur dioxide, and its source was coal. Both fossil fuels, coal and petroleum, were formed from once-living plant material. Along with the plant carbohydrates, there was sulfur-containing protein material. As a result, sulfur remains as a contaminant in fossil fuels. Low-grade coal contains about 8% of the sulfur impurity. When such fuel is burned, the sulfur also undergoes combustion. Typical reactions are

$$S + O_2 \longrightarrow SO_2$$

$$2H_2S + 3 O_2 \longrightarrow 2SO_2 + 2H_2O$$

Much of the sulfur dioxide is converted in the atmosphere to sulfur trioxide:

$$2SO_2 + O_2 \longrightarrow 2SO_3$$

This sulfur trioxide can then dissolve in the water vapour of a fog or in the water in the human body to form sulfuric acid:

$$H_2O + SO_3 \longrightarrow H_2SO_4$$

The Londoners in 1952 were actually breathing a mist of dilute sulfuric acid! This sulfuric acid irritated the lungs, resulting in oxygen starvation and heart failure. Sulfuric acid, which is produced in the atmosphere, eventually returns to earth as acid rain.

12.10 Acid Rain

In this section we shall look at a global problem, acid rain. We shall examine the origin and effects of acid rain both world-wide and in Canada. Finally, we shall examine what can be done to reduce acid rain and, in particular, how Canada proposes to deal with the problem.

Origin

The term *acid rain* originated more then a century ago. It was coined by Robert Angus Smith, who studied the atmospheric conditions in an industrialized area of England.

Acid rain consists mainly of acids formed in the atmosphere from the oxides of sulfur, SO_2 and SO_3, and to a lesser extent, the oxides of nitrogen, NO and NO_2. These acids are mainly sulfuric acid (H_2SO_4) and nitric acid (HNO_3). While sulfur dioxide has been given much publicity for its role in acid rain, it should be pointed out that nitrogen oxides are responsible for about one-third of the acidity in precipitation.

Sulfur dioxide is produced from a number of sources. These sources include the burning of fossil fuels (coal, oil, and natural gas) which contain sulfur. Coal-fired electricity generation and industrial and residential heating are examples of the ways in which fuels are burned. Other sources include petroleum refining and sulfite pulping. Another major source of SO_2 emissions is the smelting of ores to extract such metals as nickel, copper, lead, and zinc. The ores of these metals contain sulfur. When they are strongly heated in air, SO_2 is one of the products of combustion. For example, in Sudbury, nickel is extracted from its ore by the reaction

$$NiS(s) + O_2(g) \longrightarrow Ni(s) + SO_2(g)$$

Hydrogen sulfide is also a source of sulfur dioxide. Hydrogen sulfide is emitted into the atmosphere when volcanoes erupt. It is also found in "sour" natural gas and is produced when sulfur is removed from petroleum at oil refineries. In the atmosphere, hydrogen sulfide is converted into sulfur dioxide:

$$2H_2S(g) + 3\ O_2(g) \longrightarrow 2SO_2(g) + 2H_2O(g)$$

As we saw in Section 12.9, sulfur dioxide can be converted in the atmosphere to sulfur trioxide which dissolves in fog, mist, or rain, forming sulfuric acid.

The oxides of nitrogen are produced during combustion reactions at high temperatures. These oxides include emissions from automobiles, trucks, machines, power plants, and industrial sources.

A typical reaction involving nitrogen in the formation of acid rain is the direct combination of nitrogen and oxygen at the high temperature inside automobile engines:

$$N_2(g) + O_2(g) + heat \longrightarrow 2NO(g)$$

This nitrogen monoxide immediately reacts with oxygen to form nitrogen dioxide:

$$2NO(g) + O_2(g) \longrightarrow 2NO_2(g)$$

Nitrogen dioxide dissolves in fog, mist, and rain to form nitric acid:

$$3NO_2(g) + H_2O(\ell) \longrightarrow 2HNO_3(aq) + NO(g)$$

Acid rain may actually be in the form of fog, mist, rain, or snow. Normal rain has a pH of 5.6, because CO_2 in the air dissolves in water to form carbonic acid, a weak acid. Therefore, rain which we consider to be acid has a pH less than 5.6. One of the lowest pH levels in rain ever recorded was in 1974, during a storm in Scotland. That rain was said to have a pH of 2.4.

Acid rain may affect mainly the local area, or it may be transported by the atmosphere many hundreds or thousands of kilometres to affect other parts of a country or even other countries.

Effects of Acid Rain

Let us first consider the effect of acid rain on humans. It irritates the whole respiratory tract, from the mucous membranes in the nose and throat to the tissues in the lungs. As a result, it poses serious problems for people with respiratory ailments. The U.S. Council on Environmental Quality has estimated that health-related problems which are directly related to acid precipitation cost the United States $2 billion annually.

Acid rain is especially destructive to vegetation, affecting crop yield and forest productivity.

Some rivers and lakes are naturally buffered against acid rain. That is, they contain chemicals that will react with the acid precipitation and neutralize it. Such bodies of water are, therefore, not affected by acid rain as much as those that possess only limited quantities of these chemicals. In other cases, acid rain is harmful to aquatic life, since it lowers the pH of the water into which it falls. The water becomes more acidic. Most fish species die at a pH below 4.5. In addition, the reproductive process of fish is adversely affected. Salmon and trout are among the most vulnerable species. Snails and aquatic plants are also affected.

Figure 12-6
(a) Some lakes that are in areas heavily affected by acid rain, such as this one in Kilarney, Ontario, can become acidified within a decade.
(b) Taking water samples from a lake in central Ontario to test for acidity and pollutants.

Acid Rain Around the World

The occurrence of acid rain is increasing around the world. Rain and snow in many regions of the world are 5 to 30 times more acidic than normal. Thousands of lakes in Sweden and the north-eastern United States are too acid to support aquatic life. Lakes in Scotland and Norway have also been adversely affected.

Aquatic life often receives high doses of acid in the spring as acidic snow melts. The meltwater is sometimes as much as 100 times more acidic than normal. The effect of spring meltwater is especially severe, since spring is spawning time for most fish and amphibians. The effect is known as *acid shock* or *spring shock*.

An acidic atmosphere has harmful effects on most materials, especially paper, cloth, leather, and nylon. The airborne acid attacks almost everything it contacts, causing unsightly pitting and corrosion. Acid rain may significantly reduce the durability of concrete. Many statues, monuments, and buildings contain limestone or marble (calcium carbonate, $CaCO_3$). The sulfuric acid in acid rain reacts with calcium carbonate to form a crust of gypsum ($CaSO_4 \cdot 2H_2O$). As the more soluble gypsum dissolves, layers of the limestone or marble are washed away.

Acid rain can also dissolve ores and form soluble salts of metals such as manganese, cadmium, aluminum, lead, copper, and mercury. These salts can then enter lakes and streams and kill aquatic life or enter the water supply of humans and cause health problems.

In 1983, West Germany reported that 34% of the country's forested areas were damaged by acid rain. Forest damage is severe in western Czechoslovakia and south-western Sweden. Many historic buildings throughout Europe are being damaged by acid rain.

What are the countries of the world doing about acid rain? Many European countries have laws that regulate the type of fuels that may be burned. Some of them are trying to cut their annual emissions of sulfur dioxide by up to 30%.

The global nature of the acid rain problem is shown by the fact that in 1983 a number of countries including Canada, the United States, and the Soviet Union, ratified a resolution to "limit and, as far as possible, gradually reduce and prevent air pollution, including long-range transboundary air pollution." Since the effects of acid rain are often found thousands of kilometres from the sources, acid rain is a global problem.

Acid Rain in Canada

Annual emissions of sulfur dioxide in Canada are about five million tonnes. Annual emissions of nitrogen oxides are about two million tonnes. The major sources of sulfur dioxide and nitrogen oxide emissions are found east of the Manitoba-Ontario border. About 60% of the sulfur dioxide emissions originate in smelters; about 16% originate in thermal electric generating stations; and the remainder of these emissions come from other sources such as petroleum refining and sulfite pulping. About 62% of the nitrogen oxide emissions come from automobiles and other vehicles; about 13% originate in power plants; and the remainder of the nitrogen oxide emissions come from industrial and residential sources.

About 50% of the acid rain in Canada comes from the United States, while Canada is responsible for 10 to 15% of acid rain in the United States.

One area of Canada which has seen the destruction of vegetation by acid rain is Sudbury, Ontario. Sulfur dioxide is produced there during the smelting of nickel ore (Fig. 12-7). In fact, the largest point source of SO_2 emissions in the world is a smelter at Copper Cliff, near Sudbury. As a result of public protests and government pressure, the company that owns this smelter took measures to reduce ground level pollution by sulfur dioxide. It built higher stacks, including a superstack which, at 379 m, is the tallest in the world. Although the higher stacks lowered the concentration of sulfur dioxide at ground level, they did not reduce the amount of sulfur dioxide entering the atmosphere. They merely allowed the sulfur dioxide to be transported by the atmosphere to other, more distant locations. However, the company has recently made a commitment to reduce its sulfur dioxide emissions further.

Acid rain has the greatest impact upon aquatic and terrestrial ecosystems in the Canadian Shield and Atlantic Canada. The soils and granite bedrock in these areas lack the limestone that is required to neutralize the acid. Eastern Canada is downwind from large SO_2 emission sources in the northeastern United States, where many coal-burning power plants are located. On the other hand, in western Canada, the little acid rain that exists is largely of local origin.

In Ontario, 220 lakes have been found to be acidified; that is, their pH is less than 5.1 year-round. Some lakes in the Haliburton-Muskoka area have even lost the ability to neutralize the increased acidity. Hundreds of lakes in Quebec are in similar danger of destruction.

Figure 12-7
Thousands of tonnes of gases from smokestacks in Sudbury, Ontario stream into the atmosphere daily.

Between 1954 and 1977, the acidity of a number of Nova Scotia rivers increased significantly. For example, the acidity of the Mersey River increased ten-fold, and its pH changed from 6 to 5. Atlantic salmon runs have disappeared from seven rivers in Nova Scotia. By 1986 the mean annual pH in these rivers was less than 4.7. About two-thirds of the sulfuric acid in the rain falling on Nova Scotia originates from locations west of New Brunswick. About half of it comes from the densely populated, industrial areas of central Canada and the eastern United States.

The acidification of lakes and rivers could have a serious effect on employment for many people. In 1980, sport fishing in Canada generated $1.1 billion in direct revenues and $10 billion in related tourism earnings. This income will decrease as more fish are killed by high acid levels in lakes. The agriculture industry could be adversely affected also, since much of the soil in Ontario, Quebec, and Atlantic Canada is sensitive to acid rain. Another major area of concern is the forestry industry. About one-half of the forests in Canada are exposed to acid rain. The forestry industry produces $23 billion worth of shipments annually and employs, directly and indirectly, one million people, or about one in every 10 jobs in Canada.

Solving the Problem of Acid Rain

What can we do to control the effects of acid rain? On a small scale, the acidity of lakes can be reduced by adding lime or pulverized limestone. The limestone removes the hydronium ions in the following reaction:

$$CaCO_3(s) + 2H_3O^+(aq) \longrightarrow Ca^{2+}(aq) + CO_2(g) + 3H_2O(\ell)$$

This reaction also explains why lakes and streams in which the rock is limestone are able to resist the effects of acid rain.

The cost of controlling SO_2 emissions in Canada is predicted to be about $1 billion per year or $40 for each Canadian. The benefits of control may be well worth the cost. Acid rain causes about $300 million damage to buildings each year in Canada and adds hundreds of millions of dollars yearly to health costs.

The federal government and the provinces have agreed to reduce SO_2 emissions. The government of Canada has designed measures to reduce acid rain. These measures include plans to reduce SO_2 emissions in eastern Canada by 50% by 1994; to adopt tough emission standards for new cars; to provide funds for emission controls at smelters; to investigate increased use of low-sulfur coal from western Canada; and to provide money for developing new methods of reducing acid rain.

Part Three Review Questions

1. What was the source of the killer smog in London in 1952?
2. What are the main components of acid rain?
3. What are the main sources of the chemicals that make up acid rain?
4. Write the chemical equations for the formation of acid rain.

5. What is the pH of normal rain?
6. What is the pH of acid rain?
7. What are the effects of acid rain?
8. What is "acid shock"?
9. How much of the acidity in acid rain is contributed by nitrogen oxides?
10. What are the major sources of SO_2 and nitrogen oxides in eastern Canada?
11. Where is the largest single source of SO_2 emissions located?
12. Which parts of Canada are the most sensitive to acid precipitation? Why?

Chapter Summary

- Acids and bases are electrolytes.
- Acids have a sour taste, affect indicators, neutralize bases, and react with metals, carbonates, and hydrogen carbonates.
- Bases have a bitter taste, feel slippery, affect indicators, and neutralize acids.
- In an acid-base neutralization reaction the acid loses its acidic properties and the base loses its basic properties.
- An operational definition is based on the observed experimental behaviour of a category or class.
- A conceptual definition attempts to explain the behaviour of a category or class.
- An acid increases the concentration of H_3O^+ ions in water.
- A base increases the concentration of OH^- ions in water.
- The hydronium ion, H_3O^+, results from the combination of an H^+ ion with a water molecule and is responsible for the properties of acids.
- The hydroxide ion, OH^-, is responsible for the properties of bases.
- Strong acids are completely dissociated in solution; weak acids are only partially dissociated in solution. Bases can also be classified as strong or weak.
- The net ionic equation for an acid-base neutralization reaction is

$$H_3O^+(aq) + OH^-(aq) \longrightarrow 2H_2O(\ell)$$

- An acid-base neutralization reaction produces a salt and water.
- An acid-base titration is used to determine the concentration of an acid or a base.
- Indicators are dyes which indicate the acidity or basicity of a solution.
- The pH of a solution is related to the concentration of hydronium ions.
- A solution with a pH of 7 is neutral.
- A solution with a pH below 7 is acidic.
- A solution with a pH above 7 is basic.
- A Brønsted-Lowry acid donates protons.
- A Brønsted-Lowry base accepts protons.
- An amphoteric substance is capable of acting as either an acid or a base, depending upon the substance with which it reacts.

- London-type smog consists of a mixture of sulfur dioxide, smoke, and fog.
- Acid rain consists of a mixture of mainly sulfuric acid and nitric acid dissolved in rain, snow, or fog.
- The gases SO_2, NO, and NO_2 which are formed during combustion, eventually turn into the acids present in acid rain.

Key Terms

hydronium ion	titration
acid	pH
strong acid	indicator
weak acid	Brønsted-Lowry Theory
base	amphoteric substance
acid-base neutralization	London-type smog
salt	acid rain

Test Your Understanding

Each of these statements or questions is followed by four responses. Choose the correct response in each case. (Do not write in this book.)

1. Which of the following is not a property of acids?
   a) Acids taste sour.
   b) Acids react with certain metals to generate hydrogen gas.
   c) Acids turn red litmus blue.
   d) Acids are electrolytes.
2. Which of the following is not a property of bases?
   a) Bases are electrolytes.
   b) Basic solutions react with metallic carbonates to generate carbon dioxide gas.
   c) Bases have a bitter taste.
   d) Basic solutions feel slippery.
3. According to Arrhenius, an acid is
   a) a substance that produces hydrogen ions in water.
   b) a substance that produces hydroxide ions in water.
   c) a proton donor.
   d) a proton acceptor.
4. Which of the following is not a strong acid?
   a) sulfuric acid   b) nitric acid   c) acetic acid   d) hydrochloric acid
5. The products of an acid-base neutralization are
   a) a salt and water.   b) an acid and a salt.
   c) a base and a salt.   d) an acid and a base.
6. Which of the following does not occur in an acid-base neutralization?
   a) The acid loses its characteristic properties.
   b) The concentration of both the H_3O^+ and OH^- ions increases.
   c) The products of the reaction are a salt and water.
   d) The positive ion from the base is a spectator ion.

7. The volume of 0.10 mol/L NaOH that is required to neutralize 20 mL of 0.20 mol/L HCl is
   a) 20 mL.   b) 40 mL.   c) 10 mL.   d) 2.0 mL.
8. What is the concentration of HCl if 25.0 mL of the acid are needed to neutralize 15.0 mL of 0.300 mol/L KOH?
   a) 0.150 mol/L   b) 0.500 mol/L   c) 0.300 mol/L   d) 0.180 mol/L
9. What is the concentration of H_2SO_4 if 20.0 mL of the acid are needed to neutralize 10.0 mL of 0.400 mol/L NaOH?
   a) 0.800 mol/L   b) 0.400 mol/L   c) 0.200 mol/L   d) 0.100 mol/L
10. Which of the following statements concerning acid-base titrations is incorrect?
    a) There is never any difference between the theoretical endpoint and the experimental endpoint.
    b) An indicator such as phenolphthalein is often used to detect the endpoint.
    c) The percentage of acetic acid in vinegar can be determined by titration.
    d) At the endpoint the acid and the base have neutralized each other.
11. Which of the following statements is incorrect?
    a) Sulfuric acid is the acid which is produced in largest quantity by the chemical industry.
    b) Carbolic acid is present in soft drinks.
    c) Hydrochloric acid is found in the digestive juices in the stomach.
    d) Amino acids are important building blocks of proteins in our bodies.
12. Which one of the following would have the lowest pH?
    a) H_2O b) 0.1 mol/L HCl
    c) 0.01 mol/L HNO_3 d) 0.1 mol/L NH_3
13. Which one of the following would have the highest pH?
    a) milk of magnesia   b) 0.01 mol/L NH_3
    c) 0.1 mol/L NaOH   d) H_2O
14. Methyl red turns orange when mixed with a certain solution, but red litmus remains red when mixed with the same solution. Within what range does the pH of the solution fall?
    a) 4.2 – 5.0   b) 4.2 – 6.3
    c) 0.0 – 5.0   d) impossible to tell
15. According to the Brønsted-Lowry Theory, a base is
    a) a proton acceptor.
    b) a proton donor.
    c) an electron acceptor.
    d) a substance that increases the OH^- concentration.
16. The conjugate acid of water is
    a) OH^-.   b) H_2.   c) H_3O^+.   d) O_2.
17. In the following equation, which substance is a conjugate base?

$$HClO_4(\ell) + H_2O(\ell) \rightleftharpoons H_3O^+(aq) + ClO_4^-(aq)$$

    a) $HClO_4$ b) H_2O c) H_3O^+ d) ClO_4^-
18. Which of the following is likely to be an amphoteric substance?
    a) HCl   b) NaOH   c) NH_3 d) Na_2HPO_4

19. Which of the following substances is not involved in acid rain?

a) SO_2 **b)** SO_3 **c)** HCl **d)** NO_2

20. Which of the following is the greatest source of sulfur dioxide emissions?

a) smelters **b)** power stations **c)** petroleum refining **d)** paper making

Review Your Understanding

1. Why would you expect acids and bases to be electrolytes?
2. Why is pure water neither acidic nor basic?
3. What are the equilibrium processes occurring in **a)** an aqueous solution of acetic acid; **b)** an aqueous solution of ammonia?
4. Explain why HNO_3 is considered an Arrhenius acid. Why is KOH an Arrhenius base?
5. What is the difference between a weak acid and a dilute solution of a strong acid?
6. Equal volumes of acid A and acid B, both at the same concentration, were tested with a conductivity apparatus. The light bulb glowed brightly with acid B and only dimly for acid A. Comment on the relative strength of these two acids. Explain the observations.
7. Draw the Lewis structures for sulfuric acid and nitric acid.
8. What property of bases are you taking advantage of when you use antacid tablets?
9. Which of the following hypothetical definitions of a carbonate compound is an operational definition? What is a conceptual definition? Explain.
    a) Carbonate compounds produce carbon dioxide gas, a salt, and water, when in contact with hydrochloric acid.
    b) Carbonate compounds are compounds which contain the CO_3^{2-} ion.
10. Explain by a chemical equation how milk of magnesia, $Mg(OH)_2(s)$, is used to neutralize excess stomach acid, HCl(aq). Why isn't NaOH(aq) used for this purpose?
11. Write the balanced chemical equation, the ionic equation, and the net ionic equation when each of the following acid-base neutralization reactions goes to completion:
    a) KOH(aq) and HCl(aq)   **b)** H_2SO_4(aq) and $Sr(OH)_2$ (aq)
    c) HI(aq) and NaOH(aq)   **d)** $Ba(OH)_2$(aq) and HBr(aq)
12. Write the formula and give the names of the salts produced for each reaction in the previous question.
13. What is the difference between the theoretical endpoint and the experimental endpoint of an acid-base titration?
14. In an acid-base titration 25.00 mL of nitric acid were neutralized by 18.00 mL of 0.200 mol/L potassium hydroxide. What is the concentration of the nitric acid?
15. In an acid-base titration 23.50 mL of sodium hydroxide were neutralized by 16.75 mL of 0.180 mol/L sulfuric acid. Calculate the concentration of the sulfuric acid.

16. In an acid-base titration 20.00 mL of acetic acid were neutralized by 27.50 mL of 0.10 mol/L sodium hydroxide. What is the concentration of the acetic acid?

17. In an acid-base titration 15.00 mL of aqueous ammonia, $NH_3(aq)$, were neutralized by 17.25 mL of 0.250 mol/L hydrochloric acid. Calculate the concentration of $NH_3(aq)$.

18. In an acid-base titration 17.35 mL of barium hydroxide were neutralized by 35.65 mL of 0.150 mol/L nitric acid. Calculate the concentration of the barium hydroxide.

19. In an acid-base titration 32.00 mL of sulfuric acid were neutralized by 45.35 mL of 0.120 mol/L calcium hydroxide. What is the concentration of the sulfuric acid?

20. A solution of sodium hydroxide contained 3.2×10^{-3} mol of sodium hydroxide. If a solution of 0.14 mol/L hydrochloric acid was used to neutralize the base completely, how many millilitres of hydrochloric acid were used? How many moles of HCl were needed to reach the endpoint in the titration?

21. Which has a greater concentration of hydronium ions, a solution with a pH of 2 or a solution with a pH of 3?

22. Which is more acidic, a solution with a pH of 5 or a solution with a pH of 6?

23. How does the pH scale enable us to assess the basicity or acidity of a solution?

24. How could you tell if an acid and a base were completely neutralized when they were mixed?

25. To one portion of an unknown solution, a few drops of phenolphthalein indicator were added. The solution remained colourless. A few drops of bromthymol blue, added to a second portion, turned the solution blue. Was the solution acidic, neutral, or basic? What was its approximate pH?

26. In each of the following equations select the reactant which acts as a Brønsted-Lowry base and select its conjugate acid:
    a) $C_2H_3O_2^-(aq) + H_2O(\ell) \longrightarrow HC_2H_3O_2(aq) + OH^-(aq)$
    b) $NH_4^+(aq) + OH^-(aq) \longrightarrow NH_3(aq) + H_2O(\ell)$
    c) $HSO_4^-(aq) + H_2O(\ell) \longrightarrow H_3O^+(aq) + SO_4^{2-}(aq)$

27. Why is it sometimes difficult to pinpoint the exact source of SO_2 emissions?

28. Which chemical can help to protect lakes from acid rain? How does it work?

29. Which major Canadian industries are affected by acid rain? Why?

30. What steps has Canada taken to reduce acid rain?

31. How can we, as citizens, reduce emissions of nitrogen oxides?

Apply Your Understanding

1. Diethylzinc is used to neutralize the acidity in books. However, it explodes on contact with water and bursts into flames on contact with air.

Under what conditions do you think diethylzinc is used in the neutralization of the acid in books?

2. Two 20.00 mL solutions of sodium hydroxide were titrated, using two different strong acids of equal concentration. It took 15.00 mL of acid X, but 30.00 mL of acid Y, to neutralize the sodium hydroxide. Explain.

3. Compute the hydronium ion concentration in a solution prepared by mixing 25.0 mL of 0.150 mol/L NaOH(aq) and 75.0 mL of 0.105 mol/L HCl(aq).

4. A number of antacid tablets contain water-insoluble metal hydroxides such as milk of magnesia, $Mg(OH)_2$. Stomach acid is about 0.10 mol/L HCl(aq). Calculate the number of millilitres of stomach acid that can be neutralized by 500 mg of $Mg(OH)_2(s)$.

5. A 1.50 g sample of aspirin (acetylsalicylic acid) is dissolved in 100 mL of water and titrated to an endpoint with 0.180 mol/L NaOH(aq). The volume of base required is 46.2 mL. If one mole of acetylsalicylic acid reacts with one mole of NaOH, calculate the molar mass of acetylsalicylic acid.

6. Vinegar is a dilute aqueous solution of acetic acid. A 14.7 mL sample of vinegar requires 27.0 mL of 0.400 mol/L NaOH(aq) to neutralize the $HC_2H_3O_2(aq)$. If the density of vinegar is 1.060 g/mL, what is the percentage by mass of acetic acid in the vinegar?

7. The $H_2PO_4^-$ ion is an amphoteric substance. Illustrate this fact by writing balanced equations for its reactions with CN^- and with H_2SO_4 in aqueous solutions.

Investigations

1. It will cost about $150 per car for the catalytic converters needed to meet the stricter Canadian automobile emission standards. Debate the resolution, "Motorists should pay the full costs of catalytic converters for their automobiles."

2. Research the methods of reducing SO_2 emissions and write a report on your findings. Your report should include an indication of the costs involved in reducing SO_2 emissions and suggestions for sources of the necessary funds.

3. Acid rain is a global problem. When one country complains that another is polluting its atmosphere, how can the complaint be resolved? Who should pay the costs of any damage done?

4. Pretend that you are a member of a town council. The council has asked a company that is a major emitter of SO_2 in your area to cut its level of pollution. The company tells you that, in order to find money to pay for the technology to reduce SO_2 emissions, it must lay off 200 workers. Prepare a response to submit to the board of directors of the company.

5. Prepare a research report on viable methods of producing electricity that could take the place of coal-burning power plants.

6. Debate the following resolution: "It is morally acceptable to build high smokestacks to relieve local ground-level concentrations of sulfur dioxide."

7. Extract water-soluble dyes from a variety of different types of plants. Test your dyes to determine whether they are affected by acids or bases. Indicate at least three naturally-occurring dyes that could serve as acid-base indicators.

8. Select at least four brands of antacids. Determine, by titration, which of these can neutralize the most 0.10 mol/L HCl per unit cost.

9. Use indicator paper or a pH meter to determine the pH of a variety of common solutions. What is the most acidic liquid that is normally consumed in your home?

13 Gases

CHAPTER CONTENTS

Part One: **Gas Laws**
- **13.1** Early Ideas About the Nature of Matter
- **13.2** The Discovery of Air Pressure
- **13.3** Robert Boyle
- **13.4** Boyle's Law
- **13.5** Charles' Law
- **13.6** Application of Boyle's and Charles' Laws
- **13.7** Standard Temperature and Pressure
- **13.8** Dalton's Law of Partial Pressures

Part Two: **The Kinetic Molecular Theory and the Behaviour of Real Gases**
- **13.9** A Model for Gases: The Kinetic Molecular Theory
- **13.10** Further Applications of the Gas Laws
- **13.11** Behaviour of Real Gases
- **13.12** Liquefaction of Gases
- **13.13** Cooling of Gases by Expansion
- **13.14** Heating of Gases by Compression

Part Three: **Chemistry and the Gaseous Environment**
- **13.15** Technology and the Environment
- **13.16** Changes in the Air Environment
- **13.17** The Effects of Solid Particles in the Atmosphere
- **13.18** Carbon Dioxide and the Greenhouse Effect
- **13.19** Photochemical or Los Angeles-Type Smog
- **13.20** Effects of Photochemical Smog
- **13.21** What Can Be Done About Air Pollution?

What is the greenhouse effect? How might it cause the flooding of many Canadian coastal cities? How are scuba divers able to breathe under water for long periods of time even though they have such small air tanks? How can we reduce the emissions of gaseous pollutants from automobile exhaust? Why do weather balloons expand to large volumes at high altitudes? How does a drinking straw work? You will be able to answer these questions after you have studied this chapter.

We can find many examples of substances in the gaseous state in our daily lives. In fact, the physical and chemical properties of certain gases enable them to support life, to enhance our comfort, to be used in performing useful tasks, and to be of great economic importance. In this chapter we will discuss the behaviour of gases and introduce the laws that summarize their observed behaviour.

Key Objectives

When you have finished studying this chapter, you should be able to

1. use Boyle's Law and Charles' Law, singly or together, to solve gas problems involving pressure, volume and temperature.
2. calculate the volume of a gas at STP, given its volume at a specific temperature and pressure.
3. use Dalton's Law of Partial Pressures to correct for the vapour pressure of water in gas law calculations involving gases collected over water.
4. explain the compressibility of gases, Boyle's Law, Charles' Law, and Dalton's Law of Partial Pressures, using the basic principles of the kinetic molecular theory.
5. use Gay-Lussac's Law, singly or together with Boyle's Law and Charles' Law, to solve gas problems involving pressure, volume, and temperature.
6. describe the conditions under which real gases deviate most from ideal gas behaviour.
7. describe and explain the effects of solid particles in the atmosphere.
8. describe the nature and effects of Los Angeles-type smog.
9. describe the greenhouse effect.

Part One: Gas Laws

13.1 Early Ideas About the Nature of Matter

Long before Dalton's time, scientists knew of gases, liquids, and solids. They understood, in some cases, how to bring about changes from one state to another. However, they did not understand exactly what was happening within the sample of matter as it changed from one state to another.

The ancient Greeks believed that all matter was made up of four elements—earth (solid), water (liquid), air (gas), and fire (heat)—in various proportions. Their philosophers believed that matter consisted of small particles which could not be subdivided further into smaller particles. These smallest particles of matter they called atoms (Greek *atomos*, meaning "uncut"). These atoms were believed to have the characteristic properties of the substance. Thus, "atoms" of water were believed to be round and smooth so that they could slip easily over one another. "Atoms" of iron were hard and rough. "Atoms" of air were springy. The philosophers, however, did not perform any experiments to test their beliefs.

In the Middle Ages, the alchemists began their attempts to change lead into gold. They believed they could find a magical "philosopher's stone" which would cause the transformation. Eventually, the people who proposed the theories were the same people who performed experiments. Almost all of them kept their findings secret.

Figure 13-1
In 1669, the alchemist H. Brand discovered phosphorus during an attempt to find the philosopher's stone. This was just one of the many contributions made to chemical knowledge during the 17th century. Shown here is a depiction of the moment of discovery by an artist of the following century.

Thus, until the seventeenth century, there was little progress in chemistry. Suction pumps had been invented, but no one could explain how they worked, other than to say that "nature abhors a vacuum." A widespread incorrect belief called the phlogiston theory also hindered progress. The basic idea of this theory was that all combustible materials contain a common substance called phlogiston which escapes on burning.

We now know that combustion involves a reaction with oxygen gas. It was largely as a result of experiments with gases that sufficient knowledge was accumulated to discredit the phlogiston theory and allow chemical progress to be made. How do the physical properties of gases and the physical laws that describe these properties lead us to believe that matter does in fact consist of small particles? We shall look at some of the laws and at the people who discovered them.

13.2 The Discovery of Air Pressure

Galileo Galilei, an Italian physicist, discovered that if he pumped the air out of the top of a tube whose bottom was immersed in water, the water rose in the tube (Fig. 13-2a). However, it was not possible to raise the water higher than about 10 m, no matter how hard he pumped.

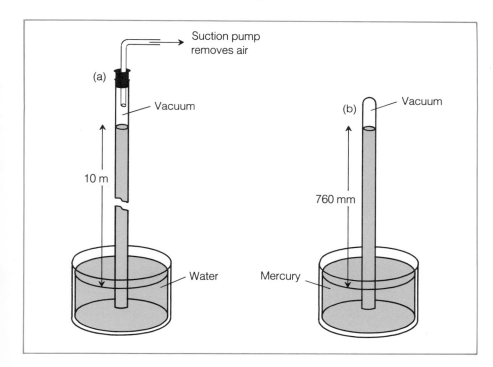

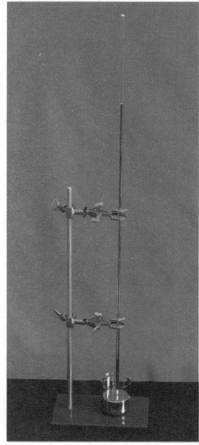

Figure 13-2
(a) Galileo's apparatus using water
(b) Torricelli's mercury barometer

A mercury barometer

Galileo called this problem to the attention of his pupil and assistant, Evangelista Torricelli. Eventually it dawned on Torricelli that it was the pressure of the atmosphere on the surface of the water outside the tube that forced the water up into the tube. At a height of about 10.3 m the pressure exerted by the column of water against the atmosphere would just balance the pressure exerted by the atmosphere against the surface of the water. He next theorized that the atmosphere should balance a much shorter column of mercury than water. Mercury is 13.6 times as dense as water, so the column ought to be only $\frac{1}{13.6}$ as high as 10.3 m, or about 0.8 m. Torricelli tried the experiment in 1643 using a closed tube and found that the mercury column stayed at a level of about 760 mm (0.76 m). This experiment led him to the invention of the mercury barometer, a device which measures the small variations in atmospheric pressure (Fig. 13-2b).

13.3 Robert Boyle

Robert Boyle was a dedicated scientist who devoted all his effort to the study of science and religion. With his new, more efficient air pump he conducted many experiments. In one experiment he placed a barometer inside a container from which he could pump the air. He found that the mercury level fell as the air was removed from the container (Fig. 13-3). Boyle felt that this experiment proved conclusively that it was the pressure of the air on the surface of the mercury which supported the column, thus confirming Torricelli's findings.

Figure 13-3
The level of mercury falls as air is pumped from the container.

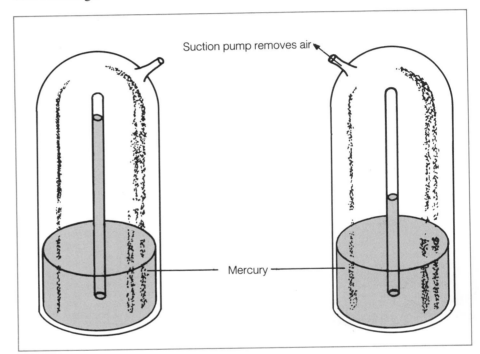

Suction pump removes air

Mercury

When Boyle published his results, his conclusions were immediately attacked by Franciscus Linus, a Jesuit priest. Linus explained the action of a barometer by supposing that the mercury column was held up by an invisible internal cord. Boyle naturally felt this was a poor hypothesis and immediately began a series of experiments to support his own ideas. In this process he formulated Boyle's Law, which stated the mathematical relationship between the pressure of a gas and its volume.

Figure 13-4
Boyle's apparatus. The longer the column of mercury, h, the more the air is compressed.

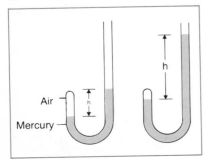

Air

Mercury

13.4 Boyle's Law

With a J-shaped tube similar to that shown in Figure 13-4, Boyle used a column of mercury to trap a sample of air in the short closed end. When he added more mercury to the longer open end, the air was compressed to a smaller volume. When he decreased the pressure by removing mercury, the air expanded to a greater volume. When he measured the air volumes corresponding to varying pressures, he noticed a simple relationship. Some of his data are given in Table 13-1.

Table 13-1   Some of Robert Boyle's Data

Volume of Air V(m)*	Pressure P(m)†	PV Product
0.305	0.74	0.226
0.254	0.90	0.229
0.203	1.12	0.227
0.152	1.49	0.226
0.102	2.23	0.227

*Boyle actually estimated the volume of the air by measuring the length of the air space in the tube. The actual volume of the air in cubic metres was unknown.

†Boyle expressed the pressure as the height (h) in metres of the mercury column plus the pressure of the atmosphere above the open end, which he estimated at 0.74 m. The metre is not an SI unit of pressure and will not be used as such in this book again. It is used here for historical reasons only.

Boyle noticed that the product of the volume of air times the pressure exerted on it was very nearly a constant (in this case 0.227), or PV = constant. This can be rearranged as

$$P = \frac{\text{constant}}{V} \quad \text{or} \quad V = \frac{\text{constant}}{P}$$

Biography

R OBERT BOYLE was the seventh son and fourteenth child of fifteen born to the Earl of Cork. He was sent to Eton at the age of eight and to Geneva three years later for completion of his studies. He returned to England in 1644 to reside in Dorset. Here he was admitted to the *invisible college*, the forerunner of the Royal Society. In the same year he moved to Oxford, where young Robert Hooke assisted Boyle in his investigation of "The Spring of the Air."

By 1659 he had constructed a new and superior air pump, which he used to prove that air pressure supports the column of a mercury barometer and to demonstrate the connection between pressure and the boiling point of water. Criticism led him to conduct further experiments, which resulted in the publication of his famous law two years later.

In addition to the formulation of his law, Boyle taught the value of experimentation. In his book *The Sceptical Chymist*, he examined the pretensions of the chemists of his time and exposed the mixture of error and imposture in most of their writings. He was the first to use the term "chemical analysis" in its modern sense. He showed that fire is not the only method to use in the analysis of compounds and that other methods may sometimes be necessary. He laid the foundations of modern analytical chemistry.

In 1669, Boyle moved to London and lived as an invalid, the victim of a lifelong kidney disorder which led to his death in 1691.

Figure 13-5
Robert Boyle (1627-1691)—Boyle's Law states the mathematical relationship between the pressure and volume of a gas.

Note that if *V* increases, *P* decreases proportionately, and vice versa. Such relations are called inverse relations or inverse proportions. These expressions are simply the mathematical expression of **Boyle's Law** which states that the volume of a fixed mass of gas varies inversely with the pressure, provided the temperature remains constant. Actually, Boyle was not really concerned about the effects of temperature. Luckily for him the temperature stayed roughly constant during his experiments.

A graph of volume versus pressure shows that, as the pressure on the gas increases, the volume of the gas decreases (Fig. 13-6).

Figure 13-6
Graph of Boyle's Law data

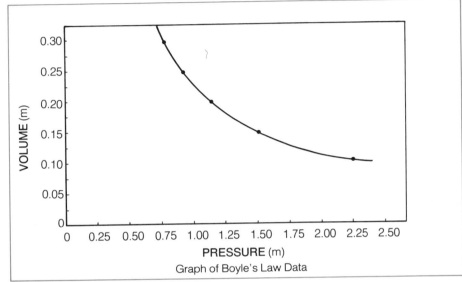

Graph of Boyle's Law Data

The cold temperature of liquid nitrogen causes the blue balloon to contract.

Gases tend to resist being squeezed into a smaller volume. For example, if you push in the plunger of a syringe with closed ends, the trapped air will cause the plunger to spring back to its original position. As a result, Boyle thought in terms of the "spring" of gases. He thought that gases were made up of very small particles similar to extremely minute coiled springs. Gases exerted a pressure because the "ether" (a mysterious "something" that was thought to fill the spaces between the gas particles) whirled the particles so violently that each particle tried to prevent all others from coming into its own neighbourhood.

13.5 Charles' Law

Boyle had investigated the relationship between the volume and the pressure of a gas in 1659. However, it was not until 1787, over a century later, that the French physicist J. A. Charles discovered the relationship between the volume and the temperature of a gas at constant pressure.

Today, however, we are all familiar with the fact that gases expand when heated and contract on cooling. A balloon becomes larger when held near a hot stove and smaller when it is removed from the heat. Hot air balloons operate on this principle. Many motorists release air from their tires to relieve the pressure caused by heat buildup from sustained high-speed driving. The problem is that they may forget to replace this air when tire temperatures return to normal, and the tire may wear excessively because of under-inflation.

Charles' experiments were rather crude and were performed more accurately in 1801 by John Dalton and in 1802 by the French chemist Joseph Louis Gay-Lussac. Nevertheless, Charles deserves the credit for the early investigations.

We can repeat Charles' early observations by using a mercury "piston," made by trapping a quantity of air in a piece of narrow-bore glass tubing with a small plug of mercury (Fig. 13-7). Since the tube is open to the atmosphere and the mercury is free to move up and down the tube, the pressure exerted on the trapped air remains constant. If the tube is immersed in hot water, the air expands, and the mercury plug moves up the tube. Typical data for a Charles' Law experiment are shown in Table 13-2.

It is obvious that the volume increases as the Celsius temperature increases. The relationship is not a direct proportionality, however, because doubling the temperature from 50°C to 100°C results in a volume increase of only $\frac{885}{766}$ or 1.16 times. A graph of volume versus temperature is a straight line, showing that the increase in volume is regular and uniform (Fig. 13-8).

What is the relationship? We notice that at 100°C the volume is $\frac{885}{648}$ or 1.366 times the volume at 0°C. An increase in temperature of 100°C causes a volume increase of 0.366 times the volume at 0°C. So an increase in temperature of 1°C causes an increase in volume of $\frac{0.366}{100}$ or 0.003 66 times its volume at 0°C. The decimal fraction 0.003 66 is equivalent to the proper fraction $\frac{1}{273}$. So another way of stating the results of the experiment is that the volume of a sample of gas at 0°C increases by $\frac{1}{273}$ of its original volume for each degree it is warmed above 0°C. We can check this statement: at 75°C the volume should be $1\frac{75}{273}$ or 1.275 times the volume at 0°C. The predicted volume is 1.275×648 mm^3 = 826 mm^3. The experimental value is 817 mm^3. The difference can be attributed to the difficulty of making sufficiently precise measurements (that is, to experimental error!).

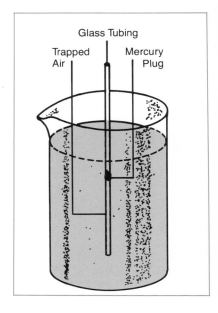

Figure 13-7
Charles' Law experiment

Table 13-2   Typical Data for Charles' Law Experiment

Temperature, T (°C)	Gas Volume, V (mm^3)
0	648
25	714
50	766
75	817
100	885

Figure 13-8
Graph of Charles' Law data

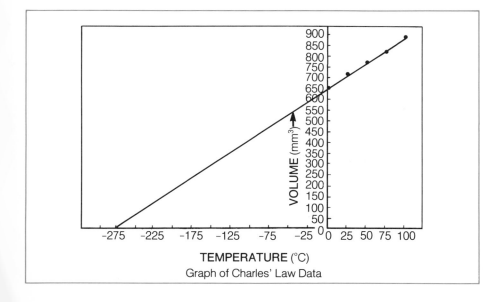

TEMPERATURE (°C)
Graph of Charles' Law Data

The Kelvin Scale

If the results of the experiment are valid, we should be able to state the results in a different way: the volume of a gas at 0°C decreases by $\frac{1}{273}$ of its original volume for each degree it is cooled below 0°C. At some temperature, the volume of the gas should become zero. What is this temperature? If we examine Figure 13-8 we can see that a gas should have a zero volume at −273°C.

In practice, this zero volume has never been observed because all gases liquefy and the majority solidify before reaching −273°C. Nevertheless, −273°C is a theoretically important temperature. About 100 years after Charles' experiments, the British physicist Lord Kelvin realized that it would be useful to define a temperature scale that took zero as its lowest value. The *Kelvin temperature scale* is based on the Celsius scale. One kelvin is the same size as one degree on the Celsius scale. In this scale, −273°C becomes zero kelvins (0 K). This temperature is called *absolute zero*. Other temperatures are obtained simply by adding 273 to the corresponding Celsius temperature. Thus, the freezing point of water is 273 K and the boiling point of water is 373 K. (Notice that the degree symbol is omitted when writing Kelvin temperatures.) The kelvin is the SI base unit for temperature.

Let us now reexamine the data of Table 13-2, expressing the temperatures in the Kelvin scale shown in Table 13-3. Now we see that increasing the temperature from 273 K to 373 K (a factor of $\frac{373}{273} = 1.366$) causes the volume to increase from 648 mm³ to 885 mm³, also a factor of $\frac{885}{648} = 1.366$. There is now a direct proportionality. We can now state **Charles' Law** in modern terms: the volume of a fixed mass of gas varies directly with the Kelvin temperature, provided the pressure remains constant:

$$V = \text{constant} \times T$$

Table 13-3   Data for Charles' Law Experiment

Temperature °C	K	Volume mm³
0	273	648
25	298	714
50	323	766
75	348	817
100	373	885

13.6 Application of Boyle's and Charles' Laws

If we know the volume and the pressure of a fixed mass of gas at a given temperature, then we can use Boyle's Law to calculate the volume at any other pressure, or the pressure at any other volume, providing the temperature remains constant.

EXAMPLE PROBLEM 13-1

A 2.00 L sample of a gas at a pressure of 1000 kPa is allowed to expand until its pressure drops to 300 kPa. If the temperature remains constant, what will be the new volume? (Note that the symbol for the SI unit of pressure, called the kilopascal, is kPa. The normal atmospheric pressure at sea level is about 101 kPa. Laboratory barometers are usually calibrated to measure atmospheric pressure in millimetres of mercury. One millimetre of mercury exerts a pressure of 133 Pa or 0.133 kPa.)

SOLUTION

To solve this problem, let us first list the quantities we know and those we are to find:

$$P_{old} = 1000 \text{ kPa} \qquad P_{new} = 300 \text{ kPa}$$
$$V_{old} = 2.00 \text{ L} \qquad V_{new} = ?$$

According to Boyle's Law, at a constant temperature, pressure affects the volume of a gas and vice versa. Therefore, we know that the new volume will be the old volume, changed as a result of the pressure change. We shall call this influence of pressure a *pressure factor*. The new volume equals the old volume multiplied by the pressure factor:

$$V_{new} = V_{old} \times (\text{pressure factor})$$

The pressure factor is the ratio of the two pressures. It is either 1000 kPa/300 kPa or 300 kPa/1000 kPa. The first of these is a number greater than one. The second is a number smaller than one. Since a gas expands as the pressure decreases, the new volume must be larger than the old volume, and the pressure factor must be a number greater than one:

$$V_{new} = 2.00 \text{ L} \times \frac{1000 \text{ kPa}}{300 \text{ kPa}} = 6.67 \text{ L}$$

The new volume will be 6.67 L.

If we know the volume and temperature of a fixed mass of gas at a given pressure, we can use Charles' Law to calculate the volume at any other temperature, or the temperature at any other volume, providing the pressure remains constant.

EXAMPLE PROBLEM 13-2

If the 6.67 L of gas in the previous example was originally at a temperature of 300 K and is now cooled to 150 K, what will be its new volume, assuming the pressure remains constant?

SOLUTION

According to Charles' Law, at a constant pressure, temperature affects the volume of a gas. Therefore, we know that the new volume will be the old volume, changed as a result of a temperature change. We shall call this influence of temperature a *temperature factor*. The new volume equals the old volume multiplied by the temperature factor:

$$V_{new} = V_{old} \times (\text{temperature factor})$$

The temperature factor is either 150 K/300 K or 300 K/150 K. Since Charles' Law tells us that gases contract on cooling, the temperature factor is a number less than one, and we can write

$$V_{new} = 6.67 \text{ L} \times \frac{150 \text{ K}}{300 \text{ K}} = 3.33 \text{ L}$$

It is possible to combine the two laws into one mathematical operation. The question might have been worded like this: A certain mass of a gas occupies a volume of 2.00 L at a pressure of 1000 kPa and a temperature of 300 K. What will be its volume at a pressure of 300 kPa and a temperature of 150 K?

According to Boyle's Law and Charles' Law, both pressure and temperature affect the volume of a gas. Thus, the new volume will be the old volume, changed as a result of both pressure and temperature changes. The new volume equals the old volume multiplied by both a pressure factor and a temperature factor:

$$V_{new} = V_{old} \times (\text{pressure factor}) \times (\text{temperature factor})$$

The pressure decrease causes a volume increase, and the pressure factor > 1. The temperature decrease causes a volume decrease and the temperature factor < 1.

$$V_{new} = 2.00 \text{ L} \times \frac{1000 \text{ kPa}}{300 \text{ kPa}} \times \frac{150 \text{ K}}{300 \text{ K}} = 3.33 \text{ L}$$

The new volume will be 3.33 L. Notice that the volume has increased since the effect of the pressure change is greater than the effect of the temperature change.

EXAMPLE PROBLEM 13-3

If a given mass of a gas occupies a volume of 8.50 L at a pressure of 95.0 kPa and 35°C, what volume will it occupy at a pressure of 75.0 kPa and a temperature of 150°C?

SOLUTION
We can summarize the data as

$$V_{old} = 8.50 \text{ L} \qquad V_{new} = ?$$
$$P_{old} = 95.0 \text{ kPa} \qquad P_{new} = 75.0 \text{ kPa}$$
$$T_{old} = 35°C = 308 \text{ K} \qquad T_{new} = 150°C = 423 \text{ K}$$

Note: We must always use Kelvin temperatures in the gas law equations.

$$V_{new} = V_{old} \times (\text{pressure factor}) \times (\text{temperature factor})$$

Since pressure decreases, volume increases, and the pressure factor > 1. Since temperature increases, volume increases, and the temperature factor > 1.

$$\therefore V_{new} = 8.50 \text{ L} \times \frac{95.0 \text{ kPa}}{75.0 \text{ kPa}} \times \frac{423 \text{ K}}{308 \text{ K}} = 14.8 \text{ L}$$

PRACTICE PROBLEM 13-1

If a given mass of a gas occupies a volume of 8.4 L at a pressure of 101 kPa, what is its volume at a pressure of 112 kPa and the same temperature? (*Answer*: 7.6 L)

PRACTICE PROBLEM 13-2

If a given mass of a gas occupies a volume of 4.2 L at a temperature of 0°C, what is its volume at a temperature of 91°C, if the pressure remains constant? (*Answer*: 5.6 L)

PRACTICE PROBLEM 13-3

If a given mass of a gas occupies a volume of 6.3 L at a pressure of 101 kPa and 0°C, what volume will it occupy at a pressure of 143 kPa and a temperature of 113°C? (*Answer*: 6.3 L)

13.7 Standard Temperature and Pressure

Since the volume of a gas varies with temperature and pressure, it is customary to choose some standard conditions of temperature and pressure so that the properties of different gases can be compared on a uniform basis. The choice is rather arbitrary, but usually the pressure of the atmosphere at sea level (101.3 kPa) is chosen as **standard pressure**. The freezing point of water (0°C or 273 K) is chosen as **standard temperature**. These two conditions together are usually referred to as **standard temperature and pressure**, or **STP**. Scientists do not usually measure gas volumes at STP, but they usually use Boyle's and Charles' Laws to "correct" their experimental values to standard conditions.

EXAMPLE PROBLEM 13-4

A chemist finds that a sample of a gas occupies a volume of 43.4 mL at a pressure of 97.3 kPa and a temperature of 27°C. What is its volume at STP?

SOLUTION
We can summarize the data as

$$V_{old} = 43.4 \text{ mL} \qquad V_{new} = ?$$
$$P_{old} = 97.3 \text{ kPa} \qquad P_{new} = 101.3 \text{ kPa}$$
$$T_{old} = 27°C = 300 \text{ K} \qquad T_{new} = 0°C = 273 \text{ kPa}$$

$$V_{new} = V_{old} \times (\text{pressure factor}) \times (\text{temperature factor})$$

Since pressure increases, volume decreases, and the pressure factor < 1. Since temperature decreases, volume decreases, and the temperature factor < 1.

$$\therefore V_{new} = 43.4 \text{ mL} \times \frac{97.3 \text{ kPa}}{101.3 \text{ kPa}} \times \frac{273 \text{ K}}{300 \text{ K}} = 37.9 \text{ mL}$$

PRACTICE PROBLEM 13-4

What is the STP volume of a sample of gas which occupies 500 mL at a temperature of 100°C and a pressure of 200 kPa? (*Answer*: 723 mL)

13.8 Dalton's Law of Partial Pressures

We have already noted that John Dalton performed experiments with gases. He had a keen interest in the weather, and his early experiments dealt with measuring the water content of the air. In Dalton's time air was generally thought to be a compound of oxygen, nitrogen, and water. Dalton showed that air always contained 21 parts of oxygen for every 79 parts of nitrogen, but the amount of water vapour in the air varied. He concluded that air was a mixture. He demonstrated that if he added water vapour to a sample of dry air, the pressure exerted by the air increased. The increase in pressure was just equal to that exerted by the water vapour alone at the same temperature. As a result of these observations and his experiments with other gas mixtures as well, Dalton formulated the law which is now called **Dalton's Law of Partial Pressures**. The total pressure exerted by a mixture of gases is the sum of the pressures of each gas when measured alone:

$$P_{total} = P_1 + P_2 + P_3 + \ldots$$

For example, let us force a 1.00 L volume of nitrogen with a pressure of 100 kPa into a rigid 1.00 L container of oxygen with a pressure of 200 kPa as shown in Figure 13-9. If the temperature is constant, the 1 L container of oxygen and nitrogen will now have a pressure of 300 kPa:

$$P_{total} = P_{nitrogen} + P_{oxygen}$$

$$= 100 \text{ kPa} + 200 \text{ kPa}$$

$$= 300 \text{ kPa}$$

Figure 13-9
Dalton's Law of Partial Pressures
(temperature remains constant)

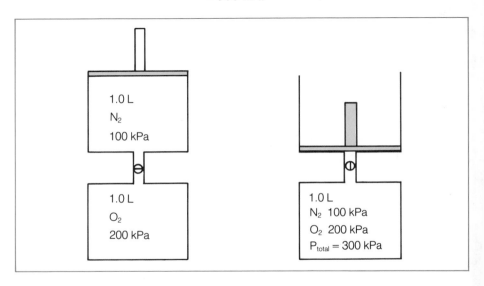

Dalton's studies led him to believe that a gas consisted of particles which could move throughout a volume already occupied by the particles of another gas, without being affected by the gas particles already there, provided the gases did not react. This was a reasonable assumption provided that gas particles were extremely small compared with the distances between them.

Oxygen is often prepared in the laboratory by heating potassium chlorate and collecting the gas over water (Fig. 13-10). In this method a bottle is filled with water in a pneumatic trough. The oxygen gas produced by heating the potassium chlorate is bubbled into the bottle of water. The oxygen is not very soluble and displaces the water from the bottle. The gas collected is not pure oxygen. It is a mixture of oxygen and water vapour. The water vapour arises from the evaporation of some liquid water. We must correct for (allow for) the presence of this water vapour whenever we wish to make calculations involving the pressure or volume of a gas collected over water. Say, for example, that we have collected 400 mL of oxygen over water at a temperature of 27°C (300 K) and at an atmospheric pressure of 100.0 kPa. The pressure exerted by the wet gas (oxygen + water vapour) is equal to the atmospheric pressure when the water level is the same inside and outside the bottle. According to Dalton's Law,

$$P_{total} = P_{oxygen} + P_{water\ vapour}$$

or

$$P_{oxygen} = P_{total} - P_{water\ vapour}$$

At 27°C the vapour pressure of the water is 3.57 kPa. (See Appendix D for a table of water vapour pressures at various temperatures.)

$$\therefore P_{oxygen} = 100.0\ kPa - 3.57\ kPa = 96.4\ kPa$$

Figure 13-10
(a) Collecting a gas over water
(b) Diagram of a pneumatic trough
(c) A pneumatic trough

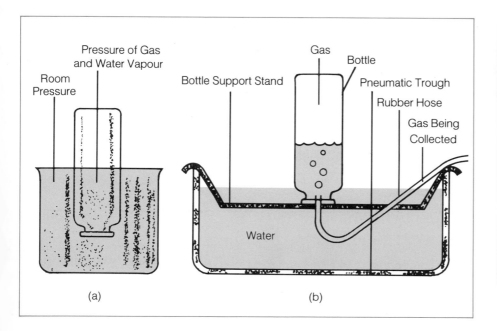

(a)

(b)

(c)

The reaction of magnesium with hydrochloric acid generates hydrogen, which is collected over water in a gas-measuring tube.

What would be the volume of the dry oxygen at 27°C and 100.0 kPa? If the water vapour were removed, the pressure exerted by 400 mL of dry oxygen would be 96.4 kPa. However, since the pressure exerted on the dry oxygen must be 100.0 kPa, the volume of the dry oxygen must decrease so that its pressure will increase from 96.4 kPa to 100.0 kPa. According to Boyle's Law,

$$V_{new} = V_{old} \times (\text{pressure factor})$$

$$V_{new} = 400 \text{ mL} \times \frac{96.4 \text{ kPa}}{100.0 \text{ kPa}} = 386 \text{ mL}$$

Dalton's Law of Partial Pressures is thus a valuable tool for calculations involving mixtures of gases. If the partial pressures of the individual gases are known, the total pressure of the mixture can be calculated. If the total pressure and the partial pressures of all but one of the gases are known, the partial pressure of the remaining gas can be calculated.

EXAMPLE PROBLEM 13-5

A volume of 75.0 mL of hydrogen gas is collected over water at a temperature of 28°C and an atmospheric pressure of 98.0 kPa. What is the volume of the dry hydrogen at STP?

SOLUTION

At 28°C, $P_{H_2O} = 3.78$ kPa

$P_{old} = (98.0 - 3.78) \text{ kPa} = 94.2 \text{ kPa}$	$P_{new} = 101.3 \text{ kPa}$
$V_{old} = 75.0 \text{ mL}$	$V_{new} = ?$
$T_{old} = 28°C = 301 \text{ K}$	$T_{new} = 273 \text{ K}$

$$V_{new} = V_{old} \times (\text{pressure factor}) \times (\text{temperature factor})$$

Since P increases, V decreases and pressure factor < 1. Since T decreases, V decreases and temperature factor < 1.

$$\therefore V_{new} = 75.0 \text{ mL} \times \frac{94.2 \text{ kPa}}{101.3 \text{ kPa}} \times \frac{273 \text{ K}}{301 \text{ K}} = 63.3 \text{ mL}$$

The dry hydrogen would occupy a volume of 63.3 mL at STP.

EXAMPLE PROBLEM 13-6

The total pressure of a mixture of carbon dioxide, oxygen, and helium is 92.5 kPa. If the partial pressure of carbon dioxide is 27.3 kPa and the partial pressure of helium is 40.5 kPa, use Dalton's Law of Partial Pressures to determine the partial pressure of oxygen.

SOLUTION

$$P_{total} = P_{CO_2} + P_{O_2} + P_{He}$$

$$P_{O_2} = P_{total} - P_{CO_2} - P_{He}$$

$$= 92.5 \text{ kPa} - 27.3 \text{ kPa} - 40.5 \text{ kPa}$$

$$= 24.7 \text{ kPa}$$

∴ The partial pressure of the oxygen is 24.7 kPa.

PRACTICE PROBLEM 13-5

A volume of 110 mL of hydrogen is collected over water at a temperature of 17°C and an atmospheric pressure of 95.0 kPa. What is the volume of the dry hydrogen at STP? (*Answer*: 95.2 mL)

This model is used to illustrate Brownian motion. Ball bearings simulate gas molecules colliding with a much larger particle (the red plastic disk).

Part One Review Questions

1. State Boyle's Law, Charles' Law, and Dalton's Law of Partial Pressures.
2. How is the Kelvin temperature scale related to the Celsius temperature scale?
3. What is meant by absolute zero?
4. A gas with a volume of 25.0 mL and a pressure of 95 kPa is allowed to expand until its pressure is 75 kPa. What is the new volume of the gas?
5. A gas with a volume of 5.5 L at 20°C is heated to 150°C. What is the new volume of the gas?
6. A gas has a volume of 35.6 mL at 30°C and 125 kPa. What will be its volume at a temperature of 95°C and a pressure of 75.0 kPa?
7. What is meant by STP?
8. A volume of 125.0 mL of oxygen is collected over water at a temperature of 24°C and an atmospheric pressure of 105.0 kPa. What is the volume of the dry oxygen at STP?

Part Two: The Kinetic Molecular Theory and the Behaviour of Real Gases

13.9 A Model for Gases: The Kinetic Molecular Theory

If a bottle of concentrated ammonia is opened in a room, the odour of the ammonia will soon be detected throughout the room. This suggests that the particles of a gas can move or diffuse through another gas. A Scottish botanist,

Robert Brown, discovered in 1827 that small particles suspended in a gas or a liquid are in a constant zigzag motion. This so-called **Brownian motion** was assumed to be due to collisions of the gas particles with the suspended material.

We can now make assumptions concerning the nature of gases. That is, we will now develop a model by making the simplest possible assumptions.

Building on the results of Boyle, Charles, and others, as well as the results of his own experiments, John Dalton in 1801 proposed a kinetic molecular model for gases. This was refined and developed by J. C. Maxwell, a British physicist, in 1859. The basic assumptions of the model are:

1. Gases consist of extremely small particles called molecules. These molecules are so small that their volume is negligible in comparison with the volume of the container.
2. The molecules of a gas are in rapid, random, straight-line motion. They collide with each other and with the walls of the container.
3. All collisions are perfectly elastic; that is, there are no energy losses due to friction.
4. There are no attractive forces between the molecules.
5. Molecules of different gases have equal average kinetic energies at the same temperature. If the temperature increases, the average kinetic energy of the molecules increases.

The last postulate (that is, the last basic principle) was added to Dalton's theory by Maxwell. These few postulates are sufficient to explain all the observations about gases that we have discussed.

Postulate 1 explains the compressibility of gases. Since the volume of a gas is mostly empty space, it should be fairly easy to force the molecules into a smaller volume.

Postulate 3 is necessary because, if molecules lost energy during nonelastic collisions, they would gradually slow down and come to rest at the bottom of the container. This has never been observed for any gas.

The first four postulates together explain Dalton's Law of Partial Pressures. Since the distance between molecules is relatively large, it is always possible to add more gas molecules; and since there are no intermolecular forces, the gases will act independently in their collisions with the walls of a container. Thus the total pressure is equal to the sum of the individual partial pressures.

The kinetic molecular theory of gases explains Boyle's Law. Pressure must be due to the collisions of the gas molecules with the walls of the container (Fig. 13-11). What would happen if the volume of the container were decreased? The gas molecules would hit the walls of the container more often, and therefore the pressure on the container walls would be increased. If the volume of the container were increased, the gas molecules would collide with the walls less frequently because they have further to travel. Therefore the pressure would be decreased.

The kinetic molecular theory also explains Charles' Law. Consider a container which can easily expand, such as a balloon. If the gas inside the balloon is heated, Postulate 5 states that the average kinetic energy of the gas molecules will increase. Therefore, the molecules will speed up. They will strike the wall of the balloon more often and with more force, and therefore in order to keep

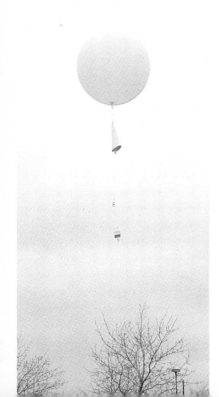

Figure 13-11
This radiosonde weather balloon is filled with hydrogen gas. Radiosonde balloons are sent up twice daily at 36 Canadian radiosonde stations to measure temperature, air pressure, and humidity.

the pressure constant, the balloon will expand. Thus the volume of a gas increases as its temperature increases.

Gay-Lussac's Law

The kinetic molecular theory of gases predicts a relationship we have not yet mentioned, the relation between pressure and temperature. Consider a gas-filled container, such as a scuba tank, that cannot easily expand. If the container is heated, the gas molecules will speed up. They will strike the walls of the container more frequently and with more force. Therefore the pressure will increase. The pressure of a gas increases as its temperature increases, and decreases with a decrease in temperature. The pressure exerted by a gas is directly proportional to its Kelvin temperature, provided the volume remains constant. This holds true experimentally and is called **Gay-Lussac's Law**. Stated mathematically, Gay-Lussac's Law can be written

$$P = (\text{constant}) \times T$$

The Meaning of Pressure and Temperature

The pressure of a gas is due to collisions of gas molecules with the walls of a container. Each collision of a gas molecule with the walls of a container can be considered to be a little "push." Pressure is thus related to the number of collisions taking place per unit area and the force of these collisions. The faster the molecules are moving, the greater is the force of the collisions and the greater is the number of collisions per unit of time.

Gas pressure is commonly measured by using a device called a manometer (Fig. 13-12). The pressure of a gas is measured by comparing the pressure it exerts on a column of mercury with the known pressure exerted by the atmosphere. The difference (h) in the height of the columns of mercury indicates the difference between the pressure of the gas and the known atmospheric pressure. A manometer is usually calibrated to indicate the pressure difference in millimetres of mercury. One millimetre of mercury exerts a pressure of 133 Pa or 0.133 kPa.

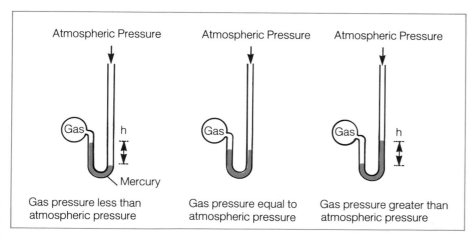

Figure 13-12
An open-end manometer

What are we measuring when we determine the temperature of a gas? The Kelvin temperature is proportional to the average kinetic energy of the gas molecules. Temperature is a measure of the random motions of the gas molecules. Thus, the greater the motions of the gas molecules, the higher the temperature.

13.10 Further Applications of the Gas Laws

We have been using Boyle's and Charles' Laws to calculate new volumes. It is possible to use these laws in conjunction with Gay-Lussac's Law to solve problems involving the pressure, volume, and temperature relationships of a gas.

EXAMPLE PROBLEM 13-7

A certain mass of a gas occupies a volume of 400 mL at a pressure of 150 kPa and a temperature of 280 K. What will be its pressure when its volume is 380 mL and its temperature is 300 K?

SOLUTION
We could write

$$P_{new} = P_{old} \times (\text{volume factor}) \times (\text{temperature factor})$$

The volume factor is the ratio of the two volumes. It is either 400 mL/380 mL or 380 mL/400 mL. The first of these is a number greater than 1. The second is a number smaller than 1. Since the pressure increases as the volume decreases, the new pressure must be larger than the old one, and the volume factor must be a number greater than 1:

$$P_{new} = 150 \text{ kPa} \times \frac{400 \text{ mL}}{380 \text{ mL}} \times (\text{temperature factor})$$

The temperature factor is either 280 K/300 K or 300 K/280 K. Since Gay-Lussac's Law tells us that the pressure increases as the Kelvin temperature increases, the temperature factor is a number greater than 1:

$$P_{new} = 150 \text{ kPa} \times \frac{400 \text{ mL}}{380 \text{ mL}} \times \frac{300 \text{ K}}{280 \text{ K}} = 169 \text{ kPa}$$

A certain mass of a gas occupies a volume of 7.50 L at a pressure of 101.0 kPa and a temperature of 27°C. What will be its Celsius temperature when its volume is 7.15 L and its pressure is 85.0 kPa?

SOLUTION
We could write

$$T_{new} = T_{old} \times (\text{pressure factor}) \times (\text{volume factor})$$

The pressure factor is either 85.0 kPa/101.0 kPa or 101.0 kPa/85.0 kPa. Since Gay-Lussac's Law tells us that the pressure decreases as the temperature decreases, the pressure factor is a number less than 1, and we can write

$$T_{new} = (273 + 27)\ K \times \frac{85.0\ kPa}{101.0\ kPa} \times (\text{volume factor})$$

The volume factor is either 7.15 L/7.50 L or 7.50 L/7.15 L. Since Charles' Law tells us that gases contract on cooling, the volume factor is a number less than 1, and we can write

$$T_{new} = 300\ K \times \frac{85.0\ kPa}{101.0\ kPa} \times \frac{7.15\ L}{7.50\ L} = 241\ K$$

$$241\ K = (241 - 273)°C = -32°C$$

PRACTICE PROBLEM 13-6

If a given mass of a gas occupies a volume of 175 mL at a pressure of 95.0 kPa and a temperature of 100°C, what will be its pressure when its volume is 200 mL and its temperature is 50°C? (*Answer*: 72.0 kPa)

PRACTICE PROBLEM 13-7

If a given mass of a gas occupies a volume of 125.0 mL at a pressure of 110.0 kPa and a temperature of 30°C, what will be its Celsius temperature when its volume is 148.0 mL and its pressure is 127.0 kPa? (*Answer*: 141°C)

13.11 Behaviour of Real Gases

A gas which obeys Boyle's Law, Charles' Law, and Gay-Lussac's Law is called an **ideal gas** and is said to behave ideally. At ordinary temperatures and pressures, most gases, such as hydrogen, nitrogen, and carbon dioxide do obey the gas laws reasonably well. Thus, they behave ideally. However, at low temperatures and high pressures, gases may deviate from ideal behaviour. Gases which deviate from ideal behaviour are called real gases.

Assume, for example, that you have one litre of carbon dioxide gas at 25°C and at a pressure of 100 kPa. What happens to the volume at 25°C as the pressure on the gas is increased to 10 000 kPa? The volume decreases. The data are given in Table 13-4. According to Boyle's Law, the product of the pressure and the volume of a gas at a given temperature should remain constant at all pressures. We see from the third column of Table 13-4 that the PV product is not constant. It decreases steadily to about one-third of its original value, then starts to increase again. Carbon dioxide does not obey Boyle's Law at high pressures and does not behave ideally.

Table 13-4   Pressure-Volume Measurements for Carbon Dioxide at 25°C

Pressure (kPa)	Volume (mL)	$P \times V$ (kPa·mL)
100	1000	100 000
2 000	46	92 000
4 000	20	80 000
6 000	10	60 000
8 000	4	32 000
10 000	4	40 000

The Behaviour of Real Gases at High Pressures

How can we explain this deviation from ideal behaviour? Recall two postulates of the kinetic molecular theory. The kinetic molecular theory assumes that the volume of the gas molecules is negligible and that there are no attractive forces between molecules.

When a high pressure is used to force molecules into a small volume, the average distances between molecules become small. Attractive forces between molecules (van der Waals forces) are greater when the molecules are close together. These attractive forces cause the volume of the gas to be smaller than it would be if there were no attractions. Because the volume is less than it would be if there were no attractive forces, the PV product decreases as the pressure on the gas increases.

Another point to notice in Table 13-4 is that it appears difficult to reduce the volume of the gas to less than 4 mL. When high pressures are used to force molecules into a small volume, the volume of the molecules themselves becomes a significant portion of the total volume. Thus, increasing the pressure on the carbon dioxide from 8000 to 10 000 kPa does not decrease its volume below 4 mL. Since V remains constant as P increases above 8000 kPa, the PV product increases.

All real gases show deviations from ideal behaviour. Figure 13-13 is a plot of PV vs. P for equal numbers of moles of hydrogen, nitrogen, and carbon

Figure 13-13
Non-ideal behaviour of real gases

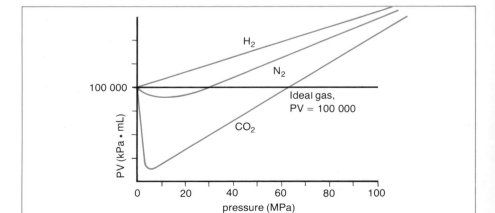

dioxide. If Boyle's Law were followed, the value of the *PV* product would always be the same for each gas and therefore would always lie on the same horizontal line. We see, however, that both nitrogen and carbon dioxide have *PV* products which are less than predicted, because van der Waals forces reduce the volumes. At higher pressures, the N_2 and CO_2 molecules occupy enough volume themselves to prevent theoretical compression, and the *PV* product is larger than expected. In contrast, the *PV* product of hydrogen does not decrease at all, but starts to increase almost immediately. This means that, although the H_2 molecules are closer together at higher pressures, the van der Waals forces between them still remain negligible.

The Behaviour of Real Gases at Low Temperatures

At lower temperatures, the molecules move more slowly and spend more time at closer distances to each other. Thus, as long as the pressure is not large enough to have forced the molecules as close together as they can get, the van der Waals forces of attraction increase at lower temperatures. The increased van der Waals forces reduce the volume of the gas below the volume that would be predicted by Charles' Law.

13.12 Liquefaction of Gases

At high pressures, the molecules of a gas are closest to each other, and the attractive forces are most important. As the temperature increases, the molecules move more rapidly, and this motion tends to keep the molecules apart. As the temperature decreases, the molecules slow down and are less able to overcome the attractive forces. When the temperature is low enough, the attractive forces can draw the molecules together to form a liquid. The temperature at which the gas molecules come together to form a liquid is called the *liquefaction temperature*. It is easier to liquefy a gas at higher pressures, because the molecules are close to each other, and the intermolecular forces are more effective. Thus, the higher the pressure, the higher the liquefaction temperature, that is, the easier it is to liquefy the gas.

Critical Temperature

For each gas there is some temperature, called the **critical temperature**, above which it is impossible to liquefy the gas, no matter how much pressure is applied. Above the critical temperature, the substance can exist only as a gas. The motion of the molecules is so vigorous that the attractive forces can no longer overcome the motions of the molecules, no matter how high the pressure.

The critical temperature depends on the strength of the intermolecular forces. A polar substance has strong attractive forces which aid in overcoming molecular motion. Thus, a polar substance may be liquefied at relatively high temperatures. A nonpolar substance has weak attractive forces. It has a low critical temperature. At higher temperatures the molecular motions are able to overcome the weak attractive forces. Table 13-5 lists the critical temperatures for some common substances. We see that water, which is highly polar and has strong attractive forces, can be liquefied at temperatures as high as 647 K. Helium, on the other hand, has such weak attractive forces that it cannot be liquefied above 5.2 K, no matter how much pressure is exerted on it.

Table 13-5   Critical Temperatures and Pressures for Some Common Substances

Substance	Critical Temperature (K)	Critical Pressure (MPa)
H_2O	647	22.1
NH_3	406	11.3
HCl	324	8.27
CO_2	304	7.40
O_2	153	5.04
N_2	126	3.39
H_2	33	1.30
He	5.2	0.229

Critical Pressure

The minimum pressure required to liquefy a gas at its critical temperature is called the **critical pressure** of the gas. Critical pressures are listed along with critical temperatures in Table 13-5.

13.13 Cooling of Gases by Expansion

The Joule-Thomson Effect

You may have noticed that air cools as it escapes rapidly through the valve of a bicycle tire. The cooling of a gas as it expands rapidly is called the *Joule-Thomson effect*. As the gas expands, the average distance between molecules increases, in spite of the intermolecular forces of attraction. Energy is required to overcome the attractive forces and separate the molecules. Since no outside energy is available, the molecules must use some of their kinetic energy. As a result, the average kinetic energy drops. Because the Kelvin temperature is directly proportional to the average kinetic energy of the molecules, the temperature also drops.

Liquid Air

The Joule-Thomson effect is used commercially in the preparation of liquid air. Before the air can be liquefied, the water vapour and carbon dioxide must be removed because both of these substances solidify on cooling and clog the pipes of the liquid air machine (Fig. 13-14). The dry air is first compressed to about 20 MPa. It is then pumped through pipes leading through a cooling bath containing a coolant such as liquid ammonia (normal boiling point −33°C). The cold compressed air is then passed through other cooling coils in the liquefier, where it escapes through an expansion valve and expands to a pressure of 2 MPa. The air cools upon expansion, because it uses up kinetic energy to overcome the intermolecular forces of attraction. The cold, expanded air is passed over the cooling coils to further cool the incoming air. As the process continues, each quantity of air that escapes from the expansion valve is colder than that which preceded it. Finally, part of the air condenses to a liquid, which is drawn off at the bottom of the expansion chamber.

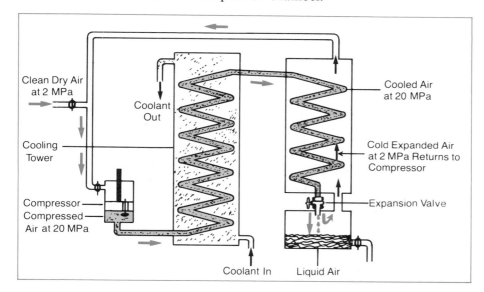

Figure 13-14
Liquefaction of air

Liquid air is a mobile liquid with a pale blue colour. It consists of about 78% nitrogen, 21% oxygen, 1% argon, and smaller amounts of ozone, hydrogen, and the noble gases. Oxygen boils at −183.0°C, argon at −185.7°C, and nitrogen at −195.8°C. Liquid air has no definite boiling point, because it is a mixture. It is often used as a low temperature coolant. Liquid oxygen, liquid nitrogen, and liquid argon are obtained commercially by fractional distillation of liquid air. Liquid oxygen is used in rockets and missiles.

13.14 Heating of Gases by Compression

We have seen that, if a gas in a container is allowed to expand rapidly, the temperature of the gas will be lowered. In the same way, when a gas in a container is compressed very rapidly, the temperature of the gas will be increased. This happens because the energy from the piston of the compressor is transferred to the gas molecules which collide with it. These molecules rebound with greater kinetic energy, and therefore the temperature of the gas increases.

The Diesel Engine

The heating of a gas by rapid compression is used in the diesel engine (Fig. 13-15). At the beginning of the compression stroke, a blower fills the cylinder with air. As the piston rises, the exhaust valves close, and the air is rapidly compressed to one-sixteenth of its original volume. The temperature of the compressed air rises above the ignition temperature of diesel fuel. At this point, fuel is injected into the cylinder, and it ignites immediately. The expanding gaseous reaction products drive the piston down in a power stroke. When the piston reaches its lowest point, the exhaust gases are blown out, and the sequence begins again.

Figure 13-15
A diesel engine

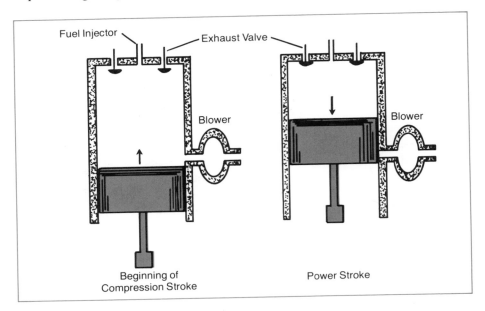

Diesel engines need no spark plugs—the hot air causes self-ignition of the fuel. Because it uses an excess of air, a diesel engine emits less carbon monoxide than an ordinary gasoline engine. However, if it is poorly maintained, a diesel engine has special smoke and odour problems, as anyone who has driven in the dark fumes behind some trucks and cars will know!

Part Two Review Questions

1. State the five basic postulates of the kinetic molecular theory.
2. Use the kinetic molecular theory to explain Dalton's Law of Partial Pressures.
3. Use the kinetic molecular theory to account for the compressibility of gases.
4. State Gay-Lussac's Law.
5. How do you explain the pressure of a gas?
6. A certain mass of gas occupies a volume of 275 mL at a pressure of 102 kPa and a temperature of 35°C. What will be its pressure when its volume is 365 mL and its temperature is 75°C?

7. A certain mass of a gas occupies a volume of 5.5 L at a pressure of 105 kPa and a temperature of 20°C. What will be its Celsius temperature when its volume is 8.5 L and its pressure is 75 kPa?

8. Explain why real gases stop behaving like ideal gases at low temperatures or high pressures.

9. Explain what is meant by **a)** critical temperature; **b)** critical pressure; and **c)** the Joule-Thomson effect.

Part Three: Chemistry and the Gaseous Environment

13.15 Technology and the Environment

In the 1970s, people became more aware of the effect of their actions on the environment and of the interrelatedness of all living things as they exist in nature. During the centuries that human beings have occupied this earth, the human population and technological advances have increased together. This has been especially true during the last several decades. To many people, technology seems to be the main reason for our problems with pollution. Some may regard technology and progress as demons responsible for the destruction of the environment. However, we should recall that technology is merely the application of scientific knowledge by people. People have misused technology to despoil the environment (Fig. 13-16). Therefore, people must use technology to clean up the environment.

Figure 13-16
A cloud of pollution hangs over a large urban centre.

Forest fires emit large amounts of carbon dioxide into the atmosphere.

One other point should be made. We cannot lay all the blame for our environmental problems at the feet of technologists. We should also consider the effect of the increasing human population. Living things have always had an effect on nature. Because of the sheer size of the global environment, nature has been able to adjust its balance. However, the disruptive effect of a highly concentrated population in a particular area may so alter the environment of that area that the balance of nature can never be achieved. The concentration of humanity in large cities means the concentration of humanity's effects on the environment in those areas. This is true regardless of the state of the technology. Would Toronto be more habitable if horses were used instead of automobiles? Even if there were room for the horses, the accumulated animal refuse would make the state of streets in Toronto unacceptable.

Although technology affects many aspects of the environment, we shall limit discussion to the air environment in the remainder of this chapter.

13.16 Changes in the Air Environment

Air is a mixture of gases such as oxygen, nitrogen, argon, and carbon dioxide. The atmosphere consists of air plus droplets of liquid (mainly water) and finely divided solid particles. The gaseous part of the atmospheric mixture is quite uniform near the earth's surface. The gas molecules are constantly in random motion and tend to mix evenly. Wind aids this mixing process. The atmosphere is a dynamic system in which the molecules are in continuous random motion. Chemical reactions occur among the substances present in the atmospheric mixture, and physical changes also take place. These physical changes include the settling out of solid and liquid particles. Water vapour condenses on dust particles to form water droplets (clouds and fog). The water droplets in clouds may become large enough to fall to earth as rain, or the drops may freeze and fall as hail or snow.

Most scientists believe that the earth has been in a stable orbit for eons and that, over the period of the earth's history, natural balances have slowly been established in the atmosphere and a wide variety of equilibria have been reached. What is an equilibrium? To learn about equilibria in general, we shall consider the cycles by which the amount of carbon dioxide in the atmosphere has been limited to a concentration of about 0.03 percent.

Carbon Dioxide and the Atmosphere

Natural events such as forest fires, animal respiration, decay of organic matter, and volcanic eruptions pour carbon dioxide into the atmosphere. Some of this carbon dioxide is used by plants for photosynthesis. When animals eat the plants and breathe, the carbon dioxide returns to the atmosphere to complete the cycle. Other carbon dioxide dissolves in the water present in the atmosphere:

$$CO_2(g) + H_2O(\ell) \longrightarrow \underset{\textit{carbonic acid}}{H_2CO_3(aq)}$$

The carbonic acid reacts with minerals to form limestone, $CaCO_3$. Carbon dioxide also dissolves in the oceans and some of it ends up as $CaCO_3$ in the shells of ocean creatures.

Thus, carbon dioxide is involved in an equilibrium process. The amount of carbon dioxide entering the atmosphere from forest fires, volcanic activity, and animal and plant respiration is equal to the amount of carbon dioxide leaving the atmosphere through photosynthesis and carbonic acid formation. An equilibrium or balance like this takes years to be reached and can be destroyed if another factor enters the picture.

The other factor is our need for heat. People have increased the carbon dioxide content of the atmosphere by the massive burning of fossil fuels such as coal and petroleum. The natural processes by which carbon dioxide is removed from the atmosphere cannot keep pace with this increase in carbon dioxide content. Thus, we have destroyed nature's carbon dioxide balance. It now appears that the carbon dioxide content of the atmosphere is increasing (Figs. 13-17 and 13-18).

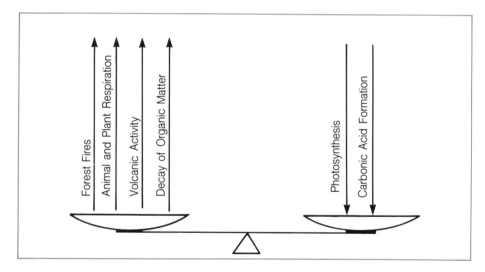

Figure 13-17
Nature's carbon dioxide balance

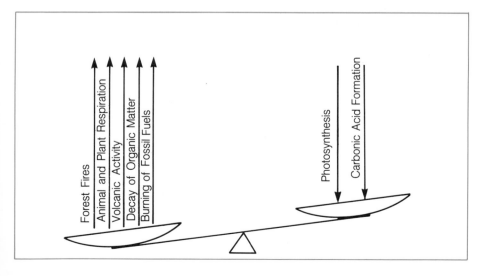

Figure 13-18
Disturbance of carbon dioxide balance through combustion of fossil fuels

Other Gases and the Atmosphere

Other gases such as carbon monoxide (CO), sulfur oxides (SO_2 and SO_3), nitrogen oxides (NO and NO_2), and hydrocarbons have always been present in the atmosphere. Volcanic action has produced sulfur oxides, and lightning produces nitrogen oxides from nitrogen and oxygen. However, we have added to the concentration of these noxious gases (Fig. 13-19). In some areas of the world, people have died because of high concentrations of these pollutants.

A **pollutant** is a substance that is present in high enough concentration to produce adverse effects on the things that human beings value. These include not only their own health, safety, and property, but also acceptable conditions for all plants and animals. We cannot control the amount of pollutants injected into the atmosphere by natural processes. However, the balancing mechanisms of nature normally act to prevent these quantities of pollutant molecules from rising to dangerous levels. What we *must* control is the quantity of pollutant molecules that enter the atmosphere as a result of our actions. In the next sections, a number of substances that are known air pollutants will be discussed.

Figure 13-19
This paper mill in Thunder Bay, Ontario, emits both SO_2 and CO_2 into the atmosphere.

APPLICATION

The Lethal Lake

In August, 1986 a deadly cloud of gas escaped from Lake Nios in the West African nation of Cameroon. Almost 1700 people were killed. Researchers believe that the cloud consisted of carbon dioxide which smothered the victims. The air we breathe normally contains about 0.03% carbon dioxide; carbon dioxide concentrations above 10% in the air can be lethal to humans.

Lake Nios is a deep lake that lies in a crater formed by a volcano. In deep lakes, a layer of warm water floats on top of a denser, cold layer. This layering traps the colder water at the bottom. In other words, the warm surface layer acts like an enormous lid.

Researchers studying the Lake Nios disaster think that carbon dioxide had gradually filtered up from molten rock below the lake and dissolved in ground water that flowed into it. For centuries, carbon dioxide had accumulated in the cold lower layer of Lake Nios. Eventually, the lake became saturated with carbon dioxide, like a giant bottle of soda water. If a bottle of soda is opened carefully, very little of the carbon dioxide bubbles out. However, if the bottle is disturbed by shaking before the cap is removed, the carbon dioxide gushes out of the water. In the same way, any disturbance could have set off a violent release of carbon dioxide from Lake Nios.

The fact that the disaster happened in August could be significant. In August, the air temperatures above lake Nios are at their lowest, and the surface layer of the lake begins to cool. If the surface layer cooled to the same temperature as the lower layer, the two layers would be able to mix. This would be equivalent to removing the warm water lid, permitting a deadly release of carbon dioxide.

Some scientists believe that gas-charged lakes such as Lake Nios can be rendered harmless. They suggest that the lower layers of such lakes could be pumped gradually through large pipes to the surface. The dissolved carbon dioxide could then escape into the atmosphere slowly and safely.

13.17 The Effects of Solid Particles in the Atmosphere

The earth is warmed by infrared radiation from the sun. The presence of particles of dust and smoke has a great effect on the ability of the atmosphere to transmit radiation from the sun. Any change in the amount of infrared radiation reaching the earth would change the temperature of the earth. Most atmospheric particles have a size ranging from 0.1 µm to 10 µm in diameter. Larger particles are too heavy to remain in the atmosphere. The smaller particles are bounced about by collisions with gas molecules and tend to remain airborne for long times. These particles create a hazy atmosphere which allows less infrared light to reach the surface of the earth. Evidence actually exists that the average global temperature fell 0.2°C between 1940 and 1967. This may have been due to a larger number of particles in the air. Continued temperature decreases would change the climate of portions of the globe.

Some solid particles, such as those in the smoke of metal smelters, contain poisons which are harmful to animals. The harmful effects of filling the lungs with soot, silica, arsenic oxides, or lead are well known. Studies suggest that respiratory diseases can be caused by breathing air which contains large quantities of particles.

Nature removes particles from the atmosphere by gravitational settling, rain, and snow. We already have technology that can aid nature in removing particles from the atmosphere. Taller smoke stacks are not the answer since they merely disperse the particles higher into the atmosphere. Instead, filter bags made of glass fibre, cotton, nylon, wool, or felt can be used to filter particles from gases. In addition, centrifugal separators can be used. These separators whirl gases and throw particles to the walls of the apparatus where they can be collected.

Electrostatic Precipitators

Figure 13-20
An electrostatic precipitator

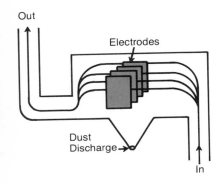

Electrostatic precipitators (Fig. 13-20) are effective in removing dust particles from industrial plant exhaust gases. The dust particles pass through a strong electric field (about 50 kV). If the particles are not already charged, they become charged. The charged particles are quickly attracted to oppositely charged plates where they are neutralized and fall to the bottom of the precipitator to be collected. Equipment like electrostatic precipitators can remove more than 99 percent of the dust particles from exhaust gases.

Another control consists of better regulation of the combustion process. If enough air is used to burn carbon-containing compounds, more carbon dioxide and less carbon soot is produced. However, as we shall see in the next section, increasing levels of CO_2 in the atmosphere pose a threat to the environment.

13.18 Carbon Dioxide and the Greenhouse Effect

The carbon dioxide content of the atmosphere is increasing. Carbon dioxide is not a poisonous gas, nor is it very corrosive; however, it is interesting to observe that even carbon dioxide can cause environmental problems if its concentration in the atmosphere becomes too great.

Light from the visible and ultraviolet regions of the spectrum passes through carbon dioxide molecules without being absorbed. However, infrared light (heat energy) causes the CO_2 molecules to vibrate more vigorously. Thus, the carbon dioxide molecules absorb and later emit infrared light.

A greenhouse

The Greenhouse Effect

When the surface of the earth cools at night or under a cloud layer, the infrared energy which the earth gives off is trapped by the carbon dioxide in the atmosphere, and part of this infrared energy is emitted back to the earth. This leads to what is called the **greenhouse effect** (Fig. 13-21). The glass roof of a greenhouse allows light and heat in, but it reflects back into the greenhouse some of the heat that tries to escape. The temperature inside a greenhouse is higher than the temperature outside it. Water vapour in the air adds to the effect of CO_2, since it also absorbs and reemits infrared energy back to earth.

It has been estimated that without the greenhouse effect the average temperature of the earth's surface would be about 45°C lower than at present. Thus, without this effect, the average surface temperature of the earth would be −30°C instead of 15°C. The greenhouse effect is even more remarkable on the planet Venus. This planet has a very dense atmosphere of carbon dioxide and a surface temperature of about 500°C.

The atmospheric CO_2 content was constant until the start of the Industrial Revolution. The natural level of CO_2 at that time is thought to have been between 260 and 290 ppm (parts per million). The atmospheric CO_2 content has risen from 316 ppm in 1958 to 344 ppm in 1984. If we consider the pre-industrial amount to be 275 ppm, the 1984 values represent a 25% increase. This change is the result of the increased burning of fossil fuels. The CO_2 level is currently estimated to be increasing by about 0.5% per year. It is thought that the CO_2 content will have doubled from the time of the Industrial Revolution (from about the middle of the eighteenth century) to the year 2050. Since it is believed that the earth should warm by 0.1°C for each 10 ppm increase in CO_2 concentration, the global temperature could rise by about 3°C by the year 2050.

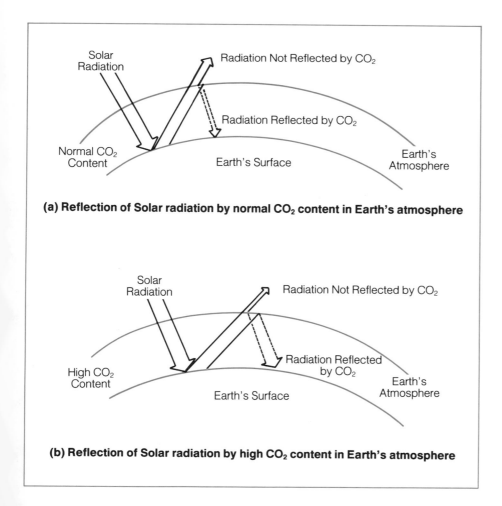

Figure 13-21
The greenhouse effect

(a) Reflection of Solar radiation by normal CO_2 content in Earth's atmosphere

(b) Reflection of Solar radiation by high CO_2 content in Earth's atmosphere

How will Canada be affected by such an increase in CO_2 levels? Northern Canada, with an average level of 344 ppm, has one of the highest CO_2 concentrations in the world. Southern Canada will be about 3°C warmer than now. The western Arctic may be 10 to 15°C warmer than it is now.

A warming trend could cause the melting of Arctic ice, higher ocean levels, lower water levels (by as much as a metre) in the Great Lakes, greater forest fire hazards, prairie droughts, increased photosynthesis and plant growth, increased agricultural capacity in cooler and wetter regions, and lower home heating costs. Thus, some regions of Canada will experience positive effects from higher CO_2 levels, while other areas will experience negative effects. Canada as a whole could benefit from the greenhouse effect. Many scientists believe that the warming effect of CO_2 emissions will continue to outweigh the cooling effect of solid particles. The net result will be a continuing warming trend.

Nobody knows the maximum permissible level of atmospheric CO_2 (the level above which the greenhouse effect would become too dangerous). A level 50% above pre-industrial values would cause a temperature increase of 1 to 2°C, which might be tolerable. In that case, our production of CO_2 could continue to increase until 2000. However, it would then have to level off rapidly between 2030 and 2080 to pre-1970 production values in order to prevent the CO_2 level from increasing to more than 50% above pre-industrial values. Clearly, we cannot continue to burn fossil fuels at the present growth rate for much longer. We have more than enough fossil fuel to supply our needs for several hundred years, but by the turn of the century we will have to find new technologies for the world-wide production of energy. If we do not find these new technologies, the time may come when CO_2 will have to be regarded as a dangerous pollutant.

13.19 Photochemical or Los Angeles–Type Smog

There are two general types of smog. London-type smog, which is largely caused by the combustion of coal and oil, is discussed in Section 12.9. This type of smog contains sulfur dioxide mixed with soot, fly ash, smoke and some organic compounds. The second type of smog is photochemical smog, or Los Angeles-type smog. It is called photochemical smog because light from the sun is important in starting the chemical process. This second type of smog is practically free of sulfur dioxide, but it contains large amounts of nitrogen oxides, ozone, ozonated hydrocarbons, organic peroxides, and hydrocarbons of varying complexities. As we shall see, the smog for which Los Angeles is so famous (and which most large Canadian cities have in varying degrees) is caused primarily by automobiles.

Los Angeles had almost no air pollution problem until around 1940, when population growth and industrial expansion in the area suddenly accelerated. It was at this time that the air pollution problem began to appear. The previously clean atmosphere frequently became transformed into smog—a brown haze of smoke and offensive odours. Several factors were responsible for the problem, especially the sudden increase in the number of automobiles in southern California and the peculiar natural climatic conditions in and around Los Angeles (including the legendary California sunshine). These factors, combined with the fact that the city lies in a mountain-rimmed depression, contributed to the build-up of smog.

Before the anti-pollution laws controlling automobile emissions were enacted, the thousands of cars around Los Angeles daily poured out enormous quantities of pollutants. The pollutants consisted mainly of nitrogen oxides and hydrocarbons (crankcase oil vapours and unburnt gasoline). These were the starting materials in the smog formation process.

Thermal Inversion

Because of its unusual geography, Los Angeles often experiences an abnormal meteorological phenomenon known as **thermal inversion** which prevents any pollutants from moving to the upper atmosphere where they can be dispersed. Normally, a layer of warmer, less dense air first forms near the earth and then rises, carrying with it pollutants such as smoke, gases, and dust. The colder, denser air from the upper atmosphere moves downward to replace it and the cycle is repeated. When thermal inversion occurs, a layer of warm air is *already* resting on top of a layer of cooler, denser air near the earth. This lower layer cannot move upward; nor can it move sideways because of the mountains ringing the city. As a result, a stagnant layer of polluted air remains trapped in and over the city. The abundant solar radiation acts on the trapped nitrogen oxides and hydrocarbons and the resultant photochemical reactions produce the smog (Fig. 13-22).

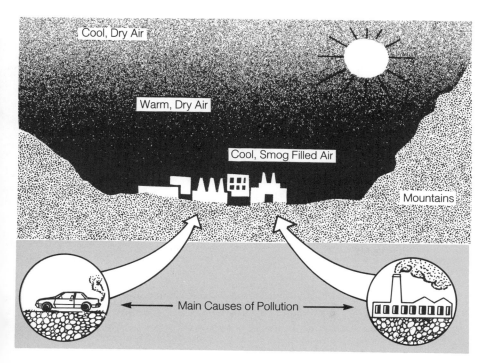

Figure 13-22
A thermal inversion. The cool, smog-filled air is trapped by a layer of warm, dry air.

Many big cities become blanketed with smog under certain weather conditions.

Nature supplies the energy and the reaction vessel in the form of sunshine and thermal inversion. People supply the chemicals in the form of emissions from gasoline combustion and evaporation and factory exhausts. When suitable meteorological conditions prevail, the concentration of contaminants in Los Angeles are enough to produce an objectionable smog during a one- or two-hour exposure to bright sunlight.

The Chemistry of Smog

Research into the series of photochemical reactions that produce smog has added to our understanding of the process. Nitrogen(II) oxide (NO) arises from a direct combination of nitrogen and oxygen which is produced by the high temperature inside the automobile engine cylinders during the ignition stroke:

$$N_2 + O_2 + heat \longrightarrow 2\ NO$$

This nitrogen(II) oxide immediately reacts with oxygen to form nitrogen dioxide:

$$2\ NO + O_2 \longrightarrow 2\ NO_2$$

At this point, the photochemical process is thought to begin. Ultraviolet light from the sun breaks up the nitrogen dioxide molecule:

$$NO_2 + UV\ light \longrightarrow NO + O$$

The oxygen atom has only six outer electrons, and it is a highly reactive species. It can react with oxygen to form ozone:

$$O + O_2 \longrightarrow O_3$$

Ozone is a form of oxygen which has three atoms of oxygen per molecule. Ozone is called an allotrope of oxygen, and it is toxic and corrosive to tissue. Even at low concentrations ozone can kill plants.

Some ozone reacts with nitrogen(II) oxide to re-form nitrogen dioxide:

$$NO + O_3 \longrightarrow NO_2 + O_2$$

Some of the reactive oxygen atoms follow another path. They can react with hydrocarbon molecules in the atmosphere to produce organic free radicals. Free radicals are organic fragments in which unpaired electrons impart a great reactivity to the species. These free radicals can react with molecular oxygen to produce even more reactive free radicals, which can then react to form secondary pollutants such as aldehydes (formaldehyde, acetaldehyde, acrolein), ketones, and peroxyacyl nitrates (PAN):

$$\underset{acrolein}{CH_2\!\!=\!\!CH\!\!-\!\!\overset{\displaystyle O}{\overset{\|}{C}}\!\!-\!\!H} \qquad \underset{a\ peroxyacyl\ nitrate}{CH_3\!\!-\!\!\overset{\displaystyle O}{\overset{\|}{C}}\!\!-\!\!O\!\!-\!\!O\!\!-\!\!NO_2}$$

It is known that formaldehyde, acrolein, and PAN are all lachrymators (tear producers). These three substances cause most of the physical discomfort produced by smog. Smog, however, contains hundreds of different molecules as a result of these free radical reactions.

13.20 Effects of Photochemical Smog

What is it like to be in Los Angeles when a combination of thermal inversion, low winds, bright sunlight, factory exhaust, and hundreds of thousands of automobiles bring on the smog? Consider the possibilities.

The day begins clear and sunny. As traffic increases, the air begins gradually to become a yellow-brown haze. The odour is sharp and pungent. The shroud of pollution obscures prominent landmarks and limits visibility. The intensity of the smog increases during the day. Around noon the pollution becomes intense. People complain of eye irritation, and many find it difficult to breathe.

At street level, the carbon monoxide emitted from automobile exhausts reaches 50 mg/kg, and traffic police complain of headaches. The carbon monoxide is combining with hemoglobin in their blood and is starving their bodies of oxygen.

The smog worsens, and school children are not allowed to have outdoor recreation classes. The ozone level has exceeded 0.35 mg/kg, and at that level fifteen minutes in the outdoors can cause respiratory irritation accompanied by choking, coughing, and severe fatigue.

The smog grows thicker. The air mass is still held stationary by the thermal inversion and by the mountains. There is no wind—not even a breeze.

The people of Los Angeles live in an atmospheric sewer, each person breathing 25 m³ of this polluted air every day, three hundred days a year. The air contains more than 50 pollutants. More than 2 million Californians (one person in eight) have some kind of respiratory problem such as emphysema, bronchitis, or asthma. There is more lung disease in California than in any other part of North America.

Although Los Angeles remains the chief centre for photochemical smog, the phenomenon has been detected in varying degrees in most major metropolitan areas. Unless pollutant emissions are controlled, the sun, which has made life on this planet possible, could make city life unbearable.

13.21 What Can Be Done About Air Pollution?

The previous sections have included discussions concerning some (but not all) of the possible air pollutants. The question is, what should be done about air pollution?

The first step is to pass laws to regulate the introduction of pollutants into the atmosphere (Fig. 13-23). Soft coal has a high sulfur content, and, when it burns, it releases large quantities of sulfur dioxide into the air. Thus, banning the use of soft coal would be helpful. Forbidding the burning of rubbish and the use of private incinerators would help to reduce air pollution further. Smokestacks should have electrostatic precipitators to cut down on the amount of particles they produce. The use of atomic energy would cut down on air pollution.

Figure 13-23
An air sampling device traps particulate matter and rainwater to allow environmentalists to study air pollution.

Figure 13-24
Devices for controlling automobile exhaust pollution

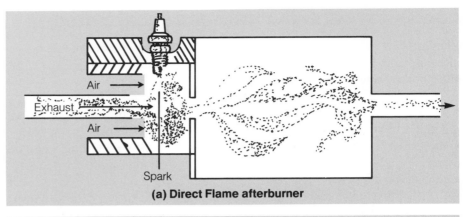

(a) Direct Flame afterburner

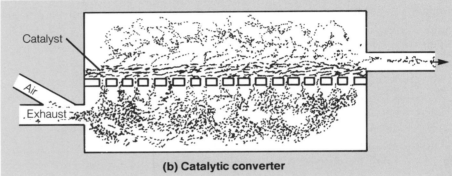

(b) Catalytic converter

We can reduce pollutants such as hydrocarbons, nitrogen(II) oxide, and carbon monoxide from automobile exhaust. Hydrocarbons and carbon monoxide can be burned more efficiently to produce carbon dioxide instead. A direct flame afterburner (Fig. 13-24a) would mix unburned pollutants from engine exhaust with additional fresh air, and then ignite the mixture with a spark or a flame. Many cars are already equipped with catalytic converters (Fig. 13-24b). These devices consume the same pollutants by passing them together with additional air over a catalyst. Most recent automobiles contain a platinum catalyst which converts hydrocarbons, CO, and NO to CO_2, H_2O and N_2. Cars equipped with catalytic converters have to use unleaded fuel (gasoline which contains no tetraethyllead) because lead renders the catalyst in the converter ineffective.

We require national and international air standards. As the world population grows, we will need tougher standards. With a combination of common sense, necessity, and research, there is no reason to believe that we will not be successful.

Part Three Review Questions

1. Describe the operation of an electrostatic precipitator.
2. Describe the greenhouse effect.
3. How might Canada be affected by the greenhouse effect if the CO_2 levels double by the year 2050?
4. How can the pollutants in automobile exhaust be reduced?

APPLICATION

Warnings from Volcanic Gases

Volcanoes emit a number of gases. These gases include carbon monoxide, carbon dioxide, sulfur dioxide, hydrogen sulfide, hydrogen, water, and hydrogen halides. The gases may be collected for analysis from aircraft flying over the volcano or by taking samples from fumaroles. Fumaroles are cracks in the side of a volcano that emit gases but not lava. The analysis of volcanic gases promises to be very useful.

Researchers have found a relationship between the amount of sulfur dioxide emitted and the volcanic activity of Mount St. Helens, in Washington. They hope that this finding might help forecast eruptions. Also, the technique of measuring changes in hydrogen gas concentration appears promising in predicting both volcanic eruptions and earthquake activity. Hydrogen sensors located at 30 stations in Hawaii and in the western United States monitor hydrogen gas concentrations. The sensors then send signals which are relayed by satellite to a computer in Maryland. Large changes in hydrogen concentration have been observed in the days before a volcanic eruption. Such changes in hydrogen concentration have also been noted before the onset of earthquakes at Long Valley, California. This hydrogen gas monitoring technique is still in the experimental stage. Perhaps one day, when collection, analysis, and interpretation have been refined, this technique could be used to give advanced warning to people living in volcanic and earthquake zones.

Chapter Summary

- Boyle's Law states that the volume of a fixed mass of a gas varies inversely with the pressure, provided the temperature remains constant.
- The Kelvin temperature scale is used in gas law equations.
- To convert a Celsius temperature to a Kelvin temperature, add 273 to the Celsius temperature.
- Charles' Law states that the volume of a fixed mass of gas varies directly with the Kelvin temperature, provided the pressure remains constant.
- Standard temperature and pressure (STP) is 273 K and 101.3 kPa.
- Dalton's Law of Partial Pressures states that the total pressure exerted by a mixture of gases is the sum of the pressures of each gas when measured alone.
- The kinetic molecular theory is a model for gases which can be used to explain the gas laws and the behaviour of gases.
- Gay-Lussac's Law states that the pressure exerted by a fixed mass of a gas is directly proportional to its Kelvin temperature, provided the volume remains constant.

- The pressure of a gas is due to the collisions of gas molecules with the walls of the container.
- The Kelvin temperature is proportional to the average kinetic energy of the gas molecules.
- A gas which obeys Boyle's Law, Charles' Law, and Gay-Lussac's Law is called an ideal gas.
- The behaviour of real gases is least ideal under conditions of low temperature or high pressure.
- Critical temperature is the temperature above which it is impossible to liquefy a gas, no matter how much pressure is applied.
- Critical pressure is the minimum pressure required to liquefy a gas at its critical temperature.
- The Joule-Thomson effect is the cooling of a gas as it expands rapidly.
- A pollutant is a substance that is present in high enough concentration to produce adverse effects on our environment.
- Solid particles in the atmosphere lead to a cooling effect.
- The greenhouse effect is caused by CO_2 levels and leads to a warming effect.
- Photochemical or Los Angeles-type smog requires ultraviolet light from the sun.
- Thermal inversion is a meteorological phenomenon in which a warm layer of air over a cooler layer prevents pollutants from being dispersed.

Key Terms

Boyle's Law	Dalton's Law of	critical temperature
Charles' Law	Partial Pressures	critical pressure
standard pressure	Brownian motion	pollutant
standard temperature	Gay-Lussac's Law	greenhouse effect
STP	ideal gas	thermal inversion

Test Your Understanding

Each of these statements or questions is followed by four responses. Choose the correct response in each case. (Do not write in this book.)

1. A certain liquid is five times denser than water. If the air is pumped out of the top of a tube whose bottom is immersed in the liquid, how high will the liquid rise in the tube at normal atmospheric pressure?
   a) 10.3 m   **b)** 51.5 m   **c)** 1.0 m   **d)** 2.1 m

2. Which of the following is a statement of Boyle's Law?
   a) VT = constant   **b)** P/constant = V
   c) PV = constant   **d)** V/constant = T

3. What will be the new volume of a fixed mass of gas whose pressure is increased from 50 kPa to 200 kPa at constant temperature?
   a) four times the original volume   **b)** twice the original volume
   c) one fourth the original volume   **d)** one half the original volume

4. Which of the following is a statment of Charles' Law?
   a) PV = constant
   b) V/T = constant
   c) PT = constant
   d) VT = constant
5. What must be the new Kelvin temperature of a fixed mass of gas whose volume has increased from 10 mL to 50 mL at constant pressure?
   a) one fifth the original temperature
   b) 40 times the original temperature
   c) 1/60 times the original temperature
   d) five times the original temperature
6. The pressure of a fixed mass of gas is doubled and the Kelvin temperature is tripled. The new volume will be
   a) 1.50 times the original volume.
   b) 0.67 times the original volume.
   c) 6.00 times the original volume.
   d) 0.17 times the original volume.
7. Which of the following are STP conditions?
   a) 273 K, 101.3 kPa
   b) 0 K, 101.3 kPa
   c) 0°C, 273 kPa
   d) 23°C, 103 kPa
8. Oxygen gas was collected over water at 24°C and an atomspheric pressure of 105.0 kPa. What is the pressure of the dry oxygen?
   a) 2.98 kPa
   b) 102.0 kPa
   c) 107.98 kPa
   d) 98.3 kPa
9. Which of the following is not a postulate of the kinetic molecular theory?
   a) Molecules of different gases at the same temperature have the same average velocity.
   b) There are no attractive forces between the molecules.
   c) Gases consist of extremely small particles called molecules.
   d) All collisions of gas molecules are perfectly elastic.
10. Which of the following is a statement of Gay-Lussac's Law?
    a) V/T = constant
    b) PV = constant
    c) PT = constant
    d) P/T = constant
11. The temperature of a fixed volume of gas changes from 200 K to 500 K. The new pressure of the gas is
    a) 0.4 times the old pressure.
    b) 1.0 times the old pressure.
    c) 2.5 times the old pressure.
    d) 300 times the old pressure.
12. Which of the following will not decrease the pressure of a gas in a balloon?
    a) reducing the volume of the balloon
    b) decreasing the temperature of the gas
    c) decreasing the atmospheric pressure
    d) removing some gas particles
13. The Kelvin temperature of a gas is inversely proportional to
    a) the pressure.
    b) the volume.
    c) the number of moles.
    d) none of the above.
14. The volume and Kelvin temperature of a fixed mass of a gas are each reduced to one third of their original values. The new pressure of the gas will be
    a) two thirds of the original pressure.
    b) one ninth of the original pressure.
    c) nine times the original pressure.
    d) the same as the original pressure.

15. Real gases deviate most from ideal gas behaviour at
    a) low pressure.
    b) low temperature.
    c) high temperature.
    d) STP.

16. Which of the following has the highest critical temperature?
    a) NH_3
    b) H_2
    c) O_2
    d) N_2

17. Solid particles in the atmosphere are responsible for
    a) the greenhouse effect.
    b) lowering the global temperature.
    c) raising the global temperature.
    d) photochemical smog.

18. Which of the following are removed from smokestack gases by electro-static precipitators?
    a) carbon dioxide
    b) sulfur dioxide
    c) dust particles
    d) nitrogen oxides

19. Which of the following molecules is most responsible for the greenhouse effect?
    a) NH_3
    b) SO_2
    c) NO_2
    d) CO_2

20. Which of the following does not contribute to photochemical smog?
    a) sunlight
    b) carbon dioxide
    c) automobiles
    d) ozone

Review Your Understanding

1. The pressure on 220 mL of a gas is 110 kPa. What will be the volume if the pressure is changed to 55.0 kPa, keeping the temperature constant?

2. A gas initially at a pressure of 300 kPa is allowed to expand at constant temperature until its volume has increased from 100 to 225 mL. What is the final pressure?

3. In a McLeod gauge a large volume of a gas at an unknown low pressure is compressed to a much smaller volume, and the new pressure is measured. In one experiment a sample of nitrogen at 20°C was compressed from 300 mL to 0.360 mL, and its new pressure was found to be 400 Pa. What was the original pressure of the nitrogen?

4. The pressure on 6.00 L of a gas is 200 kPa. What will be the volume if the pressure is doubled, keeping the temperature constant?

5. The initial pressure of a gas is 150 kPa. What will be the final pressure if the gas is compressed to one half its original volume?

6. Convert the following Celsius temperatures to Kelvin temperatures:
   a) 27°C; b) 273°C; c) −162°C; d) 727°C.

7. Convert the following Kelvin temperatures to Celsius temperatures:
   a) 0 K; b) 273 K; c) 1000 K; d) 328 K; e) 225 K.

8. Convert a) 400 K to °C; b) −150°C to K; c) 298 K to °C; d) 1 MK to °C; e) 1×10^6 °C to K.

9. If a sample of gas measures 2.00 L at 25°C, what is its volume at 50°C if the pressure remains constant?

10. A sample of a gas whose volume at 27°C is 127 mL is heated at constant pressure until its volume becomes 317 mL. What is the final Celsius temperature of the gas?

11. A gas has a volume of 225 mL at 75°C and 175 kPa. What will be its volume at a temperature of 20°C and a pressure of 100 kPa?

12. A gas has a volume of 400 mL at 50°C and 120 kPa. What will be its volume at a temperature of 80°C and a pressure of 180 kPa?

13. A 200 mL sample of gas is collected at 50.0 kPa and a temperature of 271°C. What volume would this gas occupy at 100 kPa and a temperature of −1°C?

14. The scientists of the remote oil-rich island of Salamia use units of potlins (P) for pressure, vurniks (V) for volume, and kelvins (K) for temperature. If a sample of methane gas has a volume of 13.0 V at a pressure of 4.00 P and a temperature of 150 K, what is its volume when the temperature has increased to 298 K at a pressure of 4.00 P?

15. Natural gas is usually stored in large underground reservoirs or in above-ground tanks. Suppose that a supply of natural gas is stored in an under-ground reservoir of volume 8.0×10^5 m^3 at a pressure of 360 kPa and a temperature of 16°C. How many above-ground tanks of volume 2.7×10^4 m^3 at a temperature of 6°C could be filled with the gas at a pressure of 120 kPa?

16. A gas has a volume of 55.0 mL at 45°C and 85.0 kPa. What will be its volume at STP?

17. If a sample of gas measures 500 mL at STP, what is its volume at 101.3 kPa and 34°C?

18. Correct the following volumes to STP: **a)** 24.5 L at 25°C and 104 kPa; **b)** 1000 mm^3 at 100°C and 75.0 kPa; **c)** 45.0 mL at −40°C and 140 kPa.

19. A gas mixture consists of 60.0% argon, 30.0% neon, and 10.0% krypton by volume. If the pressure of this gas mixture is 80.0 kPa, what is the partial pressure of each of the gases?

20. Air is a mixture of many gases. The partial pressure of nitrogen is 80.0 kPa. The partial pressure of oxygen is 20.3 kPa. Atmospheric pressure is 101.3 kPa. What is the partial pressure due to all the other gases present in air?

21. The total pressure of a mixture of hydrogen, helium, and argon is 99.3 kPa. If the partial pressure of helium is 42.7 kPa and the partial pressure of argon is 54.7 kPa, what is the partial pressure of hydrogen?

22. A student collects 100 mL of oxygen over water at 100.7 kPa and 21°C. What is the partial pressure of the oxygen?

23. If oxygen is collected over water at 18°C and standard pressure, what is the partial pressure of the oxygen when the water level inside the flask equals the water level outside the flask?

24. In an experiment a student collects 107 mL of hydrogen over water at a pressure of 104.8 kPa and a temperature of 31°C. What volume would this hydrogen occupy when dry at STP?

25. If 80.0 mL of oxygen are collected over water at 20°C and 95.0 kPa, what volume would the dry oxygen occupy at STP?

26. A student collects 45.0 mL of hydrogen over water at 19°C and 104.2 kPa. What would be the volume of the dry hydrogen at STP?

27. A volume of 135 mL of nitrogen is collected over water at a temperature of 24°C and an atmospheric pressure of 92.0 kPa. What is the volume of the dry nitrogen at 35°C and 120 kPa?

28. A sample of argon gas having a volume of 200 mL at 98.0 kPa and 26°C is collected over water at the same temperature. What is the new volume of the dry gas if the atmospheric pressure is 98.0 kPa?

29. What volume would 250 mL of pure, dry oxygen at STP occupy if collected over water at 20°C and at an atmospheric pressure of 98.0 kPa?

30. If 2.00 L of dry nitrogen at STP are collected over water at 20°C and 95.0 kPa, what volume will be occupied by the wet gas?

31. If 450 mL of hydrogen at STP occupy 511 mL when collected over water at 18°C, what is the atmospheric pressure?

32. Use the postulates of the kinetic molecular theory to explain why the pressure of a gas decreases when the volume of the gas increases.

33. Use the postulates of the kinetic molecular theory to explain why the volume of a gas increases when its temperature increases.

34. A certain mass of a gas has a pressure of 135 kPa at 25°C. What will be its pressure at 80°C if the volume is kept constant?

35. A certain mass of a gas occupies a volume of 325 mL at a pressure of 85.0 kPa and a temperature of 100°C. What will be its pressure when its volume is 185 mL and its temperature is 190°C?

36. On a day when the temperature is 20°C, an automobile tire has a pressure of 200 kPa. After several hours of high speed driving the tire temperature has risen to 30°C. Assuming that the volume of the tire does not change, what is the tire pressure?

37. The gaseous contents of an "empty" aerosol spray can are at a pressure of 110 kPa at room temperature (20°C). What pressure is generated inside the can if it is thrown into a fire where the temperature is 750°C?

38. One litre of a gas at 100 kPa and −20°C is compressed to half a litre at 40°C. What is its final pressure?

39. A certain mass of a gas occupies a volume of 8.50 L at a pressure of 95.0 kPa and a temperature of 55°C. What will be its Celsius temperature when its volume is 6.5 L and its pressure is 130 kPa?

40. A sample of a gas occupies a volume of 119 mL at STP. To what temperature must the sample be heated to occupy a volume of 92 mL at 225 kPa?

41. A gas storage tank is designed to hold a fixed volume of a gas at 175 kPa and 30°C. In order to prevent excessive pressure build-up by overheating, the tank is fitted with a relief valve that opens at 200 kPa. At what temperature will the relief valve open?

42. Complete Table 13-6.

Table 13-6

P_{old}(kPa)	V_{old}(L)	T_{old}(K)	P_{new}(kPa)	V_{new}(L)	T_{new}(K)
100	200	300	200	?	300
100	200	300	?	800	300
200	300	400	200	?	200
200	300	400	200	600	?
300	100	200	300	100	?
300	100	200	?	100	400

43. List five uses of gases in our daily lives.
44. What are two main causes of the increase in pollution in recent years?
45. What is the definition of a pollutant?
46. Evidence exists that the average global temperature has fallen 0.2°C between 1940 and 1967. To what do we attribute this temperature decrease? Explain your answer.
47. What is the main effect of an increasing amount of carbon dioxide in the atmosphere? Explain your answer.
48. At one time, the carbon dioxide content in the atmosphere was kept constant because the amount of carbon dioxide entering the atmosphere was equal to the amount of carbon dioxide leaving the atmosphere. What processes injected carbon dioxide into the atmosphere, and what processes removed carbon dioxide from the atmosphere? What fairly recent factor has disturbed this carbon dioxide balance?
49. What is a thermal inversion?
50. What is the main cause of photochemical smog?
51. Describe the factors that operate to give Los Angeles such a severe photochemical smog problem.
52. What are the major sources of the following air pollutants: **a)** oxides of nitrogen; **b)** ozone; **c)** carbon monoxide?
53. What would be the positive and negative effects of a continued increase in CO_2 levels over Canada? Which effects do you think would dominate?

Apply your Understanding

1. A gas is contained in a spherical vessel of 450 mL volume at a pressure of 101 kPa. When a stopcock connecting the vessel to an adjacent evacuated chamber is opened to allow the gas to flow into the chamber, the pressure stabilizes to 6.50 kPa. What is the volume of the adjacent chamber?
2. When a bottle of ammonia is opened, the odour is eventually detected across the room. Explain this observation in terms of the kinetic molecular theory. If molecules have the velocities of rifle bullets, why don't you smell the ammonia immediately after the bottle is opened?
3. The barrel of a bicycle pump usually becomes warm when it is used to inflate a tire. Explain this observation in terms of kinetic molecular theory.
4. Figure 1-8 shows a model in which ball bearings illustrate the movement of gas molecules. In what ways does the behaviour of this model differ from that of a real gas?
5. You partially fill a balloon with helium and release the balloon to the atmosphere. As the balloon ascends, will it expand, contract, or remain the same size? Explain.
6. Can the speed of a given molecule in a gas double at constant temperature? Explain your answer.
7. The volume of a fixed mass of a gas is doubled at 22°C. In order to restore the gas to its original pressure, what should be the new Celsius temperature of the gas?

8. Why do you suppose that, although the temperature of the air over deserts is so hot during the day, the nights can be very cold?

9. Suppose you are disturbed by the odour of some small animals you keep in a cage in your room. Comment on each of these possible solutions to your problem:
   a) Place a fan in the window to blow out the bad air.
   b) Spray a pleasant scent into the room to make it smell better.
   c) Clean the cage every day.
   d) Spray a disinfectant into the room to kill the germs.
   e) Install a device to pass the room air through activated charcoal.
   f) Install an air conditioner to cool and recirculate the room air.

Investigations

1. It is frequently difficult to force an industry that is polluting the atmosphere to comply with clean air standards. What arguments might a company executive make to justify continuing air pollution by the company? What arguments might a pollution control board make to justify allowing the pollution to continue?

2. Write a newspaper report on Los Angeles-type smog as it affects Canadian cities.

3. Investigate to determine whether your community has an air pollution problem. Identify the major pollutants and their sources.

4. A concentration of 1 μg/kg of ethylene (C_2H_4) in the air cannot be smelled. It has no effect on people or animals but it injures growing orchids and makes them unfit for sale. Debate the following proposition: "Ethylene is a pollutant, and its level in the atmosphere should be controlled."

5. Debate the proposition, "The use of atomic energy is an excellent method of reducing air pollution."

6. Design and conduct an experiment to determine the rates (change in volume per unit time) at which two different gases escape through the walls of a balloon. Propose an explanation for your results.

Gases and the Mole

14

Chapter Contents

Part One: **The Development of Avogadro's Hypothesis**
   14.1 Dalton's Problem
   14.2 Gay-Lussac's Law of Combining Volumes
   14.3 Avogadro's Hypothesis

Part Two: **The Ideal Gas Law**
   14.4 The Derivation of the Ideal Gas Law
   14.5 Using the Ideal Gas Law
   14.6 Molar Mass and the Ideal Gas Law

Part Three: **Calculations Involving Gases in Chemical Reactions**
   14.7 Volume-Volume Calculations
   14.8 Mass-Volume Calculations

Now that we have established the concept that the amount of matter can be measured by counting the number of particles (the mole concept), we are able to introduce laws which relate the number of particles to other physical properties such as pressure, volume, and temperature. In the case of gases, the laws are quite simple. In fact, the mole concept was arrived at largely by studying the pressure, volume, temperature, and mass relationships in gases. In this chapter, we shall see how these laws were arrived at and use the laws to expand the quantitative description of chemical reactions in some interesting ways.

Why did Dalton disagree with Gay-Lussac's conclusions about gases? Why did Berzelius disagree with Avogadro's hypothesis? Why were Avogadro's ideas rejected for 50 years even though they were correct? How were chemists first able to determine the relative masses of atoms and molecules? How can we calculate the volume of oxygen that must be breathed by a person who has consumed 100 g of sugar, knowing that oxygen is needed to react with sugar in the body to give carbon dioxide and water? The answers to these questions are found in this chapter.

Figure 14-1
Manipulating gases under pressure in a research laboratory in order to calculate their volume

Key Objectives

When you have finished studying this chapter, you should be able to

1. state Avogadro's hypothesis.
2. given any three of the variables P, V, T, and n, calculate the fourth variable using the ideal gas law.
3. calculate the molar mass and molecular mass of a fixed mass of gas, given its pressure, volume, and temperature.
4. perform mass-volume and volume-volume calculations for reactions involving gases.

Part One: The Development of Avogadro's Hypothesis

14.1 Dalton's Problem

Dalton's atomic theory of 1803 suggested that atoms of different elements should have different atomic masses. The question naturally arose, what is the mass of an atom?

Since it was impossible to determine the mass of individual atoms, Dalton and his contemporaries had to use data obtained from chemical analyses of various compounds. For example, water was found to contain hydrogen and oxygen in a ratio of one gram of hydrogen to eight grams of oxygen. Dalton had no way of knowing how many atoms of hydrogen were combined with one atom of oxygen in a molecule of water. He arbitrarily assumed that there was one atom of each element in water, and assigned water the formula HO. This gave oxygen an atomic mass eight times that of a hydrogen atom. If the formula were HO_2, then an oxygen atom had only four times the mass of a hydrogen atom, and so on. Dalton recognized the problem, but having nothing better to go on, he stuck to his assumption that equal numbers of atoms react to form compounds. He analyzed many compounds and constructed a table of relative atomic masses. Many of his numbers were wrong, of course, because of his wrong assumption. It was therefore necessary to modify the model. The key was provided several years later as the result of a study of chemical reactions of gases.

EXAMPLE PROBLEM 14-1

An analysis of water shows that it consists of 11.2% hydrogen and 88.8% oxygen by mass. If the formula of water were H_3O_2 and a hydrogen atom were assigned a mass of 1 u, what would be the atomic mass of an oxygen atom?

SOLUTION

In 100 g of water, there are 11.2 g of hydrogen and 88.8 g of oxygen. The mass ratio of oxygen to hydrogen is

$$\frac{88.8 \text{ g O}}{11.2 \text{ g H}} = \frac{7.93 \text{ g O}}{1 \text{ g H}} = \frac{7.93 \text{ u O}}{1 \text{ u H}}$$

If the formula of water were H_3O_2, and the mass of each hydrogen atom were 1 u, the mass of the three hydrogen atoms in the water molecule would be 3 u. The mass of the two oxygen atoms in the water molecule would be

$$3 \text{ u H} \times \frac{7.93 \text{ u O}}{1 \text{ u H}} = 23.8 \text{ u O}$$

Thus, the mass of each oxygen atom would be 11.9 u.

PRACTICE PROBLEM 14-1

A compound containing only carbon and hydrogen atoms is found on analysis to consist of 3.00 g of carbon for every 1.00 g of hydrogen. If a hydrogen atom is assigned a mass of 1 u, what would be the atomic mass of a carbon atom if the formula of the compound were CH? (*Answer*: 3 u)

14.2 Gay-Lussac's Law of Combining Volumes

In 1805 Joseph Louis Gay-Lussac (Fig. 14-4, page 470), a French chemist, and Alexander von Humboldt, a German scientist, worked together to measure the exact proportions by volume in which gases react with each other. They found that no matter which gas was in excess, one volume of oxygen always reacted with two volumes of hydrogen to form two volumes of water vapour. The experimental error was less than 0.1%. Gay-Lussac investigated other gas reactions and found that all gases combine in simple whole-number ratios by volume—1:1, 2:1, 1:2, 1:3, 3:1, 3:2, etc. Also, the volumes of the product gases were in simple whole-number ratios to the volumes of the reactants. We can summarize some of his results as follows:

Hydrogen   + Oxygen $\longrightarrow$ Water

| 1 vol | 1 vol | | 1 vol | | 1 vol | 1 vol | Ratio = 2 : 1 : 2 |

Nitrogen   + Oxygen $\longrightarrow$ Nitrogen(I) Oxide

| 1 vol | 1 vol | | 1 vol | | 1 vol | 1 vol | Ratio = 2 : 1 : 2 |

Nitrogen   + Oxygen $\longrightarrow$ Nitrogen(II) Oxide

| 1 vol | | 1 vol | | 1 vol | 1 vol | Ratio = 1 : 1 : 2 |

These findings can be summarized in **Gay-Lussac's Law of Combining Volumes**: when gases react, they do so in simple, whole-number ratios by volume.

Gay-Lussac knew that the particles of a gas were small in relation to the total volume of the container. He supposed that the effective volume of a gas particle (the total volume divided by the total number of particles) might be the same for all gases. If this were so, and if the gas particles were atoms, then equal volumes of gases would contain equal numbers of atoms. In this case one could determine relative atomic masses by measuring equal volumes of gases under identical conditions.

We can follow Gay-Lussac's reasoning by referring to Figure 14-2. Assume that Gay-Lussac had one-litre boxes of hydrogen, nitrogen, and oxygen, all at the same temperature and pressure. He might have determined the mass of each gas to be 0.089 g, 1.25 g, and 1.43 g, respectively. He didn't know how many atoms were in each box, but he reasoned that the same number of atoms occupied each box. Since 1 L of nitrogen had a mass which was $1.25 \div 0.089 = 14$ times the mass of 1 L of hydrogen, the mass of a nitrogen "atom" was therefore 14 times the mass of a hydrogen "atom." Also, since the mass of 1 L of oxygen was $1.43 \div 0.089 = 16$ times the mass of 1 L of hydrogen, the mass of an oxygen "atom" was 16 times the mass of a hydrogen "atom."

Figure 14-2
The relative "atomic" masses of gases can be determined by measuring the masses of equal volumes under the same conditions.

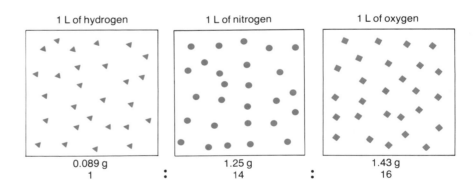

1 L of hydrogen	1 L of nitrogen	1 L of oxygen
0.089 g	1.25 g	1.43 g
1	14	16

Dalton rejected Gay-Lussac's conclusions for two reasons. He thought (erroneously) that Gay-Lussac's data were less accurate than his own. Also, he argued (correctly) that equal volumes of a gas did not contain equal numbers of atoms. For if that were the case, then the reaction between 1 volume of nitrogen "atoms" and 1 volume of oxygen "atoms" should give 1 volume of nitrogen(II) oxide "atoms." Gay-Lussac had found that 2 volumes of nitrogen(II) oxide were formed. Someone was wrong.

14.3 Avogadro's Hypothesis

In 1811 the Italian physicist Amedeo Avogadro finally solved the problem. He suggested that the smallest particles of gases were not atoms but molecules. These molecules could contain one, two, three, or more atoms, depending on the individual gases. Avogadro believed that gaseous hydrogen, oxygen, and nitrogen each consist of diatomic molecules. That is, hydrogen, oxygen, and nitrogen molecules each contain two atoms joined together. His rewording of Gay-Lussac's original guess is now called **Avogadro's hypothesis**: equal volumes of gases at the same temperature and pressure contain equal numbers of molecules. That is, a litre of nitrogen contains the same number of molecules as a litre of oxygen at the same temperature and pressure.

At this time Jöns Jakob Berzelius (Fig. 14-3), the great Swedish chemist, had developed a theory that atoms in molecules are held together by attractive forces between oppositely charged atoms. He did not believe that atoms of the same element could possibly pair up, and his influence was so great that no attention was paid to Avogadro's hypothesis for almost 50 years. It was not until 1860, 12 years after Berzelius' death and four years after Avogadro's death, when Avogadro's student Stanislao Cannizzaro made a clear and forceful presentation to a meeting of chemists at Karlsruhe, Germany, that Avogadro's hypothesis gained general acceptance.

Gay-Lussac's Law of Combining Volumes now made sense. For if one litre of nitrogen and one litre of oxygen combined to give two litres of nitrogen(II) oxide, then the observations could be explained by assuming that one molecule of nitrogen reacted with one molecule of oxygen to form two molecules of the oxide, each of which contained one atom of nitrogen and one atom of oxygen. These two nitrogen atoms must have come from a nitrogen molecule which contained two atoms, and the two oxygen atoms must have come from an

Biography

J ÖNS JAKOB BERZELIUS was the son of a Swedish school principal. He became interested in natural history and medicine at the age of 14. Two years later he began medical studies at Uppsala, Sweden. He actually failed the chemistry exam, but his professor gave him a passing grade on the basis of the strength of his knowledge in physics. Berzelius then studied chemistry in his free time, and learned how to blow glass and make barometers and thermometers.

In 1802 he received his degree and moved to Stockholm, where he took up hospital work and devoted his spare hours to experiments on the chemical action of voltaic cells.

By 1807, Berzelius had begun a lifelong attempt to determine with utmost accuracy the combining weights of the elements. In about a decade he prepared, purified, and analyzed at least 2000 compounds of 43 elements. The atomic weights calculated from his results were usually within about 0.3% of those determined a century later.

Figure 14-3
Jöns Jakob Berzelius (1779-1848)—Determined combining weights of the elements.

oxygen molecule which also consisted of two atoms. The arguments can be summarized:

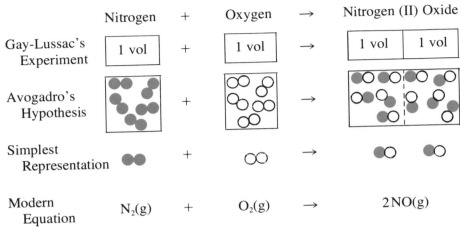

Of course, the observations could be explained equally well by the equation

$$N_4(g) + O_4(g) \longrightarrow 2N_2O_2(g)$$
1 vol   1 vol   2 vol

Other gas phase reactions involving nitrogen and oxygen contradict the assumption of tetraatomic molecules, but are consistent with an assumption of diatomic molecules. Therefore, the equation above involving N_4 and O_4 is incorrect.

In the reaction of hydrogen with oxygen to form water, similar arguments lead to the conclusion that a hydrogen molecule is diatomic and the formula for water is H_2O:

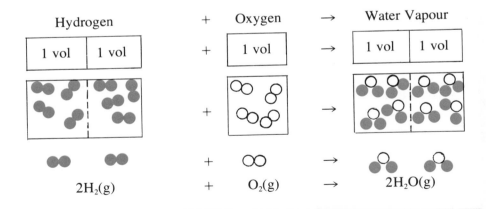

PRACTICE PROBLEM 14-2

Show how Gay-Lussac's volume data lead to the conclusion that a molecule of nitrogen(I) oxide contains two atoms of nitrogen and one atom of oxygen.

Avogadro's concept of a molecule consisting of one or more atoms marked an important turning point in the history of chemistry. After the Karlsruhe conference in 1860, chemistry developed rapidly as a science.

We should not conclude our discussion of Avogadro's hypothesis without asking why his ideas were rejected for fifty years even though they were correct. One reason may be that he offered no experimental evidence to support his hypothesis. Perhaps a more important reason is that any hypothesis or theory must not only explain a limited set of experimental data, but it must lead to the discovery of new facts and relationships. Chemists did not have the experimental techniques necessary to make further discoveries in 1811. Thus, a correct explanation was ignored for fifty years.

APPLICATION

The Depletion of the Ozone Layer

Ozone is a form of oxygen with the formula O_3. Pure ozone is a pale blue gas. It is present in air at sea level in concentrations of less that 0.05 parts per million. However, it is the concentration of the ozone in the stratosphere that has attracted recent attention.

In 1983, British scientists observed that the concentration of ozone in the stratosphere over Antarctica was falling at a dramatic rate each spring. Although the ozone concentration was partially restored by the end of each November, it appeared that this so-called "ozone hole" was growing larger each year.

Scientists are concerned that this ozone depletion could extend over more populated regions of the earth. If this were to happen, the effects would be quite detrimental to living organisms. Ozone in the stratosphere absorbs ultraviolet radiation from the sun and reduces the amount of ultraviolet radiation reaching the earth's surface. In this way, it shields us from exposure to excessive ultraviolet radiation, which can cause eye damage and an increased incidence of skin cancer. A 2 to 5% increase in skin cancer has been predicted for each 1% depletion of the ozone layer. Increased ultraviolet radiation can also affect plants, marine organisms, and synthetic materials.

What is causing this decrease in ozone? There is normally a balance in the production and depletion of ozone by natural means. However, this balance is being upset. One of the major reasons is believed to be the release of chlorofluorcarbons (CFCs) to the upper atmosphere. CFCs have been used as spray-can propellants and as refrigerator coolants. Their use is now banned in most spray-can preparations. They are also found in certain liquid cleaners and plastic foams, from which they evaporate. The CFCs are inert to most chemicals. Thus, once they leave the earth's surface, they make their way into the stratosphere. However, in the stratosphere, ultraviolet radiation causes them to decompose. This reaction releases highly reactive free chlorine atoms. These atoms then react with ozone in a series of steps which produce oxygen and a chlorine-containing compound. This compound reacts further to produce a free chlorine atom which can once again react with ozone. Thus, it has been estimated that for every chlorine atom released in the atmosphere, 100 000 molecules of ozone are destroyed.

Another chemical believed to be responsible for ozone depletion is NO, nitrogen monoxide. This substance is produced in automobile engines during the combustion process and by high-flying supersonic airplanes.

So great are the potential dangers of ozone depletion that 24 countries gathered at a conference sponsored by the United Nations in Montreal in September, 1987. They agreed in principle to a treaty that limits the production of CFCs and similar compounds.

Biography

Figure 14-4
Joseph Louis Gay-Lussac (1778-1850)—Published the Law of Combining Volumes.

JOSEPH GAY-LUSSAC was born in France in 1778. In 1795, he moved to Paris where he received an excellent education at the École Polytechnique.

Gay-Lussac was a versatile person. In 1802, he published an excellent paper on the effect of heat on gases (Charles' Law and Gay-Lussac's Law), but gave full credit for the principles to Charles. In 1804, he made a solo balloon flight to a height of 7 km and confirmed that the earth's magnetic field persists at that height. He also showed that samples of air from this altitude have the same composition as at sea level. In 1805, he collaborated with Humboldt to show that gases combine chemically in simple whole-number ratios. He published his Law of Combining Volumes in 1808, and it has since become a fundamental proposition of chemistry.

In 1808, Gay-Lussac devised a method for obtaining sodium and potassium in large quantity by the action of red-hot iron on their molten carbonates. He used the potassium to isolate boron from boracic acid. At the time the announcement of its isolation was made, Gay-Lussac was seriously ill following an explosion involving potassium which nearly cost him his sight.

In 1832, he became a professor at the Jardin des Plantes. Throughout his career, his talents were devoted entirely to the cause of science.

Part One Review Questions

1. What problem did Dalton face when he constructed a table of relative atomic masses?
2. State Gay-Lussac's Law of Combining Volumes.
3. Why did Dalton reject Gay-Lussac's conclusions?
4. State Avogadro's hypothesis.
5. Why did Berzelius reject Avogadro's hypothesis?
6. Why were Avogadro's ideas rejected for fifty years?

Part Two: The Ideal Gas Law

14.4 The Derivation of the Ideal Gas Law

Avogadro's hypothesis states that equal volumes of gases at the same temperature and pressure contain equal numbers of molecules. This implies that the volume of a gas at a fixed temperature and pressure is proportional to the number of molecules (that is, the number of moles). If the number of moles of a gas is doubled, the volume is doubled, and so on. This statement is written mathematically as

$$V \propto n$$

where n is the number of moles of the gas. According to Boyle's Law, volume is inversely proportional to pressure:

$$V \propto \frac{1}{P}$$

According to Charles' Law, volume is directly proportional to Kelvin temperature:

$$V \propto T$$

We can combine these three statements:

$$V \propto n \cdot \frac{1}{P} \cdot T \quad \textbf{or} \quad V \propto \frac{nT}{P}$$

By inserting a proportionality constant, R, the expression becomes

$$V = R \cdot \frac{nT}{P}$$

This can be rearranged as

$$PV = nRT$$

R is the **universal gas constant**.

This equation is usually known as the **ideal gas law**. It is the simplest example of an **equation of state**—an equation that gives the relationship between the pressure, volume, temperature, and mass of a substance. Any gas that obeys the ideal gas law is called an **ideal gas**. The value of R must be determined by performing an experiment on one gas. Once its value is determined, however, it can be used in calculations involving all other gases. Since equal volumes of gases contain equal numbers of molecules under identical conditions, one mole of a gas (6.02×10^{23} molecules) should occupy some definite volume under specified conditions. Experimentally, it is found that one mole of an ideal gas at 273.15 K and 101.325 kPa (STP) occupies 22.414 L. The volume 22.414 L is called the **molar volume** of a gas at STP. From the ideal gas law we find

$$R = \frac{PV}{nT} = \frac{101.325 \text{ kPa} \times 22.414 \text{ L}}{1 \text{ mol} \times 273.15 \text{ K}} = 8.314 \text{ kPa} \cdot \text{L/(K} \cdot \text{mol)}$$

Now that we know the value of R, a knowledge of any three of the variables in the ideal gas law will enable us to calculate the value of the fourth variable.

14.5 Using the Ideal Gas Law

Suppose we wanted to determine the amount of oxygen gas stored in a steel cylinder. It is easy to measure the pressure of the gas by attaching a pressure gauge to the cylinder. We would also need to know the volume of the cylinder and the temperature. A knowledge of the pressure, temperature, and volume of the oxygen allows us to calculate the amount of oxygen in the cylinder.

EXAMPLE PROBLEM 14-2

How many grams of oxygen are there in a 25.0 L tank at 20°C when the oxygen pressure is 1500 kPa?

SOLUTION

After converting the temperature to kelvins, we use the ideal gas law to calculate the number of moles of oxygen:

$$20°C = (273 + 20) \text{ K} = 293 \text{ K}$$

$$PV = nRT$$

$$n = \frac{PV}{RT} = \frac{1500 \text{ kPa} \times 25.0 \text{ L}}{8.314 \text{ kPa} \cdot \text{L/(K} \cdot \text{mol)} \times 293 \text{ K}}$$

$$n = 15.4 \text{ mol}$$

Then we convert moles of oxygen to grams of oxygen:

$$15.4 \text{ mol O}_2 \times \frac{32.0 \text{ g O}_2}{1 \text{ mol O}_2} = 493 \text{ g O}_2$$

EXAMPLE PROBLEM 14-3

What volume is occupied by 0.0330 mol of a gas at 91°C and 50.0 kPa?

SOLUTION

$$91°C = (273 + 91) \text{ K} = 364 \text{ K}$$

$$PV = nRT$$

$$V = \frac{nRT}{P} = \frac{0.0330 \text{ mol} \times 8.314 \text{ kPa} \cdot \text{L/(K} \cdot \text{mol)} \times 364 \text{ K}}{50.0 \text{ kPa}}$$

$$V = 2.00 \text{ L}$$

PRACTICE PROBLEM 14-3

What volume is occupied by 0.0273 mol of oxygen at 27°C and 107 kPa? (*Answer*: 636 mL)

PRACTICE PROBLEM 14-4

What is the pressure in a 50.0 L tank that contains 2.53 kg of oxygen, O_2, at 22°C? (*Answer*: 3880 kPa)

14.6 Molar Mass and the Ideal Gas Law

Since a mole of any gas occupies 22.4 L at STP, it now becomes easy to determine the molar mass of any substance, provided it can be converted into a gas. All that we have to do is measure the mass of a given volume of the gas at STP, and then calculate what would be the mass of 22.4 L of the gas.

EXAMPLE PROBLEM 14-4

If 2.00 g of liquid benzene can be converted into a vapour which occupies 903 mL at 150°C and 99.7 kPa, what is its approximate molar mass?

SOLUTION
We shall outline two methods of solving this problem.
Method 1
First, find the volume of the gas at STP.

$$V_{\text{old}} = 0.903 \text{ L} \qquad V_{\text{new}} = ?$$

$$P_{\text{old}} = 99.7 \text{ kPa} \qquad P_{\text{new}} = 101.3 \text{ kPa}$$

$$T_{\text{old}} = 150°C = 423 \text{ K} \quad T_{\text{new}} = 0°C = 273 \text{ K}$$

$$\therefore V_{\text{new}} = 0.903 \text{ L} \times \frac{99.7 \text{ kPa}}{101.3 \text{ kPa}} \times \frac{273 \text{ K}}{423 \text{ K}} = 0.574 \text{ L}$$

Since 0.574 L of benzene at STP has a mass of 2.00 g and since one mole of any gas occupies 22.4 L at STP, we can calculate the molar mass of benzene as

$$\frac{2.00 \text{ g}}{0.574 \text{ L}} \times \frac{22.4 \text{ L}}{1 \text{ mol}} = 78.0 \text{ g/mol}$$

Method 2

We can use the ideal gas equation to determine the number of moles of benzene:

$$PV = nRT$$

$$n = \frac{PV}{RT} = \frac{99.7 \text{ kPa} \times 0.903 \text{ L}}{8.314 \text{ kPa} \cdot \text{L/(K} \cdot \text{mol)} \times 423 \text{ K}} = 0.0256 \text{ mol}$$

Since 0.0256 mol of benzene has a mass of 2.00 g, we can calculate the molar mass of benzene as

$$\frac{2.00 \text{ g}}{0.0256 \text{ mol}} = 78.1 \text{ g/mol}.$$

PRACTICE PROBLEM 14-5

What is the molar mass of a vapour, 0.842 g of which occupies 450 mL at a pressure of 100 kPa and a temperature of 100°C? (*Answer*: 58.0 g)

EXAMPLE PROBLEM 14-5

A compound is found, on analysis, to be composed of 24.27% C, 4.07% H, and 71.66% Cl. If 1.71 g of the compound is vapourized and occupies 519 mL at a pressure of 102 kPa and a temperature of 95°C, what is the molecular formula of the compound?

SOLUTION

We can first use the ideal gas equation to determine the number of moles of the compound:

$$PV = nRT$$

$$n = \frac{PV}{RT} = \frac{102 \text{ kPa} \times 0.519 \text{ L}}{8.314 \text{ kPa} \cdot \text{L/(K} \cdot \text{mol)} \times 368 \text{ K}} = 0.0173 \text{ mol}$$

Since 0.0173 mol of the compound has a mass of 1.71 g, we can calculate the molar mass of the compound:

$$\frac{1.71 \text{ g}}{0.0173 \text{ mol}} = 98.8 \text{ g/mol}$$

We can use the percentage composition and the molar mass to determine the molecular formula. In 98.8 g of the compound there must be

$$0.2427 \times 98.8 \text{ g} = 24.0 \text{ g C}$$

$$0.0407 \times 98.8 \text{ g} = 4.02 \text{ g H}$$

$$0.7166 \times 98.8 \text{ g} = 70.8 \text{ g Cl}$$

$$\text{C}: 24.0 \text{ g C} \times \frac{1 \text{ mol C}}{12.0 \text{ g C}} = 2.00 \text{ mol C}$$

$$\text{H}: 4.02 \text{ g H} \times \frac{1 \text{ mol H}}{1.01 \text{ g H}} = 3.98 \text{ mol H}$$

$$\text{Cl}: 70.8 \text{ g Cl} \times \frac{1 \text{ mol Cl}}{35.5 \text{ g Cl}} = 1.99 \text{ mol Cl}$$

The mole ratio of C to H to Cl is 2 to 4 to 2, and the molecular formula of the compound is $C_2H_4Cl_2$.

PRACTICE PROBLEM 14-6

A compound is found, on analysis, to be composed of 45.87% C, 8.98% H, and 45.14% Cl. If 1.15 g of the compound is vapourized and occupies 401 mL at a pressure of 101.5 kPa and a temperature of 60°C, what is the molecular formula of this compound? (*Answer*: C_3H_7Cl)

Part Two Review Questions

1. What is the equation that is usually called the ideal gas law?
2. What is meant by an equation of state?
3. What is meant by an ideal gas?
4. What is meant by the molar volume? What is the numerical value of the molar volume?
5. If you wanted to determine the amount of gas stored in a steel cylinder, what quantities would you need to know?

Part Three: Calculations Involving Gases in Chemical Reactions

14.7 Volume–Volume Calculations

In some cases, a chemical equation involves several gases. If we know the volume of one of the gases, we can calculate the volume of any of the other gases at the same conditions of temperature and pressure. This type of problem is called a volume-volume calculation. Avogadro's hypothesis states that equal volumes of gases measured at the same conditions of temperature and pressure contain equal numbers of molecules. Therefore, the mole ratio and the volume ratio are the same in a chemical equation involving gases. A typical volume-volume calculation is illustrated in Example Problem 14-6.

EXAMPLE PROBLEM 14-6

Acetylene, C_2H_2, is a gas used as a fuel in welding. **a)** What volume of oxygen is required to burn 10.0 L of acetylene to carbon dioxide and water, if all gases are at the same temperature and pressure? **b)** What is the total volume of gaseous products? The balanced equation for the reaction is

$$2C_2H_2(g) + 5\ O_2(g) \longrightarrow 4CO_2(g) + 2H_2O(g)$$

SOLUTION

$$\overset{10.0\ L}{2C_2H_2(g)} + 5\ O_2(g) \longrightarrow 4CO_2(g) + 2H_2O(g)$$

mole ratio: 2 mol   5 mol   4 mol   2 mol
volume ratio: 2 L   5 L   4 L   2 L

a) $10.0\ L\ C_2H_2 \times \dfrac{5\ L\ O_2}{2\ L\ C_2H_2} = 25.0\ L\ O_2$

b) Two litres of C_2H_2 would give 4 L CO_2 and 2 L H_2O for a total of 6 L of products.

$$\therefore\ 10.0\ L\ C_2H_2 \times \dfrac{6\ L\ products}{2\ L\ C_2H_2} = 30.0\ L\ products$$

PRACTICE PROBLEM 14-7

a) How many litres of oxygen are required to burn 11 L of hydrogen?
b) How many litres of water vapour will be formed? (*Answers*: **a)** 5.5 L; **b)** 11 L)

14.8 Mass–Volume Calculations

One mole of a gas occupies 22.4 L at STP. We can use this knowledge to calculate the volumes of gases involved in chemical reactions or to calculate the numbers of moles of gases involved in chemical reactions. If any volumes are not at STP, they can be corrected to STP by using Boyle's and Charles' Laws. Problems which involve both mass and volume data are called mass-volume calculations. A number of mass-volume calculations are illustrated in the example problems that follow.

EXAMPLE PROBLEM 14-7

Calculate the volumes of hydrogen and oxygen at STP that can be obtained by electrolysis of 50.0 g of water.

SOLUTION
Balanced Equation:

$$
\begin{array}{cccc}
50.0 \text{ g} & & & \\
2H_2O(\ell) \longrightarrow & 2H_2(g) & + & O_2(g) \\
2 \text{ mol} & 2 \text{ mol} & & 1 \text{ mol} \\
2(18.0 \text{ g}) & 2(22.4 \text{ L}) & & 22.4 \text{ L} \\
36.0 \text{ g} & 44.8 \text{ L} & &
\end{array}
$$

$$
50.0 \text{ g } H_2O \times \frac{44.8 \text{ L } H_2}{36.0 \text{ g } H_2O} = 62.2 \text{ L } H_2
$$

$$
50.0 \text{ g } H_2O \times \frac{22.4 \text{ L } O_2}{36.0 \text{ g } H_2O} = 31.1 \text{ L } O_2
$$

EXAMPLE PROBLEM 14-8

How many grams of magnesium are required to react with hydrochloric acid to produce 300 mL of hydrogen gas at 25°C and 105 kPa? The equation is

$$Mg(s) + 2HCl(aq) \longrightarrow MgCl_2(aq) + H_2(g)$$

SOLUTION
We shall outline two methods of solving this problem.
Method 1
Since the volume of hydrogen gas was given at conditions other than STP, we cannot use it in our calculations until it is first corrected to STP:

$$
0.300 \text{ L } H_2 \times \frac{273 \text{ K}}{298 \text{ K}} \times \frac{105 \text{ kPa}}{101.3 \text{ kPa}} = 0.285 \text{ L } H_2
$$

$$
\begin{array}{ccc}
& & 0.285 \text{ L} \\
Mg(s) + 2HCl(aq) \longrightarrow & MgCl_2(aq) + & H_2(g) \\
1 \text{ mol} & & 1 \text{ mol} \\
24.3 \text{ g} & & 22.4 \text{ L}
\end{array}
$$

$$
0.285 \text{ L } H_2 \times \frac{24.3 \text{ g } Mg}{22.4 \text{ L } H_2} = 0.309 \text{ g } Mg
$$

Therefore, 0.309 g of magnesium is required to react with hydrochloric acid to produce 300 mL of hydrogen gas at 25°C and 105 kPa.

Method 2

Use $PV = nRT$ to first determine the number of moles of hydrogen produced during the reaction:

$$Mg(s) + 2HCl(aq) \longrightarrow MgCl_2(aq) + H_2(g)$$
$$\text{1 mol} \qquad\qquad\qquad\qquad\qquad \text{1 mol}$$

$$n = \frac{PV}{RT} = \frac{105 \text{ kPa} \times 0.300 \text{ L}}{8.314 \text{ kPa} \cdot \text{L}/(\text{K} \cdot \text{mol}) \times 298 \text{ K}} = 0.0127 \text{ mol H}_2$$

$$0.0127 \text{ mol H}_2 \times \frac{1 \text{ mol Mg}}{1 \text{ mol H}_2} \times \frac{24.3 \text{ g Mg}}{1 \text{ mol Mg}} = 0.309 \text{ g Mg}$$

EXAMPLE PROBLEM 14-9

A single cylinder of an automobile engine has a volume of 500 mL. The cylinder is full of air at 75°C and 99.0 kPa, and the amount of O_2 in dry air is 0.209 mol of O_2/mol of dry air. How many grams of octane, C_8H_{18}, could be burned by the oxygen in this amount of air, assuming complete combustion to H_2O and CO_2?

SOLUTION

Once again, we shall outline two methods of solving this problem.
Method 1
Balanced equation:

$$\qquad\qquad\qquad\quad ?\text{ L}$$
$$C_8H_{18}(\ell) + 12\tfrac{1}{2} O_2(g) \longrightarrow 8CO_2(g) + 9H_2O(g)$$
$$\text{1 mol} \qquad \text{12.5 mol}$$
$$\text{114 g} \qquad \text{12.5(22.4 L)}$$
$$\qquad\qquad \text{280 L}$$

First, correct the volume to STP:

$$0.500 \text{ L air} \times \frac{273 \text{ K}}{348 \text{ K}} \times \frac{99.0 \text{ kPa}}{101.3 \text{ kPa}} = 0.383 \text{ L air}$$

Then use the mole fraction to determine the volume of O_2 at STP. According to Avogadro's hypothesis, if the mole fraction is 0.209 mol O_2/mol of dry air, the volume fraction is also 0.209 L O_2/L of dry air.

$$0.383 \text{ L air} \times \frac{0.209 \text{ L O}_2}{1 \text{ L air}} = 0.0801 \text{ L O}_2$$

$$\qquad\qquad\qquad\quad 0.0801 \text{ L}$$
$$C_8H_{18}(\ell) + 12\tfrac{1}{2} O_2(g) \longrightarrow 8CO_2(g) + 9H_2O(g)$$
$$\text{1 mol} \qquad \text{12.5 mol}$$
$$\text{114 g} \qquad \text{280 L}$$

$$0.0801 \text{ L O}_2 \times \frac{114 \text{ g C}_8\text{H}_{18}}{280 \text{ L O}_2} = 0.0326 \text{ g C}_8\text{H}_{18}$$

Method 2

Use $PV = nRT$ to first determine the number of moles of air:

$$\text{C}_8\text{H}_{18}(\ell) + 12\tfrac{1}{2}\,\text{O}_2(\text{g}) \longrightarrow 8\text{CO}_2(\text{g}) + 9\text{H}_2\text{O}(\text{g})$$
$$\quad\quad 1 \text{ mol} \quad\quad 12.5 \text{ mol}$$

$$n = \frac{PV}{RT} = \frac{99.0 \text{ kPa} \times 0.500 \text{ L}}{8.314 \text{ kPa} \cdot \text{L}/(\text{K} \cdot \text{mol}) \times 348 \text{ K}} = 0.0171 \text{ mol air}$$

$$0.0171 \text{ mol air} \times \frac{0.209 \text{ mol O}_2}{1 \text{ mol air}} \times \frac{1 \text{ mol C}_8\text{H}_{18}}{12.5 \text{ mol O}_2} \times \frac{114 \text{ g C}_8\text{H}_{18}}{1 \text{ mol C}_8\text{H}_{18}}$$

$$= 0.0326 \text{ g C}_8\text{H}_{18}$$

EXAMPLE PROBLEM 14-10

Decomposition of potassium chlorate produces 400 mL of oxygen collected over water at 97.0 kPa and 20°C. How many grams of potassium chlorate are required?

SOLUTION

Two methods of solving this problem are outlined.
Method 1

$$\quad\quad\quad\quad\quad\quad\quad\quad\quad\quad ? \text{ L}$$
$$2\text{KClO}_3(\text{s}) \longrightarrow 2\text{KCl}(\text{s}) + 3\text{ O}_2(\text{g})$$
$$\quad 2 \text{ mol} \quad\quad\quad\quad\quad\quad 3 \text{ mol}$$
$$\quad 2(122.6 \text{ g}) \quad\quad\quad\quad\quad 3(22.4 \text{ L})$$
$$\quad\quad 245 \text{ g} \quad\quad\quad\quad\quad\quad 67.2 \text{ L}$$

Since the volume of oxygen gas was given at conditions other than STP, we cannot use it in the solution until it is corrected. Furthermore, since the oxygen was collected over water, we must determine the partial pressure of the dry oxygen before we can correct the volume to STP.

First, find the partial pressure of the oxygen. The vapour pressure of water at 20°C is 2.3 kPa.

$$\therefore P_{\text{oxygen}} = 97.0 \text{ kPa} - 2.3 \text{ kPa} = 94.7 \text{ kPa}.$$

Then convert the volume to STP:

$$0.400 \text{ L O}_2 \times \frac{94.7 \text{ kPa}}{101.3 \text{ kPa}} \times \frac{273 \text{ K}}{293 \text{ K}} = 0.348 \text{ L O}_2$$

$$0.348 \text{ L O}_2 \times \frac{245 \text{ g KClO}_3}{67.2 \text{ L O}_2} = 1.27 \text{ g KClO}_3$$

Method 2

After finding the partial pressure of the oxygen, use the ideal gas law to find the number of moles of oxygen.

$$n = \frac{PV}{RT} = \frac{94.7 \text{ kPa} \times 0.400 \text{ L}}{8.314 \text{ kPa} \cdot \text{L}/(\text{K} \cdot \text{mol}) \times 293 \text{ K}} = 0.0156 \text{ mol O}_2$$

$$2KClO_3(s) \xrightarrow{\triangle} 2KCl(s) + 3 O_2(g)$$
$$ 2 \text{ mol} \phantom{\xrightarrow{\triangle} 2KCl(s) +} 3 \text{ mol}$$

$$0.0156 \text{ mol O}_2 \times \frac{2 \text{ mol KClO}_3}{3 \text{ mol O}_2} \times \frac{122.6 \text{ g KClO}_3}{1 \text{ mol KClO}_3} = 1.28 \text{ g KClO}_3$$

PRACTICE PROBLEM 14-8

How many litres of hydrogen gas, measured at 23°C and 103 kPa, can be obtained by the reaction of 75.0 g of aluminum with excess sulfuric acid? The equation for the reaction is

$$2Al(s) + 3H_2SO_4(aq) \longrightarrow Al_2(SO_4)_3(aq) + 3H_2(g)$$

(*Answer*: 99.5 L)

PRACTICE PROBLEM 14-9

What volume of hydrogen, collected over water at 25°C and 97.0 kPa, will be formed by the reaction of 18.0 g of calcium hydride with water according to the following equation?

$$CaH_2(s) + 2H_2O(\ell) \longrightarrow Ca(OH)_2(s) + 2H_2(g)$$

(*Answer*: 22.6 L)

PRACTICE PROBLEM 14-10

What mass of aluminum is required to produce 36.3 L of hydrogen, collected over water at 22°C and 103.0 kPa, according to the following equation?

$$2Al(s) + 6HCl(aq) \longrightarrow 2AlCl_3(aq) + 3H_2(g)$$

(*Answer*: 26.7 g)

PRACTICE PROBLEM 14-11

How many grams of propane would react with the oxygen contained in a 50.0 L tank at 20°C and an internal pressure of 2000 kPa? The balanced equation is:

$$C_3H_8(g) + 5 O_2(g) \longrightarrow 3CO_2(g) + 4H_2O(g)$$

(*Answer*: 361 g C_3H_8)

PRACTICE PROBLEM 14-12

What volume of oxygen at 100 kPa and 21.0°C must be breathed by a person who has consumed 100 g of sugar, $C_{12}H_{22}O_{11}$, knowing that oxygen is needed to react with sugar in the body to give carbon dioxide and water? If the mole fraction of O_2 in dry air is 0.209 mol of O_2/mol of dry air, what volume of air would have to be breathed?
(*Answers*: 85.8 L O_2; 411 L air)

Chapter Summary

- Gay-Lussac's Law of Combining Volumes states that when gases react, they do so in simple, whole-number ratios by volume.
- Avogadro's hypothesis states that equal volumes of gases at the same temperature and pressure contain equal numbers of molecules.
- The ideal gas law equation is $PV = nRT$.
- An equation of state is an equation that gives the relationship between the pressure, volume, temperature, and mass of a substance.
- An ideal gas is any gas that obeys the ideal gas law.
- The molar volume is the volume occupied by one mole of a gas. At STP, the molar volume of an ideal gas is 22.4 L.
- The universal gas constant, R, is the constant in the ideal gas law equation. Its numerical value is 8.314 kPa · L/(K · mol).

Key Terms

Gay-Lussac's Law of Combining Volumes
Avogadro's hypothesis
universal gas constant
ideal gas law

equation of state
ideal gas
molar volume

Test Your Understanding

Each of these statements or questions is followed by four responses. Choose the correct response in each case. (Do not write in this book.)

1. A compound containing only silicon and hydrogen atoms consists of 7.0 g of silicon for every 1.0 g of hydrogen. If a hydrogen atom is assigned a mass of 1 u and the formula of the compound was thought to be SiH, what would be the atomic mass of a silicon atom?
    a) 28 u   **b)** 14 u   **c)** 7 u   **d)** 4 u

2. If the formula of the compound in the preceding question was later found to be SiH_4, what is the atomic mass of a silicon atom?

a) 28 u   **b)** 14 u   **c)** 7 u   **d)** 4 u

3. A given volume of oxygen has a mass of 0.58 g. The same volume of a second gas measured at the same temperature and pressure has a mass of 0.80 g. What is the molar mass of the second gas?

a) 23 g   **b)** 44 g   **c)** 32 g   **d)** 16 g

4. If 1.00 mol of a gas occupies 20.5 L at 26°C, what is the pressure of the gas?

a) 101 kPa   **b)** 10.5 kPa   **c)** 50 kPa   **d)** 121 kPa

5. If a sample of a gas occupies 450 mL at 105 kPa and 25°C, how many moles of gas are present in the sample?

a) 0.227 mol **b)** 0.0191 mol   **c)** 0.0185 mol   **d)** 52.4 mol

6. If 1.57 g of N_2 occupy 350 mL at 27°C, what is the pressure of the gas?

a) 101 kPa   **b)** 800 kPa   **c)** 200 kPa   **d)** 400 kPa

7. If a sample of sulfur dioxide occupies 325 mL at 102 kPa and 50°C, how many grams of sulfur dioxide are present in the sample?

a) 0.791 g   **b)** 0.756 g   **c)** 0.886 g   **d)** 5.11 g

8. If 0.150 mol of gas are collected at 105 kPa and 22°C, what is the volume of the gas?

a) 0.261 L   **b)** 3.50 L   **c)** 2.61 L   **d)** 4.50 L

9. If 0.37 g of CH_4 is collected over water at an atmospheric pressure of 104 kPa and 29°C, what is the volume of the dry gas?

a) 55.8 mL   **b)** 593 mL   **c)** 558 mL   **d)** 581 mL

10. What is the temperature of 1.10 g of CO_2 that occupies 600 mL at 99.0 kPa?

a) 286°C   **b)** 25°C   **c)** 13°C   **d)** 20°C

11. What is the density of N_2 gas at STP?

a) 1.25 g/L   **b)** 0.63 g/L   **c)** 1.08 g/L   **d)** 1.35 g/L

12. If the density of a gas is 1.96 g/L at STP, what is the molar mass of the gas?

a) 12.0 g   **b)** 32.1 g   **c)** 43.9 g   **d)** 28.0 g

13. If 1.74 g of a volatile liquid is vapourized and occupies 224 mL at STP, what is the molar mass of the compound?

a) 200 g   **b)** 186 g   **c)** 224 g   **d)** 174 g

14. If 2.38 g of a volatile liquid is vapourized and occupies 448 mL at STP, what is the molar mass of the compound?

a) 238 g   **b)** 119 g   **c)** 448 g   **d)** 60 g

15. What volume of ammonia can be produced from 18 L of hydrogen, if all gases are at the same temperature and pressure? The balanced equation for the reaction is $N_2(g) + 3H_2(g) \longrightarrow 2NH_3(g)$

a) 6 L   **b)** 18 L   **c)** 12 L   **d)** 9 L

16. What volume of carbon dioxide at STP can be obtained from the combustion of 0.50 g of carbon?

a) 0.88 L   **b)** 22.4 L   **c)** 1.00 L   **d)** 0.93 L

17. How many grams of carbon would have to be burned to produce 560 mL of carbon dioxide collected at STP?

a) 0.30 g   **b)** 0.60 g   **c)** 0.15 g   **d)** 0.45 g

18. What volume of carbon dioxide collected at 100 kPa and 27°C can be obtained from the combustion of 0.75 g of carbon?
    a) 0.14 L   b) 1.6 L   c) 0.89 L   d) 1.1 L
19. How many grams of carbon would have to be burned to produce 500 mL of carbon dioxide collected at 100 kPa and 27°C?
    a) 0.24 g   b) 2.7 g   c) 0.27 g   d) 0.22 g
20. How may grams of carbon would have to be burned to produce 448 mL of carbon dioxide collected over water at 29°C and an atmospheric pressure of 102 kPa?
    a) 0.24 g   b) 0.22 g   c) 0.21 g   d) 0.23 g

Review your Understanding

1. A compound containing only carbon and hydrogen atoms is found on analysis to consist of 3.00 g of carbon for every 1.00 g of hydrogen. If a hydrogen atom is arbitrarily assigned a mass of 1 u, what is the atomic mass of a carbon atom if the formula of the compound is CH_2? C_2H_3?
2. If you know that the atomic mass of a carbon atom is 12 u, what is a reasonable formula for the compound of Question 1?
3. An oxide of iron contains 39.3 g of iron for every 15.0 g of oxygen. Using the known atomic masses of iron and of oxygen, deduce a reasonable formula for the oxide of iron.
4. One litre of nitrogen reacts with 3 L of hydrogen to form 2 L of ammonia, all at the same conditions of temperature and pressure. Review the arguments Gay-Lussac would use to show that hydrogen and nitrogen are diatomic and that ammonia has the formula NH_3.
5. You have two identical metal containers. One is filled with hydrogen. The other is filled with helium. Both containers are at the same temperature and pressure.
    a) Which contains more molecules?
    b) Which contains the greater mass of gas?
    c) If the temperature of the hydrogen container is increased, which contains molecules with a greater average kinetic energy?
6. It is found that 50 mL of arsenic vapour react with 300 mL of hydrogen to produce 200 mL of a gaseous product. If each product molecule contains one arsenic atom, how many atoms are in a molecule of arsenic? What is the formula of the product?
7. When 2 L of a gaseous compound containing only carbon, hydrogen, and nitrogen are burned, the products consist of 2 L of carbon dioxide, 5 L of water vapour, and 1 L of nitrogen gas. What is the formula of the compound?
8. An element Z forms two gaseous oxides, ZO_2 and ZO_x. Two litres of ZO_2 react with one litre of oxygen to form two litres of ZO_x. Use Avogadro's Hypothesis and Gay-Lussac's Law of Combining Volumes to determine the formula of the unknown oxide.

9. It is found that 1.50 L of a gaseous hydrocarbon, C_xH_y, burns in oxygen to form 3.00 L of CO_2 and 4.50 L of water vapour. All volumes are measured at the same temperature and pressure. What is the formula of the hydrocarbon?

10. Calculate the volume occupied by half a mole of carbon dioxide gas at 33.0 kPa and −35°C.

11. How many moles of gas are there in a sample that occupies a volume of 570 mL at 78°C and a pressure of 103 kPa?

12. What pressure is exerted by 6.40 g of methane (CH_4) gas in a sealed 5.35 L flask at 27°C?

13. You are given 1.00 g of hydrogen gas. a) How many moles of hydrogen molecules are in this sample? b) How many molecules are there? c) How many atoms are there? d) What is the volume of this gas at STP?

14. How many grams of carbon dioxide (CO_2) are contained in a sample that occupies 1.09 L at 127°C and 40.5 kPa?

15. A 0.19 g sample of N_2 gas is collected over water at 27°C and 96.9 kPa. What would be the volume of the dry gas at STP?

16. The volume of 2.37 g of oxygen collected over water at 25°C was 1.92 L. What was the barometric pressure at the time of the experiment?

17. You are given a 1.00 mL container at 20°C and 100.0 kPa. a) How many molecules of chlorine are there? b) How many atoms are there? c) How many moles of atoms are there? d) How many moles of molecules?

18. How many moles of SO_2 gas are contained in 4.00 L at 25°C and 101 kPa? How many grams of SO_2 are in the container? What is the density (in grams per litre) under these conditions of temperature and pressure?

19. Air can be considered to have an average molar mass of 29.0 g/mol. What is the Celsius temperature of the air if it has a density of 1.2927 g/L at a pressure of 101.3 kPa?

20. What is the density of acetylene (C_2H_2) in grams per litre at STP?

21. What is the molar mass of a compound that has a density of 1.87 g/L at STP?

22. What is the molecular mass of a compound if 560 mL have a mass of 1.10 g at STP?

23. If 91.0 g of a gas occupy 14.0 L at STP, what is the molar mass of the gas?

24. If the density of a gas is 1.71 g/L at a pressure of 97.0 kPa and a temperature of 26.5°C, what is the molar mass of the gas?

25. What is the molar mass of a gas if 2.40 g of the gas occupy a volume of 2.80 L at 180°C and 50.0 kPa?

26. What is the molar mass of a gaseous compound if 0.833 g occupies a volume of 518 mL at a temperature of 17°C and a pressure of 104.0 kPa?

27. An analysis of a volatile liquid showed that it was composed of 14.14% C, 2.37% H, and 83.49% Cl. If 1.75 g of the liquid was vaporized and occupied 528 mL at 60°C and 108 kPa, what would be the molecular formula of the compound?

28. A 1.22 g sample of a liquid is vaporized at 100°C and 102 kPa. The sample occupies 617 mL. If the percentage composition of this compound is 59.96% C, 13.42% H, and 26.62% O, what is its molecular formula?

29. A 2.600 g sample of a volatile compound contains 0.644 g of carbon and 1.902 g of chlorine. The compound contains only carbon, hydrogen, and chlorine. The sample is vapourized and occupies 730 mL at 98.7 kPa and 50°C. What is the molecular formula of this compound?

30. Zirconium metal and chlorine gas react to form zirconium(IV) chloride:

$$Zr(s) + 2Cl_2(g) \longrightarrow ZrCl_4(g)$$

   a) What volume of chlorine gas must be used at 350°C and 50.0 kPa to produce 200 mL of $ZrCl_4$, under the same conditions?
   b) What mass of zirconium will be used up?

31. Methane burns according to the equation:

$$CH_4(g) + 2\ O_2(g) \longrightarrow CO_2(g) + 2H_2O(g)$$

What volume of O_2 is required to burn 5.0 L of methane when both are at the same temperature and pressure?

32. Octane, a component of gasoline, reacts with O_2 according to the equation:

$$2C_8H_{18}(g) + 25\ O_2(g) \longrightarrow 16CO_2(g) + 18H_2O(g)$$

What volume of octane will react with 500 mL of oxygen at a temperature of 50°C and a pressure of 100 kPa?

33. How many grams of $KClO_3$ are required to produce 5.00 L of oxygen gas at STP according to the following reaction?

$$2KClO_3(s) \xrightarrow{\triangle} 2KCl(s) + 3\ O_2(g)$$

34. How many grams of benzene (C_6H_6) are required to react with 25.0 L of oxygen at STP in the following reaction?

$$2C_6H_6(\ell) + 15\ O_2(g) \longrightarrow 12CO_2(g) + 6H_2O(\ell)$$

35. a) What volume of oxygen at STP is required to react with one mole of ammonia according to the following equation?

$$4NH_3(g) + 5\ O_2(g) \longrightarrow 4NO(g) + 6H_2O(\ell)$$

   b) How many litres of NO are formed at STP?

36. How many grams of zinc are required to prepare 6.00 L of hydrogen at STP? The reaction is

$$Zn(s) + H_2SO_4(aq) \longrightarrow ZnSO_4(aq) + H_2(g)$$

37. Aluminum oxide is decomposed on electrolysis, according to the equation

$$2Al_2O_3(s) \longrightarrow 4Al(s) + 3\ O_2(g)$$

What volume of oxygen at STP is obtained by electrolysis of 40.8 g of aluminum oxide?

38. Acetylene is prepared in the laboratory by dropping water on pieces of calcium carbide:

$$CaC_2(s) + 2H_2O(\ell) \longrightarrow Ca(OH)_2(s) + C_2H_2(g)$$

What volume of dry acetylene at STP can be prepared from 1.00 g of calcium carbide?

39. How many grams of calcium carbide would be required to produce 14.3 L of acetylene (C_2H_2) collected over water at 24°C and 106.0 kPa, according to the following reaction?

$$CaC_2(s) + 2H_2O(\ell) \longrightarrow C_2H_2(g) + Ca(OH)_2(s)$$

40. How many grams of sodium are required to produce 2.24 L of hydrogen gas, measured at 25°C and 110 kPa, according to the following reaction?

$$2Na(s) + 2H_2O(\ell) \longrightarrow 2NaOH(aq) + H_2(g)$$

41. What volume of oxygen at 150°C and 110 kPa is necessary to burn 14 g of methane, CH_4, to carbon dioxide and water?

42. Tungsten metal is prepared by heating tungsten trioxide with hydrogen gas according to the equation

$$WO_3(s) + 3H_2(g) \longrightarrow W(s) + 3H_2O(g)$$

What volume of hydrogen at 35°C and 95 kPa is required to prepare one mole of tungsten by this method?

43. How many grams of hydrogen are required to produce 140 L of ammonia at 55°C and 400 kPa according to the following reaction?

$$N_2(g) + 3H_2(g) \longrightarrow 2NH_3(g)$$

44. The Apollo Lunar Module was powered by a reaction similar to the following:

$$2N_2H_4(\ell) + N_2O_4(\ell) \longrightarrow 3N_2(g) + 4H_2O(g)$$

How many grams of N_2H_4 would be required to produce 450 L of exhaust gases at a temperature of 650°C and a pressure of 1.00 kPa?

45. Nitric acid is made commercially by the platinum-catalyzed air oxidation of ammonia. The first step in the sequence is the reaction given in Question 35. How many litres of air (21.0% oxygen) at 27°C and 100 kPa are needed to convert 50.0 kg of ammonia to NO by this method?

Apply Your Understanding

1. A hydrocarbon, C_xH_y, burns in oxygen to form CO_2 and H_2O. It is found that 15.0 mL of the gaseous hydrocarbon burns in 120.0 mL of oxygen (an excess!) to give a gaseous mixture of CO_2 and unreacted O_2 with a volume of 90.0 mL. The H_2O product is in the liquid state and, therefore, has a negligible volume. Removal of CO_2 by treating the mixture with sodium hydroxide solution reduces the volume to 45.0 mL. The gas volumes are all measured at STP. What is the formula of the hydrocarbon?

2. The organic compound, 1,2-dichloroethane, is a volatile liquid. How would you show experimentally that the molecular formula of this compound is $C_2H_4Cl_2$ and not CH_2Cl?

3. When 0.583 g of neon is added to an 800 mL container in which there is a sample of argon, the pressure of the gas mixture is found to be 118.5 kPa at a temperature of 22°C. Find the mass of the argon in the container.

4. A glass container fitted with a stopcock has a mass of 258.672 g when all the air is removed. When it is filled with argon, it has a mass of 260.423 g. The container is evacuated again and refilled with a mixture of neon and argon, under the same conditions of temperature and pressure. It has a mass of 259.944 g. What is the mole fraction of argon in the mixture?

5. A krypton atom is a sphere whose radius is 1.1×10^{-10} m. Find the volume of the krypton atoms in a 1.0 L container at STP. What percentage of the volume of the container is actually taken up by the krypton atoms themselves?

6. A mixture contains only KCl and $KClO_3$. When 10.53 g of the mixture is heated, 1.80 L of O_2 gas is collected over water at 21°C and an atmospheric pressure of 99.7 kPa. If all of the oxygen has been removed from the $KClO_3$ in the mixture, forming KCl, what is the percent (by mass) of $KClO_3$ in the mixture?

7. Lithium metal reacts with nitrogen according to the equation

$$6Li(s) + N_2(g) \longrightarrow 2Li_3N(s)$$

A sample of lithium was placed in a nitrogen atmosphere in a sealed 1.00 L container at a pressure of 125.0 kPa and a temperature of 23°C. After the lithium had reacted, the pressure had dropped to 90.0 kPa and the temperature was 24°C. What was the mass of the sample of lithium?

8. A sample of 0.2324 g of a compound of phosphorus and fluorine was completely vapourized in a 378 mL container. The resulting gas had a pressure of 12.96 kPa at 77°C. The gas was then mixed with calcium chloride solution, and all of the fluorine was removed by precipitation forming 0.2631 g of CaF_2. Determine the molecular formula of the original compound of phosphorus and fluorine.

Investigations

1. Real gases do not behave like ideal gases under all conditions of temperature and pressure. An equation known as the van der Waals equation has been developed to compensate for deviations from ideal behaviour. Write a report in which you outline how this equation compensates for these deviations.

2. Write a report on the Dumas method of determining the molar mass of a volatile liquid.

3. Write a report on the feasibility of raising the Titanic by the use of chemical reactions that generate gaseous products.

4. Debate the following proposition: "Chemists should devise a method of raising the Titanic by the use of chemical reactions that generate gaseous products."

5. Devise and conduct an experiment to determine the percentage of carbonate in a carbonate-containing antacid.

15 Organic Chemistry

Chapter Contents

Part One: **Saturated Hydrocarbons**
15.1 Introduction
15.2 The Alkanes
15.3 IUPAC Nomenclature of Alkanes
15.4 Sources of Alkanes
15.5 Chemical Properties of the Alkanes
15.6 Cycloalkanes
15.7 Physical Properties and Uses of Alkanes

Part Two: **Unsaturated Hydrocarbons**
15.8 Alkenes
15.9 IUPAC Nomenclature of Alkenes
15.10 Sources of Alkenes
15.11 Reactions of Alkenes
15.12 Alkynes
15.13 Ethyne—A Typical Alkyne
15.14 Chemical Reactions of Ethyne
15.15 Alkadienes
15.16 Aromatic Hydrocarbons

Part Three: **Compounds with Other Functional Groups**
15.17 Functional Groups
15.18 Alcohols
15.19 Ethers
15.20 Aldehydes
15.21 Ketones
15.22 Carboxylic Acids
15.23 Esters
15.24 Other Types of Organic Compounds

Part Four: **Organic Compounds in Industry**
15.25 Petroleum
15.26 Polymers

All chemical substances can be divided into two classes. One class includes sugar, starch, butter, rubber, paper, food, coal, petroleum, alcohol, candy, cosmetics, pesticides, ink, plastics, and penicillin. These substances are all *organic* (Fig. 15-1). The other class includes rocks, glass, air, water, salts, and metals. These are called *inorganic* substances (Fig. 15-2). This chapter will deal with organic compounds. We shall examine the structures, properties, and uses of different types of organic compounds and the chemical reactions they undergo.

Figure 15-1
Organic substances. From plastic measuring spoons to chocolate cake, all objects in this picture are made of organic compounds.

Figure 15-2
Inorganic substances. From an aluminum ingot to a piece of granite, all objects in this picture are made of inorganic compounds.

Key Objectives

When you have finished studying this chapter, you should be able to:

1. state the structural features of each of the following types of compounds and give an example of each: hydrocarbons, alkanes, alkenes, alkynes, alkadienes, cycloalkanes, aromatic hydrocarbons, alcohols, ethers, aldehydes, ketones, carboxylic acids, esters, mercaptans, amines, amides, organic halides.

2. choose from a list of structural formulas those which contain one or more of the structural groupings listed in the preceding objective.

3. write IUPAC names for simple aliphatic compounds, given their structural formulas.

4. write structural formulas for simple aliphatic compounds, given their names.

5. write equations for organic reactions such as the preparation and reactions of alkanes, alkenes, and alkynes; the preparation of ethanol; the dissociation and esterification of carboxylic acids.

6. describe some common polymers and their uses.

Part One: Saturated Hydrocarbons

15.1 Introduction

The division of substances into organic and inorganic classes was first used in 1807 by the Swedish chemist Jöns Jakob Berzelius. At that time he assumed that some substances had been present since the earth began, but other substances existed only because they had been manufactured by living things.

Berzelius called substances that were obtained from living organisms *organic*. Everything else he called inorganic.

At first this seemed to be a useful division; organic materials differ from inorganic materials in several ways. For example, organic materials are more easily decomposed than inorganic substances. You can heat salt (inorganic) until it is red hot and melts; however, the salt will remain chemically unchanged. If sugar (organic) is heated, it turns black and decomposes. Cooling does not restore its original properties.

Organic substances can be converted to inorganic substances. For example, the products of the combustion of many organic compounds are water and carbon dioxide. However, Berzelius believed that there was no way to convert an inorganic substance to an organic substance. It was believed that organic substances could be produced only by living things. At that time, chemists thought that a mysterious "vital force" was necessary to produce organic compounds. They believed that this vital force was present only in living things, and they did not expect to be able to prepare organic compounds in the laboratory.

The Synthesis of an Organic Compound

In 1827 Friedrich Wöhler, a German chemist, accidentally found that by heating an inorganic compound called ammonium cyanate he could prepare urea:

$$NH_4OCN \xrightarrow{\;\;\Delta\;\;} \begin{matrix} & & H \\ & & | \\ & H-N & \\ & & \diagdown \\ & & C=O \\ & & \diagup \\ & H-N & \\ & & | \\ & & H \end{matrix}$$

ammonium cyanate	urea
inorganic	*organic*
(*nonliving origin*)	(*living origin*)

Urea is an organic substance, one of the waste products of the body which is eliminated in urine. Wöhler had converted an inorganic compound into an organic compound. He repeated his experiment over and over before he dared to announce his results to the world. Once he had done so, it did not take long for chemists to prepare a number of other organic substances in the laboratory. Their experiments showed that atoms and molecules obey the same rules for chemical combination whether they are found in living or nonliving matter.

Chemists now know that the idea of a "vital force" is incorrect and that compounds formed in living matter can be synthesized from substances that are not found in living matter. However, they have found it convenient to retain the name *organic* to denote most compounds of carbon. The only common carbon compounds that are considered to be inorganic are the carbonates, carbon dioxide, and carbon monoxide. Thus, with very few exceptions, **organic chemistry** is the study of carbon-containing compounds. Inorganic chemistry is the study of the chemistry of the rest of the elements.

The Composition of Organic Compounds

Organic compounds contain few elements. The molecules of many organic compounds contain only carbon and hydrogen atoms. The other atoms found in organic molecules are generally limited to nitrogen, oxygen, sulfur, or halogen atoms. Despite this limitation, there are many more organic compounds than inorganic compounds. The main reason for the variety of organic compounds stems from the fact that carbon atoms are the only atoms that can effectively join together in long chains. Polyethylene molecules, for example (Fig. 15-3), can consist of chains of as many as 5000 carbon atoms. In other compounds, such as cholesterol (Fig. 15-4), the carbon atoms can be joined together in rings. Thus, organic compounds are often made up of molecules containing hundreds or even thousands of atoms.

Figure 15-3
Polyethylene

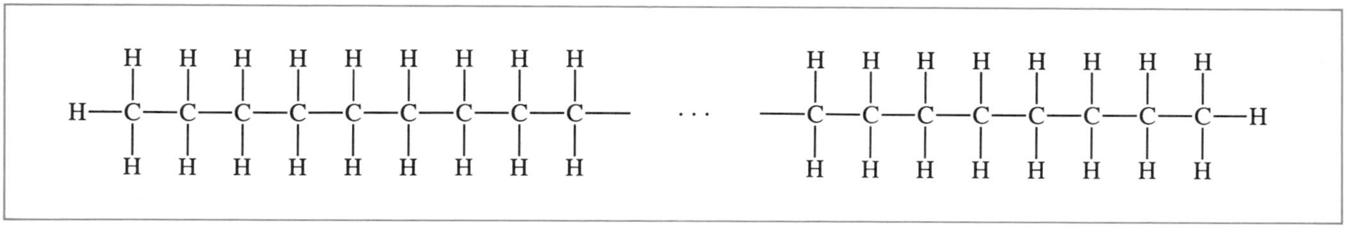

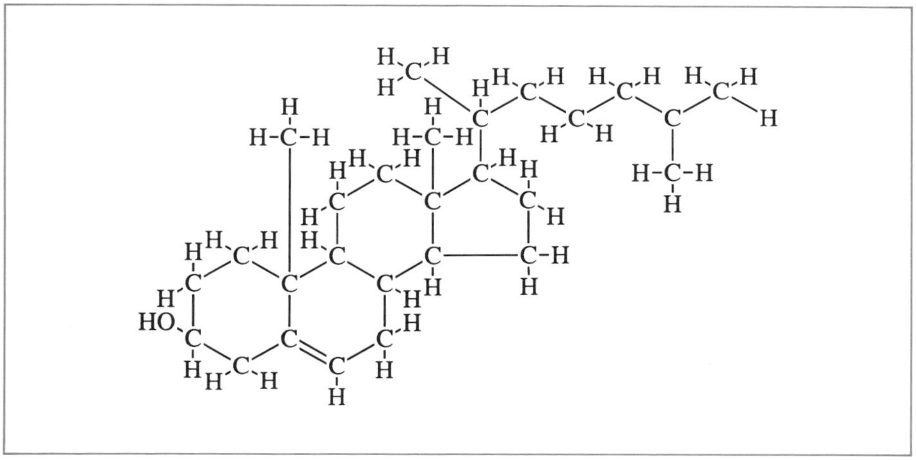

Figure 15-4
Cholesterol

15.2 The Alkanes

One of the most important classes of organic compounds is the **hydrocarbons**. These compounds are made up of carbon and hydrogen atoms only. The hydrocarbons are subdivided into a number of families. The first of the families is called the alkane family.

The structures of the alkanes are determined by the valence electrons of the carbon and hydrogen atoms. A carbon atom has four valence electrons ($\cdot \overset{\cdot}{C} \cdot$). It requires four more electrons to complete an octet, and we would expect it to form four bonds. Carbon is therefore tetravalent. A hydrogen atom has only one valence electron, and it can form only one bond. Therefore, four hydrogen atoms are required to bond with a single carbon atom:

$$H:\overset{\displaystyle H}{\underset{\displaystyle H}{\overset{..}{\underset{..}{C}}}}:H$$

This compound is the first member of the alkane family. It is called methane. Alkanes are called **saturated** hydrocarbons because the carbon atoms are bonded to as many hydrogen atoms as possible. In alkanes, the carbon atoms are saturated with hydrogen atoms.

Let us examine two carbon atoms bonded together:

$$\cdot\overset{\cdot}{C}:\overset{\cdot}{C}\cdot$$

By counting the unpaired electrons in this two-carbon system, we can see that six electrons are available to bond to six hydrogen atoms. A second member of the series of alkanes is formed when six hydrogen atoms bond to the two carbon atoms:

$$H:\overset{\displaystyle H}{\underset{\displaystyle H}{\overset{..}{\underset{..}{C}}}}:\overset{\displaystyle H}{\underset{\displaystyle H}{\overset{..}{\underset{..}{C}}}}:H$$

This compound, C_2H_6, is called ethane. In methane and ethane single bonding (two-electron bonds) is the rule. This is true for every alkane. All bonds are single bonds in alkane molecules.

The next alkane is the familiar fuel, propane. It has 3 carbon atoms per molecule, and its molecular formula is C_3H_8. The formula of every alkane differs from the alkane before it by a constant CH_2. The first three alkanes are CH_4, C_2H_6, and C_3H_8. In fact, we can write a general formula for the alkanes. The formula is C_nH_{2n+2}. The alkanes form a homologous series. A **homologous series** is one in which the formula of each member differs from that of the preceding member in a constant, regular way (in this case by a CH_2 unit).

At one time organic compounds were given their names by the people who first prepared or discovered them. However, the number of new organic compounds grew rapidly, and an efficient system of naming them had to be found. The International Union of Pure and Applied Chemistry (IUPAC) devised such a system at a meeting in Paris in 1957. The IUPAC system gives specific rules for naming organic compounds.

Many compounds have several names. For example, the compound with the IUPAC name propanone is also known as acetone or dimethyl ketone. Although all three names are correct, the one name on which all chemists will agree is the IUPAC name. Therefore, if a compound has both an IUPAC name and a non-IUPAC name, the structural formula of the compound will normally be labelled at its first use with the IUPAC name, followed by the other

Table 15-1   The First Ten Alkanes

Methane	CH_4
Ethane	C_2H_6
Propane	C_3H_8
Butane	C_4H_{10}
Pentane	C_5H_{12}
Hexane	C_6H_{14}
Heptane	C_7H_{16}
Octane	C_8H_{18}
Nonane	C_9H_{20}
Decane	$C_{10}H_{22}$

name. In this chapter, we will generally use the IUPAC name; however, for some compounds the IUPAC name is rarely used by chemists even though it is correct. For these compounds, we will use the more common non-IUPAC name.

Table 15-1 shows a list of the first ten alkanes and their molecular formulas: The name of each alkane is formed by adding the suffix -*ane* to a numerical prefix which is called the *stem* of the name. The stem indicates the number of carbon atoms in an alkane chain (Table 15-2). A glance at Table 15-2 will quickly tell you that a hydrocarbon with the stem *non-* contains a chain of nine carbon atoms.

The molecular formulas show the total number of carbon atoms and hydrogen atoms in each molecule. For methane, ethane, and propane the molecular formulas are sufficient because in each case the carbons and hydrogens can be assembled in only one way:

$$
\begin{array}{ccc}
\text{H} & \text{H H} & \text{H H H} \\
\text{H:}\overset{\cdot\cdot}{\underset{\cdot\cdot}{\text{C}}}\text{:H} & \text{H:}\overset{\cdot\cdot}{\underset{\cdot\cdot}{\text{C}}}\text{:}\overset{\cdot\cdot}{\underset{\cdot\cdot}{\text{C}}}\text{:H} & \text{H:}\overset{\cdot\cdot}{\underset{\cdot\cdot}{\text{C}}}\text{:}\overset{\cdot\cdot}{\underset{\cdot\cdot}{\text{C}}}\text{:}\overset{\cdot\cdot}{\underset{\cdot\cdot}{\text{C}}}\text{:H} \\
\text{H} & \text{H H} & \text{H H H}
\end{array}
$$

The more detailed formulas below are called **structural formulas** since they show more about the structure of the molecule. A dash often represents each electron dot pair, so that the usual structural formulas for methane, ethane, and propane are

$$
\begin{array}{ccc}
\text{H} & \text{H}\quad\text{H} & \text{H}\quad\text{H}\quad\text{H} \\
| & |\quad\ | & |\quad\ |\quad\ | \\
\text{H}-\text{C}-\text{H} & \text{H}-\text{C}-\text{C}-\text{H} & \text{H}-\text{C}-\text{C}-\text{C}-\text{H} \\
| & |\quad\ | & |\quad\ |\quad\ | \\
\text{H} & \text{H}\quad\text{H} & \text{H}\quad\text{H}\quad\text{H}
\end{array}
$$

There is only one structure for methane, one for ethane, and one for propane.

The Structure of Methane

Before going any further, we should take a closer look at the structure of methane as an example of a carbon atom with four single bonds. It appears from the structural formula that all carbon-hydrogen bonds are at 90° angles to one another. This appearance is misleading and is due to the limitations imposed by trying to draw a three-dimensional object in two dimensions. Actually, as we learned in Section 6-8, the structure of methane is three-dimensional, with the four C—H bonds pointing from the carbon atom toward the corners of a regular tetrahedron (Fig. 15-5).

In Figure 15-6 the carbon atom and the two hydrogens attached to it by solid lines are in the plane of the page. The dotted line indicates that its hydrogen atom is behind the plane of the page, and the wedge indicates that its hydrogen atom is in front of the plane of the page. All carbon-hydrogen bond angles are about 109°. The methane molecule is most stable when it has a tetrahedral rather than a planar structure. The tetrahedral structure ensures that the bonding pairs of electrons are as far from one another as possible, thus reducing the forces of repulsion.

Table 15-2	Stems Which Refer to Specific Numbers of Carbon Atoms

Stem	Number of Carbon Atoms
meth-	1
eth-	2
prop-	3
but-	4
pent-	5
hex-	6
hept-	7
oct-	8
non-	9
dec-	10

Figure 15-5
Model of methane (CH_4)

Figure 15-6
Perspective diagram of methane

Figure 15-7
Model of ethane (C_2H_6)

The Structures of Other Alkanes

In the alkanes, each carbon atom always forms four single bonds which are at 109° to one another. Usually the structural formula for ethane is drawn as

$$H-\overset{\overset{\displaystyle H}{|}}{\underset{\underset{\displaystyle H}{|}}{C}}-\overset{\overset{\displaystyle H}{|}}{\underset{\underset{\displaystyle H}{|}}{C}}-H$$

However, keep in mind that the molecule is not really planar and the bonds are not at 90° to one another (Figs. 15-7 and 15-8).

The structural formula for propane is usually drawn as

$$H-\overset{\overset{\displaystyle H}{|}}{\underset{\underset{\displaystyle H}{|}}{C}}-\overset{\overset{\displaystyle H}{|}}{\underset{\underset{\displaystyle H}{|}}{C}}-\overset{\overset{\displaystyle H}{|}}{\underset{\underset{\displaystyle H}{|}}{C}}-H$$

**Figure 15-8
Perspective diagram of ethane**

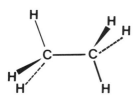

However, the following structural formula for propane is more correct and clearly shows that the three carbon atoms cannot be joined by a straight line:

In the case of butane, C_4H_{10}, an interesting possibility occurs. Two legitimate structural formulas fit the one molecular formula. They are

butane (*n-butane*)   **and**   **2-methylpropane** (*isobutane*)

**Table 15-3   Melting Points and
Boiling Points of
the Isomeric
Butanes and
Pentanes**

	m.p. (°C)	b.p. (°C)
butane	−138	0
2-methylpropane	−160	−12
pentane	−130	36
2-methylbutane	−160	28
2,2-dimethylpropane	− 17	10

We shall shortly be learning the rules for deriving the names for these compounds. For the moment, we are interested mainly in their structures. In both structures each carbon atom has four bonds, and each hydrogen atom has one bond. These two formulas represent butane molecules. The molecules are isomers. Molecules are **isomers** if they have the same molecular formula but different structural formulas. However, *isomers have different physical and chemical properties*. Table 15-3 shows that the isomeric butanes do indeed have different melting points and boiling points.

The two isomers of butane differ in the way that the four carbon atoms and ten hydrogen atoms are assembled. The first isomer is called butane or *n-*

butane. The *n* means that the carbon atoms are joined in one continuous chain. Hydrocarbons in which the carbon atoms are joined in one continuous chain are called *straight-chain* or *normal* hydrocarbons. The second isomer is called 2-methylpropane or isobutane. It is a *branched* hydrocarbon. Models of the two isomers of butane are shown in Figure 15-9.

Pentane, C_5H_{12}, has three isomers. Their structural formulas are

pentane (*n-pentane*)

2-methylbutane (*isopentane*)

2,2-dimethylpropane (*neopentane*)

Figure 15-9
Models of two butane (C_4H_{10}) isomers

The melting points and boiling points of the three isomeric pentanes are shown in Table 15-3. The different physical properties of the isomers show that they are three different compounds. Models of the three pentane isomers are shown in Figure 15-10.

There are five isomers of hexane and nine isomers of heptane. For $C_{30}H_{62}$, a large molecule, there are about four billion isomers.

Thus, there is yet another reason for the large number of organic compounds. Not only can carbon atoms join together to make long chains, but also compounds can have many isomers with the same molecular formula but with different structural formulas.

Figure 15-10
Models of the three pentane (C_5H_{12}) isomers

15.3 IUPAC Nomenclature of Alkanes

Because there are so many possible alkanes, there must be a very systematic method for generating names from structures and for determining the formula of a molecule from its name. The IUPAC rules of nomenclature provide this systematic method. The rules given here represent only a portion of the total number. They are sufficient, however, to enable us to name simple organic molecules.

As we have already seen, the names of the straight chain (normal) alkanes are formed by adding the suffix *-ane* to a numerical prefix called the stem of the name. More complicated (branched) alkanes consist of a continuous backbone of carbon atoms with one or more alkyl groups attached. Here is an example of a branched alkane:

$$CH_3CH_2CH_2CHCH_2CH_2CH_3 \longleftarrow \text{backbone}$$

$$CH \longleftarrow$$

$$CH_3 \quad CH_3 \longleftarrow \text{alkyl group}$$

An **alkyl group** is an alkane from which one hydrogen atom has been removed. It is named by adding the suffix *-yl* to an appropriate stem. The names of some common alkyl groups are given in Table 15-4.

Table 15-4   Alkyl Groups Derived from Alkanes

Parent Alkane	Formula of Parent Alkane	Formula of Alkyl Group	Name of Alkyl Group
Methane	CH_4	CH_3-	Methyl
Ethane	CH_3CH_3	CH_3CH_2-	Ethyl
Propane	$CH_3CH_2CH_3$	$CH_3CH_2CH_2-$	Propyl
Butane	$CH_3(CH_2)_2CH_3$	$CH_3(CH_2)_2CH_2-$	Butyl
Pentane	$CH_3(CH_2)_3CH_3$	$CH_3(CH_2)_3CH_2-$	Pentyl
Hexane	$CH_3(CH_2)_4CH_3$	$CH_3(CH_2)_4CH_2-$	Hexyl

Let us use the IUPAC rules given in Table 15-5 to name some of the branched alkanes we met in the previous section. For example, what is the IUPAC name of the compound we called isobutane? We can condense its structure to $CH_3-CH-CH_3$. Now let us apply the rules:

$$CH_3$$

Rule 1. The longest continuous chain contains three carbon atoms:

$$CH_3-CH-CH_3$$
$$CH_3$$

Rule 2. This chain is therefore called propane.
Rule 3. There is one alkyl group:

$$CH_3-CH-CH_3$$
$$CH_3$$

Rule 4. This is a methyl group.
Rule 5. Therefore the compound is a methylpropane.
Rule 6. The chain can be numbered as:

$$\overset{1}{CH_3}-\overset{2}{CH}-\overset{3}{CH_3}$$
$$CH_3$$

Rule 7. Since the methyl group is attached to carbon-2 of the main chain, the compound is named 2-methylpropane.

Table 15-5   IUPAC Rules for Nomenclature of Alkanes

1. Find the longest continuous chain of carbon atoms (the main chain). It is not necessary that the longest chain be written either horizontally or in a straight line.
2. Name this chain by adding *–ane* to the stem name.
3. Pick out the alkyl groups attached to the main chain.
4. Name the alkyl groups.
5. Attach the names of the alkyl groups as prefixes to the name of the main chain. If two or more alkyl groups of the same type occur, indicate how many there are by the prefixes *di-*, *tri-*, *tetra-*, etc.
6. Number the carbon atoms of the main chain consecutively from the end nearest to an alkyl group.
7. Indicate the positions of the alkyl groups according to the numbers of the carbon atoms in the main chain to which they are attached. These numbers precede the names of the alkyl groups and are connected to them by hyphens. If there are two or more alkyl groups of the same type, locate the position of each by a separate number. Use commas to separate consecutive numbers from each other.
8. If different alkyl groups are present, arrange their names in alphabetical order as prefixes to the name of the main chain. Use numbers to indicate the position of each group, with commas between numbers, and hyphens between numbers and letters. Note: the multiplying prefixes (*di-*, *tri-*, *tetra-*, etc.) are not alphabetized. The whole is written as a single word.
9. If chains of equal length are competing for selection as the main chain, choose that chain which has the greatest number of alkyl groups.
10. Other common groups are frequently found attached to hydrocarbon chains. Their names are F—*fluoro*; Cl—*chloro*; Br—*bromo*; I—*iodo*; NO_2—*nitro*.

EXAMPLE PROBLEM 15-1

What is the IUPAC name of the compound with the following formula?

$$CH_3-CH-CH_2-CH_3$$
$$|$$
$$CH_3$$

SOLUTION

Rule 1. The longest continuous chain contains four carbon atoms:

$$CH_3-CH-CH_2-CH_3$$
$$|$$
$$CH_3$$

(We could also have chosen the four carbon atoms in a horizontal row. There would be no difference in the final result.)

Rule 2. This chain is therefore called butane.
Rule 3. There is one alkyl group:

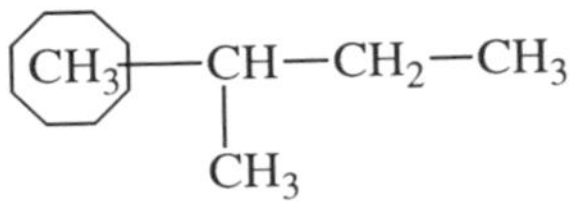

$$CH_3-CH-CH_2-CH_3$$
$$\qquad\ |$$
$$\qquad CH_3$$

Rule 4. This is a methyl group.
Rule 5. Therefore the compound is a methylbutane.
Rule 6. The chain must be numbered from the left-hand end:

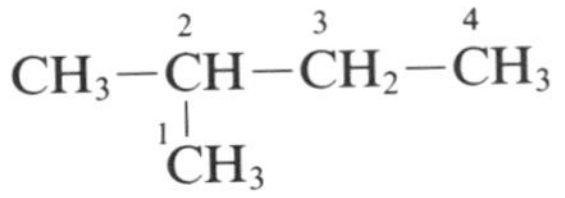

$$\overset{2}{CH_3}-\overset{3}{CH}-\overset{}{CH_2}-\overset{4}{CH_3}$$
$$\qquad\ \ \overset{1}{|}$$
$$\qquad\ \ CH_3$$

Rule 7. Since the methyl group is attached to carbon-2 of the main chain, the compound is named 2-methylbutane.

EXAMPLE PROBLEM 15-2

What is the IUPAC name of the compound with the formula shown below?

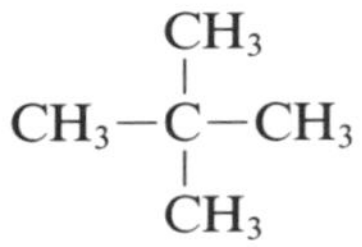

$$\begin{array}{c} CH_3 \\ | \\ CH_3-C-CH_3 \\ | \\ CH_3 \end{array}$$

SOLUTION

Rule 1. There are many different possibilities, but whichever we choose, the longest continuous chain contains only three carbon atoms. We shall choose the horizontal row:

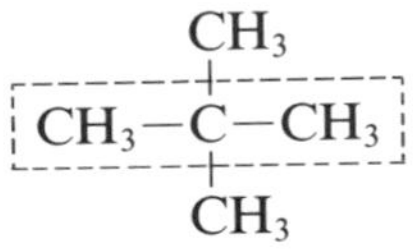

$$\begin{array}{c} CH_3 \\ | \\ CH_3-C-CH_3 \\ | \\ CH_3 \end{array}$$

Rule 2. This chain is called propane.
Rule 3. There are two alkyl groups:

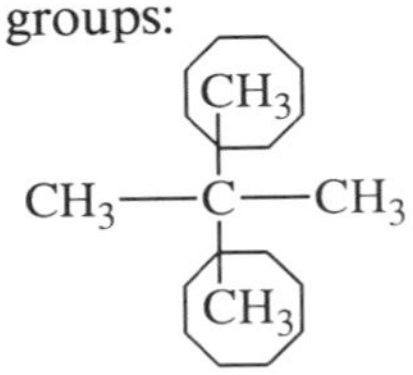

$$\begin{array}{c} CH_3 \\ | \\ CH_3-C-CH_3 \\ | \\ CH_3 \end{array}$$

Rule 4. They are both methyl groups.
Rule 5. Therefore the compound is a dimethylpropane.
Rule 6. The chain can be numbered as

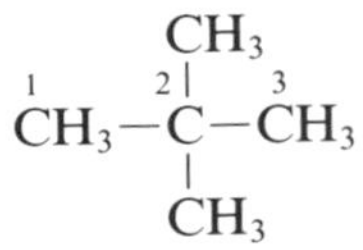

$$\begin{array}{c} CH_3 \\ \overset{1}{}\ \ \overset{2}{|}\ \ \overset{3}{} \\ CH_3-C-CH_3 \\ | \\ CH_3 \end{array}$$

Rule 7. Both methyl groups are attached to carbon-2. Therefore the name of the compound is 2,2-dimethylpropane.

PRACTICE PROBLEM 15-1

What is the name of the compound with the formula shown below?

$$CH_3$$
$$|$$
$$CH-CH_2-CH_3$$
$$|$$
$$CH_3$$

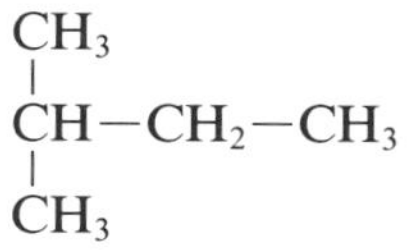

(*Answer:* 2-methylbutane).

PRACTICE PROBLEM 15-2

What is the name of the compound with the formula shown below?

$$CH_2-CH-CH_3$$
$$|\quad\quad|$$
$$CH_3 \quad CH_2-CH_3$$

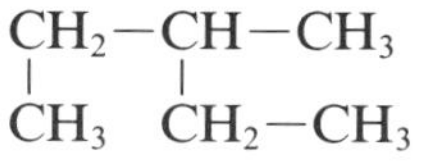

(*Answer:* 3-methylpentane).

Once you know how to derive the name of a compound from its structural formula, it becomes easy to go in the other direction, that is, to write a structural formula when you know the name of the compound. Suppose you are told that the "octane" used as a standard to determine the octane rating of gasolines is really 2,2,4-trimethylpentane. Could you write a structural formula for the compound? You would start from the end of the name and work backwards. Written as a set of steps, your procedure might look something like this:

Step 1. Determine the class of compound from the suffix. (The -*ane* ending here tells us that the compound is a saturated hydrocarbon.)
Step 2. Determine the stem (*pent*).
Step 3. Draw a chain with the number of carbon atoms indicated by this stem. (The stem *pent* indicates five carbon atoms in the main chain: C−C−C−C−C.)
Step 4. Number these carbon atoms consecutively from one end to the other:
$$\overset{1}{C}-\overset{2}{C}-\overset{3}{C}-\overset{4}{C}-\overset{5}{C}$$
Step 5. Identify the branches (methyl, ethyl, etc.) that are attached to the main chain. (This compound contains only methyl groups.)
Step 6. Use the multiplying prefixes to determine how many branches of each type are present. (The prefix *tri-* indicates that three methyl groups are present.)
Step 7. Attach the branches to the carbon atoms of the main chain having the same numbers as those immediately preceding the multiplying prefixes. (There are two methyl groups at carbon-2 and one at carbon-4.)

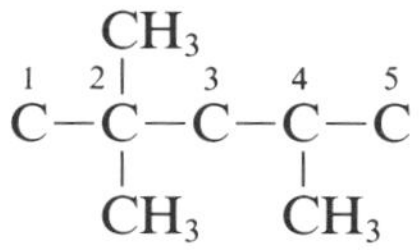

$$CH_3$$
$$\overset{1}{C}-\overset{2}{C}-\overset{3}{C}-\overset{4}{C}-\overset{5}{C}$$
$$|\quad\quad|$$
$$CH_3 \quad CH_3$$

Step 8. Add enough hydrogen atoms to give each carbon atom a total of four bonds. The complete structure is shown below.

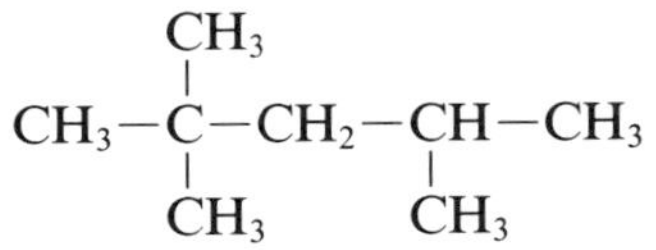

$$CH_3-\underset{\underset{\displaystyle CH_3}{|}}{\overset{\overset{\displaystyle CH_3}{|}}{C}}-CH_2-\underset{\underset{\displaystyle CH_3}{|}}{CH}-CH_3$$

EXAMPLE PROBLEM 15-3

What is the structural formula of 4-ethyl-2-methylhexane?

SOLUTION
Step 1. The -*ane* ending indicates an alkane.
Step 2. The stem is *hex*.
Step 3. This indicates a chain of six carbon atoms: C—C—C—C—C—C.
Step 4. The chain is numbered as: $\overset{1}{C}-\overset{2}{C}-\overset{3}{C}- \overset{4}{C}-\overset{5}{C}-\overset{6}{C}.$
Step 5. The branches are ethyl and methyl groups.
Step 6. There is one of each.
Step 7. The ethyl group is attached at carbon-4, and the methyl group is attached at carbon-2:

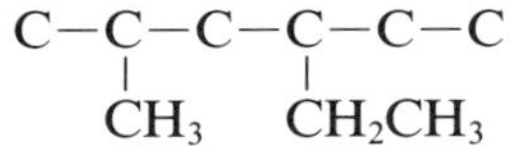

$$C-\underset{\underset{\displaystyle CH_3}{|}}{C}-C-\underset{\underset{\displaystyle CH_2CH_3}{|}}{C}-C-C$$

Step 8. The complete structure is

$$CH_3-\underset{\underset{\displaystyle CH_3}{|}}{CH}-CH_2-\underset{\underset{\displaystyle CH_2-CH_3}{|}}{CH}-CH_2-CH_3$$

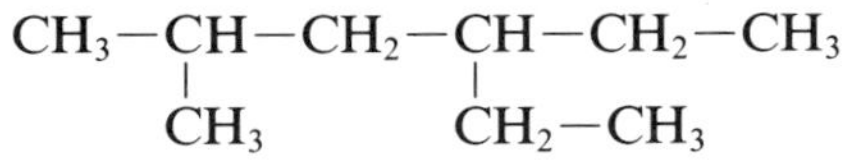

EXAMPLE PROBLEM 15-4

What is the structural formula for 5-ethyl-2,3,6-trimethyloctane?

SOLUTION
Step 1. The -*ane* ending indicates an alkane.
Step 2. The stem is *oct*.
Step 3. This indicates a chain of eight carbon atoms:
C—C—C—C—C—C—C—C.
Step 4. The chain is numbered as: $\overset{1}{C}-\overset{2}{C}-\overset{3}{C}- \overset{4}{C}-\overset{5}{C}-\overset{6}{C}-\overset{7}{C}-\overset{8}{C}.$
Step 5. The branches are ethyl and methyl groups.
Step 6. There is one ethyl group (at C-5), and there are three methyl groups (at C-2, C-3, and C-6).
Step 7. They are therefore attached as follows:

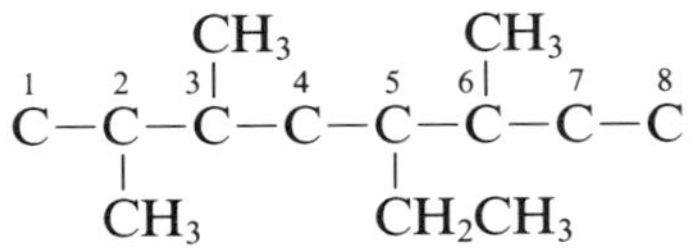

$$\overset{1}{C}-\overset{2}{C}-\overset{\overset{\displaystyle CH_3}{\overset{|}{\underset{\underset{\displaystyle CH_3}{|}}{\overset{3}{C}}}}}{}-\overset{4}{C}-\overset{5}{C}-\overset{\overset{\displaystyle CH_3}{\overset{|}{\underset{\underset{\displaystyle CH_2CH_3}{|}}{\overset{6}{C}}}}}{}-\overset{7}{C}-\overset{8}{C}$$

Notice that it does not matter whether the alkyl groups are written above or below the main chain.

Step 8. The complete structure is

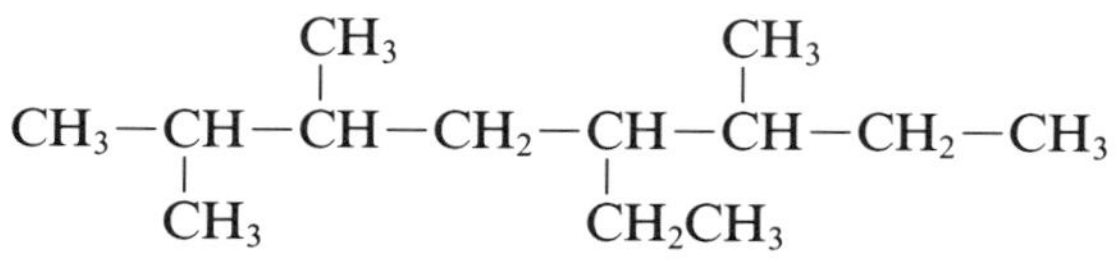

$$CH_3-CH-CH-CH_2-CH-CH-CH_2-CH_3$$

(with CH_3 groups above the second and fifth carbons, CH_3 below the second carbon, and CH_2CH_3 below the fifth carbon)

PRACTICE PROBLEM 15-3

What is the structural formula for 2,2-dimethylpentane?

(*Answer:*

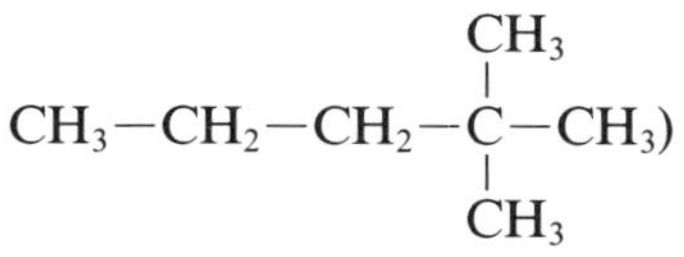

$$CH_3-CH_2-CH_2-\overset{\overset{\displaystyle CH_3}{|}}{\underset{\underset{\displaystyle CH_3}{|}}{C}}-CH_3)$$

PRACTICE PROBLEM 15-4

What is the structural formula for 5-butyl-2,6,7-trimethylnonane?

(*Answer:*

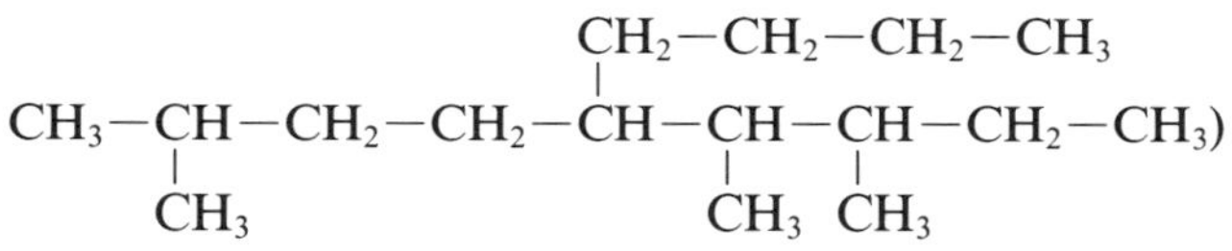

$$CH_3-CH-CH_2-CH_2-\overset{\overset{\displaystyle CH_2-CH_2-CH_2-CH_3}{|}}{CH}-CH-CH-CH_2-CH_3)$$

(with CH_3 below the second carbon, and CH_3 CH_3 below the sixth and seventh carbons)

15.4 Sources of Alkanes

All of the alkanes are obtained commercially from the refining of petroleum. Only one alkane, methane, is also available from a laboratory preparation.

Methane is a common constituent of the gas found in coal mines, where it makes up 20 to 40% of coal gas. It also constitutes about 90% of natural gas. Methane is formed in the laboratory by heating sodium acetate and sodium hydroxide:

$$NaC_2H_3O_2(s) + NaOH(s) \xrightarrow{\triangle} CH_4(g) + Na_2CO_3(s)$$

15.5 Chemical Properties of the Alkanes

Combustion

Chemically, methane and the other alkanes are relatively inert. They will, of course, burn in oxygen to form carbon dioxide and water, but they do not react easily with other compounds. Normally, methane, ethane, and propane burn quietly, but some mixtures of these gases with air are explosive. For example, many coal miners have been killed by methane explosions. When gaseous alkanes are used as heating fuels, extreme care must be used to avoid leaks which might lead to a build-up of explosive concentrations of these substances in the air. The equation for the complete combustion of methane is:

$$CH_4(g) + 2\ O_2(g) \longrightarrow CO_2(g) + 2H_2O(g)$$

When there is insufficient oxygen for complete combustion, the alkanes will still burn, but in these cases a large proportion of the product is deadly carbon monoxide gas, rather than relatively harmless carbon dioxide. The equation for the partial combustion of methane to form carbon monoxide is:

$$2CH_4(g) + 3\ O_2(g) \longrightarrow 2CO(g) + 4H_2O(g)$$

Substitution Reactions

Although alkanes are not very reactive, they will undergo substitution reactions (Section 9.6). When an alkane undergoes a substitution reaction, one or more hydrogen atoms is replaced by another atom. An example is

$$
\begin{array}{ccccccc}
& \text{H} & & & & \text{H} & \\
& | & & & & | & \\
\text{H}-\text{C}-\text{H} & + & \text{Cl}-\text{Cl} & \longrightarrow & \text{H}-\text{C}-\text{Cl} & + & \text{H}-\text{Cl} \\
& | & & & & | & \\
& \text{H} & & & & \text{H} &
\end{array}
$$

In this reaction the hydrogen atom of the C—H bond is substituted by a chlorine atom. Since only one chlorine atom has been substituted in the molecule, the product is called a monochloro substitution product. The equation is frequently written using condensed molecular formulas:

$$CH_4(g) + Cl_2(g) \longrightarrow CH_3Cl(\ell) + HCl(g)$$
chloromethane
(methyl chloride)

The substitution process could continue:

$$CH_3Cl(\ell) + Cl_2(g) \longrightarrow CH_2Cl_2(\ell) + HCl(g)$$
dichloromethane
(methylene chloride)

$$CH_2Cl_2(\ell) + Cl_2(g) \longrightarrow CHCl_3(\ell) + HCl(g)$$
trichloromethane
(chloroform)

$$CHCl_3(\ell) + Cl_2(g) \longrightarrow CCl_4(\ell) + HCl(g)$$
tetrachloromethane
(carbon tetrachloride)

In each of these reactions the same general process is occurring: a hydrogen atom in the starting material is being substituted by a chlorine atom. Since two, three, and four of the original hydrogen atoms of methane have been substituted by chlorine, these products are called dichloro, trichloro, and tetrachloro substitution products of methane.

Because all of the carbon atoms in the alkanes are bonded by single covalent bonds to four other atoms, nothing can add on to an alkane. Substitution reactions are the only possible reactions.

15.6 Cycloalkanes

Hydrocarbons can take the form of a closed ring as well as an open chain. The simplest of such cyclic compounds are the cycloalkanes:

cyclopropane   *cyclobutane*   *cyclopentane*   *cyclohexane*

These formulas are often drawn in the following manner:

At each corner of the geometric symbol a carbon atom is understood to be present. The lines from corner to corner are carbon-carbon bonds. Unless otherwise indicated, hydrogens are assumed to be present at the corner carbons in sufficient quantity to give each carbon atom four bonds. Chemically, the cycloalkanes are similar to open-chain alkanes. Five- and six-membered carbon rings are common in organic chemistry. (See cholesterol, Fig. 15-4.) They are abundant in the molecules of living things. Cyclopropane is sometimes used as an anesthetic, and cyclohexane (Fig. 15-11) is used in the manufacture of nylon.

Figure 15-11
Model of cyclohexane (C_6H_{12})

APPLICATION

A Strange-Shaped Carbon Molecule

In 1985, a new 60-carbon molecule was discovered at Rice University in Houston, Texas. It was formed during experiments which were designed to study the formation of long carbon chains in interstellar space. These experiments involved the laser vaporization of graphite.

The number of carbon atoms in the new molecule is not unusual. However, what is novel is the arrangement of the 60 carbon atoms. The atoms are grouped in the same geometric pattern as that found on the surface of a soccer ball—12 pentagons of carbon atoms. Its similarity to the geometry of R. Buckminster Fuller's geodesic dome has given the molecule its name: buckminsterfullerene.

Usually carbon atoms arrange themselves at the corners of a tetrahedron (as in diamond) or at the vertices of hexagons (as in graphite). The fact that there are 60 carbon atoms in the spherical molecule supports the belief that this number represents the largest number of identical objects that can be arranged symmetrically on the surface of a sphere. This is a stable molecule with each carbon atom bonded to three other carbon atoms.

Some scientists speculate that these C_{60} molecules are common in the universe, especially around carbon-rich stars and inside interstellar dust. They may even form the core of soot particles. Nevertheless, the method by which the C_{60} cluster arranges itself into a spherical shape remains a mystery.

15.7 Physical Properties and Uses of Alkanes

Figure 15-12
An oil rig in Noel, British Columbia

The first four members of the normal alkane series are gases at room temperature (20°C) and atmospheric pressure. From pentane, C_5H_{12}, up to $C_{16}H_{34}$, they are colourless liquids. Above $C_{16}H_{34}$, they are solids at room temperature and are referred to as paraffin waxes.

The boiling points, melting points, and densities of normal (unbranched) alkanes increase with increasing chain length. This is because the molecular surface area increases, resulting in greater attractive forces. Since the densities of alkanes are all less than that of water, alkanes float on water. For example, gasoline and oil are hydrocarbon mixtures consisting almost entirely of alkanes. Gasoline and oil float on water.

The electronegativity of carbon is 2.5. The electronegativity of hydrogen is 2.1. Such a small electronegativity difference means that a C—H bond is only slightly polar. Furthermore, in many hydrocarbons the C—H bonds are arranged so that their polarities cancel or nearly cancel each other. Thus, alkanes are nonpolar molecules. Alkanes are soluble in nonpolar solvents such as carbon tetrachloride and toluene. Alkanes are insoluble in polar solvents such as water.

Some of the uses of alkanes include the combustion of alkanes to produce energy. For example, methane is the main constituent of natural gas; propane is used in stoves and heaters; butane is used in cigarette lighters; and alkanes are the main components in kerosene, jet fuel, gasoline, home heating oil, and diesel fuel. Alkanes are also used to make waxed paper, candles, petroleum jelly, and lubricants. The chief source of alkanes is crude petroleum from oil wells such as the one shown in Figure 15-12.

The first petrochemical refinery was established in Brooklyn, N. Y., by the Nova Scotian inventor, Abraham Gesner. In 1846 Gesner invented a process for distilling a hydrocarbon lamp fuel from coal or oil shales. He called his product kerosene. He tried to set up a plant in Halifax for the commercial production of kerosene. Failing in this venture, he went to New York and obtained three United States patents for the manufacture of "kerosene burning fluid" in 1854. His refinery was established soon afterward.

Part One Review Questions

1. What is the difference between organic and inorganic compounds?
2. Describe an experiment in which an organic compound is prepared from inorganic substances.
3. What is organic chemistry?
4. Why are there so many organic compounds?
5. What is **a)** a hydrocarbon; **b)** a saturated hydrocarbon?
6. "The alkanes form a homologous series." What does this statement mean?
7. Describe the structure of a methane molecule. Use a perspective diagram to illustrate your description.
8. What are isomers?
9. What is the IUPAC name of the compound with the following structural formula?

$$CH_3 \diagdown \atop CH_3 \diagup CH-CH_2-CH-CH_2-CH_3 \atop \underset{CH_3}{|}$$

10. Write a structural formula for 2,2,4,4-tetramethylpentane.
11. What is a substitution reaction? Give an example.
12. Draw a structure for cycloheptane.
13. Would you expect kerosene to be more soluble in gasoline or in water? Explain your answer.

Part Two: Unsaturated Hydrocarbons

15.8 Alkenes

As we have seen, when two carbon atoms bond together in ethane, there is a requirement for six hydrogen atoms to bond with the two carbon atoms. However, in the compound ethene only four hydrogen atoms bond with the two carbon atoms. Ethene is an unsaturated hydrocarbon. It is not saturated with hydrogen atoms. Any hydrocarbon which contains fewer hydrogen atoms than the alkane with the same carbon skeleton is said to be **unsaturated**.

One way of writing the formula of ethene is:

$$\text{H:}\overset{..}{\underset{\overset{|}{\text{H}}}{\text{C}}}\text{:}\overset{..}{\underset{\overset{|}{\text{H}}}{\text{C}}}\text{:H}$$

In this structure, however, each carbon atom has an unpaired electron, and neither carbon atom obeys the octet rule. Carbon is tetravalent, and it forms four bonds. Thus, the two unpaired electrons form a second bond between the two carbon atoms:

$$\underset{\text{H}}{\overset{\text{H}}{}}\text{:C::C:}\overset{\text{H}}{\underset{\text{H}}{}}$$

or

$$\underset{\text{H}}{\overset{\text{H}}{}}\text{C}=\text{C}\overset{\text{H}}{\underset{\text{H}}{}}$$

Figure 15-13
Model of ethene (C$_2$H$_4$)

Each carbon atom now has four bonds: one bond to each of two hydrogen atoms and two bonds to the other carbon atom (Fig. 15-13).

Like ethane, ethene has two carbon atoms. However, ethene has one double bond in the molecule. Hydrocarbons which contain one carbon-carbon double bond are called **alkenes**. All alkenes have names ending in the suffix *-ene*, indicating the presence of a carbon-carbon double bond. The general formula of the alkenes is C$_n$H$_{2n}$.

Ethene is a colourless gas with a sweet odour. It is sometimes used as an anesthetic in dentistry and surgery. It is produced naturally by fruits as they ripen and is used commercially to speed the ripening of green fruits.

Below are some other simple alkenes:

$$\underset{\text{H}}{\overset{\text{H}}{}}\text{C}=\text{C}\overset{\text{H}}{\underset{\text{CH}_3}{}}$$

propene, C$_3$H$_6$

$$\underset{\text{H}}{\overset{\text{H}}{}}\text{C}=\text{C}\overset{\text{CH}_3}{\underset{\text{CH}_3}{}}$$

$$\underset{\text{H}}{\overset{\text{H}}{}}\text{C}=\text{C}\overset{\text{H}}{\underset{\text{CH}_2\text{CH}_3}{}}$$

$$\underset{\text{H}}{\overset{\text{CH}_3}{}}\text{C}=\text{C}\overset{\text{H}}{\underset{\text{CH}_3}{}}$$

isomers of **butene**, C$_4$H$_8$

We notice that in addition to the arrangement of the carbon atoms in butene being varied, the position of the double bond can also be varied.

15.9 IUPAC Nomenclature of Alkenes

The rules for naming alkenes are similar to those for naming alkanes, with the following modifications:

a) The longest continuous chain must be chosen to contain both carbon atoms of the double bond, even if a longer chain could be found by choosing a different path.

b) The characteristic family suffix is *-ene*.

c) The chain must be numbered from the end closest to a double-bonded carbon atom. If there is no difference, then it is numbered from the end closest to an alkyl group.

d) The position of the double bond is indicated by the lower-numbered carbon atom in the double bond, and this number is placed immediately before the stem. (The number is omitted if there is only one possible location for the double bond.)

EXAMPLE PROBLEM 15-5

What is the IUPAC name for the butene isomer with structural formula $CH_2=CH-CH_2-CH_3$?

SOLUTION

The steps to use in developing the name are:

Rule 1. A four-carbon chain with a double bond.

Rule 2. ∴ *But-* + *-ene* ⟶ butene.

Rules 3 to 5. No alkyl groups.

Rule 6. Numbering is $\overset{1}{C}=\overset{2}{C}-\overset{3}{C}-\overset{4}{C}$.

∴The name is 1-butene.

EXAMPLE PROBLEM 15-6

What is the name of the butene isomer with the following structural formula?

$$CH_2=\underset{\underset{CH_3}{|}}{C}-CH_3$$

SOLUTION

The steps to use in developing the name are:

Rule 1. A three-carbon chain with a double bond.

Rule 2. ∴ *Prop-* + *-ene* ⟶ propene.

Rules 3 and 4. One alkyl group (methyl).

Rule 5. ∴ A methylpropene.

Rule 6. Numbering is $\overset{1}{C}=\underset{\underset{CH_3}{|}}{\overset{2}{C}}-\overset{3}{C}$.

∴The name is 2-methylpropene. *Note*: No number is used to locate the double bond in propene, because there is no such thing as 2-propene.

EXAMPLE PROBLEM 15-7

What is the structural formula for 2-butene?

SOLUTION

The steps to use in developing a structural formula are:

Step 1. The suffix -*ene* tells us the compound is an alkene.

Step 2. The stem *but*- tells us the main chain contains four carbon atoms.

Step 3. Write the skeleton main chain: C—C—C—C.

Step 4. Number the carbon atoms, and insert the double bond between the indicated carbon atom and the one with the next higher number. In this case, the double bond goes between carbon atoms 2 and 3: $\overset{1}{C}-\overset{2}{C}=\overset{3}{C}-\overset{4}{C}$.

Step 5. Since there are no alkyl groups, add enough hydrogen atoms to give each carbon atom a total of four bonds. The complete structure is

$$CH_3-\underset{\underset{H}{|}}{C}=\underset{\underset{H}{|}}{C}-CH_3$$

EXAMPLE PROBLEM 15-8

What is the structural formula for 2-methyl-1-butene?

SOLUTION

The steps to use in developing a structural formula are:

Step 1. The suffix -*ene* tells us the compound is an alkene.

Step 2. The stem *but*- tells us the main chain contains four carbon atoms.

Step 3. Write the skeleton main chain: C—C—C—C.

Step 4. Insert the double bond between carbon atoms 1 and 2: $\overset{1}{C}=\overset{2}{C}-\overset{3}{C}-\overset{4}{C}$.

Step 5. Attach the methyl group to carbon-2: $C=\underset{\underset{CH_3}{|}}{C}-C-C$.

Step 6. Add enough hydrogen atoms to give each carbon atom a total of four bonds: $CH_2=\underset{\underset{CH_3}{|}}{C}-CH_2-CH_3$.

PRACTICE PROBLEM 15-5

What is the IUPAC name for the following?

$$CH_3-CH_2-\underset{\underset{H}{|}}{C}=\underset{\underset{H}{|}}{C}-CH_2-\underset{\underset{CH_3}{|}}{CH}-CH_2-CH_3$$

(*Answer:* 6-methyl-3-octene).

PRACTICE PROBLEM 15-6

What is the IUPAC name for the following?

$$CH_3-\underset{\underset{CH_3}{|}}{C}=CH-CH_2-\underset{\underset{CH_3}{|}}{\overset{\overset{CH_3}{|}}{CH}}$$

(*Answer:* 2,5-dimethyl-2-hexene).

PRACTICE PROBLEM 15-7

Give the structural formula for 3-ethyl-2-pentene.

$$(Answer: CH_3-CH=\underset{\underset{CH_2-CH_3}{|}}{C}-CH_2-CH_3)$$

PRACTICE PROBLEM 15-8

Give the structural formula for 4-ethyl-2-methyl-2-hexene.

$$(Answer: CH_3-\underset{\underset{CH_3}{|}}{C}=CH-\underset{\underset{CH_2CH_3}{|}}{CH}-CH_2-CH_3)$$

15.10 Sources of Alkenes

Almost all alkenes are obtained commercially by the cracking of petroleum. **Cracking** is a process in which hydrocarbons are broken into smaller molecules by using heat and a catalyst. Although a hydrocarbon molecule is relatively stable, in the presence of a catalyst at 600°C it breaks up or "cracks" to give hydrocarbons of lower molecular mass. Many of these hydrocarbons are alkenes.

One alkene, ethene, is frequently prepared in the laboratory by the action of hot, concentrated sulfuric acid on ethanol:

$$C_2H_5OH(\ell) \xrightarrow[\Delta]{H_2SO_4} C_2H_4(g) + H_2O(\ell)$$

$$\textit{ethanol}$$
$$(\textit{ethyl alcohol})$$

15.11 Reactions of Alkenes

Because all alkenes behave similarly, we shall use ethene as a representative example in the following discussion.

Addition Reactions

The double bonds in alkenes are chemically reactive, and the alkenes will undergo what is called an **addition reaction**. For example, a bromine molecule will add to an ethene molecule to form 1,2-dibromoethane:

$$\begin{array}{c} H \\ \diagdown \\ \diagup \\ H \end{array} C{=}C \begin{array}{c} H \\ \diagup \\ \diagdown \\ H \end{array} + Br_2 \longrightarrow \begin{array}{c} H\ \ H \\ | \ \ | \\ H{-}C{-}C{-}H \\ | \ \ | \\ Br\ Br \end{array}$$

Notice that the suffix changes from *-ene* to *-ane*. Since 1,2-dibromoethane has no double bonds, its name ends in *-ane*. In this reaction the bromine is not substituted for anything. It actually adds to the ethene. This is why we call this type of reaction an addition reaction.

In the presence of a suitable catalyst such as platinum, it is possible to add hydrogen to alkenes to form the corresponding alkanes.

$$\begin{array}{c} H \\ \diagdown \\ \diagup \\ H \end{array} C{=}C \begin{array}{c} H \\ \diagup \\ \diagdown \\ H \end{array} + H_2 \xrightarrow{\text{Pt, pressure}} \begin{array}{c} H\ \ H \\ | \ \ | \\ H{-}C{-}C{-}H \\ | \ \ | \\ H\ \ H \end{array}$$

ethene   ethane

Ethene may be converted to ethanol by the acid-catalyzed addition of water:

$$\begin{array}{c} H \\ \diagdown \\ \diagup \\ H \end{array} C{=}C \begin{array}{c} H \\ \diagup \\ \diagdown \\ H \end{array} + H_2O \xrightarrow{\text{acid}} \begin{array}{c} H\ \ H \\ | \ \ | \\ H{-}C{-}C{-}H \\ | \ \ | \\ H\ \ OH \end{array}$$

Polymerization

Alkenes have the important property of being able to add to themselves to form hydrocarbon chains up to several thousand carbon atoms long. This reaction is important in the manufacture of polymers such as polyethylene (Section 15.26).

$$n \begin{array}{c} H \\ \diagdown \\ \diagup \\ H \end{array} C{=}C \begin{array}{c} H \\ \diagup \\ \diagdown \\ H \end{array} \xrightarrow{\text{catalyst}} \left(\begin{array}{c} H\ \ H \\ | \ \ | \\ {-}C{-}C{-} \\ | \ \ | \\ H\ \ H \end{array} \right)_n$$

where *n* is a large number

ethene   polyethylene

Combustion

Ethene, like any hydrocarbon, will burn in air to form carbon dioxide and water vapour:

$$C_2H_4(g) + 3\ O_2(g) \longrightarrow 2CO_2(g) + 2H_2O(g)$$

In the presence of limited amounts of air, ethene burns to produce poisonous carbon monoxide:

$$C_2H_4(g) + 2\ O_2(g) \longrightarrow 2CO(g) + 2H_2O(g)$$

15.12 Alkynes

We have seen examples of two carbon atoms bonding to six hydrogen atoms, and two carbon atoms bonding to four hydrogen atoms. In the compound ethyne (acetylene), only two hydrogen atoms are bonded to the two carbon atoms:

$$H:\overset{\cdot}{\underset{\cdot}{C}}:\overset{\cdot}{\underset{\cdot}{C}}:H$$

In this structure, however, each carbon atom has two unpaired electrons, and neither carbon atom obeys the octet rule. However, you will recall that carbon is tetravalent and forms four bonds. Thus the four unpaired electrons form a second and a third bond between the two carbon atoms:

$$H:C::C:H \quad \textbf{or} \quad H-C\equiv C-H$$

Now each carbon atom has four bonds: one bond is to a hydrogen atom, and the other three bonds are to the other carbon atom (Fig. 15-14). Any hydrocarbon that has one triple bond is called an **alkyne**. Because an alkyne contains less than the maximum possible number of hydrogen atoms, it is unsaturated.

Alkynes are named in the same way as alkenes. The major difference is that the characteristic suffix used to indicate the presence of an alkyne is *-yne*. The general formula of the alkynes is C_nH_{2n-2}. Some of the other simple alkynes are

propyne	C_3H_4	$H-C\equiv C-CH_3$
1-butyne	C_4H_6	$H-C\equiv C-CH_2-CH_3$
		and
2-butyne		$CH_3-C\equiv C-CH_3$

Figure 15-14
Model of ethyne (C_2H_2)

15.13 Ethyne—A Typical Alkyne

By far the most important alkyne is ethyne, which is commonly called acetylene. (In this common name, the *-ene* ending does not indicate a carbon-carbon double bond.) Acetylene is obtained both commercially and in the laboratory by the action of water on calcium carbide:

$$CaC_2(s) + 2H_2O(\ell) \longrightarrow C_2H_2(g) + Ca(OH)_2(aq)$$

Calcium carbide was accidentally discovered in 1892 by T. J. Willson, a native of Woodstock, Ontario. Willson was attempting to obtain aluminum from its ore by heating it with coke and lime (CaO) in an electric furnace. Instead of aluminum, he obtained a gray powder which reacted with water to form a flammable gas. The gray powder was apparently formed by the reation

$$CaO(s) + 3C(s) \overset{\triangle}{\longrightarrow} CaC_2(s) + CO(g)$$

This reaction was the basis on which the carbide and acetylene industries were subsequently built.

Acetylene will decompose explosively when compressed to about 20 times atmospheric pressure. Even liquid acetylene (b.p. −83°C) must be handled with extreme care. Thus, the acetylene cylinders used for transportation and storage of acetylene do not contain pure acetylene. Instead, the gas is dissolved at a moderate pressure (about 12 times atmospheric pressure) in acetone (Section 15.21). The solution is prepared in the cylinder, which also contains a porous material that absorbs the liquid and prevents it from splashing about during transit.

15.14 Chemical Reactions of Ethyne

Addition

Because it is an unsaturated hydrocarbon, ethyne will undergo addition reactions. Thus, it will add two molecules of a halogen such as bromine or chlorine to form a saturated product:

$$H-C\equiv C-H + 2H_2 \xrightarrow{Pt} H-\underset{\underset{H}{|}}{\overset{\overset{H}{|}}{C}}-\underset{\underset{H}{|}}{\overset{\overset{H}{|}}{C}}-H$$

In the presence of a catalyst such as platinum it will add two molecules of hydrogen and form ethane:

$$H-C\equiv C-H + 2H_2 \xrightarrow{Pt} H-\underset{\underset{H}{|}}{\overset{\overset{H}{|}}{C}}-\underset{\underset{H}{|}}{\overset{\overset{H}{|}}{C}}-H$$

Combustion

Like any hydrocarbon, acetylene undergoes combustion. Acetylene burns in air with a highly luminous flame, probably because at the combustion temperature the hydrocarbon is partly broken down into finely divided carbon particles that become incandescent. The incomplete combustion of acetylene in the presence of limited amounts of oxygen results in the formation of carbon monoxide and soot (carbon particles):

$$2C_2H_2(g) + 5\,O_2(g) \longrightarrow 4CO_2(g) + 2H_2O(g)$$

$$2C_2H_2(g) + 3\,O_2(g) \longrightarrow 4CO(g) + 2H_2O(g)$$

$$2C_2H_2(g) + O_2(g) \longrightarrow 4C(s) + 2H_2O(g)$$

Because of its luminous flame, acetylene was used as an illuminating gas before the advent of the electric light, especially in automobile headlamps. Several decades ago, the usual bicycle lamp and the miner's lamp were small acetylene generators, each of which consisted of a calcium carbide canister into which water could be dripped as desired through a valve.

When acetylene is burned in pure oxygen, the flame temperature is very high (2800°C to 3000°C). As a result, an important use of acetylene is as a fuel in the oxyacetylene torches used for the welding and cutting of metals (Fig. 15-15).

Figure 15-15
This welder is using an oxyacetylene torch to cut through a piece of sheet steel.

15.15 Alkadienes

The alkadienes or dienes are molecules which have just what their name implies—two (*di-*) double bonds (*-ene*) per molecule. The most important diene is 1,3-butadiene. The stem *but-* tells us that the molecule has four carbon atoms. The numbers one and three tell us that the double bonds are attached to the first and the third carbon atoms:

Figure 15-16
Car tires made of synthetic rubber

1,3-Butadiene is prepared by cracking petroleum samples which contain butane. Butadiene is used in making synthetic rubber, which is widely used in car tires (Fig. 15-16). During World War II, the world's main provider of natural rubber, the Malayan Peninsula, became an unreliable source. Therefore, the Canadian government set up Polymer Corporation in Sarnia, Ontario. This corporation, now known as Polysar, first produced Buna-S in 1942. This synthetic rubber is made from 1,3-butadiene and styrene (Section 15.26).

15.16 Aromatic Hydrocarbons

In 1825, Michael Faraday purified and analyzed an oily liquid given to him by the Portable Gas Company of London. Because of his careful work, he is credited with the discovery of benzene. Several of the aromatic, pleasant-smelling oils obtainable from plant sources were also found to consist of molecules similar in some ways to benzene. Later the term **aromatic compound** was applied to any substance whose molecules contained the benzene structural skeleton, even if the substance did not have a pleasant smell. The properties of aromatic hydrocarbons are so different from those of other hydrocarbons that we have a special name for the others. **Aliphatic compounds** are hydrocarbons that contain no aromatic or other rings. These compounds were called aliphatic (from Greek *aleiphatos*, "fat") because many of them that contain chains of 12 to 30 carbon atoms—such as paraffin wax—are fat-like in their physical properties. Thus, ethene, propyne, and 1,3-butadiene are all aliphatic hydrocarbons. Hydrocarbons which contain rings but are not aromatic (for example, cyclohexane) are called alicyclic (*ali*phatic + *cyclic*) compounds.

Since benzene is the parent aromatic hydrocarbon, we shall examine its structure and properties first. The molecular formula of benzene is C_6H_6. Any compound which has only six hydrogen atoms for six carbon atoms must be highly unsaturated. However, unlike the other unsaturated families (for example, alkenes and alkynes) benzene undergoes substitution reactions rather than addition reactions. Thus, C_6H_6 reacts with chlorine in the presence of a catalyst to form a monosubstituted product with the formula C_6H_5Cl:

$$C_6H_6(\ell) + Cl_2(g) \longrightarrow C_6H_5Cl(\ell) + HCl(g)$$

August Kekulé (Fig. 15-17), a German chemist, attempted to explain some of these experimental results by proposing structure 1 for benzene and structure 2 for the monosubstituted product:

(1)   **(2)**

Structure 1 does not account for all the properties of benzene. For example, it appears that benzene still has three carbon-carbon double bonds. These should readily *add* molecules of bromine, for example, but benzene undergoes only substitution, not addition, reactions.

It is now known that the carbon-carbon bonds in benzene are neither single bonds nor double bonds. Rather each bond is half-way between a single bond and a double bond in character. This characteristic is shown by drawing a dotted line instead of a solid line for one of the two bonds:

$$C === C$$

Biography

AUGUST KEKULÉ was born in 1829 in Darmstadt, Germany. He attended the University of Giessen where he studied architecture before the lectures of Justus von Liebig won him over to chemistry. He obtained his doctorate in 1852. Following study in Paris and London, Kekulé returned to Germany in 1856 and became a lecturer at the University of Heidelberg. In 1858, he became a full professor of pure chemistry at the Belgian State University of Ghent. In 1867, he became a professor of chemistry at Bonn. He held that post until his death in 1896.

Kekulé made important contributions to chemical theory, especially with regard to the structures of carbon compounds. He published a famous paper in which he stated that carbon is tetravalent. In another paper, he explained the large number of carbon compounds in terms of the ease with which carbon atoms bond to one another to form chains.

In 1865, as the result of a dream about benzene, Kekulé made the most brilliant prediction to be found in the whole of organic chemistry. He dozed off while writing his textbook and dreamed that chains of carbon atoms appeared as snakes. "And look, what was that? One snake grabbed its own tail, and mockingly the shape whirled before my eyes." He had hit upon the idea of assuming that the six carbon atoms of benzene were arranged in a ring. In this way, he opened up the field of aromatic chemistry.

Kekulé was a great promoter of chemistry. He was a good lecturer, combining the abilities of a great researcher with those of an impressive speaker. In order to help his students understand his theories of chains and rings of carbon atoms, Kekulé invented atomic models. These consisted of coloured wooden spheres which could be attached to one another. Similar models are still used in classrooms today.

Figure 15-17
August Kekulé (1829-1896)—
Suggested ring structure for benzene.

It is this special type of bond that makes benzene and its derivatives so different from aliphatic compounds. Thus, the structure of benzene is best written as

H—C≷C—H ... **or** ⬡ **or** ⬡
H—C≷C—H

Figure 15-18
Model of benzene (C_6H_6)

Every carbon-carbon bond in benzene is identical, and is best considered as a bond and a half. The six identical carbon-carbon bonds are seen in Fig. 15-18.

Benzene is a colourless, highly flammable liquid which is obtained from coal tar and is used as a starting material for the manufacture of drugs, dyes, and explosives. It is an excellent solvent for nonpolar solutes, but it is quite toxic. An oral dose of 3.8 mL/kg of body mass kills 50% of the rats to which it is given. The sudden ingestion of a comparable amount of benzene by humans causes irritation of the mucous membranes, restlessness, convulsions and excitement, followed by depression and death from respiratory failure. The ingestion of smaller amounts of benzene over a longer period (orally, by inhalation, or by absorption through the skin) can cause bone marrow depression and arrest development of selected parts of the body. It occasionally causes leukemia and other types of cancer. For these reasons, benzene is no longer used in most high school laboratories.

The compound which is normally used to replace benzene in a high school laboratory is toluene, which is less toxic than benzene. In toluene a methyl group replaces one hydrogen atom of benzene. It is used as a gasoline additive; as a solvent for paints, lacquers, gums, and resins; and as a starting material in the manufacture of benzoic acid (Section 15.22), benzaldehyde (Section 15.20), dyes, and explosives such as trinitrotoluene (TNT):

toluene   *2,4,6-trinitrotoluene*

The xylenes have two CH_3 groups replacing two hydrogen atoms:

1,2-dimethylbenzene   *1,3-dimethylbenzene*   *1,4-dimethylbenzene*
(*ortho xylene*)   (*meta xylene*)   (*para xylene*)

The terms *ortho*, *meta*, and *para* are used in naming disubstituted aromatic substances to indicate the relative position of the two substituted groups. The term *ortho* indicates that the two groups are on adjacent carbon atoms; the term *para* indicates that the two groups are on opposite carbon atoms; and the term *meta* indicates that the two groups are on carbon atoms that are separated by one carbon atom of the ring.

Part Two Review Questions

1. Describe the structure of an alkene. Give the structural formulas of two alkenes.
2. What is an addition reaction? Using ethene as an example, write equations for three different addition reactions.
3. Describe the structure of an alkyne. Give the structural formulas of two alkynes.
4. Write balanced chemical equations for the reaction of propyne with **a)** excess bromine; **b)** excess oxygen; **c)** a limited amount of oxygen (assume the only carbon-containing product is carbon monoxide).
5. Write the structures for **a)** 1,2-butadiene; **b)** 2,4-pentadiene.
6. What are the structural differences between aliphatic, alicyclic, and aromatic compounds? Give one example of each type of compound.
7. Describe the structure of benzene.
8. In what ways do the properties of aromatic compounds differ from those of nonaromatic compounds?

Part Three: Compounds With Other Functional Groups

15.17 Functional Groups

A large number of hydrocarbons of all kinds can be made. We shall now consider organic compounds that contain other atoms in addition to carbon atoms and hydrogen atoms. Here the emphasis will be on the properties of the groups of atoms which can be attached to the hydrocarbon part of the molecule. The symbol R— will be used to designate that hydrocarbon part. The R— groups are often alkyl groups and they are derived from alkanes.

The atom or reactive group of atoms attached to the alkyl group is called a **functional group**. The term *functional group* is applied to any reactive atom or group of atoms in an organic compound to distinguish these atoms from those which do not react. You have already seen two functional groups. The carbon-carbon double bond and the carbon-carbon triple bond are both functional groups because they are groups of atoms which act as sites of reactivity within a molecule. Other examples of functional groups are:

$$-OH \text{ (hydroxyl)} \quad -NH_2 \text{ (amino)} \quad \overset{\displaystyle O}{\overset{\displaystyle \|}{-C-}} \text{(carbonyl)} \quad \overset{\displaystyle O}{\overset{\displaystyle \|}{-C-}}OH \text{ (carboxyl)}$$

It is the functional group that is responsible for the type of chemical reaction shown by a particular class of compounds. During a reaction, the hydrocarbon part usually remains unchanged while the functional group is chemically reactive.

15.18 Alcohols

The **alcohols** are compounds which contain a hydroxyl group attached to an alkyl group. A **hydroxyl** (*hydr*ogen + *ox*ygen + *yl*) **group** is an $-OH$ group. Thus, the general formula for an alcohol is $R-OH$.

The IUPAC names of alcohols are formed from the names of the corresponding alkanes. All you do is drop the final *e* of the alkane name and replace it with the suffix *-ol*. Thus CH_3OH is methanol, and CH_3CH_2OH is ethanol. For other alcohols, the main chain must be chosen to contain the carbon atom bearing the OH group, and the chain is numbered from the end that gives the lowest possible number to the carbon atom bearing the OH group. Thus, $CH_3CH_2CH_2CHCH_3$ is 2-pentanol.
$$\qquad\qquad\qquad\qquad \underset{\displaystyle OH}{|}$$

The simplest member of the family is methanol or methyl alcohol, CH_3OH. Methanol is a colourless liquid with a pleasant odour. It is a good fuel and is used in antifreeze. Unfortunately, some people drink it in ignorance of its effects. Since it is poisonous, even small quantities can bring about blindness or death. Often methanol is sold under the names wood alcohol, methyl hydrate, or methylated spirits. Just under 300 kt are used in Canada each year in the production of formaldehyde (Section 15.20), acetic acid (Section 15.22), methanol-blended fuels, windshield de-icers, pipeline dehydrating agents, and solvents for dyes, resins, shellac, and cellulose.

The most familiar member of the alcohol family is ethanol, also called ethyl alcohol or grain alcohol, C_2H_5OH. It is colourless, has a pleasant odour and a sharp taste, and burns with a colourless flame. Ethanol is used as a laboratory and industrial solvent, in the manufacture of pharmaceuticals, in perfumes, and as a starting material in the synthesis of other organic compounds.

Despite its many industrial and laboratory uses, most ethanol is used in alcoholic beverages. Since early times, people have known that certain fruit juices spoiled (fermented) when left to stand. Drinking these fermented liquids (wine) produced a euphoric state. About one thousand years ago it was discovered that the active substances in the liquid mixture could be concentrated by distillation. Since then many types of liquors have been developed. All of these alcoholic beverages contain ethanol in varying concentrations, from 5% to as high as 75% by volume. Like the other lower members of the alcohol family (those with relatively few C-atoms in the hydrocarbon chain), ethanol is poisonous. A concentration of only 0.40% in the blood of an individual can cause death. Because alcoholic beverages are taxed in almost all areas of the world, most ethanol sold for laboratory or industrial purposes (and not taxed as liquor) is denatured; that is, small amounts of toxic impurities are added so that the alcohol will not be diverted from the laboratory or factory into illegal beverages. Among the substances used to make denatured alcohol are methanol, camphor, gasoline, 2-propanol, benzene, castor oil, acetone, nicotine, ether, pyridine, sulfuric acid, and kerosene.

Ethanol is often prepared by fermentation of sugars by yeast:

$$C_6H_{12}O_6(aq) \xrightarrow[\text{fermentation}]{} 2C_2H_5OH(aq) + 2CO_2(g)$$

glucose   ***ethanol***

$$C_{12}H_{22}O_{11}(aq) + H_2O(\ell) \xrightarrow[\text{fermentation}]{} 4C_2H_5OH(aq) + 4CO_2(g)$$

sucrose (sugar)   ***ethanol***

Ethanol is also made industrially from ethene:

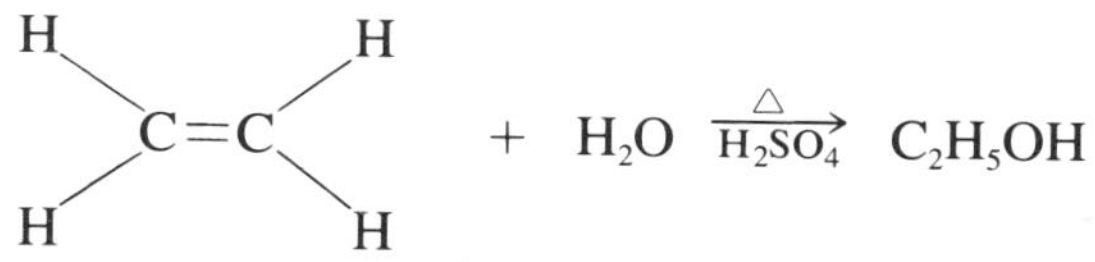

Other alcohols are derived from propane, butane, and pentane. They are the propanols (propyl alcohols), butanols (butyl alcohols), and pentanols (pentyl alcohols). Isopropyl alcohol is used as rubbing alcohol because it kills germs better than the other simple alcohols and has a temporary lubricating effect during the rubbing process. Just over 5 kt of 1-butanol (*n*-butyl alcohol) are used each year in Canada as a solvent for paints and coatings, gums, resins, vegetable oils and dyes. 1-Butanol is also used in the production of pharmaceuticals, glues, safety glass, artificial leather, films, perfumes, and in the fabrication of plastics.

We should also look at two alcohols which have more than one hydroxyl group per molecule. 1,2-Ethanediol (ethylene glycol) is a two-carbon, saturated molecule which has two hydroxyl groups:

$$\begin{array}{l} CH_2-OH \\ | \\ CH_2-OH \end{array}$$

It is used as an antifreeze in the ratio of 1:1 by volume with water. The resultant mixture has a lower freezing point and a higher boiling point than water.

1,2,3-Propanetriol (glycerol) is a three-carbon, saturated molecule which has three hydroxyl groups:

$$\begin{array}{l} CH_2-OH \\ | \\ CH-OH \\ | \\ CH_2-OH \end{array}$$

Glycerol is a colourless, odourless, viscous (thick) liquid with a sweet taste. It is used in cigarette tobacco to keep it moist, in the manufacture of hand lotions and cellophane, and in the manufacure of the explosive, nitroglycerin.

The cyclic and branched alcohol, menthol, is used as a flavouring agent in products such as cigarettes and cough drops.

$$
\begin{array}{c}
CH_3 \\
| \\
CH \\
\diagup \quad \diagdown \\
CH_2 \qquad CH_2 \\
| \qquad\qquad | \\
CH_2 \qquad CH-OH \\
\diagdown \quad \diagup \\
CH \\
| \\
CH \\
\diagup \quad \diagdown \\
CH_3 \qquad CH_3
\end{array}
$$

The alcohols containing fewer than five carbon atoms tend to be soluble in water due to hydrogen bonding. The alcohols themselves are good solvents for many organic compounds.

APPLICATION

New Drug with Sobering Effects

A new drug has shown promise in blocking the effects of alcohol. The drug, known as Ro15-4513, has been used to study the effect of alcohol on the brain. The data indicate that the drug will prevent the effects of ethanol (ethyl alcohol). Ethanol is known to stimulate the uptake of chloride ions in the brain. These chloride ions block the transmission of messages by certain nerve cells, which results in a reduction of tension or anxiety.

Research with the new drug on rats has shown that its use will block all the outward indications of drunkenness, from a staggering gait to the release of tension. Chemists are now developing a more potent drug with fewer side effects. Eventually we may have a drug which could be used to sober dangerously drunk individuals or to treat alcoholism.

15.19 Ethers

An **ether** is a compound which has two alkyl (or aromatic) groups attached to a single oxygen atom. For this reason, ethers have been called organic oxides. Their general formula is $R-O-R'$. R can be the same as R' as in CH_3-O-CH_3 (dimethyl ether) and $C_2H_5-O-C_2H_5$ (diethyl ether), or R can be different from R′ as in $CH_3-O-C_2H_5$ (methyl ethyl ether).

The most common ether is diethyl ether, which is normally called ether. It is used as an organic solvent and has been used as an anesthetic. It is rarely used for the latter purpose today because of its unpleasant side-effects.

Ethers, like the alkanes, are relatively unreactive. However, they make excellent solvents for waxes, fats, oils, perfumes, and gums. Because of its low boiling point (35°C) ether is frequently used to extract organic compounds from plant and animal tissues. The ether can then be removed by boiling at such a low temperature that heat-sensitive substances are not harmed. The low boiling point of ether makes it useful as a starter for diesel engines in cold weather.

The "petroleum ether" found in many laboratories is not an ether at all. It is a mixture of alkanes, mostly isomers of pentanes and hexanes, with an approximate boiling range of 30 to 60°C. The "ether" part of the name is used because diethyl ether has a similarly low boiling point.

15.20 Aldehydes

Aldehydes are compounds which contain an alkyl (or aromatic) group and a hydrogen atom attached to a carbonyl group. A **carbonyl group**, $-\overset{\displaystyle O}{\overset{\|}{C}}-$, has a carbon atom joined to an oxygen atom by a double bond. Thus, the general formula of an aldehyde is $R-\overset{\displaystyle O}{\overset{\|}{C}}-H$.

The IUPAC names of aliphatic aldehydes are formed from the names of the corresponding alkanes. All you do is drop the final *e* of the alkane name and replace it with the suffix *-al*. Thus $H-\overset{\displaystyle O}{\overset{\|}{C}}-H$ is methanal and $CH_3-\overset{\displaystyle O}{\overset{\|}{C}}-H$ is ethanal. For branched aldehydes, the main chain must be chosen to contain the carbonyl carbon, even if a longer chain could be found by choosing a different path. The chain is always numbered starting from the carbonyl carbon as number 1, but the 1 is not written in the name. Thus

$$CH_3-CH_2-\underset{\underset{\displaystyle CH_2CH_3}{|}}{CH}-\overset{\displaystyle O}{\overset{\|}{C}}-H \quad \text{is 2-ethylbutanal.}$$

Formaldehyde (methanal), $H-\overset{\displaystyle O}{\overset{\|}{C}}-H$, is a gas with a sharp, suffocating odour. A solution of formaldehyde in water is called formalin. Formalin is used for perserving biological specimens because it reacts with the proteins in the specimens, making them harder and less susceptible to putrefaction (rotting).

Acetaldehyde (ethanal), $CH_3-\overset{\displaystyle O}{\overset{\|}{C}}-H$, is another simple aldehyde. It is used largely as an intermediate for the synthesis of other organic compounds.

15.21 Ketones

A **ketone** contains two alkyl (or aromatic) groups attached to a carbonyl group.

Thus, ketones have the general formula $R-\overset{\overset{\displaystyle O}{\|}}{C}-R'$. As in ethers, R and R' can

be the same in ketones such as $CH_3-\overset{\overset{\displaystyle O}{\|}}{C}-CH_3$ (propanone), or they can be

different in ketones such as $CH_3-\overset{\overset{\displaystyle O}{\|}}{C}-C_2H_5$ (butanone).

In the IUPAC system, aliphatic ketones are named after the longest chain containing the carbonyl group. The final *e* of the alkane name is replaced with the suffix *-one*. This chain is numbered to give the carbonyl carbon the lowest

possible number. For example, $CH_3-CH_2-CH_2-\overset{\overset{\displaystyle O}{\|}}{C}-CH_3$ is 2-pentanone. Since the carbonyl group can be only at carbon-2 in either propanone or butanone, the 2 is not included in those two names.

Biography

Figure 15-19
Raymond Lemieux—Developed methods for synthesizing complex organic molecules.

RAYMOND U. LEMIEUX was born in Lac La Biche, Alberta. He obtained his B.Sc. from the University of Alberta and his Ph.D. from McGill University. In 1961, after holding posts in several Canadian and American universities, he took a position at the University of Alberta. There he made major contributions to the development of methods for synthesizing complex organic molecules containing carbohydrate structures. His more recent research has focussed on the synthesis of carbohydrate-containing substances that are important in blood typing and in the removal of offending antibodies from blood prior to tissue transplantation.

Dr. Lemieux's scientific achievements are matched by his dedication to the development of a viable Canadian manufacturing industry in the field of bioscience.

Raymond Lemieux is the author of over 200 scientific publications. As one of Canada's outstanding scientists, he has received numerous awards, including honorary doctorates from 11 universities. In 1967, he was elected Fellow of the prestigious Royal Society of London and made an Officer of the Order of Canada the following year. In 1982, he became the first recipient of the prestigious Izaak Walton Killam Memorial Prize which recognizes distinguished lifetime achievement and outstanding contribution to the advancement of knowledge in the natural sciences. In 1985, he was awarded the Canadian Medical Association's Medal of Honour.

The most common ketone is propanone, which is usually called acetone. Acetone is used as a solvent for paint, varnish, and lacquer; for cleaning precision machinery; and in the production of lubricating oils, chloroform, pharmaceuticals, and pesticides. About 25 kt are used in Canada each year. Diabetics produce abnormally large quantities of acetone in their metabolism. In severe cases of diabetes, an odour of acetone can be detected in the breath.

Another industrially important ketone is 4-methyl-2-pentanone. About 8 kt of this ketone are used in Canada each year as a solvent for coatings, for use in dewaxing of lube oils, and in the rubber industry.

15.22 Carboxylic Acids

A **carboxylic acid** contains an alkyl (or aromatic) group attached to a **carboxyl** (*carb*onyl + hydr*oxyl*) **group**, $-\overset{\overset{\displaystyle O}{\|}}{C}-OH$. Thus, the general formula for a carboxylic acid is $R-\overset{\overset{\displaystyle O}{\|}}{C}-OH$.

Acids are substances which dissociate in water to generate hydronium (H_3O^+) ions (Section 12.2). A carboxylic acid molecule can dissociate in the following way:

$$R-\overset{\overset{\displaystyle O}{\|}}{C}-O-H(aq) + H_2O(\ell) \rightarrow R-\overset{\overset{\displaystyle O}{\|}}{C}-O^-(aq) + H_3O^+(aq)$$

When the oxygen-hydrogen bond breaks, the electron pair remains with the oxygen atom, which now becomes negative. The hydrogen atom loses its electron and an H_3O^+ ion is formed.

Carboxylic acids are named after the longest carbon chain containing the carboxyl group. The carbon of the carboxyl group is always numbered 1. The name of the acid ends in *-oic acid*, which replaces the final *e* of the name of the alkane. Thus $H-\overset{\overset{\displaystyle O}{\|}}{C}-OH$ is methanoic acid, and $CH_3-\overset{\overset{\displaystyle O}{\|}}{C}-OH$ is ethanoic acid. Despite the existence of perfectly logical IUPAC names, the carboxylic acids are still commonly named according to an older system in which the name is derived from an original source of the acid. For example, methanoic acid is more commonly called formic acid (from Latin *formica*, "ant"), because in medieval times alchemists obtained it by distilling red ants.

Carboxylic acids are found in vinegar, fruits, and rancid butter. The first member of the family is formic acid (methanoic acid), $H-\overset{\overset{\displaystyle O}{\|}}{C}-OH$. Formic acid is an acrid-smelling liquid found in the sting of bees and the "bite" of many types of ants. Formic acid is used in the dying of wool; for the dehairing and tanning of animal hides; in electroplating; in the coagulation of rubber latex; and as an aid in the regeneration of old rubber.

Acetic acid (from Latin *acetum*, "vinegar") or ethanoic acid,

$$CH_3-\overset{\overset{\displaystyle O}{||}}{C}-OH$$, is found in vinegar, which is a 5% solution of acetic acid in water. In its pure form it is called glacial acetic acid because at 16.7°C it freezes to form an icy-looking solid. It is used in the manufacture of cellulose acetate, acetate rayon, plastics, and rubber; as an acidifier and preservative in foods (for example, pickles); and as a solvent for gums and resins. It is widely used in commercial organic syntheses. Like the other carboxylic acids, it can react with a base such as sodium hydroxide to form a salt (sodium acetate) and water:

$$CH_3\overset{\overset{\displaystyle O}{||}}{C}-OH + NaOH \rightarrow CH_3\overset{\overset{\displaystyle O}{||}}{C}-ONa + H_2O$$

Butyric acid (from Latin *butyrum*, "butter") or butanoic acid,

$$CH_3-CH_2-CH_2-\overset{\overset{\displaystyle O}{||}}{C}-OH$$, is found in rancid butter and gives it its characteristic disagreeable odour. It is also responsible for the characteristic locker-room odour of stale perspiration.

Some acids contain more than one functional group. For example, oxalic acid (from Latin *oxalis*, "sorrel") consists of two carboxyl groups joined together: $HO-\overset{\overset{\displaystyle O}{||}}{C}-\overset{\overset{\displaystyle O}{||}}{C}-OH$. Its IUPAC name is ethanedioic acid. It is present in low concentrations in many vegetables (such as rhubarb) and in varying concentrations in many plants (such as wood sorrel). This acid is poisonous. It is an ingredient in many products used for the cleaning of copper and brass, for the removal of rust, and in the preparation of blueprints.

Lactic acid (from Latin *lac*, "milk") has a hydroxyl group as well as a carboxyl group: $CH_3-\underset{\underset{\displaystyle OH}{|}}{CH}-\overset{\overset{\displaystyle O}{||}}{C}-OH$. Its IUPAC name is 2-hydroxypropanoic acid. Lactic acid is an acidic component of sour milk. It is used for acidifying beverages, in the manufacture of lactates which are used in food products, in medicine, and as a plasticizer for certain resins.

Many carboxylic acids are important in metabolic processes. For example, acetic acid is used by living organisms as a building block for the manufacture of larger molecules such as fats and steroids. Many molds convert sugar into oxalic acid, some with a 90% efficiency. Lactic acid is one of the end products in the series of biochemical reactions that converts the chemical energy of carbohydrates into useful work. Lactic acid accumulates in the muscles during strenuous activity, especially if a person is out of shape, and causes the painful reminders of unaccustomed exertion the following day.

Some acids contain aromatic rings. A typical example is salicylic acid (from Latin *salix*, "willow") which can be obtained from the willow tree. It consists of a benzene molecule with a carboxyl group and a hydroxyl group on adjacent carbon atoms:

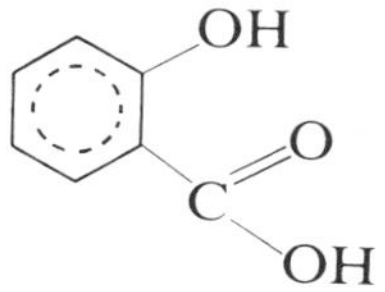

In some countries salicylic acid is used as a preservative of food products, but its major use is in the manufacture of certain esters (Section 15.23).

15.23 Esters

An **ester** is a compound in which the acidic hydrogen atom of a carboxylic acid molecule has been substituted by an alkyl (or aromatic) group. Thus, the general formula for an ester is $R-\overset{\overset{\displaystyle O}{\|}}{C}-OR'$. Organic acids react with alcohols to form esters and water. A general equation is:

$$R-\overset{\overset{\displaystyle O}{\|}}{C}-OH + R'-OH \xrightarrow[H_2SO_4]{\triangle} R-\overset{\overset{\displaystyle O}{\|}}{C}-OR' + H_2O$$

Carboxylic Acid + Alcohol $\longrightarrow$ Ester   +   Water

Esters are named after the carboxylic acid and the alcohol from which they were formed. In the name of the ester, the alkyl group of the alcohol is named first, followed as a separate word by the name of the carboxylic acid with the ending *-ic acid* replaced by *-ate*. Thus the IUPAC name of $CH_3-\overset{\overset{\displaystyle O}{\|}}{C}-OCH_3$ is methyl ethanoate. Unfortunately, the IUPAC names of esters are rarely used. Fortunately, however, the common names of esters are obtained by applying exactly the same rules, but to the common names of the acid and the alcohol. Thus, the common name of methyl ethanoate is methyl acetate.

Esters have many commercial uses ranging from cosmetics and perfumes to synthetic fibres. Many vegetables oils are esters, and the odours of fruits such as apples and oranges are due to the presence of esters. The odour of bananas is due to pentyl acetate, which is an ester of acetic acid and pentanol:

$$C_5H_{11}OH + CH_3\overset{\overset{\displaystyle O}{\|}}{C}-OH \xrightarrow[H_2SO_4]{\triangle} CH_3\overset{\overset{\displaystyle O}{\|}}{C}-OC_5H_{11} + H_2O$$

Ethyl acetate is formed by the reaction between ethanol and acetic acid:

$$CH_3\overset{\overset{\displaystyle O}{\|}}{C}-OH + CH_3CH_2OH \xrightarrow[H_2SO_4]{\triangle} CH_3\overset{\overset{\displaystyle O}{\|}}{C}-OCH_2CH_3 + H_2O$$

It is a colourless liquid with an odour similar to that of nail polish remover. About 6 kt are used each year. It is used as a solvent in inks, lacquers, and adhesives, and in the production of cellulose acetate, artificial flavours, and perfumes.

Salicylic acid contains both a carboxyl group and a hydroxyl group. The carboxyl group can react with an alcohol such as methanol to form the ester methyl salicylate:

salicylic acid   *methyl salicylate*

Methyl salicylate is the flavouring constituent of oil of wintergreen and a common ingredient in liniments.

On the other hand, the hydroxyl group of salicylic acid can react with acids such as acetic acid to form acetylsalicylic acid (aspirin) which is the most common of all analgesics (pain killers):

salicylic acid   *aspirin*

Esters can also be prepared by reacting inorganic acids with alcohols. One well-known ester is the explosive nitroglycerin. Glycerol reacts with nitric acid to give nitroglycerin and water:

nitroglycerin

Nitroglycerin was first prepared by the Italian chemist Ascanio Sobrero in 1846. It was left to Alfred Nobel (Fig. 15-19), a Swedish chemist, to discover a way to make its use less dangerous.

15.24 Other Types of Organic Compounds

Alcohols can be related to water ($H-O-H$) by considering that an alkyl group has been substituted for one of the two hydrogen atoms ($R-O-H$). The *mercaptans* can be related to hydrogen sulfide ($H-S-H$) by considering that an alkyl group has been substituted for a hydrogen atom ($R-S-H$).

The mercaptans have an offensive rotten egg odour. A mercaptan is the principal part of the well-known fluid that is the defensive weapon of the skunk. In the same way that alcohols relate to water and mercaptans relate to hydrogen sulfide, *amines* relate to ammonia (NH_3). Amines have the formula $R-NH_2$ and are often detected by their fish-like odour.

A compound in which the $-OH$ group of an acid

$$R-\overset{\displaystyle \overset{O}{\|}}{C}-OH$$

is replaced by an $-NH_2$ is called an *amide* ($R-\overset{\displaystyle \overset{O}{\|}}{C}-NH_2$). Amides are of special importance because the amide group $-\overset{\displaystyle \overset{O}{\|}}{C}-\underset{\displaystyle \underset{H}{|}}{N}-$ is the basic structural group in proteins and enzymes. The amide bond links amino acids in proteins.

Organic halides contain the halogens fluorine, chlorine, bromine, and iodine. They have physical properties similar to those of the corresponding hydrocarbons.

Biography

ALFRED NOBEL was born in Stockholm, Sweden, in 1833. Nobel attended school in Stockholm from 1841 to 1842. When his family moved to St. Petersburg in Russia, he was given private instruction by Russian and Swedish tutors from 1843 to 1850. He then made a two year trip to study chemistry in Italy, France, Germany, and North America.

Nobel became interested in the explosive nitroglycerin, and in 1863 he made his first important invention, a detonator for nitroglycerin. In 1865, he set up the world's first nitroglycerin factory. Because of the large number of accidents involving the explosive he experimented with ways of reducing the danger of handling it. He found that it could be adsorbed on (mixed with) a powder called kieselguhr (diatomaceous earth), which is made up of the remains of marine organisms called diatoms. Mixing nitroglycerin with kieselguhr made the explosive harder to detonate and enabled it to be transported much more safely. This mixture is called dynamite.

Nobel went on to invent other types of explosives. He also did work in the areas of optics, electrochemistry, biology, and physiology. He held over 350 patents throughout the world. As a result of his many inventions and factories, he became quite wealthy. Nobel died in 1896 in Italy. He left his whole fortune of millions of dollars to be used to establish the Nobel Prizes in literature, science, and international understanding.

Figure 15-20
Alfred Nobel (1833-1896)—Inventor of dynamite

Carbon tetrachloride is a colourless, sweet-smelling liquid. Over 20 kt are used in Canada each year, mainly in the preparation of chlorofluoromethanes (freons). It is also used as a metal degreaser, a fumigant, and a solvent for fats, oils, rubber, waxes, and resins. Carbon tetrachloride is highly toxic. Moderate exposure to it causes headaches, nausea, dizziness, and confusion. It damages the gastrointestinal tract, central nervous system, liver, and lungs. It irritates the eyes and dries the skin. Severe exposure has a narcotic effect, leading to unconsciousness and death. Furthermore, carbon tetrachloride is a suspected carcinogen—a cancer causing agent—in humans. For these reasons, it has been banned from most high school laboratories.

A common substitute for carbon tetrachloride in industry is 1,1,1-trichloroethane, CH_3CCl_3. It is a colourless liquid with a sweetish, chloroform-like odour, and is much less toxic than carbon tetrachloride. About 15 kt are used each year in Canada in metal cleaning and as a solvent in dry cleaning and degreasing operations.

Perhaps the most widely used organic halide in Canada is 1,2-dichloroethane (ethylene dichloride), $Cl-CH_2CH_2-Cl$. It is a colourless, oily liquid with a chloroform-like odour and a sweet taste. Over 600 kt are consumed each year in the production of vinyl chloride (Section 15.26) and 4 kt in anti-knock gasolines. Other uses include fumigation, degreasing, paint removing, and ore flotation.

Trichlorofluoromethane, $CFCl_3$, commonly known as Freon-11, is a highly volatile, colourless liquid. About 20 kt are used in Canada each year in aerosol propellants, refrigerants, cleaning solvents, and to sterilize medical supplies.

Biography

Figure 15-21
Paul de Mayo (b.1924)—Developed the use of ultraviolet light for organic synthesis.

PAUL DE MAYO was born in London, England, in 1924. He obtained his B.Sc. in 1944 and Ph.D. in 1954 from the University of London, where he helped to determine the structures of complex natural products. He was a lecturer in organic chemistry at Imperial College, London, and later spent a year in the Harvard University laboratory of Nobel Prize-winner R. B. Woodward. In 1959 he joined the faculty at the University of Western Ontario, where he became a driving force in revitalizing the undergraduate and graduate chemistry programs.

Dr. de Mayo developed organic photochemistry (the reactions of organic compounds under the influence of ultraviolet light) into a useful tool for the syntheses of complex organic molecules. His numerous publications include two books on natural products and texts on molecular rearrangements that have become standard reference works for chemists. Paul de Mayo is also known as an excellent teacher.

Dr. de Mayo has received, among many honours, the Centennial Medal in 1967, the Chemical Institute of Canada Medal in 1982, and the E. W. R. Steacie Award in Photochemistry in 1985.

In this chapter, you learned the structures, properties, and uses of many organic compounds. In particular, you learned how to identify different organic families by their characteristic functional groups. The characteristics of these families are summarized in Table 15-6.

Table 15-6   Summary of the Families of Organic Compounds

Family	General Formula	Distinguishing Feature	Example Formula	Name
Alkane	C_nH_{2n+2}	all single bonds	CH_3CH_3	**ethane**
Alkene	C_nH_{2n}	one C=C bond	$CH_2=CH_2$	**ethene**
Alkyne	C_nH_{2n-2}	one C≡C bond	HC≡CH	**ethyne**
Alkadiene	C_nH_{2n-2}	two C=C bonds	$CH_2=CH—CH=CH_2$	**1,3-butadiene**
Cycloalkane	C_nH_{2n}	carbon ring	▢	**cyclobutane**
Aromatic		benzene-like structure	⬡	**benzene**
Alcohol	R—OH	hydroxyl group (OH)	C_2H_5OH	**ethanol**
Ether	R—O—R′	—O—	$CH_3OC_2H_5$	*methyl ethyl ether*
Aldehyde	R—C(=O)—H	—C(=O)—H	$CH_3C(=O)—H$	**ethanal** *acetaldehyde*
Ketone	R—C(=O)—R′	carbonyl —C(=O)— group	$CH_3C(=O)—C_2H_5$	**butanone**
Carboxylic acid	R—C(=O)—OH	carboxyl —C(=O)—OH group	$CH_3C(=O)—OH$	**ethanoic acid** *acetic acid*
Ester	R—C(=O)—OR′	—C(=O)—O—	$CH_3C(=O)—OC_5H_{11}$	*pentyl acetate*
Mercaptan	R—S—H	—S—H	C_4H_9SH	*butyl mercaptan*
Amine	R—NH_2	amino -NH_2 group	$C_2H_5NH_2$	*ethylamine*
Amide	R—C(=O)—NH_2	—C(=O)—NH_2	$CH_3C(=O)—NH_2$	*acetamide*
Organic halide		one or more of F, Cl, Br, I	CCl_4	*carbon tetrachloride*

Part Three Review Questions

1. What is a functional group? Write formulas for five specific compounds, each containing a different functional group.
2. What is **a)** an alcohol; **b)** a hydroxyl group? Write structural formulas for two different alcohols.

3. What is an ether? Write formulas for three different ethers.
4. What is **a)** a carbonyl group; **b)** an aldehyde; **c)** a ketone? Write formulas for two different aldehydes and two different ketones not mentioned in the text. Give the names for the aldehydes and ketones you have chosen.
5. What is **a)** a carboxyl group; **b)** a carboxylic acid? Write formulas for four different carboxylic acids.
6. What is an ester? Choose two different alcohols and two different acids. Write the equations for the formation of the four different esters that can be formed from your chosen alcohols and acids. Name all organic compounds appearing in your equations.
7. What is the general formula of **a)** a mercaptan; **b)** an amine; **c)** an amide?

Part Four: Organic Compounds in Industry

15.25 Petroleum

Products of the chemical industry such as gasoline, rubber, plastics, drugs, and colouring and flavouring agents are important in our modern society. We have already learned about many of them. Where do all of these substances come from?

Ultimately, the primary sources of most organic compounds are coal and petroleum. Coal originated about 250 000 years ago when vegetation was especially lush. Dead plant matter became covered by earth. Some of the material dissolved in water and some was destroyed by microorganisms. The remaining materials, containing about 50% carbon, were heated and compressed by earth movements to become successively peat, brown coal, soft coal, and hard coal, which is about 94% carbon.

When a tonne of soft coal is strongly heated, it forms 800 kg of coke and about 40 kg of volatile by-products called coal tar. From the coal tar are obtained various aromatic compounds such as benzene (10 L from 1 t of soft coal), toluene, xylene, ethylbenzene, naphthalene, biphenyl, anthracene, and creosote oil. Figure 15-22 shows the structures of some of these compounds. The aromatic compounds are well suited to provide starting materials for the chemical transformations that ultimately lead to synthetic dyes. The dye industry was one of the first major organic chemical industries.

Figure 15-22
Representative compounds from the destructive distillation of coal

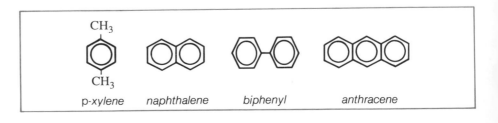

p-xylene   naphthalene   biphenyl   anthracene

Crude petroleum is mainly a mixture of hydrocarbons averaging 86% carbon by mass. Its empirical formula is approximately CH_2. Aromatic substances are present in petroleum (Fig. 15-23), but they are only a few among the hundreds of compounds identified. In an oil refinery, the components are separated by distillation into "fractions" which are mixtures of compounds characterized by boiling range and carbon number, that is, the number of carbon atoms per molecule. Common descriptions of the fractions are given in Table 15-7.

Table 15-7   Composition of Petroleum Fractions

Boiling Range (°C)	Carbon Number	Fraction
< 30	$C_1 - C_7$	light gas
30–60	$C_5 - C_8$	petroleum ether
60–100	$C_6 - C_7$	naphtha
40–205	$C_5 - C_{12}$	natural gasoline
175–320	$C_9 - C_{15}$	kerosene
275–350	$C_{14} - C_{18}$	heating and fuel oils
> 350	C_{18} up	lubricating oils and greases
tarry residue	high	asphalt; sludge

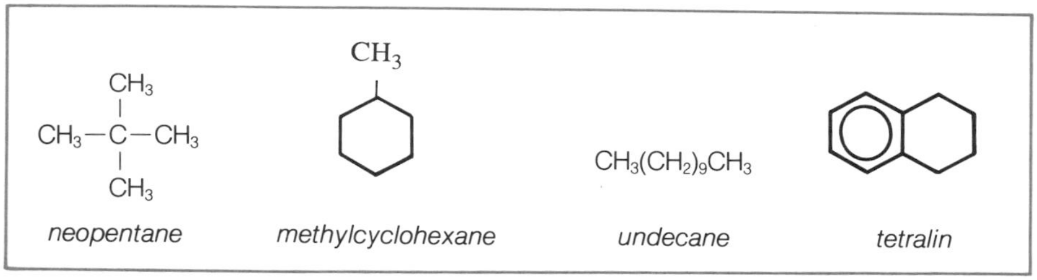

neopentane   methylcyclohexane   undecane   tetralin

Figure 15-23
Representative compounds from the distillation of petroleum

The larger molecules are converted to smaller, more useful ones by heating them with a catalyst in large catalytic "crackers" (Section 16.4). For example,

$$CH_3CH_2CH_2CH_2CH_2-CH_2CH_2CH_2CH_2CH_3 \xrightarrow[\text{catalyst}]{\Delta}$$
decane

$$CH_3CH_2CH_2CH_2CH_3 + CH_3CH_2CH_2CH=CH_2$$
pentane   1-pentene

Alkenes are converted to alcohols by the acid-catalyzed addition of water:

$$R-CH=CH_2 + H_2O \xrightarrow{acid} RCHCH_3$$
$$|$$
$$OH$$

an alkene   an alcohol

Alcohols are easily converted to aldehydes, ketones, and carboxylic acids, and these are in turn converted to other functional groups. In ways such as these, hundreds and even thousands of different compounds are made from petroleum and coal. Examples of some of the uses of ethylene, a typical petrochemical, are shown in Figure 15-24.

Figure 15-24
Ethylene in the petrochemical industry

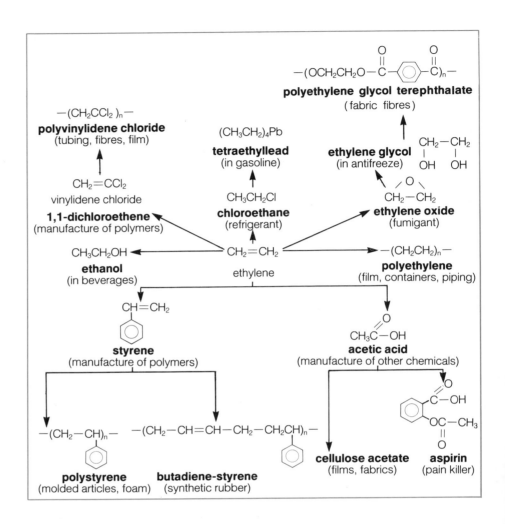

15.26 Polymers

Many useful organic compounds, including some of those in Figure 15-24, are polymers. A **polymer** is a substance which has a large molecular mass and which is composed of many fundamental building block molecules or *monomer* units. These monomer units have chemically united to form the polymer.

The well-known polymer, polyethylene, is made from the polymerization of the monomer ethylene:

many $H_2C=CH_2$ (ethylene) $\longrightarrow$ $-CH_2CH_2(CH_2CH_2)_nCH_2CH_2-$
polyethylene

Each $-CH_2CH_2-$ unit in the polyethylene is derived from an ethylene molecule. The double bond in ethylene has been converted to a single bond as the polymer is formed. A single polyethylene molecule can contain many thousands of these ethylene monomer units.

There are both naturally occurring polymers and synthetic polymers. Proteins, nucleic acids, polysaccharides, and natural rubber are examples of naturally occurring polymers. The synthetic polymers include nylon, Dacron®, Lucite®, synthetic rubber, and polystyrene.

The two principal types of synthetic polymers are condensation polymers and addition polymers. Condensation polymers form slowly and require heating of the reaction mixture for several hours before the polymer reaches the proper chain lengths. Condensation polymers have relatively low molecular masses in comparison with addition polymers. The former have molecular masses which are generally less than 100 000 u, while addition polymer molecules can have masses of as much as 10 million atomic mass units. Addition polymers also form much more rapidly than condensation polymers. The formation of condensation polymers involves the elimination of small molecules such as H_2O and NH_3. In addition polymerization, the monomer units merely add on to one another.

Condensation Polymers

When ethylene glycol, $HO-CH_2-CH_2-OH$, reacts with terephthalic acid,

$$HO-\overset{\overset{O}{\|}}{C}-\langle\bigcirc\rangle-\overset{\overset{O}{\|}}{C}-OH$$

terephthalic acid

a polymer of high molecular mass is produced:

$$\overline{H}{+}O-CH_2CH_2-O{+}\overline{H + HO}{+}\overset{\overset{O}{\|}}{C}-\langle\bigcirc\rangle-\overset{\overset{O}{\|}}{C}{+}\overline{OH + H}{+}O-CH_2CH_2-O{+}\overline{H}$$

$$\xrightarrow{\triangle} ----O-CH_2CH_2-O-\overset{\overset{O}{\|}}{C}-\langle\bigcirc\rangle-\overset{\overset{O}{\|}}{C}-O-CH_2CH_2-O---- + H_2O$$

Because an acid has reacted with an alcohol to form a polymeric substance containing many ester groups, the product is called a polyester. When this particular terephthalate ester is made into thin films, it is known as Mylar. When it is made into fibres for fabrics, it is known as Dacron or Terylene.

Nylons are a well-known type of condensation polymer. One of the best known of the nylons is made by reacting hexamethylenediamine with adipic acid:

$$H_2N-(CH_2)_6-NH_2 + HO\overset{\overset{O}{\|}}{C}-(CH_2)_4-\overset{\overset{O}{\|}}{C}OH \longrightarrow$$

hexamethylenediamine   *adipic acid*

$$------\overset{\overset{H}{|}}{N}-(CH_2)_6-\overset{\overset{H}{|}}{N}-\overset{\overset{O}{\|}}{C}-(CH_2)_4-\overset{\overset{O}{\|}}{C}-------- + H_2O$$

nylon

In the polymer, nylon, the hexamethylenediamine units alternate with adipic acid units. The $-\overset{\overset{\displaystyle H}{|}}{N}-\overset{\overset{\displaystyle O}{||}}{C}-$ segments are amide groups, and nylon is a polyamide. Nylon has a molecular mass of around 25 000 u. It is used in textiles. If nylon is heated to 250°C and forced through a plate in which tiny holes have been drilled, nylon fibres are produced. This nylon in fibre form is subjected to mechanical stretching called cold drawing, and a strong fibre results. Nylon fibre is used in clothing, ropes, tire cord, and brushes, while solid nylon is formed into gears or bearings.

Addition Polymers

Many polymerization reactions proceed by means of a free-radical mechanism. An atom, ion, or molecule which contains one or more unpaired electrons is a **free radical**. Unpaired electrons have a tendency to pair with other unpaired electrons of opposite spins. Thus free radicals are *reactive*.

Addition polymerization involves monomers which have double bonds. During the process of forming a polymer by a free-radical mechanism, the double bond becomes a single bond. This releases two electrons, allowing two new bonds to form. Styrene is a monomer which has a double bond and can be made to form an addition polymer:

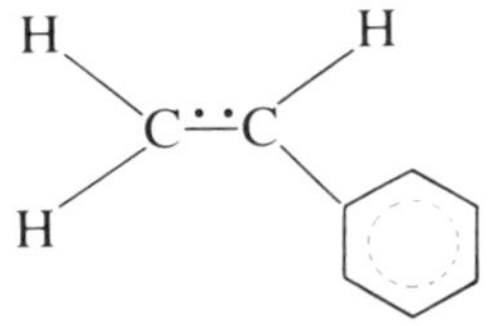

The second bond of the double bond between the two carbon atoms in styrene is shown as an electron pair. The reason for this will become clear as we proceed.

A small amount of an *initiator* (a substance which generates free radicals) is mixed with the styrene monomer. Heating the mixture causes the initiator to break down into free radicals, as we see in the case of benzoyl peroxide:

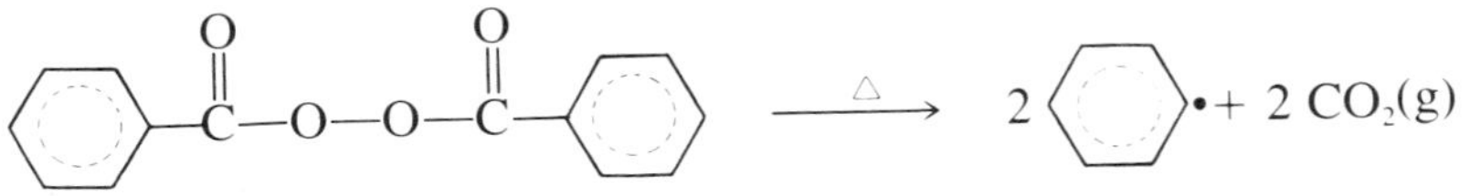

benzoyl peroxide
(initiator)

phenyl radical

The phenyl radical then reacts with one end of the carbon-carbon double bond of the styrene molecule:

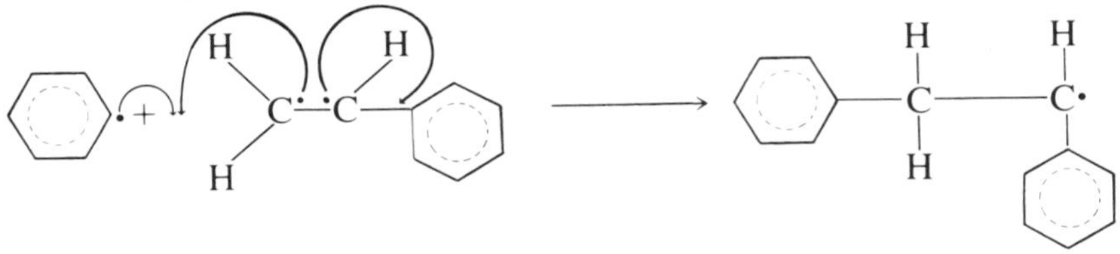

Now an unpaired electron appears on one of the styrene carbon atoms. This new free radical then reacts with a second styrene molecule to form a new free radical which reacts with a third styrene, then a fourth, a fifth, and so on:

polystyrene

Finally, the growth of the addition polymer chain is stopped by a termination step. For example, the growing chain could react with a phenyl free radical. However, a more likely possibility is for two different growing chains to react together. When two free radicals come together, the unpaired electrons of the two species pair up and form a stable covalent bond.

Polymers are used by manufacturers to produce commercial plastics. The first plastic was produced in 1869 in answer to a newspaper contest which offered a reward of $10 000 to the first person to produce a synthetic substitute for ivory billiard balls. There was concern that the large demand for ivory billiard balls would endanger the existence of elephant herds. The name of the first plastic was Celluloid®. It was made by treating a natural polymer, cellulose, with nitric acid, and then adding camphor. Since that time the amount and variety of plastics have grown phenomenally. Today the plastics industry plays an important part in the economy and annually produces billions of kilograms of plastics.

Polymers are classified according to their method of formation: addition or condensation. They are also classified according to their structure and properties. Some polymers may be heated and melted without being destroyed. They return to their original condition upon cooling. Such polymers are **thermoplastics**. Thermoplastics can be melted and moulded into different shapes. Thermoplastics are made up of straight chains with few cross-links between chains. Polymers that cannot be heated or melted without being destroyed are called **thermosetting polymers**. They will not regain their original structure on cooling. Thermosetting polymers are made up of thoroughly cross-linked chains in a three-dimensional network. Polyethylene, polystyrene, and polyamides (nylon) are all thermoplastics. Other members of this class include acrylics, Teflon®, polypropylene, and polyvinyl chloride (PVC). Thermosetting polymers include polyesters, phenol-formaldehyde resins, epoxy resins, and polyurethanes.

Polymer chains of different lengths can be made by varying such conditions as pressure, temperature, heating time, and amount of reactants. Polymers that are used in the production of useful end products are generally mixed with other chemicals to give them special properties. The basic polymer is referred to as the resin. This resin often comes in the form of powder or pellets. The polymerization reaction is not completed in the resin; rather, it is completed when other chemicals are added to the resin, and the polymer is made into its desired form.

The plastics industry has grown so rapidly because so many plastics with special properties can be made, and the raw materials have been both cheap and abundant. Table 15-8 summarizes some types of polymers and their uses.

Table 15-8 Common Polymers and Their Uses

Monomer	Polymer	Name	Some Uses
CH_3 $\ \ \ $ H $\diagdown$ C=C $\diagup$ H $\diagup\ \ \diagdown$ H *propylene*	$\left(\begin{array}{cc} CH_3 & H \\ -C & -C- \\ H & H \end{array}\right)_n$	polypropylene	packaging, moulded articles, carpet fibres
H $\ \ \ $ H $\diagdown$ C=C $\diagup$ H $\diagup\ \ \diagdown$ Cl *vinyl chloride*	$\left(\begin{array}{cc} H & H \\ -C & -C- \\ H & Cl \end{array}\right)_n$	polyvinyl chloride (PVC)	piping, toys, records
F $\ \ \ $ F $\diagdown$ C=C $\diagup$ F $\diagup\ \ \diagdown$ F *tetrafluoroethylene*	$\left(\begin{array}{cc} F & F \\ -C & -C- \\ F & F \end{array}\right)_n$	Teflon®	gaskets, seals, coating for cooking utensils
O ‖ H $\ \ \ $ C—OCH_3 $\diagdown$ C=C $\diagup$ H $\diagup\ \ \diagdown$ H *methyl methacrylate*	$\left(\begin{array}{cc} H & \overset{O}{\overset{\|}{C}}-OCH_3 \\ -C & -C- \\ H & H \end{array}\right)_n$	Lucite®, Plexiglas®	moulded articles, coatings, windows
H $\ \ \ $ CN $\diagdown$ C=C $\diagup$ H $\diagup\ \ \diagdown$ H *acrylonitrile*	$\left(\begin{array}{cc} H & CN \\ -C & -C- \\ H & H \end{array}\right)_n$	polyacrylonitrile (Orlon®, Acrilan®)	textiles, carpets

Part Four Review Questions

1. What are the ultimate sources of most organic compounds?
2. What is a polymer? Give a structural formula for a typical polymer.
3. What is a condensation polymer? Write an equation for the formation of a condensation polymer from its monomers.
4. What is **a)** a free radical; **b)** an addition polymer; **c)** a thermoplastic; **d)** a thermosetting polymer? Give an example of each.

Chapter Summary

- Organic chemistry is the study of carbon-containing compounds.
- Hydrocarbons are compounds made up of carbon and hydrogen atoms only.
- Saturated hydrocarbons contain the maximum number of hydrogen atoms bonded to each carbon atom.
- Alkanes are saturated hydrocarbons.
- A homologous series is one in which the formula of each member of the series differs from that of the preceding member in a constant regular way.
- A structural formula shows the way in which the atoms are attached to each other.
- Isomers are molecules which have the same molecular formula but different structural formulas.
- An alkyl group is an alkane from which a hydrogen atom has been removed.
- A substitution reaction is a reaction in which one atom is replaced by another atom.
- Cracking is a method of breaking hydrocarbons into smaller molecules by the action of heat and a catalyst.
- An alkene is a hydrocarbon which contains a carbon-carbon double bond.
- An unsaturated hydrocarbon is one which has one or more carbon-carbon double or triple bonds.
- In an addition reaction with an alkene, the two parts of the reagent add to the double-bonded carbon atoms and convert the double bond to a single bond.
- An alkyne is a hydrocarbon which contains a carbon-carbon triple bond.
- An alkadiene is a hydrocarbon which contains two carbon-carbon double bonds.
- An aromatic compound is one which contains the benzene structural skeleton.
- An aliphatic compound is one which contains no aromatic or other rings.
- An alicyclic compound is one which contains one or more rings but which is not aromatic.
- A functional group is an atom or group of atoms that serves as a site of reactivity in a molecule.
- A hydroxyl group is an —OH group.
- An alcohol is a compound which contains a hydroxyl group attached to an alkyl group.
- An ether is a compound which contains two alkyl or aromatic groups attached to an oxygen atom.
- A carbonyl group has a carbon atom attached to an oxygen atom by a double bond.
- An aldehyde is a compound which contains an alkyl or aromatic group and a hydrogen atom attached to a carbonyl group.
- A ketone is a compound which contains two alkyl or aromatic groups attached to a carbonyl group.
- A carboxyl group has a hydroxyl group attached to a carbonyl group.

- A carboxylic acid is a compound which contains an alkyl or aromatic group attached to a carboxyl group.
- An ester is a compound in which the acidic hydrogen atom of a carboxylic acid molecule has been replaced by an alkyl or aromatic group.
- Mercaptans have the general formula $R-SH$.
- Amines have the general formula $R-NH_2$.
- Amides have the general formula $R-\overset{\overset{\displaystyle O}{\|}}{C}-NH_2$.
- Organic halides have the general formula RX, where X = F, Cl, Br, or I.
- A polymer is a substance which has a large molecular mass and is composed of many monomer units.
- A condensation polymer is one which is formed by the reaction between two substances in such a way that many small molecules (such as H_2O or NH_3) are also formed.
- A free radical is an atom, ion, or molecule which contains one or more unpaired electrons.
- An addition polymer is one which is formed by the addition of its monomers to the end of a growing polymer chain.
- A thermoplastic is a polymer which melts or softens on heating and returns to its original state on cooling.
- A thermosetting polymer is one that cannot be heated or melted without being destroyed.

Key Terms

organic chemistry
hydrocarbon
saturated
homologous series
structural formula
isomer
alkyl group
unsaturated
alkene
cracking
addition reaction
alkyne
aromatic compound
aliphatic compound

functional group
alcohol
hydroxyl group
ether
aldehyde
carbonyl group
ketone
carboxylic acid
carboxyl group
ester
polymer
free radical
thermoplastic
thermosetting polymer

Test Your Understanding

Each of these statements or questions is followed by four responses. Choose the correct response in each case. (Do not write in this book.)

1. The general formula of an alkane is
   a) C_nH_{2n-4}.   b) C_nH_{2n-2}.   c) C_nH_{2n}.   d) C_nH_{2n+2}.

2. The $CH_3CH_2CH_2-$ group is called
   a) a butyl group.
   b) an ethyl group.
   c) a propyl group.
   d) a pentyl group.

3. A student incorrectly named a compound 3,4-dimethylpentane. Its correct IUPAC name is most likely to be
   a) 2,3-dimethylpentane.
   b) 3,4-methylpentane.
   c) 3-methyl-4-methylpentane.
   d) 2-methyl-3-methylpentane.

4. Give the IUPAC name of

$$\begin{array}{ccccccc} & CH_3 & & & CH_3 & & \\ & | & & & | & & \\ CH_2-CH_2-CH_2-CH-CH-C-CH_2-CH_2 \\ & & & | & | & | & | \\ & & & CH_3 & CH_3 & CH_3 & CH_3 \end{array}$$

   a) 4,4,5,6-tetramethyldecane
   b) 5,6,7,7-tetramethyldecane
   c) 1,4,5,6,6,8-hexamethyloctane
   d) 1,3,3,4,5,8-hexamethyloctane

5. Of the reagents H_2SO_4, Br_2, NaOH, and O_2, the only one which will react with propane is
   a) H_2SO_4.
   b) Br_2.
   c) NaOH.
   d) O_2.

6. A compound whose formula is abbreviated as

   is

   a) an alkene.
   b) an alkyne.
   c) an aromatic hydrocarbon.
   d) a cycloalkane.

7. Alkenes
   a) are saturated hydrocarbons.
   b) contain only carbon-carbon single bonds.
   c) do not contain double bonds.
   d) are unsaturated hydrocarbons.

8. The general formula of an alkene is
   a) C_nH_{2n-4}.
   b) C_nH_{2n-2}.
   c) C_nH_{2n}.
   d) C_nH_{2n+2}.

9. When bromine reacts with 2-butene, the product is

   a) $CH_3CH-CHCH_3$
         $\ \ \ \ \ |\ \ \ \ |$
         $\ \ \ \ Br\ \ \ Br$

   b) $CH_3CH_2CH-CH_2$
         $\ \ \ \ \ \ \ \ \ \ |\ \ \ \ |$
         $\ \ \ \ \ \ \ \ \ Br\ \ \ Br$

   c) $CH_3CH_2-\overset{\overset{\displaystyle Br}{|}}{C}-CH_3$
         $\ \ \ \ \ \ \ \ \ \ |$
         $\ \ \ \ \ \ \ \ \ Br$

   d) $CH_3C=CCH_3$
         $\ \ \ \ \ \ |\ \ |$
         $\ \ \ \ Br\ Br$

10. The product formed by the acid-catalyzed addition of water to 2-butene is

   a) $CH_3CH_2CHCH_3$
         $\ \ \ \ \ \ \ \ \ \ |$
         $\ \ \ \ \ \ \ \ \ OH$

   b) $CH_3-CH-CH-CH_3$
         $\ \ \ \ \ \ \ \ |\ \ \ \ |$
         $\ \ \ \ \ \ OH\ \ OH$

   c) $CH_3CH_2CH_2CH_2OH$

   d) $CH_3-C=CHCH_3$
         $\ \ \ \ \ \ \ |$
         $\ \ \ \ \ OH$

11. A hydrocarbon which contains a carbon-carbon triple bond is called an
   a) alkane.
   b) alkene.
   c) alkyne.
   d) aromatic hydrocarbon.

12. The general formula of an alkyne is
   a) C_nH_{2n-4}.
   b) C_nH_{2n-2}.
   c) C_nH_{2n}.
   d) C_nH_{2n+2}.

13. When excess Br_2 reacts with 2-butyne, the product is

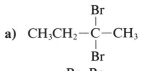

a)
$$CH_3CH_2-\overset{\overset{\displaystyle Br}{|}}{\underset{\underset{\displaystyle Br}{|}}{C}}-CH_3$$

b)
$$CH_3\underset{\underset{\displaystyle Br}{|}}{CH}-\underset{\underset{\displaystyle Br}{|}}{CH}CH_3$$

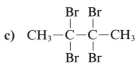

c)
$$CH_3-\overset{\overset{\displaystyle Br}{|}}{\underset{\underset{\displaystyle Br}{|}}{C}}-\overset{\overset{\displaystyle Br}{|}}{\underset{\underset{\displaystyle Br}{|}}{C}}-CH_3$$

d)
$$CH_3-\underset{\underset{\displaystyle Br}{|}}{C}=\underset{\underset{\displaystyle Br}{|}}{C}-CH_3$$

14. Which of the following does not contain a carbonyl group?
  a) a carboxylic acid   **b)** an alcohol   **c)** an ester   **d)** an aldehyde

15. The compound which is an aldehyde is

a) $CH_3\overset{\overset{\displaystyle O}{||}}{C}-OH$ b) $CH_3\overset{\overset{\displaystyle O}{||}}{C}-H$ c) $CH_3\overset{\overset{\displaystyle O}{||}}{C}CH_3$ d) $CH_3\overset{\overset{\displaystyle OH}{|}}{C}HCH_3$

16. The compound $CH_3-\overset{\overset{\displaystyle O}{||}}{C}-CH_3$ is
  a) an alcohol.   **b)** a ketone.   **c)** an ester.   **d)** an aldehyde.

17. The $-\overset{\overset{\displaystyle O}{||}}{C}-OH$ group is called
  a) an aldehyde group.   **b)** a carbonyl group.
  c) a carboxyl group.   **d)** a hydroxyl group.

18. An alcohol reacts with a carboxylic acid to yield
  a) an ether and water.   **b)** a ketone and a new alcohol.
  c) an aldehyde and a new acid.   **d)** an ester and water.

19. Which of the following is an ester?

a) $CH_3\overset{\overset{\displaystyle O}{||}}{C}-H$ b) $CH_3O\overset{\overset{\displaystyle O}{||}}{C}CH_3$ c) $CH_3\overset{\overset{\displaystyle O}{||}}{C}CH_3$ d) $CH_3\overset{\overset{\displaystyle O}{||}}{C}-OH$

20. The reaction of 2-propanol (isopropyl alcohol) with acetic acid will yield
  a) isopropyl acetate.   **b)** acetyl isopropionate.
  c) 2-propanal.   **d)** diisopropyl ether.

Review Your Understanding

1. What is organic chemistry?
2. Table salt is an example of an inorganic compound, while table sugar is an example of an organic compound. How do their properties differ?
3. Why are there so many organic compounds?
4. Describe the experiment which indicated that a "vital force" is not necessary to produce organic compounds.
5. What is the molecular formula of an alkane that has 16 carbon atoms?
6. Silicon is just below carbon in the periodic table. Predict the molecular shape of silane, SiH_4.
7. Why are the alkanes considered to be a homologous series?

8. Would it be possible to have a homologous series of hydrocarbons in which each member differs from the preceding member by one carbon and three hydrogen atoms? Give reasons for your answer.

9. Draw the structural formulas of the isomers of hexane.

10. Draw all the isomers having the formula $C_4H_8Br_2$.

11. Give IUPAC names for all the isomers of $C_4H_8Br_2$.

12. Give IUPAC names for the following:

a) $CH_3CH_2CH_2CH_2CH_2CH_2CH_2CH_2CH_3$

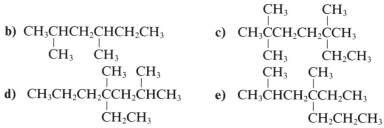

13. Draw structural formulas for each of the following compounds:
a) 4-bromo-3-methylheptane;   b) 2,4,5-trimethylheptane;
c) 3-iodo-2,2,4-trimethylpentane;   d) 3-chloro-2,3,4-trimethylhexane;
e) 4-bromo-2,2-dimethylpentane.

14. Arrange the following compounds in order of increasing boiling point: pentane, 2-methylpentane, 2,2-dimethylpentane, ethane.

15. Write the equation for the combustion of butane.

16. Write the equation for the substitution reaction involving one molecule each of ethane and bromine.

17. Why are the alkenes said to be unsaturated hydrocarbons?

18. Distinguish between structural, empirical, and molecular formulas for butene.

19. Draw the structural formulas for all the isomers of C_4H_8.

20. Give the IUPAC names for the aliphatic isomers of C_4H_8.

21. Draw the structural formulas for all the isomers having the formula C_5H_{10}.

22. Give the IUPAC names for the aliphatic isomers of C_5H_{10}.

23. Draw the structural formulas for all four isomers having the formula C_3H_5Cl.

24. Give the IUPAC names for the aliphatic isomers of C_3H_5Cl.

25. Complete each of the following equations by drawing the structural formulas of the products.

a) propene + H_2 $\xrightarrow{\text{Pt}}$

b) [cyclohexene structure]CH_3 + Cl_2 $\longrightarrow$

c) cyclohexene + H_2O $\xrightarrow{H_2SO_4}$

26. Although both alkanes and alkenes each contain only carbon and hydrogen, why are the alkenes more reactive than the alkanes?

27. Write the equation for the addition reaction involving propene and bromine.

28. What is the molecular formula of an alkyne that has 20 carbon atoms?

29. Write structural formulas for each of the following compounds:
- **a)** 2,3-dimethyl-2-butene;
- **b)** 2,4,4-trimethyl-1-pentene;
- **c)** 2,3-dibromopropene;
- **d)** 4-octene;
- **e)** 1,5-hexadiene;
- **f)** 2-bromo-2-butene;
- **g)** 1,1-dichloro-1,2-propadiene;
- **h)** 2,4-pentadiene.

30. Write the equation for the reaction of one molecule of propyne and **a)** one molecule of bromine; **b)** 2 molecules of bromine. Name the products in each case.

31. Provide an acceptable name for each of the following structures:

a) $CF_2{=}CCl_2$ **b)** $CH_3CH_2-C{\equiv}C-CH_3$ **c)** $CH_3-\underset{\underset{\displaystyle CH_2}{\|}}{C}-CH_3$

d) $CH_3CH{=}\underset{\underset{\displaystyle Br}{|}}{C}-CH_2Br$ **e)** $CH_3\underset{\underset{\displaystyle CH_3}{|}}{C}{=}CHCH_3$ **f)** $CH_3\underset{\underset{\displaystyle CH_3}{|}}{CH}-C{\equiv}CH$

32. What does the dotted circle in ⬡ represent?

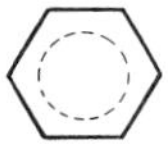

33. Why is ⬡ a better representation of the benzene molecule than ⬡ ?

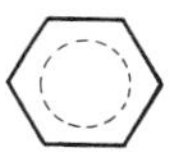

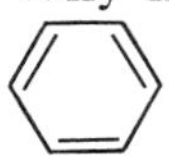

34. Draw the structural formulas of all isomers made by substituting one hydrogen atom in ⬡ (with Cl, Cl) with a chlorine atom.

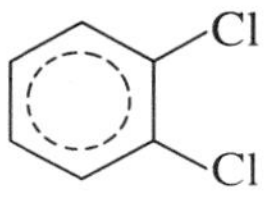

35. Write the structures of the following compounds: **a)** tetrapropylmethane; **b)** trimethylmethanol; **c)** dibutylmethane; **d)** butylmethylmethane; **e)** triethylmethanol. What are the correct IUPAC names for these compounds?

36. Write the structures of the following compounds: **a)** 2,2-dimethylpropane; **b)** 3-ethyl-2,3,5,5-tetramethylheptane; **c)** 2,2,5,5-tetramethylhexane; **d)** hexabromoethane; **e)** 1,2,3-propanetriol. What is the common name of the last compound?

37. Write IUPAC names for the following compounds:

a) $CH_3-\underset{\underset{\displaystyle CH_3}{|}}{\overset{\overset{\displaystyle CH_3}{|}}{C}}-OH;$ **b)** $CH_3CH_2CH_2CH_2\underset{\underset{\displaystyle CH_3CH_2CH_2CH_2}{|}}{\overset{\overset{\displaystyle CH_3CH_2CH_2CH_2}{|}}{C}}-OH;$

c) $CH_3-\underset{\underset{\displaystyle CH_3}{|}}{\overset{\overset{\displaystyle CH_3}{|}}{C}}CH_2OH;$ **d)** $CH_3\underset{\underset{\displaystyle CH_3}{|}}{CH}CH_2CH_2CH_2Cl$

38. Name the following compounds by the IUPAC system:

a) $CH_3CHCH_2CH_2OH$;
   $\overset{|}{CH_3}$

b) $CH_3CH-CHCH_2-\overset{\overset{\displaystyle CH_3}{|}}{\underset{\underset{\displaystyle CH_3}{|}}{C}}-CH_3$;
   $\overset{|}{OH}$ $\overset{|}{Br}$

c) $CH_3-\overset{\overset{\displaystyle Cl}{|}}{\underset{\underset{\displaystyle Cl}{|}}{C}}-CH_2-\overset{\overset{\displaystyle CH_3}{|}}{\underset{\underset{\displaystyle OH}{|}}{C}}-CH_3$;

d) $CH_3CH_2CH_2CH_2\overset{}{\underset{\underset{\displaystyle CH_2Br}{|}}{CH}}-OH$

39. What is denatured alcohol? Why is it denatured?

40. Write the structural formula for ethyl propyl ether.

41. Draw the structural formula for an ether having a total of two carbon atoms.

42. Draw the three structural formulas for the compounds having the molecular formula C_3H_8O.

43. Write structures that correspond to the following names: a) cyclohexyl methyl ether; **b)** 2-butanol; **c)** 2-methyl-2-propanol; **d)** 1,3-propanediol; **e)** 4,4-dimethyl-3-heptanol; **f)** butyl ethyl ether.

44. What functional group is found in both aldehydes and ketones?

45. Write the structures of the organic molecules that correspond to the following names: **a)** 2-bromopropanal; **b)** heptanal; **c)** 2-methyl-3-pentanone; **d)** 1,1,1-trichloro-3-pentanone.

46. Write the structural formula of a carboxylic acid which has four carbon atoms.

47. How is dynamite related to nitroglycerin?

48. Draw the structural formula for any ester having a total of five carbon atoms. Name the ester.

49. Write the equation for the reaction of 1-propanol with acetic acid. This type of reaction is called an esterification.

50. Write the structural formulas of the alcohol and the carboxylic acid from which the following ester can be made. Name the ester.

$$H-\overset{\overset{\displaystyle H}{|}}{\underset{\underset{\displaystyle H}{|}}{C}}-\overset{\overset{\displaystyle H}{|}}{\underset{\underset{\displaystyle H}{|}}{C}}-\overset{\overset{\displaystyle H}{|}}{\underset{\underset{\displaystyle H}{|}}{C}}-\overset{\overset{\displaystyle O}{||}}{C}-O-\overset{\overset{\displaystyle H}{|}}{\underset{\underset{\displaystyle H}{|}}{C}}-\overset{\overset{\displaystyle H}{|}}{\underset{\underset{\displaystyle H}{|}}{C}}-H$$

51. Write IUPAC names for the following compounds:

a) $CF_3\overset{\overset{\displaystyle O}{||}}{C}CF_3$;

b) $CH_3\overset{}{\underset{\underset{\displaystyle CH_2CH_3}{|}}{CH}}CH_2\overset{\overset{\displaystyle O}{||}}{C}-OH$;

c) $CH_3-\overset{\overset{\displaystyle CH_3}{|}}{\underset{\underset{\displaystyle CH_3}{|}}{C}}-\overset{\overset{\displaystyle O}{||}}{C}-H$;

d) $CH_3CH_2CH_2CH_2O-\overset{\overset{\displaystyle O}{||}}{C}-CH_3$.

52. Tetrahydrocannabinol (THC), shown below, is the major active component in marijuana. What functional groups are present in THC? Predict the product that would be obtained by treating THC with **a)** bromine; **b)** acetic acid containing a trace of sulfuric acid.

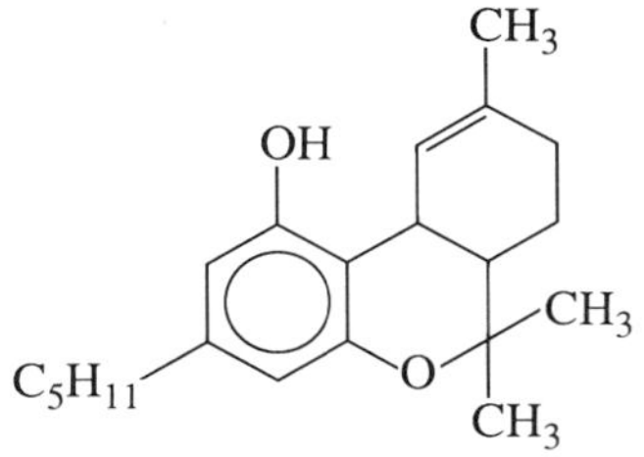

THC

53. The structure of cholesterol is shown in Figure 15-4. What functional groups are present in cholesterol? Write the structural formula of the product that would be obtained by treating cholesterol with **a)** bromine; **b)** acetic acid containing a trace of sulfuric acid.

54. Match the compounds in Column A with the appropriate type of compound listed in Column B.

A	B
$\begin{array}{c} \quad\ O \\ \quad\ \| \\ \textbf{1. } CH_3C-H \end{array}$	**a)** alkane
2. CH_3CH_2Br	**b)** alkene
3. CH_3CH_2OH	**c)** alkyne
4. $CH_3C\equiv CCH_3$	**d)** aromatic hydrocarbon
5. $CH_3CH_2CH_3$	**e)** alcohol
6. 	**f)** ether
7. $CH_3CH=CHCH_3$	**g)** organic halide
8. $CH_3CH_2NH_2$	**h)** aldehyde
$\begin{array}{c} \quad\ O \\ \quad\ \| \\ \textbf{9. } CH_3C-NH_2 \end{array}$	**i)** ketone
10. $CH_3CH_2OCH_2CH_3$	**j)** carboxylic acid
$\begin{array}{c} \quad\ O \\ \quad\ \| \\ \textbf{11. } CH_3C-CH_3 \end{array}$	**k)** amide
$\begin{array}{c} \quad\ O \\ \quad\ \| \\ \textbf{12. } CH_3C-OH \end{array}$	**l)** amine

55. Name the group of compounds (for example, alkane, alkyne, ketone, etc.) to which each of the following belongs:

a)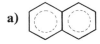

b) $CH_2{=}C{=}CH_2$

c) $CH_3CH_2CH_2OH$

d) $CH_3-C{\equiv}C-CH_3$

e) $C_{16}H_{30}$

f) $CH_3CH_2-O-CH_2CH_3$

g)
$$CH_3CH_2\overset{\displaystyle O}{\overset{\|}{C}}-H$$

h)
$$CH_3\overset{\displaystyle O}{\overset{\|}{C}}CH_3$$

i)  ⬠

j)
$$H-\overset{\displaystyle O}{\overset{\|}{C}}-OH$$

k) $CH_3CH_2CH_2CH_3$

l) $CH_2{=}CH-CH_2-CH_3$

m)
$$CH_3CH_2\overset{\displaystyle O}{\overset{\|}{C}}-OCH_2CH_3$$

n) C_2H_5-S-H

o) $CHCl_3$

56. Why are plastics so useful?

57. Why are free radicals so reactive?

58. Name ten items found around a household that are made of plastic.

59. Show how ethylene glycol reacts with adipic acid to form a polyester.

60. Give three examples of natural polymers and three examples of synthetic polymers.

61. How do condensation polymers differ from addition polymers? Give an example of each type.

62. What is the difference between a thermoplastic and a thermosetting polymer? Give two examples of each type.

63. Complete each of the following equations by drawing the structural formulas of the products. If no reaction occurs, write NR.

a) $CH_3CH_3 + Br_2 \longrightarrow$

b) ethane + Na $\longrightarrow$

c) $C_3H_8 + O_2 \longrightarrow$

d)  ⬡ $+ Cl_2 \longrightarrow$

Apply Your Understanding

1. There are four isomeric butenes, even though the structures of only three of them are given in Section 15.8. What is the structure of the missing isomer?

2. A solution of bromine in CCl_4 is brown. Substitution reactions are slow at room temperature, while addition reactions are almost instantaneous. In each case, the products of the bromine-containing reaction are colourless. How would you use this information to decide whether an unknown gas was butane or 1-butene? Explain your reasoning.

3. Occasionally, you can identify an unknown compound simply by counting the number of isomeric products it forms in different reactions. For example, in the chlorination of propane, four different isomeric products with the formula $C_3H_6Cl_2$ were isolated and designated by the letters $A,B,C,$ and D. Each was separated and reacted with more chlorine to give one or more trichloropropanes, $C_3H_5Cl_3$. A gave one trichloropropane, B gave two, and C and D each gave three trichloro compounds. One of the trichloro compounds from C was identical to the trichloro compound from A. Deduce structures for $A,B,C,$ and D.

4. Draw the correct structural formula for each of the following alkanes:

Alkane	Molecular mass (u)	Monochloro substitution products
A	16	1
B	56	1
C	58	2
D	72	1
E	72	3

5. The general formula for an alkane is C_nH_{2n+2}; for an alkene it is C_nH_{2n}; and for an alkyne it is C_nH_{2n-2}. Deduce the general formulas for the following families of hydrocarbons: a) cycloalkanes; b) alkadienes; c) bicycloalkanes (containing two rings). Explain how your answers lead to the rule, "To determine the total number of rings plus double bonds in a hydrocarbon, subtract the number of hydrogen atoms in a molecule of the compound from the number of hydrogen atoms in a molecule of the alkane with the same number of carbon atoms. Then divide by two." Apply the rule to naphthalene, $C_{10}H_8$. What is the total number of rings plus double bonds in naphthalene?

6. Two compounds, X and Y, react in the presence of sulfuric acid to form water and a compound Z, which has a strong rum-like odour. $X,Y,$ and Z contain only carbon, hydrogen, and oxygen. The elemental analyses are as follows: Compound X—52.1% C and 13.1% H; Compound Y—26.1% C and 4.38% H; Compound Z—48.6% C and 8.17% H. Identify the three compounds and write the chemical equation for the reaction.

7. Would you expect the structural isomers dimethyl ether and ethanol to differ significantly in their boiling points? Explain your answer.

8. A 230 mg sample of an organic compound containing only carbon, hydrogen, and oxygen was burned in excess oxygen to give 461 mg of CO_2 and 188 mg of water. In a separate experiment the molar mass of the

compound was found to be 44.0 g. Suggest a suitable structure for the compound. What is the functional group in the compound?

9. An unknown gas which contained 85.6% C and 14.4% H added chlorine to form a product which contained 31.9% C, 5.35% H, and 62.8% Cl. Using structural formulas, write a balanced equation for the reaction.

Investigations

1. Write a report on the laboratory experiment that produced the first organic compound from inorganic starting materials.
2. Write a report on the development of organic nomenclature.
3. Debate the following proposition: "Given that drug abuse can lead to employee health problems, accidents, and absenteeism, all workers should undergo mandatory drug testing."
4. In consultation with your teacher, obtain a laboratory procedure for the extraction of a natural product such as caffeine from tea, or trimyristin from nutmeg. Perform this extraction in the laboratory.

16

The Chemical Industry, Technology, and Society

Chapter Contents

Part One: **The Relation of Science and Technology to Industry and Society**
  16.1 Science, Technology, and Industry
  16.2 Technology and Society

Part Two: **The Chemical Industry**
  16.3 Features of the Chemical Industry
  16.4 Sources of Raw Materials for the Chemical Industry
  16.5 Production of Important Industrial Chemicals
  16.6 The Chemical Industry in Canada
  16.7 Factors that Determine the Location of an Industrial Plant
  16.8 Household Chemicals

Part Three: **Canadian Technology**
  16.9 Canadian Contributions to Technology
  16.10 Citizens of a Technologically-Based Society

Pure science helps to expand our knowledge and understanding of the universe. It is an essential basis for technology, which is the application of scientific discoveries to the solution of practical problems. Technology, therefore, has enhanced our comfort, convenience, health, and standard of living.

Many industries have been created as a result of technology. The resulting rise in the standard of living has caused society to expect a continually expanding use of technology. However, the benefits of industrialization are often accompanied by social and environmental problems.

After studying this chapter, you will be able to make informed decisions about some of the complex science-related issues that affect society and to consider your responsibilities as a citizen in a technological society. You will be able to discuss questions such as: What factors determine the location of an

industrial plant? Why was the discovery of the process of making ammonia so important in prolonging World War I? How are computers aiding the pharmaceutical industry? Why is sulfuric acid considered to be the most important chemical? Which chemicals affect our daily lives?

Figure 16-1
A mainframe university computer. Computer technology is playing an important role in chemical research.

Key Objectives

When you have finished studying this chapter, you should be able to:
1. explain the relation of science and technology to both industry and society.
2. describe the features of the chemical industry.
3. explain the importance of chemical industries.
4. list some of the most important chemicals produced by industry and state their uses.
5. list the factors that determine the location of an industrial plant.
6. list common household chemicals and state their uses.
7. describe the contributions of Canadians to technological development.
8. discuss the responsibilities of citizens in a technological society.

Part One: The Relation of Science and Technology to Industry and Society

16.1 Science, Technology, and Industry

Through pure scientific research, **science** attempts to satisfy curiosity and to provide us with a better understanding of our universe. It may, therefore, try to answer questions that may not have direct practical applications. Scientists involved in pure research may ask questions such as: What type of bonding is involved in a particular molecule? How could we synthesize a particular new compound? What would be the physical and chemical properties of such a compound? Would this compound be similar in physical and chemical properties to other known compounds?

The answers to such questions contribute to the vast body of scientific knowledge. In the process of answering these questions, discoveries may be made which could have practical applications. For instance, scientists asking questions about bonding would gain insights about the forces holding atoms together. Their findings could lead to cheaper and simpler methods of chemical production. Perhaps, in the synthesis of a new compound, scientists might learn how to prepare other useful compounds. For example, a new compound may have properties that would make it useful in the pharmaceutical industry. When we mention practical applications, we are referring to technology. **Technology** is the application of scientific principles in order to solve a problem or to create a product that will enhance our health, comfort, convenience, or leisure time.

Figure 16-2
Laser spectroscopy in which information is gathered on chemical reactions. Laser beams have important applications in fields ranging from industry to medicine.

Products such as lasers, television, Teflon®, nylon, and computers are the result of technology. Although pure scientific research may lead eventually to applications, we must realize that most pure scientific research has no immediate application.

Most industries are the result of technological breakthroughs. Further technological advances keep them in operation. For example, the development of the internal combustion engine led eventually to the creation of the automobile industry. A number of further technological breakthroughs have improved the production and efficiency of automobiles. The discovery of the transistor and of integrated circuits led to the development of computers. Computers, in turn, are finding a use in the design of automobiles, in controlling the robots that build the cars, and as important components of the vehicles themselves. The motion picture industry owes its beginning to the production of celluloid in 1870 which led to the invention of celluloid film. The chemical industry owes its development in part to the application of electrolysis to produce important chemicals such as sodium hydroxide and chlorine. More recently, computers have allowed the pharmaceutical industry to save much time and money in the development of new drugs. A large, complex molecule can be created on the screen of a computer monitor so as to appear three dimensional.

Biography

JACQUELINE BARTON received her Ph.D. in 1979 from Columbia University, where she is now a professor of inorganic and biophysical chemistry. Her research has focussed on clarifying the precise locations of specific genes (complex chemical units) which make up DNA.

The strands of DNA consist of sequences of genes strung together like the words of an enormously long sentence. Researchers have looked for years to identify the punctuation marks that signify the beginning and the end of an individual gene. Dr. Barton has developed a completely new approach to locating individual genes. She uses metals such as rhodium, ruthenium, copper, and cobalt to build families of inorganic compounds which bind to specific locations on a strand of DNA. Dr. Barton has discovered that the locations where her compounds attach to the DNA appear to serve as punctuation marks for the ends of genes. Her work therefore helps biologists to pinpoint genes more precisely.

Dr. Barton's achievements have been recognized with several major honours. In 1985 she became the first woman to receive the National Science Foundation's Alan T. Waterman Award, given to the outstanding young scientist in the United States. In 1987 she received the American Chemical Society's Eli Lilly Award in biological chemistry.

Figure 16-3
Jacqueline Barton—Locating genes in DNA.

Canadarm—Canadian technology in space.

The computer-generated molecule can be rotated and its structure viewed from many different angles. The active sites of this molecule, as well as its ability to bind to specific sites on biological molecules in the body, can be studied by simulating on the screen its interaction with these molecules. This type of computer simulation can help a company decide whether to synthesize a compound for further testing.

Thus, technology may lead to the start of a new industry, may improve the quality or efficiency of industrial products, and may result in more economical production methods. Advances in technology can even aid pure scientific research. This happens, for example, when sophisticated measuring instruments are developed.

APPLICATION

Voyager: High-Tech Aircraft

On December 14, 1986, the experimental aircraft Voyager left California and travelled non-stop around the world in nine days. The remarkable aspect of this trip is that the aircraft never stopped for refuelling.

In order to achieve this feat, the pilots, Dick Rutan and Jeana Yeager, had to use innovative technology. The plane's self-supporting skin was made of a honeycombed paper core placed between layers of an advanced composite material consisting of carbon-fibre tape impregnated with an epoxy resin. This composite material is stronger than steel, yet lighter than aluminum. Unlike metal, it does not expand or contract from exposure to heat or cold. Since the plane's composite skin was smooth, with no joints or rivets, air was able to flow more smoothly over the wings. Drag effects were, therefore, reduced.

The aircraft's two engines were separated from the fuel tanks (which contained 4500 L of fuel) by firewalls made of advanced lightweight ceramic materials which could withstand temperatures of 1100°C. In fact, most of the airplane was made of advanced materials. Except for the two engines and a few nuts and bolts, no traditional metals were used.

Voyager's wingspan was 34 m, longer than that of a Boeing 727, and its mass was a mere 843 kg. The aircraft travelled at an average speed of 185 km/h. When it landed, only 40 L of fuel were left.

16.2 Technology and Society

There is a close relationship between technology and society. Sometimes technology evolves because of a direct need in society. Technology is, therefore, influenced by society. In other cases a technological development may lead to new products. Technology, therefore, influences society.

Nitrogen is an essential plant nutrient making up about 78% of the atmosphere. However, before it can be useful to plants, it must be converted to a soluble compound. The process of converting atmospheric nitrogen to nitrogen-containing compounds is called nitrogen fixation. When the world population was small and a smaller food supply was required, certain forms of bacteria were able to fix enough nitrogen to satisfy world needs. In the latter part of the nineteenth century, as the world population began to increase greatly, it became apparent that there would eventually be a shortage of food. This shortage would have been the result of the limited amount of nitrogen fixed by bacteria. It would not have been sufficient to support the production of enough food. Chemists sought a solution to the problem. Although there was much nitrogen available in the atmosphere, it was difficult to convert it to soluble nitrogen-containing compounds, because of the strong triple bond holding the nitrogen molecule together. This problem was solved in 1908 when a German chemist named Fritz Haber developed a process in which elemental hydrogen and nitrogen are combined at high temperatures and pressures in the presence of a catalyst to yield ammonia.

$$N_2(g) + 3H_2(g) \rightleftharpoons 2NH_3(g)$$

Haber was awarded the Nobel Prize in chemistry in 1918 for developing the process named after him. Since World War I was imminent, Germany was interested in ammonia not as a fertilizer, but as a source of explosives such as ammonium nitrate. Germany normally relied on Chile for potassium nitrate to make explosives but this source was cut off by the British forces at the start of the war. However, Germany was still able to produce explosives by using the Haber process to make ammonia which was then converted to ammonium nitrate. The availability of ammonia certainly helped to prolong the war. The application of the Haber process met a military need of the German government. Later, through the use of ammonia and ammonium compounds, it would meet the agricultural needs of the world.

Another technological breakthrough that served the purposes of both the military and society at large was the development of synthetic rubber. During World War II, it appeared that the Allies would face a shortage of natural rubber which they needed for airplane and automobile tires. Within a short time, the program set up to develop synthetic rubber was successful.

Today, society requires alternative fuels, more efficient automobiles, improved fertilizers, cures for diseases through new drugs, and a cleaner environment. Technology will attempt to meet the challenge of these demands.

New developments in technology do not always occur because a serious need exists in society. Sometimes technology merely enhances our convenience or comfort with such products as electric can-openers, microwave ovens, electric heaters, and nylon carpets. In the future, robots may do household chores. Technology also affects clothing styles and the way we spend our leisure time. Home computers, televisions, radios, stereos, and sports equipment have changed our lifestyles. Changes ranging from television to the technology used to reach the moon have influenced the way we view our world and our place in the universe.

APPLICATION

Canadian Biotechnology

Canadian researchers are at the forefront of developments in biotechnology, which is the application of technology to living organisms. The Canadian government has formed a special committee to advise it on ways to foster commercial use of these developments. This committee has defined areas of commercial importance as human and animal health care, development of new plant strains, fixation of nitrogen, new uses of cellulose, waste treatment, mining, and leaching of minerals. For basic research, the committee identified priorities including protein engineering, bioprocess engineering, and improving the genetics of plants, animals, and cells.

Canadian farmers will be the major beneficiaries of new developments in biotechnology. Future advances will be based largely on the genetic engineering of plants, animals, and microorganisms. Food production will be enhanced by the development of plants with greater nutritional value, earlier maturation, disease resistance, and stress tolerance. Development of biocontrol agents and biofertilizers will reduce adverse environmental impact and production costs. Crops will be designed to meet specific food product or processing requirements.

Canada has a well-established forest-product industry. It produces about $25 billion worth of products each year and contributes substantially to the country's foreign exchange earnings. But the industry is facing stiffer foreign competition and higher trade barriers. As a result, the techniques of biotechnology, such as tissue culture, development of fast-growing trees, and use of bioreactors for pulp and paper waste treatment, will be of immense value. The industry could also adopt sophisticated forest management strategies such as biological pest control. Trials of new pulp treatment methods using microorganisms and enzymes to improve the whiteness and strength of paper are already under way.

The importance of our biotechnology industry is reflected in the fact that the Canadian health-care market is the eighth largest in the world. Spending on drugs is expected to increase to about $3.3 billion by 1990. In addition, a growing market for medical equipment accounts for another billion dollars in annual sales. Several Canadian companies have become world leaders in the production of vaccines. Some companies are now selling technology-based diagnostic systems. Canadian biotechnology has applications in other medical fields. Preliminary trials have suggested that the drug azidothymidine (AZT) may be effective in the treatment of AIDS. The price of thymidine, its principal ingredient, has already doubled, and if AZT is indeed effective, current suppliers will be unable to meet the expected demand. As a result, an Alberta company has geared up to produce thymidine in quantity, at a lower cost. It has already developed a new synthetic process that uses less expensive starting materials, and is testing the process in a pilot plant. With the new process in large-scale production, this company will increase the world supply of thymidine 10 to 20 times.

Part One Review Questions

1. State whether each of the following is mainly pure science or technology. Explain your answers.
   a) Developing catalytic converters to reduce pollution from automobiles.
   b) Studying the effect of acidity on marine life.
   c) Sending a space probe to study our solar system.
   d) Studying the effect of concentrations of reactants on the speed of a chemical reaction.
   e) Experimenting with ways to reduce the emission of sulfur dioxide pollution from smokestacks.
2. What is the main difference in the type of scientific research normally carried out by universities and research carried out by industrial companies?
3. How does the work of a chemist differ from that of a chemical engineer?
4. Why is the development of new and improved technologies so important to industry?
5. What technologies are evident in the modern automobile?

Part Two: The Chemical Industry

16.3 Features of the Chemical Industry

The chemical industry produces many chemical substances. Chemical substances are elements or compounds in a reasonably pure state. Sodium chloride can be found in fairly pure deposits, so it is not produced by the chemical industry. However, sodium chloride is a valuable raw material for the production of other chemicals. Petroleum is not produced by the chemical industry either, but the components of petroleum also serve as raw materials. When sodium hydroxide, chlorine, and hydrogen are produced from sodium chloride and water, or when the components of petroleum are used to produce such chemicals as ethene (ethylene) and 1,3-butadiene, the chemical industry is involved. One of the characteristics of the chemical industry is that it does not normally manufacture products that go directly into the hands of the consumer. Another characteristic of the chemical industry is that large amounts of money are spent on research.

A small number of fundamental chemicals serve as the backbone of the chemical industry. These are referred to as **heavy chemicals** because they are produced in large amounts. They are valuable because they are used to produce many other chemicals which ultimately lead to useful finished products. Another branch of the chemical industry produces **fine chemicals**, which are produced in smaller amounts. Fine chemicals, such as dyes and pharmaceuticals, are more expensive than heavy chemicals. The chemical industry uses a great deal of automated equipment, so production can be handled by a few

Figure 16-4
The chemical industry is a vital part of Canada's economy.

workers. More money is invested in plants and equipment per worker than in most other industries. The industry enjoys manufacturing flexibility because some chemicals can be prepared by a number of different methods. Economic conditions determined by the availability and price of raw materials dictate which manufacturing path is taken.

16.4 Sources of Raw Materials for the Chemical Industry

Raw materials are needed to produce both heavy chemicals and fine chemicals. The chief sources of these raw materials are petroleum, coal, natural gas, and mineral deposits. The atmosphere supplies nitrogen and oxygen, which are also valuable raw materials.

Petroleum

The chemical industry produces many organic compounds. One of the chief sources of raw materials for some of them is petroleum (Fig. 16-5). Chemicals which are produced from components of petroleum are called petrochemicals.

Figure 16-5
Drilling for oil on Issungnak, an artificial island created for this purpose in the Beaufort Sea.

Crude oil, as it comes from the ground, consists of a highly complex mixture of compounds. Most of these are saturated hydrocarbons, but there are a number of aromatic hydrocarbons present as well. At the oil refinery, crude oil is separated into a number of fractions by distillation. The division into fractions is largely determined on the basis of boiling point and the number of carbon atoms present in the hydrocarbon molecules. The gas fraction is made up of those compounds that boil at less than 30°C. It is really natural gas dissolved in the liquid petroleum. Natural gas is mainly methane which is mixed with some ethane and some propane. The naphtha or petroleum-ether

Table 16-1 Major Uses of Petroleum Fractions

Number of Carbon Atoms	Major Uses
$C_6 - C_9$	gasoline
$C_{10} - C_{18}$	kerosene jet fuel diesel fuel heating oil
$C_{18} - C_{21}$	lubricating oil
$C_{22} - C_{40}$	asphalt paraffin wax

fraction consists of low boiling (30-65°C) hydrocarbons which have five or six carbon atoms per molecule. It is used primarily as a solvent or cleaning material. Gasoline boils from 60 to 150°C and contains six to nine carbon atoms per molecule. Kerosene boils from 170 to 230°C and is used as a jet fuel and as a light diesel fuel. The molecules in the kerosene fraction have 10 to 13 carbon atoms. Light oils have 14 to 18 carbon atoms per molecule and are used as a diesel fuel and for oil-fired furnaces. The heavy oils (18 to 21 carbon atoms per molecule) are used for lubrication. The C_{22} to C_{40} fraction is paraffin wax. After the distillation process, some nondistillable residues remain. These are the asphalts which are used in road surfacing and roofing materials. Table 16-1 summarizes the major uses of the petroleum fractions.

Cyclic compounds in petroleum, such as cyclohexane, can be converted into benzene (Section 15.16) with the aid of a catalyst at high temperature and pressure:

cyclohexane   *benzene*

Benzene is manufactured in this manner because not enough of it is produced from coal to meet the demands of industries that use benzene to manufacture plastics, detergents, drugs, dyes, insecticides, and synthetic fibres.

Petroleum products are also used to produce cigarette lighter fluid, enamel, weed-killers and fertilizers, synthetic rubber, photographic film, candles, waxed paper, polish, floor coverings, protective paints, and other goods.

Petroleum often contains undesirable quantities of organic sulfur and nitrogen compounds. One way of removing them is to treat the petroleum with high-pressure hydrogen, which converts the sulfur and nitrogen to H_2S and NH_3. These two gases are removed from the petroleum by dissolving them in a sodium hydroxide solution.

The automobile gets its power from the controlled ignition of a mixture of air and vaporized fuel. The gasoline fraction is the most useful fuel fraction since the naphtha fraction would evaporate in the gasoline tank, and the kerosene fraction is not volatile enough to evaporate sufficiently for efficient ignition. Only a small fraction of the original crude oil is actually useful in automobile engines without further treatment.

Large molecules in the lubricating oil range ($C_{18}-C_{21}$) can be converted into more useful molecules in the gasoline range (C_6-C_9) by the **cracking** process. This process occurs under the influence of heat and appropriate catalysts. A typical reaction in the process is

hexadecane   *octane*   *1-octene*

In this way smaller molecules in the gasoline range are made from larger, less useful molecules. The cracking process is carried out in large "cat crackers" (catalytic crackers, Fig. 16-6).

Branched-chain hydrocarbons burn more evenly and are less susceptible to premature ignition (engine knock) in an engine cylinder than straight-chain hydrocarbons. The knock is due to the explosion taking place in the cylinder before the piston has reached its position of minimum displacement.

A gasoline is rated for performance in a test engine where its performance is compared with that of a known mixture of two pure hydrocarbons. Isooctane (2,2,4-trimethylpentane) has a branched structure and knocks very little. Pure isooctane has an octane rating of 100. Heptane has a straight-chain structure and knocks very badly. It has an octane rating of 0. A gasoline is rated at 91-octane if it behaves the same in a test engine as a mixture of 91% isooctane and 9% heptane would. Poor-quality gasoline has a low octane rating.

There are two ways of improving the octane rating of gasoline. One is to alter the structure of the molecules to increase the proportions of branched-chain hydrocarbons. During the 1930s it was discovered that a variety of catalysts such as aluminum chloride and sulfuric acid would promote the "reforming" of branched-chain molecules in the cracking process. Also, it became possible to combine molecules in the volatile C_4 to C_6 range to produce the larger gasoline range molecules.

The second way to improve the octane rating is to use additives. Tetraethyl-lead is chief among the additives. A small quantity of this compound improves the anti-knock quality of the gasoline. However, about 0.1 g of lead per litre of gasoline burned finds its way into the atmosphere. Most of this lead eventually ends up in the soil. Like most heavy metals, lead is a poison, even in trace amounts. It is true that lead compounds already occur naturally in the soil; however, any extra amount deposited there by human activity could be dangerous. An important start on pollution control has involved the removal of lead from gasoline. The octane rating of gasoline without lead has been maintained by structural modification of hydrocarbon molecules. This structural modification of hydrocarbons may be accomplished by cracking, isomerization, alkylation, and aromatization.

Isomerization is the conversion of a straight chain hydrocarbon to its branched chain isomer. For example, hexane is changed to neohexane (2,2-dimethylbutane) by the use of a platinum catalyst in the presence of hydrogen gas at high temperatures. The hydrogen gas is used to prevent unwanted side reactions:

hexane   *2,2-dimethylbutane*
                              (*neohexane*)

Alkylation is the process which combines a branched alkane with an alkene to produce a large branched alkane. Concentrated sulfuric acid may be used as a catalyst. Reaction vessels constructed of a special steel are used to prevent corrosion by the acid. The reaction must be carried out at low temperatures (2 to 7°C); heat must be removed efficiently because of the large amount of heat released during the reaction:

$$
\begin{array}{ccc}
\underset{\substack{\text{2-methylpropane}\\(\textit{isobutane})}}{\text{CH}_3\!\!-\!\!\overset{\displaystyle\text{CH}_3}{\underset{\displaystyle\text{CH}_3}{\text{C}}}\!\!-\!\!\text{H}} \; + \; \underset{\substack{\text{2-methylpropene}\\(\textit{isobutylene})}}{\text{CH}_3\!\!-\!\!\overset{\displaystyle\text{CH}_3}{\text{C}}\!\!=\!\!\text{CH}_2} & \xrightarrow[\substack{\text{low}\\\text{temperature}}]{\text{H}_2\text{SO}_4} & \underset{\substack{\text{2,2,4-trimethylpentane}\\(\textit{isooctane})}}{\text{CH}_3\!\!-\!\!\overset{\displaystyle\text{CH}_3}{\underset{\displaystyle\text{CH}_3}{\text{C}}}\!\!-\!\!\overset{\displaystyle\text{H}}{\underset{\displaystyle\text{H}}{\text{C}}}\!\!-\!\!\overset{\displaystyle\text{CH}_3}{\underset{\displaystyle\text{H}}{\text{C}}}\!\!-\!\!\text{CH}_3}
\end{array}
$$

Figure 16-6
A catalytic cracker (left foreground) at the oil refinery in Sarnia, Ontario.

Aromatization is the conversion of straight-chain alkanes to cycloalkanes, followed by the removal of hydrogen from the cycloalkane to produce an aromatic hydrocarbon. The reaction is carried out at high temperatures with a catalyst:

$$CH_3CH_2CH_2CH_2CH_2CH_3 \xrightarrow[\triangle, \text{ catalyst}]{-H_2}$$

hexane　　　　　　　　　　　**cyclohexane**　　　　　　　　**benzene**

Coal

When soft coal (Fig. 16-7) is heated to about 1200°C in the absence of air, a number of chemicals are obtained. Among these products are gases such as carbon monoxide, hydrogen, ammonia, methane, and low molecular mass hydrocarbons. Most of the coal is converted to coke, which consists of approximately 95% pure carbon. In the manufacture of iron and steel, coke is first heated to produce carbon monoxide, which is then reacted with iron ore:

$$Fe_2O_3(s) + 3CO(g) \longrightarrow 2Fe(s) + 3CO_2(g)$$

Table 16-2　**Some Chemicals Found in Coal Tar**

Chemical	Use
Aniline	dye industry
Naphthalene	dyes and agricultural chemicals
Benzene	manufacture of styrene for polystyrene; detergents, drugs, dyes
Cresols	in wood preservatives (creosote)
Toluene	component in motor fuel; also used to produce other materials such plastics
Phenol	disinfectants, plastics

Figure 16-7
This Nova Scotia coal mine supplies coal for various industrial uses. Here an armour-plated conveyor carries coal away from the face.

A small quantity of a gummy, black residue, known as coal tar, is also produced from coal. When the coal tar is further distilled and chemically treated, a number of substances are produced (Table 16-2).

Coal products are also used to produce drugs, explosives, medicines, rubber, flavourings, perfume, and roofing materials. Ammonium salts, also found in coal, are used as fertilizers.

Natural Gas

Natural gas is often found during the drilling of wells for petroleum (Fig. 16-8). Its composition ranges from approximately 50 to 94% methane. Some ethane, propane, butane, nitrogen, and helium are also present. Natural gas is used as a fuel to produce heat for homes and industries.

Natural gas is also used as a source of raw materials to produce many other chemicals, one of the most important being ammonia. Alkenes such as ethylene, propylene, and butylene, along with the alkadiene, 1,3-butadiene, can be made from natural-gas components and used to produce plastic toys, vinyl flooring, phonograph records, artificial leather, auto tires, and rubber hoses. Ethanol for industrial purposes is made from natural gas. Chlorine reacts with natural gas to produce such compounds as chloroform and carbon tetrachloride. Natural gas is also used to produce fertilizers, carbon black (finely powdered carbon), synthetic fibres, paints, insecticides, cleaning fluids, solvents, and many other chemicals.

Figure 16-8
Natural gas plant at Judy Creek, Alberta

Mineral Deposits

Deposits of table salt, sodium chloride, provide important raw materials for the chemical industry. Electrolysis of a sodium chloride solution produces hydrogen, chlorine, and sodium hydroxide. Sodium chloride is also used to produce the heavy chemicals hydrochloric acid and sodium carbonate.

Another mineral, potash, is also useful. Potash deposits are minerals that contain potassium salts, such as potassium chloride and potassium carbonate. Some of the largest potash deposits in the world are located in Saskatchewan. Potash provides chemicals for the fertilizer industry and for the production of soap and glass.

The mineral, calcium phosphate, is also found in the earth's crust. Elemental phosphorus, obtained from this mineral, is used to manufacture phosphoric acid.

Another important source of raw materials for the chemical industry is limestone. The chemical name for limestone is calcium carbonate. Limestone, when heated, produces quicklime (calcium oxide) and carbon dioxide gas:

$$CaCO_3(s) \xrightarrow{\triangle} CaO(s) + CO_2(g)$$
$$\text{limestone} \qquad \text{quicklime}$$

Quicklime has a wide variety of uses. It is used in manufacturing glass, in poultry feeds, in the treatment of organic wastes, in food processing, and in making mortar (when mixed with sand and water). Quicklime is also converted into calcium hydroxide:

$$CaO(s) + H_2O(\ell) \longrightarrow Ca(OH)_2(s)$$

Calcium hydroxide is used in the rubber industry, for making cements and plasters, as a sanitizing agent, as a disinfectant, and for removing hair from hides in the leather industry. It finds its main use in agriculture, where it is added to the soil to neutralize undesirable acids.

Sulfur is used mainly in the manufacture of sulfuric acid, the most important industrial chemical. There are a number of sources of sulfur. The largest amounts of sulfur produced in the world are obtained from solid sulfur deposits by the Frasch process (Fig. 16-9). In this process, water which has been superheated under pressure to about 165°C is forced down a pipe into the sulfur deposit. The hot water melts the sulfur, which is then forced out through another pipe to the surface by compressed air.

In Canada, the chief source of sulfur is *sour gas*, natural gas containing hydrogen sulfide. The Athabasca tar sands are another source. Canada is one of the world's largest producers of sulfur.

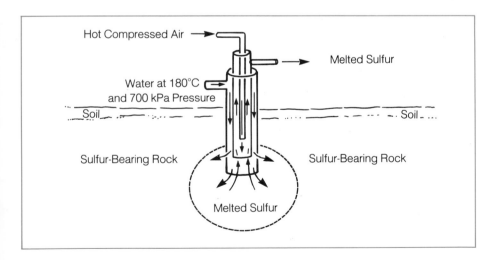

Figure 16-9
Frasch process for mining sulfur

The Atmosphere

Nitrogen, which makes up about 78% of the atmosphere, is used as a raw material in the production of ammonia. Ammonia may then be used to produce fertilizers and also nitric acid, another important chemical.

Table 16-3   Heavy Chemicals Produced by the Chemical Industry

sulfuric acid
hydrochloric acid
nitric acid
sodium hydroxide
ammonia
ammonium nitrate
ammonium phosphate
sodium carbonate
sodium chlorate
chlorine
hydrogen
benzene
toluene
xylene
methanol
ethanol
urea
ethylene (ethene)
propylene (propene)
1,3-butadiene
polyethylene resin

Sulfuric acid removes the water from sugar, leaving a charred mass of carbon.

16.5 Production of Important Industrial Chemicals

The chemical industry is important, not only because it provides employment and contributes to the economy of a nation, but also because it produces many useful products. These include fertilizers, pharmaceuticals, insectides, herbicides, food additives, explosives, adhesives, plastics, paints and dyes. This list is by no means complete, but it does show the wide variety of products of the chemical industry. Try to visualize a day in your life without them.

Some of the heavy chemicals produced by the chemical industry are listed in Table 16-3. Many more products can be produced from these heavy chemicals. Let us examine how some of these heavy chemicals are produced.

Sulfuric Acid

The most important chemical produced by an industrial society is sulfuric acid, H_2SO_4. This chemical is so important that a country's industrial prosperity and state of development is often gauged by the amount of sulfuric acid consumed yearly. Pure sulfuric acid is a colourless, odourless, oily liquid, which freezes at 10.5°C. Concentrated sulfuric acid contains 98.33% H_2SO_4 and 1.67% water. Sulfuric acid is a good dehydrating agent, removing water from organic compounds. It is a strong acid and therefore dissociates readily into ions. It reacts with most metals and neutralizes bases such as sodium hydroxide. It is very corrosive, causing severe burns as it reacts with organic compounds in the skin.

Most of the sulfuric acid manufactured in the world today is produced by the contact process. The contact process receives its name from the process in which sulfur dioxide is converted to sulfur trioxide through contact with a catalyst. The first step in the process occurs when nearly pure sulfur is burned in air to produce sulfur dioxide:

$$S_8(s) + 8\ O_2(g) \longrightarrow 8SO_2(g)$$

Sulfur dioxide is then reacted with oxygen in air at 400°C in the presence of a catalyst such as V_2O_5 (vanadium(V) oxide) to produce sulfur trioxide:

$$2SO_2(g) + O_2(g) \xrightarrow[400°C]{V_2O_5} 2SO_3(g)$$

The sulfur trioxide is cooled to 100°C and passed into a tower to be absorbed in 97% sulfuric acid, producing pyrosulfuric acid:

$$H_2SO_4(\ell) + SO_3(g) \longrightarrow H_2S_2O_7(s)$$
$$\textit{pyrosulfuric acid}$$

Sulfur trioxide is absorbed into concentrated sulfuric acid instead of water, because sulfur trioxide mixed with air does not dissolve rapidly in water. A concentration of 97% sulfuric acid is used because lower concentrations produce a mist which is difficult to remove from the waste gases leaving the plant. The addition of water to pyrosulfuric acid gives sulfuric acid:

$$H_2S_2O_7(s) + H_2O(\ell) \longrightarrow 2H_2SO_4(\ell)$$

During the process, two moles of H_2SO_4 have been produced from one mole of SO_3 and one mole of H_2SO_4. Since the sulfur dioxide used in the process must be pure (to avoid damage to the catalyst), the sulfuric acid produced by the contact process is also pure. Highly concentrated acid is produced. The production and uses of sulfuric acid are summarized in Figure 16-10.

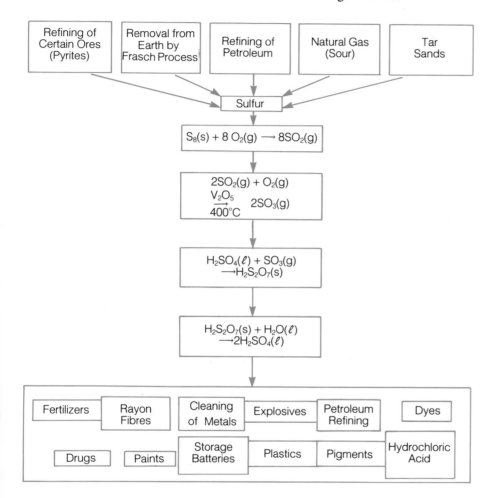

Figure 16-10
Production and uses of sulfuric acid

Ammonia

Ammonia is a colourless gas with a very pungent odour. As we have seen, it is produced by the Haber process. Hydrogen and nitrogen react to form ammonia:

$$N_2(g) + 3H_2(g) \rightleftharpoons 2NH_3(g)$$

This is a reversible reaction, and a state of chemical equilibrium is attained. In order to produce the maximum yield of ammonia, the hydrogen and nitrogen gases are reacted at a temperature of 400 to 450°C and pressures which are 200 to 600 times higher than normal atmospheric pressure. This reaction is carried out in the presence of a special catalyst which consists of a mixture of potassium oxide, iron, and aluminum oxide. The catalyst causes equilibrium to be reached faster. The ammonia is separated from the equilibrium mixture by condensing it to form a liquid, since its boiling point (−33°C) is much higher than that of hydrogen (b.p. −253°C) or nitrogen (b.p. −196°C). The unreacted hydrogen and nitrogen are recycled to produce more ammonia.

Ammonia has many uses. These include being used to produce nitric acid, rayon, nylon, fertilizers, sodium carbonate, and hydrogen cyanide. It is used as a refrigerant and as a household cleaning agent. It is the active substance in smelling salts. It is used in metallurgy, in the plastics industry as a catalyst, in the pharmaceutical industry to produce vitamins and sulfa drugs, and in the petroleum industry as a neutralizing agent. It is also used in the rubber industry to prevent the suspended particles in latex from coming out of suspension (coagulating) during shipment.

Nitric Acid

Nitric acid is a colourless liquid. It fumes in moist air and has a choking odour. Nitric acid is manufactured by the Ostwald process. Ammonia is reacted with oxygen in the presence of a platinum catalyst at about 800°C to produce nitrogen monoxide:

$$4NH_3(g) + 5\ O_2(g) \xrightarrow[800°C]{Pt} 4NO(g) + 6H_2O(g)$$

The nitrogen monoxide reacts readily with atmospheric oxygen to form nitrogen dioxide:

$$2NO(g) + O_2(g) \longrightarrow 2NO_2(g)$$

Nitrogen dioxide is cooled and reacted with water to form nitric acid:

$$3NO_2(g) + H_2O(\ell) \longrightarrow 2HNO_3(aq) + NO(g)$$

The nitric acid solution is drawn off and concentrated. The nitrogen monoxide is recycled to produce more nitrogen dioxide and nitric acid. Some of the nitric acid is neutralized with additional ammonia to produce ammonium nitrate:

$$NH_3(g) + HNO_3(aq) \longrightarrow NH_4NO_3(aq)$$

Much ammonia and nitric acid is used for this purpose. Ammonium nitrate is used as a fertilizer and explosive.

Nitric acid is also used for making other explosives such as nitroglycerin, TNT, and picric acid. In addition to the production of fertilizers and explosives, nitric acid is used in the manufacture of plastics, drugs, dyes, and pharmaceuticals. It is used in metallurgy to clean the surface of stainless steel and other metals. Engravers use nitric acid to make tracings on copper plates.

Chlorine, Hydrogen, and Sodium Hydroxide

Three important chemicals of the chemical industry may be produced by the same process, the electrolysis of a concentrated solution of sodium chloride in water (Fig. 16-11):

$$2NaCl(aq) + 2H_2O(\ell) \xrightarrow{electrolysis} Cl_2(g) + H_2(g) + 2NaOH(aq)$$

The following reactions occur:

At the positive electrode: $2 Cl^-(aq) \longrightarrow Cl_2(g) + 2e^-$ (oxidation)

At the negative electrode: $2H_2O(\ell) + 2e^- \longrightarrow H_2(g) + 2 OH^-(aq)$ (reduction)

Overall reaction: $2 Cl^-(aq) + 2H_2O(\ell) \longrightarrow Cl_2(g) + H_2(g) + 2 OH^-(aq)$

Chloride ions are oxidized to chlorine atoms at the positive electrode. At the negative electrode water is reduced, yielding hydrogen gas and hydroxide ions. (Even though the sodium ions are attracted to the negative electrode, there is less tendency for them to gain electrons than there is for the water molecules to gain electrons.) Since the solution contains sodium ions, the products are chlorine gas, hydrogen gas, and an aqueous solution of sodium hydroxide.

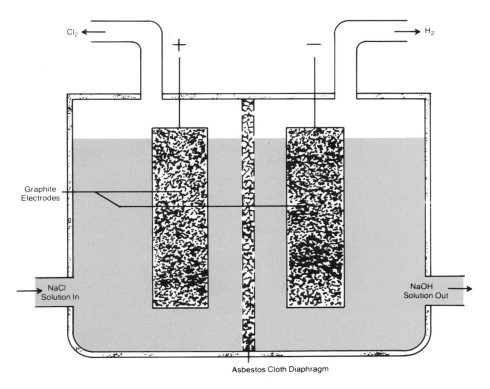

Figure 16-11
Electrolysis cell for the preparation of sodium hydroxide, chlorine, and hydrogen

Chlorine is a greenish-yellow gas which has a suffocating odour. It can react explosively with the hydrogen that is formed as a by-product. The two gases are kept apart by asbestos cloth diaphragms which, when wet, prevent the passage of gases but allow aqueous ions to pass through to reach the electrodes.

Chlorine is used in enormous quantities in the bleaching of paper and textiles, where the unwanted coloured substances are oxidized to colourless compounds. It is used in the purification of drinking water to kill bacteria and remove undesirable tastes and odours. Chlorine is also used to produce insecticides, plastics, and the solvent carbon tetrachloride.

Hydrogen is a colourless, odourless, tasteless gas. It is flammable and can form explosive mixtures with oxygen. Hydrogen is used in the production of ammonia, methanol, soap, and margarine, in petroleum refining, in welding, and in weather balloons.

Sodium hydroxide is a white waxy solid, which melts at 318°C and boils at 1390°C. It is a very corrosive compound, which is frequently called caustic soda, lye, or soda lye. It is soluble in water, producing a strongly basic solution. Sodium hydroxide reacts with atmospheric carbon dioxide to produce sodium carbonate and water. An aqueous solution of sodium hydroxide etches glass by attacking the surface silicate minerals.

Sodium hydroxide (caustic soda or lye) is also used in the paper industry, in petroleum refining, and in the manufacture of cellophane, rayon, soap, textiles, oven cleaners, and drain cleaners.

Hydrochloric Acid

Hydrochloric acid is a solution of hydrogen chloride in water. It is a clear, colourless liquid with a pungent odour. The method that is most often used to produce hydrochloric acid is the treatment of sodium chloride with sulfuric acid:

$$NaCl(s) + H_2SO_4(aq) \xrightarrow{\triangle} NaHSO_4(aq) + HCl(g)$$

The $HCl(g)$ is dissolved in water to produce hydrochloric acid.

Hydrochloric acid is used to clean metal surfaces before they are plated or enamelled, to remove excess mortar from brick buildings, and in the petroleum and pharmaceutical industries.

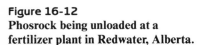

Figure 16-12
Phosrock being unloaded at a fertilizer plant in Redwater, Alberta.

Phosphoric Acid

Phosphoric acid is made by reacting concentrated sulfuric acid with pulverized phosphate rock, also known as phosrock (Fig. 16-12), which is composed largely of calcium phosphate, $Ca_3(PO_4)_2$:

$$Ca_3(PO_4)_2(s) + 3H_2SO_4(\ell) \longrightarrow 2H_3PO_4(\ell) + 3CaSO_4(s)$$

Water is added to the mixture, and the solid calcium sulfate is filtered off. The filtrate (liquid) that results is evaporated until a clear, syrupy liquid containing 85% H_3PO_4 is obtained.

Most phosphoric acid is used to make phosphate fertilizers such as ammonium phosphate, calcium phosphate, and calcium hydrogen phosphate. Phosphoric acid is used to produce detergents, soil conditioners, and metal-cleaning preparations. It is used in the pharmaceutical and food industries.

Ethylene

Ethylene is produced in great quantities by the cracking of ethane gas from petroleum:

$$C_2H_6(g) \xrightarrow{\triangle} C_2H_4(g) + H_2(g)$$

ethane   *ethylene*

Ethylene is the most important industrial organic chemical. It is a colourless gas with a sweet odour. Like all hydrocarbons it is flammable, burning in air to produce carbon dioxide and water vapour:

$$C_2H_4(g) + 3\ O_2(g) \longrightarrow 2CO_2(g) + 2H_2O(g)$$

It also undergoes addition reactions in which the double bond is broken and each carbon atom picks an additional atom:

$$CH_2{=}CH_2(g) + Br_2(\ell) \longrightarrow \underset{\underset{Br}{|}}{CH_2}{-}\underset{\underset{Br}{|}}{CH_2}(\ell)$$

Ethylene may be converted to ethanol by the acid-catalyzed addition of water:

$$CH_2{=}CH_2(g) + H_2O(\ell) \longrightarrow CH_3CH_2OH(\ell)$$

Ethylene molecules have the important property of being able to join to one another, forming long chains. This reaction is of importance in the plastics industry (Section 15.26).

Ethylene is used as a fuel in welding, as an anesthetic, as a refrigerant, and in the ripening of fruit. Ethylene is also used to produce ethylene dichloride (1,2-dichloroethane), a starting material in the manufacture of vinyl chloride (chloroethene), used in plastics. Ethylene may also be converted to other chemicals which are used to produce rubber, plastics, solvents, antifreeze, degreasers, and fumigants.

A summary of the properties and uses of some heavy chemicals is given in Table 16-4.

Table 16-4   Properties and uses of some heavy chemicals

Chemical	Properties	Uses
Sulfuric acid	clear, colourless, odourless, viscous, oily liquid; very corrosive	fertilizers, drugs, paints, rayon fibres; cleaning of metals; storage batteries, explosives, plastics, dyes, pigments, petroleum refining, production of hydrochloric acid.
Ammonia	colourless gas; very pungent odour	production of nitric acid, rayon, nylon, fertilizers; plastics industry; cleaning agent; pharmaceuticals; refrigerant.
Nitric acid	colourless liquid; fumes in moist air; choking odour	production of fertilizers, explosives, plastics, drugs, dyes; engraving copper plates; metallurgy, pharmaceuticals.
Sodium hydroxide	white, waxy solid; very corrosive	paper industry; petroleum refining; manufacture of cellophane, rayon, soap, textiles, drain cleaners, oven cleaners.
Chlorine	greenish-yellow gas; suffocating odour	bleaching of paper and textiles; purification of drinking water; production of insecticides and plastics.
Hydrogen	colourless, odourless, tasteless gas; flammable and explosive	production of ammonia, methanol, soap, and margarine; petroleum refining; welding; weather balloons.
Hydrochloric acid	clear, colourless liquid; pungent odour	cleaning metal surfaces; removing excess mortar from brick buildings; in the petroleum and pharmaceutical industries.
Phosphoric acid	clear, syrupy liquid	fertilizers, detergents, metal-cleaning preparations; in the pharmaceutical and food industries.
Ethylene	colourless, flammable gas; sweet odour	plastics; welding; as an anesthetic; as a refrigerant; production of rubber, solvents, antifreeze; ripening of fruit; degreaser; fumigant.

Table 16-5   Production of Industrial Chemicals and Synthetic Resins in Canada, 1985

Chemical	Quantity (kilotonnes)
Sulfuric acid	3900
Ammonia	3600
Urea	2000
Ethylene	1700
Sodium hydroxide (caustic soda)	1600
Chlorine	1400
Ammonium phosphate	1300
Nitric acid	1100
Ammonium nitrate	1100
Polyethylene (resins)	960
Benzene	690
Propylene	650
Phosphoric acid	630
Toluene	470
Xylene	410
Sodium chlorate	340
Synthetic rubber	210
Carbon black	170
Hydrochloric acid	160
Butylene	130
Butadiene	130
Polyesters (unsaturated)	36
Acetone	23

Source: "Industrial Chemicals and Synthetic Resins," Catalogue 46-002 Monthly, Statistics Canada, December 1985.

16.6 The Chemical Industry in Canada

A total of 147 plants manufacturing inorganic and/or organic industrial chemicals in Canada was listed for 1983. Ontario had the largest number of these plants, followed by Quebec, Alberta, and then British Columbia. Over 12 000 people worked in the chemical industry, and many others were involved in producing fine chemicals and chemical-related consumer products.

Only one plant employed over 1000 workers; the majority of the plants employed fewer than 100 workers. Most plants employed few workers because the industry is highly automated. The total value of manufacturing by the Canadian chemical industry was approximately six billion dollars in 1983.

An indication of the amounts of some of the heavy chemicals produced in Canada in 1985 is given in Table 16-5. Sulfuric acid is produced in the largest quantity. Seven of the ten chemicals produced in greatest quantity are inorganic. Urea and ethylene are the organic chemicals produced in the largest quantities. The materials and supplies used for the production of heavy chemicals in Canada in 1983 are listed in Table 16-6. A wide variety of chemicals is required by the manufacturers.

As a result of recent growth and diversification, Canada now makes almost all major organic and inorganic compounds. However, the small domestic market in Canada has been a deciding factor in limiting the progress of diversification of manufacture.

Canadian industry faces the problems of being in a resource-rich, developing country located close to a highly developed, wealthy nation. The technology necessary for the exploitation of Canada's natural resources initially came from foreign sources. Foreign capital, investment, and control usually follow this transfer of technology. Thus, many of the companies involved in the chemical industry are foreign-owned. Such companies prefer to spend their research dollars in the company's home country. This fact, along with the limited support of research by the Canadian government, has caused Canada's chemical research and development program to be small. Nevertheless, despite facing many problems, the Canadian chemical industry will continue to play an important role in our economy.

16.7 Factors that Determine the Location of an Industrial Plant

Suppose that, as an executive of a large corporation, you were asked to set up an industrial plant in Canada. What factors would influence your choice of location? Factors to be considered would include the cost of energy in a particular area, the availability of any raw materials required, any special tax concessions for locating in a certain area, the cost of labour, the stability and skill of the labour force in a particular region, the cost of transportation of products to intended markets, and the effect on the local environment. Some of these factors may favour a proposed location; others may not. Usually the location that will yield the lowest cost for the delivered product will be chosen.

16.8 Household Chemicals

One does not have to look to industry or research laboratories to find useful chemicals. Our homes contain many important chemicals. We will examine some of the more common ones.

Sodium hydrogen carbonate, $NaHCO_3$, is also known as sodium bicarbonate or baking soda. Acids react with baking soda to produce carbon dioxide:

$$NaHCO_3(aq) + H_3O^+(aq) \longrightarrow Na^+(aq) + CO_2(g) + 2H_2O(\ell)$$

This reaction occurs during the leavening, or rising, process in baking. Carbon dioxide is trapped in the dough and expands when it is warmed. Thus, it produces small holes in the bread or cake, giving the quality of lightness. Baking soda is also used in refrigerators to absorb food odours and as a mild abrasive cleanser.

Baking power contains a mixture of baking soda, a solid acid, and starch or flour to keep the solid mixture dry. When water is added the acid reacts with the sodium hydrogen carbonate to produce the carbon dioxide responsible for making the batter rise.

Vinegar contains 3-6% acetic acid, $HC_2H_3O_2$, by volume. This solution is used in salad dressing and for seasoning and preserving food.

Drain cleaners usually contain a mixture of a strong base such as sodium hydroxide and aluminum filings. This solid mixture reacts with the water in the clogged drain to produce much heat and hydrogen gas. The heat released serves to melt the drain blockage which usually contains fats. Mechanical agitation by the bubbles of hydrogen gas also help to remove the blockage.

Teflon®, an organic polymer, has the formula

$$\begin{bmatrix} \overset{\displaystyle F}{\underset{\displaystyle F}{\vphantom{|}}} \\ -C \\ \end{bmatrix}$$

$$\left[\begin{array}{cc} F & F \\ | & | \\ -C - & C - \\ | & | \\ F & F \end{array}\right]_n$$

where n is a large number. It has the special property of being resistant to chemicals and to high temperatures. Thus, it finds uses in gaskets, electrical insulation and cooking-pan coatings.

Other chemicals found around the house are used for cleaning purposes. Examples are sodium hydroxide in oven cleaners and aqueous ammonia, $NH_3(aq)$, in window cleaners.

Sodium iodide, NaI, is added to salt to produce "iodized" salt. Using iodized salt helps to prevent iodine deficiencies which may produce goitre, a thyroid condition. Hydrogen peroxide, H_2O_2, in the form of a 3% aqueous solution is used as an antiseptic to treat minor cuts. Sodium hypochlorite, $NaOCl$, is the active ingredient in household bleach. Benzoyl peroxide is used in ointments for treating the skin condition referred to as acne.

Many other chemicals are found around the home. The list includes waxes, polishes, insecticides, spot removers, toothpaste, detergents, paints, shampoos, medicines, fats, oils, and soaps.

Many household chemicals must be used with care. Some produce harmful vapours and must be used with proper ventilation; others are flammable or corrosive. Many are harmful if accidentally ingested.

Table 16-6 **Materials and Supplies Used by the Manufacturers of Industrial Chemicals in Canada, 1983**

Chemical	Quantity (kilotonnes)
Salt (sodium chloride)	1200
Ethylene	590
Sulfuric acid	320
Benzene	250
Sulfur (excluding pharmaceuticals)	210
Propylene	140
Fluorspar, crude	84
Sodium carbonate	58
Sodium hydroxide	54
Bauxite ore	35
Isopropyl alcohol	21
Ethylene oxide	13
Toluene	7
Xylene	1
	Millions of Cubic Metres
Natural gas	1900
Acetylene	2
	Millions of Litres
Methyl alcohol	43

Source: "Industrial and Agricultural Chemical Products," Catalogue 46-224 Annual, Statistics Canada, 1983.

Part Two Review Questions

1. List two of the features of the chemical industry.
2. What does the term *heavy chemicals* mean?
3. What process is used to separate the components of crude oil?
4. List four of the chemicals obtained by refining crude petroleum. What are their uses?
5. What is a cat cracker?
6. Explain what is meant by octane rating.
7. What chemicals are present in natural gas?
8. What are four of the most important sources of raw materials for the chemical industry?
9. In what ways are the chemical industries important?
10. What is the most important chemical produced by the chemical industry? Why is it so important?
11. Name five of the most important chemicals produced by the chemical industry. For each chemical, give the formula, physical properties, and major uses.
12. Ethylene is one of the most important organic chemicals produced. Why is it so important?
13. List the factors that govern the location of an industrial plant.
14. Name five common household chemicals. For each chemical, give its major uses.

Part Three: Canadian Technology

16.9 Canadian Contributions to Technology

Canada has made significant contributions to the advancement of technology in a number of areas.

Nuclear Energy

In the early 1900s, Sir Ernest Rutherford experimented with radioactivity at McGill University. In 1940, experiments involving uranium and graphite were begun in the Ottawa laboratories of the National Research Council. Had the materials been purer, Canada might have had the world's first controlled nuclear chain reaction.

In 1945, the Zero Energy Experimental Pile (ZEEP) at Chalk River, Ontario, became the first nuclear reactor outside of the United States. In 1962, Canada's first trial nuclear power plant was set up at Rolphton, Ontario. The project was successful, and a full-size plant was constructed at Douglas Point in 1967. This reactor was built by Atomic Energy of Canada Limited and was operated by Ontario Hydro. It was a CANDU (Canada Deuterium Uranium) reactor.

The Canadian nuclear power program is based on our experience with CANDU reactors. A number of countries have bought Canada's CANDU reactors which are considered to be among the most efficient and safest in the world. By 1982 the Pickering nuclear power station near Toronto had produced more electricity than any other nuclear power station in the world. In 1986 there were 15 nuclear power reactors in Canada which accounted for 13% of the electricity generated nation-wide.

There are two nuclear research establishments in Canada. One is at Chalk River and one is the Whiteshell research plant in Manitoba. The NRX reactor at Chalk River produced radioisotopes which were marketed as early as 1949. The world's first cobalt radiotherapy units, used in the treatment of cancer, were produced in 1951 using radioactive cobalt from the NRX reactor. These units have saved many lives and have been exported to over 80 countries.

This power station at Rolphton, Ontario produced the first commercial nuclear-generated electricity in Canada.

APPLICATION

Food Irradiation

The irradiation of food has many beneficial effects. It produces sterile products that can be stored without refrigeration. It kills microorganisms, parasites, and insects in stored foods. It delays ripening in some vegetables and fruits and stops sprouting in stored root crops such as onions and potatoes. As a result, it extends the shelf life of foods. The irradiation of food could eliminate the use of pesticides and preservatives that may have dangerous side effects.

Food is irradiated by exposing it to X-rays, high-speed electrons from an electron accelerator, or gamma rays emitted by radioactive isotopes such as cobalt-60 or cesium-137. How does this process provide the benefits described above? When the ionizing energy passes through food, it breaks apart bonds between the atoms. This chemical change stops normal cell division in insects and microorganisms, sterilizing or killing them. The process does not leave any radioactive materials.

Like any proposed advance in technology, food irradiation has some negative aspects. Some public interest groups have voiced concerns about the safety of workers at irradiation facilities, and about the supply, transportation, and disposal of radioactive isotopes. These groups are also concerned about the nutritional quality of irradiated food. In addition, they point out that we still do not know if new compounds resulting from

**Table 16-7   Canada as a Pro-
ducer of Minerals
(1981)**

Mineral	World Rank
Nickel	1
Zinc	1
Asbestos	2
Gypsum	2
Potash	2
Sulfur	2
Uranium concentrate	2
Molybdenum	2
Titanium concentrate	2
Aluminum	3
Platinum	3
Gold	4
Copper	4
Lead	4
Silver	4
Cadmium	4
Iron ore	6
Antimony	7

irradiation will have toxicological or mutagenic (that is, causing muta-
tion) effects.

In 1986, a standing committee of the House of Commons was set up to
study the irradiation of food. In 1987, this committee recommended that
Canada regulate irradiation of food by classifying ionizing radiation as a
food additive, subject to special federal rules. The committee also recommended that a panel be organized to examine food irradiation and the
safety of eating irradiated food more closely. At present, irradiated
onions, potatoes, wheat flour, spices, and dehydrated seasonings are
allowed to be sold in Canada. However, the process is not being carried
out commercially in this country.

Canada is in an excellent position to market its technology for irradiation of food. Canada is the world leader in using gamma irradiation to
sterilize medical products, and has provided 85 of the 135 gamma irradiators used around the world. It supplies 80% of the world's cobalt-60, the
main source of gamma rays for irradiators. This technology could be
applied to food irradiation.

Mining

Canada has many mineral resources. Technical expertise has allowed Canadians to exploit these resources, making Canada third in the world for the value
and diversity of mineral production. On a per capita basis, Canada is ranked
first in value of minerals produced. Minerals account for about one-third of
Canada's total commodity exports. Table 16-7 shows Canada's world ranking
as a producer of various minerals.

Canada is the world's largest producer of zinc. One of the largest zinc-
producing plants is located at Trail, B.C. At this plant, the impure ore, containing sphalerite (ZnS) is crushed, and the ZnS is separated from other metallic
sulfides. The ZnS is heated in air or oxygen (that is, roasted) to produce ZnO:

$$2ZnS(s) + 3\ O_2(g) \longrightarrow 2ZnO(s) + 2SO_2(g)$$

The ZnO is purified by adding sulfuric acid to dissolve it from the solid
mixture in a process known as leaching:

$$ZnO(s) + H_2SO_4(aq) \longrightarrow ZnSO_4(aq) + H_2O(\ell)$$

The $ZnSO_4$ is purified by crystallization and electrolyzed in aqueous solution.
Metallic zinc is deposited on the negative electrode. Zinc is used to make
galvanized iron and brass. It is used in the production of rubber, paint pigments, Ni-Zn batteries, and skin ointments.

Canada is also the world's largest producer of nickel. Large deposits of nickel
occur in the Sudbury Basin of Northern Ontario. This ore deposit contains
sulfides of nickel, iron, and copper. The ore is crushed, and NiS is separated
from the other sulfides. Nickel sulfide is then roasted to produce nickel oxide:

$$2NiS(s) + 3\ O_2(g) \longrightarrow 2NiO(s) + 2SO_2(g)$$

The oxide is then reduced by heating with carbon:

$$2NiO(s) + C(s) \longrightarrow 2Ni(s) + CO_2(g)$$

The nickel produced in this way is about 96% pure. It is further purified by one of two methods, either electrolytically or by the Mond process in which the impure nickel is warmed to 60°C in an atmosphere of carbon monoxide gas. Nickel carbonyl (b.p. 43°C) forms and distills off:

$$Ni(s) + 4CO(g) \longrightarrow Ni(CO)_4(g)$$

The nickel carbonyl is heated to 200°C. It decomposes and deposits pure nickel:

$$Ni(CO)_4(g) \longrightarrow Ni(s) + 4CO(g)$$

Nickel is a component in many alloys. It is used to make coins and magnets. Nickel is also used as a catalyst in petroleum refining and in the conversion of vegetable oils to solid fats.

Pulp and Paper

Canada has developed new technologies in wood harvesting and in the production of pulp and paper. Research in the making of pulp and paper is carried out by the Pulp and Paper Institute of Canada, by the Forest Engineering Research Institute of Canada, and by some manufacturers.

Canada supplies much of the world's newsprint, following only the United States in pulp and paper manufacture. Canada is ranked first in pulp and paper exports. This industry is larger than any other manufacturing industry in Canada. Canada has about 140 pulp, pulp and paper, and paper mills, employing about 85 000 workers in total.

The essential component of paper is cellulose fibre. In order to produce paper from wood, the brown resinous component, called lignin, must be removed. The lignin holds the cellulose fibres together. The bark is removed from the tree, which is then broken into chips. In the Kraft process, the chips are placed in a digester with a basic solution of NaOH and Na_2S at 170°C and 800 to 1000 kPa. During this procedure, called pulping, the lignin and most of the impurities dissolve in the solution, leaving cellulose fibres (that is, pulp). The brown pulp is filtered, washed, and bleached with chlorine. Chemicals such as alum, starch, clay, titanium dioxide, and barium sulfate are added to the pulp to produce a paper of good quality and weight and to prepare the surface for printing. The pulp is then fed into the paper-making machinery. One machine can produce a continuous sheet of paper about six metres wide and three hundred kilometres long in a single day. There is, unfortunately, an environmental problem related to pulp and paper plants. The gases, SO_2 and H_2S, may be released to the atmosphere. The H_2S gas is highly poisonous, while the SO_2 gas contributes to the production of acid rain.

The Petroleum Industry

The petroleum and related industries employ approximately 250 000 people in Canada. In 1982 there were about 1500 people engaged in petroleum research and development in Canada. Some worked for the Alberta Research Council;

others worked for large industrial oil companies. The largest petroleum research centre in Canada is operated by Imperial Oil in Sarnia, Ontario.

Canada has developed technology for both the drilling and production of crude oil. Canada produces a large amount of sulfur. Some is removed from sour natural gas which contains H_2S and some is processed from petroleum. In both cases the sulfur is produced by burning H_2S:

$$2H_2S(g) + 3 O_2(g) \longrightarrow 2SO_2(g) + 2H_2O(g)$$

The sulfur dioxide which is produced is then reacted with additional H_2S to form elemental sulfur:

$$SO_2(g) + 2H_2S(g) \longrightarrow 3S(\ell) + 2H_2O(g)$$

Canada has had to develop special lubricants to allow work in the petroleum industries to continue in extremely cold conditions. Ice-resistant drilling platforms and artificial islands have been developed to cope with the harsh conditions experienced offshore and in northern Canada. Some of this technology has been adopted by other countries.

Raw material for the chemical and petroleum industries is found in the deposits of quartz sand (SiO_2) in northern Alberta known as the Athabasca oil sands. Each sand grain in these deposits is surrounded by a film of water, which in turn is surrounded by a layer containing oil. In order to convert the oil sand into a usable form, the oil-bearing layer must be separated from the water and sand. This is a several-stage process which produces a mixture of hydrocarbons called *bitumen*. The bitumen is processed further in different ways to give synthetic crude oil. Another by-product of these processes is sulfur.

Two major plants are involved in this method of extracting oil in Athabasca. They have a combined output of about 200 000 barrels of oil a day. This represents 15% of Canada's production of oil. The larger of the plants, the Syncrude plant (Fig. 16-13), uses the largest open-pit mine in the world. This pit is operated in temperatures which range from −40°C in winter to +32°C in summer. Chemical engineers have played a vital role in the operation of this large plant under such extreme climatic conditions. Much research is being done on methods of extracting oil from Alberta's oil sands.

Figure 16-13
(a) Aerial view of the Syncrude upgrading area. Here the bitumen is converted to naptha and gas oil, the components of synthetic crude oil.
(b) A bucketwheel reclaimer feeds oil sand onto conveyor belts. There are 19 km of conveyors to carry oil sand to the plant.

APPLICATION

Synthetic Fuels in Canada

Attempts to develop the oil sands of northern Canada date back to the mid-1920s, when there were a number of small-scale attempts to obtain the oil by using a hot-water extraction process. In 1962, the Alberta government approved the Great Canadian Oil Sands proposal. The plant built as part of the scheme came on stream in 1967, with the capability of producing 5000 m³/d (cubic metres per day). In the following decade it expanded its production to 10 000 m³/d. In 1973 the Syncrude project was approved. The plant for this project was completed in 1978, and by 1983 it was producing an average of 17 000 m³/d. These two plants represent the Canadian synthetic fuel industry, with a combined output of about 15% of Canada's domestic production of petroleum.

Syncrude leases 400 km² of land containing oil sands sufficient to produce a total of 25 000 to 28 000 m³ of synthetic crude oil per day. The Syncrude plant has the largest open pit mine in the world. It has to operate in temperatures ranging from +32°C in summer to −40°C in winter, and the oil sands that it handles are extremely abrasive and corrosive. It contains the world's largest extraction plant, a utilities complex producing enough electricity for a city of 300 000 people, and a refinery-type upgrader capable of processing 24 000 m³ of heavy oil feedstock daily.

In recent years, both the Alberta and the Canadian governments have introduced policies to make oil sands and heavy oil plants attractive to investors. As a result, Alberta now has a large number of small pilot projects. The development of Canada's northern oil resources is not cheap. The direct costs of producing synthetic oil are approximately $125/m³. However, with this investment come significant spin-offs such as new jobs, markets for domestic goods and services, new government revenue, and new community development.

Chemical Technology

Canada's contribution to chemical technology began in the 1850s when a Nova Scotian, Abraham Gesner, invented kerosene oil. Gesner also obtained patents for distilling the heavy component of petroleum, bitumen. As a result, he is considered a founder of the modern petroleum industry.

Canadians have contributed to other technological advances since that time. Here are just a few examples. Canadians helped the Allies during World War II by producing synthetic rubber at Sarnia in 1942. In the same year, ammonia, which is important in producing fertilizers and explosives, was obtained from natural gas in Calgary. During the 1960s, research at DuPont Canada in Kingston and Sarnia resulted in the production of linear low-density polyethylene (plastic resin), a material which is useful in packaging. Research is also taking place in the area of petrochemicals—that is, chemicals made from the components of oil and natural gas. These chemicals can be used to produce such end-products as solvents, dyes, fertilizers, rubber, plastics, and paints.

The largest number of people developing and using chemical technology work in and around Sarnia, Ontario. This region is often referred to as Can-

Biography

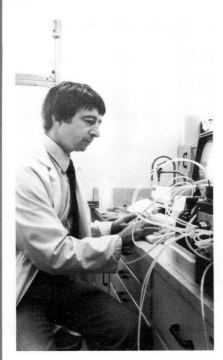

Figure 16-14
Kelvin Ogilvie—Developed the gene machine.

KELVIN K. OGILVIE was born and raised in Nova Scotia. He received his B.Sc. from Acadia University in 1964. He went on to obtain his Ph.D. from Northwestern University in 1968 and then joined the faculty at the University of Manitoba as an assistant professor. He moved to McGill University in 1974 and was promoted to full professor in 1978. In 1987, he became a Vice-President of Acadia University.

Dr. Ogilvie has concentrated his research on the chemistry of DNA and RNA. These are large molecules which carry the genetic information during cell division. They direct the synthesis of substances which control a host of reactions in the cells of living organisms. His work led to the development of the "gene machine," an instrument which uses an automated process to synthesize DNA and RNA fragments in a matter of hours, rather than in the weeks or months taken by the manual process. He has successfully synthesized small RNA molecules which carry out numerous functions essential to living cells and are the key components of many viruses. In the process, he has succeeded in preparing a new class of compounds that are highly active against a wide range of viruses, particularly Herpes Simplex I and II (which are responsible for cold sores and some sexually transmitted diseases).

Kelvin Ogilvie has received many awards and fellowships. In 1982, he was given an E.W.R. Steacie Fellowship. This award is made annually by the National Sciences and Engineering Research Council of Canada to the most promising young researchers in Canadian universities to enable them to devote their time entirely to research for a period of up to two years.

ada's "Chemical Valley." Some of the largest plants located in this region are Dow Chemical Canada, C-I-L, Du Pont Canada and Polysar Limited. The first of these, Dow Chemical Canada, has the largest and most diversified chemicals and plastics complex in Canada. Some of the chemicals produced are chlorine, sodium hydroxide, ammonia, styrene, latex, vinyl chloride, and polyethylene.

C-I-L operates one of the largest fertilizer plants in Canada which includes two world-scale anhydrous ammonia plants. Ammonia can be used directly as a fertilizer or made into other fertilizers. C-I-L's high-nitrogen fertilizer chemical plant, which produces a sulfur-coated urea, was the first of its kind in the world. This company is Canada's largest producer of chemicals and allied products. These products include plastics, paints, explosives, and agricultural and industrial chemicals.

Du Pont Canada operates a world-scale polyethylene plant near Sarnia. It produces more than 235 million kilograms of polyethylene annually. The main raw material, ethylene, is obtained from a nearby Polysar Limited refinery. The end-products of polyethylene include food packages, car parts, and swimming pool covers.

Polysar Limited is the world's largest producer of synthetic rubber and latex. It operates a butyl rubber plant which is one of the world's most advanced synthetic rubber facilities. Polysar's products are sold in more than 90 countries.

16.10 Citizens of a Technologically-Based Society

We have responsibilities as citizens of one of the most technologically advanced countries in the world. One of these is to educate ourselves about the interrelationships among science, technology, industry and society. We must understand the role played by the chemical industry. We must also have the scientific and technological knowledge required to enable us to make informed decisions about technological issues. We must be prepared to voice our opinions. Scientists and technologists did not create our industrial society alone. Citizens receive the benefits of industrialization, so they must share responsibility for helping to diminish the problems that accompany technological change.

We must realize that decisions affecting scientific development are not always made by scientists alone. In many cases, the opinion of citizens is sought and acted upon. A number of commissions have been set up to seek public input on various issues, for example, acid rain and potential hazards of uranium mining. Meetings are held on such issues and citizens are invited to make presentations on both sides. Citizens do have the power to influence ultimate decisions.

Companies which explore for oil and gas are required to carry out environmental impact studies. Citizens should try to get involved in these studies by obtaining copies of the reports and responding to them at public meetings where the reports are presented. It is not always easy to make an informed decision. We must look at all sides of each issue and consider scientific, technical, geographic, social, and economic factors.

CAREER

The Chemical Engineer

Figure 16-15
A chemical engineer gives advice on handling chemicals in many situations.

Chemical engineers are the people who transform the discoveries of chemists into commercially profitable processes. They design and supervise the construction of industrial chemical plants. In operating plants, chemical engineers supervise the manufacturing of chemicals as well as the storage and unloading of raw materials. They also find ways to improve existing manufacturing processes to make better products and increase safety and efficiency. They deal with products and equipment problems, controlling losses and keeping the plant running on schedule. Many chemical engineers are responsible for the proper disposal of the hazardous wastes generated by their companies.

Not all chemical engineers are directly involved in the production of chemicals. Some sell chemicals, others buy them. Some study markets for various products, and some arrange financing for new chemical plants. Many chemical engineers become managers in their companies because their jobs offer opportunities to show leadership skills.

The chemical engineer needs to know how chemical substances behave in an actual factory situation: how they will flow through pipes and valves, how they will behave when subjected to high temperatures and pressures, what kinds of containers can be used to store them, what effects they have on valves, and so on. For this reason, a university degree in engineering is essential for anyone who wants to become a chemical engineer. An engineering course would cover subjects such as engineering drawing, strength of materials, heat and mass transfer, thermodynamics, fluid dynamics, process engineering, and chemical engineering design, but would also stress basic areas of science such as chemistry, physics, and mathematics. A chemical engineering program would also include courses in the use of computers and communication skills.

Chemical engineers are expected to accept shift work since most large factories operate 24 hours a day in order to use plant and production capacity most efficiently.

Part Three Review Questions

1. What is the largest manufacturing industry in Canada?
2. What are petrochemicals? Why are they so valuable?
3. What contribution did Abraham Gesner make to science?
4. Where is the centre of Canada's chemical industry located?
5. Name the two industries that contribute the most to Canada's exports, in terms of volume of material exported.
6. List four areas of technology in which Canada has made contributions. State what these contributions were.
7. What are the responsibilities of citizens in a technological society? Why do we have these responsibilities?

Chapter Summary

- Pure science helps to expand our knowledge of the universe.
- Technology is the application of scientific discoveries to the solution of practical problems.
- Many industries have been created as a result of technology.
- Technology influences society, and society influences technological developments.
- The chemical industry produces chemicals which are then used to produce many other chemicals for the manufacture of useful finished products.
- The sources of raw materials for the chemical industry include petroleum, coal, natural gas, mineral deposits, and the atmosphere.
- Chemical industries are important because they affect the economy of a nation, provide employment, and produce many useful products.
- Some important chemicals produced by the chemical industry are sulfuric acid, ammonia, nitric acid, sodium hydroxide, and ethylene.
- A number of geographical, economic, environmental, and labour-related factors determine the location of an industrial plant.
- Many important, useful chemicals are found around the household. These chemicals should be used with care.
- Canada has made significant contributions to the development of technology in areas such as nuclear energy, mining, pulp and paper, the petroleum industry, and chemical production.
- We have special responsibilities as citizens in a technological society.

Key Terms

science
technology
heavy chemical

fine chemical
cracking

Test Your Understanding

Each of these statements or questions is followed by four responses. Choose the correct response in each case. (Do not write in this book.)

1. Which one of the following is best considered to be an example of pure science?
   a) the development of a faster computer
   b) determining why leaves are green
   c) putting salt on the roads in winter
   d) the launching of a new space satellite

2. Which one of the following is best considered to be an example of technology?
   a) measuring the density of carbon dioxide
   b) observing a burning candle
   c) developing a new antibiotic drug
   d) explaining the blue colour of the sky

3. Which of the following is not a general feature of the chemical industry?
   a) The chemical industry produces a wide variety of chemicals.
   b) The chemical industry uses a great deal of automated equipment.
   c) The chemical industry spends large amounts of money on research.
   d) There is usually only one method to produce a particular chemical.

4. The octane rating of gasoline can be improved by
   a) cracking.   b) isomerization. c) alkylation.   d) all of these.

5. Which of the following is not a source of raw materials for the chemical industry?
   a) coal   b) petroleum   c) silicon   d) the atmosphere

6. The most important chemical produced by an industrial society is
   a) sulfuric acid. b) ammonia.   c) nitric acid.   d) ethylene.

7. Which of the following is not a "heavy chemical"?
   a) hydrogen   b) aspirin   c) phosphoric acid   d) sodium hydroxide

8. Canada is ranked first in the world as a producer of
   a) aluminum.   b) gold.   c) lead.   d) nickel.

9. "Sour" natural gas owes its name to the presence of
   a) CO_2.   b) H_2.   c) H_2S.   d) NH_3.

10. Canada's "Chemical Valley" is located in and around
    a) Sarnia.   b) Montreal.   c) Edmonton.   d) Toronto.

Review Your Understanding

1. Which of the following are best considered as examples of science, and which as examples of technology? Explain.
   a) It has been found that the carbon dioxide molecule contains two double covalent bonds.
   b) Monosodium glutamate is added to foods to enhance their flavour.
   c) Trisodium phosphate is a water-softening agent.
   d) Zinc reacts with hydrochloric acid to form hydrogen gas.
   e) Aureomycin is used as an antibiotic.
   f) Silver melts at 961°C.

2. Give three examples of the influence of technology on our daily lives.
3. Why is it important to have scientific and technological information when making a decision concerning a science-related issue?
4. Why do chemical plants, in general, employ so few workers?
5. Why are catalysts so important to the chemical industry?
6. Why are heavy chemicals so important?
7. If a particular chemical can be produced from a number of different raw materials, what determines which raw materials are used?
8. List four characteristics of the chemical industry.
9. How are hydrocarbons modified structurally by the following methods: **a)** cracking; **b)** isomerization; **c)** alkylation; and **d)** aromatization?
10. Name two ways of improving the octane rating of gasoline.
11. What effect does a lower oil price have on the chemical industry?
12. What is coal tar?
13. Why would some scientists argue that burning coal and petroleum is a misuse of these materials?
14. Why are sodium chloride, limestone, and potash useful?
15. What is sour natural gas?
16. List two major sources of sulfur.
17. What is the chief source of sulfur in Canada?
18. What is the chief use of sulfur?
19. List three industries that are dependent upon the chemical industry.
20. Using appropriate chemical equations, describe the production and uses of the following heavy chemicals: **a)** sulfuric acid; **b)** ammonia; **c)** sodium hydroxide; **d)** chlorine; and **e)** ethylene.
21. List five inorganic heavy chemicals and five organic heavy chemicals.
22. What are the three chemicals produced in greatest quantity by the Canadian chemical industry?
23. List three possible dangers associated with the use of household chemicals.
24. As an advanced technological society, what responsibilities does Canada have towards underdeveloped countries?
25. Write a brief description of Canada's contribution to nuclear technology.
26. Describe the method used to produce zinc.
27. Canada is the world's largest producer of nickel. Describe the method used to produce nickel in Canada.
28. Briefly describe the Kraft process for making paper.
29. Describe the production of sulfur from sour natural gas.
30. List four end-products obtained from petrochemicals.

Apply Your Understanding

1. Suppose a substance is found to have the following properties: it melts at an extremely high temperature, is insoluble in water, has a very low density, and is not reactive towards most chemicals. Suggest possible uses for such a substance.
2. What would be the effect on society if a new inexpensive drug which greatly increased intelligence were made readily available?

3. Rubber and nylon were developed to meet a military need. The technology surrounding these products eventually led to peacetime benefits to society. Describe another example.

4. Is it possible to have technology without pure scientific research? Is it possible to have pure scientific research without technology?

5. Comment on the following statement: "If our environmental problems are caused by the misuse of technology, it will take technology to correct these problems."

6. Suppose a plant, employing many people, is causing much air pollution. When asked to reduce the emission of pollutants, company officials say it will mean loss of jobs, because of the costs involved for anti-pollution devices. Discuss the issues.

7. Should the provincial and federal governments have stricter pollution laws? Give reasons for your answer.

8. Discuss the effect on society of the development of the hydrogen bomb.

9. Should a government which grants money to universities for scientific research specify the particular areas of the research? Explain.

10. What are the environmental concerns associated with offshore oil drilling?

Investigations

1. Write a library research report on the effect of technology on one of the following:
   a) the sports equipment industry   b) the automobile industry
   c) the music industry   d) the computer industry

2. Debate the following statement: "The development of technology causes the loss of jobs."

3. If a chemical company were considering building a large chemical plant in your area, what positive and negative factors would the company have to consider?

4. Debate the following statement: "Canada should spend less money on pure scientific research and use the money saved for social welfare programs."

5. What arguments could be used for and against the establishment of a nuclear power plant near your home?

6. What arguments could be used for and against the establishment of a large oil refinery near your home?

7. Select a technologically-based industry located near you. Determine the impact of this industry on the area in terms of employment, pollution, and economic benefits.

8. Debate the following statement: "Scientific research should not be permitted on live animals."

Nuclear Chemistry

17

CHAPTER CONTENTS

Part One: **The Principles of Nuclear Chemistry**
  17.1 Radioactivity
  17.2 Properties of a Radioactive Element—Radium
  17.3 The Lead Block Experiment
  17.4 Stability of the Nucleus
  17.5 Balancing Nuclear Equations
  17.6 Types of Nuclear Reactions
  17.7 Radioactive Decomposition
  17.8 Applications of Natural Radioactivity

Part Two: **Splitting the Atom**
  17.9 Artificial Transmutation
  17.10 Particle Accelerators
  17.11 Neutrons Are Better Bullets
  17.12 Fission
  17.13 The Atomic Bomb

Part Three: **Canada's Nuclear Industry**
  17.14 Nuclear Reactors
  17.15 Canada and Nuclear Chemistry
  17.16 AECL Power Projects
  17.17 CANDU Reactors
  17.18 Production of Heavy Water
  17.19 AECL Commercial Products—Uses of Radioactive Isotopes

Part Four: **The Nuclear Age**
  17.20 Breeder Reactors
  17.21 Fusion
  17.22 The Hydrogen Bomb
  17.23 Ionizing Radiation
  17.24 Problems of the Nuclear Age

What is a radioactive element? How does a nuclear reactor operate? How safe are nuclear reactors? What is ionizing radiation? What is a safe level of ionizing radiation? What are some of the problems associated with living in the nuclear age? Answering these questions requires a knowledge of nuclear chemistry.

In this chapter, we examine radioactivity and its relationship to the nucleus of the atom. We study the properties of the particles and rays that can be emitted from nuclei. We learn the types of nuclear reactions that can occur. A detailed discussion of the operation of a CANDU reactor is given. The role of Canada in the development of nuclear chemistry is described.

Key Objectives

When you have finished studying this chapter, you should be able to
1. describe the properties of radioactive substances such as compounds of radium.
2. list the properties of alpha, beta, and gamma rays.
3. list the four types of nuclear reactions and give an example of each.
4. balance nuclear equations.
5. describe the operation of the CANDU nuclear reactor.
6. discuss the benefits and the hazards associated with the production of nuclear energy for commercial and military purposes.

Part One: The Principles of Nuclear Chemistry

Up to this point we have paid little attention to the nucleus of an atom, other than to note that it is the number of protons in the nucleus that determines the identity of an atom. There are nuclear phenomena, however, that have applications in chemistry. In this chapter we shall learn about these phenomena and the experiments that led to their discovery.

17.1 Radioactivity

Radioactivity was discovered by the French physicist Henri Becquerel in 1896. He had been studying the properties of metals and was especially interested in those metals that fluoresce (emit light). He found accidentally that exposure to a uranium ore caused a covered photographic plate to become fogged. He wrapped photographic plates in thin sheets of copper and aluminum and placed uranium salts on the covered plates. Again the photographic plates were blackened when they were developed. Becquerel decided that uranium salts emit rays which are capable of penetrating black paper and thin sheets of metal. The elements which give off these invisible rays are said to be **radioactive**. The property of giving off these rays is called **radioactivity**.

The French wife and husband team, Marie and Pierre Curie, took up Becquerel's research. They found that pitchblende, an ore which is partially made up of uranium, was more radioactive than uranium itself. The Curies believed that this ore contained a new element which was more radioactive than uranium.

The Curies obtained several tonnes of pitchblende and attempted to concentrate the material. For the next two years they laboured. The work was tedious, painstaking, and unexpectedly dangerous since they were ignorant of the detrimental effects of exposure to radioactive materials. Nevertheless, they finally succeeded in concentrating the sources of the radioactivity. One of the sources was a new element resembling nickel which Marie Curie (Fig 17-1) called *polonium*. Polonium was many times more radioactive than uranium.

Biography

MARIE SKLODOWSKA was born in Poland. She dreamed of a career in science, and at the age of 24 she moved to Paris to study and teach. After graduating from the Sorbonne, she met Pierre Curie in the Municipal School of Physics and Chemistry and worked beside him in the laboratory. They were married in 1895.

In 1896, Becquerel found that pitchblende, a uranium ore, darkened a photographic plate more than could be accounted for by the amount of uranium present. He asked Madame Curie, a trained and gifted experimenter, to undertake the search for the element responsible.

The Curies boiled and cooked a tonne of the ore. They filtered and separated impurity after impurity. When the poisonous fumes threatened to stifle them under the leaky roof of their improvised lab, Mme. Curie herself moved large vats of liquid to the adjoining yard. For hours at a time she stood beside the boiling pots stirring the thick liquids with a great iron rod. In the bitter winter of 1896 Mme. Curie caught pneumonia. It was three months before she could return to work. Later, she became pregnant, but continued working. During the later stages of their work, both the Curies were continually ill. Finally, after two years, they had a small quantity of material about 300 times as potent as uranium. From this material Mme. Curie isolated a new element which she named polonium in honour of her native land.

Figure 17-1
Marie Sklodowska Curie
(1867-1934)—Discovered polonium and radium.

Mme. Curie kept working with the residues, which seemed to be even more radioactive than polonium. After a long series of purification steps, she isolated a very small amount of radium salt.

Mme. Curie studied every property of this strange new element. After five more years of research she presented her thesis in 1902. Her examining committee unanimously agreed that it was the greatest single contribution of any doctoral thesis in the history of science. The Curies became world famous and, along with Becquerel, were awarded the Nobel prize.

In 1906, Pierre Curie was killed in a traffic accident. Marie Curie was then chosen to succeed him in the chair in physics he had held at the Sorbonne, where no woman had ever held a professorship. She also retained the professorship at the school in Sèvres which she had held since 1900. Mme. Curie continued to work with radium. Finally, in 1910 she isolated pure radium by electrolysis of radium chloride. For this she was awarded a second Nobel Prize in 1911. She was the first person to win the Nobel Prize twice. Marie Curie died in 1934 of leukemia, evidently caused by her exposure to radioactivity.

They continued the tedious separation process and isolated a new element which was even more radioactive than polonium. The new metal was named *radium*. It is about a million times more radioactive than uranium. They separated only a few milligrams of radium chloride from the large quantity of pitchblende. Marie Curie later isolated pure radium from radium chloride. She found that it had the same chemical properties as the Group 2 element, calcium. However, it also possessed a new set of properties due to its strong radioactivity.

17.2 Properties of a Radioactive Element—Radium

The invisible rays given off from radioactive elements affect the light-sensitive emulsion on a photographic plate in the same way that ordinary light affects a photographic plate. These rays from a radioactive element are able to penetrate paper, wood, flesh, and thin metal sheets.

The invisible rays from a radioactive element will also discharge a charged electroscope (Fig. 17-2). When a charged rod touches the metal disc of the electroscope, the charge is distributed throughout the electroscope, including the two pieces of metal foil. The pieces of foil separate because like charges repel each other. The rays from a radioactive source ionize the gas (air) molecules in the electroscope by knocking off the outer electrons of the gas molecules. These charged ions are able to conduct electricity, and thus the charge is removed from the foil pieces which then fall back together.

The invisible rays can be measured in a Geiger-Müller tube (Fig. 17-3). The rays enter the gas contained in the tube through a thin glass envelope. Energy from the rays ionizes the gas molecules present in the tube, and electrons and positive ions which are formed are attracted to the charged electrodes. When they hit an electrode, a pulse of current is created. These electrical pulses are counted in a Geiger counter.

Figure 17-2
A charged electroscope

Figure 17-3
A Geiger-Müller tube

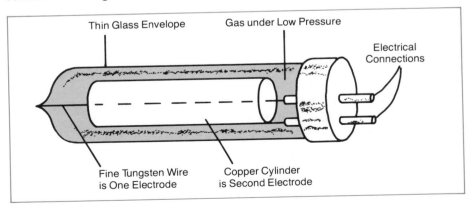

Radium compounds mixed with another compound such as zinc sulfide, cause the latter to glow in the dark. This type of mixture has been used in making the luminous paint for instrument dials. Most manufacturers have stopped making this type of paint because of the radiation hazard to the workers who prepare and apply the paint.

Radium salts are constantly emitting energy. Part of this energy is in the form of light. Radium salts glow in the dark; however, they do not give off enough light for this glow to be visible in the daylight. Radium salts also give off heat energy. In one hour radium emits enough heat to melt 1.5 times its own mass of ice.

A portable Geiger counter.

17.3 The Lead Block Experiment

The British scientist Lord Rutherford (Section 4.5) placed a few grains of radium salts in the bottom of a hole in a block of lead. The radiation escaped from the hole and was allowed to strike a zinc sulfide screen. The zinc sulfide flashed when it was bombarded with the radiation, and Rutherford observed a small luminous patch on the screen. By placing a magnet near the path of the rays, Rutherford found three patches of light were produced. The magnet had an effect on the charged particles. Since charged particles in motion have a magnetic field surrounding them, the magnetic fields of the magnet and of the charged particles interacted to cause a deflection of the charged particles. Some streams of particles, or rays, were bent in one direction, some in the opposite direction, while others were not bent at all. From the direction in which the rays were bent, Rutherford concluded that some were positively charged, some were negatively charged, and others were uncharged. He named these *alpha*, *beta*, and *gamma rays*, respectively (Fig 17-4).

Figure 17-4
Lead block experiment

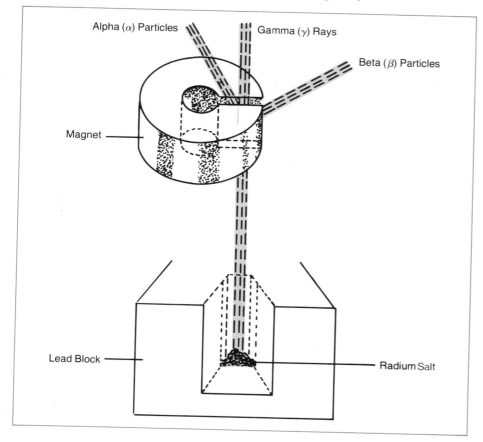

Alpha particles were found (after Rutherford's experiment) to be helium nuclei. They have two protons and two neutrons. Their positive charge is two units, and they have a mass of 4 u. They travel at high speeds (around 0.05 times the speed of light), but they have weak penetrating power. They have a range of only a few centimetres in air, and they are stopped by thin sheets of aluminum or even paper. However, they do ionize the molecules of any gas through which they pass.

Beta particles are high-energy electrons, and they are negatively charged. They travel at speeds which approach the speed of light (around 0.3 to 0.99 times the speed of light) and have a greater penetrating power than alpha particles. They can pass through several millimetres of aluminum. However, they do not ionize a gas through which they pass as well as alpha particles do.

Gamma rays are not deflected by a magnet. They are not charged particles. In fact, they are not particles at all. Gamma rays are high-energy, high-frequency electromagnetic radiations, similar to X-rays. They travel at the speed of light and are very penetrating. They can pass through 5 cm of lead or 30 cm of steel. They have little ionizing effect on gases.

In experiments like the one described above, the alpha particles are too heavy to be deflected very much. The beta particles are light and negatively charged. They are deflected in the opposite direction and to a greater extent than alpha particles. Because they have no charge, the gamma rays are not deflected at all.

17.4 Stability of the Nucleus

In order for an atom to be radioactive, its nucleus must be unstable. Thus, certain unstable nuclei spontaneously break down into more stable nuclei, emitting particles and rays as they do so. Three factors can be considered when we are trying to decide whether or not a nucleus is stable.

Mass

The atomic mass of a helium nucleus is 4.0015 u. This nucleus is made up of two protons (1.0073 u each) and two neutrons (1.0087 u each). The sum of the masses of the two protons and the masses of the two neutrons is 4.0320 u:

$$2(1.0073) + 2(1.0087) = 4.0320 \text{ u}$$

Apparently, 0.0305 atomic mass units were lost when two protons and two neutrons combined to form a helium nucleus. This is called the mass defect. The **mass defect** is the mass converted into energy when the helium nucleus is made from its components. This energy is called **binding energy**. Whenever the binding energy is large, the nucleus will be stable. However, it has been found that the binding energy is smaller for the very heavy and very light nuclei than it is for nuclei of intermediate mass (Fig. 17-5). Thus, the nuclei of the very light and the very heavy atoms are least stable because they have the smallest binding energies. In fact, all nuclei heavier than bismuth (mass, 209 u) are unstable. The binding energy per nucleon plotted in Fig. 17-5 is a measure of the average energy required to remove a proton or neutron from the nucleus. This graph shows that the binding energy per nucleon is greatest for elements with mass numbers between 50 and 60. An isotope of iron, $^{56}_{26}$Fe, is the most stable nucleus in nature.

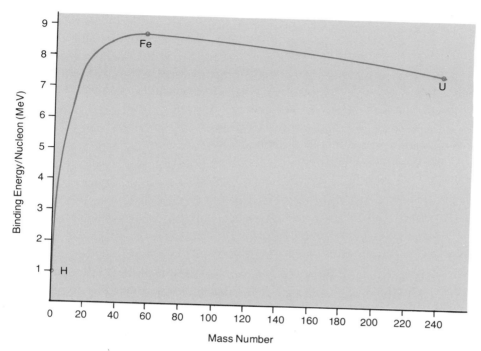

Figure 17-5
Relationship between binding energy and mass number

p/n Ratio

The stability of a nucleus also depends on the proton to neutron ratio. For any given number of protons, there is a small range of numbers of neutrons that permit a stable nucleus. In the case of the first 20 elements of the periodic table, the preferred ratio is one proton to every one neutron.

Beyond the first 20 elements, nuclei have more neutrons than protons. A proton to neutron ratio of 1 : 1 is not necessary to ensure stability for elements beyond the first 20. For example, the most stable nucleus, $^{56}_{26}$Fe, has a p/n ratio of 1 : 1.15. However, $^{238}_{92}$U has a p/n ratio of 1 : 1.59 and is unstable. There is a favourable "belt" of p/n ratios for nuclear stability. If the p/n ratio of a nucleus lies outside this belt, the nucleus is radioactive. To provide for a more favourable p/n ratio, radioactive nuclei emit particles. In doing so, they achieve a p/n ratio that provides for stability. The p/n ratio is decreased by emission of alpha particles and is increased by emission of beta particles.

Even-Odd Rule

We can also predict the stability of a nucleus by looking at the number of protons and neutrons in it. A nucleus with an even number of protons and an even number of neutrons has a better chance of being stable than a nucleus with an odd number of protons and an even number of neutrons, or a nucleus with an even number of protons and an odd number of neutrons. There are

Table 17-1   Combinations of Protons and Neutrons in Stable Nuclei

Number of Protons	Number of Neutrons	Number of Stable Nuclei
Even	Even	164
Odd	Even	50
Even	Odd	55
Odd	Odd	4

very few stable nuclei that have both an odd number of neutrons and an odd number of protons. The four nuclei which are stable in spite of an odd number of both protons and neutrons are ^2_1H, ^6_3Li, $^{10}_5B$, and $^{14}_7N$. Table 17-1 shows the combinations of protons and neutrons in stable nuclei. The known number of stable nuclei decreases when the number of either protons or neutrons is odd.

These three factors (mass, p/n ratio, and the even-odd rule) can be used to predict the stability of a nucleus. However, in some cases prediction is difficult because the rules contradict one another. In the case of nitrogen (seven protons and seven neutrons), the p/n ratio is 1 : 1, but the nucleus has odd numbers of protons and neutrons. In the case of boron (five protons and six neutrons), the p/n ratio is not 1 : 1, but there is an even number of neutrons. However, in the case of carbon (six protons and six neutrons), both rules are satisfied. One would readily predict that its nucleus is stable and, in fact, it is.

It has been found that nuclei are unusually stable if they have certain "magic numbers" of protons or neutrons. Nuclei with the following magic number of protons or neutrons have been found to be exceptionally stable and abundant in nature:

$$\text{number of protons} = 2, 8, 20, 28, 50, \text{ and } 82$$

$$\text{number of neutrons} = 2, 8, 20, 28, 50, 82, \text{ and } 126$$

Nuclei with magic numbers of either protons or neutrons tend not to take part in nuclear reactions, just as the noble gases tend not to take part in chemical reactions.

Several models have been proposed to describe the organization of nucleons in atomic nuclei. One such model, the shell model, uses the principles of quantum mechanics to describe the arrangement of nucleons in a nucleus in a manner similar to the description of the arrangement of electrons in the orbitals. An important result of this model is that only certain nuclear energy levels are allowed. Each energy level can accommodate a fixed number of nucleons. The magic numbers may be the numbers of nucleons that fill successive nuclear energy shells.

17.5 Balancing Nuclear Equations

Before we can study nuclear reactions, we need to learn how to balance a nuclear equation. There are a number of concepts to be remembered.

First, only nuclei are represented. The carbon isotope, with six protons and six neutrons, is represented by the symbol $^{12}_6C$. The subscript (6) is the number of protons in the nucleus (atomic number). The superscript (12) represents the total number of protons and neutrons in the nucleus (mass number). Thus, $^{80}_{35}Br$ indicates that bromine has 35 protons and 45 (80 − 35 = 45) neutrons. The symbols for some important particles are shown in Table 17-2. Notice that the beta particle is $^0_{-1}e$ because its charge is −1 and its mass number is 0. The deuteron is the deuterium nucleus. Deuterium is an isotope of hydrogen. It has one proton and one neutron, and its mass number is two. It is often called heavy hydrogen.

Table 17-2   Names and Symbols of Some Important Particles

Name of Particle	Symbol
Alpha	^4_2He or α
Beta	$^0_{-1}e$ or β^-
Proton	^1_1H or p
Neutron	^1_0n or n
Deuteron	^2_1H

When we attempt to balance a nuclear equation, we must remember to make sure that the sum of the superscripts on the left equals the sum of the superscripts on the right (conservation of mass). We must also ensure that the sum of the subscripts on the left equals the sum of the subscripts on the right (conservation of charge):

For example,

$$^{238}_{92}U \longrightarrow \, ^{234}_{90}Th + \, ^{4}_{2}He$$
$$238 = 234 + 4$$
$$92 = 90 + 2$$

$$^{13}_{7}N + \, ^{0}_{-1}e \longrightarrow \, ^{13}_{6}C$$
$$13 + 0 = 13$$
$$7 + (-1) = 6$$

$$^{24}_{11}Na \longrightarrow \, ^{24}_{12}Mg + \, ^{0}_{-1}e$$
$$24 = 24 + 0$$
$$11 = 12 + (-1)$$

$$^{235}_{92}U + \, ^{1}_{0}n \longrightarrow \, ^{90}_{38}Sr + \, ^{143}_{54}Xe + 3 \, ^{1}_{0}n$$
$$235 + 1 = 236 = 90 + 143 + (3 \times 1)$$
$$92 + 0 = 92 = 38 + 54 + (3 \times 0)$$

The total positive charge of the reacting nuclei must equal the total positive charge of the product nuclei. The mass of the reactants must equal the mass of the products.

17.6 Types of Nuclear Reactions

There are four types of nuclear reactions. The first type is radioactive decomposition. If a nucleus is altered by the loss of an alpha particle or a beta particle, the reaction is a **radioactive decomposition**.

If a nucleus is bombarded by alpha particles, protons, or neutrons, an unstable nucleus may result. The unstable nucleus can emit a proton or a neutron in order to gain stability. This process is called **artificial transmutation**.

If a heavy nucleus splits to form nuclei of intermediate mass, the process is called **fission**. If two light nuclei combine to form a heavier, more stable nucleus, the process is called **fusion**. Examples of these types of nuclear reactions appear in the following pages of this chapter.

17.7 Radioactive Decomposition

Alpha Decay

Certain heavy nuclei spontaneously break down into lighter nuclei. Alpha particles (helium nuclei) are emitted. This process is known as alpha decay:

$$^{238}_{92}U \longrightarrow \, ^{234}_{90}Th + \, ^{4}_{2}He$$

The uranium nucleus has been changed to a thorium nucleus because two protons have left the uranium nucleus. A **transmutation** is a change in the identity of a nucleus because of a change in the number of protons in the nucleus. Uranium has been transmuted to thorium.

When uranium is transmuted to thorium, the above equation describes 77% of the decay events. However, for 23% of the decay events, the following reaction is more correct:

$$^{238}_{92}U \longrightarrow {}^{234}_{90}Th^* + {}^{4}_{2}He$$

then $\quad {}^{234}_{90}Th^* \longrightarrow {}^{234}_{90}Th + Gamma Rays$

The ${}^{234}_{90}Th^*$ is an excited (high energy) thorium nucleus which loses its extra energy by giving off gamma rays.

If a sample initially contained a certain number of nuclei, then after a period of time called the **half-life** ($t_{\frac{1}{2}}$), only half of the original number of nuclei will remain (Fig. 17-6). The half-life for uranium-238 is 4.6×10^9 years. After 4.6×10^9 years, only one half of the original uranium will be left.

Figure 17-6
Half-life curve for radioactive decay of an element with $t_{1/2} = 20$ h

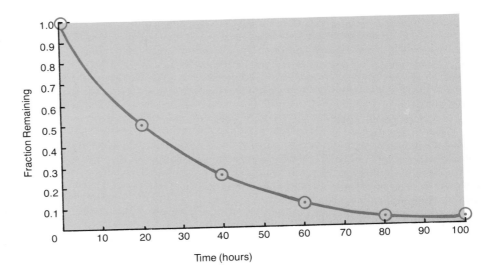

Beta Decay

The emission of a negatively charged electron (beta particle) from a nucleus causes the positive nuclear charge to increase by 1 [since − (−1) is equal to + 1]. Since the mass of an electron is very small, loss of one electron causes no change in the mass number. The thorium-234 nucleus formed during the alpha decay of uranium-238 is itself unstable and is transmuted to a protactinium-234 nucleus by beta decay:

$$^{234}_{90}Th \longrightarrow {}^{234}_{91}Pa + {}^{0}_{-1}e$$

Since nuclei are supposed to contain only protons and neutrons, we might wonder where the electron (beta particle) originated. The Italian nuclear scientist Enrico Fermi suggested that an electron is created at the moment of beta decay. A neutron is believed to be transformed into a proton and an electron at the moment of decay:

$$^{1}_{0}n \longrightarrow {}^{1}_{1}H + {}^{0}_{-1}e$$

Just as in alpha decay, some beta decays produce an excited state of protactinium-234. Gamma rays are produced as the excited protactinium decays.

The half-life of thorium-234 is 24 days. Thus, as the uranium slowly decays to thorium by alpha decay, the thorium more rapidly decays to protactinium by beta decay.

This radioactive decay series began with uranium-238, decaying to thorium-234, which in turn decayed to protactinium-234. The protactinium-234 decays by beta decay:

$$^{234}_{91}\text{Pa} \longrightarrow \,^{234}_{92}\text{U} + \,^{0}_{-1}\text{e}$$

The uranium-234 then decays by five alpha decays in a row to $^{214}_{82}\text{Pb}$. The lead-214 decays by two beta, then one alpha, then two more beta, and finally one alpha decay to $^{206}_{82}\text{Pb}$. Lead-206 is a stable nucleus and the series ends. This series is one of three naturally occurring radioactive decay series. All naturally occurring radioactive elements belong to one of the three series.

Alpha decays produce more stable nuclei by decreasing the mass of a heavy nucleus. The loss of an alpha particle lowers the mass of a heavy nucleus by four mass units. You will recall that nuclei of intermediate mass have more stability than heavy nuclei.

Beta decays produce more stable nuclei by converting a neutron into a proton. Since many nuclei have more neutrons than protons, the conversion of a neutron into a proton brings the proton-to-neutron ratio closer to one to one.

17.8 Applications of Natural Radioactivity

A reliable method of determining the age of an old object depends on the presence of natural radioactivity. Consider the chemical composition of a very old crystal of pitchblende (U_3O_8). If we assume that the U_3O_8 crystal precipitated from molten rock when the earth's crust first cooled, the resulting crystal would tend to be pure U_3O_8. However, analysis shows that every deposit of pitchblende contains some lead. Scientists assume that this lead came from uranium by natural radioactive decay. The quantity of lead is proportional to the length of time from the solidification of the rock to the present. The rate at which uranium disintegrates is known, and the age of the rock can be calculated from the amounts of lead and uranium in the rock. Minerals dated this way have been found to be at least 5×10^9 years old.

The age of minerals can also be determined by using the decay of $^{40}_{19}\text{K}$ to form $^{40}_{18}\text{Ar}$. By determining the amounts of the two nuclei, it is possible to determine the age of the mineral in which they are present. Such a determination has resulted in establishing that the solar system solidified about four and half billion years ago.

Another type of dating is called **carbon-dating**. Radioactive carbon-14 is produced from the bombardment of atmospheric nitrogen by neutrons from cosmic rays:

$$^{14}_{7}\text{N} + \,^{1}_{0}\text{n} \longrightarrow \,^{14}_{6}\text{C} + \,^{1}_{1}\text{H}$$

The concentration of radioactive $^{14}CO_2$ in the atmosphere is kept in balance by the constant production of carbon-14 and decay of carbon-14 back to nitrogen-14.

Living plants and animals establish a balance with the atmospheric $^{14}CO_2$, and thus they contain a constant amount of carbon-14. When these plants and animals die, the balance is upset as the carbon-14 decays without being replaced. Carbon-14 has a half-life of 5570 years, and measurement of the carbon-14 radioactivity in a specimen allows the calculation of how long ago death took place. Carbon-dating has shown, for example, that the Dead Sea Scrolls are about two thousand years old.

Part One Review Questions

1. Describe Rutherford's lead block experiment which showed that three different kinds of rays can be emitted from radioactive substances.
2. Explain the meanings of the terms *binding energy*, *p/n ratio*, and *even-odd rule*. Show how they are involved in determining the relative stability of a nucleus.
3. Complete the following equations:
   a) $^{27}_{13}Al + ^{2}_{1}H \longrightarrow ^{4}_{2}He + ?$ b) $^{7}_{3}Li + ? \longrightarrow 2\,^{4}_{2}He$
4. What is the difference between nuclear fusion and nuclear fission?
5. What is meant by the half-life of a radioactive substance? How can a knowledge of the half-life of a substance be used to determine the age of a material in which the substance is present?

Part Two: Splitting the Atom

Not all radioactive elements occur naturally. Many elements undergo artificially induced nuclear reactions with the formation of unstable nuclei, which then undergo radioactive disintegration. The process can be used to generate isotopes which are useful and to generate huge amounts of energy. In this part we shall see some of the ways in which these aims are accomplished.

17.9 Artificial Transmutation

Lord Rutherford produced the first artificial transmutation in 1919. He allowed alpha particles from a radium source to bombard nitrogen atoms. Protons and an isotope of oxygen were produced:

$$^{14}_{7}N + ^{4}_{2}He \longrightarrow ^{17}_{8}O + ^{1}_{1}H$$

Rutherford repeated this experiment with other elements. Nearly all the light elements behaved like nitrogen and emitted protons. However, for some reason the alpha particles would not penetrate the nuclei of elements heavier than potassium.

Since alpha particles have two positive charges each, they are repelled by the positively charged nuclei. A fast-moving alpha particle can overcome the reasonably small repulsive forces of the light elements. However, the repulsive force is greater when heavier elements with more protons in their nuclei are used as targets, and the alpha particles are repelled by these heavier nuclei.

If we wish to use a more efficient "bullet," a particle with a smaller positive charge should be used. A proton has only one half of the positive charge of an alpha particle. Therefore, protons should be able to enter the nucleus of an element more readily than alpha particles can.

In 1932, two English scientists, John Cockcroft and Ernest Walton, bombarded a target of lithium with high-energy protons obtained by ionizing hydrogen gas. They accelerated the protons with a voltage of 250 000 V. Alpha particles were obtained:

$$^{7}_{3}Li + ^{1}_{1}H \longrightarrow ^{4}_{2}He + ^{4}_{2}He + energy$$

Also in 1932, the English scientist James Chadwick bombarded beryllium with alpha particles. A particle with no charge and a mass of 1 u was obtained. This particle was called the neutron:

$$^{9}_{4}Be + ^{4}_{2}He \longrightarrow ^{12}_{6}C + ^{1}_{0}n$$

17.10 Particle Accelerators

The energy required for a transmutation reaction is usually provided as kinetic energy by one of the reactants. For example, the energy required for the reaction of alpha particles with $^{9}_{4}Be$ atoms comes from the kinetic energy of the alpha particles which are travelling at speeds of 2 to 3% of the speed of light.

Generally, the charged particles used to produce transmutation reactions are accelerated to the necessary kinetic energies by machines (accelerators) that use magnetic and electric fields to accelerate the particles. Such machines include cyclotrons (Fig. 17-7) and linear accelerators (Fig. 17-8). Cyclotrons

Figure 17-7
Canada's TRIUMF cyclotron under construction on the campus of the University of British Columbia

accelerate particles in a spiral path (Fig. 17-9). In linear accelerators, particles pass down a series of charged cylinders so that they are continuously accelerated (Fig. 17-10). In all accelerators, the particles move in a vacuum so as to avoid collisions with gas molecules.

Figure 17-8
The Stanford Linear Accelerator Centre (SLAC). This is the largest linear accelerator in the world. It is about 3.2 km long.

Figure 17-9
The path of charged particles in a cyclotron

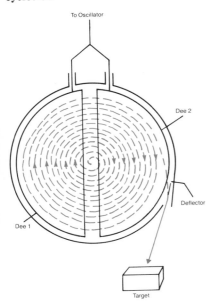

Figure 17-10
Linear accelerator

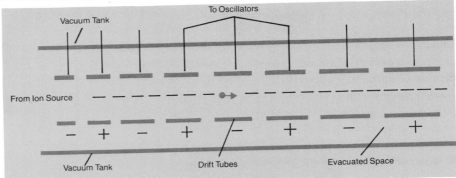

APPLICATION

Canada's TRIUMFant Particle Accelerator

TRIUMF—the *TRI-U*niversity *M*eson *F*acility—on the University of British Columbia campus, is one of only three establishments in the world that specialize in producing huge quantities of pi-mesons, or pions.

Pions are subatomic particles that are responsible for most nuclear forces between protons and neutrons in atomic nuclei. TRIUMF uses a cyclotron 18 m in diameter to accelerate protons to 75% of the speed of light (fast enough to travel from Vancouver to Halifax in 30 ms) and then smash them into target nuclei. The collisions produce a variety of subatomic particles that can be used in research and medical applications.

These particles can be detected by sophisticated electronic instruments that record their speeds, masses, charges, directions, and types. Scientists then analyze the records for clues to the nature of matter and the forces that hold the nucleus together.

The standard model of matter assumes that everything in the universe is made of atoms, which in turn are made of protons, neutrons, and electrons. Protons and neutrons are composed of even smaller particles called quarks, which are held together by gluons. The model does not reveal whether subatomic particles such as quarks and electrons are divisible into even more basic units, although it does allow for predictions of the ways in which unstable particles decay. Studies such as those carried out at TRIUMF enable scientists to probe for the existence of new particles and to test the assumptions of the standard model.

TRIUMF supports many applied research programs. The best known of these is the use of an intense beam of pions to treat deep-seated tumors. The pions are deposited inside the tumor, where they give up all their energy, thereby destroying cancer cells. Several pharmaceutical companies use TRIUMF particles in the production of radioactive isotopes for medical diagnosis. TRIUMF neutron beams are also used to analyze ore samples and detect valuable elements such as gold and platinum in quantities as small as a few micrograms per kilogram of ore. TRIUMF has contributed to the growth of companies that build its equipment. In 1975, a small company won the contract for TRIUMF's vacuum chamber and resonators. The knowledge gained on the job enabled them to expand internationally, and the value of their annual exports now exceeds TRIUMF's annual budget.

17.11 Neutrons Are Better Bullets

Before the discovery of neutrons in 1932, alpha particles, deuterons, and protons were used for studying atomic nuclei. These positively charged particles must be hurled against nuclear targets at tremendous speeds because they are repelled by the positive nuclear targets. However, neutrons have no charge, and they are more readily able to penetrate the nuclei of target atoms. As a result, neutrons are often used in the study of atomic nuclei.

Usually neutrons produced by nuclear reactions must be slowed down before they can be absorbed by other atoms, causing their nuclei to disintegrate. Fast neutrons lose their kinetic energy when they collide with the atoms, which are able to slow the neutrons without absorbing them or reacting with them. Materials commonly used to slow neutrons include heavy water (D_2O, Section 8.10), graphite, carbon dioxide, and ordinary water.

A convenient laboratory source of neutrons is obtained by bombarding beryllium with alpha particles:

$$^{9}_{4}Be + ^{4}_{2}He \longrightarrow ^{12}_{6}C + ^{1}_{0}n$$

Neutron bombardment may produce artificial elements. When slow neutrons strike uranium-238, the product, uranium-239, is unstable. This nucleus undergoes beta decay and is converted to a new element, neptunium-239. The neptunium is also unstable and undergoes beta decay to form another new element, plutonium-239.

$$^{238}_{92}U + ^{1}_{0}n \longrightarrow ^{239}_{92}U$$

$$^{239}_{92}U \longrightarrow ^{239}_{93}Np + ^{0}_{-1}e$$

$$^{239}_{93}Np \longrightarrow ^{239}_{94}Pu + ^{0}_{-1}e$$

Neptunium and plutonium are two **transuranium elements**: that is, they have more than 92 protons in their nuclei. Other transuranium elements may be prepared by bombardment with different particles:

$$^{238}_{92}U + ^{12}_{6}C \longrightarrow ^{246}_{98}Cf + 4\,^{1}_{0}n$$

$$^{238}_{92}U + ^{14}_{7}N \longrightarrow ^{247}_{99}Es + 5\,^{1}_{0}n$$

Stable nuclei can be made artificially radioactive by neutron bombardment. Cobalt-59 is not radioactive, but it can be converted to radioactive cobalt-60:

$$^{59}_{27}Co + ^{1}_{0}n \longrightarrow ^{60}_{27}Co$$

Sulfur-32 is not radioactive but can be converted to radioactive phosphorus-32:

$$^{32}_{16}S + ^{1}_{0}n \longrightarrow ^{32}_{15}P + ^{1}_{1}H$$

It has been possible to prepare radioactive isotopes of all the elements. Many of the radioactive isotopes are useful, as we shall see in Section 17.19.

17.12 Fission

At the time that Enrico Fermi was demonstrating that uranium-238 could be transmuted to neptunium and plutonium, a number of unexplained results indicated that occasionally a process which released tremendous energy took place. In 1939 two Germans, Otto Hahn and Fritz Strassmann, proved that whenever this large energy release occurred, atoms of intermediate atomic number were produced. Lise Meitner, a refugee from Hitler's Germany, suggested that the uranium nucleus absorbed a neutron and then split into two roughly equal fragments. The smaller masses associated with these fragments allow for a great release of energy. In fact, as the uranium nucleus splits, about 0.1 percent of its mass is converted into a huge amount of energy. Meitner passed her idea to Niels Bohr, who was on his way to the United States. This idea caused great excitement.

The splitting of a uranium nucleus into two smaller nuclei was called fission. It was found that uranium-235 was responsible for most of the fission and that, in the process, two or three neutrons were set free:

$$^{235}_{92}U + ^{1}_{0}n \longrightarrow ^{141}_{56}Ba + ^{92}_{36}Kr + 3\ ^{1}_{0}n + energy$$

The fact that neutrons were released in the reaction meant that if these neutrons encountered other uranium-235 nuclei, they would cause additional fissions. A rapidly growing chain reaction would occur (Fig. 17-12). Since nuclear reactions are very fast, the almost instantaneous release of huge quantities of energy with every fission could produce an explosion of unprecedented size. The fission of 1 kg of nuclear fuel such as uranium-235 releases as much energy as the burning of 2 700 000 kg of coal or 2 600 000 L of fuel oil.

Biography

LISE MEITNER was born in Vienna in 1878. At a young age, she displayed an interest in science by reading newspaper accounts of the discovery of radioactivity and the isolation of radium by Marie Curie. In 1906, she was one of the first women to acquire a doctorate from the University of Vienna. From 1908 to 1911, she served as an assistant to Max Planck at the University of Berlin.

Meitner became a professor at the Kaiser Wilhelm Institute for Chemistry in Berlin, where she became famous for her studies on radioactive radium, thorium, and actinium. For three decades, she co-operated with the German chemist, Otto Hahn. In 1917, she and Hahn discovered element 91, protactinium.

In 1938, Meitner had to move to Stockholm, Sweden to escape the Hitler regime. In the same year, Hahn and another associate, Fritz Strassmann, found that the bombardment of uranium with neutrons changed some of the uranium to barium. In 1939, Meitner explained that the nuclei of uranium atoms had split into smaller fragments. She developed a mathematical theory to explain this process which she called nuclear fission, and she calculated the energy released during the reaction.

Lise Meitner continued to work on nuclear physics at the University of Stockholm until 1960, when she retired in England. In 1966, she shared the Enrico Fermi Award with Hahn and Strassmann for their joint work which led to the discovery of nuclear fission.

Figure 17-11
Lise Meitner (1878-1968)—
Explaining nuclear fission

Figure 17-12
A nuclear chain reaction

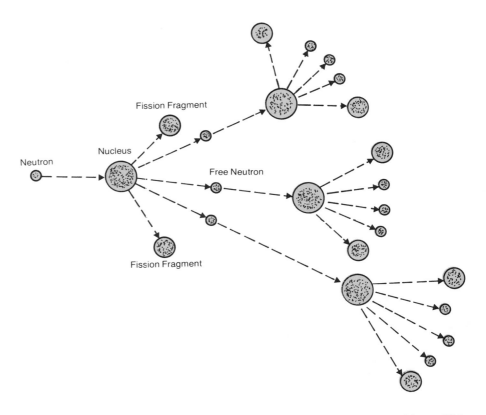

Scientists in the United States could easily foresee a weapon of incredible destructiveness. The fact that German scientists knew as much about fission as Americans did was very disturbing to the American scientists. In 1939, Albert Einstein urged President Roosevelt to increase research in the area of nuclear fission. The atomic bomb was developed from a research effort (the Manhattan Project) which cost 2.2 billion dollars.

17.13 The Atomic Bomb

Natural uranium is 99.3% uranium-238 and 0.7% uranium-235. Uranium-235 is fissionable when it is struck by neutrons. The fission of uranium-235 produces a variety of pairs of fission products such as barium and krypton. Uranium-238 does not split. It absorbs slow neutrons without fission. To isolate enough uranium-235 or plutonium (which will also undergo rapid fission in a chain reaction) is one of the difficult tasks in making an atomic bomb.

In order for it to explode, the bomb must contain a certain minimum quantity of fissionable material. This minimum quantity is called the **critical mass**. If the amount of fissionable material is too small, too many neutrons leave the mass without striking other nuclei. Neutrons from nuclei on the surface escape into space. The bomb, in the safety position, consists of two or more separate quantities of fissionable material, all of which contain less than the critical mass. If such masses are jammed together to form a larger mass which has less surface area per unit volume, the critical mass is exceeded and an explosive chain reaction results. The chain reaction is started by stray neutrons.

An atomic bomb was first used in war on August 6, 1945, when the United States dropped a bomb containing uranium-235 on the Japanese city of Hiroshima (Fig. 17-13). This bomb released a quantity of energy equivalent to 18 kt of TNT. Sixty-six thousand people were killed and 69 000 injured as a result of the explosion. Two days later, an atomic bomb containing fissionable plutonium-239, which requires a smaller critical mass than does uranium, was dropped on another Japanese city, Nagasaki. As a result of this explosion, 39 000 were killed and 25 000 were injured. The explosion of an atomic bomb produces extremely high temperatures, a severe shock wave, and gamma rays. Gamma rays are dangerous because, even at great distances from the centre of the explosion, they can cause radiation sickness and genetic damage.

Figure 17-13
Hiroshima's main street after the atomic bomb blast in 1945. This view is from about 2.5 km from the centre of the explosion.

Part Two Review Questions

1. Write an equation for a reaction which involves the artificial transmutation of one element into another.
2. Why are neutrons better bullets for studying atomic nuclei than alpha particles or protons?
3. Describe in words a chain reaction such as the one that is involved in the fission of uranium nuclei.
4. What is meant by the critical mass of a fissionable material? Why is critical mass important?

Part Three: Canada's Nuclear Industry

Canada has long been regarded as a world leader in the development of nuclear energy for peaceful purposes. Peaceful uses of nuclear energy include the harnessing of nuclear reactions to provide the energy required by an industrial society and the development of useful radioactive isotopes.

17.14 Nuclear Reactors

In 1942, in a converted squash court at the University of Chicago, a group of scientists led by Fermi showed that a chain reaction could be controlled in a device called a **nuclear reactor**. Their reactor was called an atomic pile. It was constructed by building up layers of uranium oxide and uranium metal separated by graphite bricks. Graphite is a **moderator**; that is, it has the ability to slow down neutrons. When they are first thrown out of a fissioning nucleus, the neutrons have enormously high speeds. Fermi believed that since uranium-235 is present in such small proportions in natural uranium the neutrons must be slowed down so that they can be more easily captured by the uranium-235 nuclei that are present. The neutrons can be slowed down by bouncing them off the carbon atoms in the graphite blocks. The original pile also contained control rods of cadmium metal. When these rods were inserted in the reactor, they absorbed neutrons and slowed down the chain reaction. The first pile contained more than the amount of uranium needed to maintain the chain reaction. The control rods were in place and, as they were removed, fissions were able to occur. At the critical point, the pile operated on its own. The pile could be shut down by replacing the control rods.

Fermi's pile was cooled by allowing air to circulate between the graphite blocks. All reactors have a fuel, a moderator, a control, and a cooling system.

Reactors have three main functions. They produce radioactive isotopes which have many applications; they release vast amounts of energy in the form of heat from the fission of small quantities of matter; and they produce new nuclear fuels from nonfissionable elements.

17.15 Canada and Nuclear Chemistry

Much of the fundamental research on radioactivity has been done in Canada, and it is appropriate that Canadians are very interested in the peaceful uses of nuclear energy. The Atomic Energy Research Laboratories at Chalk River, Ontario, began operation in the summer of 1944. In 1947, they were taken over as a division of the National Research Council. In 1952, the Crown corporation Atomic Energy of Canada Limited (AECL) took over the job of running the Chalk River Laboratories. Today, besides the nuclear research laboratories at

Chalk River, AECL has a reactor design group in the Sheridan Park Laboratories, Toronto, a group with a reactor at Pinawa, Manitoba, and a commercial products group in Ottawa.

Canadian scientists have concentrated on using natural uranium as fuel and heavy water as moderator in their reactors. Canadians made the first reactor built outside the United States. In 1945, the ZEEP (Zero Energy Experimental Pile) was made operational at Chalk River. The ZEEP is an aluminum tank (2.4 m in diameter and 3.0 m deep) which is filled with heavy water and surrounded with a graphite layer (0.9 m thick) which tends to reflect departing neutrons back into the heavy water. Movable rods of uranium cased in aluminum are hung in the heavy water. The distance which gives the best thickness of heavy water between rods is found experimentally. If the rods are too close together, the fast neutrons are not slowed down enough to cause sufficient fission reactions. If the rods are too far apart, the neutrons are scattered without causing fission reactions. The ZEEP is used to find the best distances between fuel rods for various fuels. The distance differs for metallic uranium, uranium oxide, and uranium carbide. The best distance between fuel rods also depends on the kind of metal container in which the fuel is placed.

The heavy water NRX reactor (Fig. 17-14) began operation at Chalk River in 1947. It is now operating to release 42 MW (megawatts) of power. NRX has automatic means of irradiating materials for the production of isotopes. NRX produces a great quantity of neutrons from the fission of uranium-235. Any element placed in NRX is subjected to high neutron bombardment and will become a radioactive isotope of the same or some other element. For example:

$$\underset{\text{nonradioactive}}{{}^{59}_{27}\text{Co}} + {}^{1}_{0}\text{n} \longrightarrow \underset{\text{radioactive}}{{}^{60}_{27}\text{Co}}$$

The cobalt nucleus has picked up a neutron and has become unstable.

NRX was unsurpassed as a research and fuel testing reactor until the construction of the 200 MW reactor, NRU. NRU began operation in 1957. It has the finest research facilities of any reactor. NRU is similar to NRX except that it is both cooled and moderated by heavy water. It is fueled with natural uranium.

Figure 17-14
NRX reactor

17.16 AECL Power Projects

The AECL Power Projects group is responsible for the design of nuclear power plants and for the project management of nuclear power stations built by AECL. Since the fission of 1 kg of a nuclear fuel such as uranium-235 releases as much heat as the burning of 2 700 000 kg of coal or 2 600 000 L of fuel oil, it is natural that scientists should be interested in using this source of energy for the production of electric power.

The Canadian nuclear power program is based on experience with the heavy water-moderated, natural uranium type of reactors which have been so successful at Chalk River. Canadian scientists have had more experience than any other group with this type of reactor. Southern Ontario requires more electricity than can be supplied by hydro stations. In the past coal-burning stations were built. It was decided to develop a nuclear power system in order to see if it could compete with a coal-burning system.

A 20 MW nuclear power demonstration plant (NPD) was built in 1962 at Rolphton, Ontario. NPD was successful, and a full-size plant was constructed at Douglas Point, Ontario, in 1967. The Douglas Point Power Reactor is similar to NPD, except that it is bigger and produces 200 MW. This reactor was built by AECL and is operated by Ontario Hydro. It is a CANDU (Canada Deuterium Uranium) reactor.

17.17 CANDU Reactors

The CANDU reactor has proven to be one of the most efficient types of nuclear reactors in the world. The Bruce Nuclear Generating Station at Douglas Point has a capacity of 3400 MW. There are eight operating units on the Bruce site. The four largest CANDU reactors in the world have been built at Darlington, Ontario. Darlington is one of the world's largest power stations. Another CANDU project has been completed at Gentilly, Quebec, and one more has been completed at Point Lepreau, New Brunswick.

Another plant operated by Ontario Hydro is located at Pickering, about 30 km east of Toronto, Ontario (Fig. 17-15). It is one of the most successful

Figure 17-15
The Pickering nuclear power station

commercial nuclear power stations in the world. Pickering A produces about 2000 MW and is made up of four CANDU reactors. Pickering B, which is located next to it, provides an additional four CANDU reactors. The eight reactors have the capability of lighting up a city of three million people.

We will now examine the operation of a CANDU reactor.

Fuel

The fuel for a CANDU reactor is prepared from uranium ore (U_3O_8). The ore is converted into pellets of UO_2 which are placed in zircaloy tubes whose ends are welded shut. Twenty-eight of these tubes are assembled to form a fuel bundle (Fig. 17-16). Each fuel bundle contains 22.2 kg of UO_2.

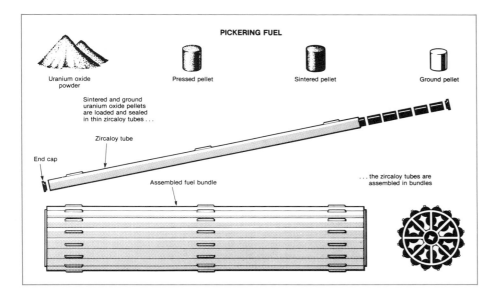

Figure 17-16
Fuel for the CANDU reactor

The fuel bundles are then placed in fuel channels in the reactor vessel, called a calandria (Fig. 17-17). A full fuel charge for one reactor consists of 4680 fuel bundles having a total mass of 105 t of fuel. The heat produced by this much fuel is equal to the heat that could be provided from about 2 700 000 t of coal.

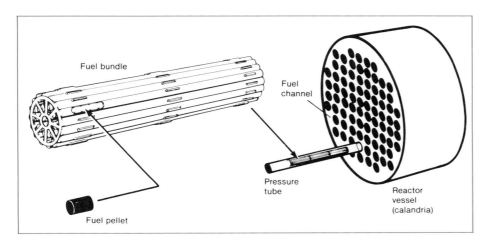

Figure 17-17
Fuel bundle and fuel channel relationship

A calandria during construction

The natural uranium in the UO_2 fuel pellets consists of 99.3% $^{238}_{92}U$ and 0.7% $^{235}_{92}U$. The more plentiful uranium-238 cannot be used directly for producing electricity. Rather, it is the small amount of uranium-235 which actually undergoes fission and produces energy in the reactor.

There are two types of reactions in a reactor which is fueled with natural uranium: the fission reaction and the neutron capture reaction. The fission reaction involves the splitting of uranium-235 into smaller products. An example of the fission reaction is

$$\underset{\text{slow}}{^{235}_{92}U + {}^{1}_{0}n} \longrightarrow {}^{90}_{38}Sr + {}^{143}_{54}Xe + \underset{\text{fast}}{3 \, {}^{1}_{0}n} + Energy$$

The fast neutrons that are produced are slowed down by a moderator so that they can bombard other uranium-235 nuclei efficiently. Some of the fast neutrons, however, hit the more abundant uranium-238 nuclei before they get slowed down by the moderator. These fast neutrons are captured by the uranium-238 nuclei, which are converted to plutonium-239 nuclei:

$$^{238}_{92}U + {}^{1}_{0}n \longrightarrow {}^{239}_{92}U \longrightarrow {}^{239}_{93}Np + {}^{0}_{-1}e$$

$$\text{then} \quad {}^{239}_{93}Np \longrightarrow {}^{239}_{94}Pu + {}^{0}_{-1}e$$

Notice that the fission reaction is a chain reaction because neutrons are used up and they are also produced. The neutron capture reaction uses up neutrons, but it does not produce them.

The plutonium-239 that is produced is also capable of undergoing fission. About half the plutonium-239 undergoes fission to contribute more than one-third of all the heat produced by a fuel bundle.

As the uranium fuel is used up, spent fuel is removed from one end of each fuel channel while fresh fuel is added to the other end. This process is done automatically under the direction of a computer and takes place while the reactor is in full operation. A typical 600 MW CANDU plant consumes 95 t of UO_2 fuel every 17 months.

Moderator

In order for uranium-235 to be able to undergo fission its nuclei must absorb slow-moving neutrons. The neutrons released as a result of fission of uranium-235 are moving much too fast (up to 42 000 km/s). The neutrons must be slowed down to about 3 km/s so that they may be absorbed by uranium-235 nuclei. They are slowed by having them collide with particles of similar mass. CANDU reactors use the moderator heavy water (D_2O) to slow neutrons down. The fast moving "fission" neutrons bounce off deuterium atoms a number of times and are slowed down.

Heavy water is by far the best moderating material. It is a better moderator than light water or graphite because the deuterium in heavy water has a much smaller probability of actually capturing these neutrons than does ordinary hydrogen in light water or carbon in graphite.

Any capture of "fission" neutrons by the moderator would interfere with the fission chain reaction. A higher concentration of uranium-235 would then be needed in the reactor in order to sustain a chain reaction. Thus, CANDU reactors using heavy water as a moderator are able to use natural uranium (containing only 0.7% $^{235}_{92}U$) fuel while reactors in the United States and elsewhere using ordinary water as a moderator must use enriched uranium (3 to 4% $^{235}_{92}U$) fuel.

The efficiency of heavy water as a moderator makes it possible to carry natural uranium to a very high value of burn-up. Burn-up indicates the amount of energy extracted from the fuel during its stay in the reactor. A higher power output is obtained from a smaller amount of fuel when the burn-up is high. The disadvantage of heavy water is its high cost. Each reactor at the Pickering plant uses 301 000 kg of heavy water.

Controls

The rate of energy output from a reactor is controlled by adjusting the level of ordinary water in one or more of the 14 compartments (liquid zone control units) placed among the fuel channels in the calandria. Ordinary water is a good absorber of neutrons. If the level of water is raised in one of these units, more neutrons in that zone of the calandria are captured by the water and prevented from colliding with uranium-235. Thus, the chain reaction (and energy output) is reduced in that zone of the calandria. If the level of ordinary water in all 14 liquid zone control units is raised enough, the chain reaction can be slowed or even halted everywhere in the calandria.

In addition to methods for controlling the reactor, there are three types of shut-down mechanisms. The first type involves shutoff rods (Fig. 17-18). Shutoff rods are made of a neutron absorbing material such as a cadmium-stainless steel alloy. When these rods are dropped into the core of the reactor, they capture the fission neutrons and stop the chain reaction.

The second type involves the injection of a liquid "poison" directly into the moderator (Fig. 17-18). This "poison" is a neutron absorber such as a concentrated solution of gadolinium nitrate. Once again, a loss of fission neutrons stops the chain reaction.

The third type of shut-down mechanism involves dumping the heavy water moderator into a special dump tank (Fig. 17-19) located below the calandria. Without the moderator, the fast neutrons are not slowed sufficiently for fission to continue.

Coolant

The products of the fission reactions move apart with great energy. They jostle the atoms in the fuel and thus cause the fuel to heat up. Heavy water is used as a coolant in the CANDU system and flows through tubes which enclose the fuel bundles. This coolant is under high pressure so that it will not boil. The hot heavy water coolant is circulated to a steam generator where it causes ordinary water to boil. The steam produced is fed to turbines which drive electrical generators. The heavy water coolant is pumped back through the reactor core. A typical reactor uses 143 t of heavy water as coolant. The operation of a CANDU reactor is schematically represented in Figure 17-20.

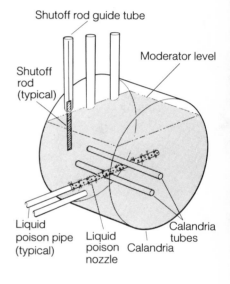

**Figure 17-18
Shutdown mechanisms for a CANDU reactor**

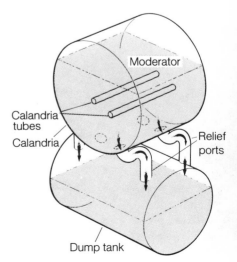

**Figure 17-19
Location of the heavy water dump tank**

Figure 17-20
The CANDU-PHW System
(Canada Deuterium Uranium—
Pressurized Heavy Water)

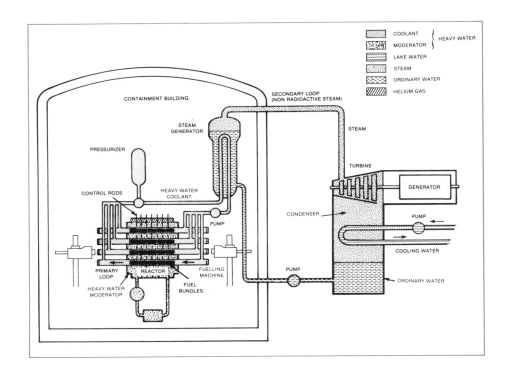

Used Fuel

A fuel bundle produces heat for about 17 months and is removed almost unchanged in appearance. About 70% of the uranium-235 is used up. In addition to the remaining uranium, radioactive fission products are present, along with plutonium produced from uranium-238. A used fuel bundle thus contains 98.6% uranium-238, 0.2% uranium-235, 0.4% plutonium, and 0.8% of other isotopes.

Figure 17-21
(a) Temporary storage of spent fuel
(b) A spent fuel storage bay at a
CANDU generating station.

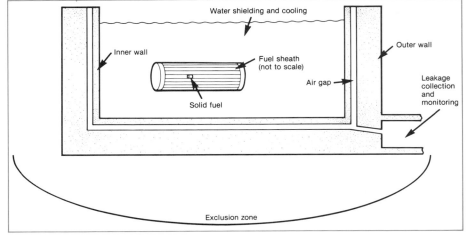

All the radioactive fission products are locked into the used fuel bundles, and natural decay eliminates 99.8% of the radiation in ten years. However, plutonium, which is extremely dangerous if it enters the body, has a half-life of 24 000 years.

Most of Canada's used fuel is temporarily stored in thick-walled, reinforced concrete pools at the nuclear power stations (Fig. 17-21). Water must be present for the initial few years to remove heat from the spent fuel. The concrete and water ensure protection from the penetrating radiation of the fission products.

It is not yet certain what will happen to the radioactive fuel bundles when they can no longer be kept in temporary storage. There are two suggestions for permanent disposal. First, the entire fuel bundle could be sealed in corrosion-proof containers and disposed of. Second, the useful materials such as plutonium could be extracted from the fuel bundles, leaving waste materials which could be chemically reacted with glass-forming substances to form insoluble solids for permanent disposal.

In either event, the disposable materials would be placed in a deep vault carved out of rock in the Canadian Shield. Certain geological formations in the Canadian Shield are known to have been stable for millions of years. The nuclear waste deep vault would resemble a large hard-rock mine, with the wastes stored under 500 to 1000 m of rock (Fig. 17-22). A single vault would be able to store all the nuclear fuel wastes produced in Canada for the next fifty years. After a period of monitoring, the vault would be sealed. The question of how to dispose of used fuel has been one of the most controversial issues related to the use of nuclear power.

Figure 17-22
Deep vault for permanent disposal of radioactive waste

17.18 Production of Heavy Water

Heavy water (D_2O, deuterium oxide) is present in ordinary water to the extent of one molecule of D_2O for every 7000 molecules of H_2O. This deuterium oxide must be extracted for the use of the nuclear industry. The process used in Canadian plants located at the Bruce Nuclear Power Development is based on the exchange of deuterium between liquid water and hydrogen sulfide gas at two different temperatures. The deuterium migrates to the hydrogen sulfide gas at a temperature of 128°C:

$$D_2O(\ell) + H_2S(g) \longrightarrow H_2O(\ell) + D_2S(g)$$

The D_2S is then dissolved in cooler (32°C) water, and the deuterium migrates to the water, enriching its content of D_2O:

$$D_2S(g) + H_2O(\ell) \longrightarrow D_2O(\ell) + H_2S(g)$$

The process is repeated many times, until eventually the concentration of D_2O reaches 20 to 30%. At this point, use is made of the fact that D_2O has a higher boiling point (b.p. 101.4°C) than has H_2O (b.p. 100.0°C). The lower boiling H_2O is removed in a very efficient distillation process to yield reactor grade D_2O that is 99.75% pure.

A facility for preserving food by using gamma radiation.

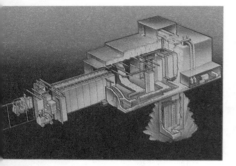

17.19 AECL Commercial Products—Uses of Radioactive Isotopes

The Commercial Products group of AECL is responsible for the processing and distribution of radioactive isotopes and for the design and manufacture of associated equipment. It is a unique organization, being the only one of its kind in the world to offer a complete "package" comprising the preparation and supply of radioisotopes; the manufacture and installation of equipment; and the provision of a variety of consulting and technical services.

Radioactive isotopes have application in medical therapy. Canadians and Canadian reactors have played an important role in this field. Through the effort of Dr. Cipriani at Chalk River, the first radioactive cobalt-60 sources were made for the treatment of malignant tumors. These cobalt-60 sources produce intense beams of gamma rays which destroy tumors. In addition to the cobalt-60 cancer therapy machines (*cobalt bombs*), cesium-137 *Caesatrons* have been made and sold by AECL Commercial Products. Over 700 cancer therapy units have been made by the Commercial Products group of AECL and shipped to 51 different countries.

The gamma radiation from cobalt-60 is used to sterilize vegetables, fruit, and grain, as well as medical supplies and wool. Potatoes sterilized by the radiation from cobalt-60 will not sprout, fruit can have a longer shelf life in stores, and medical instruments can be put in packages and sterilized by gamma radiation, ready for use when unpacked. Commercial Products supplied the first commercial units for this type of sterilization to companies in Canada, the United States, India, and New Zealand.

Radioisotopes are used in industry to study the wearing of machine parts. For example, the efficiency of lubricants to diminish wear on moving parts of machines is determined by incorporating radioisotopes into the metal surfaces. After even the slightest amount of metal wear, radioactive atoms will be detected in the lubricants. Radioactive isotopes may be added to a liquid flowing in a pipeline. By means of a Geiger counter, one can determine the position of a leak or obstruction in the pipeline.

Part Three Review Questions

1. What are the three main functions of a nuclear reactor?
2. What is the fuel used in a CANDU reactor?
3. Write the equations for the chain-reaction fission of uranium-235 and for the neutron capture reactions of uranium-238.
4. Describe the function of a moderator. What is the moderator used in a CANDU reactor?
5. Describe briefly the three types of shut-down mechanisms used in a CANDU reactor.
6. How is heavy water obtained from ordinary water?
7. State four uses of radioactive isotopes.

Part Four: The Nuclear Age

In almost all fields of human endeavour, the discovery of one method of accomplishing a task frequently leads to other improved methods of achieving the same ends. The nuclear industry is no exception. Thus, improved nuclear reactor design has led to more efficient production of energy. There are, however, dangers associated with the production of energy by nuclear reactors. There is always a small possibility of exposure to radioactivity accidentally leaked from a nuclear power plant (as happened at Chernobyl in the Ukraine in 1986) or to radiation from mishandling of isotopic products. In the next several sections we shall discuss these and other problems associated with living in the nuclear age.

17.20 Breeder Reactors

As we noted in Section 17.17, when uranium-235 undergoes fission in a reactor, a second reaction also takes place:

$$^{238}_{92}U + ^{1}_{0}n \longrightarrow ^{239}_{92}U \longrightarrow ^{239}_{93}Np + ^{0}_{-1}e$$

$$\text{then} \quad ^{239}_{93}Np \longrightarrow ^{239}_{94}Pu + ^{0}_{-1}e$$

In fact, nonfissionable uranium-238 is converted to fissionable plutonium-239. Since an average of 2.5 neutrons are released during each fission of a uranium-235 nucleus (two neutrons are produced in some fissions, and three are produced in others; the average is 2.5), it is possible to make a reactor that produces more fissionable fuel than it consumes.

A breeder reactor is operated by deliberately reducing the amount of moderator so that more fast neutrons are produced to convert uranium-238 to plutonium-239. In this way, the more plentiful component of natural uranium becomes usable for energy production. Some of the latest breeder reactors, called *fast breeders*, use virtually no moderators so that a maximum of plutonium-239 is produced. The plutonium-239 is recovered by shutting down a breeder reactor from time to time and reprocessing the unspent fuel. The plutonium-239 can then be used to produce fuel cores for other reactors.

17.21 Fusion

Energy is released when light atoms combine to form heavier atoms. These reactions are called fusion reactions. A simple fusion is the combination of hydrogen nuclei (protons) to form helium nuclei:

$$4\,^{1}_{1}H + 2\,^{0}_{-1}e \longrightarrow ^{4}_{2}He + \text{Energy}$$

This is the type of reaction that occurs in the sun. The reactants have more mass than the products. Thus, some mass has been converted into a tremendous amount of energy. In fact, a fusion reaction liberates more energy than a fission reaction.

The fusion of light elements is a difficult operation. All nuclei are positively charged. To make them combine, they must be shot at one another with high enough speeds to overcome their mutual repulsion. As a result, fusion will take place only at high temperatures. The fusion of hydrogen nuclei to form helium nuclei occurs only in the sun and other stars where the interior temperatures are around 20 million degrees. However, research in the field of constructing fusion reactors is being done in a number of countries.

17.22 The Hydrogen Bomb

The heavier isotopes of hydrogen (deuterium and tritium) fuse more rapidly than ordinary hydrogen and at a lower temperature and can be used to construct a new type of bomb. Hydrogen bombs have already been tested. The main reaction is the fusion of deuterium and tritium. Tritium is prepared from lithium-6, and the compound lithium deuteride supplies both the tritium and deuterium. An atomic bomb is the detonator for a hydrogen bomb. The atomic bomb produces the high temperature necessary for the fusion to take place.

There are probably three different reactions in the hydrogen bomb. The explosion of the atomic bomb is the first. In addition to high temperature, it also liberates neutrons. The lithium-6 is split into tritium and helium by the neutrons:

$$^6_3\text{Li} + ^1_0\text{n} \longrightarrow ^3_1\text{H} + ^4_2\text{He} + \text{Energy}$$

The tritium fuses with the deuterium:

$$^3_1\text{H} + ^2_1\text{H} \longrightarrow ^4_2\text{He} + ^1_0\text{n} + \text{Energy}$$

The hydrogen bomb releases much more energy than the atomic bomb. Consequently, the use of fusion for non-peaceful purposes is even more dangerous than the use of fission.

17.23 Ionizing Radiation

Ionizing radiation is radiation which is capable of interacting with atoms or molecules to produce ions. Such radiation can be particularly harmful to humans if it comes into contact with cells and causes chemical reactions. Such an interaction may lead to damage of cells, resulting in cancer and possible birth defects.

A number of units are used for measuring ionizing radiation. The sievert is the SI unit of dose equivalent. However, the **rem** (radiation equivalent man) is much more commonly used as a measure of the different effects of various types of radiation on people. It does not correspond to a fixed amount of radiation energy. Rather, it takes into account the varying effects of different types of radiation on biological matter. For example, alpha particles are twice as damaging as protons with the same energy, and these in turn are five times as damaging as X-rays.

Cosmic rays, X-rays, gamma rays, alpha and beta particles, and high speed neutrons are all forms of ionizing radiation. The sources of the ionizing radiation are both natural (radiation from space, from radioactive elements in the earth and in our bodies) and the result of human activity (medical X-rays, radiation therapy, nuclear power plants, and weapons-testing fallout). A small amount of radiation comes from sources such as bricks, television, and certain foods. Table 17-3 shows the estimated annual public exposure to radiation.

An international committee of scientists and medical doctors recommends that 500 mrem/year be considered a safe maximum dose for a member of the general public. Workers at nuclear power plants may be exposed to annual maximum doses of 5000 mrem. However, doses up to 25 000 mrem produce no detectable physiological effects. A dose of 500 000 mrem causes an estimated death rate of half an exposed population within 30 days after exposure. A comparison of these figures with those listed in Table 17-3 suggests that the annual public exposure to radiation is well below the estimated danger level.

Table 17-3   Estimated Average Annual Public Exposure to Radiation

Source	Radiation Exposure (mrem)
From artificial sources:	
Nuclear power plant (at the boundary)	4-5
Weapons–testing fallout	4-5
Industrial/Commercial	1
Medical diagnoses	65-75
From natural sources:	
Elements in the body	20-30
Earth and space	70-80

17.24 Problems of the Nuclear Age

The first problem of the nuclear age is what to do with the radioactive wastes from a reactor. Some of these radioactive wastes have short half-lives and "cool off" rapidly. However, some of the radioactive wastes, such as strontium-90 and cesium-137, have long half-lives.

These wastes must be stored in safe places for many years. Strontium-90 and cesium-137 have half-lives of about 30 years. It has been suggested that it takes 20 half-lives before a decaying isotope is safe. Thus, it would require 600 years for cesium-137 and strontium-90 to decay to a safe radiation level. This means that the storage facilities for radioactive wastes must be safe and impervious to geological action for up to a thousand years.

The total problem of radioactive waste disposal, not only of fission products but also of the slightly contaminated water or even trash from reactor plants, is one on which no compromises can be allowed. This is one area where neither industry nor government agencies should be allowed to take shortcuts.

A second problem involves the hazards associated with uranium exploration and mining. Some of these hazards are the possible contamination of drinking water by radioactive substances and the spreading of radioactive materials such as radon-222 into the air. There is the potential threat to uranium miners of having radioactive dust lodged in their lungs. Another danger arises from residual radioactivity remaining in the piles of spent uranium ore (tailings) after the uranium has been removed.

A third problem is the possible release of large amounts of radiation into the atmosphere from nuclear reactors. Although nuclear reactors are built with many safety features, accidents are still possible. This is shown by the accident that occurred in 1979 at the Three Mile Island nuclear power plant in Pennsylvania. This accident, in which radiation was released into the atmosphere, was the result of a combination of equipment malfunction and human error. The disastrous explosion at the Chernobyl nuclear power plant in 1986 was also the result of human error—an unauthorized test of the reactor's emergency systems by the Chernobyl operators.

A fourth problem is the fallout from the explosion of atomic and hydrogen bombs. The explosions of these bombs blow radioactive materials such as strontium-90 high into the stratosphere. Eventually these radioactive particles fall back to earth. A shower of intensely radioactive particles can produce severe burns. These particles can also contaminate our food. Strontium-90 gets into plants and then into animals where it accumulates. People consume animal products and may take in a sizable dose of strontium-90. Because strontium-90 is chemically similar to calcium, it collects in our bones. There it can cause cancer; it can destroy skin tissue; it can ultimately cause death.

The last and probably the greatest problem of the nuclear age is the fact that several countries have at their disposal a large number of atomic and hydrogen bombs. Ironically, the prospect of an all-out nuclear war is so terrifying that it has been considered by some to be an effective deterrent against even the use of smaller tactical atomic weapons. Still, the threat will remain as long as "conventional" wars continue.

APPLICATION

Nuclear Disaster

What caused the explosion at the Chernobyl nuclear power plant? In a report given to the International Atomic Energy Agency, Soviet officials stated that plant operators were responsible for the 1986 explosion at the nuclear power plant which is located near Kiev in the Soviet Union. The accident occurred while the operators were conducting an unauthorized test of the reactor's emergency systems.

In a nuclear reactor, so much heat is generated by nuclear fission that it is vital to have cooling systems to remove the heat, and control rods to slow the fission to a manageable level. Nevertheless, Chernobyl operators shut down two vital systems in order to do their test. First, they turned off the emergency cooling system. This system is designed to prevent the reactor from overheating in the event of a failure of the main cooling system. Second, they turned off the emergency shut-down system. This system is designed to turn off the reactor if the rate of fission becomes too rapid.

During the course of the test, the power level of the reactor dropped to 1% of its capacity. The operators decreased the flow of coolant water and removed the control rods. A sudden power surge resulted. The power level jumped to 50% of the reactor's capacity in only three seconds. Since the emergency shut-down system had been turned off, the only way to stop the reactor was to reinsert the control rods. However, the operators did not have enough time to do this, and the reactor became so hot that the coolant water vapourized rapidly, causing an explosion that blew the 900 t lid right off the reactor.

The effects of this accident have been severe. The topsoil from a 2600 km² area around the Chernobyl power station was so contaminated by radioactive particles from the explosion that it had to be removed and stored in a nuclear waste dump. In addition, Soviet officials predicted that, during the next 30 years, 6500 to 40 000 Soviet citizens would die of cancer caused by radiation from the explosion.

Part Four Review Questions

1. How does a breeder reactor differ from a conventional reactor (such as a CANDU reactor)?
2. What characteristic distinguishes nuclear fusion from nuclear fission? Write nuclear equations that illustrate the two processes.
3. List five natural sources of ionizing radiation and four sources that result from human activity.
4. State five major problems of the nuclear age.

Chapter Summary

- Radioactivity is the ability of a substance to emit invisible rays which will fog a covered photographic plate.
- Alpha (α) particles are helium nuclei.
- Beta (β) particles are high-energy electrons.
- Gamma (γ) rays are high-energy, high-frequency electromagnetic radiation similar to X-rays.
- The mass defect is the amount of mass that is lost and converted into energy when a nucleus is constructed from separate protons and neutrons.
- The binding energy of a nucleus is the energy equivalent to the mass defect.
- A nuclear equation balances the neutrons, protons, and electrons that are involved in nuclear reactions.
- Radioactive decomposition is the alteration of a nucleus by the loss of an alpha particle or a beta particle.
- Transmutation is a change in the identity of a nucleus because of a change in the number of protons in the nucleus.
- Artificial transmutation is the conversion of one nucleus to another by bombarding it with alpha particles, beta particles, or neutrons.
- Fission is the splitting of a nucleus to form nuclei of intermediate mass.
- Fusion is the combination of two nuclei to form a heavier, more stable nucleus.
- The half-life of a radioactive substance is the time required for half of the nuclei to undergo radioactive decomposition.

- Carbon-dating is a technique for measuring the age of a substance by measuring the carbon-14 level of radioactivity remaining in the substance.
- A transuranium element is one with more than 92 protons in its nuclei.
- The critical mass of a fissionable material is the minimum mass required for an explosive chain reaction to occur.
- A nuclear reactor is a device in which a nuclear chain reaction can be carried out in a controlled fashion.
- A moderator is a substance which can slow down fast neutrons.
- Ionizing radiation is radiation which is capable of interacting with atoms or molecules to produce ions.

Key Terms

radioactive
radioactivity
alpha particle
beta particle
gamma ray
mass defect
binding energy
radioactive decomposition
artificial transmutation
fission

fusion
transmutation
half-life
carbon-dating
transuranium element
critical mass
nuclear reactor
moderator
ionizing radiation
rem

Test Your Understanding

Each of the following statements or questions is followed by four responses. Choose the correct response in each case. (Do not write in this book.)

1. Marie Curie suspected that pitchblende contained an element that was more radioactive than uranium because:
   a) pitchblende fogs a covered photographic plate.
   b) pitchblende is more radioactive than uranium.
   c) pitchblende emits rays which can penetrate thin sheets of metal.
   d) the rays emitted by pitchblende will discharge an electroscope.
2. The most highly radioactive element among the following is:
   a) Ni.   b) Po.   c) Ra.   d) U.
3. Rutherford's lead block experiment showed that
   a) radium emits alpha, beta, and gamma rays.
   b) beta particles are high-energy radiation similar to X-rays.
   c) gamma rays are high-energy electrons.
   d) the nucleus of an atom contains most of the mass of the atom.
4. The energy that is released when a nucleus is made from the isolated protons and neutrons is called
   a) the mass defect.   b) transmutation energy.
   c) fission energy.   d) binding energy.

5. A nucleus is more likely to be stable when it has
 a) an even number of protons and an even number of neutrons.
 b) an even number of protons and an odd number of neutrons.
 c) an odd number of protons and an even number of neutrons.
 d) an odd number of protons and an odd number of neutrons.
6. The question mark in the nuclear equation $^{226}_{88}Ra \longrightarrow ^{222}_{86}Rn + ?$ stands for
 a) $4\,^1_0n$.　　　b) $2\,^1_1H$.　　　c) ^4_2He.　　　d) $2\,^0_{-1}e$.
7. When two light nuclei combine to form a heavier more stable nucleus, the process is called
 a) fission.　b) fusion.　c) radioactive decomposition.　d) beta decay.
8. Beta decay consists of the emission of
 a) a helium nucleus from an atom.
 b) an electron from a nucleus.
 c) high-energy electromagnetic radiation from a nucleus.
 d) a proton from a nucleus.
9. If the half-life of a radioactive element is 1 h, its level of radioactivity will have dropped to 25% of its original value after a period of:
 a) 0.5 h.　　b) 1 h.　　c) 2 h.　　d) 3 h.
10. A device which allows a nuclear chain reaction to be controlled is called
 a) a cyclotron.　　　　b) a reactor.
 c) an accelerator.　　　d) a radioactive isotope.
11. The fuel for a CANDU reactor consists of
 a) pellets of UO_2.　　　b) pellets of U_3O_8.
 c) rods of uranium.　　　d) bricks of graphite.
12. The large reaction vessel which contains the fuel bundles of a CANDU reactor is called a
 a) cooling tank.　b) dump tank.　c) calandria.　d) moderator.
13. A substance which slows down neutrons sufficiently to allow their capture by nuclei is called a
 a) poison.　　b) decelerator.　c) coolant.　　d) moderator.
14. The rate of energy output from a reactor can be increased by
 a) dumping the heavy water into a dump tank.
 b) removing some ordinary water from the liquid zone control units.
 c) dropping cadmium alloy rods into the reactor core.
 d) adding a solution of gadolinium nitrate to the moderator.
15. The nuclear fuel produced by a breeder reactor is
 a) neptunium-239.　　　b) plutonium-239.
 c) uranium-238.　　　d) uranium-235.

Review Your Understanding

1. How does the presence of a radioactive substance affect a charged electroscope?
2. What information regarding radioactivity was gained from Rutherford's lead block experiment?
3. How do α, β, and γ rays differ from each other in their properties?
4. From what part of a radioactive atom do the alpha or beta particles come?

5. Why is $^{35}_{15}P$ likely to be an unstable nucleus?

6. Why is $^{18}_{9}F$ likely to be an unstable nucleus?

7. Would you expect the following nuclei to be stable or unstable? State your reason in each case. a) $^{3}_{1}H$; b) $^{20}_{10}Ne$; c) $^{58}_{29}Cu$; d) $^{210}_{83}Bi$

8. What is the relationship between nuclear binding energy and the stability of a nucleus?

9. Write the equation for the emission of an alpha particle from $^{226}_{88}Ra$.

10. Write the equation for the emission of a beta particle from $^{214}_{82}Pb$.

11. Write the equation for successive emissions of an alpha particle and a beta particle from $^{214}_{84}Po$.

12. What is meant by the half-life of a radioactive element?

13. The half-life of Fm-253 is 4.5 days. What fraction of 1 g of Fm would remain after 13.5 days?

14. Why are neutrons better particles for bombarding atomic nuclei than protons or alpha particles?

15. What can happen when a neutron is fired at the nucleus of an atom?

16. What is meant by the term *transmutation*?

17. What is a fission reaction?

18. What nucleus is left out of the following equation?

$$^{235}_{92}U + ^{1}_{0}n \longrightarrow ^{147}_{59}Pr + ? + 3\,^{1}_{0}n$$

19. Describe a chain reaction and state how the fission of $^{235}_{92}U$ can produce a chain reaction.

20. Describe the parts of a nuclear reactor and their functions. What are the main uses of a reactor?

21. In the CANDU system, what happens when the heavy water moderator is released into the dump tank? Why?

22. What is a transuranium element?

23. For what purposes can cobalt-60 be used?

24. What is the purpose of a breeder reactor?

25. What type of nuclear reaction is the following?

$$^{2}_{1}H + ^{3}_{1}H \longrightarrow ^{4}_{2}He + ^{1}_{0}n$$

26. Complete the following nuclear equations:

a) $^{31}_{15}P + ^{1}_{1}H \longrightarrow ^{28}_{14}Si + ?$ b) $^{27}_{13}Al + ^{4}_{2}He \longrightarrow ? + ^{1}_{1}H$

c) $^{23}_{11}Na + ^{2}_{1}H \longrightarrow ^{1}_{1}H + ?$ d) $^{235}_{92}U + ^{1}_{0}n \longrightarrow ^{90}_{37}Rb + ? + 2\,^{1}_{0}n$

e) $^{3}_{1}H + ^{3}_{1}H \longrightarrow ^{4}_{2}He + ?$ f) $^{228}_{90}Th \longrightarrow ^{4}_{2}He + ?$

g) $^{12}_{5}B \longrightarrow ^{0}_{-1}e + ?$

27. State which type of nuclear reaction is illustrated by each equation in question 26.

28. What is the half-life of a radioactive isotope if 1.20 mg of it decays to 0.30 mg in 40 min?

29. Describe how heavy water is produced for use in CANDU reactors.

30. If ionizing radiation from all sources is added up, what is the approximate annual dosage received by an individual? Is this dosage considered dangerous by health authorities?

31. How is spent nuclear fuel stored?

Apply Your Understanding

1. A one gram sample of wood unearthed from an excavation produced 1.9 beta particles per minute due to the decay of carbon-14. If a one gram sample of new wood produces 15.2 beta particles per minute, what is the age of the old wood?

2. The left-over ore (*tailings*) from uranium processing plants has been used to make concrete for the construction of houses. These tailings contain radium (half-life 1620 years) and its decay product radon (half-life 3.8 days). The radon gas escapes from the concrete into the surrounding air. State whether you agree or disagree with the following statements, and give your reasons in each case.

   a) Old tailings do not constitute a health hazard because radon has such a short half-life that it disappears quickly.

   b) Radon is a noble gas and does not enter into biochemical reactions, so it should not constitute a health hazard.

   c) In any case, any health hazard could be decreased by using a ventilation system to blow the radon gas outdoors.

3. The uranium-238 decay series begins with $^{238}_{92}U$ and ends with $^{206}_{82}Pb$. The particles given off are, in succession: α, β, β, α, α, α, α, α, β, β, α, β, β, α. Write the balanced nuclear equations for each step in the process.

4. Why would you expect $^{12}_{6}C$ to behave the same chemically as radioactive $^{14}_{6}C$?

5. Why is there no such thing as a pure radioactive isotope?

6. Why does a nuclear reaction usually involve the conversion of one element into another?

7. What is the final product in the uranium-235 natural radioactivity series in which there are seven alpha and four beta decays?

8. Natural uranium contains uranium-235 which can undergo fission. Why doesn't an explosive chain reaction take place in uranium ores?

9. How many alpha particles and how many beta particles are emitted when $^{241}_{94}Pu$ decays to $^{209}_{83}Bi$?

Investigations

1. Why do environmentalists complain about thermal pollution from nuclear power stations when it is well known that thermal generating stations (often the only alternative source of power) also give off heat into the environment?

2. Many nations now have the atomic bomb. In your opinion, does this make world peace more or less likely?

3. If you were asked to vote on whether or not a nuclear power plant should be built 50 km from your home, how would you vote? What would your reasons be?

4. Write a research report on the operation of particle accelerators.

5. Debate the following proposition: "In light of the fact that used fuel rods from CANDU reactors can be reprocessed to obtain plutonium (an ingredient in nuclear weapons), Canada should not sell reactors to other countries."

Appendix

Appendix A
International Atomic Masses

Name	Symbol	Atomic Number	Atomic Mass*	Name	Symbol	Atomic Number	Atomic Mass
Actinium	Ac	89	227.0	Californium	Cf	98	(251)
Aluminum	Al	13	27.0	Carbon	C	6	12.01
Americium	Am	95	(243)	Cerium	Ce	58	140.1
Antimony	Sb	51	121.8	Cesium	Cs	55	132.9
Argon	Ar	18	39.9	Chlorine	Cl	17	35.5
Arsenic	As	33	74.9	Chromium	Cr	24	52.0
Astatine	At	85	(210)	Cobalt	Co	27	58.9
Barium	Ba	56	137.3	Copper	Cu	29	63.5
Berkelium	Bk	97	(247)	Curium	Cm	96	(247)
Beryllium	Be	4	9.01	Dysprosium	Dy	66	162.5
Bismuth	Bi	83	209.0	Einsteinium	Es	99	(252)
Boron	B	5	10.8	Erbium	Er	68	167.3
Bromine	Br	35	79.9	Europium	Eu	63	152.0
Cadmium	Cd	48	112.4	Fermium	Fm	100	(257)
Calcium	Ca	20	40.1	Fluorine	F	9	19.0

*A value given in parentheses denotes the mass of the isotope with the longest known half-life.

Name	Symbol	Atomic Number	Atomic Mass
Francium	Fr	87	(223)
Gadolinium	Gd	64	157.2
Gallium	Ga	31	69.7
Germanium	Ge	32	72.6
Gold	Au	79	197.0
Hafnium	Hf	72	178.5
Helium	He	2	4.00
Holmium	Ho	67	164.9
Hydrogen	H	1	1.008
Indium	In	49	114.8
Iodine	I	53	126.9
Iridium	Ir	77	192.2
Iron	Fe	26	55.8
Krypton	Kr	36	83.8
Lanthanum	La	57	138.9
Lawrencium	Lr	103	(260)
Lead	Pb	82	207.2
Lithium	Li	3	6.94
Lutetium	Lu	71	175.0
Magnesium	Mg	12	24.3
Manganese	Mn	25	54.9
Mendelevium	Md	101	(258)
Mercury	Hg	80	200.6
Molybdenum	Mo	42	95.9
Neodymium	Nd	60	144.2
Neon	Ne	10	20.2
Neptunium	Np	93	237.0
Nickel	Ni	28	58.7
Niobium	Nb	41	92.9
Nitrogen	N	7	14.01
Nobelium	No	102	(259)
Osmium	Os	76	190.2
Oxygen	O	8	16.00
Palladium	Pd	46	106.4
Phosphorus	P	15	31.0
Platinum	Pt	78	195.1
Plutonium	Pu	94	(244)
Polonium	Po	84	(209)
Potassium	K	19	39.1

Name	Symbol	Atomic Number	Atomic Mass
Praseodymium	Pr	59	140.9
Promethium	Pm	61	(145)
Protactinium	Pa	91	231.0
Radium	Ra	88	226.0
Radon	Rn	86	(222)
Rhenium	Re	75	186.2
Rhodium	Rh	45	102.9
Rubidium	Rb	37	85.5
Ruthenium	Ru	44	101.1
Samarium	Sm	62	150.4
Scandium	Sc	21	45.0
Selenium	Se	34	79.0
Silicon	Si	14	28.1
Silver	Ag	47	107.9
Sodium	Na	11	23.0
Strontium	Sr	38	87.6
Sulfur	S	16	32.1
Tantalum	Ta	73	180.9
Technetium	Tc	43	(98)
Tellurium	Te	52	127.6
Terbium	Tb	65	158.9
Thallium	Tl	81	204.4
Thorium	Th	90	232.0
Thulium	Tm	69	168.9
Tin	Sn	50	118.7
Titanium	Ti	22	47.9
Tungsten	W	74	183.8
Unnilennium	Une	109	—
Unnilhexium	Unh	106	(263)
Unnilpentium	Unp	105	(262)
Unnilquadium	Unq	104	(261)
Unnilseptium	Uns	107	(262)
Uranium	U	92	238.0
Vanadium	V	23	50.9
Xenon	Xe	54	131.3
Ytterbium	Yb	70	173.0
Yttrium	Y	39	88.9
Zinc	Zn	30	65.4
Zirconium	Zr	40	91.2

Appendix B Periodic Table and

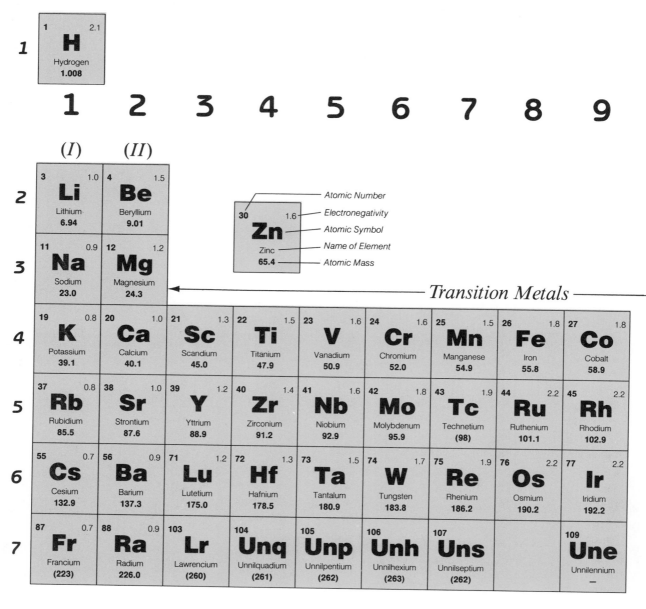

		Atomic Number
30	1.6	Electronegativity
Zn		Atomic Symbol
Zinc		Name of Element
65.4		Atomic Mass

Transition Metals

Lanthanide Series

Actinide Series

Table of Electronegativities

18

(VIII)

10	11	12	13 *(III)*	14 *(IV)*	15 *(V)*	16 *(VI)*	17 *(VII)*	
								2 **He** Helium 4.00
			5 2.0 **B** Boron 10.8	**6** 2.5 **C** Carbon 12.01	**7** 3.0 **N** Nitrogen 14.01	**8** 3.5 **O** Oxygen 16.00	**9** 4.0 **F** Fluorine 19.0	**10** **Ne** Neon 20.2
			13 1.5 **Al** Aluminum 27.0	**14** 1.8 **Si** Silicon 28.1	**15** 2.1 **P** Phosphorus 31.0	**16** 2.5 **S** Sulfur 32.1	**17** 3.0 **Cl** Chlorine 35.5	**18** **Ar** Argon 39.9
28 1.8 **Ni** Nickel 58.7	**29** 1.9 **Cu** Copper 63.5	**30** 1.6 **Zn** Zinc 65.4	**31** 1.6 **Ga** Gallium 69.7	**32** 1.8 **Ge** Germanium 72.6	**33** 2.0 **As** Arsenic 74.9	**34** 2.4 **Se** Selenium 79.0	**35** 2.8 **Br** Bromine 79.9	**36** **Kr** Krypton 83.8
46 2.2 **Pd** Palladium 106.4	**47** 1.9 **Ag** Silver 107.9	**48** 1.7 **Cd** Cadmium 112.4	**49** 1.7 **In** Indium 114.8	**50** 1.8 **Sn** Tin 118.7	**51** 1.9 **Sb** Antimony 121.8	**52** 2.1 **Te** Tellurium 127.6	**53** 2.5 **I** Iodine 126.9	**54** **Xe** Xenon 131.3
78 2.2 **Pt** Platinum 195.1	**79** 2.4 **Au** Gold 197.0	**80** 1.9 **Hg** Mercury 200.6	**81** 1.8 **Tl** Thallium 204.4	**82** 1.8 **Pb** Lead 207.2	**83** 1.9 **Bi** Bismuth 209.0	**84** 2.0 **Po** Polonium (209)	**85** 2.2 **At** Astatine (210)	**86** **Rn** Radon (222)

64 1.1 **Gd** Gadolinium 157.2	**65** 1.1 **Tb** Terbium 158.9	**66** 1.1 **Dy** Dysprosium 162.5	**67** 1.1 **Ho** Holmium 164.9	**68** 1.1 **Er** Erbium 167.3	**69** 1.1 **Tm** Thulium 168.9	**70** 1.1 **Yb** Ytterbium 173.0
96 1.3 **Cm** Curium (247)	**97** 1.3 **Bk** Berkelium (247)	**98** 1.3 **Cf** Californium (251)	**99** 1.3 **Es** Einsteinium (252)	**100** 1.3 **Fm** Fermium (257)	**101** 1.3 **Md** Mendelevium (258)	**102** 1.3 **No** Nobelium (259)

Appendix C
The First Twenty Elements

C-1 Hydrogen (H)

Element 1, hydrogen (Greek *hydro*, water; *genes*, former) burns in oxygen, forming water. It is the most abundant element in the universe and is found in the sun and most stars. On earth, hydrogen is mainly found combined with oxygen as water; however, it is also present in organic matter such as living plants and petroleum. Hydrogen is the least dense of all known substances.

Molecular hydrogen (H_2) is a colourless, odourless, tasteless gas. At room temperature, it is rather inert. At elevated temperatures, it reacts with nonmetals to form compounds such as H_2O, H_2S and NH_3 and with active metals to form metallic hydrides such as NaH and LiH. Hydrogen reduces some metallic oxides such as CuO, forming the free metal and water.

Hydrogen is prepared by the electrolysis of water or by the action of HCl or H_2SO_4 on Fe or Zn. It is used in the Haber process for producing NH_3, in the hydrogenation of fats and oils, as a fuel in welding, and for filling weather balloons.

C-2 Group 1 (Alkali Metals): Lithium (Li), Sodium (Na), and Potassium (K)

Element 3, lithium (Greek *lithos*, stone), element 11, sodium (Medieval Latin *sodanum*, headache remedy), and element 19, potassium (English *potash*, pot ashes) each have one easily-lost valence electron. Thus, they are very active metals. In fact, they each react with water at room temperature, forming a metallic hydroxide ($LiOH$, $NaOH$, and KOH) and H_2. Because they are active metals, they are never found free in nature; rather, they are found in mineral deposits. Sodium is the most abundant of the alkali metals, and its most common mineral is $NaCl$.

Lithium, sodium, and potassium are soft, silvery metals. They have low melting points, and their densities are quite low for metals. Lithium is the least dense of all metals. These metals and their compounds impart colours to flames. A lithium flame is crimson; a sodium flame is yellow; and a potassium flame is violet. In addition to reacting with water, the alkali metals react with hydrogen to form hydrides, and with oxygen to form Li_2O, Na_2O_2, or KO_2. They also react with the elements of Group 17, forming metallic halides such as $LiCl$, $NaCl$, and KCl.

Lithium and sodium are prepared by the electrolysis of their molten chlorides, and potassium is prepared by the electrolysis of molten KOH. Lithium is used in the manufacture of alloys. Some lithium salts are useful; lithium chloride is used in industrial drying systems because it absorbs water so well. Sodium is used in the manufacture of many important compounds such as NaOH and NaCN. Sodium compounds are important to the paper, glass, soap, textile, and petroleum industries. Because potassium is essential for plant growth, potassium compounds are used in fertilizers.

C-3 Group 2 (Alkaline Earth Metals): Beryllium (Be), Magnesium (Mg), and Calcium (Ca)

Element 4, beryllium (Greek *beryllos*, beryl) occurs in the mineral beryl $(Be_3Al_2(SiO_3)_6)$. The precious forms of beryl are emerald and aquamarine. Beryllium and its compounds are toxic. Element 12, magnesium (*Magnesia*, district in Thessaly, Greece) is found in a number of minerals such as dolomite $(CaCO_3 \cdot MgCO_3)$, and in seawater. Element 20, calcium (Latin *calx*, lime) occurs abundantly in minerals such as limestone $(CaCO_3)$ and gypsum $(CaSO_4 \cdot 2H_2O)$. The sulfate and hydrogen carbonate salts of magnesium and calcium are the main components of hard water.

Beryllium is one of the least dense metals. Magnesium and calcium are relatively soft metals. They are all good conductors of heat and electricity. Beryllium is the least metallic of the three. The activity of these metals increases with increasing atomic number. Calcium is the most active and is similar in activity to the alkali metals. Calcium reacts with cold water to give $Ca(OH)_2$ and H_2. Magnesium reacts with steam to give $Mg(OH)_2$ and H_2. Beryllium does not react with water. Calcium reacts with hydrogen to form the hydride (CaH_2). All three metals react directly with nitrogen, oxygen, sulfur, and the halogens to give compounds in which they exhibit the +2 oxidation state.

Magnesium and calcium are prepared by the electrolysis of their molten chlorides. Beryllium is prepared by the reduction of BeF_2 with Mg to form Be and MgF_2. Beryllium is used as an alloying agent in beryllium copper, which is used for springs, electrical contacts, and spot-welding electrodes. Finely divided magnesium, such as magnesium powder or wire, burns with a dazzling white flame and is used in flash bulbs and flares. Magnesium is often used to make alloys for aircraft construction. When used as an alloying agent, it improves the fabrication and welding characteristics of aluminum. Magnesium is also important for plant and animal life; it is the metal ion present in chlorophyll. Milk of magnesia $(Mg(OH)_2)$ is a common home remedy for acid indigestion. Calcium is used as a reducing agent for preparing other metals such as uranium and thorium. Quicklime (CaO) is used for making mortar and plaster.

C-4 Group 13: Boron (B) and Aluminum (Al)

Element 5, boron (Arabic *buraq*; Persian *burah*) was known to ancient peoples as borax ($Na_2B_4O_7 \cdot 4H_2O$). Element 13, aluminum (Latin *alumen*, alum) is the most abundant metal in the earth's crust. Neither element is found free in nature. The most important source of boron is a mineral deposit in the Mojave Desert of California. Aluminum is found in minerals such as bauxite ($Al_2O_3 \cdot nH_2O$) and cryolite (Na_3AlF_6).

Boron has nonmetallic properties and resembles silicon and carbon more than it does aluminum. It is brittle and is not a good conductor of heat or electricity; however, it does conduct electricity at elevated temperatures. Aluminum is an active metal. It is a silvery-white metal with a low density and a pleasing appearance. It is the second most malleable metal (after gold). Boron and aluminum react with the most electronegative elements such as oxygen and the halogens; they both exhibit only the +3 oxidation state.

Boron is prepared by reducing B_2O_3 with powdered Mg, and aluminum is prepared by the electrolysis of Al_2O_3. Boron is used in small quantities in steel. The most important compounds of boron are boric acid ($B(OH)_3$), used as a mild antiseptic, and borax, used as a cleaning flux in welding and as a water softener in washing powders. Aluminum is used for making kitchen utensils, and alloys of aluminum are vital for the construction of aircraft.

C-5 Group 14: Carbon (C) and Silicon (Si)

Element 6, carbon (Latin *carbo*, charcoal) was known in prehistoric times. Carbon is widely distributed throughout the universe. On earth, carbon is found free in nature as diamond and graphite. Diamond is one of the hardest substances; graphite is one of the softest. Carbon is also found combined with other elements in compounds such as carbon dioxide, limestone ($CaCO_3$), and the hydrocarbons. Element 14, silicon (Latin, *silex, silicis*, flint) is present in the sun and the stars. Silicon does not exist in the free state on the earth; however, silica (SiO_2) and metallic silicates are found everywhere. Silicon is second only to oxygen in its natural abundance in the earth's crust.

Carbon and silicon are the two nonmetallic elements of Group 14. Carbon is fairly inert at room temperature; however, at higher temperatures, it combines directly with oxygen, the halogens, and most other nonmetals. Silicon is ordinarily rather inert, but it does react with oxygen to form SiO_2 and with the halogens to form tetrahalides such as $SiCl_4$.

Carbon is used as a reducing agent in the preparation of many metals. It is unique among the elements with respect to the vast number and variety of its compounds. There are millions of known carbon compounds, thousands of which are vital to life processes. In fact, one branch of chemistry (organic chemistry) is devoted wholly to the study of carbon compounds. Silicon is prepared by heating SiO_2 and C in an electric furnace. Silicon is an important ingredient in steel. Silica, in the form of sand, is a principal ingredient in the manufacture of glass.

C-6 Group 15: Nitrogen (N) and Phosphorus (P)

Element 7, nitrogen (Greek *nitron*, native soda and *genes*, former) makes up about 78% of the atmosphere by volume. Element 15, phosphorus (Greek *phosphoros*, light-bearing) is not found free in nature, but it is widely distributed in mineral deposits, chiefly as calcium phosphate.

Nitrogen and phosphorus are nonmetals. Molecular nitrogen (N_2) is a colourless, odourless, tasteless, generally inert gas. Phosphorus exists in two forms. White phosphorus is reactive, volatile, and toxic. It ignites spontaneously in air, but changes to red phosphorus when it is heated in the absence of air. Red phosphorus is relatively unreactive, nonvolatile, and nontoxic. Nitrogen reacts with hydrogen to from NH_3, with active metals to form nitrides, and with oxygen to form a variety of oxides such as N_2O, NO and NO_2. Phosphorus reacts with hydrogen to form PH_3(phosphine), with chlorine to form PCl_3 and PCl_5, and with oxygen to form P_2O_5 and P_4O_6.

Nitrogen is obtained by the distillation of liquid air. White phosphorus can be made by heating $Ca_3(PO_4)_2$ in the presence of C and SiO_2 in an electric furnace. Because nitrogen and phosphorus are essential for plant growth, their compounds are used in fertilizers. Nitrogen is used in large quantities to form important compounds such as NH_3 and HNO_3. Red phosphorus is used in the manufacture of matches, incendiary shells, smoke bombs, and tracer bullets.

C-7 Group 16: Oxygen (O) and Sulfur (S)

Element 8, oxygen (Greek *oxys*, sharp, acid and *genes*, former) makes up about 21% of the atmosphere by volume. Oxygen is the most abundant element in the earth's crust. Element 16, sulfur (Latin *sulphurium*) was known to ancient peoples and is referred to in the Bible as *brimstone*. In nature, sulfur occurs both combined with other elements in minerals, such as metallic sulfides and sulfates, and also in deposits of free sulfur. Sulfur also occurs in natural gas and crude oil, and large amounts of sulfur are now being recovered from Alberta gas fields.

Oxygen and sulfur are nonmetals. Molecular oxygen (O_2) is a colourless, tasteless, odourless gas. Oxygen is capable of reacting with most elements. Sulfur is a pale yellow, odourless, brittle solid. Sulfur reacts, on heating, with most metals and nonmetals.

Oxygen is obtained by the distillation of liquid air or by the electrolysis of water. It is used during the manufacture of steel in blast furnaces and for oxyacetylene welding. It is also used in hospitals to aid the respiration of patients. Oxygen is essential for plant and animal respiration and for almost all combustions. Ozone (O_3) is a highly reactive form of oxygen. Its presence in the atmosphere prevents excesses of deadly ultraviolet radiation from reaching the earth's surface. Sulfur is mined by the Frasch process. In this process, superheated water is forced into the sulfur deposit to melt the sulfur; then compressed air is used to force the molten sulfur to the surface. Sulfur is a component of black gunpowder and is added to rubber to strengthen it in the process called vulcanization. A tremendous quantity of sulfur is used to produce sulfuric acid, the most important manufactured chemical. Sulfur dioxide is a dangerous contributor to air pollution and is the main cause of acid rain. Hydrogen sulfide has a rotten egg odour and is an insidious poison that quickly deadens the victim's sense of smell.

C-8 Group 17 (Halogens): Fluorine (F) and Chlorine (Cl)

Element 9, fluorine (Latin *fluere*, to flow) is found only in the combined state, chiefly in fluorspar (CaF_2) and cryolite (Na_3AlF_6). Fluorine is the most electronegative and reactive of all elements. Element 17, chlorine (Greek *chloros*, greenish-yellow) is also found only in the combined state, chiefly as common salt (NaCl).

Fluorine and chlorine are members of the most active family of elements—the halogens. The halogens each have seven valence electrons and can achieve a noble gas electron configuration by gaining one electron. Fluorine and chlorine readily form compounds in which they have a −1 oxidation state. Molecular fluorine (F_2) is a pale yellow, pungent, corrosive gas. It reacts with almost all organic and inorganic substances. In fact, compounds of fluorine with krypton, xenon, and radon (three noble gases) have been prepared. Molecular chlorine (Cl_2) is a greenish-yellow gas with a sharp, irritating odour. It reacts directly with nearly all of the elements.

Fluorine is obtained by the electrolysis of molten fluorides. Hydrofluoric acid is used to etch the glass of light bulbs. Compounds of fluorine are added to drinking water and some brands of toothpaste to prevent tooth decay in children. Fluorchlorohydrocarbons are used in refrigerators and air conditioners. Chlorine is obtained by the electrolysis of aqueous sodium chloride, and is used in the manufacture of compounds for sanitation, pulp bleaching, disinfectants, and textile processing.

C-9 Group 18 (Noble Gases): Helium (He), Neon (Ne), and Argon (Ar)

Element 2, helium (Greek *helios*, the sun) is the second most abundant element in the universe (after hydrogen). Alpha particles are helium nuclei, and helium is a decay product present in radioactive minerals. However, most of our helium comes from natural gas wells in the southern United States, which yield an average helium content of about 2%. Element 10, neon (Greek *neos*, new) is present in the atmosphere in small concentrations. Element 18, argon (Greek *argos*, inactive) makes up nearly 1% of the atmosphere by volume.

These elements are members of the noble gas family. Their electron configurations are such that they neither gain nor lose electrons readily. Helium, neon and argon are inert gases and do not exist in compounds. However, compounds of the other noble gases have been prepared, and it is possible that compounds of any, or all, of these three elements will eventually be prepared.

Helium has the lowest melting point of any element. Because its boiling point is close to absolute zero (0 K), liquid helium is used in cryogenic (low temperature) research. Helium has a low density and is a safer gas than hydrogen for filling balloons. A mixture of 80% helium and 20% oxygen is used as an artificial atmosphere for deep-sea divers. Neon and argon are obtained by the distillation of liquid air. In gas discharge tubes, neon gives off a reddish-orange light. This property has made is useful for making neon lights. Helium and neon are used for making gas lasers. Liquid neon is used as a cryogenic refrigerant. Argon and helium and used as inert gas shields for arc welding. Argon is used as an inert gas in lightbulbs.

Appendix D
Vapour Pressure of Water at Different Temperatures

Temperature (°C)	Vapour Pressure (kPa)	Temperature (°C)	Vapour Pressure (kPa)
0	0.61	26	3.36
5	0.87	27	3.57
10	1.23	28	3.78
15	1.71	29	4.00
16	1.82	30	4.24
17	1.94	35	5.62
18	2.06	40	7.38
19	2.20	45	9.58
20	2.34	50	12.33
21	2.49	60	19.92
22	2.64	70	31.16
23	2.81	80	47.34
24	2.98	90	70.10
25	3.17	100	101.32

Glossary

Accuracy The closeness of a measurement to its true value.

Acid Any substance which increases the number of hydronium ions present in an aqueous solution.

Acid rain Rain which contains acids formed in the atmosphere from the oxides of sulfur, SO_2 and SO_3, and to a lesser extent, the oxides of nitrogen, NO and NO_2.

Acid-base neutralization The process whereby hydronium ions and hydroxide ions react to form water molecules.

Actual yield The amount of product that is actually obtained from chemical reaction.

Addition reaction A reaction in which no substitution occurs; rather, atoms and molecules join together directly to form larger molecules.

Alcohol A compound that contains a hydroxyl group attached to an alkyl group.

Aldehyde A compound that contains an alkyl or aromatic group and a hydrogen atom attached to a carbonyl group.

Aliphatic compound A hydrocarbon that contains no aromatic or other rings.

Alkali metals The elements found in Group 1 of the periodic table: lithium, sodium, potassium, rubidium, cesium, and francium. They all have metallic properties and are highly reactive.

Alkane A saturated hydrocarbon.

Alkene A hydrocarbon that contains one carbon-carbon double bond.

Alkyl group An alkane from which one hydrogen atom has been removed.

Alkyne A hydrocarbon that contains one carbon-carbon triple bond.

Alloy A solution of one metal, or several, in another metal. A substance which has metallic properties and which contains more than one element.

Alpha particle A helium nucleus; a particle with two protons and two neutrons.

Amphoteric substance A compound which is capable of acting as either an acid or a base, depending on the substance with which it is reacting.

Anion A negative ion.

Applied science The application of scientific discoveries to practical problems. Same as Technology.

Aqueous solution A solution made by dissolving a solute in water.

Aromatic compound Any substance whose molecules contain the benzene structural skeleton.

Artificial transmutation The process of bombarding a nucleus with alpha particles, protons, or neutrons to give an unstable nucleus, which can then emit a proton or neutron in order to gain stability.

Atom The smallest particle of an element that has all the chemical properties of that element. An atom is electrically neutral.

Atomic mass The mass of an atom expressed in atomic mass units.

Atomic mass unit One-twelfth the mass of a carbon-12 atom.

Atomic number The number of protons in the nucleus of an atom.

Aufbau principle The principle which states that in going from a hydrogen atom to a larger atom, you add protons to the nucleus and electrons to orbitals having the same shape as those derived for the hydrogen atom.

Avogadro's hypothesis The hypothesis which states that equal volumes of gases at the same temperature and pressure contain equal numbers of molecules.

Base Any substance which increases the number of hydroxide ions present in an aqueous solution.

Base units The seven fundamental units from which the SI is constructed: metre, kilogram, second, kelvin, mole, ampere, and candela.

Beta particle A high-energy electron emitted in radioactive decay.

Binding energy The energy produced when a nucleus is made from its components.

Biochemical oxygen demand (BOD) The oxygen take-up by the microorganisms that decompose organic materials in the water.

Boiling The process wherein vapour bubbles reach the surface of a liquid which is being heated, break, and allow the vapour to escape into the air.

Boiling point The temperature at which vapour bubbles in a liquid which is being heated reach the surface, break, and allow the vapour to escape into the air.

Boyle's Law The law which states that the volume of a fixed mass of a gas varies inversely with the pressure, provided the temperature remains constant.

Brønsted-Lowry Theory The theory which states that an acid is a proton donor and a base is a proton acceptor.

Brownian motion The constant zigzag motion of small particles suspended in a gas or a liquid.

Carbon-dating A dating technique which uses the amount of radioactive carbon-14 in organic matter to measure the age of the organic matter.

Carbonyl group An organic functional group that has a carbon atom attached to an oxygen atom by a double bond.

Carboxyl group An organic functional group that has a hydroxyl group attached to a carbonyl group.

Carboxylic acid A compound that contains an alkyl or aromatic group attached to a carboxyl group.

Catalyst A chemical which speeds up a reaction and can be recovered unchanged when the reaction is complete.

Cathode rays The invisible rays which originate at the cathode and travel to the anode in a discharge tube.

Cation A positive ion.

Charles' Law   The law which states that the volume of a fixed mass of gas varies directly with the Kelvin temperature, provided the pressure remains constant.

Chemical bonds   The attractions between the atoms or ions of a substance.

Chemical change   A change which converts a substance into a different kind or different kinds of matter each with a different composition and new properties.

Chemical energy   The potential energy stored in chemical bonds.

Chemical equation   A shorthand description of a chemical change that uses symbols and formulas to indicate the substances involved in the change.

Chemical formula   A formula in which the symbols of the elements are used to represent the composition of a substance.

Chemical nomenclature   The assignment of names to compounds.

Chemical properties   The properties of a substance which can be observed only when the substance undergoes a change in composition.

Chemical reactions   The changes or transformations of various kinds of matter into other substances with different compositions, different structures, and different properties.

Chemistry   The science that deals with the composition, structure, and properties of matter.

Chromatography   A technique used to separate the components of a mixture by using differences in the degree to which various substances are attracted to the surface of a nonreactive substance.

Combination reaction   A reaction in which atoms and molecules join together directly to produce larger molecules. Also called an addition reaction or a synthesis.

Compound   A pure substance consisting of two or more elements in combination. May be decomposed into two or more simpler substances by ordinary chemical methods.

Concentrated solution   A solution which contains a relatively large amount of solute.

Concentration   The quantity of solute dissolved in a certain amount of solution.

Condensation   The change of state from a gas to a liquid.

Continuous spectrum   A spectrum which consists of all the colours or frequencies of visible light: red, orange, yellow, green, blue, indigo, and violet.

Coordinate covalent bond   A covalent bond in which both electrons of the shared pair come from only one of the bonded atoms.

Covalent bond   The type of chemical bonding resulting from the sharing of valence electrons between atoms.

Cracking   A method by which organic molecules are broken down into simpler molecules by the action of heat and a catalyst.

Critical mass   The minimum mass of fissionable material required for an explosive chain reaction.

Critical pressure   The minimum pressure required to liquefy a gas at its critical temperature.

Critical temperature   The temperature above which it is impossible to liquefy a gas, no matter how much pressure is applied.

Crystallization   A technique which may be used to separate one solid from the other components in a solid mixture. This solid must be much more soluble in hot solvent than in cold solvent. The mixture is dissolved in the hot solvent. As the solution cools, the solid which is insoluble in cold solvent precipitates and is collected by filtration.

Dalton's Law of Partial Pressures   The law which states that the total pressure exerted by a mixture of gases is the sum of the pressures of each gas when measured alone.

Decomposition reaction   A reaction in which molecules break apart to form atoms or smaller molecules.

Density   The mass of an object per unit volume of the object.

Dilute solution   A solution which contains a relatively small amount of solute.

Dimensional analysis   A method of calculation in which both the numerals and the SI units (dimensions) are carried through the algebraic operations.

Dipole   An atom or molecule in which one side has a slightly positive charge, while the other side has a slightly negative charge, because of the distribution of the electrons.

Displacement reaction   A reaction in which one atom or group of atoms in a molecule is replaced by another atom or group of atoms. Also called substitution reaction.

Dissociation   The separation or breaking apart of solute particles, resulting in ions in solution.

Distillation   A method of separating a solution into its components by heating the solution in a distilling apparatus. One substance vapourizes at a lower temperature than the others and boils off, leaving the remaining substances in the distilling flask. The vapour is liquefied in a condenser and is collected.

Double covalent bond   The sharing of two pairs of electrons between two atoms.

Double-displacement reaction   A reaction which involves a joint exchange of partners between two reactants. An atom or group of atoms in each of two new compounds gains a new partner. Also called a metathetic reaction.

Dynamic equilibrium   The state in which two opposing processes, such as dissolving and crystallizing, occur at equal rates.

Electrolyte   The solute in a solution which conducts electricity.

Electron   A particle which has a relative charge of -1 and a mass of $0.000\ 55$ u.

Electron affinity   A measure of the energy given off when an electron is added to a neutral atom. Each element has a characteristic electron affinity.

Electron configuration   The distribution of electrons among the various orbitals of an atom.

Electron dot symbol   The symbol developed by chemists to represent an atom and its valence electrons, the electrons which participate in chemical reactions.

Electronegativity   A quantitative measure of the electron-attracting ability of the atoms in a molecule.

Element   A pure substance which cannot be broken down into simpler substances by ordinary chemical methods.

Empirical formula   The simplest formula, which shows only the relative number of moles of each type of atom in a compound.

Empirical knowledge   Knowledge gained by using the senses; that is, by seeing, tasting, hearing, feeling, or smelling.

Endothermic reaction   A reaction during which energy is absorbed. Energy must be supplied continuously to keep the reaction going.

Energy   The ability to do work; that is, to move matter against a force which opposes the motion.

Energy levels   Certain states, each with definite, fixed energies, in which an electron is allowed to move.

Energy level population   The number of electrons occupying each of the lowest available energy levels in an atom.

Equation of state   An equation which gives the relationship between the pressure, volume, temperature, and mass of a substance.

Ester   A compound in which the acidic hydrogen atom of a carboxylic acid molecule has been replaced by an alkyl or aromatic group.

Ether   A compound that contains two alkyl or aromatic groups attached to an oxygen atom.

Eutrophication   The natural aging process of a lake or pond in which aquatic organisms live, die, and are decomposed by microorganisms in the water.

Evaporation   The escape of molecules from the surface of a liquid to enter the gas phase above the liquid.

Exothermic reaction   A reaction during which energy is given off.

Experiment   A planned series of observations.

Filtration   A method of separating insoluble substances from the soluble components in a liquid mixture by passing the mixture through a filter paper. The solid particles are trapped in the filter paper, and the soluble components pass through the filter paper.

Fine chemicals   Chemicals which are more expensive than heavy chemicals, and are produced in smaller amounts.

Fission   The process wherein a heavy nucleus splits to form nuclei of intermediate mass.

Free radical   An atom, ion, or molecule which contains one or more unpaired electrons.

Freezing   The process of changing phase from liquid to solid.

Freezing point   The temperature at which a liquid releases heat energy and changes phase to become a solid.

Functional group   Any reactive atom or group of atoms in an organic compound.

Fusion   The process wherein two light nuclei combine to form a heavier, more stable nucleus.

Gamma ray   High-energy, high-frequency electromagnetic radiation.

Gas   Matter which has neither a definite shape nor a definite volume, but which takes the shape and volume of the container in which it is placed.

Gay-Lussac's Law   The law which states that the pressure exerted by a gas is directly proportional to its Kelvin temperature, provided the volume remains constant.

Gay-Lussac's Law of Combining Volumes   The law which states that when gases react, they do so in simple, whole-number ratios by volume.

Greenhouse effect   The raising of the surface temperature of the earth by infrared energy trapped by atmospheric carbon dioxide and transmitted back to the surface of the earth.

Group   Elements with similar properties arranged in a vertical column of the periodic table.

Half-life   The length of time required for half of the original number of nuclei in a radioactive sample to decay.

Halogens   The elements found in Group 17 of the periodic table: fluorine, chlorine, bromine, iodine, and astatine. They all have nonmetallic properties and are highly reactive.

Hard water   Water containing dissolved magnesium and calcium salts.

Heavy chemicals   The fundamental chemicals which are the backbone of the chemical industry and are used to produce many other chemicals which lead to finished products. Produced in large amounts.

Heavy water   Water containing the deuterium isotope rather than the regular hydrogen isotope; deuterium oxide.

Heterogeneous   Consisting of more than one phase.

Homogeneous   Consisting of only one phase. Uniform throughout.

Homologous series   A formula series in which the formula of each member differs from that of the preceding member in a consistent, regular way.

Hund's rule   The rule which states that electrons in the same sublevel will not pair up until all the orbitals of the sublevel are at least half-filled.

Hydrate   A compound containing water of hydration.

Hydrocarbons   Organic compounds made up of carbon and hydrogen atoms only.

Hydrogen bond   A dipole-dipole attraction between a highly electronegative atom (such as fluorine, oxygen, or nitrogen) and a hydrogen atom which is covalently bonded to another highly electronegative atom.

Hydronium ion   The $H_3O^+(aq)$ ion.

Hydroxyl group   An –OH group.

Hypothesis   A tentative explanation of the regularities observed in nature.

Ideal gas   A gas which obeys Boyle's Law, Charles' Law, and Gay-Lussac's Law. Any gas that obeys the ideal gas law.

Ideal gas law   The equation which gives the relationship between the pressure, volume, temperature, and number of moles of a gas. Only an ideal gas will obey the ideal gas law.

Indicator   A dye which indicates the acidity or basicity of a solution.

Inertia   The tendency for a moving body to remain in motion and for a stationary body to remain stationary. Resistance to change in state of motion.

Ionic bonding   The type of chemical bonding resulting from attraction between oppositely charged ions formed when metallic atoms transfer electrons to nonmetallic atoms.

Ionic equation   An equation in which soluble compounds that dissociate are written as ions.

Ionic solid   A solid which is hard, has a high melting point, and is a poor conductor of electricity. Ionic solids are held together by ionic bonding.

Ionization   The process of changing an atom or molecule into a ion.

Ionization energy   The energy required to remove an electron completely from an atom.

Ionizing radiation   Radiation which is capable of interacting with atoms or molecules to produce ions.

Isoelectronic   Having the same number of electrons. Particles having the same energy level populations are isoelectronic.

Isomers   Molecules which have the same molecular formula, but different structural formulas.

Isotopes   Atoms of the same element which have the same number of protons but different numbers of neutrons; that is, the same atomic number but different mass numbers.

Kernel   The nucleus and the inner electrons of an atom. It is these which are represented by the letters of an electron dot symbol.

Ketone   A compound that contains two alkyl or aromatic groups attached to a carbonyl group.

Kinetic energy   The energy of motion.

Law   A statement of regularity found in observations made on a system.

Law of Conservation of Mass   The law which states that during a chemical reaction matter is neither created nor destroyed.

Law of Conservation of Mass and Energy   The law which states that the total quantity of matter and energy in the universe is constant.

Law of Constant Composition   The law which states that the percentage composition by mass of a particular compound never varies.

Law of Multiple Proportions   The law which states that when the proportions by mass in which two elements combine to form two different compounds are divided by each other, the result is a fraction composed of small whole numbers.

Lewis structure   A formula in which single lines are used to represent shared pairs of electrons and dots are used to represent valence electrons not involved in bonding.

Limiting reagent   The first reactant in a chemical reaction to be completely consumed, before the other reactant(s) is (are).

Line spectrum   A spectrum which consists of only certain frequencies or colours of visible light. Each element has a characteristic line spectrum.

Liquid   Matter which has a definite volume, but does not have a definite shape.

London-type smog   A type of air pollution consisting of a mixture of sulfur dioxide, smoke, and fog.

Magnetic quantum number (m_l)   A quantum number which identifies the direction of orientation (pointing) of an orbital in relation to an external magnetic field.

Main-group elements   The elements of Groups 1 and 2, on the left-hand side of the periodic table, and groups 13 to 17, on the right-hand side.

Mass   A measure of the amount of matter in an object; the quantity of inertia possessed by an object.

Mass defect   The mass converted into energy when a nucleus is made from its components.

Mass number   The total number of protons and neutrons in an atom.

Matter   Anything that has mass and takes up space. The material which makes up the universe.

Metal   An element which is a good conductor of electricity, is malleable and ductile, and has a characteristic lustre, a high density, and a high boiling point.

Metalloid   An element which has some of the properties of metals and some of the properties of nonmetals.

Mixture   A combination of two or more pure substances each of which retains its own physical and chemical properties.

Model   A visualization useful in helping us to understand a theory.

Moderator   A material which slows down neutrons in a nuclear reactor so that they can be more easily captured by the uranium-235 nuclei present.

Molar mass   The mass of one mole of any substance.

Molar volume   The volume occupied by one mole of a gas.

Molarity   The number of moles of solute dissolved per litre of solution.

Mole   The quantity of a substance which contains the same number of chemical units (atoms, molecules, formula units, ions) as there are atoms in exactly 12 g of carbon-12; Avogadro's number of chemical units of a substance.

Molecular formula   The formula which shows the actual number of atoms of each element in a molecule of a compound.

Molecule   The smallest particle of a compound which has all the chemical properties of that compound; any electrically neutral group of atoms which is held together tightly enough to be considered a single particle.

Monomer   A single molecular building block from which a polymer is constructed.

Net ionic equation   An ionic equation from which the spectator ions have been cancelled.

Neutralization   The process whereby hydronium ions and hydroxide ions react to form water molecules.

Neutron   A particle which has no charge and a mass of 1.0087 u.

Noble gases   The elements found in Group 18 of the periodic table: helium, neon, argon, krypton, xenon, and radon. They are the least reactive of all the elements.

Nonelectrolyte   The solute in a solution which does not conduct electricity.

Nonmetal   An element whose properties are not characteristic of metals. Nonmetals do not have a lustre, and they are poor conductors of heat and electricity.

Normal boiling point   The temperature at which a liquid boils when the atmospheric pressure is standard pressure (101.3 kPa).

Nuclear reactor   A device in which nuclear reactions take place in a controlled manner.

Nucleon   A particle which resides in the nucleus of an atom. Protons and neutrons are both nucleons.

Nucleus   The centre of an atom in which the positive charge and most of the mass is located.

Objectivity   Objectivity requires an unbiased reporting of all observations.

Octet   A grouping of eight electrons in an outer energy level of an atom. Atoms with such a grouping are especially stable.

Octet rule   A rule which states that when atoms combine, the bonds are formed in such a way that each atom finishes with an octet of valence electrons.

Orbit   A circular path in which an electron can move around the nucleus; corresponds to an energy level.

Orbital   A region of space where an electron is most likely to be found.

Organic chemistry   The study of carbon-containing compounds.

Oxidation   The loss of electrons from an atom, with a consequent increase in oxidation number.

Oxidation-reduction reaction   A reaction which involves electron transfer. Oxidation and reduction always occur together during a chemical reaction. Also known as a Redox reaction.

Oxidized   Refers to the substance which loses the electrons in an oxidation.

Oxidizing agent   The substance which removes electrons from an atom during an oxidation.

Pauli exclusion principle   The principle which states that an orbital can be empty, have one electron, or at most have two electrons. No two electrons in the same atom can have the same set of values for the four quantum numbers.

Percent yield   The actual yield of a product of a chemical reaction expressed as a percentage of the theoretical yield.

Periodic law   The law which states that when the elements are arranged in the periodic table in order of increasing atomic number, elements with similar properties occur at regular intervals.

Periods   The horizontal rows in which the elements are placed in a periodic table.

pH   The negative logarithm of the molar concentration of the hydronium ion in a solution. $pH = -\log[H_3O^+]$.

Phase   Any portion of material with a uniform set of properties.

Physical change   A change which alters one or more of the properties of a substance with no change in its composition or identity.

Physical properties   Those properties of a substance which can be determined without changing the composition or make-up of that substance.

Polar covalent bond   A covalent bond in which there is unequal sharing of electrons because of greater attraction for the shared pair of electrons by one atom than by another.

Pollutant   A substance that is present in high enough concentration to produce adverse effects on the things that human beings value.

Polymer   The large molecule formed as a result of the bonding of many monomers.

Potential energy   The energy of position.

Precipitate   The solid that is formed as a result of mixing two aqueous solutions of ionic compounds during a precipitation reaction.

Precipitation   The process of forming a solid during a precipitation reaction.

Precipitation reaction   A reaction in which two aqueous solutions are mixed, and one of the products is an insoluble solid.

Precision   The term used to describe how well a group of measurements made of the same object or event under the same conditions actually agree with each other.

Principal quantum number (n)   A number which identifies the energy possessed by an electron in any orbital under study.

Product   Any substance formed as a result of a chemical reaction. Differs in properties from any of the reactants.

Property   A quality or characteristic.

Proton   A particle which has a relative charge of +1 and a mass of 1.0073 u.

Pure science   The search for knowledge by people who want to improve our understanding of the way the universe operates.

Pure substance   A homogeneous or uniform material which consists of only one particular kind of matter and has the same properties throughout. It always has the same composition.

Qualitative analysis   The identification of the chemicals present in a substance or a mixture.

Qualitative properties   Physical or chemical properties that cannot be measured and so cannot be expressed in terms of numbers.

Quantitative properties   Physical or chemical properties that can be measured. Numbers are used in expressing quantitative properties.

Quantized   Limited to a number of specific energy levels.

Quantum   A discrete package of energy; a photon.

Quantum mechanics   The study of the laws which describe the motion of particles of very small mass.

Radioactive   Giving off invisible rays capable of penetrating various substances.

Radioactive decomposition   The altering of a nucleus by the loss of an alpha particle or a beta particle.

Radioactivity   The name of the property of any element which gives off invisible rays capable of penetrating various substances.

Radioisotope   A radioactive isotope.

Reactant   Any starting material in a chemical reaction.

Real gas   A gas which deviates from ideal behaviour. Real gases behave least like ideal gases under conditions of low temperature or high pressure.

Redox reaction   See Oxidation-reduction reaction.

Reduced   Refers to the substance which gains the electrons during reduction.

Reducing agent   The substance which supplies electrons to an atom during reduction.

Reduction   The gain of electrons by an atom, with a consequent decrease in oxidation number.

rem (radiation equivalent man)   The unit used to give a measure of the different effects of various types of radiation on people.

Reproducible   Measurements must be reproducible; that is, all measurements made of the same object or event under the same conditions must agree regardless of when or by whom the measurements are made.

Salt   A substance which consists of a positive ion of a base combined with the negative ion of an acid.

Saturated hydrocarbon   A hydrocarbon in which the carbon atoms are bonded to as many hydrogen atoms as possible.

Saturated solution   A solution which contains the maximum amount of solute that can remain dissolved in a given amount of solvent at a particular temperature.

Science   A human activity which is directed toward increasing our knowledge about the composition and behaviour of matter, both living and nonliving.

Secondary quantum number (ℓ)   The quantum number which identifies the shape of the orbital.

SI   The International System of Units, a single set of standards against which all measurements are made.

Significant digits   The number of digits in a measurement needed to give an idea of the degree of uncertainty in the measurement. The number of digits in a measurement that are known with certainty plus one that is uncertain.

Single covalent bond   The type of covalent bond which is found when only one pair of electrons is shared between atoms.

Solid   Matter which has a definite shape and volume.

Solubility   The maximum amount of solute that can be dissolved in a given amount of solvent at a particular temperature and pressure.

Solubility curve   A graph of the solubility of a solute versus temperature.

Solute   The substance in a solution which is present in smaller quantity and which is dissolved by the solvent.

Solution   A homogeneous or one-phase mixture. It is uniform in properties throughout. Particles in a solution are of molecular size, are scattered randomly throughout the solution, and are in continuous motion.

Solvent   The substance in a solution which is present in larger quantity and which dissolves the solute.

Spectator ions   Ions that do not participate directly in a chemical reaction.

Spin quantum number (m_s)   The quantum number which identifies the orientation of the spin of an electron with respect to an external magnetic field. Can have only two values, $+\frac{1}{2}$ or $-\frac{1}{2}$.

Stable octet   A grouping of eight electrons in the outer energy level of an atom. Atoms having an octet are particularly stable and unreactive.

Standard pressure   The pressure of the atmosphere at sea level, 101.3 kPa.

Standard temperature   The freezing point of water, 0°C or 273 K.

STP (standard temperature and pressure)   The combination of the two conditions, standard temperature and standard pressure.

Strong acid   An acid that dissociates completely in solution.

Strong electrolyte   An electrolyte that readily furnishes ions in solution.

Structural formula   A detailed formula which shows the structure of a molecule.

Sublimation   The change of a solid directly to its vapour.

Supersaturated solution   A solution which contains more dissolved solute than it would if it were saturated.

Technology   The application of scientific discoveries to practical problems. Same as Applied science.

Theoretical yield   The maximum amount of a product that can form in a chemical reaction.

Theory   An attempt to explain an observed regularity. It is a guess at the underlying principles which can explain a group of related observations.

Thermal inversion   An abnormal meteorological phenomenon in which a layer of warm air resting on top of a layer of cool air prevents pollutants near the earth's surface from moving to the upper atmosphere where they can be dispersed.

Thermoplastic   A polymer which may be heated and melted without being destroyed; it returns to its original condition upon cooling.

Thermosetting polymer   A polymer which cannot be heated or melted without being destroyed; it does not regain its original structure on cooling.

Titration   The procedure of carefully measuring the volume of a solution required to react completely with a known amount of another chemical.

Transition metals   The elements of Groups 3 to 12 in the middle of the periodic table.

Transmutation   A change in the identity of a nucleus because of a change in the number of protons in the nucleus.

Transuranium elements   The elements which have more than 92 protons in their nuclei.

Triple covalent bond   The sharing of three pairs of electrons between two atoms.

Units   The dimensions of the quantity of a measurement.

Universal gas constant   The constant of proportionality in the ideal gas law.

Universality   The idea that, for a scientific communication to be effective, it must be interpreted in the same way everywhere by every person.

Unsaturated hydrocarbon   A hydrocarbon that contains fewer hydrogen atoms than the alkane with the same carbon skeleton.

Unsaturated solution   A solution which contains less than the maximum amount of solute that can dissolve in a given amount of solvent at a particular temperature.

Valence electrons   The electrons in the outer energy levels of atoms. It is these which participate in chemical reactions.

Valence orbitals   The outer s and p orbitals which contain the valence electrons, the electrons which participate in chemical reactions.

Valence shell   The occupied energy level with the highest principal quantum number. The valence electrons of an atom occupy the valence shell.

Valence-Shell Electron-Pair Repulsion Theory   The theory which proposes that the arrangement of atoms around a central atom in a molecule depends upon the repulsions between all of the electron pairs in the valence shell of the central atom.

Vapour pressure   The pressure exerted by a vapour in equilibrium with its liquid; a measure of the tendency for molecules to leave the liquid and enter the vapour phase.

VSEPR Theory   See Valence-Shell Electron-Pair Repulsion Theory.

Water of hydration   The water retained when aqueous solutions of many of the soluble salts are evaporated and the ions form crystals. The water becomes part of the crystal.

Wave equation   A mathematical expression describing the energy and motion of an electron around a nucleus.

Wave function   A mathematical function related to the probability of finding an electron in a certain volume of space.

Wave mechanics   A type of mathematics devised in 1924 by Schrödinger in order to describe the motion of electrons in atoms, taking into account both their wave nature and their particle nature.

Weak acid   An acid that dissociates only partially in solution.

Weak electrolyte   An electrolyte that does not readily furnish ions in solution.

Weight   A measurement of the gravitational force of attraction between the earth and an object. This force depends upon the mass of the object and its distance from the centre of the earth.

Work   The moving of matter against a force which opposes the motion.

Index

A

Absolute zero,   426
Accelerator
  linear,   597-8
  particle,   597-8
Accuracy,   14
Acetaldehyde,   521
Acetic acid,   524
Acetone,   523
Acetylene,   511-3
Acetylsalicylic acid,   526
Acid,   385ff.
  Arrhenius,   387
  binary,   227
  Brønsted-Lowry,   402-4
  carboxylic,   523
  conjugate,   403
  nomenclature,   227
  strong,   387
  weak,   388-9
Acid rain,   405ff., 575
  effects of,   407
  origin,   406
Acid-base neutralization,   390
Actinides,   139
Activity series of metals,   296-8
Actual yield,   363
Addition polymers,   534
Addition reaction,   510
Adipic acid,   533
Agent Orange,   260
Air pollution,   443ff.
Alchemists,   127
Alcohols,   518
  IUPAC names,   518
Aldehydes,   521
  IUPAC names,   521
Aliphatic compound,   514
Alkadienes,   513-4
Alkali metals,   139
Alkanes,   491ff.
  combustion,   501-2
  IUPAC nomenclature,   495ff.
  physical properties,   504
  sources of,   501

substitution reactions,   502
  uses of,   504
Alkenes,   505ff.
  addition reactions,   510
  combustion,   510-1
  IUPAC nomenclature,   507-9
  polymerization,   510
  reactions of,   509-11
  sources of,   509
Alkyl groups,   496
  names of,   496
Alkylation,   559-60
Alkynes,   511
  addition reactions,   512
  combustion,   512-3
Alloy,   60
Alpha decay,   593-4
Alpha particle,   89-90, 590
Aluminum hydroxide,   258
Amides,   527
Amines,   527
Amino group,   518
Ammonia,   565-6
Ampere,   26
Amphoteric substance,   403
Analogy,   24-5
Analytical chemist,   12
Anode,   84
Anomaly,   13, 16
Applied science,   7-8
Aqueous ion,   320
Aromatic compound,   514
Aromatic hydrocarbon,   514ff.
Aromatization,   561
Arrhenius acid,   387
Arrhenius base,   387
Arrhenius, S.,   317, 388
Artifical blood,   315
Artificial transmutation,   593, 596-7
Asphalts,   558
Aspirin,   526
Atom,   81
  nuclear,   91-2
Atomic bomb,   602-3
Atomic mass,   83
Atomic mass unit,   83

Atomic number,   92
Atomic orbital,   195
Atomic radius,   150
Atomic size,   150
Atomic theory, Dalton's,   80ff.
Aufbau principle,   109
Authoritative knowledge,   6
Avogadro's hypothesis,   466ff.
Avogadro's number,   341
  determining,   342
Avogadro, A.,   341

B

Bacteriological contamination
  of water,   254
Baking powder,   571
Baking soda,   571
Balanced chemical equation,   230
Balancing chemical equations,   284ff.
Balancing nuclear equations,   592-3
Barometer,   421
Bartlett, N.,   167
Base,   385ff.
  Arrhenius,   387
  Brønsted-Lowry,   402-4
  conjugate,   403
  strong,   389-90
  weak,   389-90
Base units,   26
Basicity,   401
Becquerel, H.,   88, 586
Benzene,   514-6
Benzoyl peroxide,   534, 571
Berzelius, J.,   467
Beta decay,   594-5
Beta particle,   590
Binary acid,   227
Binary compound,   222-3
Binding energy,   590
Biochemical oxygen demand,   256
Biochemist,   12
Biodegradability,   10
Biodegradable detergents,   256
Biotechnology, Canadian,   554-5
BOD,   256
Bohr's model of the atom,   96ff.

Bohr, N., 98
Boiling, 242
Boiling point, 242
　normal, 242
Bomb
　atomic, 602-3
　cobalt, 612
　hydrogen, 614
Bond, 180
　chemical, 164
　coordinate covalent, 192-3
　covalent, 172ff.
　double covalent, 185
　hydrogen, 239-40
　ionic, 167ff., 202-3
　polar covalent, 175-6, 197ff.
　single covalent, 173
　triple covalent, 186
Boyle's Law, 422-4
Boyle, R., 423
Branched hydrocarbon, 495
Breeder reactor, 613
Brønsted-Lowry theory, 402
Brownian motion, 434
Butadiene, 514
Butane, 494
Butanoic acid, 524
Butanone, 522
Butene, 506
Butyne, 511
Butyric acid, 524

C

Caesatron, 612
Calandria, 607
Calcium carbide, 511
Calcium hydroxide, 563
Calcium phosphate, 568
Candela, 26
CANDU reactor, 573, 606ff.
　controls, 609
　coolant, 609-10
　fuel, 607-8
　moderator, 608-9
　used fuel, 610-1
Cannizzaro, S., 467
Carbon dioxide
　atmospheric, 444-5
　greenhouse effect, 448-9
Carbon monoxide, 3
Carbon tetrachloride, 502, 528
Carbon-dating, 595-6
Carbonic acid, 386
Carbonyl group, 518, 521
Carboxyl group, 518, 523

Carboxylic acids, 523
　IUPAC names, 523
Catalyst, 277
Catalytic cracker, 531
Cathode, 84
Cathode rays, 86
Chadwick, J., 91, 597
Chain reaction, 601-2
Change
　chemical, 57
　physical, 57
Changes of state, 57
Charge-to-mass ratio, 86
Charles' Law, 424-6
Chemical bond, 164
Chemical change, 57
Chemical energy, 51, 272
Chemical engineer, 12. 580
Chemical equation, 230, 284
　balanced, 230
　balancing, 284ff.
　information obtained from, 288-9
　mass calculations, 359ff.
　symbols used in, 289
Chemical explosive, 275-6
Chemical formula, 216
Chemical industry, 555ff.
　features of, 555-7
　sources of raw materials, 557ff.
Chemical nomenclature, 210, 220ff.
Chemical property, 56
Chemical reaction, 56, 270ff.
　factors affecting rate, 276ff.
　recognizing, 279-80
　reversible, 279
　types, 282-4
Chemist
　analytical, 12
　inorganic, 12
　organic, 12
　physical, 12
　quality control, 70
Chemistry, 9
Chernobyl, 615-7
Chlorine, 3, 566-7
Chloroform, 502
Chloromethane, 502
Chromatography, 62
Cisplatin, 17
Coal, 561
Coal tar, 561
Cobalt bomb, 612
Cockcroft, J., 597
Colloidal dispersion, 315
Combination reaction, 282
Communication, scientific, 23-4

Compound, 58-9
　binary, 222-3
　naming, 220ff.
Concentrated, 368
Concentrated solution, 308, 368
Concentration, 368
　molarity, 368-9
　percentage by mass, 372
Condensation, 242
Condensation polymers, 533-4
Conjugate acid, 403
Conjugate base, 403
Conservation of Energy, Law of, 52
Conservation of Mass and Energy,
　Law of, 52
Conservation of Mass, Law of, 66, 82
Constant Composition, Law of, 67, 82
Contact process, 564
Continuous spectrum, 97
Conversion factor, 38-9
Coordinate covalent bond, 192-3
Covalent bond, 172ff.
Covalent compounds, properties, 178
Cracking, 509, 558
Critical mass, 602
Critical pressure, 440
Critical temperature, 439-40
Crude oil, 557
Crystallization, 62, 311
Curie, M., 587
Cycloalkanes, 503
Cyclobutane, 503
Cyclohexane, 503
Cyclopentane, 503
Cyclopropane, 503
Cyclotron, 597-8

D

Dacron, 533
Dalton's atomic theory, 80ff.
Dalton's Law of Partial Pressures, 430-2
Dalton, J., 81
Davy, H., 129
DDT, 4
Decomposition reaction, 282-3
Decomposition, radioactive, 593-5
Degenerate, 107
DeMayo, P., 528
Denatured alcohol, 519
Density, 56
Derived units, 26
Desalination, 64
Detergent action of soap, 323
Deuterium, 251
Deuterium oxide, 251

Diatomic molecule, 173
Dichloroethane, 528
Dichloromethane, 502
Diene, 513-4
Diesel engine, 442
Diethyl ether, 521
Digits, significant, 13, 14, 32ff.
    mathematical operations, 34
Dilute, 368
Dilute solution, 308, 368
Dimensional analysis, 37ff.
Dimensions, 24
Dipole-dipole forces, 239
Discharge tube, 84, 87
Displacement reaction, 283
Dissociation, 250, 318
Dissolved oxygen, 255
Dissolving, 310
Distillate, 244
Distillation, 62, 64, 244
    vacuum, 244
Double covalent bond, 185
Double displacement reaction, 283-4
Drain cleaners, 571
Dynamic equilibrium, 243, 311
Dynamite, 527

E

Electrical energy, 51
Electrolyte, 316-8
    strong, 318
    weak, 318
Electrolytes, modern theory of, 317
Electron, 84ff.
    charge, 86
    mass, 86
    valence, 145
Electron affinity, 155
Electron configuration, 113
Electron dot formula, 173
Electron dot symbol, 145
Electron transfer reaction, 291ff.
Electronegativity, 176ff.
    and bond type, 180
Electrostatic precipitation, 62
Electrostatic precipitator, 448
Elements, 58, 126ff.
    discovery of, 126
    symbols of, 133
Empirical formula, 352
Empirical knowledge, 6
Endothermic, 271, 311
Endpoint
    experimental, 395
    theoretical, 395

Energy, 27, 50
    binding, 590
    chemical, 51, 272
    electrical, 51
    ionization, 152
    kinetic, 50
    mechanical, 51
    potential, 50
    quantization, 98
    rotational, 55
    translational, 54-5
    vibrational, 55
Energy level, 99ff., 103
Energy level diagram, 107ff.
    hydrogen, 107
    many-electron atom, 110
Energy level population, 103
Energy sublevel, 142-4
Engine knock, 559
Engineer, chemical, 12, 580
Enzyme, 277
Equation of state, 471
Equation
    chemical, 230, 284
        balancing, 284ff.
    ionic, 325-7
    net ionic, 325-7
    nuclear, 592-3
        balancing, 592-3
Equilibrium
    dynamic, 243, 311
    liquid-vapour, 242-3
    physical, 242
Esters, 525
    IUPAC names, 525
Ethanal, 521
Ethane, 494
Ethanedioic acid, 524
Ethanoic acid, 524
Ethanol, 509, 518-9
Ethene, 510
Ethers, 520-1
Ethyl acetate, 525
Ethyl alcohol, 509, 518-9
Ethylene, 568-9
Ethylene glycol, 519
Ethyne, 511-3
Eutrophication, 257
Evaporation, 241
Even-odd rule, 591-2
Exothermic, 271, 311
Experiment, 11
Experimentation, rules of, 11
Exponential notation, 29-30
Exponential numbers, 29
    adding, 31

    dividing, 30
    multiplying, 30
    subtracting, 31

F

Fast breeder, 613
Fermi, E., 604
Filtration, 61
Fine chemicals, 556
Fireworks, 226
Fission, 593, 600-2
Food irradiation, 573-4
Force, 26
Forces of attraction, 239
    dipole-dipole, 239
    London dispersion, 239
    van der Waals, 239
Formaldehyde, 521
Formalin, 521
Formic acid, 523
Formula mass, 346
Formula unit, 231
Formula writing, 216ff.
Formula
    chemical, 216
    empirical, 352
    molecular, 354
    structural, 493
Frasch process, 563
Free radical, 534
Freeze drying, 244
Freezing, 241
Freezing point, 241
Freon-11, 528
Fuels, 274-5
Functional group, 517-8
Fusion, 593, 613-4

G

Galileo, G., 421
Gamma rays, 590
Gas, 53, 418ff.
    ideal, 437, 471
    real, 437-9
Gas constant, universal, 471
Gas Law, Ideal, 471-2
Gas laws, 419ff.
Gasoline, 558
Gay-Lussac's Law, 435
Gay-Lussac's Law of Combining
    Volumes, 465-6
Gay-Lussac, J., 470
Geiger counter, 588

Gesner, A., 505
Gillespie, R., 190
Glycerol, 519-20
Gold foil experiment, 90
Greek model of the atom, 80
Greenhouse effect, 448-9
Ground state, 113
Group, 136
Gypsum, 251

H

Haber process, 553, 565
Haber, F., 553
Hahn, O., 600
Half-life, 140, 594
Halogens, 140
Hard water, 252-3
Hardness
  permanent, 252
  temporary, 252
Hazardous waste, 262-3
Heavy chemicals, 556
Heavy oils, 558
Heavy water, 251-2, 611
Henry's Law, 312-3
Herzberg, G., 16
Heterogeneous, 59
Heterogeneous mixture, 59
Hexachlorophene, 3
Hexamethylenediamine, 533
Hodgkin, D., 194
Homogeneous, 58
Homogeneous mixture, 59
Homologous series, 492
Household chemicals, 571
Household detergents in water, 256
Hund's rule, 111
Hybridization, 201
Hydrate, 251
Hydrated ion, 320
Hydrocarbon, 491
  aromatic, 514ff.
  branched, 495
  normal, 495
  saturated, 492
  straight-chain, 495
  unsaturated, 506
Hydrochloric acid, 568
Hydrogen, 140, 566-8
Hydrogen bomb, 614
Hydrogen bond, 239-40
Hydrogen peroxide, 571
Hydrogen sulfide, 3
Hydronium ion, 321, 387
Hydroxide ion, 387

Hydroxyl group, 518
Hypothesis, 11

I

Ideal gas, 437, 471
Ideal Gas Law, 471-2
Immiscible, 309
Indicator, 385, 402
Industrial wastes in water, 255
Inert gases, 139
Inertia, 48
Initiator, 534
Inorganic chemist, 12
Inorganic substances, 488
Insoluble, 324
International System of Units, 25ff.
Intuitive knowledge, 7
Iodized salt, 571
Ion exchange, 253
Ion
  aqueous, 320
  hydrated, 320
  hydronium, 321, 387
  hydroxide, 387
  polyatomic, 215-6
  spectator, 326
Ionic bond, 167ff., 202-3
Ionic compounds
  naming, 171
  properties, 171
Ionic equation, 325-7
Ionization, 152
Ionization energy, 152
Ionizing radiation, 614
Isoelectronic, 167
Isomer, 494-5
Isomerization, 559
Isooctane, 559-60
Isopropyl alcohol, 519
Isotope, 83, 92ff.
IUPAC names of organic
  compounds, 492-3
  alcohols, 518
  aldehydes, 521
  alkanes, 495ff.
  alkenes, 507-9
  carboxylic acids, 523
  esters, 525
  ketones, 522

J

Joule, 27
Joule-Thomson effect, 440

K

Kekulé, A., 515
Kelvin, 26
Kelvin scale, 426
Kenney-Wallace, G., 28
Kernel, 145
Kerosene, 558
Ketones, 522
  IUPAC names, 522
Kilogram, 26
Kinetic energy, 50
Kinetic molecular theory, 15, 433ff.
Knock, engine, 559
Knowledge
  authoritative, 6
  empirical, 6
  intuitive, 7
  rational, 6
  scientific, 6
Kraft process, 575

L

Lactic acid, 524
Lanthanides, 139
Lavoiser, A., 67
Law, 14-5
  Boyle's, 422-4
  Charles', 424-6
  Gay-Lussac's, 435
  Henry's, 312-3
  Ideal Gas, 471-2
  Periodic, 134ff.
Law of Combining Volumes,
  Gay-Lussac's, 465-6
Law of Conservation of Energy, 52
Law of Conservation of Mass, 66, 82
Law of Conservation of Mass and
  Energy, 52
Law of Constant Composition, 67, 82
Law of Multiple Proportions, 69, 82
Law of Partial Pressures, Dalton's, 430-2
LD_{50}, 260
Lead block experiment, 589
Lemieux, R., 522
Length, 26
Lewis structure, 173-4
Lewis symbol, 145
Lewis, G., 165
Light oils, 558
Limestone, 563
Limiting reagent, 361
Line spectrum, 97, 100, 102
Linear accelerator, 597-8
Liquefaction temperature, 439

Liquids, 52
  physical properties, 240-1
  structure, 238
Liquid air, 440-1
Liquid-vapour equilibrium, 242-3
Litre, 26
London dispersion forces, 239
London-type smog, 404-5
Los Angeles-type smog, 450-1
Lucite, 536

M

Magic numbers, 592
Magnetic quantum number, 115
Main-group elements, 139
Manometer, 435
Mass, 26, 49
  atomic, 83
  formula, 346
  molar, 342
  molecular, 345
Mass calculations involving chemical
  equations, 359ff.
Mass defect, 590
Mass number, 92
Mass-volume calculations, 476ff.
Matter, 48
  states of, 52
Maxwell, J., 434
Mechanical energy, 51
Meitner, L., 601
Mendeleev's Periodic Law, 136
Mendeleev's periodic table, 135-6
Mendeleev, D., 135
Menten, M., 278
Menthol, 520
Mercaptans, 526
Meta, 517
Metalloids, 133
Metals, 131
  activity series, 296-8
Metathetic reaction, 283-4
Methanal, 521
Methane, structure of, 493
Methanoic acid, 523
Methanol, 518
Method, scientific, 10-1
Methyl alcohol, 518
Methyl chloride, 502
Methyl methacrylate, 536
Methyl salicylate, 526
Methylene chloride, 502
Metre, 26
Millikan, R., 86
Mineral deposits, 562

Miscible, 309
Mixture, 59
  heterogeneous, 59
  homogeneous, 59
Model, 15-6
Model of the atom
  Bohr's, 96ff.
  Greek, 80
  nuclear, 92
  plum pudding, 88
  Rutherford's, 88
  Thomson's, 88, 90
Moderator, 604
Modern Periodic Law, 138
Modern periodic table, 137-8
Molar mass, 342
  and the Ideal Gas Law, 473-5
Molar volume, 472
Molarity, 368-9
Mole, 26, 342
Mole triangle, 343
Molecular formula, 354
Molecular mass, 345
Molecular orbital, 195
Molecule, 9, 164, 180
  nonpolar, 176
  polar, 175
Molecules, shapes of, 187ff.
Mond process, 575
Monomer, 532
Moseley, H., 137
Motion
  rotational, 55
  translational, 54-5
  vibrational, 55
Multiple Proportions, Law of, 69, 82
Mylar, 533

N

Naming acids, 227
Naming binary compounds, 222-3
Naming compounds, 220ff.
Naming ionic compounds, 171
Naphtha, 557-8
Natural gas, 557, 562
Net ionic equation, 325-7
Neutralization, acid-base, 390
Neutron, 91
Newton, 26
Nitric acid, 566
Nitroglycerin, 275, 526
Nobel, A., 527
Noble gases, 139
Nomenclature, 210, 220ff.
  of acids, 227

Nonelectrolyte, 316-8
Nonmetals, 132
Nonpolar molecule, 176
Normal boiling point, 242
Normal hydrocarbon, 495
Notation
  exponential, 29-30
  scientific, 29-30
NPD reactor, 606
NRU reactor, 605
NRX reactor, 573, 605
Nuclear atom, 91-2
Nuclear energy, Canadian, 572-3
Nuclear equations, 592-3
  balancing, 592-3
Nuclear reactions, types of, 593
Nuclear reactor, 604
Nucleon, 92
Nucleus, 90
Number
  atomic, 92
  Avogadro's, 341
  mass, 92
  oxidation, 211ff.
Numbers, exponential, 29
  adding, 31
  dividing, 30
  multiplying, 30
  subtracting, 31
Nylon, 533-4

O

Objective, 11
Observation
  qualitative, 7
  quantitative, 7
Octane rating, 559
Octet, 146
  stable, 167
Octet rule, 182
Ogilvie, K., 578
Oil of wintergreen, 526
Orbit, 99-100
Orbital, 105
  atomic, 195
  molecular, 195
  valence, 148
Orbital diagram, 195
Orbital hybridization, 201
Organic chemist, 12
Organic chemistry, 490
Organic halides, 527
Organic substances, 488
Orlon, 536
Ortho, 517

Osmosis,   64
Ostwald process,   566
Outer electron,   145
Oxalic acid,   524
Oxidation,   293
Oxidation number,   211ff.
Oxidation-reduction reaction,   293
Oxidized,   293
Oxidizing agent,   293
Oxygen depletion of water,   255-6
Ozone layer,   469-70

P

p/n Ratio,   591
Para,   517
Paraffin wax,   558
Partial pressure,   430
Particle accelerators,   597-8
Particle, subatomic,   91
Pascal,   27
Pauli exclusion principle,   111
Pauling, L.,   177
PCBs,   260-1
Pentane,   495
Pentyl acetate,   525
Percent yield,   363
Percentage composition,   346
Percentage by mass,   372
Period,   136
Periodic Law,   134ff.
   Mendeleev's,   136
   modern,   138
Periodic table,   135ff.
   Mendeleev's,   135-6
   modern,   137-8
   trends,   150ff.
Periodicity,   138
Permanent hardness,   252
Petroleum,   530-2, 557ff.
Petroleum ether,   557-8
pH,   400-2
pH of common substances,   401
Phase,   59
Phenol,   3
Phosphor,   85
Phosphoric acid,   568
Photochemical smog,   450-1
   effects of,   453
Photodegradability,   10
Photon,   97
Physical change,   57
Physical chemist,   12
Physical equilibrium,   242
Physical property,   55
Planck, M.,   97

Plaster of Paris,   251
Plastics,   10
Plexiglas,   536
Plum pudding model of the atom,   88
Polanyi, J.,   280
Polar covalent bond,   175-6, 197ff.
Polar molecule,   175
Pollutant,   446
Polonium,   587
Polyacrylonitrile,   536
Polyatomic ion,   215-6
Polyester,   533
Polyethylene,   532
Polymerization,   510
Polymers,   532ff.
   addition,   534
   condensation,   533-4
   table of,   536
   thermoplastic,   535
   thermosetting,   535
Polypropylene,   536
Polystyrene,   535
Polyvinyl chloride,   536
Potash,   562
Potential energy,   50
Precipitate,   325
Precipitation,   284, 324-5
   electrostatic,   62
Precipitation reaction,   325
Precision,   13
Prefixes, SI,   27
Pressure,   27
   critical,   440
   effect on solubility,   312-3
Principal quantum number,   106, 114
Product,   270
Propane,   494
Propanone,   522
Propene,   506
Property,   55
   chemical,   56
   physical,   55
   qualitative,   57
   quantitative,   56
Propylene,   536
Propyne,   511
Proton,   87
   charge,   87
   mass,   87
Proton-to-neutron ratio,   591
Pure covalent bond,   180
Pure science,   7
Pure substance,   58-9
Purifying water,   258-9
PVC,   536
Pyrolysis,   263

Q

Qualitative analysis,   328
Qualitative observation,   7
Qualitative property,   57
Quality control chemist,   70
Quantitative observation,   7
Quantitative property,   56
Quantized,   99
Quantum,   97
Quantum mechanics,   102
Quantum number,   114ff.
   magnetic,   115
   principal,   106, 114
   secondary,   115
   spin,   116
Quicklime,   563

R

Radiation, ionizing,   614
Radical, free,   534
Radioactive,   586
Radioactive decomposition,   593-5
Radioactivity,   586
Radioisotope,   95-6, 612
Radium,   588
Radius, atomic,   150
Rational knowledge,   6
Reactant,   270
Reaction
   addition,   510
   chain,   601-2
   chemical,   56, 270ff.
   combination,   282
   decomposition,   282-3
   displacement,   283
   double displacement,   283-4
   electron transfer,   291ff.
   factors affecting the rate,   276ff.
   metathetic,   283-4
   nuclear,   593
   oxidation-reduction,   293
   precipitation,   284, 325
   recognizing a,   279-80
   redox,   293
   reversible,   279
   substitution,   283, 502
   types of,   282-4
Reactor
   breeder,   613
   CANDU,   573, 606ff.
   nuclear,   604
Real gas,   437-9
   behaviour of,   437-9
Recrystallization,   315

Redox reaction, 293
Reduced, 293
Reducing agent, 293
Reduction, 293
Rem, 614
Reproducible, 13
Reverse osmosis, 64
Reversible reaction, 279
Rotational energy, 55
Rotational motion, 55
Rounding off, 36-7
Rutherford's model of the atom, 88
Rutherford, E., 89

S

Salicylic acid, 524-5
Salt, 391
Saturated hydrocarbon, 492
Saturated solution, 311
Schrödinger, E., 105
Science, 6, 8, 550
    applied, 7-8
    pure, 7
Scientific communication, 23-4
Scientific knowledge, 6
Scientific law, 14
Scientific method, 10-1
Scientific notation, 29-30
Second, 26
Secondary quantum number, 115
Shapes of molecules, 187ff.
Shared pair, 173
Shell, 100
SI, 25ff.
    prefixes, 27
    style conventions, 28
Side reaction, 363
Significant digits, 13, 14, 32ff.
    mathematical operations, 34
Single covalent bond, 173
Smog
    chemistry of, 452
    effects of photochemical, 453
    London-type, 404-5
    Los Angeles-type, 450-1
    photochemical, 450-1
Soap, detergent action, 323
Sodium bicarbonate, 571
Sodium hydrogen carbonate, 571
Sodium hydroxide, 566-8
Sodium hypochlorite, 571
Solid, 52
Solubility, 312
    factors affecting, 312ff.
Solubility curve, 314

Soluble, 324
Solute, 308
Solution, 59, 308
    aqueous, 230
    concentrated, 308, 368
    dilute, 308, 368
    saturated, 311
    supersaturated, 311
    types of, 308-10
    unsaturated, 311
Solution process, 310
Solvent, 308
Sour gas, 563
Spectator ion, 326
Spectrum
    continuous, 97
    line, 97, 100, 102
Spin quantum number, 116
Stable octet, 167
Standard pressure, 242, 429
Standard temperature, 429
Standard temperature and
    pressure, 429
States of matter, 52
    changes in, 54
Stock system, 221
STP, 429
Straight-chain hydrocarbon, 495
Strassmann, F., 600
Strong acid, 387
Strong base, 389-90
Strong electrolyte, 318
Structural formula, 493
Style, SI, 28
Styrene, 534
Subatomic particle, 91
Sublevel, energy, 142-4
Sublimation, 245
Substitution reaction, 283, 502
Sulfur, 563
Sulfur dioxide, 404-5, 564-5
Sulfur trioxide, 405, 564-5
Sulfuric acid, 564-5
Superconductivity, 178-9
Supersaturated solution, 311
Symbol
    electron dot, 145
    Lewis, 145
Symbols of the elements, 133
Synthetic fuels, 577

T

TCDD, 260-1
Technology, 7-8, 550
Technology and society, 552-3

Technology, Canadian, 572ff.
    chemical, 578-9
    mining, 574-5
    nuclear energy, 572-3
    pulp and paper, 575
Teflon, 571
Temperature
    critical, 439-40
    effect on solubility, 313
Temporary hardness, 252
Terephthalic acid, 533
Terylene, 533
Tetrachloromethane, 502
Tetraethyllead, 559
Tetrafluoroethylene, 13
Theoretical yield, 362
Theory, 11, 15, 16
Thermal inversion, 451
Thermal pollution of water, 255
Thermoplastics, 535
Thermosetting polymers, 535
Thomson's model of the atom, 88, 90
Thomson, J. J., 86
Titration, 395-6
TNT, 275, 516
Toluene, 516
Torricelli, E., 421
Tracer, 95
Transition metals, 139
Translational energy, 54-5
Translational motion, 54-5
Transmutation, 593-4
Transuranium elements, 600
Trends in the periodic table, 150ff.
Trichloroethane, 528
Trichlorofluoromethane, 528
Trichloromethane, 502
Trinitrotoluene, 275, 516
Triple covalent bond, 186
TRIUMF, 598
Trivial name, 220
Truncation, 36

U

Uncertainty, 32-3
Units, 24
    base, 26
    derived, 26
Units of measurement, 26
Universal gas constant, 471
Universality, 24
Unsaturated hydrocarbon, 506
Unsaturated solution, 311
Uranium, 601ff.

V

Vacuum distillation, 244
Valence bond theory, 195
Valence electron, 145
Valence orbital, 148
Valence shell, 188
Valence-shell electron-pair repulsion
  theory, 187ff.
Van der Waals forces, 239
Vapour, 242
Vapour pressure, 243
Vibrational energy, 55
Vibrational motion, 55
Vinegar, 571
Vinyl chloride, 536
Vital force, 490
Vitamin A, 322
Vitamin C, 322
Volume, 26
  molar, 472
Volume-volume calculations, 475-6
Volumetric flask, 368
VSEPR theory, 187ff.

W

Walton, E., 597
Water, 247ff.
  bacteriological contamination, 254
  chemical properties, 249-51
  dissociation of, 250
  hard, 252-3
  heavy, 251-2
  household detergents in, 256
  industrial wastes in, 255
  occurrence of, 247
  oxygen depletion of, 255-6
  physical properties, 247-9
  purifying, 258-9
  reaction with metals, 250
  reaction with nonmetals, 250
  reaction with oxides, 250-1
  thermal pollution, 255
Water of hydration, 251
Water pollution, 254ff.
Water vapour pressure, 431-2
Wave equation, 105
Wave function, 105
Wave mechanics, 105
Weak acid, 388-9
Weak base, 389-90
Weak electrolyte, 318
Weight, 50
Willson, T., 511
Wöhler, F., 490
Work, 27, 50

X

X-rays, 137
Xylene, 516

Y

Yield
  actual, 363
  percent, 363
  theoretical, 362

Z

ZEEP reactor, 573, 605
Zeolite, 253
Zero sum rule, 216
Zircaloy, 607

Periodic Table and Tabl

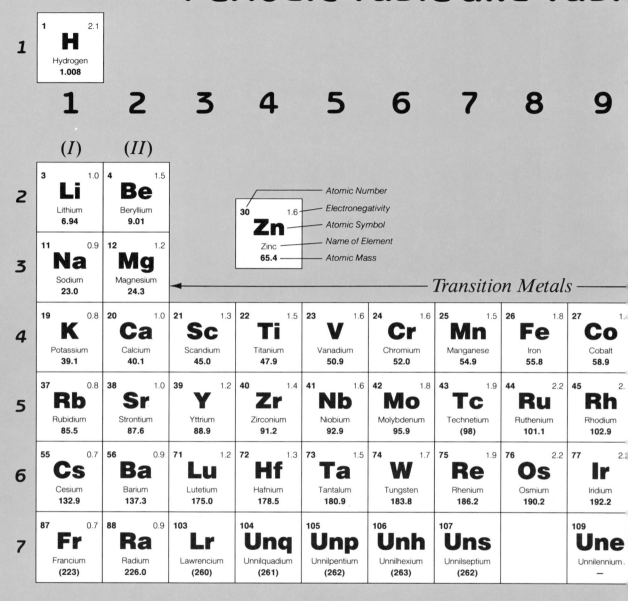

	1 (I)	2 (II)	3	4	5	6	7	8	9
1	**H** 1, 2.1, Hydrogen, 1.008								
2	**Li** 3, 1.0, Lithium, 6.94	**Be** 4, 1.5, Beryllium, 9.01							
3	**Na** 11, 0.9, Sodium, 23.0	**Mg** 12, 1.2, Magnesium, 24.3							
4	**K** 19, 0.8, Potassium, 39.1	**Ca** 20, 1.0, Calcium, 40.1	**Sc** 21, 1.3, Scandium, 45.0	**Ti** 22, 1.5, Titanium, 47.9	**V** 23, 1.6, Vanadium, 50.9	**Cr** 24, 1.6, Chromium, 52.0	**Mn** 25, 1.5, Manganese, 54.9	**Fe** 26, 1.8, Iron, 55.8	**Co** 27, 1., Cobalt, 58.9
5	**Rb** 37, 0.8, Rubidium, 85.5	**Sr** 38, 1.0, Strontium, 87.6	**Y** 39, 1.2, Yttrium, 88.9	**Zr** 40, 1.4, Zirconium, 91.2	**Nb** 41, 1.6, Niobium, 92.9	**Mo** 42, 1.8, Molybdenum, 95.9	**Tc** 43, 1.9, Technetium, (98)	**Ru** 44, 2.2, Ruthenium, 101.1	**Rh** 45, 2., Rhodium, 102.9
6	**Cs** 55, 0.7, Cesium, 132.9	**Ba** 56, 0.9, Barium, 137.3	**Lu** 71, 1.2, Lutetium, 175.0	**Hf** 72, 1.3, Hafnium, 178.5	**Ta** 73, 1.5, Tantalum, 180.9	**W** 74, 1.7, Tungsten, 183.8	**Re** 75, 1.9, Rhenium, 186.2	**Os** 76, 2.2, Osmium, 190.2	**Ir** 77, 2., Iridium, 192.2
7	**Fr** 87, 0.7, Francium, (223)	**Ra** 88, 0.9, Radium, 226.0	**Lr** 103, Lawrencium, (260)	**Unq** 104, Unnilquadium, (261)	**Unp** 105, Unnilpentium, (262)	**Unh** 106, Unnilhexium, (263)	**Uns** 107, Unnilseptium, (262)		**Une** 109, Unnilennium, —

Atomic Number — 30
Electronegativity — 1.6
Atomic Symbol — **Zn**
Name of Element — Zinc
Atomic Mass — 65.4

Transition Metals

Lanthanide Series

La 57, 1.1, Lanthanum, 138.9	**Ce** 58, 1.1, Cerium, 140.1	**Pr** 59, 1.1, Praseodymium, 140.9	**Nd** 60, 1.1, Neodymium, 144.2	**Pm** 61, 1.1, Promethium, (145)	**Sm** 62, 1.1, Samarium, 150.4	**Eu** 63, 1., Europium, 152.0

Actinide Series

Ac 89, 1.1, Actinium, 227.0	**Th** 90, 1.3, Thorium, 232.0	**Pa** 91, 1.5, Protactinium, 231.0	**U** 92, 1.7, Uranium, 238.0	**Np** 93, 1.3, Neptunium, 237.0	**Pu** 94, 1.3, Plutonium, (244)	**Am** 95, 1., Americium, (243)